Antibiotics

Current Innovations and Future Trends

Edited by

Sergio Sánchez

Departamento de Biología Molecular y Biotecnología
Instituto de Investigaciones Biomédicas
Universidad Nacional Autónoma de México
México, DF
México

and

Arnold L. Demain

Research Institute for Scientists Emeriti (RISE)
Drew University
Madison, NJ
USA

Caister Academic Press

Caister Academic Press
Norfolk, UK

www.caister.com

British Library Cataloguing-in-Publication Data
A catalogue record for this book is available from the British Library

ISBN: 978-1-908230-54-6 (hardback)
ISBN: 978-1-908230-55-3 (ebook)

Cover design adapted from Figures 2.1, 2.2, 2.3 and 2.4

Contents

Contributors

Joan Wennstrom Bennett
Department of Plant Biology and Pathology
Rutgers University
New Brunswick, NJ
USA

profmycogirl@yahoo.com

János Bérdy
H-1124 Budapest
Thoman I. u. 6
Hungary

jberdy@t-online.hu

Naowarat Cheeptham
Department of Biological Sciences
Faculty of Science
Thompson Rivers University
Kamloops, BC
Canada

ncheeptham@tru.ca

Brian Conlon
Department of Biology
Northeastern University
Boston, MA
USA

brian.patrick.conlon@gmail.com

Arnold L. Demain
Research Institute for Scientists Emeriti (RISE)
Drew University
Madison, NJ
USA

ademain@drew.edu

Stefano Donadio
Naicons Srl
Milan
Italy

sdonadio@naicons.com

Allan Espinosa-Gómez
Departamento de Biología Molecular y Biotecnología
Instituto de Investigaciones Biomédicas
Universidad Nacional Autónoma de México
México D.F.
México

espinosa_allan@hotmail.com

Prabhavathi Fernandes
Cempra Inc.
Chapel Hill, NC
USA

prabha@fernandes-domain.com

Luis B. Flores-Cotera
Departamento de Biotecnología
Centro de Investigación y de Estudios Avanzados
Instituto Politécnico Nacional
México D.F.
México

lfcotera@cinvestav.mx

Octavio L. Franco
Universidade Católica de Brasília
Centro de Análises Proteômicas e Bioquímicas
Pós-Graduação em Ciências Genômicas e Biotecnologia UCB
Brasília-DF
Brazil

ocfranco@gmail.com

Ponnampalam Gopalakrishnakone
Department of Anatomy
Yong Loo Lin School of Medicine
National University Health System (NUHS)
National University of Singapore
Singapore

antgopal@nus.edu.sg

Silvia Guzmán-Trampe
Departamento de Biología Molecular y Biotecnología
Instituto de Investigaciones Biomédicas
Universidad Nacional Autónoma de México
México D.F.
México

noencontre@gmail.com

Jianying Han
Chinese Academy of Sciences Key Laboratory of Pathogenic Microbiology and Immunology
Institute of Microbiology
Chinese Academy of Sciences; and
University of Chinese Academy of Sciences
Beijing
China

jyhan0524@163.com

Vladimir Havlicek
Institute of Microbiology
Academy of Sciences of the Czech Republic
Prague
Czech Republic

vlhavlic@biomed.cas.cz

Wenni He
Chinese Academy of Sciences Key Laboratory of Pathogenic Microbiology and Immunology
Institute of Microbiology
Chinese Academy of Sciences
Beijing
China

winner0047@163.com

Pei Huang
Chinese Academy of Sciences Key Laboratory of Pathogenic Microbiology and Immunology
Institute of Microbiology
Chinese Academy of Sciences; and
University of Chinese Academy of Sciences
Beijing
China

fenglin4620@163.com

Nelson Kardos
International Technology Development Corporation
Coatesville, PA
USA

drnkardos@gmail.com

Michael LaFleur
Arietis Corporation
Boston, MA
USA

m.lafleur@neu.edu

Kim Lewis
Department of Biology
Northeastern University
Boston, MA
USA

k.lewis@neu.edu

Paloma Liras
Microbiology Section
Departament of Molecular Biology
Faculty of Biological and Environmental Sciences
University of León
León
Spain

paloma.liras@unileon.es

Xueting Liu
Chinese Academy of Sciences Key Laboratory of Pathogenic Microbiology and Immunology
Institute of Microbiology
Chinese Academy of Sciences
Beijing
China

liuxt2010@gmail.com

Martha L. Macías-Rubalcava
Departamento de Productos Naturales
Instituto de Química
Universidad Nacional Autónoma de México
México D.F.
México

mlmacias@unam.mx

Girish B. Mahajan
Piramal Enterprises Limited
Mumbai
Maharashtra
India

girish.mahajan@piramal.com

Juan F. Martín
Microbiology Section
Departament of Molecular Biology
Faculty of Biological and Environmental Sciences
University of León
León
Spain

jfmarm@unileon.es

Satoshi Ōmura
Graduate School of Infection Control Sciences and Kitasato Institute for Life Sciences
Kitasato University
Shirokane
Tokyo
Japan

omuras@insti.kitasato-u.ac.jp

Hyunjun Park
Department of Bacteriology
University of Wisconsin
Madison, WI
USA

hpark67@wisc.edu

Sung Ryeol Park
Life Sciences Institute
University of Michigan
Ann Arbor, MI
USA

sungpark@umich.edu

Karol Rodríguez-Peña
Departamento de Biología Molecular y Biotecnología
Instituto de Investigaciones Biomédicas
Universidad Nacional Autónoma de México
México D.F.
México

karol_rodriguez@hotmail.com

Daniel Piso
Control Division
European Spallation Source (ESS)
Sweden

daniel.pisofernandez@esss.se

Steven J. Projan
Infectious Diseases and Vaccines Innovative Medicines Unit
MedImmune
Gaithersburg, MD
USA

projansj@medimmune.com

Biao Ren
Chinese Academy of Sciences Key Laboratory of Pathogenic Microbiology and Immunology
Institute of Microbiology
Chinese Academy of Sciences; and
University of Chinese Academy of Sciences
Beijing
China

phoenix.renbiao@gmail.com

Cesareo Saiz-Jimenez
Instituto de Recursos Naturales y Agrobiologia
IRNAS-CSIC
Sevilla
Spain

saiz@irnase.csic.es

Ramar Perumal Samy
Departments of Anatomy, Microbiology and Orthopaedic Surgery
Yong Loo Lin School of Medicine
National University Health System (NUHS)
National University of Singapore
Singapore

rperumalsamy@yahoo.co.uk

Sergio Sánchez
Departamento de Biología Molecular y Biotecnología
Instituto de Investigaciones Biomédicas
Universidad Nacional Autónoma de México
México D.F.
México

sersan@biomedicas.unam.mx

Rosa E. Sánchez-Fernández
Departamento de Productos Naturales
Instituto de Química
Universidad Nacional Autónoma de México
México D.F.
México

resf0311@yahoo.com.mx

Angeles Sánchez-Pérez
Discipline of Physiology and Bosch Institute
School of Medical Sciences
University of Sydney
Australia

angeles.sanchez-perez@sydney.edu.au

Seetharama Satyanarayanajois
Department of Basic Pharmaceutical Sciences
School of Pharmacy
University of Louisiana at Monroe
Monroe, LA
USA

jois@ulm.edu

Kazuro Shiomi
Graduate School of Infection Control Sciences and Kitasato Institute for Life Sciences
Kitasato University
Shirokane
Tokyo
Japan

shiomi@lisci.kitasato-u.ac.jp

Lynn L. Silver
LL Silver Consulting, LLC
Springfield, NJ
USA

silverly@comcast.net

Margherita Sosio
Naicons Srl
Milan
Italy

msosio@naicons.com

Jaroslav Spizek
Institute of Microbiology
Academy of Sciences of the Czech Republic
Prague
Czech Republic

spizek@biomed.cas.cz

Bradley G. Stiles
Integrated Toxicology Division
US Army Medical Research Institute of Infectious Diseases
Fort Detrick, MD; and
Department of Biology
Wilson College
Chambersburg, PA
USA

bstiles@wilson.edu

Michael Thomas
Department of Bacteriology
University of Wisconsin
Madison, WI
USA

thomas@bact.wisc.edu

Patricia Veiga-Crespo
Department of Microbiology
Faculty of Pharmacy; and
School of Biotechnology
University of Santiago de Compostela
Spain

pvcrespo@gmail.com

Tomás G. Villa
Department of Microbiology
Faculty of Pharmacy; and
School of Biotechnology
University of Santiago de Compostela
Spain

tomas.gonzalez@usc.es

Miquel Viñas
Department of Pathology and Experimental Therapeutics, IDIBELL
Schools of Medicine and Dentistry
University of Barcelona
Spain

mvinyas@ub.edu

Yeo Joon Yoon
Department of Chemistry and Nano Sciences
Ewha Womans University
Seoul
Republic of Korea

joonyoon@ewha.ac.kr

Jingyu Zhang
Chinese Academy of Sciences Key Laboratory of Pathogenic Microbiology and Immunology
Institute of Microbiology
Chinese Academy of Sciences
Beijing
China

zhangjingyu1985@126.com

Lixin Zhang
Chinese Academy of Sciences Key Laboratory of Pathogenic Microbiology and Immunology
Institute of Microbiology
Chinese Academy of Sciences; and
University of Chinese Academy of Sciences
Beijing
China

lzhang03@gmail.com

Current Books of Interest

Advanced Vaccine Research Methods for the Decade of Vaccines	2015
Antifungals: From Genomics to Resistance and the Development of Novel Agents	2015
Bacterial-Plant Interactions: Advanced Research and Future Trends	2015
Aeromonas	2015
Leishmania: Current Biology and Control	2015
Acanthamoeba: Biology and Pathogenesis (2nd edition)	2015
Microarrays: Current Technology, Innovations and Applications	2014
Metagenomics of the Microbial Nitrogen Cycle: Theory, Methods and Applications	2014
Pathogenic *Neisseria*: Genomics, Molecular Biology and Disease Intervention	2014
Proteomics: Targeted Technology, Innovations and Applications	2014
Biofuels: From Microbes to Molecules	2014
Human Pathogenic Fungi: Molecular Biology and Pathogenic Mechanisms	2014
Applied RNAi: From Fundamental Research to Therapeutic Applications	2014
Halophiles: Genetics and Genomes	2014
Molecular Diagnostics: Current Research and Applications	2014
Phage Therapy: Current Research and Applications	2014
Bioinformatics and Data Analysis in Microbiology	2014
The Cell Biology of Cyanobacteria	2014
Pathogenic *Escherichia coli*: Molecular and Cellular Microbiology	2014
Campylobacter Ecology and Evolution	2014
Burkholderia: From Genomes to Function	2014
Myxobacteria: Genomics, Cellular and Molecular Biology	2014
Next-generation Sequencing: Current Technologies and Applications	2014
Omics in Soil Science	2014
Applications of Molecular Microbiological Methods	2014
Mollicutes: Molecular Biology and Pathogenesis	2014
Genome Analysis: Current Procedures and Applications	2014
Bacterial Toxins: Genetics, Cellular Biology and Practical Applications	2013
Bacterial Membranes: Structural and Molecular Biology	2014
Cold-Adapted Microorganisms	2013
Fusarium: Genomics, Molecular and Cellular Biology	2013
Prions: Current Progress in Advanced Research	2013
RNA Editing: Current Research and Future Trends	2013
Real-Time PCR: Advanced Technologies and Applications	2013

Full details at www.caister.com

Preface

Antibiotics have played a most crucial role in the improvement of human health and have doubled life expectancy since the start of the 20th century. Between 1940 and 1970, there was an explosion of antibiotic discoveries and their use in human medicine. However, since 1970, there has been a rapid drop in the introduction of new antibiotics and a sad departure by many large pharmaceutical companies from antibiotics as commercial products. This has been accompanied by a dangerous rise in resistance to present antibiotics by pathogenic microbes, which is especially evident in hospitals. Thus, we are now experiencing a crisis in the antibiotic field and something must be done to correct this problem. We hope this book will aid in this noble effort. The authors of the chapters define what an antibiotic is, describe the producing microorganisms, explain why we need new antibiotics and point out applications for antibiotics, both old and new. Also provided are chapters on antibiotic resistance, dormant pathogens, toxicity, overuse and new sources, including animal venoms, endophytes, hot springs and caves. Recent developments are described, such as new targets and technologies for finding new antibiotics. These include the use of silent gene clusters and combinatorial biosynthesis. Compounds such as peptides, antiviral agents and core structures are described, as well as the current pipeline of antibiotics. We hope that this book will reignite the much needed efforts to discover new antibiotics, especially by biotechnology companies and academic laboratories which constitute the new hope for solving the predicament in which the antibiotic field is in today. An extra bonus would be the re-entry of some pharmaceutical companies into the hunt for new antibiotics.

What is an Antibiotic?

1

Joan Wennstrom Bennett

Abstract

The word 'antibiotic' was originally used in the English language as an adjective. In 1947, Selman A. Waksman published a definition of 'antibiotic' as a noun:

> An antibiotic is a chemical substance, produced by micro-organisms, which has the capacity to inhibit the growth of and even to destroy bacteria and other micro-organisms.

The clinical efficacy of penicillin and streptomycin brought public attention to this group of life-saving drugs and the word 'antibiotic' soon became commonplace in scientific and ordinary language. Waksman stressed that antibiotics were not only antimicrobial in action but also that they were products of microbial metabolism. He excluded compounds made by plants, animals or synthetic chemists. Contemporary usage makes no such distinction. Because most of the early antibiotics were effective only against bacteria, and because responsible health care workers warned patients against the asking for antibiotics to treat viral infections, nowadays 'antibiotic' is often used as a synonym for 'antibacterial' in order to simplify communication. Curiously, Waksman's original definition and most of the energy he subsequently devoted to defending his definition rarely included mention of the single most important property of a good antibiotic, namely that it does not harm the host. Many of the currently published definitions continue in the Waksman tradition and focus on the antimicrobial activity of the compounds, not on their selective action. The published literature can be an important resource in the search for new efficacious compounds with antimicrobial action. Thus, it is more important than ever to have an understanding of how the word 'antibiotic' is used in different contexts so as to guide contemporary information scientists to mine the vast scientific and patent literature.

Introduction

The discovery of the first antibiotic was among the most far-reaching events in the history of medicine. In 1940, despite the development of phage therapy, salvarasan for syphilis, quinine for malaria, serum antibodies, and the sulfa drugs, there was little that healthcare workers could do to treat most infectious diseases. No less of an authority than Sir Almroth Wright, who was Alexander Fleming's boss at St. Mary's Hospital in London, believed that 'antibacterial drugs are a delusion' (Wilson, 1976).

Penicillin, discovered by Alexander Fleming, and researched as a systemic clinical agent by Howard Florey, Ernest Chain and colleagues, was developed and introduced during the Second World War. It was the first of the major antibiotics, capable of killing Gram-positive bacteria including the organisms that caused gonorrhoea, syphilis, and puerperal infections. It was widely heralded as a miracle drug but it was not called an antibiotic until after the discovery of streptomycin by Selman A. Waksman, Albert Schatz, and Elizabeth Bugie at the New Jersey Agricultural Experiment Station at Rutgers University. Streptomycin proved to be useful for the treatment of tuberculosis and several other diseases that were not helped by penicillin; it was Waksman who championed the use of the term 'antibiotic' to describe penicillin, streptomycin and several other antimicrobial compounds that had been isolated from microbial sources. Waksman is generally credited with having developed the modern definition of antibiotic. Here is how he described its genesis, in a book written towards the end of his life:

> At the beginning of the antibiotic era the very word 'antibiotic' for designating antimicrobial substances produced by microbes did not exist. It was introduced soon afterward, however. Dr J. E. Flynn, editor of *Biological Abstracts*, asked me in July, 1942[1], to suggest a word that could be used for designating a type of chemical substance which would include those compounds and preparations that are produced by microbes and that possess antimicrobial properties. A proper generic word was needed for indexing purposes to embrace such compounds as pyocyanase, pyocyanin, penicillin, gramicidin, tyrothricin and actinomycin ... I suggested the word 'antibiotic,' and began to use it in my own lectures and scientific papers ... The word almost immediately became very popular both in the scientific literature and in everyday usage. Unfortunately, there soon appeared a tendency to change and often to confuse its meaning, so that in 1947 I found it advisable to publish a comprehensive definition of it, which has now been almost universally accepted ...
>
> (Waksman, 1975, p. 28)

Here is Waksman's comprehensive definition:

> An antibiotic is a chemical substance, produced by micro-organisms, which has the capacity to inhibit the growth of and even to destroy bacteria and other micro-organisms. The action of an antibiotic against micro-organisms is selective in nature, some organisms being affected and others not at all or only to a limited degree; each antibiotic is thus characterized by a specific antimicrobial spectrum. The selective action of an antibiotic is also manifested against microbial vs. host cells. ... some antibiotics have remarkable chemotherapeutic potentialities and can be used for the control of various microbial infections in man and animal.
>
> (Waksman, 1947)

In a subsequent paper published four years later, he refined the definition to include the fact that antibiotics act in low concentration:

> An antibiotic is a chemical substance, produced by microorganisms, which has the capacity to inhibit the growth and even to destroy bacteria and other microorganisms, in dilute solutions.
>
> (Waksman, 1951)

Waksman shaped his definition to exclude naturally occurring, non-microbial substances that kill bacteria such as gastric juice. His definition also excluded synthetic antibacterial compounds such as the sulfonamides. It is noteworthy that Waksman's abbreviated definition put the emphasis on the compound itself and on its target, but not on its therapeutic potential. 'Antibiotic' has become a commonplace word in popular and scientific discourse to describe anti-infectious drugs. It usually is used nowadays without bothering to give it a definition. The assumption that everyone agrees on what an antibiotic is, however, unwarranted. As antibiotics have transformed agriculture, aquaculture, and public health, denotations and connotations of the word have shifted, sometimes leading to a breakdown in communication between scientists, healthcare workers and patients. Many physicians and scientists now assume that 'antibiotic' is synonymous with 'antibacterial' and use the word accordingly. Others continue to use 'antibiotic' as Waksman originally defined it, i.e. broadly to mean 'antimicrobial' and thereby encompassing antiviral, antifungal and antiprotozoal drugs. There are even those who expand Waksman's definition to include anti-helminthic agents.

This essay will give a brief review of the etymology of the word 'antibiotic;' summarize something of the changing connotations of the

[1] In other published sources, Waksman said the request from Flynn came in 1941.

term; and then address the inconvenient fact that Waksman's famous definition, which is still widely encountered in dictionaries, glossaries, and textbooks, does not emphasize the single most important attribute of a good antibiotic drug, namely that it should exhibit selective toxicity. Finally, I will comment on the way in which the emergence of microorganisms that are resistant to antimicrobial drugs has brought new focus on the need for a word that means the same thing to all the people who use it.

The etymology of the word antibiotic

The word 'antibiotic' has a rich pedigree. It is generally accepted that 'antibiotic,' as used in microbiology, was derived from the French words *antibiose* and *antibiotique* introduced by Vuillemin in the late nineteenth century to describe antagonistic effects between microorganisms. Vuillemin was a mycologist and he used it to describe any compound that was 'injurious to or destructive of living matter, especially microorganisms' (Vuillemin, 1890). Still earlier, Tyndall, Pasteur, and Koch and others observed antagonisms between microbes and theorized that there could be therapeutic potential in some of these interactions. In 1928, a French textbook, *Les Associations Microbiennes, Leurs Applications Thérapeutiques* contained an 80-page section titled 'L'Antibiose,' and the literature review had approximately 600 citations concerning antibiotic actions (Papacostas and Gaté, 1928).

In *Biological Abstracts*, 'antibiosis' (noun) first appeared as an index heading in 1928 (volume 2). From 1929–1932, there were no 'antibiosis' entries but in 1933 there were two, in 1935 there were three, in 1938 there was one, and in both 1939 and 1940 there were three. Articles indexed under 'antibiosis' increased to seven each in 1941 and 1942. By 1943, there were three headings under the noun 'antibiosis' (five abstracts) and the adjective form appeared in two ways: 'antibiotic action' (24 abstracts) and 'antibiotic substances' (listing the names of 18 specific substances). The supplement to the Oxford English Dictionary (OED) published in 1933 gave a definition for 'antibiotic' as an adjective: 'injurious to or destructive of living matter' (OED, 1933). In the USA, Constantine J. Alexopoulos, a prominent mycologist, used the noun 'antibiosis' in three papers dealing with the possible inhibition of fungi by common species of bacteria under the generic title, 'Studies in antibiosis between bacteria and fungi' (Alexopoulos and Herrick, 1942).

In general, in the English language, the suffix '-ic' is used to form an adjective from other parts of speech. Thus, in proposing his definition, Waksman had taken the adjective as used in 'antibiotic substance' or 'antibiotic compound' and turned the adjective into a noun. Such noun forms are called substantives by grammarians. Over the course of several decades, Waksman devoted considerable energy towards taking credit for having coined the word 'antibiotic,' vigorously defended certain aspects of his definition and widely promulgated its use. In particular, Waksman felt strongly that the term 'antibiotic' should be limited to products of microbial metabolism. At a 1956 symposium, he made the point explicitly:

> As I understand antibiotics, we are dealing here entirely with microbial products. We are not dealing with plant products; we are not dealing with animal products. Blood has antibacterial properties. Quinine has antibacterial properties. Therefore, we would be lost if we consider the whole system of life, because we will find that all plants, all animals, insects and what not, produce in one way or another antimicrobial activity. Therefore, we must adopt the concept that has been said time and again by some of the greatest philosophers: 'In order to understand one another, let us define our terms first of all'
>
> (Waksman, 1956)

In summary, starting in the early 1940s, in his published 1947 definition, and in later papers that reiterated and refined the 1947 definition, Waksman championed the use of 'antibiotic' as a noun and emphasized that according to his proscribed definition, antibiotics should be products of microbial metabolism that were active against microbes. He clearly recognized the need for a single term with a carefully stipulated definition to encompass the new class of compounds that were transforming medical practice and during his lifetime, was vigilant in defending the way he felt that 'antibiotic' should be used. For example, when some people started using the term to describe agents produced by microbes and found to be worthwhile in cancer chemotherapy, he rejected this usage. For a colourful summary of Waksman attempts to control the definition of 'antibiotic', see Pramer (1988). Nevertheless, the word and its definition were not without critics. Howard Florey was opposed to a word that meant, literally, 'against life' to describe lifesaving drugs and seems to have preferred the word 'bacteriostat' (Wainwright, 1990). Others suggested alternatives that included 'antibiotin' (Cohen and Spegelman, 1981) and 'mycoine' (Vonkennel *et al.*, 1943). And, as is the nature of language, as time passed, others started using 'antibiotic' in ways different from that which Waksman had sanctioned (see below).

For a more detailed summary of emergence of the terms 'antibiotic' (used as both an adjective and as a noun) see Waksman (1975), Wainwright (1990), Bentley and Bennett (2003), and Bryskier (2005b). The citations within these sources will lead to a huge number of papers in which different definitions of 'antibiotic' are given. In the sections below, I highlight the ways in which 'antibiotic' has come to have multiple meanings. Because the literature on antibiotics is so vast, in most cases I cite sources from the secondary literature. The references in these general sources lead to citations in the primary literature. In cases where specific definitions are quoted, they are meant to illustrate the ways in which different experts have used different meanings for 'antibiotic' and are not meant to be restrictive, proscriptive, or even close to exhaustive.

Disinfectants and related compounds

Many terms are encountered in relation to antimicrobial effects. Physicians had been describing and treating infections long before the scientific discovery of pathogenic microorganisms and had found empirically that infection could be prevented or allayed by certain treatments. Various disinfectants, fumigants, antiseptics, sanitizing substances and procedures had been developed to purify materials that had come in contact with the sick and their corpses. Treatments were developed to treat wounds (Forrest, 1982) and a vocabulary was developed to describe agents that would prevent putrefaction or sepsis. In particular, Joseph Lister, a British surgeon, was a pioneer of antiseptic medical techniques, and the use of antiseptic substances to exclude microorganisms from wounds is sometimes called 'Listerism' (Pennington, 1995; Bankston, 2004)

A list of words that have to do with agents/processes that kill disease-causing organisms is listed in Table 1.1. All of the definitions are taken from the *Oxford English Dictionary.* For readers whose first language is not English, it should be pointed out that this illustrious unabridged dictionary of the English language is constructed on historical principles and employs citations to illustrate the way in which words have been used in language (Murray, 1977). In the table, only the first illustrative citation is reproduced and figurative definitions are excluded.

Nowadays, disinfectants are generally viewed as substances that eliminate microbes found on inanimate objects ('fomites') and antiseptics as substances to be used on the outside of the body. Well-known antiseptics include alcohol, phenol, iodine and certain dyes. Collectively, both groups are often subsumed under the heading 'biocide' or 'germicide' and distinguished from antibiotics which can be used systemically. Nevertheless, there are many compounds called 'antibiotics' that have toxic effects when taken internally but that are nevertheless used topically. In these cases, it sometimes is hard to draw a clear line with antiseptics. In general, however, natural products that fall under the Waksman definition are labelled 'topical antibiotics' while the earlier, non-specific agents like alcohol are called 'antiseptics.' For good general coverage of antisepsis, disinfection and sterilization see Russell *et al.* (1992), Ascenzi (1996), Block (2001) and McDonnell (2007).

In order to create a linguistic label that avoids some of the confusion around 'antibiotic,' one excellent recent monograph on antisepsis and disinfection uses the term 'anti-infective' and distinguishes these agents from 'biocides.'

> Anti-infectives are substances (or drugs) capable of inhibiting or inactivating microorganisms (particularly pathogens) that are associated with various infections in animals, plants, and humans. The term is used to encompass drugs that act specifically on certain types of microorganism including antibacterials (antibiotics), antifungals, antivirals and antiprotozoal agents. In contrast, biocides are chemical or physical agents that are used on inanimate surfaces or on the skin and mucous membranes.
>
> (McDonnell, 2007, p. 217)

In further refining the distinction between anti-infectives and biocides, McDonnell (2007) points out that anti-infectives have a specific, targeted mechanism of action with minimal toxicity while biocides generally have multiple targets and are broadly toxic.

In recent years, a flood of germ-fighting products has been introduced to the consumer market. Many soaps, household cleaners, medicated shampoos, and skin care products now incorporate antimicrobial substances and use labels that imply that they will help to avoid disease. For the lay public, this non-medical use of antimicrobial products has blurred the distinction between 'germicides', 'antibacterial substances', 'sanitizers', 'disinfectants' and 'antibiotics'. Marketers encourage and exploit an emotional germophobia in order to sell their products. For example, triclosan is an antiseptic substance that has shown activity in the control of *Staphylococcus aureus*, other Gram-positive bacteria as well as yeasts. It has been used as a surgical wash because of its low skin irritation. The incorporation of triclosan into personal hygiene products has been used as a marketing tool, promising to reduce body odour and improve health. Unfortunately, it may cause adverse health effects and triclosan-containing products promote the emergence of bacteria resistant to antibiotic medications (Chuanchuen *et al.*, 2001; Levy, 2001). It has been shown that using soaps with triclosan is no more effective than hand washing with soap and water (Aiello *et al.*, 2007). Both the American Medical Association and the American Public Health Association have recommended against its use (AMA, 2000; APHA, 2009).

In summary, antibiotics undoubtedly are antibacterial and/or antimicrobial in action. However, it does not follow that all antibacterial and antimicrobial substances are antibiotics. For the lay public in contemporary times, the proliferation of personal hygiene products that incorporate disinfectants and antiseptics, and the number of over-the-counter medications for treating infections, often blurs the linguistic distinctions between the different kinds of microbe-killing substances and may be contributing to the emergence of antibiotic-resistant microbes.

Chemotherapeutic agents, magic bullets and selective toxicity

It was not until the nineteenth and twentieth centuries that we needed a word that was specifically focused on substances that could be administered systemically to kill microbes *without* killing us or making us sick (at least not usually). One such term is 'chemotherapeutic agent.'

The discovery of microorganisms as the cause of infectious disease in the middle of the nineteenth century stimulated new efforts to find ways in which to treat and cure these scourges. Famously, Paul Ehrlich used bacteria-specific dyes in his histological research and realized that certain of them stained bacteria but not animal cells (Krantz, 1974; Mann, 1999). He felt that the discriminatory property of dyes could be exploited for medicinal purposes and described his goal as a search for a 'magic bullet.'

> Picture an organism as infected by a certain species of bacterium, it will ... be easy to effect a cure if substances have been discovered which have a specific affinity for these bacteria and act ... on these alone ... while they possess no affinity for the normal constituents of the body... such substances would then be magic bullets.
>
> (Paul Ehrlich, translated from the German in Levy, 2011, p. 201)

Ehrlich, working with Sahachiro Hara and other colleagues, hypothesized that scientists could develop specific drugs to target specific

Table 1.1 Definitions of terms used to describe agents that kill or inhibit microorganisms from the Oxford English Dictionary

antimicrobial

Med. and Pharmacol.

A. adj. That kills or inhibits the growth of microorganisms.
1910 Lippincott's New Med. Dic. 67/1 Antimicrobial, antagonistic to microbes.

B. n. A substance active against microorganisms.
1957 Dorland's Med. Dict. (ed. 23) Antimicrobial, an agent which destroys microorganisms or suppresses their multiplication or growth.

antipyretic

Med.

A. adj. Tending to prevent or allay fever.
1875 H. C. Wood Treat. Therapeutics (1879) 74 All antipyretic remedies appear to act more strongly on children.

B. n. (sc. medicine.)
1681 in S. Pordage tr. T. Willis Remaining Med. Wks. Table, Antipyreticks, medicines against burning feavers.

antiseptic

A. adj. 1. Counteracting putrefaction; antiputrescent.
1751 Gentleman's Mag. 557 Myrrh in a watery menstruum was 12 times more antiseptic than sea salt.

B. n. 1. (sc. agent.)
1751 Gentleman's Mag. Dec. 557 Acids per se are most powerful antiseptics.

bacteriostasis

n. Inhibition of the growth of bacteria without destroying them.
1936 Nature 15 Aug. 295/1 Selective bacteriostasis (the inhibitory action of such substances as dyestuffs and the products of their own metabolism on the growth of bacteria and fungi).

Derivatives:

bacˈteriostat n. an agent that causes bacteriostasis.
1944 Lancet 17 June 792/2 Cultures whose growth is delayed or prevented by one of the substances called bacteriostats.

bacteriostatic adj. (and n.)/-ˈstætɪk/
1912 Jrnl. Exper. Med. 16 229 Its bacteriostatic action on the violet positive organisms is evident enough.

disinfectant

A. adj. Having the property of disinfecting.
1875 Ure's Dict. Arts III. 1192 The disinfectant liquor of Sir W. Burnett is chloride of zinc.

B. n. Something having this property; an agent used for disinfecting or destroying the germs of infectious disease.
1837 Penny Cycl. VII. 107/2 The hypochlorite of lime, usually called chloride of lime..is a compound of great importance, both in the arts, and as a disinfectant.

disinfection

n. The action of disinfecting or purifying from infection; destruction of the germs of infectious diseases.
1803 Duncan's Ann. Med. II. ii. 35 On the influence of Oxygen in the process of disinfection.

fumigant

A. adj. That fumes. Obs.
1727–36 N. Bailey Universal Etymol. Eng. Dict., Fumigant, smoaking (sic), fuming.

B. n. That which fumigates.
1890 Sc. Leader 7 Feb. 7 The production of the fashionable little fumigant [cigarette] has trebled in the last two years.
1940 Chambers's Techn. Dict. 359/1 Examples of fumigants are hydrogen cyanide and ethylene oxide.
1952 Oxf. Junior Encycl. VI. 327/1 The farmer and gardener have a number of methods of control they can use against pests, apart from..spraying with..Fumigants.

fumigate

A. trans. To apply smoke or fumes to; esp. to disinfect or purify by exposure to smoke or fumes.
1781 W. Cowper Let. 18 Sept. (1979) I. 521 You never fumigate the Ladies, or force them out of company.

Table 1.1 (Continued)

B. To scent with fumes; to perfume.
1530 J. Palsgrave Lesclarcissement 559/2, I fumygate a place with a swete fumygacion, je enfume or je parfume. Let the place be well fumygate, or ever they come.

germicide

A. n. An agent that kills microorganisms, esp. bacteria (cf. germ n. 4); a microbicide; a disinfectant.
1870 Br. Med. Jrnl. 15 Oct. 407/2 These doubts arose from some cases of wounds in which the 'germs' had full access for some time, in which no germicide was used to the surface of the wound, and yet in which no pus was formed.

B. adj. (attrib.) = germicidal adj. Now rare.
1871 Med. Press & Circular 9 Aug. 127/1 It forms an exhaustive exposition of what has been said in praise of carbolic acid in its germicide aspect.

microbicide

adj. Now historical. That kills microbes; of or relating to a microbicide or its action. 1885 Br. Med. Jrnl. 2 1097/1 The sulphates of copper and zinc..have an effective microbicide power.

sanitizer

n. A substance which sanitizes: a disinfectant, or a preservative of food.
1950 Jrnl. Milk & Food Technol. 13 63/2 The most recent trend has been toward the use of the quaternary ammonium detergent sanitizer.

sterilize*

v. 5. To render free from micro-organisms.
1878 J. Tyndall Fragm. Sci. (1879) II. 297 Schwann..sterilised the flask by boiling.

Derivatives:

sterilized adj.
1877 J. Tyndall Ess. Floating Matter of Air (1881) 133 A sterilized infusion..remained sterile.

*Note: most of the definitions under this entry have to do with rendering infertile.

microorganisms, thereby directly attacking the cause of the disease, rather than merely treating the symptoms of the disease. His strategy involved using systemic chemical modifications to synthesize a large number of related compounds, followed by pharmacological screening to evaluate efficacy. Ehrlich's group applied this approach in an exhaustive fashion and isolated arsphenamine, a compound that was active against syphilis and trypanosomiasis. They followed the German practice of designating synthesized chemicals by a development number. Hence, it was called 'Compound 606' which Hoechst marketed under the name Salvarsan, 'the arsenic that saves' (Winau *et al.*, 2007; Bosch and Rosich, 2008). Arsphenamine is generally viewed as the first modern chemotherapeutic agent because it was discovered as a result of a rational screening programme. Although it had serious side-effects, it was more toxic to bacteria than to the host and was better than anything else available. Salvarsan was used widely and helped to inspire Paul Gerhard Domagk in his successful search for sulfa drugs (Hager, 2006). Ehrlich is generally credited for having introduced the concept of 'chemotherapy' whereby clinicians prescribe chemical agents that selectively target the organism causing disease while not harming, or only minimally harming, the host. He later applied the same theory to the study of tumour cells.

In summary, 'chemotherapy' refers to the use of a physiologically active chemical (the 'chemotherapeutic agent') to selectively treat the disease while not damaging the patient. In the broadest sense, chemotherapy refers to the use of any drug, including antibiotics like penicillin, to treat any disease. In contemporary use, 'chemotherapeutic agent' usually refers to a drug that is cytotoxic and that targets malignant tumours. Nowadays, when adopted for systemic cancer treatment, the process is called chemotherapy, often abbreviated as 'chemo' (DeVita and Chu, 2008).

Although the word 'chemotherapeutic agent' has a different meaning in contemporary times than the way it was originally used by Paul Ehrlich, the understanding that therapeutically acceptable drugs have to be selective is a major part of his legacy. It is curious that this legacy did not have greater impact on the first formal definitions of 'antibiotic.'

The long shadow of Selman Waksman and synthetic antibiotics

After the introduction of penicillin and streptomycin, other clinically useful antibiotics soon entered the market. Aureomycin was introduced by Lederle Laboratories in 1948; Chloromycetin (chloramphenicol) by Parke, Davis and Company in 1949; Terramycin by Pfizer in 1950; and Nystatin, the first anti-fungal antibiotic, was discovered at a public health laboratory in Albany, New York, and by 1951 E. R. Squibb was producing it for the treatment of yeast infections. In the middle of the twentieth century, the pharmaceutical industry grew rich and expanded on profits made from the treatment of infectious disease (Parascandola, 1980). Looking back, it has become a cliché to call it the Golden Age of Antibiotics. Millions and millions of lives were saved. Medical historians have written scores of books about the era, especially about the discovery of penicillin (Hare, 1970; Macfarlane, 1979; Hobby, 1985; Lax, 2004) and the controversies surrounding the discovery of streptomycin (Pringle, 2012).

During the Golden Age of Antibiotics, several major classes of synthetic compounds with powerful antibacterial activity were approved for clinical use. For example, the fluoroquinolones were discovered in 1949. In order to avoid calling them antibiotics, labels such as 'synthetic antibacterial agents' were used. Even now, many writers are careful to avoid use of the word 'antibiotic' when writing about quinolones, using an alternative like 'potent antibacterial chemotherapeutic agents.'

Over the decades, numerous people have asked the same question we are asking here: 'What is an antibiotic?' Because of the way in which Waksman had proscribed a definition for his substantive noun, and because so many non-microbial compounds with anti-microbial activity were discovered, the question has often been followed by other questions: 'Do antibiotics have to be microbial products?' 'Do antibiotics have to be natural products?' 'Can a synthetic molecule that is effective against a microbial infectious agent be called an antibiotic?'

Waksman, who clearly had understood the need for a single, useful word to describe these wonderful antimicrobial drugs, had promulgated – and then vigorously defended – a definition that specifically stipulated a microbial origin for the compounds he felt could legitimately be called 'antibiotics.' The sulfonamides, which had been introduced during the 1930s before penicillin and streptomycin, provided antibacterial activity with low toxicity in humans, but were excluded from his naming, as were Ehrlich's earlier chemotherapeutic agents for the treatment of syphilis. Waksman's insistence that the term 'antibiotic' be limited to products of microbial metabolism, excluding both synthetic drugs and active natural products from other sources, has generated decades of scientifically unproductive debate about whether or not a given antimicrobial compound can legitimately be called one.

For example, in 1990, the *Journal of Antibiotics* convened a six-person 'Antibiotic Definition Committee' which was charged with deciding what kinds of compounds could legitimately be called 'antibiotics.' The majority of the members of the committee felt that an antibiotic could be defined 'as a natural product produced by any type of cell, microbial, plant or animals' but one member felt that only products produced by microbes should be called antibiotics. In order to resolve the impasse, the Committee delivered a majority and a minority report (Demain, 2002, personal communication). Both the majority and minority report agreed that antibiotics were natural products (Bentley and Bennett, 2003).

Another example comes from an authoritative, encyclopaedic volume (over 1400 pages) of antimicrobial agents entitled *Antimicrobial Agents, Antibacterials and Antifungals* and published by the American Society for Microbiology in 2005. This book follows the Waksman tradition, as seen by the following quote from the section on 'Definition of antibiotics.'

> An antibiotic is a derivative produced by the metabolism of microorganism that possesses antibacterial activity at low concentrations and is not toxic to the host...The antibacterial agents obtained by total synthesis, such as the quinolones, benzylpyrimidines ... and penems, *are synthetic antibacterial agents and not antibiotics.*
>
> (Bryskier, 2005b, p. 4, author's emphasis)

In the very next sentence, Bryskier modifies this statement as follows:

> However, usage has meant that any substance possessing antibacterial activity that is not toxic to the host is referred to as an antibiotic
>
> (Bryskier, 2005b, p. 4)

Although Bryskier prefers not to use the word 'antibiotic,' to describe selective anti-bacterial agents, he clearly understands why so many other people do. Namely, most scientists, physicians, and lay people focus on what an antibiotic does, not where it comes from.

In particular, research on the penicillin family has blurred the difference between natural and synthetic compounds. Amoxicillin (α-amino-p-hydroxybenzylpenicillin), a semisynthetic material, was readily accepted as a genuine antibiotic because it was derived from a fermentation product. Aztreonam, the first wholly synthetic monocyclic β-lactam, was also accepted as an antibiotic. These compounds all contained the β-lactam structure, were strongly antibacterial, and had few negative side-effects in human or veterinary use (Sheehan, 1982). It seemed silly to distinguish them by whether they were synthesized by batch fermentation, laboratory bioconversion, or chemical synthesis.

With few exceptions, current meanings of 'antibiotic' are extremely inclusive when it comes to drug origin. Most people simply have expanded and re-defined 'antibiotic' to include therapeutically suitable antimicrobials whether they are synthetic, semisynthetic or natural non-microbial compounds. Here is the entry from the second edition of the Oxford English Dictionary:

> As adjective: injurious to or destructive of living matter, esp. microorganisms. As substantive: an antibiotic substance: one of a class of substances produced by living organisms and capable of destroying or inhibiting the growth of microorganisms, spec. any of those substances used for therapeutic purposes. Also used of synthetic organic compounds having similar properties.
>
> (OED, 1997)

The commodious definition has had a retroactive effect. Now it is fairly commonplace to read that Salvarsan, the sulfa drugs, or other active agents were 'the first antibiotics.' (see, for example, Levy, 2002; Hager, 2006). On the other hand, many contemporary writers are still careful to use other terms such as 'antimicrobial' in speaking about therapeutically useful pre-penicillin compounds (see, for example, Sepkotitz, 2011).

In summary, it is a testament to the force of Waksman's intellect, his international stature, his generous philanthropy, his love of microbiology, and perhaps his inchoate and unstated recognition that in the 1940s there was something very different about a drug manufactured using biotechnological methods, that so many decades after his death some scientists still feel it is important to limit the word 'antibiotic' to products of microbial metabolism, and even believe that Waksman's promulgated definition is the single 'right' definition.

Antibiotic resistance and the emergence of 'antibacterial' as a synonym for 'antibiotic'

In the early days of antibiotic research, the dream of the medical conquest of infectious diseases seemed about to come true. Nevertheless, from the very beginning of the antibiotic era, bacterial strains that were resistant to penicillin, streptomycin and other antibiotic substances were discovered in clinical practice. Physicians responded by increasing the dosage of antibiotic used to treat infections. Thus began a huge

Darwinian experiment whereby humans created situations in which only resistant strains could survive to produce offspring. The speed at which resistance to antibiotics emerged was increased by reckless and inappropriate use of these lifesaving drugs.

Now, in the twenty-first century, infectious disease experts are reporting increasing morbidity and mortality from bacteria that are resistant to all approved antibiotic options (Amyes, 2001). The emergence of bacterial resistance in turn has had a feedback on patterns of drug usage.

An aspect of the antibiotic revolution not foreseen by Waksman and the other early pioneers was that the vast majority of the clinically significant antibiotics that were isolated during the twentieth century were active only against bacteria. The antifungal, antiprotozoal and antiviral agents isolated generally had unacceptably high levels of negative side-effects; their lack of selective action made them too toxic for therapeutic use. Unfortunately, although clinically efficacious antibiotics were largely restricted to drugs against bacterial pathogens, many physicians succumbed to patient requests for antibiotics even when they were ineffective (Bud, 2007).

There is increasing awareness that physicians should do a better job in the management of patient expectations and deny patient requests for antibiotic treatment when it is not warranted, especially against viral conditions such as the common cold and the flu. Health professionals now commonly advise patients that antibiotics are effective only against bacteria. As a result, there is an increasing tendency to use the word 'antibiotic' as a synonym for 'antibacterial agent.' Moreover, since the discovery of effective agents for the treatment of human immunodeficiency virus, the trend often has been to come up with new labels such as 'antiviral antibiotic,' 'antiviral agent' or simply 'antiviral' to distinguish them from the broad class of effective antibacterial drugs. Cognate terms are in use for other classes of antimicrobial agents, for example, 'antifungal antibiotics' (also called 'antifungals' or 'antimycotics') and 'antiprotozoal agents'. Some experts have taken to use the term 'antimicrobial agent' as the new umbrella to cover antibacterial, antiviral, antifungal, antiprotozoal and antimycotic agents, thereby trying to turn 'antimicrobial' into the usage that Waksman had originally claimed for 'antibiotic.' No less an authority than the World Health Organization has gone this route and stated, 'the term antibiotic is used as a synonym for antibacterials used to treat bacterial infections in both people and animals' (WHO 2011) and has defined 'antimicrobial' as 'any substance of natural, synthetic or semisynthetic origin which at low concentrations kills or inhibits the growth of micro-organisms but causes little or no host damage' (WHO 2000).

Communication of health risks is necessarily reductive. It is understandable that the public health community has chosen to emphasize that most of the anti-microbial therapeutic drugs are active only against bacteria. This semantic shift makes it easier for physicians to deny patients when they request an unsuitable medication to treat a viral infection. Nevertheless, the terms 'antimicrobial' and 'anti-infective' are not necessarily the best of substitutes for Waksman's portmanteau word.

This point has been made by a blogger named Collin Purrington who has a number of witty and opinionated posts. He has a personal 'quest to show that 'antibiotic' is a word that does more harm than good to public health outreach.' In his blogs, Purrington regularly points out that when people ask for an antibiotic to treat a viral infection it is because they believe that these drugs will help them. Or as he puts it: 'People believe that antibiotics treat viruses because on Earth we regularly use a word's roots to infer meaning' (Purrington, 2013a) and that 'When an average person hears "antibiotics," they envision a wonder drug that can kill viruses in addition to bacteria' (Purrington, 2013b).

Unlike the World Health Organization, the American Medical Association seems to understand the linguistic conundrum. The following recommendation was published in 2000. The alternate use of 'antibiotic' and 'antimicrobial' is highlighted by using bold.

> The AMA: (a) Encourages the federal government, the World Health Organization, the World Medical Association, and the International Federation of Pharmacists to promote more effective education

concerning the appropriate use of **antibiotics**, (b) strongly urges physicians to educate their patients about their **antimicrobial therapy**, the importance of compliance with the prescribed regimen, and the problem of **antimicrobial resistance**; (c) will continue to educate physicians and physicians-in-training about the appropriate prescribing of **antimicrobial agents**; (d) encourages the use of **antibiotic resistance management programmes**; these education-based programmes should be multidisciplinary and cooperative (i.e. including infectious disease physicians, infection-control specialists, microbiology laboratory personnel, and clinical pharmacists); and (e) encourages continued scientific research on the issue of **antibiotic resistance**.

(http://www.ama-assn.org/ama1/pub/upload/mm/443/csaa-00.pdf)

In summary, many medical professionals have narrowed the definition of 'antibiotic' to become synonymous with 'antibacterial'. Simultaneously, it has become commonplace to use 'antimicrobial agent' as the broad rubric to subsume antifungal and antiviral antibiotics. Unfortunately, 'antimicrobial agent' is a very commodious term. For example, the Wikipedia entry for 'antimicrobial' opens with the following sentence: 'An "antimicrobial" or "antibiotic" is an agent that kills microorganisms or inhibits their growth.' The article then goes on to present information not only on drugs considered as traditional antibiotics, but also on heavy metals, certain acids, essential oils and other biocides, as well as disinfectant treatments like heat, ozone and radiation (http://en.wikipedia.org/wiki/Antimicrobial).

Selective toxicity, lexicography and the evolving definition of antibiotic

It is ironic that a word that literally means 'against life' has come to be synonymous with the saving of life. The incongruity stems from what was underemphasized in Selman Waksman's original definition, namely differential toxicity.

His definition focused on the killing action of antibiotics against microbes and gave little attention to the selective aspects of the killing. We have seen how Waksman's definition interpreted the world of biologically active natural products through his professional lens as a soil microbiologist and how his proscribed definition stressed the importance of a *microbial* origin of these drugs. In the semantic debates that continue even to this day, the notion of the microbe as the target for the action of the drug is usually stressed, but mention of discriminatory toxicity is often lacking. In contrast, Ehrlich had focused on finding a compound that would impede the microbe with minimal damage to healthy body tissue, and clearly understood the need for finding a way to combat infectious diseases by designing selective drugs. It is unfortunate that the intellectual legacies of Waksman and Ehrlich have not come closer together.

To repeat: Waksman's original definition focused on the producing organism (microbes) and the target organism (again microbes): *An antibiotic is a chemical substance, produced by micro-organisms, which has the capacity to inhibit the growth of and even to destroy bacteria and other micro-organisms.* The second sentence of his comprehensive 1947 definition focused on the spectrum of antimicrobial action: *The action of an antibiotic against micro-organisms is selective in nature, some organisms being affected and others not at all or only to a limited degree; each antibiotic is thus characterized by a specific antimicrobial spectrum.* Only in the third sentence did he address the fact that a clinically useful antibiotic has to function without killing host tissues: *The selective action of an antibiotic is also manifested against microbial vs. host cells* (Waksman, 1947).

A definition is merely a statement that explains the meaning of a given term. The term itself may have multiple meanings and a good dictionary will list all of these definitions. Lexicography refers to the practices and principles of dictionary making. The vast majority of the definitions of 'antibiotic' that have been published in dictionaries, encyclopaedias, textbooks and monographs follow the Waksman tradition, usually paraphrasing the first part of proscribed definition, emphasizing the antimicrobial activity of antibiotics. For example, the *Academic Press Dictionary of*

Science and Technology defines 'antibiotic' as: 'A chemical substance that is important in the treatment of infectious diseases produced either by a micro-organism or semisynthetically, having the capacity in dilute solution to either kill or inhibit the growth of certain other harmful microorganisms' (Morris, 1992). Interestingly, the same dictionary alludes to selective toxicity in its definition of 'antimicrobial:' 'Any substance that either destroys or inhibits the growth of a micro-organism at concentrations tolerated by the infected host' (Morris, 1992).

A representative, but far from exhaustive list of definitions of 'antibiotic' is given in Table 1.2.

Semantic ambiguity is a reality across language, including the language of science. Words and their definitions change over time, in a kind of etymological evolution. That said, most scientists do not give a lot of thought to the origin of our technical words. Moreover, not everyone recognizes that the definitions given in dictionaries are descriptive, not proscriptive, i.e. the definitions appearing in dictionaries reflect the way in which a word is actually being used rather than the way someone thinks it should be used. Waksman's proscribed, big umbrella definition of 'antibiotic' covered compounds that showed activity against algae, bacteria, fungi, protozoa, and viruses, and encompassed many compounds now considered far too noxious to be labelled as antibiotics. As we have seen, during the Golden Age of Antibiotics almost all of the clinically significant antibiotics that were discovered were active only against bacteria. Antifungal, antiprotozoal and antiviral agents were isolated, but with few exceptions, were too toxic for therapeutic use. As time went on, some of the compounds originally called 'antibiotics' in the scientific literature were re-categorized. A good case in point is patulin, a fungal metabolite heralded during the 1940s as an antiviral agent capable of providing relief from the common cold. This fungal

Table 1.2 Selected published definitions of antibiotics

An antibiotic or antibiotic substance is a substance produced by microorganisms, which has the capacity of inhibiting the growth and even of destroying other microorganisms (Waksman, 1949)
Substances with diverse chemical structures and biological activities. They range in their action from those which inhibit the growth of certain strains of bacteria in a highly selective manner to those which are relatively toxic to all living cells (Abraham, 1949)
A bacteriocidal or bacteriostatic substance produced by certain microorganisms (King, 1968)
Substances produced by some microorganisms that destroy or arrest the growth of others (Medawar and Medawar, 1983)
A natural substance of relatively low molecular weight, produced by a microorganisms, which in dilute solutions inhibits growth or destroys other organisms. Toxicity is generally selective. Most natural antibiotics, whose structures vary widely, are derived from the Streptomyces, an exception being penicillin (Combs, 1992)
Substances isolated from microorganisms, especially moulds, that destroy or inhibit the growth of other microorganisms, particularly disease-producing bacteria and fungi (Daintith, 1996)
Antibiotics are a collection of natural products and synthetic compounds that kill bacteria (Creighton *et al.*, 1999)
Any of numerous substances of relatively low M_r produced by living microorganisms (and also certain plants) that are able selectively and at low concentrations to destroy or inhibit the growth of other organisms, especially microorganisms. Also included are the many semi- or wholly synthetic organic compounds with similar antimicrobial properties. Many are useful chemotherapeutic agents (Smith *et al.*, 2000)
Antibiotics are drugs of natural or synthetic origin that have the capacity to kill or to inhibit the growth of microorganisms (FAO, 2005)
A drug that is used to kill harmful bacteria and to cure infections' and 'a substance produced by or a semisynthetic substance derived from a microorganism and able in dilute solution to inhibit or kill another microorganism. http://www.merriam-webster.com/dictionary/antibiotic (accessed 2 February 2014)
A substance, such as penicillin or streptomycin, produced by or derived from certain fungi, bacteria, and other organisms, that can destroy or inhibit the growth of other microorganisms. Antibiotics are widely used in the prevention and treatment of infectious diseases. http://www.thefreedictionary.com/antibiotic (accessed 2 February 2014)

secondary metabolite also had anti-fungal and anti-cancer properties. As research data accumulated, however, it became apparent that its strong anti-microbial and anti-neoplastic activities were accompanied by severe adverse health effects. It is a compound that is broadly 'against life.' Now patulin is firmly classified as a mycotoxin (Ciegler *et al.*, 1971; Wainwright, 1990; Bennett and Klich, 2003).

Certain other compounds originally called antibiotics had another fate. Actinomycin A was isolated in Waksman's laboratory before streptomycin and was powerfully antimicrobial. However, it was also toxic and caused an unacceptably high level of adverse side-effects such as diarrhoea, fatigue, hair loss, and mouth ulcers and could not be used as an antibiotic drug. Later, however, it was shown to have anti-tumour activity and re-emerged for medical use as a chemotherapeutic anti-cancer agent, while actinomycin D became a widely used laboratory tool in molecular biology (Waksman and Woodruff, 1941; Tan *et al.*, 1959; Hollrich, 1974).

It will be interesting to see what crowd sourcing will do to lexicography. Such bottom-up dictionary writing is alive and well at urbandictionary.com, where terms such as 'virus' and 'infect' are defined entirely in terms of their relationship to computers and hackers, and where the entire entry under 'antibiotics' says only 'see cocaine.'

Why worry about definitions?

Semiotic theory highlights the relationship between a term and what the term describes. Each individual has to negotiate the verbal message. Dictionaries and glossaries provide guidance in stabilizing meaning. They are 'anchors for definition' (Garrity, quoted in a personal communication 2014).

The scientists who develop, refine, and study the compounds we call 'antibiotics' generally do not give much thought to lexicography. The novelists, journalists, bloggers and other communicators who write about antibiotics are usually from outside of the healthcare professions and rarely worry about the formal definition of 'antibiotic.' While an antibiotic might be a clinical tool for a physician, it can also be a growth-promoting agent for a chicken farmer, a miracle medicine for an impoverished mother with a sick child in Africa, an additive to a cosmetic product for a teenager with acne, and an intellectual challenge for a synthetic chemist. For those physicians attempting to curb the inappropriate use of antibiotics, the most important thing is to draw patient's attention to the fact that the vast majority of clinically effective antimicrobials are useful only against bacteria, hence the synonymy with 'antibacterial.'

Apart from the need for public communication, defining the word 'antibiotic' is not merely a semantic exercise because new research on antibiotic discovery is urgently needed. During the twentieth century, many physicians, pharmaceutical companies and public health workers became complacent about the threat of infectious disease, choosing to focus on other killers like cancer, cardiovascular disease and neurological conditions such as Alzheimer's and Parkinson's disease. There has been less emphasis on hygiene as a defence against infectious disease when antibiotics seem to do the trick. However, with each passing year, more and more cases of untreatable bacterial infections occur. In an attempt to draw attention to the growing heath threat posed by antibiotic-resistant bacteria, the Infectious Disease Society of American (IDSA) has been publicizing the so-called 'ESKAPE pathogens, *Enterococcus faecium*, *Staphylococcus aureus*, *Klebsiella pneumoniae*, *Acinetobacter baumannii*, *Pseudomonas aeruginosa* and *Enterobacter* spp., that are posing new challenges in bacterial pathogenesis, resistance and transmission (Boucher *et al.*, 2007; Pendelton *et al.*, 2013). IDSA tracks the status of antimicrobial drugs and has set a goal 'to develop ten new, safe, and efficacious systemically administered antibiotics' in the near future. In particular, new agents are needed to treat infections caused by Gram-negative bacilli resistant to currently available antibiotics. They warn that progress remains 'alarmingly elusive' (Boucher *et al.*, 2014). It will behove the scientific community to be more precise in the way we talk about anti-infectious drugs, the nature of antibiotic resistance, and the responsible stewardship of the drugs we have. As antibiotic resistance becomes an ever-growing threat to public health, we need a new focus in order to find novel antibiotics that function with hitherto undiscovered or underutilized modes of action (Spellberg *et al.*, 2013).

One possible resource is the knowledge embedded in scientific research that has already been conducted. Thousands of compounds were isolated, screened and discarded during the Golden Age of Antibiotics. There is a high probability that useful compounds already have been isolated and exist in the vast natural products libraries that were amassed during the twentieth century. Contemporary information technology is developing better and better tools to mine language. Such searches of the published scientific literature, and of US and international patents, require that words be defined carefully. Yet, as we have seen, 'antibiotic' has multiple, albeit related, meanings.

Our ability to mine data banks, patents, and the scientific literature depends on having stable labels that can link concepts with compounds. As algorithms are developed for mining texts, there will be an increasing reliance on the verbal taxonomies that have been developed to classify antibiotics. These include their mechanism of action, chemical structures, mode of production (fermentation, synthetic, or semisynthetic) or spectrum of activity. Some antibiotics work on bacterial cell walls, others inhibit bacterial protein synthesis. Antibiotics from natural sources are often subclassified according to the group of producing organisms (fungal-derived, actinomycete-derived). They can be classified by mode of administration: oral, intravenous, and topical. They also can be classified by target organism: antibacterial (Gram-positive, Gram-negative, broad spectrum), antiviral, antifungal, antiprotozoal and so forth. So-called narrow-spectrum antibiotics target specific types of bacteria, such as Gram-negatives, whereas broad-spectrum antibiotics affect a wider range. There is an enormous literature; see, for example, Florey *et al.* (1949), Gale *et al.* (1972), Walsh (2003) and Bryskier (2005a).

It is impossible to resolve all of the nomenclatural ambiguity between drugs and the names we use to describe them. Nevertheless, it makes sense to develop linguistic databases that would allow us to exploit the vast amount of published and patented knowledge about antimicrobial drugs. Doing so will require that information technologists develop more precise ontologies based on stipulated definitions of the term 'antibiotic,' and on the many ways in which antibiotics, their actions, and their side-effects have been classified. Such ontologies will have to take into account the fact that some people use 'antibiotic' as a synonym for 'antimicrobial' while many others, especially within the public health community, use it as a synonym for 'antibacterial'. Similarly, we have to recognize that sometimes 'antimicrobial' is used broadly in the sense of Waksman's original definition of 'antibiotic' or even more broadly to include disinfectants and antiseptics. All good definitions attempt to cover the criteria that are essential to the designation of things as members of the relevant group. For Waksman, the important features of antibiotics were its producing species (microbes) and its target species (microbes). For Paul Ehrlich, the important features of a chemotherapeutic agent were its selective action. For our future linguistic searches, we need to devise search tools that bring both philological-scientific traditions together.

Summary

Words that end in '-ic' tend to be adjectives in the English language, and before Waksman the word 'antibiotic' was used as an adjective to modify a noun such as 'activity' or 'agent.' Waksman proposed a proscriptive definition for antibiotic as a generic noun to describe anti-infective agents such as penicillin and streptomycin. During his lifetime, he worked hard to ensure that microbiologists, chemists and healthcare professionals used his definition, i.e. antibiotics were antimicrobial agents made by microbes. Waksman's definition excluded non-microbial substances that kill microbes such as germicides (e.g. gastric juices) and it also excluded chemically synthesized drugs such as the sulfonamides. However, as time went on, scientists found it useful to expand the definition of 'antibiotic' to encompass synthetic and semisynthetic substances, as well as substances produced by plants and animals, as long at these compounds were active against the microbes that caused disease. Waksman's 'big umbrella' concept, while emphasizing broad anti-microbial activity, downplayed the importance of selective toxicity. An agent that does not discriminate between killing a microbial cell and a human cell is no longer considered an 'antibiotic.' We usually call that kind of compound a 'toxin' or a 'poison.'

Contemporary usage has expanded the possible sources of antibiotic substances, while often limiting 'antibiotic' to antibacterial substances. Nowadays, the broader terms 'antimicrobial' or 'anti-infective' are used more in the way Waksman originally defined 'antibiotic' to include antiviral and antifungal drugs. Nonetheless, there is no universal agreement on the definition of 'antibiotic,' or 'antimicrobial' and not everyone limits use of 'antibiotic' to antibacterial agents. In general, scientists try to make a serious effort to agree on definitions and to use carefully stipulated language. Therefore, it is now common to find specific modifiers to describe drug action, e.g. 'antibacterial drug,' 'antiviral compound' or 'antifungal antibiotic.'

As bacterial, fungal and viral drug resistance become an increasing problem in modern medicine, and as public health workers try to communicate the importance of not overusing the drugs we have, the precise definition of antibiotic becomes more than an academic exercise. It is important to be able to use simple words in warning patients of inappropriate drug use so it is understandable that medical professionals often use 'antibiotic' as a synonym for 'antibacterial' in order to simplify medical communication. Nevertheless, as more and more marketers, bloggers and others use the word 'antibiotic' without paying much, if any, attention to its scientific denotation, it becomes obvious that Waksman's attempts to proscribe a single, universal definition were simultaneously prescient and futile. Language, including scientific language, changes. Nowadays, when used without a modifier, 'antibiotic' usually – but not always – means 'antibacterial,' comfortably encompasses synthetic compounds, and is often confused with disinfectants, germicides and antiseptics in the public mind. To grasp fully the many ways in which 'antibiotic' has been and is being used requires a broad understanding of microbiology, clinical medicine, natural products chemistry, and public health. It does not hurt to know something of the history of pharmacology and to have a sense that words and their definitions are not fixed in time.

Why is it that we still need to review the definition of 'antibiotic?' Perhaps it is because scientists like facts, hard data and reproducible experiments. As a result, some scientists may have come to think that it is a 'fact' that antibiotics are antimicrobial products produced by microbes, and that there is a single, 'true' definition of antibiotic. While it is a fact that nature and synthetic chemists have made many compounds with antimicrobial activity that can be used in medical therapy, sometimes we need to be reminded that the way we collectively label these drugs – the way in which we construct a definition of 'antibiotic' – is not the same thing as a piece of factual information. Rather, definitions are a human linguistic convenience that can and do change over time. We are faced with a paradox: although words with more or less stable definitions are essential to communication, outside mathematics, it is impossible to stabilize meaning. Thus, while information technologists need carefully constructed definitions of words in order to mine the information embedded in the published scientific and patent literature, context, nuance and human interpretations always will be required.

In conclusion, 'antibiotic' is a word with multiple meanings and connotations. Those who seek to formalize a single, unwavering definition will not succeed. Therefore, it is appropriate to end this essay with two definitions of lexicographer. The first is from Dr Samuel Johnson, who in putting together the first major dictionary of the English language included this entry: '**Lexicographer**: A writer of dictionaries; a harmless drudge that busies himself in tracing the original, and detailing the signification of words' (Johnson, 1755; Hitchings, 2006).

The second is an irreverent quote from Ambrose Bierce who defined a lexicographer as 'A pestilent fellow who, under the pretence of recording some particular stage in the development of a language, does what he can to arrest its growth, stiffen its flexibility and mechanize its methods' (Bierce, 1911).

Acknowledgements

I am indebted to the late Ronald Bentley (University of Pittsburgh), whose scholarship and wide grasp of the literature surrounding antibiotics inspired my interest in understanding the etymology of the term. George Garrity (Michigan State University) provided insights into the use of contemporary information technology to maximize our ability to access research data that are embedded in words. Shannon Morath (Rutgers University) and Amy Vollmer (Swarthmore College) reviewed the manuscript. Over the years, many colleagues at the American Society for Microbiology, the Society for Industrial Microbiology and Biotechnology and the International Union of Microbiological Sciences have provided stimulating discussions about the nature and economics of antibiotics. Among them, I give special thanks to Arny Demain (Drew University) and Prakash Masurekar (Rutgers University). Finally, I thank Rutgers University for providing an administrative leave that has given me time to ponder the multiple and changing definitions of 'antibiotic.'

References

Abraham, E. (1949). The action of antibiotics on bacteria. In Antibiotics. A Survey of Penicillin, Streptomycin, and Other Antimicrobial Substances from Fungi, Actinomycetes, Bacteria and Plants, W.H. Florey, E. Chain, N.G. Heatley, M.A. Jennings, A.G. Sanders, E.P. Abraham, and M.E. Florey, eds. (London, UK: Oxford University Press), Vol. II, p. 1438–1496.

Aiello, A.E., Larson, E.L., and Levy, S.B. (2007). Consumer antibacterial soaps: effective or just risky? Clin. Infect. Dis. *45*, 137–147.

Alexopoulos, J., and Herrick, J.A. (1942). Studies in antibiosis between bacteria and fungi-III. Inhibitory activity of some Actinomycetes on various species of fungi in culture. Bull. Torrey Botan. Club. *69*, 257–261.

AMA (2000). Use of Antimicrobials in Consumer Products. Report 2 of the Council on Scientific Affairs (American Medical Association, A-00).

Amyes, S.G.B. (2001). Magic Bullets, Lost Horizons: the Rise and Fall of Antibiotics (Taylor & Francis, New York, NY).

APHA (2009). B4: Proposed resolution to ban non-medical uses of triclosan (submitted 17 February 2009) (American Public Health Association). Available online: http://www.apha.org/NR/rdonlyres/1129B888-4088-4D58-9AFA-25FC9E1A15A3/0/B42_17Resolution2.pdf (accessed 14 July 2014).

Ascenzi. J.M. (1996). Handbook of Disinfectants and Antiseptics (Marcel Dekker, New York, NY).

Bankston, J. (2004). Joseph Lister and the Story of Antiseptics (Uncharted, Unexplored, and Unexplained) (Mitchell Lane Publishers, Bear, DE).

Bennett, J.W., and Klich, M. (2003). Mycotoxins. Clin. Microbiol. Rev. *16*, 497–516.

Bentley, R., and Bennett, J.W. (2003). What is an antibiotic? – revisited. Adv. Appl. Microbiol. *52*, 303–331.

Bierce, A. (1911). The Devil's Dictionary. Published as an EBook by The Project Gutenberg. Release Date: 26 July 2008 [EBook #972] http://www.gutenberg.org/files/972/972-h/972-h.htm#link2H_4_0014

Block, S.S. (2001). Disinfection, Sterilization, and Preservation. (5th edn) (Lippincott Williams & Wilkins, Philadelphia, PA).

Bosch, F., and Rosich, L. (2008). The contributions of Paul Ehrlich to pharmacology: a tribute on the occasion of the centenary of his Nobel Prize. Pharmacology *82*, 171–179.

Boucher, H.W., Talbot, G.H., Bradley, J.S., Edwards, J.E., Gilbert, D., Rice, L.B., Scheld, M., Spellberg, B., and Bartlett, J. (2009). Bad bugs, no drugs: no ESKAPE! An update from the Infectious Diseases Society of America. Clin. Infect. Dis. *48*, 1–12.

Boucher, H.W., Talbot, G.H., Benjamin Jr, D.K., Bradley, J., Guidos, R.J., Jones, R.N., Murray, B.E., Bonomo, R.A., and Gilbert, D. (2013). Infectious Diseases Society of America. '10 × '20 Progress – development of new drugs active against Gram-negative Bacilli: An update from the Infectious Diseases Society of America'. Clin. Infect. Dis. *56*, 1685–1694.

Bryskier, A. (2005a). Antimicrobial Agents: Antibacterials and Antifungals (ASM Press, Washington, DC).

Bryskier, A. (2005b). Historical reviews of antibacterial chemotherapy. In Antimicrobial Agents: Antibacterials and Antifungals, A. Brysker, ed. (ASM Press, Washington, DC), pp. 1–12.

Bud, R. (2007). Penicillin. Triumph and Tragedy (Oxford, UK: Oxford University Press).

Ciegler, A., Detroy, R.W., and Lillehoj, E.B. (1971). Patulin, penicillic acid, and other carcinogenic lactones. In Microbial Toxins, A. Ciegler, S. Kadis, and S. Ajl, eds. (Academic Press, New York, NY), Vol. VI, pp. 409–434.

Chuanchuen, R., Beinlich, K., Hoang, T.T., Becher, A., Karkhoff-Schweizer, R.R., and Schweizer, H.P. (2001). Cross-resistance between triclosan and antibiotics in *Pseudomonas aeruginosa* is mediated by multidrug efflux pumps: exposure of a susceptible mutant strain to triclosan selects *nfx*B mutants overexpressing MexCD-OprJ. Antimicrob. Agents Chemother. *45*, 428–432.

Combs, J. (1992). Dictionary of Biotechnology 2nd edn (The Macmillan Press, Ltd., London, UK).

Cohen, D.L., and Segelman, A.D. (1981). Antibiotics in Historical Perspective (Merck, Sharp and Dohme, Rahway, NJ).

Creighton, T.E., Kerr, K.M., Sauna, Z.E., and Ambudkar, S.V. (1999). Encyclopedia of Molecular Biology (John Wiley & Sons, New York, NY).

Daintith, J. (ed.) (1996). A Dictionary of Chemistry 3rd. edn (Oxford University Press, Oxford, UK).
DeVita Jr., V.T., and Chu, E. (2008). A history of cancer chemotherapy. Cancer Res. *68*, 8643–8653.
Florey, H.W., Chain, E., Heatley, N.G., Jennings, M.A., Sanders, A.G., Abraham, E.P., and Florey, M.E. (1949). The antibiotics. (Oxford University Press, London, UK).
Food and Agriculture Organization of the United Nations (FAO) (2005). Technical Paper 469. Responsible use of antibiotics in aquaculture. http://ftp.fao.org/docrep/fao/009/a0282e/a0282e00.pdf
Forrest, R.D. (1982). The early history of wound treatment. J. R. Soc. Med. *75*, 198–205.
Gale, E.F., Cundliffe, E., Reynolds, P.E., Richmond, M.H., and Waring, M.J. (1972). The Molecular Basis of Antibiotic Action (Wiley, New York, NY).
Gibaud, S., and Jaouen, G. (2010). Arsenic-based drugs: from Fowler's solution to modern anticancer chemotherapy. Top. Organomet. Chem. *32*, 1–20.
Hager, T. (2006). The Demon Under the Microscope: From Battlefield Hospitals to Nazi Labs, One Doctor's Heroic Search for the World's First Miracle Drug (Harmony Books, New York, NY).
Hare, R. (1970). The Birth of Penicillin and the Disarming of the Microbes (George Allen & Unwin, London, UK).
Hitchings, H. (2005). Defining the World. The Extraordinary Story of Dr. Johnson's Dictionary (George Allen & Unwin, London, UK)
Hobby, G. (1985). Penicillin: Meeting the Challenge (Yale University Press, New Haven, CT).
Hollstein, U. (1974). Actinomycin, chemistry and mechanism of action. Chem. Rev. *74*, 625–652.
Johnson, S. (1755). Dictionary of the English Language. http://www.samueljohnson.com/definitions.html
Kaufmann, S.H. (2008). Paul Ehrlich: founder of chemotherapy. Nat. Rev. Drug. Discovery *7*, 373.
King, R.C. (1968). A Dictionary of Genetics (Oxford University Press, New York, NY).
Krantz, J.C. Jr. (1974). Paul Ehrlich and 'The Magic Bullet', Arsphenamine. Historical Medical Classics Involving New Drugs (Williams & Wilkins, Baltimore, MD).
Lax, E. (2004). The Mold in Dr. Florey's Coat. The Story of the Penicillin Miracle (Henry Holt and Company, New York, NY).
Levy, J. (2011). Poison. An Illustrated History (Lyons Press, Guilford, CT).
Levy, S.B. (1992). The Antibiotic Paradox: How Miracle Drugs are Destroying the Miracle (Plenum Press, New York, NY).
Levy, S.B. (2001). Antibacterial household products: cause for concern. Emerg. Infectious Dis. 7 (Suppl 3), 512–515.
McDonnell, G.E. (2007). Antisepsis, Disinfection, and Sterilization. Types, Action, and Resistance (ASM Press, Washington, DC).
McDonough, J.T. (1994). Stedman's Concise Medical Dictionary (Williams & Wilkins, Baltimore, MD).
Macfarlane, G. (1979). Howard Florey. The Making of a Great Scientist (Oxford University Press, Oxford, UK).
Mann, J. (1999). The Elusive Magic Bullet. The Search for the Perfect Drug (Oxford University Press, New York, NY).
Medawar, P.B., and Medawar, J.S. (1983). Aristotle to Zoos: A Philosophical Dictionary of Biology (Harvard University Press, Cambridge, MA).
Murray, K.M.E. (1977). Caught in the Web of Words: James Murray and the Oxford English Dictionary (Yale University Press, New Haven, CT).
Norris, S., and Mandell, G.L. (1988). The quinolones: history and overview. In The Quinolones, V.T. Andriole, ed. (Academic Press Inc., San Diego, CA), pp. 1–22.
OED (1933). The Oxford English Dictionary (Oxford University Press, Oxford, UK).
OED (2014). The Oxford English Dictionary. The Definitive Record of the English Language (accessed Feb/2014)/http://www.oed.com/
Papacostas, G., and Gaté, J. (1928). Les Associations Microbiennes. Leurs Applications Thérapeutiques (Librairie Octave Doin, Paris, France).
Parascandola, J. (1980). The History of Antibiotics: A Symposium. (American Institute of the History of Pharmacy, Madison, WI).
Pendleton, J.N., Gorman, S.P., and Gilmore, B.F. (2013). Clinical relevance of the ESKAPE pathogens. Expert Rev. Anti Infect. Ther. *11*, 297–308.
Pennington, T.H. (1995). Listerism, its decline and its persistence: the introduction of aseptic surgical techniques in three British teaching hospitals, 1890–99. Med. Hist. *39*, 35–60.
Pramer, D. (1988). Some choice words from Waksman. The Scientist. April 4, p. 17.
Pringle, Peter. (2012). Experiment Eleven: Dark Secrets Behind the Discovery of a Wonder Drug (Walker & Co., New York, NY).
Purrington, C. (2013a). Antibiotics are antiviral. Available at: http://colinpurrington.com/2013/evidence-based-antibiotic-usage/(accessed 17 February 2014).
Purrington, C.(2013b). Evidence-based antibiotic usage. Available at: http://colinpurrington.com/2013/antibiotics-antiviral/ (accessed 17 February 2014).
Richard, E., Reese, Robert, F.B., and Bora, G. (2000). Handbook of Antibiotic (Lippincott Williams & Wilkins, Philadelphia, PA).
Russell, A.D., Hugo, W.B., and Auliffe, G.A.J. (2011). Principles and Practice of Disinfection, Preservation and Sterilization 2nd ed. (Blackwell Science, Cambridge, MA).
Sepkowitz, K.A. (2011). One hundred years of Salvarsan. N. Engl. J. Med. *365*, 291 –293.
Sheehan, J.C. (1982). The Enchanted Ring. The Untold Story of Penicillin (The MIT Press, Cambridge, MA).
Simpson, J.A., and Weiner. E.S.C. (1997). The Oxford English Dictionary 2nd edn (Clarendon Press/Oxford University Press, Oxford, UK).
Smith, A.D., Datta, S.P., Smith, G.H., Campbell, P.N., Bentley, R., and McKenzie, H.A. (2000). Oxford Dictionary of Biochemistry and Molecular Biology Revised Edn (Oxford University Press, Oxford, UK).

Spellberg, B, Barlett, J.G., and Gilbert, D.N. (2013). The future of antibiotics and resistance. N. Engl. J. Med. *368*, 299–302.
Tan, C.T.C., Dargeon, H.W., and Burchenal, J.H. (1959). The effect of actinomycin D on cancer in childhood. Pediatrics *24*, 544–561.
Vonkennel, J., Kimmig, J., and Lembke, A. (1943). Die mycoine, eine neue gruppe therapeutisch wirksamer substanzen aus pilzen. Klin. Wochenschrift *22*, 321.
Vuillemin, P. (1890). Antibiose et symbiose. Association Française pour L'avancement des Sciences. Compte Rendu de la 18ME Session, Seconde Partie, Notes et Mémoires *11*, 525–543.
Wainwright, M. (1990). Miracle Cure. The Story of Penicillin and the Golden Age of Antibiotics (Blackwell, Oxford, UK).
Waksman, S.A. (1947). What is an antibiotic or an antibiotic substance? Mycologia *39*, 565–569.
Waksman, S.A. (1951). Antibiotics – a new field of science of life-saving drugs. Antibiot. Chemother. *1*, 1.
Waksman, S.A. (1956). The role of antibiotics in natural processes. Giorn. Microbiol. *2*, 1–14.
Waksman, S.A. (1975). The Antibiotic Era. (The Waksman Foundation of Japan, Inc. Tokyo, Japan)
Waksman, S.A., and Woodruff, H.B. (1941). *Actinomyces antibioticus,* a new soil organism antagonistic to soil organism antagonistic to pathogenic and nonpathogenic bacteria. J. Bacteriol. *42*, 231–234.
Walsh, C. (2003). Antibiotics: Actions, Origins, Resistance (ASM Press, Washington, DC).
WHO (2000). Global principals for the containment of antimicrobial resistance in animals intended for food (World Health Organization). Available online: http://whalibdoc.who.int/ha/2000/who_cds_csr_aph-2000–4.pdf
WHO (2011). Tackling antibiotic resistance from a food safety perspective in Europe (World Health Organization Regional Office for Europe, Copenhagen, Denmark).
Wilson, D. (1976). In Search of Penicillin (Alfred A. Knopf Publishers, New York, NY).
Winau, F., Westphal, O., and Winau, R. (2004). Paul Ehrlich – In search of the magic bullet. Microbes Infect. *6*, 786–789.
Winchester, S. (2003). The Meaning of Everything: The Story of the Oxford English Dictionary (Oxford Univ. Press, Oxford, UK).

Main Applications of Antibiotics

2

Biao Ren, Pei Huang, Jingyu Zhang, Wenni He, Jianying Han, Xueting Liu and Lixin Zhang

Abstract

Antibiotics have been developed for more than 80 years since the investigation of the sulfonamides and β-lactams in 1930s. Meanwhile, the pathogenic bacteria developed resistance to antibiotics rapidly and have caused urgent threat to public health. In addition, many antibiotics are used commercially, or are potentially useful, in medicine for activities other than their antimicrobial action. They are used as anti-tumour agents, enzyme inhibitors including powerful hypocholesterolemic agents, immunosuppressive agents, and anti-migraine agents, etc. The purpose of this chapter is to focus on the application of antibacterials, antifungals, and anti-cancers with their clinical use to date, including the development history, side-effects, etc. The antibiotics summarized herein were classified by their uses, structure types, and molecular mechanisms.

Introduction

Antibiotics are small molecules used for treatment or prevention of bacterial infections. Some antibiotics may be used to treat cancer and other infections like protozoa infections as well. Since the discovery of the sulfonamides and β-lactams in the 1930s, antibiotics have contributed more than 80 years to medicine and agriculture (Butler *et al.*, 2013; Demain and Sanchez, 2009). The majority of antibiotic drug class in use today was discovered in the 'golden era' of antibiotic research from the 1930s to the 1970s. However, since 1970, most of the newly approved antibiotics were originated from the known drugs, except that the discovery of five new classes of antibiotics launched: linezolid, daptomycin, retapamulin, fidaxomicin, and bedaquiline (Butler *et al.*, 2013; Butler and Cooper, 2011).

Microorganisms play a crucial role in the production of antibiotics and other drugs for the treatment of serious diseases. Of all the reported bioactive natural products, approximately 10% were obtained from microbes (Berdy, 2005). After the invention of penicillin, many more naturally occurring antibiotics, such as chloramphenicol and streptomycin, were identified, most of which were developed as drug leads and were optimized by chemical synthesis (Cragg *et al.*, 1997).

This chapter reviews the main applications of antibiotics in clinical use at present. The classification schemes for antibiotics are normally based on bacterial spectrum (broad vs. narrow), or route of administration (injectable vs. oral versus topical), or type of activity (bactericidal vs. bacteriostatic), or chemical structure. The antibiotics outlined in this chapter were divided to be three parts by their biological activities, anti-infective activity, anticancer activity, and other activities. Antibiotics in each section were grouped by their mode of action (MoA). Meanwhile, their development background, structure characteristics, clinical applications, and side-effects were documented as well. The purpose of this chapter is to provide a brief outline and index for currently used antibiotics and afford a reference for further drug discovery for resistant pathogens.

Antibiotics for treatment of infectious diseases

Infectious diseases are known as transmissible diseases or communicable diseases and the clinically evident illness resulting from the infection,

Table 2.1 The antibacterial drugs and their molecular mechanisms

Targets	Antibiotic class (example)	Mechanism
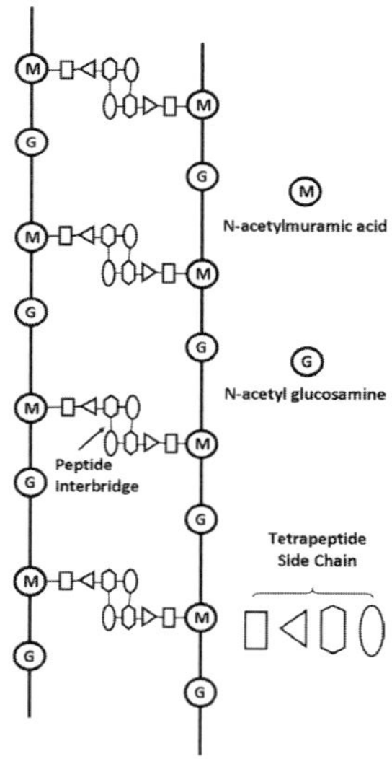	β-lactams (penicillin G, **1**)	Inhibition of the formation of peptidoglycan cross-links
	Polypeptide (bacitracin, **2**)	Interfering with the dephosphorylation of the C55-isoprenyl pyrophosphate
	Glycopeptides (vancomycin, **3**)	Preventing the synthesis of the long polymers of N-acetylmuramic acid and N-acetylglucosamine and inhibiting the cell wall cross-linking
	Phosphonic acid derivatives (fosfomycin, **4**)	Inhibition of bacterial cell wall biogenesis by inactivating the enzyme UDP-N-acetylglucosamine-3-enolpyruvyltransferase
Inhibition of cell wall biosynthesis	Analogues of D-alanine (cycloserine, **5**)	Inhibition of the enzyme alanine racemase and D-alanine ligase
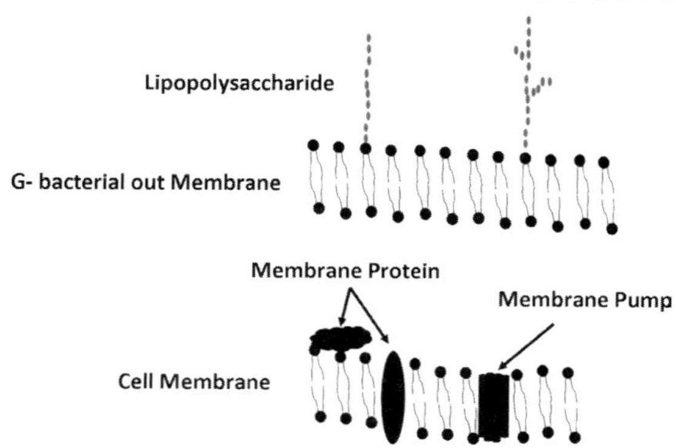	Lipopetides (daptomycin, **6**)	Altering the curvature of the membrane, which creates holes that leak ions
	Diarylquinolines (bedaquiline, **7**)	Affecting the proton pump for ATP synthase
Interference with cell membrane integrity	Polymyxins (colistin, **8**)	Binding to lipopolysaccharide

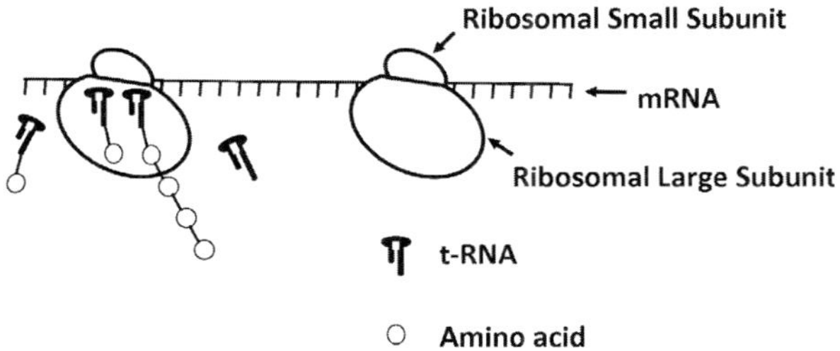 Inhibition of protein synthesis	Aminoglycosides (kanamycin, **9**)	Binding to the small S12 rRNA of the 30S subunit of the bacterial ribosome, interfering with the binding of formyl-methionyl-tRNA to the 30S subunit
	Chloramphenicols (chloramphenicol, **10**)	Inhibition of the peptidyl transferase activity of the bacterial ribosome
	Macrolides (erythromycin, **11**)	Interfere with aminoacyl translocation
	Ketolides (telithromycin, **12**)	Binding to the bacterial ribosome at two sites and modified the structures that mediated resistance
	Tetracyclines (chlortetracycline, **13**)	Prevent the attachment of charged aminoacyl-tRNA to the A site on the ribosome
	Fusidane (fusidic acid, **14**)	Inhibiting the transfer of amino acid from aminoacyl-sRNA to protein on the ribosomes
	Oxazolidinones (linezolid, **15**)	Interfering with the binding of N-formylmethionyl-tRNA to the ribosome
	Streptogramins (Pristinamycin IA, **16**)	Preventing the transfer of the tRNA bound at the A site of the rRNA complex
	Monoxycarbolic acid (mupirocin, **17**)	Inhibition of bacterial isoleucyl-tRNA synthetase
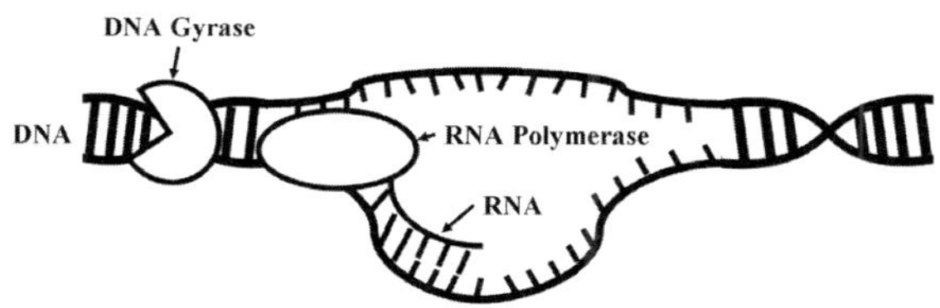 Inhibition of nucleic acid synthesis	Rifamycins (Rifampicin, **18**)	Inhibiting the bacterial DNA-dependent RNA polymerase
	Quinolones (nalidixic, acid, **19**)	Inhibition of DNA gyrase
	Macrocyclic (fidaxomicin, **20**)	Inhibition of RNA polymerase
	Nitrofurans (nitrofurazone, **21**)	Damaging bacterial DNA
	Nitroimidazoles (metronidazole, **22**)	Causing the strand breakage of DNA and the extent of DNA
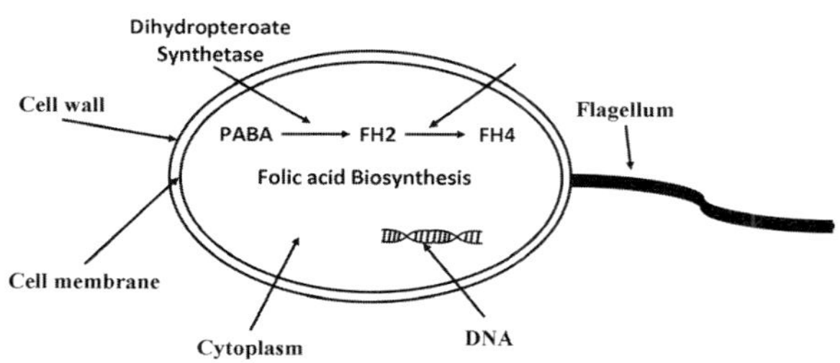 Inhibition of metabolic pathways	Sulfadrugs (prontosil, **23**)	Competitively inhibiting the enzyme dihydropteroate synthetase
	Trimethoprim (Trimethoprim, **24**)	Binding to dihydrofolate reductase and inhibit the reduction of dihydrofolic acid (DHF) to tetrahydrofolic acid (THF)

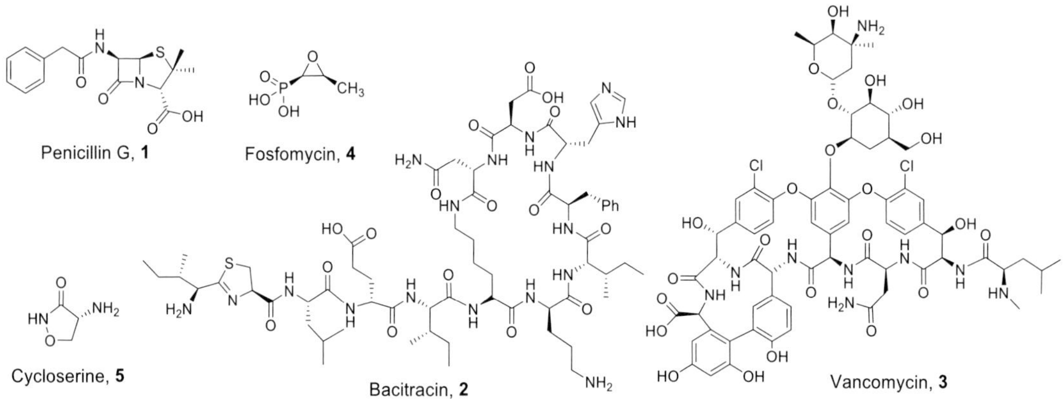

Figure 2.1 Structure of antibacterial targeting the cell wall (**1–5**).

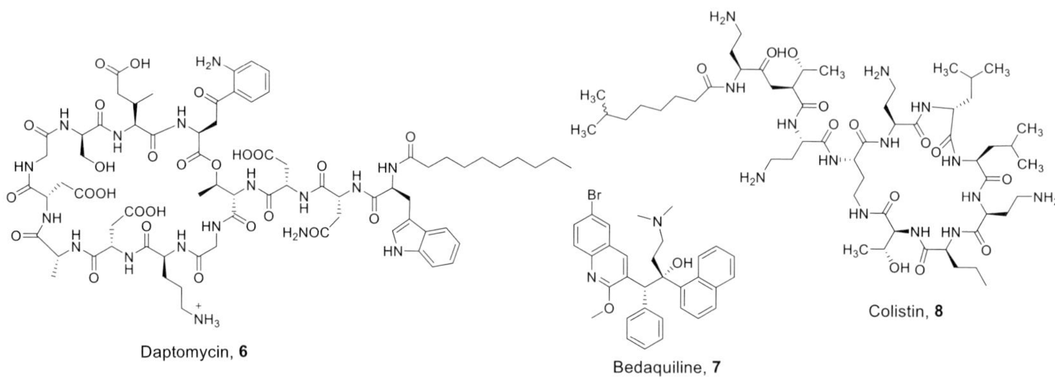

Figure 2.2 Structure of antibacterial targeting the cell membrane (**6–8**).

on-going (Cohen, 2013). The typical side-effects of this drug are nausea, joint and chest pain, and headache; it can also induce black-box warning for arrhythmias and long QT syndrome by blocking the hERG channel (Deoghare, 2013).

Polymyxins were discovered in 1940–1950 (Ainsworth *et al.*, 1947; Benedict and Langlykke, 1947; Koyama *et al.*, 1950; Stansly *et al.*, 1947), they were derived from various species of *Paenibacillus polymyxa* with a cyclic peptide in structure, consisting of polymyxins A–E. While polymyxin M was recognized in 2003 (Martin *et al.*, 2003). Polymyxins show antibacterial activity by binding to lipopolysaccharide (LPS) in the outer membrane and disrupting both the outer and inner membranes due to the function of the hydrophobic tail in the structure (Aboltins *et al.*, 2007). Polymyxins have been considered as an antibiotic niche for the clinical treatment of Gram-negative bacterial infections, among which polymyxin B and E (colistin, **8**) are most widely applied for clinical use (Evans *et al.*, 1999) (Fig. 2.2), and now hold an important role in the antibiotic armamentarium when other antibiotics are ineffective or are contraindicated, such as infections caused by multidrug-resistant *Pseudomonas aeruginosa*, *Acinetobacter baumannii*, *Klebsiella pneumoniae* and *Stenotrophomonas maltophilia* (Giamarellou and Poulakou, 2009). Although the significant activity of polymyxins on Gram-negative organisms, they are relatively neurotoxic and nephrotoxic (Falagas and Kasiakou, 2006). Further studies showed that optimized use of them could minimize those side-effects (Falagas and Kasiakou, 2006).

Antibiotics targeting protein synthesis

Aminoglycosides are a class of antibiotics with amino-modified sugars in their structures (Padilla and Burgos, 2010). Kanamycin (**9**) is an aminoglycoside bactericidal antibiotic, isolated from the bacterium *Streptomyces kanamyceticus* (Umezawa, 1958) (Fig. 2.3). It binds the 30S subunit of the bacterial ribosome and inhibits the protein synthesis by interfering with the binding of formylmethionyl-tRNA to the 30S subunit. (Misumi and Tanaka, 1980; Pestka, 1974). Kanamycin is a broad spectrum antibiotic, active against both Gram-positive and Gram-negative organisms (Prigot *et al.*, 1958). So its used in the treatment of surgical, urinary and respiratory infections caused by various Gram-positive and Gram-negative pathogens, especially staphylococci, and its effective in the treatment of some pathogens that are resistant to other antibiotics (Koota *et al.*, 1960). In addition, kanamycin is always used as a selective agent in molecular biology (Bardini *et al.*, 2003; Wilmink and Dons, 1993). Side-effects related to this drug include loss of hearing, tinnitus, toxicity to kidneys, anaphylactic shock and hypersensitivity skin reactions (Finegold, 1966).

Chloramphenicol (**10**) originally isolated from *Streptomyces venezuelae* is used in the clinic since 1949 (Fig. 2.3). It was an active drug against Gram-positive bacteria (as most strains of MRSA), Gram-negative bacteria and anaerobes, however, not against *P. aeruginosa*, chlamydiae or *Enterobacter* species (Neu and Gootz, 1996). Chloramphenicol was originally used to treat typhus. Since it could penetrate the blood–brain barrier, chloramphenicol was good for the treatment of staphylococcal brain abscesses and meningitis. In the USA, it was used to treat children with fever and a petechial rash. Because of its resistance and safety concerns, it ceased to be a first-line agent in developed countries, but in developing countries, chloramphenicol is still used broadly on account of its low price. Aplastic anaemia, which was rare and fatal, was a most serious effect of chloramphenicol (Rich *et al.*, 1950). Besides, bone marrow suppression (Yunis, 1989) and increased risk of childhood leukaemia were all side-effects of chloramphenicol (Shu *et al.*, 1987).

Erythromycin (**11**), a macrolide antibiotic, was isolated from a strain of *Saccharopolyspora erythraea* in 1949 (Fig. 2.3). Erythromycin in high concentrations inhibits the growth of bacteria, but its mechanism is not completely understood. It has been reported that erythromycin suppresses

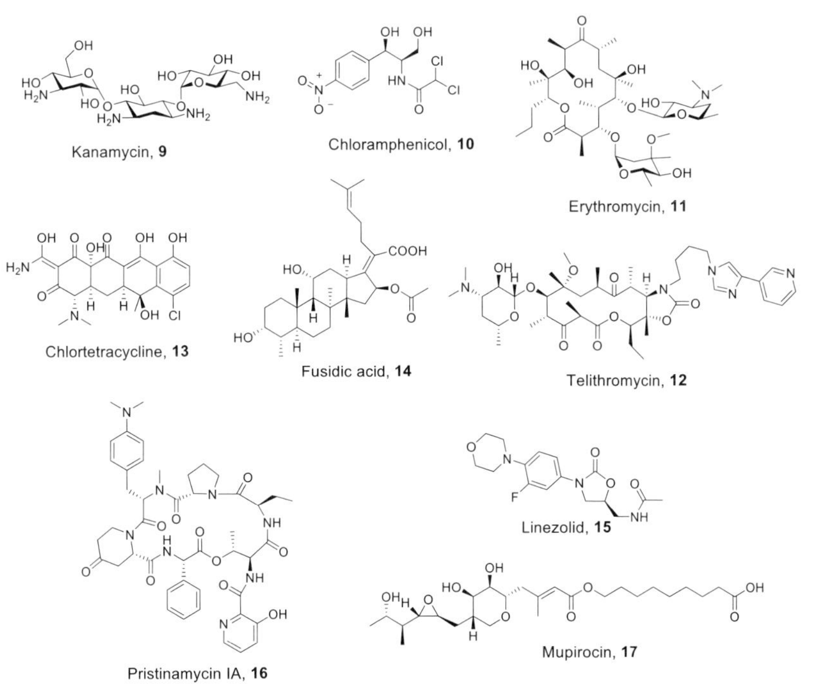

Figure 2.3 Structure of antibiotics targeting the protein synthesis (**9**–**17**).

protein synthesis and subsequent structure and function processes which are important for life (Fraschini *et al.*, 1986). Its antimicrobial spectrum is similar to that of penicillin and thus it is suitable for people allergic to penicillin. It is also occasionally used as a prokinetic agent, because it can induce gastrointestinal disturbances during the treatment (Weber Jr *et al.*, 1993). According to the report, arrhythmia with prolonged QTc intervals including torsades de pointes and reversible deafness are more serious side-effects of erythromycin (Maheshwai, 2007).

Ketolides are derivatives of erythromycin (Scheinfeld, 2004). Telithromycin (**12**) is the first ketolide antibiotic to enter clinical use (Felmingham, 2001) (Fig. 2.3). Ketolides are effective against macrolide-resistant bacteria, because they bind to the bacterial ribosome at two sites and modify the structures that mediate resistance (Lonks and Goldmann, 2005).

Tetracyclines are a group of broad-spectrum antibiotics. Chlortetracycline (**13**) is the first tetracycline identified in 1945 (Jukes, 1985) (Fig. 2.3). Tetracycline antibiotics inhibit protein synthesis by binding to the 30S ribosomal subunit in the mRNA translation complex (Guignard *et al.*, 2013). Tetracyclines are usually used to treat infections of the urinary tract and the intestines. Furthermore, for patients who are allergic to β-lactams and macrolides, they are better for the treatment of *Chlamydia* (Di Francesco *et al.*, 2013). Commonly, tetracyclines are used to treat moderately severe acne and rosacea (Chang *et al.*, 2012). It was tested that tetracyclines and their modified derivatives were effective drugs for human cancers; the promising results were confirmed by the use of CMT-3 for patients with Kaposi sarcoma (Richards *et al.*, 2011). Side-effects of tetracyclines include cramps or burning of the stomach, diarrhoea, sore mouth or tongue, skin photosensitivity, and rarely headache and vision problems. Tetracyclines cause teeth discoloration of the fetus in infancy if the mother takes these drugs during pregnancy. Expired drugs lead to kidney damage (Archer and Archer, 2002; Dréno *et al.*, 2004).

Fusidic acid (**14**) is a lipophilic steroid antibiotic, with strong activity against various Gram-positive bacteria such as *Staphylococcus* species and *Streptococcus* species *in vitro* (Fig. 2.3). Fusidic acid inhibits the transfer of amino acids from aminoacyl-sRNA to protein on the ribosomes (Leclercq *et al.*, 2000). Clinically, fusidic acid is active against MRSA (Vanderhelst *et al.*, 2013). It has been used to treat acne vulgaris and was often partially effective at improving acne symptoms (Spelman, 1999). Fusidic acid is being tested for indications beyond skin infections. In addition, fusidic acid was effective for prosthetic joint-related chronic osteomyelitis (Wolfe, 2011). The side-effects of fucidin tablets and suspension were liver upsets, which would produce jaundice, dark urine and lighter than usual faeces (Long, 2008).

Oxazolidinones are a class of compounds containing 2-oxazolidone in the structure, which was first developed in the late 1980s (Diekema and Jones, 2000). Oxazolidinones inhibit the protein synthesis by interfering with the binding of N-formylmethionyl-tRNA to the ribosome (Shinabarger, 1999). Oxazolidinones are active against many Gram-positive pathogens, including multiresistant staphylococci, vancomycin-resistant enterococci, penicillin-resistant pneumococci, and anaerobic organisms. Linezolid (**15**), the only currently available oxazolidinone (Fig. 2.3), has been used in the treatment of complicated skin infections, nosocomial pneumonia, concurrent bacteraemia, endocarditis and tuberculosis (Bozdogan and Appelbaum, 2004). The most common side-effects include nausea, diarrhoea, and vomiting. Other side-effects include headache, tongue discoloration and oral moniliasis (Birmingham *et al.*, 2003). More serious side-effects include thrombocytopenia and reversible myelosuppression; neuropathy has also been reported (Attassi *et al.*, 2002; Corallo and Paull, 2002; Green *et al.*, 2001).

Streptogramins were a class of antibiotics, first isolated by Charney *et al.* in 1953 (Charney *et al.*, 1953). Streptogramins were divided into two groups, streptogramins A (e.g. pristinamycin IIA) and B (e.g. pristinamycin IA, **16**) (Fig. 2.3). The binding of streptogramin A to ribosome resulted in a conformational change of 50S subunit, which multiplied the activity of streptogramin B 100-fold. Streptogramin B displayed the restraint of the protein chains elongation and leaded to the

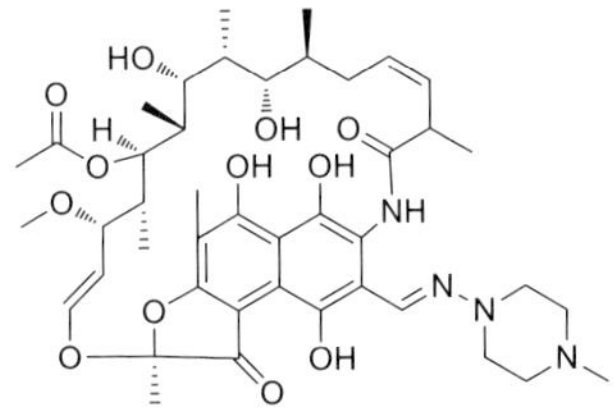

Rifampicin, **18**

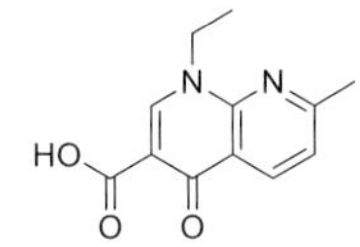

Nalidixic acid, **19**

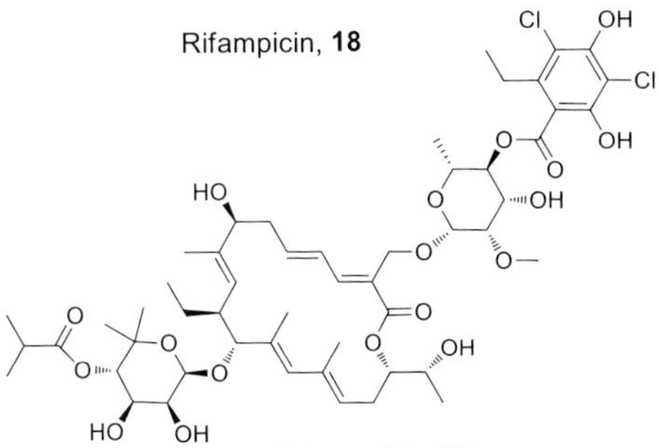

Fidaxomicin, **20**

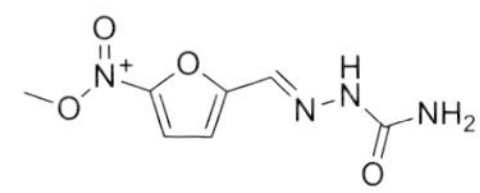

Nitrofurazone, **21**

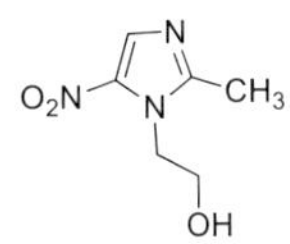

Metronidazole, **22**

Figure 2.4 Structure of antibiotics targeting at the nucleotides (**18–22**).

release of incomplete peptides. Streptogramins were active against vancomycin-resistant *S. aureus* (VRSA) and vancomycin-resistant *Enterococcus* (VRE), but these antibiotics were commercially used as feed additives in agriculture (Mukhtar and Wright, 2005). The adverse effect of streptogramins included infusion-site reactions (e.g. inflammation, pain, and oedema), thrombophlebitis, and non-specified reactions. Otherwise, nausea, diarrhoea, vomiting, rash, headache, pain, and pruritus were also typical side-effects of streptogramins (Allington and Rivey, 2001).

Mupirocin (**17**) is an antibiotic of the monoxycarbolic acid class and was first derived from *Pseudomonas fluorescens* by Fuller and his collaborators (Fuller *et al.*, 1971) (Fig. 2.3). Mupirocin inhibits the protein synthesis by binding to the isoleucyl t-RNA synthetase, and the formation of DNA and cell wall are also reduced to a lesser degree (Hughes and Mellows, 1978b). Mupirocin is active against *Staphylococci*, *Streptococci* and some Gram-negative pathogens, including *Haemophilus influenzae* and *Neisseria gonorrhoeae* (Parenti *et al.*, 1987). It is used as a topical treatment for bacterial skin infections and is effective against Gram-positive bacteria, including MRSA (Hughes and Mellows, 1978a). Side-effects of mupirocin include burning, stinging, itch and rash (Ward and Campoli-Richards, 1986).

Antibiotics targeting nucleotides

Rifampicin (**18**) is a bactericidal antibiotic drug isolated in 1957, which belongs to the rifamycin group (Sensi, 1982) (Fig. 2.4). Rifampicin inhibits the bacterial DNA-dependent RNA polymerase resulting in the inhibition of bacterial DNA-dependent RNA synthesis (Calvori *et al.*, 1965). Rifampicin was introduced to treat tuberculosis (Rybacki, 2004), along with pyrazinamide, isoniazid, ethambutol, and streptomycin ('PIERS'). In combination with fusidic acid, is also used in the treatment of MRSA (Aboltins *et al.*, 2007). Other applications of rifampicin include treat the infections caused by *Neisseria meningitidis,* the tick-borne disease pathogens *Anaplasma phagocytophilum* (Thomas *et al.*, 2009; Wormser *et al.*, 2006), and vaccinia virus (Charity *et al.*, 2007; Sodeik *et al.*, 1994). The use of rifampicin may cause side-effects including hepatotoxicity, gastrointestinal reactions, thrombocytopenic purpura, febrile reactions, shortness of breath, shock, acute haemolytic anaemia, and acute renal failure (Grosset and Leventis, 1983).

Quinolones are a class of broad-spectrum antibiotics (Andersson and MacGowan, 2003). Nalidixic acid (**19**) was the first generation that introduced for the treatment of urinary tract infections in 1962 (Lesher *et al.*, 1962) (Fig. 2.4). The subset fluoroquinolones used mostly in clinical, there inhibit the bacterial growth by trapping the DNA gyrase and/or DNA topoisomerase IV on DNA as drug/enzyme/DNA complexes (Drlica, 1999). The clinical function of fluoroquinolones focuses on the treatment of hospital-acquired infections and those in which resistance to older antibacterial orders is suspected (Liu and Mulholland, 2005). It is also used to treat intracellular pathogens due that can enter cells easily via porins (Pocidalo, 1989). In general, fluoroquinolones have mild to moderate side-effects, which are well tolerated. Occasionally, serious adverse effects happened, including peripheral neuropathy (Owens and Ambrose, 2005; Sarro and Sarro, 2001), central nervous system (CNS) and tendon toxicity (Iannini, 2007; Owens and Ambrose, 2005). Sometimes SIADH (Babar, 2013), convulsions, tendon rupture, torsade de pointes, and hypoglycaemia (Rouveix, 2006), and disturbed blood sugar levels (Lewis and Mohr III, 2008; Mehlhorn and Brown, 2007).

Fidaxomicin (**20**) is a narrow spectrum macrocyclic antibiotic drug (Fig. 2.4), derived from the fermentation products of the actinomycete *Dactylosporangium aurantiacum* subspecies *hamdenesis* (Louie *et al.*, 2009; Revill *et al.*, 2006) and it was approved by FDA in 2011 to target treat *Clostridium difficile* infection (CDI) (Venugopal and Johnson, 2012). Fidaxomicin can selectively eradicate pathogenic *Clostridium difficile* with minimal disruption to normal bacteria. It inhibits the bacterial enzyme RNA polymerase by binding to the DNA template-RNA polymerase (RNAP) complex, then block the transcription through inhibit the formation of the open RNAP–DNA complex (Srivastava *et al.*, 2011). Fidaxomicin showed relative poor activity against Gram-negative bacteria compared to significant activity against *Clostridium difficile* (Mullane and Gorbach, 2011). Up to now, no evidence shows cross-resistance between fidaxomicin and any other antibiotics and the most common adverse reactions reported is anaemia, gastrointestinal haemorrhage, nausea, neutropenia, and vomiting (Golan and Epstein, 2012).

Nitrofurans are a class of typical antibiotics with a furan ring and a nitro group in their structure. Their antimicrobial activity was detected in 1944, and since then, they were widely used as topical antibacterial or antifungal agents in medicine, as urinary tract antiseptics particularly for long-term treatments or as food additives (Herrlich and Schweiger, 1976). Nitrofurans work by damaging bacterial DNA, since its reduced form is highly reactive (McCalla *et al.*, 1971). In this class of antibiotics, two agents commonly prescribed are furazolidone (**21**) and nitrofurantoin (Fig. 2.4). Furazolidone is used to treat bacterial and protozoal diarrhoea and enteritis (Phillips and Hailey, 1986). Nitrofurantoin is used to treat urinary tract infections caused by *Escherichia coli* and other enteric pathogens of the urinary tract. And it is still a first-line therapy for acute uncomplicated cystitis in women today. Systemic administration of nitrofurans is associated with many side-effects, nausea, headache and flatulence are the most typical ones, and less common reactions include diarrhoea, dizziness, pruritus, alopecia and fever (Guay, 2001).

Nitroimidazoles are serial imidazole derivatives and have broad spectrum of activity, including Gram-positive and Gram-negative bacteria, the occasional helminth, protozoa and hypoxic tumours, and low incidence of resistance (Edwards, 1993). They work through causing strand breakage of DNA to kill bacteria (Edwards, 1980). The first nitroimidazole derivative, metronidazole (**22**), was discovered in 1959 (Cosar *et al.*, 1959) (Fig. 2.4). It is effective against anaerobic bacteria and certain parasites, and it can selectively block some of the functions within the bacterial cells and the parasites resulting in their death (Lamp *et al.*, 1999). The common side-effects of nitroimidazoles include encephalopathy, peripheral neuropathy nausea, anorexia, vomiting, metallic or bitter taste, in addition, dizziness, ataxia and headache have also been reported (Andersson, 1981).

Antibiotics that inhibit metabolic pathways

Sulfonamide drugs were the first antimicrobial drugs, of which prontosil (**23**) was the first sulfonamide (Fig. 2.5), developed in the 1930s. The antibacterial activity of prontosil was firstly found by Josef Klarer (Hager, 2006). This drug

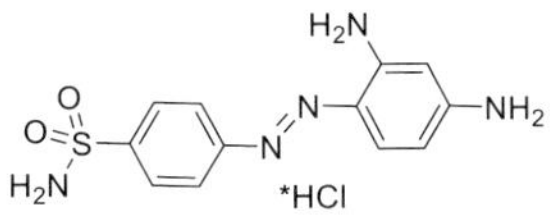

Prontosil, **23**

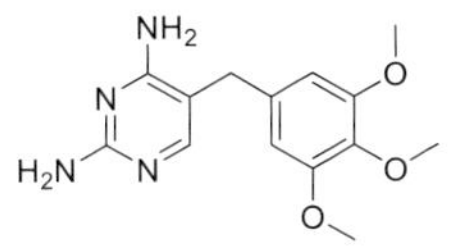

Trimethoprim, **24**

Figure 2.5 Structure of antibiotics inhibit of the metabolic pathways (**23** and **24**).

was metabolized into an active compound called sulfanilamide inside the body (Petri Jr, 2007). The antibacterial mechanism of sulfonamides is to competitively inhibit the enzyme dihydropteroate synthetase (DHPS), an enzyme involved in the folate synthesis (Sköld, 2000). Sulfonamides are effective in the treatment of chancroid, gas gangrene, lymphogranuloma, arthritis, bacillary dysentery, and the infections diseases cause by haemolytic streptococcal, meningococcal, streptococcus viridans, urinary tract, gonococcal, pneumococcal, staphylococcal, *E. coli* tissue and gonococcal (Long, 1941). Sulfonamide drugs can cause a variety of serious side-effects, including hypersensitivity reactions, nausea, vomiting, diarrhoea, anorexia, dizziness, cyanosis, rash and fever (Connor, 1998; Long, 1941). In addition, high-dose use of Sulfonamides has the potential to cause Stevens–Johnson syndrome and toxic epidermal necrolysis (Rubin, 1977; Schopf *et al.*, 1991).

Trimethoprim (**24**) is a bacteriostatic antibiotic first described by Roth and his collaborators in 1962 (Roth *et al.*, 1962) (Fig. 2.5). It interfering with the folate synthesis by binding to the dihydrofolate reductase and inhibits the reduction of dihydrofolic acid to tetrahydrofolic acid (Lacey, 1982). Combined with sulphamethoxazole (SMX), sulphadiazine or sulfamoxole as co-trimoxazole (Co-T), trimethoprim has been used widely in bronchitis, pneumonia, enteric, protozoa, urinary tract, ear, nose, throat and serious systemic infections (Wormser *et al.*, 1982) since 1969, so trimethoprim was just considered a sulfonamide potentiator (Lacey, 1982). The use of the combination may cause side-effects in the following aspects: gastrointestinal, dermatological, renal, haematological and psychiatric problems (Masters *et al.*, 2003). When used alone, trimethoprim was mainly used in the treatment of chronic bronchitis, staphylococcal pneumonia, Gram-negative septicaemia and urinary tract infections (Darrell *et al.*, 1968).

Antifungal activity

Fungi are a kind of good resource for the antibiotics productions, such as the 'wonder drug' penicillin (Kauffman, 1979). However, fungal pathogens cause one of the leading infectious diseases in clinical practice, especially in the immunocompromised population: HIV-infected patients, organ transplant recipients, radiation-treated cancer patient s(Menzin *et al.*, 2009), etc. The mortality rate of fungal infection has increased during these years. *Aspergillus, Candida* and *Cryptococcus* are the major fungal pathogen especially the species *Aspergillus fumigatus, Candida albicans* and *Candida neoformans* (Posteraro *et al.*, 2013). The candidiasis, which was considered the most serious fungal infection, caused about 40% of mortality rate in clinical practice (Menzin *et al.*, 2009).

Antifungal drug development started in the 1930s (Fig. 2.6). In 1939, the first oral antifungal drug, griseofulvin, was discovered, and it was later marketed under the name Grifulvin V by Ortho-neutrogena Labs. It was isolated from *Penicillium griseofulvum* and used to treat the skin and nails in animals and humans (Goldman *et al.*, 1960). After griseofulvin, antifungal drug development entered the 'golden age' from the 1950s to the 1980s, when many new classes of antifungals emerged, such as polyenes, flucytosine, echinocandin, imidazole, triazoles, allyamines and other classes. Amphotericin B, isolated from *Streptomyces nodosus,* is still the 'gold standard' antifungal drug (Stiller *et al.*, 1955). The emergency challenge is that sincethe 1990s, no new class of antifungals has been developed. All new drugs were the result of modification of azoles or echinocandins. The lack of new antifungals caused serious drug resistance in hospital.

The main targets for antifungal were cell membrane, cell wall and nucleotides (Table 2.2). Some rare antifungals can also target other pathways, such as microtubules, oxygen uptake, lipid protein and so on (Odds *et al.*, 2003).

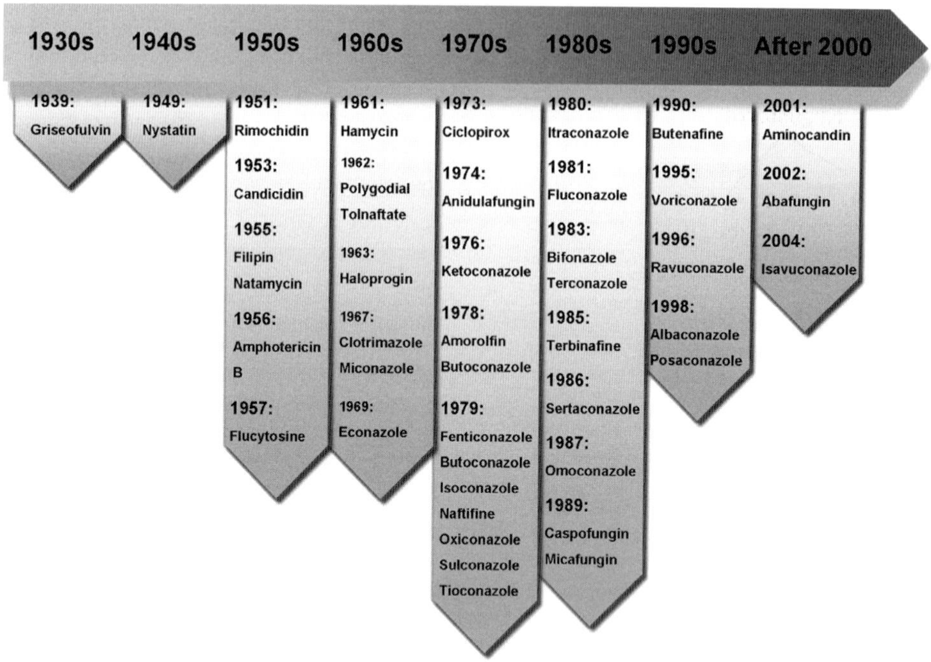

Figure 2.6 The discovered timeline of antifungal drugs.

Antifungals targeting the cell membrane

One of the most important different cell membrane components between fungal pathogens and humans is ergosterol (or the human cholesterol). Totally, one of these antifungals target at the biosynthesis pathway of ergosterol (Francois *et al.*, 2006), and the others are binding to the ergosterol to disrupt the cell membrane (Baginski *et al.*, 2005).

Polyene antifungal drugs

The polyene compounds have a macrolide ring structure conjugated with double bonds (Fig. 2.7) (Hamilton-Miller, 1973). The main mechanism of the polyene antibiotics is to bind the ergosterol of fungal cell membrane and form holes to cause the cell leaky (Bossche *et al.*, 1987; Oroshnik and Mebane, 1963). The polyene antifungal agents include amphotericin B, candicidin, filipin, hamycin, natamycin, nystatin, and rimochidin.

Take amphotericin B (**25**) as an example. The application of amphotericin B is for treatment of life-threatening mycoses and most other mycoses, and also for dermatophytosis (Kafetzis *et al.*, 2005). (Fig. 2.7). Amphotericin B has been in common clinical use since the 1960s (Abu-Salah, 1996) to treat most of the medically important moulds and yeasts pathogens such as *Candida* species, *Cryptococcus neoformans*, and *Aspergillus* species (Hay, 1999; Hector, 2005). Resistance to this agent is rare, but is noteworthy that the azole-resistant *C. albicans* with high level of ABC transporters has become sensitive to amphotericin B (Biao Ren *et al.*, 2013), and *Candida tropicalis* with dysfunctional *CtERG3* and *CtERG11* showed azole and amphotericin B cross-resistance (Eddouzi *et al.*, 2013). The challenge of application of amphotericin B is its toxicity or side-effect due to the binding with sterols from mammalian cell membrane. Currently, amphotericin B is used in liposomal form to increase the host tolerability and reduce its toxicity (Hsuchen and Feingold, 1973; Juliano *et al.*, 1985; Kinsky *et al.*, 1962; Kotler-Brajtburg *et al.*, 1974).

Azole antifungal drugs

The azole antifungal agents contain either two or three nitrogen molecules (the imidazoles, triazoles and thiazoles, respectively) (Saag and Dismukes, 1988). The clinically useful imidazoles

Table 2.2 The antifungal drugs and their molecular mechanisms

Targets	Antibiotic class (example)	Mechanism
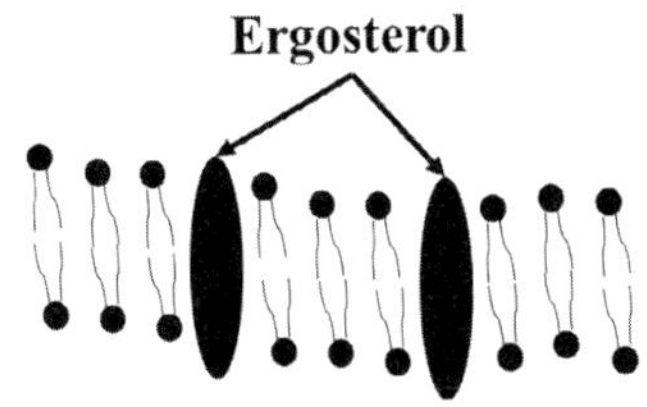 Cell membrane	Polyenes (amphotericin B, **25**)	Binding to the ergosterol to disrupt the cell membrane
	Azoles (ketoconazole, **26**)	Inhibiting the lanosterol 14-alpha-demethylase (*ERG11*) of the ergosterol biosynthesis pathway
	Allylamines (butenafine, **27**)	Inhibiting the squalene epoxidase (ERG1) of the ergosterol biosynthesis pathway
	Thiocarbamates (tolnaftate, **28**)	
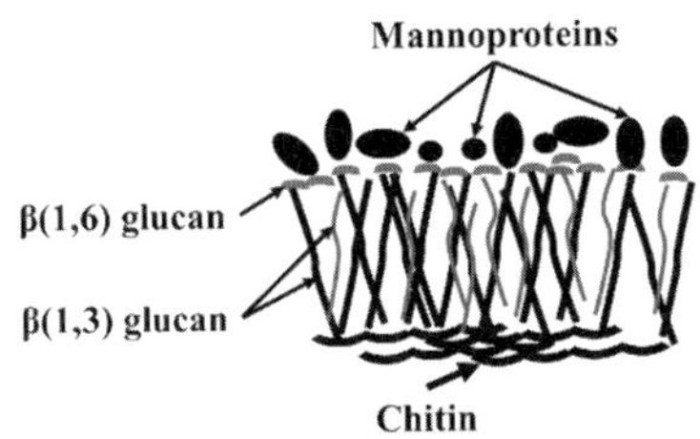 Cell wall	Echinocandins (caspofungin, **29**)	Inhibiting the 1,3-β-glucan synthase of the cell wall biosynthesis pathway
 Nucleotides	Pyrimidine analogues (5-fluorocytosine, **30**)	Inhibiting the nucleotides biosynthesis
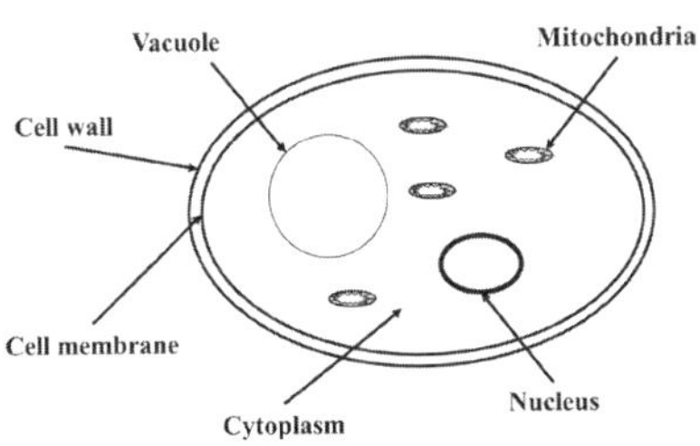 Cell	hydroxypyridones (ciclopirox, **31**)	Inhibiting the membrane transfer system by interrupting the Na^+/K^+-ATPase
	Benzofurans (griseofulvin, **32**)	Binding to the tubulin to inhibit the mitosis
	Phenyl ethers (haloprogin, **33**)	Inhibiting the oxygen uptake and disrupting yeast membrane structure and function
	Sesquiterpenes (polygodial, **34**)	Disrupting the lipid–protein interface of integral proteins and the membrane

are bifonazole, butoconazole, clotrimazole, econazole, fenticonazole, isoconazole, ketoconazole (**26**), miconazole, omoconazole, oxiconazole, sertaconazole, sulconazole, and tioconazole (Fig. 2.7). The triazoles are including albaconazole, fluconazole, isavuconazole, itraconazole, posaconazole, ravuconazole, terconazole, and voriconazole. The abafungin, a thiazole, is in the phase III stage now. In general, the azole antifungal agents inhibit cytochrome P450-dependent enzymes involved in the biosynthesis of cell membrane key component: ergosterol (Francois *et al.*, 2006; Odds *et al.*, 2003). The emergent challenge is the drug resistance to azole agents (Vanden *et al.*, 1998). The mechanism mainly due to the drug efflux pumps overexpression and enzymes mutations of the ergosterol biosynthesis pathway (Perea *et al.*, 2001; Sanglard *et al.*, 1998; Sanglard *et al.*, 1995).

Three typical medically used azoles are ketoconazole, fluconazole, and itraconazole (Dixon and Walsh, 1996; Wood *et al.*, 1994). Ketoconazole has been used to treat fungal infections of the skin and mucous membranes since the 1970s (approved in 1981 in USA) (Armstrong *et al.*, 1985; Thienpont *et al.*, 1980). It can be taken both orally and topically, and has a range of activity including infections due to mucosal candidiasis, histoplasmosis and blastomycosis, and a variety of skin mycoses, but not for treatment of

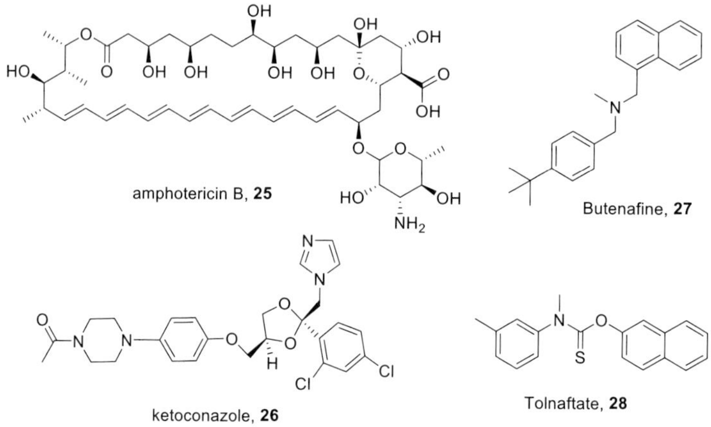

Figure 2.7 Structure of antifungal targeting the cell membrane (**25–28**).

aspergillosis or of systemic infections caused by yeasts (Ernest, 1992; Heeres *et al.*, 1979). Ketoconazole is also used to treat dandruff (Peter and Richarz, 1995). The hepatotoxicity limited the application of ketoconazole (Duarte *et al.*, 1984; Zöllner *et al.*, 2001). The mechanism of the liver injury is unknown, but it may be correlated with the ability of ketoconazole to inhibit the human sterols biosynthesis (Rozman and Monostory, 2010; Tyle, 1984).

The less toxic and more effective triazoles, fluconazole and itraconazole, are preferred in clinical use. They have also replaced amphotericin B for treating certain forms of the systemic mycoses (Chen and Sorrell, 2007). Fluconazole is now used to treat candidemia, cryptococcosis and coccidioidomycosis (Rossi, 2003). Itraconazole is effective for histoplasmosis, blastomycosis, sporotrichosis, coccidioidomycosis, cryptococcosis, and certain forms of aspergillosis (Grant and Clissold, 1989; Restrepo *et al.*, 1986). Fluconazole can be used either orally, or intravenously, while itraconazole is oral (Varhe *et al.*, 1994).

Allylamines and tolnaftate

Allylamines and tolnaftate inhibit the first enzyme of ergosterol biosynthesis pathway: Squalene epoxidase (*ERG1*) (Barrett-Bee and Dixon, 1995; Ryder *et al.*, 1986). The allylamines are including amorolfin, butenafine (**27**), naftifine and terbinafine. Tolnaftate (28) is a synthetic thiocarbamate (Fig. 2.7).

Terbinafine is mainly effective on the dermatophyte (skin plant) (Petranyi *et al.*, 1987). It also is used for the treatment of onychomycosis and *Candida* species (Balfour and Faulds, 1992). Terbinafine may cause hepatotoxicity by increasing the enzymes of liver (Iverson and Uetrecht, 2001; Rosso, 1998). Other side-effects, such as gastrointestinal problems, neurological problems, immune system problems, skin problems, etc., are also reported (Abruzzo *et al.*, 1987; Carstens *et al.*, 1994; O'Reardon *et al.*, 2002). Tolnaftate is used to treat jock itch, athlete's foot and ringworm etc. but not candidiasis (Crawford and Hollis, 2007; Ishii *et al.*, 1983; Ryder *et al.*, 1986). When using tolnoftate, the skin irritation and serious allergic reaction may occurred (Gupta *et al.*, 1994; Hashimoto *et al.*, 1966).

Antifungals targeting at the cell wall

The most obvious different character between fungal pathogen and human is the fungal cell wall. Echinocandins inhibit the synthesis of glucan in the cell wall, via inhibition of the enzyme 1,3-β-glucan synthase (Debono and Gordee, 1994; Odds *et al.*, 2003) and are thus called 'penicillin of antifungals' (van Zanten, 2010). β-Glucans are carbohydrate polymers, which are cross-linked with other fungal cell wall components. Echinocandins are including anidulafungin, caspofungin, micafungin, and aminocandin.

Caspofungin (**29**), the first inhibitor of 1,3-β-glucan synthase, is a lipopeptide antifungal agent (Fig. 2.8) (Onishi *et al.*, 2000; Saravolatz

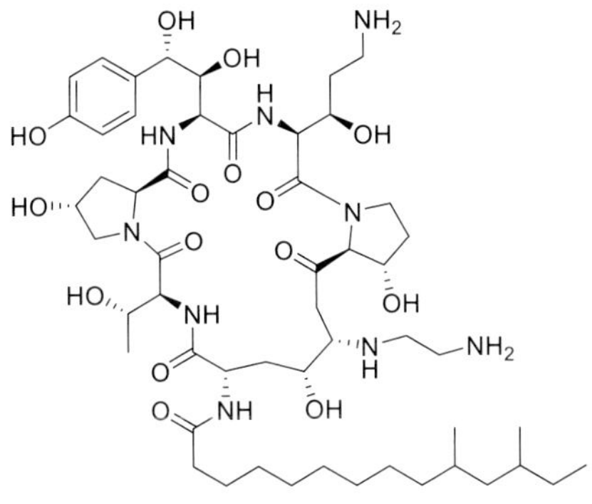

Caspofungin, **29**

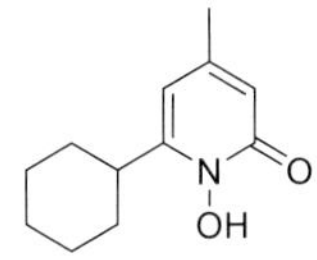

Ciclopirox, **31**

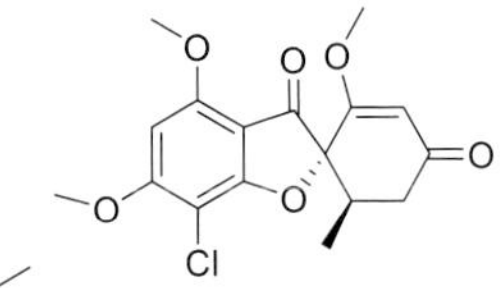

Griseofulvin, **32**

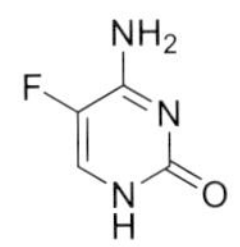

5-fluocytosine, **30**

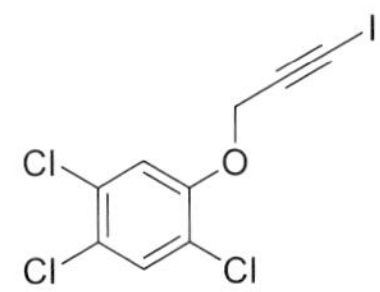

Haloprogin, **33**

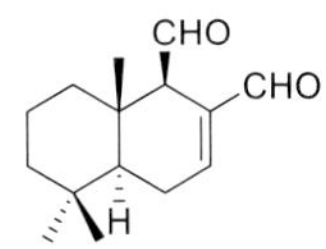

Polygodial, **34**

Figure 2.8 Structure of antifungals targeting at the cell wall, nucleotides, and other antifungal agents (29–34).

et al., 2003). Caspofungin has broad-spectrum antifungal activities, such as *Candida*, *Cryptococcus* and *Aspergillus* species (Maertens *et al.*, 2004; Odds *et al.*, 2003; Pfaller *et al.*, 2003). The resistance is rare but still reported from *Candida* or *Cryptococcus*, due to the mutations of 1,3-β-glucan synthase (Baixench *et al.*, 2007; Maligie and Selitrennikoff, 2005). The side-effect is very low for caspofungin. Sometimes, gastrointestinal problem, haematological problem, liver problem and respiratory problem occurred (Maertens *et al.*, 2004; Ullmann, 2003).

Antifungals targeting at nucleotides

5-Fluorocytosine (**30**), a fluorinated pyrimidine analogue (Fig. 2.8), inhibits both DNA and RNA synthesis and further causes the protein synthesis inhibition (Polak and Scholer, 1975). 5-Fluorocytosine is used against *Candida* and *Cryptococcus* (Shadomy, 1969; Tassel and Madoff, 1968). Resistance is common and it is seldom used alone (Polak and Scholer, 1975). In combination with amphotericin B it used for the treatment of cryptococcal meningitis and other mycoses (Fass and Perkins, 1971; Medoff *et al.*, 1971; Medoff *et al.*, 1972). The side-effects are including liver/renal problem, central nervous system problem, skin effect and anaphylaxis (Baley *et al.*, 1990; Diasio *et al.*, 1978; Kauffman and Frame, 1977).

Other antifungal agents

Ciclopirox (**31**) (Fig. 2.8) is a synthetic antifungal agent for superficial mycoses especially tinea versicolor (Abrams *et al.*, 1991; Gupta, 2001; Niewerth *et al.*, 2003; Subissi *et al.*, 2010). It is noteworthy that ciclopirox can eradicate HIV in cell cultures (Hanauske-Abel *et al.*, 2013; Hoque *et al.*, 2009).

Griseofulvin (**32**) (Fig. 2.8) is an antifungal antibiotic produced by *Penicillium griseofulvum* (Oxford *et al.*, 1939). It is effective on most dermatophytes (Artis *et al.*, 1981; GOH *et al.*, 1994).

Haloprogin (**33**) (Fig. 2.8) is an antifungal drug used to treat athlete's foot and other fungal infections (Harrison *et al.*, 1970; Hermann, 1972b; Katz and Cahn, 1972). Haloprogin had side-effects including: itching, irritation, vesiculation, scaling, and burning (Al-Nakib *et al.*, 1997; Hermann, 1972a; Weikel Jr and Bartek, 1972).

Polygodial (**34**) (Fig. 2.8) is a drimane-type sesquiterpene dialdehyde with broad spectrum of bioactivities: antifungal (Kubo *et al.*, 2001), antimicrobial (Anke and Sterner, 1991), antihyperalgesia (Mendes *et al.*, 1998), insect antifeedant activity (Caprioli *et al.*, 1987), antinociception (Mendes *et al.*, 2000), anti-inflammatory (da Cunha *et al.*, 2001) and antiallergic activities (Tratsk *et al.*, 1997).

Further discussion

The development of antifungal drugs is far lagged behind the antibacterial drugs. No novel actions of antifungal drugs were developed nearly 35 years. The widespread and overused the known drugs caused serious drug resistance in clinical, especially the multidrug resistance (Alexander and Perfect, 1997; Pfaller, 2012; Vanden *et al.*, 1998). Currently, the costly and time-consuming pipeline of drug discovery lacks the ability to overcome the rapidly increased drug-resistant pathogens (DiMasi *et al.*, 2003; Smith, 2002; Talbot *et al.*, 2006). An alternative approach is to prospect the new applications of known, clinical approved drugs (Ashburn and Thor, 2004; Campàs, 2009) and employ the synergistic screening model to sample a previously untapped expanse of bioactive chemical space including the antibiotics which even quit the market due to the drug resistance (Zhang *et al.*, 2007). Another way to speed up the discovery new antifungal drugs is to predict the agents or the combinations via computing science (Banfi *et al.*, 2006; Giordani *et al.*, 2002; Meletiadis *et al.*, 2007; Soltani *et al.*, 2009). This can be reduced the screening scale but increase the success rate (Song *et al.*, 2009; Streich *et al.*, 1988).

Antibiotics for cancer therapy

Most of the clinically effective anti-tumour agents from microbes were developed in the 1960s by testing against tumour cell lines the active compounds which were too toxic for use as anti-infective drugs (Lancini and Demain, 2013). Only a new family of antibiotics, epothilones, which could stabilize microtubulins and inhibit the overexpression of P-glycoprotein efflux pumps, was discovered by target-based screening (Bollag *et al.*, 1995). The most important anticancer drugs among the approved products could be classified to be several types of scaffolds, including anthracycline, mitosane, bleomycin, actinomycins, anthracenones, enediynes, and epothilones (Fig. 2.9 and Table 2.3).

Actinomycin D (**35**), a class of polypeptide antibiotics isolated from soil bacteria of the genus *Streptomyces*, is the most significant member of Actinomycetes and the first antibiotic with anti-cancer activity identified by Waksman in 1940 (Hollstein, 1974; Waksman and Woodruff, 1940). It has been used as a highly effective apoptosis inducer and inhibits DNA-dependent RNA synthesis by binding to guanine residues, and is clinically approved for treating rare tumours such as Wilms tumour and rhabdomyosarcoma (EL-AAL *et al.*, 2005; Khatua *et al.*, 2004; Sobell, 1985). The clinical use of Actinomycin D cause the following side-effects which determine the amount and schedule of administration, such as low white blood counts, low platelet count, anaemia, hair loss, soreness of the mouth, difficulty swallowing, diarrhoea, and skin discoloration.

Anthracyclines are isolated from *Streptomyces peucetius* var. *caesius* and they are used to treat many cancers, including leukaemias, lymphomas, breast, uterine, ovarian, and lung cancers (Bertram and Anthony, 1993; Fujiwara *et al.*, 1985). The anthracyclines are among the most effective anticancer treatments ever developed and the first anthracycline discovered was daunorubicin (**36**) (Minotti *et al.*, 2004; Peng *et al.*, 2005; Weiss, 1992). Anthracycline has four mechanisms of action: (1) inhibits DNA and RNA synthesis by intercalating between base pairs of the DNA/RNA strand, thus preventing the replication of rapidly growing cancer cells; (2) inhibits topoisomerase II enzyme, preventing the relaxing of supercoiled DNA and thus blocking DNA transcription and replication. This leads to topo II mediated DNA-cleavage, producing DNA breaks. The binding of topo II inhibitor prevents DNA repair by ligase (Jensen *et al.*, 1993; Osheroff, 1986); (3) creates iron-mediated free oxygen radicals that damage the DNA, proteins and cell membranes; (4) induces histone eviction from chromatin that deregulates DNA damage response, epigenome and transcriptome (Pang *et al.*, 2013). Their main side-effect is cardiotoxicity, which considerably limits their application. Use of anthracyclines has also been shown to be significantly associated with cycle 1 severe or febrile neutropenia (Lyman *et al.*, 2011).

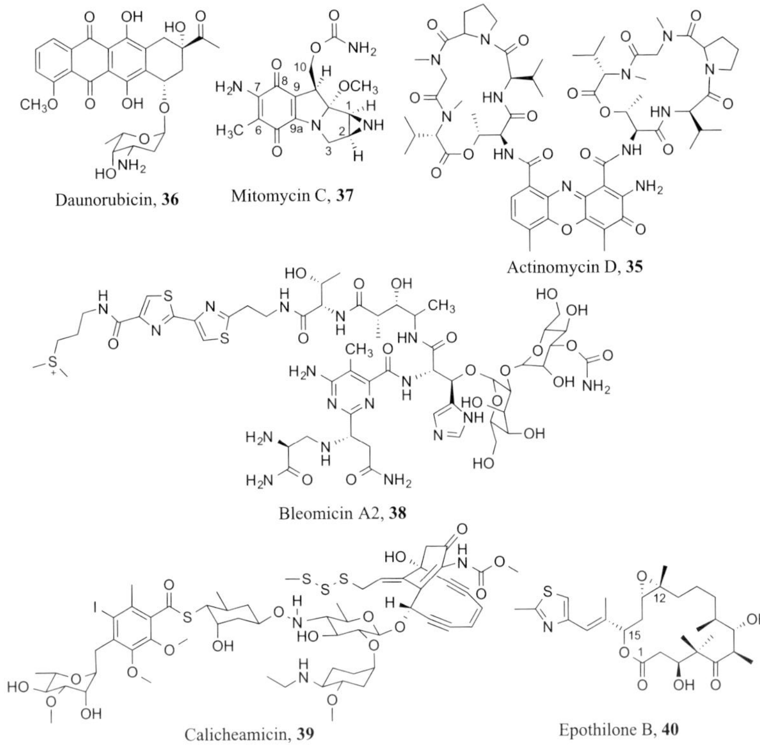

Figure 2.9 Structure of anticancer antibiotics (**35–40**).

Table 2.3 Anti-tumour antibiotics and their molecular mechanisms

Antibiotic type	Example	Structural skeleton	Mode of action	Producing strain
Actinomycin	Actinomycin D (**35**)	Polypeptide	Inhibiting RNA polymerase	*Streptomyces* sp.
Anthracycline	Daunorubicin (**36**)	Polyketide	Inhibiting DNA and RNA synthesis	*S. peucetius* var. *caesius*
Mitosane	Mitomycin C (**37**)	Aziridine-containing natural products	A potent DNA crosslinker	*S. achromogenes*
Bleomycin	Bleomycin (**38**)	Glycopeptide	Induction of DNA strand breaks and inhibiting incorporation of thymidine into DNA strands	*S. verticillus*
Enediynes	Calicheamicin (**49**)	Enediyne	A highly potent DNA-cleaving agent	*Micromonospora echinospora* ssp. *calichensis*
Epothilones	Epothilone B (**40**)	Macrolide	Stabilizing microtubulins and inhibit the overexpression of P-glycoprotein efflux pumps.	*Sorangium cellulosum*

Mitomycins are a family of aziridine-containing natural products isolated from *S. caespitosus* or *S. lavendulae* (Danshiitsoodol *et al.*, 2006). Among this type of compounds, mitomycin C (**37**) is used as a chemotherapeutic agent and to treat upper gastro-intestinal (e.g. oesophageal carcinoma), anal cancers, and breast cancers, as well as by bladder instillation for superficial bladder tumours. Mitomycin is also used as a chemotherapeutic agent in glaucoma surgery. The study on the

MoA of Mitomycin C revealed that it is a potent DNA cross-linker. A single crosslink per genome has shown to be effective in killing bacteria. This is accomplished by reductive activation followed by two N-alkylations. Both alkylations are sequence specific for a guanine nucleoside in the sequence 5′-CpG-3′ (Tomasz, 1995). Potential bis-alkylating heterocylic quinones were synthetized in order to explore their antitumoral activities by bioreductive alkylation (Renault *et al.*, 1981). Since it causes delayed bone marrow toxicity and the prolonged use may result in permanent bone-marrow damage, it is usually administered at 6-weekly intervals. In addition, it may also cause lung fibrosis and renal damage (Mao *et al.*, 1999).

Bleomycins (**38**), the glycopeptide antibiotics produced by *Streptomyces verticillus*, are used as chemotherapeuticals and the most popular forms in clinical are bleomycins A2 and B2. They are used in the treatment of Hodgkin's lymphoma, squamous cell carcinomas, testicular cancer, and in the treatment of plantar warts (Lewis and Nydorf, 2006). Bleomycin showed the induction of DNA strand breaks (Takimoto and Calvo, 2008). Some studies suggest bleomycin also inhibits incorporation of thymidine into DNA strands. DNA cleavage by bleomycin depends on oxygen and metal ions *in vitro*. The exact mechanism of DNA strand scission is unknown, but it has been suggested that bleomycin chelates metal ions (primarily iron), producing a pseudoenzyme that reacts with oxygen to produce superoxide and hydroxide free radicals that cleave DNA. An alternative hypothesis states that bleomycin may bind at specific sites in the DNA strand and induce scission by abstracting the hydrogen atom from the base, resulting in strand cleavage as the base undergoes a Criegee-type rearrangement, or forms an alkali-labile lesion (Hecht, 2000). In addition, these complexes also mediate lipid peroxidation and oxidation of other cellular molecules.

Enediynes are a class of bacterial natural products characterized by either nine- and ten-membered rings containing two triple bonds separated by a double bond and show significant anticancer activity. Three classes of enediynes exist: (1) calicheamicin/esperamicin; (2) dynemicin; (3) chromoprotein types. The first representative of enediyne antibiotic, calicheamicin Y1 (**39**), was isolated from *Micromonospora echinospora* subsp. *calichensis* in 1987 and showed potent activity against murine tumours such as P388 and L210 leukaemias and solid neoplasms such as colon 26 and B16 melanoma with optimal doses of 0.15–5.0 μg/kg (Golik *et al.*, 1987a,b; Lee *et al.*, 1987a,b). The study on the MoA of calicheamicin Y1 revealed that it is a highly potent DNA-cleaving agent giving rise primarily to sequence-selective double-strand cuts. The rigid 10-membered bridged ring of the aglycon of enediynes is capable of undergoing Bergman reaction and the trisulfide serves as the trigger, which ultimately leads to DNA cleavage (Fig. 2.10).

Epothilones are a family of new anticancer agents produced by *Sorangium cellulosum* (Reichenbach and Höfle, 2008). The study on their MoA revealed that they act by stabilizing microtubulins, like taxanes, and inhibit the overexpression of P-glycoprotein efflux pumps as well (Bollag *et al.*, 1995). In addition, in early trials epithilones have better efficacy and milder adverse effects than taxanes (Rosenberg *et al.*, 2005). Epothilones showed promising potentials by their better water solubility, cremophors (solubilizing agents used for paclitaxel which can affect cardiac function and cause severe hypersensitivity) are not needed (Julien and Shah, 2002). The semisynthetic derivative of epothilone B (**40**) was approved by the FDA in 2007 to treat advanced breast cancer and is also undergoing clinical trials for other indications (Goodin *et al.*, 2004). The endotoxin-like properties known from paclitaxel, like activation of macrophages synthesizing inflammatory cytokines and nitric oxide, are not observed for epothilone B (Mühlradt and Sasse, 1997).

Conclusion and future perspectives

Ensuring the continued availability of novel antibiotics to combat existing and emerging pathogens, especially pathogens showed resistance to currently available therapies, is a critical public health issue (Lancini and Demain, 2013). Microbial metabolites, started from the discovery of actinomycin around 1940, play a crucial role

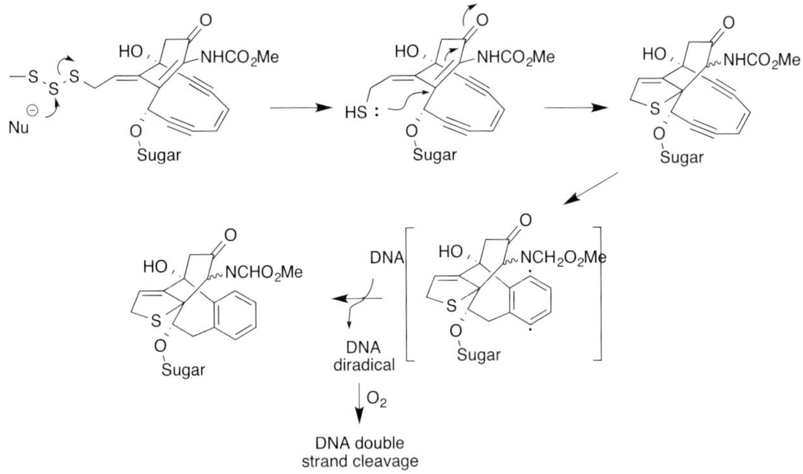

Figure 2.10 Mechanism by which calicheamicin Y_1 cleaves DNA. Nu, nucleophile. (A nucleophile, such as glutathione, attacks the central sulfur atom of the trisulfide group, causing the formation of a thiolate or a thiol which adds intramolecularly to the adjacent α,β-unsaturated ketone embedded within the framework of the aglycon. This reaction causes a significant change in structural geometry and completely relieved by the enediyne undergoing the Bergman reaction to generate a highly reactive benzenoid diradical. The calicheamicin diradical abstracts hydrogen atoms from duplex DNA at the C-5′ position of the cytidine in 5′-TCCT-3′ and the C-4′ position of the nucleotide three base pairs removed on the 3′ side of the complementary strand, leading to cleavage of both strands of DNA). (De Voss *et al.*, 1990; Hangeland *et al.*, 1992; Nicolaou *et al.*, 1993).

as the cancer chemotherapeutic agents and there are more than 60% of the current antineoplastic agents were originally isolated as natural products or their derivatives (Demain and Sanchez, 2009).

Antibiotic resistance in bacteria is caused by vertically transmitted innate genetic and physiology through species and the remarkable propensity of bacteria to exchange genetic material horizontally across species and genera (Wright and Sutherland, 2007). The major mechanisms of antibiotic resistance include enzymatic transformation, modification of the molecular target, sequestration of the drug, active efflux from the cell interior, and prevention of entry of the compound in to the cell (Wright, 2005). The growing antibiotic resistance among Gram-positive and Gram-negative pathogens has led to untreatable infection, but the pace of antibiotics discovery to combat these pathogens has slowed (Lewis, 2013).

Most antibiotics were developed during the golden era of antibiotics discovery in the 1940s and 1960s. Over the last 40 years, only five new classes of antibiotics have been introduced: the oxazolidinone linezolid (Zyvox; Pfizer) in 2000, the cyclic lipopeptide daptomycin (Cubicin; Cubist) in 2003, retapamuliin (Altabax; GlaxoSmithKline) in 2007, fidaxomicin (Dificid; Cubist) in 2010, and the diarylquinoline bedaquiline (Sirturo; Janssen) in 2012 (Butler *et al.*, 2013). The lack of new antibiotics is due to the emergence of multidrug-resistant bacteria and the economic and regulatory challenges of antibiotic research (Jones, 2011; Moellering Jr, 2011; Silver, 2011). Although there have been a number of significant developments in improving the atmosphere for antibiotic research (Policy, 2010; Cooper and Shlaes, 2011; Laxminarayan and Powers, 2011; Cain, 2012; Spellberg *et al.*, 2012), novel antibiotics, especially those with new molecular mechanisms, are urgently needed to combat the severe pathogen resistance situation. Altman indicated that using nucleic acids as part of a novel compound could be promising method to develop new drugs against drug resistance (Wesolowski, 2011; Altman, 2013). Meanwhile, Steckbeck *et al.* (2014) proposed that antimicrobial peptides might act as promising new drugs. Demain and Sanchez (2009) suggested a platform for new antibiotics discovery including investigating the uncultivated microorganisms for novel metabolites, applying metagenomic strategy to discover the untapped reservoir of genetic and metabolic diversity, and performing the chemical and biological modification of old antibiotics are the important aspects for future antibiotics discovery (Demain and Sanchez, 2009; Wesolowski, 2011).

Acknowledgements

This work was supported in part by the National Program on Key Basic Research Project (973 program, 2013CB734000), the National Natural Science Foundation of China (81102369, 30901849, 30911120483, 81102356, 30973665, 30911120484), the CAS Pillar Program (XDA04074000), the Ministry of Science and Technology of China (2013ZX10005004, 2011ZX11102-011-11, and 2007DFB31620), Genzyme, a Sanofi company, the Institute for Molecular Bioscience, the European Union's Seventh Framework Programme (FP7/2007–2013) under grant agreement no. 312184. L. Z. is an Awardee of the National Distinguished Young Scholar Program in China.

References

Abd El-Aal, H.H., Habib, E.E., and Mishrif, M.M. (2005). Wilms' tumor: the experience of the pediatric unit of Kasr El-Aini center of radiation oncology and nuclear medicine (NEMROCK). J. Egypt. Natl. Cancer Inst. *17*, 308–314.

Aboltins, C., Page, M., Buising, K., Jenney, A., Daffy, J., Choong, P., and Stanley, P. (2007). Treatment of staphylococcal prosthetic joint infections with debridement, prosthesis retention and oral rifampicin and fusidic acid. Clin. Microbiol. Infect. *13*, 586–591.

Abrams, B.B., Hänel, H., and Hoehler, T. (1991). Ciclopirox olamine: a hydroxypyridone antifungal agent. Clin. Dermatol. *9*, 471–477.

Abruzzo, G.K., Fromtling, R.A., Turnbull, T.A., and Giltinan, D.M. (1987). Effects of bifonazole, fluconazole, itraconazole, and terbinafine on the chemiluminescence response of immune cells. J. Antimicrob. Chemother. *20*, 61–68.

Abu-Salah, K. (1996). Amphotericin B: an update. Br. J. Biomed. Sci. *53*, 122.

Ainsworth, G., Brown, A.M., and Brownlee, G. (1947). Aerosporin, an antibiotic produced by Bacillus aerosporus Greer. Nature *160*, 878.

Alexander, B.D., and Perfect, J.R. (1997). Antifungal resistance trends towards the year 2000. Drugs *54*, 657–678.

Allington, D.R., and Rivey, M.P. (2001). Quinupristin/dalfopristin: a therapeutic review. Clin. Ther. *23*, 24–44.

Al-Nakib, T., Perjési, P., Varghese, R., and Meegan, M. (1997). Benzylideneindanones and benzylidenebenzosuberones: relationship between structure, antimycotic activity and acute toxicity. Med. Prin. Pract. *6*, 14–21.

Altman, S. (2013). Antibiotics present and future. Febs. Lett. *588*, 1–2.

Andersson, K. (1981). Pharmacokinetics of nitroimidazoles. Spectrum of adverse reactions. Scand. J. Infect. Dis. Suppl. *26*, 60.

Andersson, M.I., and MacGowan, A.P. (2003). Development of the quinolones. J. Antimicrob. Chemother. *51*, 1–11.

Andries, K., Verhasselt, P., Guillemont, J., Göhlmann, H.W., Neefs, J.-M., Winkler, H., Van Gestel, J., Timmerman, P., Zhu, M., and Lee, E. (2005). A diarylquinoline drug active on the ATP synthase of *Mycobacterium tuberculosis*. Science *307*, 223–227.

Anke, H., and Sterner, O. (1991). Comparison of the antimicrobial and cytotoxic activities of twenty unsaturated sesquiterpene dialdehydes from plants and mushrooms. Planta Med. *57*, 344–346.

Arbex, M.A., Varella, M.de C.L., de Siqueira, H.R., and de Mello, F.A.F. (2010). Antituberculosis drugs: drug interactions, adverse effects, and use in special situations-part 2: second line drugs. J. Bras. Pneumol. *36*, 641–656.

Archer, J.S., and Archer, D.F. (2002). Oral contraceptive efficacy and antibiotic interaction: a myth debunked. J. Am. Acad. Dermatol. *46*, 917–923.

Armstrong, D., Gold, J.W., Dryjanski, J., Whimbey, E., Polsky, B., Hawkins, C., Brown, A.E., Bernard, E., and Kiehn, T.E. (1985). Treatment of infections in patients with the acquired immunodeficiency syndrome. Ann. Intern. Med. *103*, 738–743.

Artis, W.M., Odle, B.M., and Jones, H.E. (1981). Griseofulvin-resistant dermatophytosis correlates with *in vitro* resistance. Arch. Dermatol. *117*, 16.

Ashburn, T.T., and Thor, K.B. (2004). Drug repositioning: identifying and developing new uses for existing drugs. Nat. Rev. Drug Discov. *3*, 673–683.

Attassi, K., Hershberger, E., Alam, R., and Zervos, M.J. (2002). Thrombocytopenia associated with linezolid therapy. Clin. Infect. Dis. *34*, 695–698.

Babar, S.M. (2013). SIADH associated with ciprofloxacin. Ann. Pharmacother. *47*, 1359–1363.

Baginski, M., Sternal, K., Czub, J., and Borowski, E. (2005). Molecular modelling of membrane activity of amphotericin B, a polyene macrolide antifungal antibiotic. Acta. Biochim. Pol. *52*, 655.

Baixench, M.-T., Aoun, N., Desnos-Ollivier, M., Garcia-Hermoso, D., Bretagne, S., Ramires, S., Piketty, C., and Dannaoui, E. (2007). Acquired resistance to echinocandins in *Candida albicans*: case report and review. J. Antimicrob. Chemother. *59*, 1076–1083.

Baley, J.E., Meyers, C., Kliegman, R.M., Jacobs, M.R., and Blumer, J.L. (1990). Pharmacokinetics, outcome of treatment, and toxic effects of amphotericin B and 5-fluorocytosine in neonates. J. Pediatr. *116*, 791–797.

Balfour, J.A., and Faulds, D. (1992). Terbinafine. Drugs *43*, 259–284.

Baltz, R.H. (2009). Daptomycin: mechanisms of action and resistance, and biosynthetic engineering. Curr. Opin. Chem. Biol. *13*, 144–151.

Banfi, E., Scialino, G., Zampieri, D., Mamolo, M.G., Vio, L., Ferrone, M., Fermeglia, M., Paneni, M.S., and Pricl, S. (2006). Antifungal and antimycobacterial activity of new imidazole and triazole derivatives. A combined experimental and computational approach. J. Antimicrob. Chemother. *58*, 76–84.

Bardini, M., Labra, M., Winfield, M., and Sala, F. (2003). Antibiotic-induced DNA methylation changes in calluses of *Arabidopsis thaliana*. Plant Cell Tissue Organ Cult. *72*, 157–162.

Barrett-Bee, K., and Dixon, G. (1995). Ergosterol biosynthesis inhibition: a target for antifungal agents. Acta Biochim. Pol. *42*, 465.

Benedict, R., and Langlykke, A. (1947). Antibiotic activity of *Bacillus polymyxa*. J. Bacteriol. *54*, 24.

Berdy, J. (2005). Bioactive microbial metabolites. J. Antibiot. *58*, 1–26.

Bertram, G.K., and Anthony, J. (1993). Pharmacology Examination and Board Review (Typopress, Lebanon).

Birmingham, M.C., Rayner, C.R., Meagher, A.K., Flavin, S.M., Batts, D.H., and Schentag, J.J. (2003). Linezolid for the treatment of multidrug-resistant, Gram-positive infections: experience from a compassionate-use program. Clin. Infect. Dis. *36*, 159–168.

Bollag, D.M., McQueney, P.A., Zhu, J., Hensens, O., Koupal, L., Liesch, J., Goetz, M., Lazarides, E., and Woods, C.M. (1995). Epothilones, a new class of microtubule-stabilizing agents with a taxol-like mechanism of action. Cancer Res. *55*, 2325–2333.

Bonomo, R.A., Van Zile, P.S., Li, Q., Shermock, K.M., McCormick, W.G., and Kohut, B. (2007). Topical triple-antibiotic ointment as a novel therapeutic choice in wound management and infection prevention: a practical perspective. Expert Rev. Anti. Infect. Ther. *5*, 773–782.

Bossche, H.V., Willemsens, G., and Marichal, P. (1987). Anti-candida drugs-the biochemical basis for their activity. Crit. Rev. Microbiol. *15*, 57–72.

Boucher, H.W., Talbot, G.H., Bradley, J.S., Edwards, J.E., Gilbert, D., Rice, L.B., Scheld, M., Spellberg, B., and Bartlett, J. (2009). Bad bugs, no drugs: no ESKAPE! An update from the Infectious Diseases Society of America. Clin. Infect. Dis. *48*, 1–12.

Bozdogan, B., and Appelbaum, P.C. (2004). Oxazolidinones: activity, mode of action, and mechanism of resistance. Int. J. Antimicrob. Ag. *23*, 113–119.

Brown, E.D., Vivas, E.I., Walsh, C.T., and Kolter, R. (1995). MurA (MurZ), the enzyme that catalyzes the first committed step in peptidoglycan biosynthesis, is essential in *Escherichia coli*. J. Bacteriol. *177*, 4194–4197.

Butler, M.S., and Cooper, M.A. (2011). Antibiotics in the clinical pipeline in 2011. J. Antibiot. *64*, 413–425.

Butler, M.S., Blaskovich, M.A., and Cooper, M.A. (2013). Antibiotics in the clinical pipeline in 2013. J. Antibiot. (Tokyo) *66*, 571–591.

Cain, C. (2012). Rediscovering antibiotics. Sci. B. X. *5*, doi:10.1038/scibx.2012.1198.

Calvori, C., Frontali, L., Leoni, L., and Tecce, G. (1965). Effect of rifamycin on protein synthesis. Nature *207*, 417–418.

Campàs, C. (2009). Drug repositioning summit: finding new routes to success. Drug News Perspect. *22*, 126–128.

Caprioli, V., Cimino, G., Colle, R., Gavagnin, M., Sodano, G., and Spinella, A. (1987). Insect antifeedant activity and hot taste for humans of selected natural and synthetic 1, 4-dialdehydes. J. Nat. Prod. *50*, 146–151.

Carstens, J., Wendelboe, P., Søgaard, H., and Thestrup-Pedersen, K. (1994). Toxic epidermal necrolysis and erythema multiforme following therapy with terbinafine. Acta Derm-venereol. *74*, 391.

Chang, R.K., Raoufinia, A., and Shah, N. (2012). Once daily formulations of tetracyclines (US Patent 8,206,740).

Charity, J.C., Katz, E., and Moss, B. (2007). Amino acid substitutions at multiple sites within the vaccinia virus D13 scaffold protein confer resistance to rifampicin. Virology *359*, 227–232.

Charney, J., Fisher, W., Curran, C., Machlowitz, R., and Tytell, A. (1953). Streptogramin, a new antibiotic. Antibiot. Chemother. *3*, 1283–1286.

Chen, S.C., and Sorrell, T.C. (2007). Antifungal agents. Med. J. Austral. *187*, 404.

Cohen, J. (2013). Approval of novel TB drug celebrated—with restraint. Science *339*, 130–130.

Connor, E.E. (1998). Sulfonamide antibiotics. Prim. Care Update Ob/Gyns. *5*, 32–35.

Cooper, M.A., and Shlaes, D. (2011). Fix the antibiotics pipeline. Nature *472*, 32–32.

Corallo, C.E., and Paull, A.E. (2002). Linezolid-induced neuropathy. Med. J. Aust. *177*, 332.

Cosar, C., Julou, L., and Bénazet, M. (1959). Activité de l'(hydroxy-2'éthyl)–1 méthyl-2 nitro-5 imidazole (8.823 RP) vis-à-vis des infections expérimentales à Trichomonas vaginalis. Ann. Inst. Pasteur (Paris) *96*, 238–241.

Cragg, G.M., Newman, D.J., and Snader, K.M. (1997). Natural products in drug discovery and development. J. Nat. Prod. *60*, 52–60.

Crawford, F., and Hollis, S. (2007). Topical treatments for fungal infections of the skin and nails of the foot. Cochrane Database Syst. Rev., 1–118.

da Cunha, F.M., Fröde, T.S., Mendes, G.L., Malheiros, A., Yunes, R.A., and Calixto, J.B. (2001). Additional evidence for the anti-inflammatory and anti-allergic properties of the sesquiterpene polygodial. Life Sci. *70*, 159–169.

Dai, T., Huang, Y.-Y., Sharma, S.K., Hashmi, J.T., Kurup, D.B., and Hamblin, M.R. (2010). Topical antimicrobials for burn wound infections. Recent Pat. Antiinfect. Drug Discov. *5*, 124.

Danshiitsoodol, N., de Pinho, C.A., Matoba, Y., Kumagai, T., and Sugiyama, M. (2006). The mitomycin C (MMC)-binding protein from MMC-producing microorganisms protects from the lethal effect of bleomycin: crystallographic analysis to elucidate the

binding mode of the antibiotic to the protein. J. Mol. Biol. *360*, 398–408.
Darrell, J., Garrod, L., and Waterworth, P.M. (1968). Trimethoprim: laboratory and clinical studies. J. Clin. Pathol. *21*, 202.
Davies, Y.K., Cox, K.M., Abdullah, B.A., Safta, A., Terry, A.B., and Cox, K.L. (2008). Long-term treatment of primary sclerosing cholangitis in children with oral vancomycin: an immunomodulating antibiotic. J. Pediatr. Gastr. Nutr. *47*, 61–67.
Debono, M., and Gordee, R.S. (1994). Antibiotics that inhibit fungal cell wall development. Annu. Rev. Microbiol. *48*, 471–497.
Debono, M., Barnhart, M., Carrell, C., Hoffmann, J., Occolowitz, J., Abbott, B., Fukuda, D., Hamill, R., Biemann, K., and Herlihy, W. (1987). A21978C, a complex of new acidic peptide antibiotics: isolation, chemistry, and mass spectral structure elucidation. J. Antibiot. *40*, 761–777.
Demain, A.L., and Sanchez, S. (2009). Microbial drug discovery: 80 years of progress. J. Antibiot. *62*, 5–16.
Deoghare, S. (2013). Bedaquiline: A new drug approved for treatment of multidrug-resistant tuberculosis. Indian J. Pharmacol. *45*, 536.
De Voss, J.J., Townsend, C.A., Ding, W.D., Morton, G.O., Ellestad, G.A., Zein, N., Tabor, A.B., and Schreiber, S.L. (1990). Site-specific atom transfer from DNA to a bound ligand defines the geometry of a DNA-calicheamicin.gamma.1I complex. J. Am. Chem. Soc. *112*, 9669–9670.
Diacon, A.H., Pym, A., Grobusch, M., Patientia, R., Rustomjee, R., Page-Shipp, L., Pistorius, C., Krause, R., Bogoshi, M., and Churchyard, G. (2009). The diarylquinoline TMC207 for multidrug-resistant tuberculosis. N. Engl. J. Med. *360*, 2397–2405.
Diasio, R.B., Lakings, D.E., and Bennett, J.E. (1978). Evidence for conversion of 5-fluorocytosine to 5-fluorouracil in humans: possible factor in 5-fluorocytosine clinical toxicity. Antimicrob. Agents Chemother. *14*, 903–908.
Diekema, D.J., and Jones, R.N. (2000). Oxazolidinones. Drugs *59*, 7–16.
Di Francesco, A., Favaroni, A., and Donati, M. (2013). Host defense peptides: general overview and an update on their activity against *Chlamydia* spp. Expert Rev. Anti. Infect. Ther. *11*, 1215–1224.
DiMasi, J.A., Hansen, R.W., and Grabowski, H.G. (2003). The price of innovation: new estimates of drug development costs. J. Health Econ. *22*, 151–185.
Dixon, D., and Walsh, T. (1996). Antifungal Agents. In Baron's Medical Microbiology, 4th edn. S. Baron, ed. (University of Texas Medical Branch, Galveston, USA) (via NCBI Bookshelf).
Dréno, B., Bettoli, V., Ochsendorf, F., Layton, A., Mobacken, H., and Degreef, H. (2004). European recommendations on the use of oral antibiotics for acne. Eur. J. Dermatol. *14*, 391–399.
Drlica, K. (1999). Mechanism of fluoroquinolone action. Curr. Opin. Microbiol. *2*, 504–508.
Duarte, P.A., Chow, C.C., Simmons, F., and Ruskin, J. (1984). Fatal hepatitis associated with ketoconazole therapy. Arch. Intern. Med. *144*, 1069.
Eddouzi, J., Parker, J.E., Vale-Silva, L.A., Coste, A., Ischer, F., Kelly, S., Manai, M., and Sanglard, D. (2013). Molecular mechanisms of drug resistance in clinical *Candida* species isolated from Tunisian hospitals. Antimicrob. Agents Chemother. *57*, 3182–3193.
Edwards, D.I. (1980). Mechanisms of selective toxicity of metronidazole and other nitroimidazole drugs. Br. J. Vener. Dis. *56*, 285–290.
Eliopoulos, G., Willey, S., Reiszner, E., Spitzer, P., Caputo, G., and Moellering, R. (1986). *In vitro* and *in vivo* activity of LY 146032, a new cyclic lipopeptide antibiotic. Antimicrob. Agents Chemother. *30*, 532–535.
Enoch, D.A., Bygott, J.M., Daly, M.-L., and Karas, J.A. (2007). Daptomycin. J. Infection. *55*, 205–213.
Epstein, I., Nair, K., and Boyd, L. (1955). Cycloserine, a new antibiotic, in the treatment of human pulmonary tuberculosis: a preliminary report. Antibiotic Med. Clin. Ther. *1*, 80.
Ernest, J. (1992). Topical antifungal agents. Obstet. Gyn. Clin. N. Am *19*, 587–607.
Evans, M.E., Feola, D.J., and Rapp, R.P. (1999). Polymyxin B sulfate and colistin: old antibiotics for emerging multiresistant Gram-negative bacteria. Ann. Pharmacother. *33*, 960–967.
Falagas, M.E., and Kasiakou, S.K. (2006). Toxicity of polymyxins: a systematic review of the evidence from old and recent studies. Crit. Care *10*, R27.
Farber, B.F., and Moellering, R. (1983). Retrospective study of the toxicity of preparations of vancomycin from 1974 to 1981. Antimicrob. Agents Chemother. *23*, 138–141.
Fass, R.J., and Perkins, R.L. (1971). 5-fluorocytosine in the treatment of cryptococcal and candida mycoses. Ann. Intern. Med. *74*, 535–539.
Felmingham, D. (2001). Microbiological profile of telithromycin, the first ketolide antimicrobial. Clin. Microbiol. Infect. *7*, 2–10.
Finegold, S.M. (1966). Toxicity of kanamycin in adults. Ann. N.Y. Acad. Sci. *132*, 942–956.
Fisher, J.F., Meroueh, S.O., and Mobashery, S. (2005). Bacterial resistance to β-lactam antibiotics: compelling opportunism, compelling opportunity. Chem. Rev. *105*, 395–424.
Francois, I.E.J.A., Cammue, B.P.A., Borgers, M., Ausma, J., Dispersyn, G.D., and Thevissen, K. (2006). Azoles: mode of antifungal action and resistance development. effect of miconazole on endogenous reactive oxygen species production in *Candida albicans*. Antiinfect. Agents Med. Chem. *5*, 3–13.
Fraschini, F., Scaglione, F., Ferrara, F., Marelli, O., Braga, P., and Teodori, F. (1986). Evaluation of the immunostimulating activity of erythromycin in man. Chemotherapy *32*, 286–290.
Fujiwara, A., Hoshino, T., and Westley, J.W. (1985). Anthracycline antibiotics. Crit. Rev. Biotechnol. *3*, 133–157.
Fuller, A., Mellows, G., Woolford, M., Banks, G., Barrow, K., and Chain, E. (1971). Pseudomonic acid: an antibiotic produced by *Pseudomonas fluorescens*. Nature *234*, 416–417.

Giamarellou, H., and Poulakou, G. (2009). Multidrug-resistant Gram-negative infections. Drugs *69*, 1879–1901.

Giordani, R., Buc, J., and Regli, P. (2002). Mathematical modelling of antifungal action. Mycoses *45*, 482–487.

Goh, C.L., Tay, Y.K., Ali, K., Koh, M.T., and Seow, C.S. (1994). *In vitro* evaluation of griseofulvin, ketoconazole, and itraconazole against various dermatophytes in Singapore. Int. J. Dermatol. *33*, 733–737.

Golan, Y., and Epstein, L. (2012). Safety and efficacy of fidaxomicin in the treatment of *Clostridium difficile*-associated diarrhea. Therape. Adv. Gastroenterol. *5*, 395–402.

Goldman, L., Schwarz, J., Preston, R.H., Beyer, A., and Loutzenhiser, J. (1960). Current status of griseofulvin: Report on one hundred seventy-five cases. J. Am. Med. Assoc. *172*, 532–538.

Goldman, R.C., Fesik, S., and Doran, C. (1990). Role of protonated and neutral forms of macrolides in binding to ribosomes from Gram-positive and Gram-negative bacteria. Antimicrob. Agents Chemother. *34*, 426–431.

Golik, J., Clardy, J., Dubay, G., Groenewold, G., Kawaguchi, H., Konishi, M., Krishnan, B., Ohkuma, H., Saitoh, K., and Doyle, T.W. (1987a). Esperamicins, a novel class of potent antitumor antibiotics. 2. Structure of esperamicin X. J. Am. Chem. Soc. *109*, 3461–3462.

Golik, J., Dubay, G., Groenewold, G., Kawaguchi, H., Konishi, M., Krishnan, B., Ohkuma, H., Saitoh, K., and Doyle, T.W. (1987b). Esperamicins, a novel class of potent antitumor antibiotics. 3. Structures of esperamicins A1, A2, and A1b. J. Am. Chem. Soc. *109*, 3462–3464.

Goodin, S., Kane, M.P., and Rubin, E.H. (2004). Epothilones: mechanism of action and biologic activity. J. Clin. Oncol. *22*, 2015–2025.

Grant, S.M., and Clissold, S.P. (1989). Itraconazole. Drugs *37*, 310–344.

Green, S.L., Maddox, J.C., and Huttenbach, E.D. (2001). Linezolid and reversible myelosuppression. J. Am. Med. Assoc. *285*, 1291–1291.

Grosset, J., and Leventis, S. (1983). Adverse effects of rifampin. Rev. infect. Dis. *5*, S440-S446.

Guay, D.R. (2001). An update on the role of nitrofurans in the management of urinary tract infections. Drugs *61*, 353–364.

Guignard, B., Vouillamoz, J., Giddey, M., and Moreillon, P. (2013). A positive interaction between inhibitors of protein synthesis and cefepime in the fight against methicillin-resistant *Staphylococcus aureus*. Eur. J. Clin. Microbiol *32*, 899–907.

Gupta, A.K. (2001). Ciclopirox: an overview. Int. J. Dermatol. *40*, 305–310.

Gupta, A.K., Sauder, D.N., and Shear, N.H. (1994). Antifungal agents: an overview. Part I. J. Am. Acad. Dermatol. *30*, 677–698.

Hager, T. (2006). The Demon Under the Microscope: From Battlefield Hospitals to Nazi Labs, One Doctor's Heroic Search for the World's First Miracle Drug (Random House, New York, USA).

Hamilton-Miller, J. (1973). Chemistry and biology of the polyene macrolide antibiotics. Bacteriol. Rev. *37*, 166.

Hanauske-Abel, H.M., Saxena, D., Palumbo, P.E., Hanauske, A.-R., Luchessi, A.D., Cambiaghi, T.D., Hoque, M., Spino, M., Gandolfi, D.D.A., and Heller, D.S. (2013). Drug-induced reactivation of apoptosis abrogates HIV-1 infection. PloS One *8*, e74414.

Hangeland, J.J., De Voss, J.J., Heath, J.A., Townsend, C.A., Ding, W.D., Ashcroft, J.S., and Ellestad, G.A. (1992). Specific abstraction of the 5′S- and 4′-deoxyribosyl hydrogen atoms from DNA by calicheamicin.gamma-1. J. Am. Chem. Soc. *114*, 9200–9202.

Hare, R. (1982). New light on the history of penicillin. Med. Hist. *26*, 1–24.

Harrison, E., Zwadyk, P., Bequette, R., Hamlow, E., Tavormina, P., and Zygmunt, W. (1970). Haloprogin: a topical antifungal agent. Appl. Microbiol. *19*, 746–750.

Hashimoto, Y., Noguchi, T., Kitagawa, H., and Ohta, G. (1966). Toxicologic and pharmacologic properties of tolnaftate, an antitrichophyton agent. Toxicol. Appl. Pharm. *8*, 380–385.

Hay, R. (1999). Therapeutic potential of terbinafine in subcutaneous and systemic mycoses. Br. J. Dermatol. *141*, 36–40.

Hecht, S.M. (2000). Bleomycin: new perspectives on the mechanism of action 1. J. Nat. Prod. *63*, 158–168.

Hector, R.F. (2005). An overview of antifungal drugs and their use for treatment of deep and superficial mycoses in animals. Clin. Tech. Small. An. P *20*, 240–249.

Heeres, J., Backx, L.J.J., Mostmans, J.H., and Van Cutsem, J. (1979). Antimycotic imidazoles. Part 4. Synthesis and antifungal activity of ketoconazole, a new potent orally active broad-spectrum antifungal agent. J. Med. Chem. *22*, 1003–1005.

Hermann, H. (1972a). Clinical evaluation of haloprogin. Toxicol. Appl. Pharm. *23*, 598–605.

Hermann, H.W. (1972b). Clinical efficacy studies of haloprogin, a new topical antimicrobial agent. Arch. Dermatol. *106*, 839.

Herrlich, P., and Schweiger, M. (1976). Nitrofurans, a group of synthetic antibiotics, with a new mode of action: discrimination of specific messenger RNA classes. Proc. Natl. Acad. Sci. U.S.A. *73*, 3386–3390.

Hollstein, U. (1974). Actinomycin. Chemistry and mechanism of action. Chem. Rev. *74*, 625–652.

Hoque, M., Hanauske-Abel, H.M., Palumbo, P., Saxena, D., D'Alliessi Gandolfi, D., Park, M.H., Pe'ery, T., and Mathews, M.B. (2009). Inhibition of HIV-1 gene expression by ciclopirox and deferiprone, drugs that prevent hypusination of eukaryotic initiation factor 5A. Retrovirology *6*, 1742–4690.

Hsuchen, C.-C., and Feingold, D.S. (1973). Selective membrane toxicity of the polyene antibiotics: studies on lecithin membrane models (liposomes). Antimicrob. Agents Chemother. *4*, 309–315.

Hughes, J., and Mellows, G. (1978a). Inhibition of isoleucyl-transfer ribonucleic acid synthetase in *Escherichia coli* by pseudomonic acid. Biochem. J *176*, 305–318.

Hughes, J., and Mellows, G. (1978b). On the mode of action of pseudomonic acid: inhibition of protein synthesis in *Staphylococcus aureus*. J. Antibiot. *31*, 330.

Iannini, P.B. (2007). The safety profile of moxifloxacin and other fluoroquinolones in special patient populations. Curr. Med. Res. Opin. *23*, 1403–1413.

Ishii, M., Hamada, T., and Asai, Y. (1983). Treatment of onychomycosis by ODT therapy with 20% urea ointment and 2% tolnaftate ointment. Dermatology *167*, 273–279.

Iverson, S.L., and Uetrecht, J.P. (2001). Identification of a reactive metabolite of terbinafine: insights into terbinafine-induced hepatotoxicity. Chem. Res. Toxicol. *14*, 175–181.

Jensen, P.B., Sørensen, B.S., Sehested, M., Demant, E.J., Kjeldsen, E., Friche, E., and Hansen, H.H. (1993). Different modes of anthracycline interaction with topoisomerase II: separate structures critical for DNA-cleavage, and for overcoming topoisomerase II-related drug resistance. Biochem. Pharmacol. *45*, 2025–2035.

Johnson, B.A., Anker, H., and Meleney, F.L. (1945). Bacitracin: a new antibiotic produced by a member of the *B. subtilis* group. Science *102*, 376–377.

Jones, D. (2011). A guiding hand for antibiotics. Nat. Rev. Drug Discov. *10*, 161–162.

Jukes, T.H. (1985). Some historical notes on chlortetracycline. Rev. Infect. Dis. *7*, 702–707.

Juliano, R., Lopez-Berestein, G., Hopfer, R., Mehta, R., Mehta, K., and Mills, K. (1985). Selective toxicity and enhanced therapeutic index of liposomal polyene antibiotics in systemic fungal infections. Ann. Ny. Acad. Sci. *446*, 390–402.

Julien, B., and Shah, S. (2002). Heterologous expression of epothilone biosynthetic genes in *Myxococcus xanthus*. Antimicrob. Agents Chemother. *46*, 2772–2778.

Kafetzis, D.A., Velissariou, I.M., Stabouli, S., Mavrikou, M., Delis, D., and Liapi, G. (2005). Treatment of paediatric visceral leishmaniasis: amphotericin B or pentavalent antimony compounds? Int. J. Antimicrob. Ag. *25*, 26–30.

Kaltenis, P. (1986). Cycloserine as a urinary tract antiseptic. Int. Urol. Nephrol. *18*, 125–130.

Katz, R., and Cahn, B. (1972). Haloprogin therapy for dermatophyte infections. Arch. Dermatol. *106*, 837.

Kauffman, C.A., and Frame, P.T. (1977). Bone marrow toxicity associated with 5-fluorocytosine therapy. Antimicrob. Agents Chemother. *11*, 244–247.

Kauffman, G.B. (1979). The discovery of penicillin: twentieth century wonder drug. J. Chem. Edu. *56*, 454.

Khatua, S., Nair, C.N., and Ghosh, K. (2004). Immune-mediated thrombocytopenia following dactinomycin therapy in a child with alveolar rhabdomyosarcoma: the unresolved issues. J. Pediatr. Hematol. Oncol. *26*, 777–779.

Kinsky, S., Avruch, J., Permutt, M., Rogers, H., and Schonder, A. (1962). The lytic effect of polyene antifungal antibiotics on mammalian erythrocytes. Biochem. Biophys. Res. Commun. *9*, 503–507.

Koota, G.M., Schweinburg, F.B., and Rutenburg, A.M. (1960). Role of kanamycin in the management of infections. N. Engl. J. Med. *263*, 629–633.

Kotler-Brajtburg, J., Price, H., Medoff, G., Schlessinger, D., and Kobayashi, G.S. (1974). Molecular basis for the selective toxicity of amphotericin B for yeast and filipin for animal cells. Antimicrob. Agents Chemother. *5*, 377–382.

Koyama, Y., Kurosasa, A., Tsuchiya, A., and Takakuta, K. (1950). A new antibiotic 'colistin'produced by spore-forming soil bacteria. J. Antibiot. *3*, 457–458.

Kubo, I., Fujita, K.i., and Lee, S.H. (2001). Antifungal mechanism of polygodial. J. Agr. Food Chem. *49*, 1607–1611.

Lacey, R. (1982). Do sulphonamide–trimethoprim combinations select less resistance to trimethoprim than the use of trimethoprim alone? J. Med. Microbiol. *15*, 403–427.

Lambert, M.P., and Neuhaus, F.C. (1972). Mechanism of D-cycloserine action: alanine racemase from *Escherichia coli* W. J. Bacteriol. *110*, 978–987.

Lamp, K.C., Freeman, C.D., Klutman, N.E., and Lacy, M.K. (1999). Pharmacokinetics and pharmacodynamics of the nitroimidazole antimicrobials. Clin. Pharmacokinet. *36*, 353–373.

Lancini, G., and Demain, A.L. (2013). Bacterial pharmaceutical products. In The Prokaryotes, E. Rosenberg, E.F. DeLong, S. Lory, E. Stackebrandt, and F. Thompson, eds. (Springer, Berlin Heidelberg), pp. 257–280.

Laxminarayan, R., and Powers, J.H. (2011). Antibacterial R&D incentives. Nat. Rev. Drug Discov. *10*, 727–728.

Leclercq, R., Bismuth, R., Casin, I., Cavallo, J., Croizé, J., Felten, A., Goldstein, F., Monteil, H., Quentin-Noury, C., and Reverdy, M. (2000). *In vitro* activity of fusidic acid against streptococci isolated from skin and soft tissue infections. J. Antimicrob. Chemother. *45*, 27–29.

Lee, M.D., Dunne, T.S., Chang, C.C., Ellestad, G.A., Siegel, M.M., Morton, G.O., McGahren, W.J., and Borders, D.B. (1987a). Calichemicins, a novel family of antitumor antibiotics. 2. Chemistry and structure of calichemicin gamma.1I. J. Am. Chem. Soc. *109*, 3466–3468.

Lee, M.D., Dunne, T.S., Siegel, M.M., Chang, C.C., Morton, G.O., and Borders, D.B. (1987b). Calichemicins, a novel family of antitumor antibiotics. 1. Chemistry and partial structure of calichemicin.gamma-1 J. Am. Chem. Soc. *109*, 3464–3466.

Lesher, G.Y., Froelich, E.J., Gruett, M.D., Bailey, J.H., and Brundage, R.P. (1962). 1, 8-Naphthyridine derivatives. A new class of chemotherapeutic agents. J. Med. Chem. *5*, 1063–1065.

Levine, D.P. (2006). Vancomycin: a history. Clin. Infect. Dis. *42*, S5-S12.

Lewis, K. (2013). Platforms for antibiotic discovery. Nat. Rev. Drug Discov. *12*, 371–387.

Lewis, R.J., and Mohr III, J.F. (2008). Dysglycaemias and fluoroquinolones. Drug Saf. *31*, 283–292.

Lewis, T.G., and Nydorf, E.D. (2006). Intralesional bleomycin for warts: a review. J. Drugs Dermatol. *5*, 499–504.

Liu, H., and Mulholland, S.G. (2005). Appropriate antibiotic treatment of genitourinary infections in hospitalized patients. Am. J. Med. *118*, 14–20.

Long, B.H. (2008). Fusidic acid in skin and soft-tissue infections. Acta Derm-venereol. *88*, 14–20.

Long, P.H. (1941). The clinical use of sulfanilamide, sulfapyridine, sulfathiazole, sulfaguanidine, and sulfadiazine in the prophylaxis and treatment of infections. Can. Med. Assoc. J. *44*, 217–227.

Lonks, J.R., and Goldmann, D.A. (2005). Telithromycin: a ketolide antibiotic for treatment of respiratory tract infections. Clin. Infect. Dis. *40*, 1657–1664.

Louie, T.J., Emery, J., Krulicki, W., Byrne, B., and Mah, M. (2009). OPT-80 eliminates *Clostridium difficile* and is sparing of bacteroides species during treatment of *C. difficile* infection. Antimicrob. Agents Chemother. *53*, 261–263.

Lyman, G.H., Kuderer, N.M., Crawford, J., Wolff, D.A., Culakova, E., Poniewierski, M.S., and Dale, D.C. (2011). Predicting individual risk of neutropenic complications in patients receiving cancer chemotherapy. Cancer *117*, 1917–1927.

Mühlradt, P.F., and Sasse, F. (1997). Epothilone B stabilizes microtubuli of macrophages like taxol without showing taxol-like endotoxin activity. Cancer Res. *57*, 3344–3346.

McCalla, D., Reuvers, A., and Kaiser, C. (1971). Breakage of bacterial DNA by nitrofuran derivatives. Cancer Res. *31*, 2184–2188.

MacGowan, A., and Albur, M. (2013). Frontline antibiotic therapy. Clin. Med. *13*, 263–268.

Maertens, J., Raad, I., Petrikkos, G., Boogaerts, M., Selleslag, D., Petersen, F.B., Sable, C.A., Kartsonis, N.A., Ngai, A., and Taylor, A. (2004). Efficacy and safety of caspofungin for treatment of invasive aspergillosis in patients refractory to or intolerant of conventional antifungal therapy. Clin. Infect. Dis. *39*, 1563–1571.

Maheshwai, N. (2007). Are young infants treated with erythromycin at risk for developing hypertrophic pyloric stenosis? Arch. Dis. Child. *92*, 271–273.

Maligie, M.A., and Selitrennikoff, C.P. (2005). *Cryptococcus neoformans* resistance to echinocandins: (1, 3) β-glucan synthase activity is sensitive to echinocandins. Antimicrob. Agents Chemother. *49*, 2851–2856.

Mandell, G.L., Bennett, J., and Dolin, R. (2000). Principles and Practice of Infectious Diseases. (Churchill Livingstone Press, Philadelphia, PA).

Mao, Y., Varoglu, M., and Sherman, D.H. (1999). Molecular characterization and analysis of the biosynthetic gene cluster for the antitumor antibiotic mitomycin C from *Streptomyces lavendulae* NRRL 2564. Chem. Biol. *6*, 251–263.

Martin, N.I., Hu, H., Moake, M.M., Churey, J.J., Whittal, R., Worobo, R.W., and Vederas, J.C. (2003). Isolation, structural characterization, and properties of mattacin (Polymyxin M), a cyclic peptide antibiotic produced by *Paenibacillus kobensis* M. J. Biol. Chem. *278*, 13124–13132.

Masters, P.A., O'Bryan, T.A., Zurlo, J., Miller, D.Q., and Joshi, N. (2003). Trimethoprim-sulfamethoxazole revisited. Arch. Intern. Med. *163*, 402.

Medoff, G., Comfort, M., and Kobayashi, G. (1971). Synergistic action of amphotericin B and 5-fluorocytosine against yeast-like organisms. Exp. Biol. Med. *138*, 571–574.

Medoff, G., Kobayashi, G., Kwan, C., Schlessinger, D., and Venkov, P. (1972). Potentiation of rifampicin and 5-fluorocytosine as antifungal antibiotics by amphotericin B. Proc. Natl. Acad. Sci. U.S.A. *69*, 196.

Mehlhorn, A.J., and Brown, D.A. (2007). Safety concerns with fluoroquinolones. Ann. Pharmacother. *41*, 1859–1866.

Meletiadis, J., Stergiopoulou, T., O'Shaughnessy, E.M., Peter, J., and Walsh, T.J. (2007). Concentration-dependent synergy and antagonism within a triple antifungal drug combination against *Aspergillus* species: analysis by a new response surface model. Antimicrob. Agents Chemother. *51*, 2053–2064.

Mendes, G.L., Santos, A.R., Campos, M.M., Tratsk, K.S., Yunes, R.A., and Calixto, J.B. (1998). Anti-hyperalgesic properties of the extract and of the main sesquiterpene polygodial isolated from the barks of *Drymis winteri* (Winteraceae). Life Sci. *63*, 369–381.

Mendes, G.L., Santos, A.R., Malheiros, A., Cechinel Filho, V., Yunes, R.A., and Calixto, J.B. (2000). Assessment of mechanisms involved in antinociception caused by sesquiterpene polygodial. J. Pharmacol. Exp. Ther *292*, 164–172.

Menzin, J., Meyers, J.L., Friedman, M., Perfect, J.R., Langston, A.A., Danna, R.P., and Papadopoulos, G. (2009). Mortality, length of hospitalization, and costs associated with invasive fungal infections in high-risk patients. Am. J. Health-Syst. Pharm. *66*, 1711–1717.

Michalopoulos, A.S., Livaditis, I.G., and Gougoutas, V. (2011). The revival of fosfomycin. Int. J. Infect. Dis. *15*, e732-e739.

Minotti, G., Menna, P., Salvatorelli, E., Cairo, G., and Gianni, L. (2004). Anthracyclines: molecular advances and pharmacologic developments in antitumor activity and cardiotoxicity. Pharmacol. Rev. *56*, 185–229.

Misumi, M., and Tanaka, N. (1980). Mechanism of inhibition of translocation by kanamycin and viomycin: a comparative study with fusidic acid. Biochem. Biophys. Res. Commun. *92*, 647–654.

Moellering, R.C. (2006). Vancomycin: a 50-year reassessment. Clin. Infect. Dis. *42*, S3-S4.

Moellering Jr, R.C. (2011). Discovering new antimicrobial agents. Int. J. Antimicrob. Ag. *37*, 2–9.

Mukhtar, T.A., and Wright, G.D. (2005). Streptogramins, oxazolidinones, and other inhibitors of bacterial protein synthesis. Chem. Rev. *105*, 529–542.

Mullane, K.M., and Gorbach, S. (2011). Fidaxomicin: first-in-class macrocyclic antibiotic. Expert Rev. Anti. Infect. Ther. *9*, 767–777.

Mullard, A. (2013). 2012 FDA drug approvals. Nat. Rev. Drug. Discov. *12*, 87–90.

Neu, H.C., and Gootz, T.D. (1996). Antimicrobial chemotherapy: antimicrobial inhibitors of ribosome function. In Baron's Medical Microbiology, 4th edn. S. Baron, ed. (University of Texas Medical Branch, Galveston, USA) (via NCBI Bookshelf).

Nicolaou, K., Smith, A., and Yue, E. (1993). Chemistry and biology of natural and designed enediynes. Proc. Natl. Acad. Sci. U.S.A. *90*, 5881–5888.

Niewerth, M., Kunze, D., Seibold, M., Schaller, M., Korting, H.C., and Hube, B. (2003). Ciclopiroxolamine treatment affects the expression pattern of *Candida albicans* genes encoding virulence factors, iron metabolism proteins, and drug resistance factors. Antimicrob. Agents Chemother. *47*, 1805–1817.

O'Reardon, J., Hetzmecker, J., Rynn, M., Baldassano, C., and Szuba, M. (2002). Desipramine toxicity with terbinafine. Am. J. Psychiat. *159*, 492–492.

Odds, F.C., Brown, A.J., and Gow, N.A. (2003). Antifungal agents: mechanisms of action. Trends Microbiol. *11*, 272–279.

Onishi, J., Meinz, M., Thompson, J., Curotto, J., Dreikorn, S., Rosenbach, M., Douglas, C., Abruzzo, G., Flattery, A., and Kong, L. (2000). Discovery of novel antifungal (1, 3)-β-D-glucan synthase inhibitors. Antimicrob. Agents Chemother. *44*, 368–377.

Oroshnik, W., and Mebane, A. (1963). The polyene antifungal antibiotics. In Progress in the Chemistry of Organic Natural Products/Progrès Dans La Chimie Des Substances Organiques Naturelles, L. Zechmeister, ed. (Springer Vienna, Wien, Austria), pp. 17–79.

Osheroff, N. (1986). Eukaryotic topoisomerase II. Characterization of enzyme turnover. J. Biol. Chem. *261*, 9944–9950.

Owens, R.C., and Ambrose, P.G. (2005). Antimicrobial safety: focus on fluoroquinolones. Clin. Infect. Dis. *41*, S144-S157.

Oxford, A.E., Raistrick, H., and Simonart, P. (1939). Studies in the biochemistry of micro-organisms: Griseofulvin, C17H17O6Cl, a metabolic product of Penicillium griseo-fulvum Dierckx. Biochem. J. *33*, 240.

Padilla, I., and Burgos, L. (2010). Aminoglycoside antibiotics: structure, functions and effects on *in vitro* plant culture and genetic transformation protocols. Plant Cell Rep. *29*, 1203–1213.

Pang, B., Qiao, X., Janssen, L., Velds, A., Groothuis, T., Kerkhoven, R., Nieuwland, M., Ovaa, H., Rottenberg, S., and van Tellingen, O. (2013). Drug-induced histone eviction from open chromatin contributes to the chemotherapeutic effects of doxorubicin. Nat. Commun. *4*, 1908.

Parenti, M., Hatfield, S., and Leyden, J. (1987). Mupirocin: a topical antibiotic with a unique structure and mechanism of action. Clin. Pharm. *6*, 761.

Patel, S.S., Balfour, J.A., and Bryson, H.M. (1997). Fosfomycin tromethamine. Drugs *53*, 637–656.

Payne, D.J., Gwynn, M.N., Holmes, D.J., and Pompliano, D.L. (2006). Drugs for bad bugs: confronting the challenges of antibacterial discovery. Nat. Rev. Drug Discov. *6*, 29–40.

Peng, X., Chen, B., Lim, C.C., and Sawyer, D.B. (2005). The cardiotoxicology of anthracycline chemotherapeutics: translating molecular mechanism into preventative medicine. Mol. Interv. *5*, 163–171.

Penicillins, O. (2000). Appropriate prescribing of oral β-lactam antibiotics. Am. Fam. Physician *62*, 611–620.

Perea, S., López-Ribot, J.L., Kirkpatrick, W.R., McAtee, R.K., Santillán, R.A., Martínez, M., Calabrese, D., Sanglard, D., and Patterson, T.F. (2001). Prevalence of molecular mechanisms of resistance to azole antifungal agents in *Candida albicans* strains displaying high-level fluconazole resistance isolated from human immunodeficiency virus-infected patients. Antimicrob. Agents Chemother. *45*, 2676–2684.

Pestka, S. (1974). The use of inhibitors in studies of protein synthesis. Method. Enzymol. *30*, 261–282.

Peter, R., and Richarz, B. (1995). Successful treatment and prophylaxis of scalp seborrhoeic dermatitis and dandruff with 2% ketoconazole shampoo: results of a multicentre, double-blind, placebo-controlled trial. Br. J. Dermatol. *132*, 441–445.

Petranyi, G., Meingassner, J.G., and Mieth, H. (1987). Antifungal activity of the allylamine derivative terbinafine *in vitro*. Antimicrob. Agents Chemother. *31*, 1365–1368.

Petri Jr, W.A. (2007). The first miracle drugs: how the sulfa drugs transformed medicine. JAMA. *297*, 1494–1495.

Pfaller, M.A. (2012). Antifungal drug resistance: mechanisms, epidemiology, and consequences for treatment. Am. J. Med. *125*, S3-S13.

Pfaller, M., Diekema, D., Messer, S., Hollis, R., and Jones, R. (2003). *In vitro* activities of caspofungin compared with those of fluconazole and itraconazole against 3,959 clinical isolates of *Candida* spp., including 157 fluconazole-resistant isolates. Antimicrob. Agents Chemother. *47*, 1068–1071.

Phillips, K., and Hailey, F. (1986). The use of furoxone: a perspective. J. Int. Med. Res. *14*, 19–29.

Pocidalo, J.-J. (1989). Use of fluoroquinolones for intracellular pathogens. Rev. infect. Dis. *11*, S979-S984.

Pogliano, J., Pogliano, N., and Silverman, J.A. (2012). Daptomycin-mediated reorganization of membrane architecture causes mislocalization of essential cell division proteins. J. Bacteriol. *194*, 4494–4504.

Polak, A., and Scholer, H. (1975). Mode of action of 5-fluorocytosine and mechanisms of resistance. Chemotherapy *21*, 113–130.

Policy, I.P. (2010). The 10 '20 initiative: pursuing a global commitment to develop 10 new antibacterial drugs by 2020. Clin. Infect. Dis. *50*, 1081–1083.

Posteraro, B., Posteraro, P., and Sanguinetti, M. (2013). Update on antifungal resistance and its clinical impact. Curr. Fungal Infect. Rep. *7*, 224–234.

Prigot, A., Shidlovsky, B.A., and Campbell, E.A. (1958). Intraperitoneal use of kanamycin as an adjunct in the therapy of established peritonitis and peritoneal contamination. Ann. N.Y. Acad. Sci. *76*, 204–211.

Prosser, G.A., and Carvalho, L.P.S. (2013). Kinetic mechanism and inhibition of *Mycobacterium tuberculosis* D-alanine: D-alanine ligase by the antibiotic D-cycloserine. FEBS J. *280*, 1150–1166.

Reichenbach, H., and Höfle, G. (2008). Discovery and development of the epothilones: a novel class of antineoplastic drugs. Drugs R. D. *9*, 1–10.

Ren B., Dai, H.-Q., Pei, G., Tong, Y.-J., Zhuo, Y., Yang, N., Su, M.-Y., Huang, P., Yang, Y.-Z., and Zhang, L.-X. (2013). ABC transporters coupled with the elevated ergosterol contents contribute to the azole resistance and amphotericin B susceptibility. Appl. Microbiol. Biot. *98*, 2609–2616.

Renault, J., Baron, M., and Mailliet, P. (1981). Heterocyclic quinones. 2. Quinoxaline-5, 6-(and

5–8)-diones–potential antitumoral agents. Eur. J. Med. Chem. *16*, 545–550.

Restrepo, A., Robledo, J., Gómez, I., Tabares, A.M., and Gutiérrez, R. (1986). Itraconazole therapy in lymphangitic and cutaneous sporotrichosis. Arch. Dermatol. *122*, 413–417.

Revill, P., Serradell, N., and Bolos, J. (2006). Tiacumicin B: macrolide antibiotic treatment of *C. difficile*-associated diarrhea. Drugs of the Future *31*, 494–497.

Rich, M.L., Ritterhoff, R.J., and Hoffmann, R.J. (1950). A fatal case of aplastic anemia following chloramphenicol (chloromycetin) therapy. Ann. Intern. Med. *33*, 1459–1467.

Richards, C., Pantanowitz, L., and Dezube, B.J. (2011). Antimicrobial and non-antimicrobial tetracyclines in human cancer trials. Pharmacol. Res. *63*, 151–156.

Rosenberg, S., DeVita, V.T., and Hellman, S. (2005). Cancer: Principles & Practice of Oncology. (Lippincott Williams & Wilkins, Hagerstwon, MD).

Rossi, S. (2003). Australian Medicines Handbook. In Encyclopedia of Clinical Pharmacy, T.D. Joseph, ed. (Marcel Dekker Inc., New York, USA), pp. 75–77.

Rossi, S. (2006). Australian Medicines Handbook 2006. Adelaide, Australia: Australian Medicines Handbook (ISBN 0–9757919–2–3).

Rosso, D. (1998). Hepatitis associated with terbinafine therapy: three case reports and a review of the literature. Clin. Exp. Dermatol. *23*, 64–67.

Roth, B., Falco, E.A., Hitchings, G.H., and Bushby, S. (1962). 5-Benzyl-2, 4-diaminopyrimidines as antibacterial agents. I. Synthesis and antibacterial activity *in vitro*. J. Med. Chem. *5*, 1103–1123.

Rouveix, B. (2006). Clinically significant toxicity and tolerance of the main antibiotics used in lower respiratory tract infections. Med. Maladies. Infect. *36*, 697–705.

Rozman, D., and Monostory, K. (2010). Perspectives of the non-statin hypolipidemic agents. Pharmacol. Therapeut. *127*, 19–40.

Rubin, Z. (1977). Ophthalmic sulfonamide-induced Stevens–Johnson syndrome. Arch. Dermatol. *113*, 235–236.

Rybacki, J.J. (2004). The Essential Guide to Prescription Drugs 2005. Everything You Need to Know for Safe Drug Use (HarperCollins Publishers Inc., New York).

Ryder, N., Frank, I., and Dupont, M. (1986). Ergosterol biosynthesis inhibition by the thiocarbamate antifungal agents tolnaftate and tolciclate. Antimicrob. Agents Chemother. *29*, 858–860.

Saag, M.S., and Dismukes, W.E. (1988). Azole antifungal agents: emphasis on new triazoles. Antimicrob. Agents Chemother. *32*, 1–8.

Sanglard, D., Kuchler, K., Ischer, F., Pagani, J., Monod, M., and Bille, J. (1995). Mechanisms of resistance to azole antifungal agents in *Candida albicans* isolates from AIDS patients involve specific multidrug transporters. Antimicrob. Agents Chemother. *39*, 2378–2386.

Sanglard, D., Ischer, F., Koymans, L., and Bille, J. (1998). Amino acid substitutions in the cytochrome P450 lanosterol 14α-demethylase (CYP51A1) from azole-resistant *Candida albicans* clinical isolates contribute to resistance to azole antifungal agents. Antimicrob. Agents Chemother. *42*, 241–253.

Saravolatz, L.D., Deresinski, S.C., and Stevens, D.A. (2003). Caspofungin. Clin. Infect. Dis. *36*, 1445–1457.

Sarro, A., and Sarro, G. (2001). Adverse reactions to fluoroquinolones. An overview on mechanistic aspects. Curr. Med. Chem. *8*, 371–384.

Scheinfeld, N. (2004). Telithromycin: a brief review of a new ketolide antibiotic. J. Drugs Dermatol. *3*, 409–413.

Schopf, E., Stuhmer, A., Rzany, B., Victor, N., Zentgraf, R., and Kapp, J.F. (1991). Toxic epidermal necrolysis and Stevens–Johnson syndrome: an epidemiologic study from West Germany. Arch. Dermatol. *127*, 839–842.

Sensi, P. (1982). Rifampicin. In Chronicles of Drug Discovery, Bindra, J.S. and Lednicer, D. eds. (Wiley Publishing, New York, USA:), pp. 201–221.

Shadomy, S. (1969). *In vitro* studies with 5-fluorocytosine. Appl. Microbiol. *17*, 871–877.

Shinabarger, D. (1999). Mechanism of action of the oxazolidinone antibacterial agents. Expert Opin. Investig. Drugs *8*, 1195–1202.

Shnayerson, M., and Plotkin, M.J. (2003). The Killers Within: the deadly rise of drug-resistant bacteria (Back Bay Books, Boston, USA).

Shu, X., Linet, M., Gao, R., Gao, Y., Brinton, L., Jin, F., and Fraumeni JR, J. (1987). Chloramphenicol use and childhood leukaemia in Shanghai. The Lancet *330*, 934–937.

Silver, L.L. (2011). Challenges of antibacterial discovery. Clin. Microbiol. Rev. *24*, 71–109.

Sköld, O. (2000). Sulfonamide resistance: mechanisms and trends. Drug Resist. Update. *3*, 155–160.

Smith, A. (2002). Screening for drug discovery: the leading question. Nature *418*, 453–459.

Smith, J. (1986). The mode of action of 4-quinolones and possible mechanisms of resistance. J. Antimicrob. Chemother. *18*, 21–29.

Sobell, H.M. (1985). Actinomycin and DNA transcription. Proc. Natl. Acad. Sci. U.S.A. *82*, 5328–5331.

Sodeik, B., Griffiths, G., Ericsson, M., Moss, B., and Doms, R.W. (1994). Assembly of vaccinia virus: effects of rifampin on the intracellular distribution of viral protein p65. J. Virol. *68*, 1103–1114.

Soltani, S., Dianat, S., and Sardari, S. (2009). Forward modeling of the coumarin antifungals; SPR/SAR based perspective. Avicenna J. Med. Biotechnol. *1*, 95–103.

Song, C.M., Lim, S.J., and Tong, J.C. (2009). Recent advances in computer-aided drug design. Brief. Bioinform. *10*, 579–591.

Spellberg, B., Sharma, P., and Rex, J.H. (2012). The critical impact of time discounting on economic incentives to overcome the antibiotic market failure. Nat. Rev. Drug Discov. *11*, 168–168.

Spelman, D. (1999). Fusidic acid in skin and soft tissue infections. Int. J. Antimicrob. Ag. *12*, S59-S65.

Spratt, B. (1980). Biochemical and genetical approaches to the mechanism of action of penicillin. Philos. Trans. R. Soc. Lond. B Biol. Sci. *289*, 273–283.

Srivastava, A., Talaue, M., Liu, S., Degen, D., Ebright, R.Y., Sineva, E., Chakraborty, A., Druzhinin, S.Y., Chatterjee, S., and Mukhopadhyay, J. (2011). New target for

inhibition of bacterial RNA polymerase: 'switch region'. Curr. Opin. Microbiol. *14*, 532–543.
Stansly, P.G., Shepherd, R., and White, H. (1947). Polymyxin: a new chemotherapeutic agent. Bull. Johns Hopkins Hosp. *81*, 43–54.
Steckbeck, J.D., Deslouches, B., and Montelaro, R.C. (2014). Antimicrobial peptides: new drugs for bad bugs? Expert. Opin. Biol. Th., 1–4.
Stiller, E., Vandeputte, J., and Wachtel, J. (1955). Amphotericins A and B, antifungal antibiotics produced by a streptomycete. II. The isolation and properties of the crystalline amphotericins. Antibiot. Annu. *3*, 587–591.
Stone, K.J., and Strominger, J.L. (1971). Mechanism of action of bacitracin: complexation with metal ion and C55-isoprenyl pyrophosphate. Proc. Natl. Acad. Sci. U.S.A. *68*, 3223–3227.
Streich, W., Hübel, S., and Franke, R. (1988). The use of mathematical logics in drug design. Math. Comput. Model. *11*, 647–651.
Subissi, A., Monti, D., Togni, G., and Mailland, F. (2010). Ciclopirox. Drugs *70*, 2133–2152.
Takimoto, C.H., and Calvo, E. (2008). Principles of oncologic pharmacotherapy. In Cancer Management: A Multidisciplinary Approach, R. Pazdur, L.R. Coia, W.J. Hoskins, and L.D. Wagman, eds. (CMPMedica, Lawrence, KS), pp. 23–42.
Talbot, G.H., Bradley, J., Edwards, J.E., Gilbert, D., Scheld, M., and Bartlett, J.G. (2006). Bad bugs need drugs: an update on the development pipeline from the Antimicrobial Availability Task Force of the Infectious Diseases Society of America. Clin. Infect. Dis. *42*, 657–668.
Tan, J.S., and File Jr, T.M. (1995). Antipseudomonal penicillins. Med. Clin. North Am. *79*, 679–693.
Tassel, D., and Madoff, M.A. (1968). Treatment of *Candida* sepsis and *Cryptococcus* meningitis with 5-fluorocytosine a new antifungal agent. JAMA. *206*, 830–832.
Thienpont, D., Van Cutsem, J., and Borgers, M. (1980). Ketoconazole in experimental candidosis. Rev. Infect. Dis. *2*, 570–577.
Thomas, R.J., Dumler, J.S., and Carlyon, J.A. (2009). Current management of human granulocytic anaplasmosis, human monocytic ehrlichiosis and Ehrlichia ewingii ehrlichiosis. Expert Rev. Anti. Infect. Ther. *7*, 709–722.
Tomasz, M. (1995). Mitomycin C: small, fast and deadly (but very selective). Chem. Biol. *2*, 575–579.
Trapnell, B.C., McColley, S.A., Kissner, D.G., Rolfe, M.W., Rosen, J.M., McKevitt, M., Moorehead, L., Montgomery, A.B., and Geller, D.E. (2012). Fosfomycin/tobramycin for inhalation in patients with cystic fibrosis with pseudomonas airway infection. Am. J. Resp. Crit. Care *185*, 171–178.
Tratsk, K., Campos, M., Vaz, Z., Schlemper, V., Yunes, R., and Calixto, J. (1997). Anti-allergic effects and oedema inhibition caused by the extract of Drymis winteri. Inflamm. Res. *46*, 509–514.
Turkington, C., and Ashby, B. (2007). The encyclopedia of infectious diseases (Infobase Publishing, New York, USA).
Tyle, J.H. (1984). Ketoconazole: mechanism of action, spectrum of activity, pharmacokinetics, drug interactions, adverse reactions and therapeutic use. Pharmacotherapy *4*, 343–373.
Ullmann, A.J. (2003). Review of the safety, tolerability, and drug interactions of the new antifungal agents caspofungin and voriconazole. Curr. Med. Res. Opin. *19*, 263–271.
Umezawa, H. (1958). Kanamycin: its discovery. Ann. N.Y. Acad. Sci. *76*, 20–26.
Vanden, B.H., Dromer, F., Improvisi, I., Lozano-Chiu, M., Rex, J., and Sanglard, D. (1998). Antifungal drug resistance in pathogenic fungi. Med. Mycol. *36*, 119–128.
Vanderhelst, E., De Wachter, E., Willekens, J., Pierard, D., Vincken, W., and Malfroot, A. (2013). Eradication of chronic methicillin-resistant *Staphylococcus aureus* infection in cystic fibrosis patients. An observational prospective cohort study of 11 patients. J. Cyst. Fibros. *12*, 662–666.
Varhe, A., Olkkola, K.T., and Neuvonen, P.J. (1994). Oral triazolam is potentially hazardous to patients receiving systemic antimycotics ketoconazole or itraconazole. Clin. Pharmacol. Ther. *56*, 601–607.
Venugopal, A.A., and Johnson, S. (2012). Fidaxomicin: a novel macrocyclic antibiotic approved for treatment of *Clostridium difficile* infection. Clin. Infect. Dis. *54*, 568–574.
Waksman, S.A., and Woodruff, H.B. (1940). Bacteriostatic and bactericidal substances produced by a aoil Actinomyces. Exp. Biol. Med. *45*, 609–614.
Walsh, C. (2000). Molecular mechanisms that confer antibacterial drug resistance. Nature *406*, 775–781.
Ward, A., and Campoli-Richards, D.M. (1986). Mupirocin. Drugs *32*, 425–444.
Weber Jr, F., Richards, R., and McCallum, R. (1993). Erythromycin: a motilin agonist and gastrointestinal prokinetic agent. Am. J. Gastroenterol. *88*, 485–490.
Wehrli, W., Knüsel, F., Schmid, K., and Staehelin, M. (1968). Interaction of rifamycin with bacterial RNA polymerase. Proc. Natl. Acad. Sci. U.S.A. *61*, 667–673.
Weikel Jr, J., and Bartek, M. (1972). Toxicologic properties and metabolic fate of haloprogin, an antifungal agent. Toxicol. Appl. Pharm. *22*, 375–386.
Weiss, R.B. (1992). The anthracyclines: will we ever find a better doxorubicin? Semin. Oncol. *19*, 670–686.
Wesolowski, D., Tae, H.S., Gandotra, N., Llopis, P., Shen, N., and Altman, S. (2011). A basic peptide–morpholino conjugate that is very effective in killing bacteria by gene specific and non-specific modes. Proc. Natl. Acad. Sci. U.S.A. *108*, 16582– 16587.
Wilmink, A., and Dons, J. (1993). Selective agents and marker genes for use in transformation of monocotyledonous plants. Plant Mol. Biol. Rep. *11*, 165–185.
Winkelman, W., and Gratton, D. (1989). Topical antibacterials. Clin. Dermatol. *7*, 156–162.
Wolfe, C.R. (2011). Case report: treatment of chronic osteomyelitis. Clin. Infect. Dis. *52*, S538-S541.
Wood, A.J., Como, J.A., and Dismukes, W.E. (1994). Oral azole drugs as systemic antifungal therapy. N. Engl. J. Med. *330*, 263–272.

Wormser, G.P., Keusch, G.T., and Heel, R.C. (1982). Co-trimoxazole (trimethoprim–sulfamethoxazole). Drugs *24*, 459–518.
Wormser, G.P., Dattwyler, R.J., Shapiro, E.D., Halperin, J.J., Steere, A.C., Klempner, M.S., Krause, P.J., Bakken, J.S., Strle, F., and Stanek, G. (2006). The clinical assessment, treatment, and prevention of Lyme disease, human granulocytic anaplasmosis, and babesiosis: clinical practice guidelines by the Infectious Diseases Society of America. Clin. Infect. Dis. *43*, 1089–1134.
Wright, G.D. (2005). Bacterial resistance to antibiotics: enzymatic degradation and modification. Adv. Drug Delivery Rev. *57*, 1451–1470.
Wright, G.D., and Sutherland, A.D. (2007). New strategies for combating multidrug-resistant bacteria. Trends Mol. Med. *13*, 260–267.
Yunis, A. (1989). Chloramphenicol toxicity: 25 years of research. Am. J. Med. *87*, 44N–48N.
van Zanten, A. (2010). New candidates for Candida: Echinocandins. Neth. J. Crit. Care. *14*, 242–243.
Zöllner, E., Delport, S., and Bonnici, F. (2001). Fatal liver failure due to ketoconazole treatment of a girl with Cushing's syndrome. J. Pediatr. Endocrinol. Metab. *14*, 335–338.
Zhang, L., Yan, K., Zhang, Y., Huang, R., Bian, J., Zheng, C., Sun, H., Chen, Z., Sun, N., and An, R. (2007). High-throughput synergy screening identifies microbial metabolites as combination agents for the treatment of fungal infections. Proc. Natl. Acad. Sci. U.S.A. *104*, 4606–4611.

Microorganisms Producing Antibiotics

3

János Bérdy

Abstract

The definition of antibiotics, that is bioactive secondary microbial metabolites, and specific features of the history of antibiotic research in the last 80 years shows continuous change. The real role of microbial metabolites in various life processes on Earth is still to be totally explored. Their importance, in addition to the antibiotic effect, seems to be more significant than it was believed. They may be general regulatory and signalling compounds in most life processes and in the environment.

The main producers of the microbial metabolites, i.e. the actinobacteria, fungi and other filamentous bacteria, represent inexhaustible sources for the future. The increasing taxonomy, chemistry and biodiversity of microbial metabolites will certainly give numerous new leads and new drugs and other products, not only for chemotherapy, but also in other fields of human therapy and in agriculture to cure and feed the increasing population.

The numbers of known and possibly new secondary metabolite-producing microorganisms, the produced as well as new chemical compounds, and their extremely wide range of bioactivities are discussed in detail. Several new trends in research, the full exploration of the living world of oceans, and the discovery of endophytic microbes are also presented.

The antibiotic producers and antibiotic functions

Definitions

The term "antibiotic" itself, their origins, i.e. the real antibiotic-producing organisms, and the significance of their effects, i.e. the ecological functions in the lower and higher forms of life of the compounds called antibiotics, are far from clear today.

The original term used by Waksman is derived from the words: "anti" and "biosis". However, the term "antibiosis" was used earlier. Over time, the definition continuously changed. For example: "... a chemical substance of microbial origin possessing antibiotic effect", "...secondary microbial metabolites which inhibit the growth of other microorganisms or kill them", "a chemical compound used in the chemotherapy", etc. Definitions were given at different times. The two main features in all definitions are (1) the origin, i.e. the producer organism and (2) the product with the biological effect. In all definitions, the key expressions were the *producer* and the *product(s)*. The *producer* is a living organism, belonging to the Plant or Animal Kingdom (prokaryotes or eukaryotes, i.e. microbe or higher organism), producing by biosynthesis the given metabolite(s), the *products,* which are low molecular weight organic compounds, i.e. the secondary metabolite(s) exhibiting some kind of biological effect.

The current meaning of the word antibiotic in various contexts represents somewhat different things in the minds of microbiologists, chemists, physicians and everyday people. For microbiologists, the antibiotics are secondary metabolites isolated from microorganisms. The organic chemists include the small natural antibiotics and the synthetic compounds derived as mimics of the natural products. Physicians, include all chemical compounds, sometimes including primary metabolites; also macromolecular compounds, vaccines and all chemically synthesized compounds (e.g. sulfonamides) which have antibiotic (chemotherapeutic) or similar effects. In the veterinary, agriculture and food industries, the "antibiotics" are used in numerous ways and represent one type of substance among the many chemicals used. For most people, the antibiotics still are the "magic bullets".

The meaning of "antibiotic effect" is also different from various points of view. First, besides the antibacterial metabolites, the antifungal, antiprotozoal, antiviral and antitumour natural metabolites (most of which also inhibit bacteria) are included. The antiprotozoals, antiparasitics, phytotoxics, insecticides, and similar agents are also covered and called antimicrobial/antibiotic compounds. According to some new definitions, antibiotics are derived from living things and inhibit or modify the life-cycle of other organisms or kill them. They are considered to cover all physiologically, pharmacologically, and immunologically active agents produced by the pharmaceutical industry, including all microbial products used as chemotherapeutic agents, enzyme inhibitors, antimetabolites, and many other intimately linked terms, such as plant growth modulators, feed additives, or inhibitors (insecticides, miticides, antiparasitics, phytotoxins, herbicides, etc.). Some are referred to as *agricultural antibiotics*. (Davies and Ryan, 2012). I prefer calling all small molecular weight chemical compounds derived from the living world, mainly from microbes, *bioactive secondary metabolites*.

Historical remarks

The discovery and use of the antibiotics in the last century represented one of the greatest benefits of mankind. Until the 1960s and 1970s, everything seemed to be blossoming and well controlled. It was the Golden Era of antibiotics. Unfortunately, this situation dramatically changed. About 15 to 20 years ago, and to a lesser extent nowadays, it seems that the antibiotics became victims of their own success. The reasons for the declining tendencies were the dangers of increasingly acquired multiresistant strains and the re-emergence of opportunistic old pathogens. Of importance were the continuous human irresponsibilities such as unnecessary administrations, the emerged economic, political and regulatory difficulties, the increasing cost of research, the relatively short application times of these drugs, together with decreasing profits. Additionally, the strict regulations, licensing problems, and several scientific failures (e.g. overestimating the benefits of high throughput screening (HTS) and combinational chemistry) led to a major decrease in the antibiotic drug pipeline (Bérdy, 2012; Butler *et al.*, 2013).

Humankind, with the mass production and uncontrolled applications of antibiotics, had rushed in a very aggressive way into the untouched world of the microbes which had existed in variable environments for 3.5 billion years under continuous evolutionary pressure. Fighting with other microbes and with the changing outside world, the microbes acquired tremendous biosynthetic experience. The answer of the microbes to the humans within a few decades was similarly aggressive ("The Empire Strikes Back"). They were originally and became even more incredibly adaptable, replicating and mutating rapidly and developing, in addition to simple resistance, numerous additional survival mechanisms via transcription modulation. Many of these were undesirable; for example, virulence (Appelbaum, 2012).

People, including scientists, for a long time, did not realize the dangers. In the last decades of the 20th century, most of the great pharmaceutical companies in the world, occupying at least 70% of the market, simply stopped their research activities in this area. Indeed, it is easier today to find new antibiotics or promising new lead compounds, than to find a company to develop one! Numerous opinions support the concept that the antibiotic era is close to its end and we are going back to the pre-antibiotic era. Really, we are in a public health crisis situation. To save public health and to reopen the pipeline of antibiotic discovery, we must reorganize the whole antibiotic research effort. After the long innovation gap, there is a pressing need to discover new effective antibiotics with new chemical scaffolds and new types of activities. We have to find new, untouched natural resources from unique environmental surroundings and use biochemical engineering and genome mining to wake up the undiscovered cryptic biosynthetic pathways in order to discover the hidden treasures of microbes.

The great efforts of the scientific community in the recent past, including those who remain working in the antibiotic area, are mainly in academic institutions and in several new small companies. In contrast to pessimistic, myopic opinions, I feel optimistic for the future. The partial results of these recent efforts are well known: daptomycin, retapamulin, the tetracycline analogue tigecycline, telavancin, fidaxomicin and some totally new promising natural product scaffolds such as platensimycin, kibdellomycin, krisynomycin, actinocarbasin, etc. (Walsh and Wencewitz, 2014). Interestingly, most of these compounds are actinobacterial products. A new trend is the return of the lipopeptide compounds. Besides daptomycin, and many others, polymixn (colistin) is being used in the therapy of tuberculosis. In recent years, new antibiotic drugs are mainly synthetic, including modified products such quinolones, vancomycin, macrolide analogues, etc. (Wright, 2012).

Some features of antibiotic research – diversity increasing

The first step of antibiotic research is to find and develop *new producer strains*, or recently, to find the hidden diversity of the known old and new microbes and culture the so-called unculturable strains. The next step is to isolate new compounds or rediscover old metabolites by advanced screening methods, and to explore all of their possible existing bioactivities to obtain promising *new lead compounds* (Fernandes, 2006). Additionally, there are numerous possibilities to modify them by genetic and/or chemical methods, in order to discover their analogues and obtain new medical agents. After finding promising lead compounds, the development of useful drugs unfortunately remains the most time- and investment-requiring step. The random chemical synthesis of new antimicrobial rugs unfortunately remains less effective (Singh and Palaez, 2008).

In each phase of the research on bioactive microbial metabolites, the increasing diversity of microbiological, chemical and biochemical activities is inevitable. Taxonomic diversity represents increasing numbers of new types of producing strains. This leads to increased chemical (structural) diversity and more unique chemical compounds, new types of bioactivities and new areas of possible utilization (Singh and Palaez, 2008; Cragg and Newman, 2013; O'Connel *et al.*, 2013). All of these areas fortunately are inexhaustible.

The possible *taxonomic diversity*, especially the inexhaustible mass of natural product resources from the antibiotic producing organisms, is one guarantee of the continuous development in the whole area. There exist in nature, besides the great series of known organisms, billions of marine, lower plant, and mainly microbial species which remain unknown. The known range of the frequent producer species are presently restricted to only a few easily detectable and culturable taxa which produce only a relatively narrow diversity of possible molecular structures (Davies, 2011).

Increasing *chemical diversity* is appearing in the continuously discovered unique, sometimes fantastic, almost incredibly complex chemical structures, isolated from the major terrestrial and marine taxa. Their already known and possible structural diversity is far beyond one's perception. The pressing need for new antibiotic frameworks and new scaffolds, may be reached by the culturing of the so-called "unculturable" species, and by precursor-directed biosynthesis, in order to discover the hidden biosynthetic capacity of the microbes. Also useful could be co-cultures and other biological and semi-chemical modifications, various genetic and genomic methods, and last but not least, preparation of synthetic analogues, homologues, and mimics by chemical methods (Walsh and Wencewicz, 2014).

The majority of the world's *biodiversity* remains to be explored. The old theories about the role of secondary metabolites, for example, as "waste materials" are totally absurd. The range of possible and multiple types of bioactivities (interactions with other non-microbial organisms) of natural products is extremely wide. The fermentation industry continues to produce more and more new and "old" products (statins, cyclosporins, avermectins, etc.) for use in non-chemotherapeutic and non-pharmaceutical areas. Presently, the greater part of the world's antibiotic production is used for non-pharmaceutical purposes, mainly in the agricultural and food industry areas.

According to "Perlman's law", "the microbes and their products will do anything". Microbes may influence, regulate and direct the life processes of all organisms (cross-kingdom signalling, endophytes, symbionts) in the world, both in terrestrial and marine environments. Understanding the essence of Life on the Earth depends on the total understanding of the precise working mechanisms of the microbial world!

It is sure, in this century, that the increasing diversity in all areas is the most important factor in all of natural product research. The continuously developing isolation, dereplication, identification, and structural determination techniques, plus new basic principles in the discovery process, using modern physical methods in the genetic fields (besides the solving of economic and political problems) are the main factors for renewing natural product research and reopening the antibiotic drug pipeline (Bérdy, 2005, 2012).

The real role of microbial metabolites: "bacteria without borders"

What is the role of these natural compounds in the life in the whole Earth? Their importance in communication inside the microbial world as inter-microbial signalling agents (sociomicrobiology) is clear (Greenberg, 2010). My personal opinion about the real significance of secondary metabolites in nature is that it is much wider than was supposed. They are indeed the essence, the main moving force, in the regulation of most, if not all, life processes. I think that there is no inevitable argument against my strong belief that *all* secondary metabolites have some kind of inherent biochemical, biophysical, molecular biological, and/or pharmacological, effect playing determinative roles in the total environment and in most life processes. Perhaps some of these interactions, and the methods for their detection, are still unknown (Smith *et al.*, 2013). To illustrate the whole situation, remember the case of mycophenolic acid. It was the first crystalline compound isolated from a *Penicillium* species in 1896. It was re-isolated in 1933 as an inactive compound. Its structure was elucidated in 1948 and its antimicrobial activity was described. Over time, it was isolated and identified from more than twenty different fungi, i.e. *Penicillium*, *Septoria* and *Verticicladiella* species. Later, besides its broad spectrum of chemotherapeutic, antitumour and antiviral activities, it was found to exhibit numerous other activities, for example, immunosuppressive, antidiabetic, antipsoriatic, and various enzyme- inhibitory effects. In present days, it has been "rediscovered" as a possible pharmaceutical agent (Cellcept). In the history of antibiotic research, other similar examples exist.

One must keep in mind that in publications describing the isolation and identification of a new microorganism, usually no data are given about any "bioactivity" of the microbes isolated. In some cases, the data refer only to one or a few *in vitro* tests which are not exclusive for other bioactivities. Similarly, in publications describing the isolation of a new microbial metabolite, sometimes no data are presented, and it has either never or very rarely been tested for its other activities. especially by using new specific methods.

The bioactive secondary metabolites are an open, but in great part, a hidden treasure for humankind. Most known activities may be interactions between the life cycle of various prokaryotic or eukaryotic cells, regulating the growth processes by inhibitory or stimulatory action, affecting the replication, and various responding actions on the life-cycle of living organisms at the biochemical level. Consider the role of endophytic and gut microbiota in plants and animals. Their beneficial effects ensure ecological fitness both in plants and humans. I mean that there is no secondary metabolite without some kind of biological/biochemical function. All the known "inactive" metabolites and those that will be isolated sooner or later, will become bioactive, discovering their still unknown role in the regulation of life processes.

The antibiotics, the bioactive microbial metabolites, are not only weapons against competitors, but also represent the connections by intermicrobial regulation or as signalling agents, not only inside the microbial world, but also in wide-scale interactions as interspecies regulators in the total biosphere. They must reflect, in some aspect, by cross-kingdom activity, the fundamental biochemical and biological mechanisms of the higher organisms. "Life would not exists with macromolecules alone".

The working of the total biosphere also depends on the microbial world. How, and by what mechanisms, and where do the secondary metabolites act exactly in the whole living world? These are mainly unanswered questions. Their cytotoxicity itself represents their connection with the animal world. The ecological stability of the soils and terrestrial waters would be impossible without microbes living in these habitats. My vision is that the whole Earth's life, and its fitness is greatly dependent on the undisturbed work of the Microbial World.

The advantages of natural products

The most important and promising basis for the future is the unquestionable *inexhaustibility of nature*, especially the microbial world. The countless new natural resources, and our expanding ability to discover additional new agents by genetic approaches, from the simple precursor- directed fermentation to metabolomics, using new and more selective effective methods and techniques, will ensure the discovery of the fully encrypted and dormant diversity of Mother Nature.

In one gram of a soil sample, millions of bacteria live. According to Krypides, 5×10^{30} microbes live on the Earth, while the number of cells in one human is 10^{13} and the number of stars in the known Universe is only 7×10^{21} (Krypides, 2009). According to DNA analytical investigations and mathematical modelling, actual microbial diversity is much higher than expected. Presently, less than 1% of the microorganisms seen microscopically have been cultivated. Billions of microbial species may exist in hidden or encrypted form in the Metabolome. In the living world, at least billions of microbial species exist, but only a small part, i.e. about 250,000 of these species have been isolated, investigated, identified and deposited. Up to now, only 1.5% of existing actinobacteria and about 5% of known fungi are cultivable by present methods and exactly identified by these methods. Do not forget that in one species, there are at least 10 to 20 unexpressed biosynthetic clusters. In the *Streptomyces coelicolor* species, already 25–30 independent clusters have been identified (Ochi and Hosaka, 2013).

In general, in pharmaceutical drug research, two main methods exist to obtain new lead compounds. Historically, the first is the pure chemical/synthetic approach, i.e. the random synthesis of new chemicals dating back to the 19th century. The second main method is based on the isolation of new natural products, i.e. the biological/biotechnological approach, which is going back to ancient times. The *combination* of these approaches may be the most successful (Clardy and Walsh, 2004).

Despite the clear advantages of chemical synthesis (e.g. simplicity, cost efficiency, easy scale up, etc.), there are several inherent advantages of natural products which make them promising for lead discovery for the whole of drug research. These are as follows:

1 Higher structural diversity and many chiral centres. These are not easily accessible by chemical synthesis. The functional diversity derived from unique steric structures provide selective activities on the targets due to diverse multicentre ligand–protein interactions. Clearly, their variable specificity and selectivity are higher.
2 Their action on different targets and on whole enzyme families. The variable effects, the possible "cross-talking" between microbes and their metabolites and cells of higher organisms, may lead to the discovery of non-antibiotic drugs. Several metabolites, mainly fungal ones, exhibit multiple, i.e. 10 to 15 different activities. The statins, the world's best selling anticholesterolemic agents, were originally derived from fungi.
3 Host compatibility. The drug-like natural product structures guarantee good correlation with human targets. These privileged structures, derived from the millions of years of evolutionary pressure are pre-eminent as lead compounds for successful modifications.
4 Lower risk factor for toxicity to the environment. Natural products generally meet the environmental requirements of "green chemistry".

According to the opinions of numerous scientists: "Nature is the best, most sophisticated, versatile and energetic combinatorial chemist". It seems that natural product libraries have more advantages over random or combinatorial chemistry libraries. The hit rate to find useful drugs from the known millions of synthetic compounds is about 0.01%, while from all known natural products, there is a two magnitude higher rate. About 0.1% of them became marketed drugs. From the total known microbial products, 0.5% are used as practically utilizable drugs. From the actinomycete metabolites, the hit rate is close to 1.0%, i.e. three magnitudes higher than from synthetic compounds. From all fungal products, 0.6%, and from all eubacterial metabolites, some 0.05%, are human drugs. The semi-synthetic antibiotics and mimics represents about 0.05% efficiency. As seen, the creativity of nature is far beyond the imagination and possibilities of synthetic chemists.

Preferring the natural product approach in intelligent combination with chemical or genetic modifications, hopefully, we will not revert back to the pre-antibiotic era. The clue is to find new types of producing microbes, and new chemical frameworks for direct use or modification. This is not an easy task. The well known phrase: "only the lower hanging fruits were picked (harvested)", is more valid than ever. Especially hard work is waiting for us (Capon, 2012).

The numbers of antibiotic-producing microbes

Taxonomic diversity

The antibiotic-producing organisms almost exclusively belong to the Plant Kingdom including various bacteria such as actinobacteria (filamentous bacteria), other eubacteria (unicellular bacteria), cyanobacteria, and myxobacteria; fungi (microscopic fungi, basidiomycetes, slime moulds and yeasts); algae, lichens, the lower plants (Bryophyta: liverworts, ferns, mosses) and higher (green) plants (Spermatophyta: gymnosperms and angiosperms). The antibiotic producers in the Animal Kingdom include the eukaryotic unicellular protozoa, and the series of prokaryotic animals, such as worms, insects and other invertebrates (amphibians), as well as a series of terrestrial and marine animals, invertebrates and vertebrates, alike. The marine invertebrates (sponges, tunicates, molluscs, echinoderms, coelenterates, etc.) are relatively common, while the terrestrial animals (mammals) are very rare producers. The number of antibiotic compounds isolated from animals is low and their source is very uneven. From the fishes and birds, only a few compounds have been isolated, but from marine invertebrates, a relative greater number of compounds were obtained; however, they are of uncertain origin.

In general, the existing overlap in production of chemical entities between producing taxa, except among the close relative taxonomical types, has not been extended seriously. Only between *Streptomyces* and rare actinomycetes; and all the actinobacteria and other eubacteriales, does there exist some (3 to 5%) overlapping. The number of antibiotic compounds isolated simultaneously from the microbial producer and from various sources of higher life, i.e. the overlapping, is very low. Other than the symbiotic metabolites, there are 245 marine animal, 38 terrestrial animal and 88 algal products, produced simultaneously by microbes.

Fig. 3.1 illustrates the relative number and taxonomic place of the main producers of bioactive secondary metabolites.

Modern chemotherapy since Koch, Pasteur and Ehlich's "magic bullets", from the end of the 19th century, has resulted in the discovery of more than 200,000 natural and synthetic antimicrobial (chemotherapeutic) compounds, semisynthetic derivatives of antibiotics, and active chemical analogues of natural products derived by chemical synthesis. The number of human drugs used in medicine based on synthetic or natural products is presently about 3500. According to most views, 50% to 65% of our pharmaceutical drugs are direct natural products, or biologically or chemically modified natural products (semisynthetic derivatives), or compounds obtained by chemical synthesis on the basis of natural product structures: mimics, and analogues (Butler *et al.*, 2013).

The favourite trends and efforts dedicated to the research has changed several times in the last sixty years or more. In the first years of antibiotic research, from the nineteen forties to the sixties of the 20th century, antibiotics were isolated mainly from microorganisms, chiefly from various bacteria, including actinobacteria, other eubacteria, and fungi, including microscopic fungi and basidiomycetes. They totally numbered by 1960 about 2500 bioactive compounds. This number,

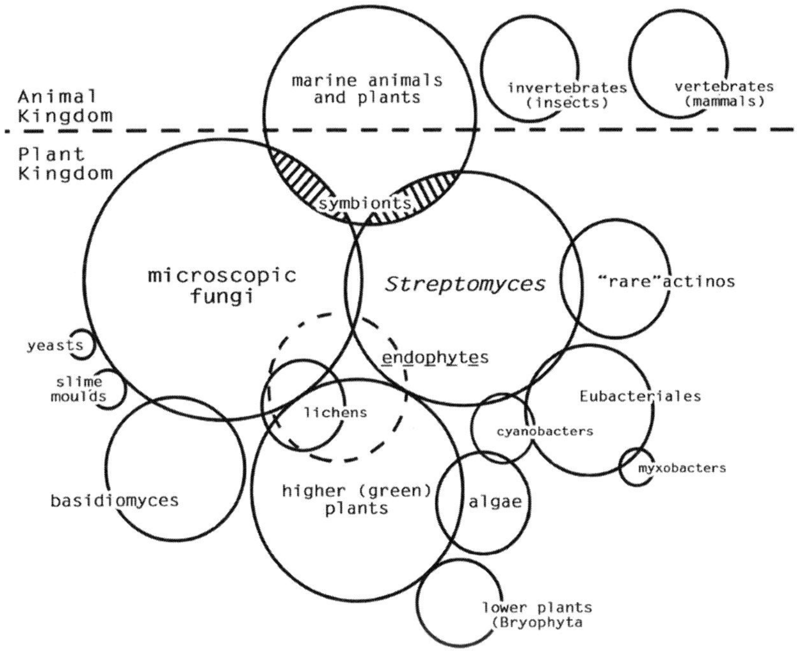

Figure 3.1 Bioactive secondary metabolites in nature.

by the year 2000, increased to 15,500 and by today is over 35,000. The producer species were, in the last century, derived mainly from terrestrial sources, in most cases from soil samples. The number of marine-derived bioactive compounds in 1960 was only 140. These numbers today are over 30,000 marine microbially derived and over 20,000 marine animal products. It is interesting that cephalosporin C really was derived from the marine *Cephalosporium* sp., isolated from a marine source in 1945.

The significance and the frequency of investigated microbes in recent decades has changed from time to time. By the end of the "Golden Era", i.e. around 1970, most antibiotics were isolated from *Streptomyces* species, their share increasing by over 70%. In the 1980s, the share from the "rare" actinomycetes had increased close to 25%, the same rate as the fungal products in that period. However, in the 21st century, the share of isolated fungal derivatives has increased to 62–65%. and the share of actinomycete products decreased almost to 30%. The increase in the discovery of cyanobacterial metabolites is more significant. In 1975, there were only 15 products, but by the year 2000, 50 cyanobacterial metabolites, and today, over 1500 bioactive metabolites have been isolated from cyanobacterial strains.

Presently close to half a million, or perhaps more, natural products are known. This number is increasing almost exponentially, due to new isolation and identification methods and techniques. The known bioactive plant metabolites (alkaloids, antitumour agents, antimicrobial agents, etc.) number today 60,000 to 70,000. From terrestrial animals, about 2500 bioactive secondary metabolites have been isolated.

To estimate the exact number of existing *bioactive* compounds in the area of natural products, i.e. the microbial secondary metabolites, is very difficult. The majority of known (identified) living species have never been tested for any kind of bioactivity. The isolated natural products have either never or very rarely been investigated for a wide range of bioactivities, especially by recently introduced sensitive and specific methods. I believe that the existing millions of microbial strains are possible antibiotic and human drug sources, and we have to find ways to reveal their full biosynthetic capacities. The majority of known synthetic compounds also have never or have rarely been tested for all possible bioactivities.

Table 3.1 summarizes the numbers of natural products derived from different taxa of the living world which are antibiotics in the strict sense (first column), all bioactive products (second column) and all the known natural products (third column) (Bérdy, private data).

In the following, the antibiotics (antimicrobial metabolites), all the known bioactive metabolites, including pharmacologically active agents, and the approximate number of all isolated secondary metabolites (including active and inactive ones), are listed according to their main producer species, basic chemical structures, and types of bioactivities. The values, due to many inherent reasons, of course, are only approximate.

Until today, some 75,000–80,000 secondary metabolites, including about 38,000 "bioactive" compounds and almost 29,000 "classical antibiotics", have been isolated from microorganisms. The microbial compounds include over two hundred "direct" natural drugs which have been or are used in medical and agricultural practices. Additionally, from the many thousands of semisynthetic antibiotics, analogues and mimics, at least 300 to 400 compounds are in use today in human medicine, veterinary medicine, agriculture and animal husbandry as marketed products.

Terrestrial plant-derived compounds today number over 300,000 (according to other views, over 500,000). From higher plants, approximately 35,000 compounds, with known biological activities, are known. Most of them are alkaloids, flavonoids, phenolic derivatives, including 15,000 natural products exhibiting antibiotic, chemotherapeutic, antitumour, and other activities.

Table 3.1 Numbers of natural products

Producer organisms	Antibiotics	All bioactive	Total known
Actinobacteria	14,500	16,000	20,000–21,000
Streptomyces sp.	11,000	12,400	17,000
Other actinobacteria	3400	3600	4400
Eubacteriales	3500	4700	11,000–12,000
Bacillus sp.			1400
Pseudomonas sp.			950
Myxobacteriales	450	750	1200
Cyanobacteria	400	1800	4500
Other bacteria including proteobacteria, etc.	800	2000	5000
Fungi	10,500	18,000	40,000–45,000
Microscopic fungi	9000	14,000	32,000
Aspergillus			2200
Penicillium			1650
Basidiomycetes	2900	4100	9000
Other fungi including yeasts, slime moulds, etc.	110	340	3000
Total microbial	~28,500	~38,000	75,000–80,000
Higher forms of life			
Higher plant products	15,000–16,000	30,000–32,000	Over 500,000
Lower plants including macroalgae, lichens	1500–2000	10,000–15,000	60,000–70,000
Terrestrial animal products	1000–1500	2500–3000	?
Marine products (plants and animals)	6000–8000	20,000–25,000	?

The data are based on the BMIC database which covers the numbers of all types of microbial secondary metabolites, terrestrial, marine, plant and animal products known by the end of 2013 (Bérdy, 2012). The data are based on close to half a million references, close to 100,000 producing species and a similar number of data covered in my private database.

From marine sources, a continuously increasing number, presently over 20,000 bioactive compounds including approximately 7000 antibiotics, have been identified. It is possible that a part of the marine and higher plant products may really belong to the group of microbial metabolites produced by symbiotic and endophytic microbes.

The total number of *chemically synthesized products*, thanks to combinatorial chemistry, is at least eight to ten million, including about three thousand practically used drugs. After random approaches, taking lessons from nature by inspiration of the specific features and complex structural cores of secondary metabolites, synthetic chemists may access more and more new drugs and drug-like compounds. Their number presently is about 2500 to 3000 compounds. Our medical armamentarium today contains about 3500 to 4000 drugs, both natural and synthetic, including close to one thousand chemotherapeutic drugs and other pharmaceutical products of natural origin, as direct drugs or analogues. From these numbers, it is clear that the wide variety of known and future-to-be-discovered new microbes are the most promising sources for new drugs and other useful agents in pharmacy and agriculture.

Numbers of producing species

Independent of the decreasing medical success of antibiotics research, the absolute number of known microbial secondary metabolites is almost exponentially increasing up to 40,000 to 80,000. To understand the reason for this wide range in numerical data, do not forget some unique features of the natural product and antibiotic research. One strain or organism is able to produce numerous, totally chemically different secondary metabolites; on the other hand, one secondary metabolite may be produced by taxonomically different organisms. The multiple production of bioactive secondary metabolites is illustrated by the following data. The existing 38,000 bioactive microbial metabolites appear to be produced by approximately 82,000 different microbes, but this number may actually be over 100,000 identified species. All microbial metabolites have, on average, at least two producers. There are, however, still numerous hidden chemical identities in the published compounds. Besides the microbes presently known to produce only a single metabolite, there are microbes, first of all fungal species, producing many, sometimes ten to fifteen or more independent compounds. The 15,000 known actinobacterial metabolites are derived from approximately 27,000 different actinobacter species, while the 17,000 fungal products are derived from over 46,000 fungal species. There are frequently five to ten different fungal species, from totally different taxa, producing the same chemical metabolite.

The real numbers and the origin of metabolites produced by a given type of organism, due to the possible endophytic and symbiotic lifestyle of numerous microbes, may be the subject of some uncertainty. In fact, several marine animal products are possibly of symbiotic microbial origin. Also, many higher plant products are endophytic fungal or bacterial products. The situation in the determination of numbers of producer species and isolated compounds today is somewhat complicated. This is one reason why all of the numbers are, in general, only approximate values. In scientific papers, it would be advantageous to carefully use the following terms: origin, isolated from, derived from, produced by, and biosynthesized by.

In reality, the situation is more complicated due to some factors. Numerous secondary metabolites are produced in chemical complexes, a series of chemically similar compounds with identical framework, differing only in one or two substituents. Actinobacteria often produces the so called minor components of the same chemical family. Most minor components are normally produced only in very small amounts. The number of these minor components, in a special case, may reach 50 isolated and identified compounds. This phenomenon, in the strict sense, may decrease the number of really independent metabolites.

According to our present knowledge in the area of the microbial world, most of the known taxa (80%–90%) are not producers or only very poor producers. The privileged types of microorganisms are the actinobacterial, some specific fungal, and other non-actinobacterial eubacterial strains. They are the producers of over 96% of all the known (about 37,500) microbial antibiotics. From about 140 genera of actinobacteria, a single *Streptomyces* species produces over 12,000 bioactive metabolites, and the "rare" actinobacterial species, produce 3500 bioactive metabolites. Altogether, the 15,500 compounds represent today 43% of the all microbial metabolites. Other privileged types, i.e. the fungi and several eubacteria, produce almost all the rest of the known metabolites. Microscopic fungi and basidiomycetes produce about 17,000 (13,000 and 4000), the eubacteriales (firmiculates) some 4500, while cyanobacteria produce about 1500 products. The smaller producer types are myxobacteria (700), yeasts (250), slime moulds (140), and other unique organisms, e.g. proteobacteria (120). In contrast, the common *Saccharomyces* yeasts are able to produce only a few (15–20) secondary metabolites. The above facts raise additional speculations and provide tasks for the future (Davies, 2006).

In the genus *Streptomyces*. there are 543 *Streptomyces hygroscopicus* strains. Its mutants, variants, recombinants, etc. are producers of chemically different types of bioactive metabolites. Of the other *Streptomyces* strains, including over 1500 known species, the major producers (making over one hundred products) are: *S. griseus* (190), *S. albus* (150), *S. antibioticus* (148), *S. platensis* (138), *S. fradiae* (136), *S. lavendulae* (119) and *S. venezuelae* (110). In the area of rare actinomycetes, there are about 140 known genera. These genera include *Micromonospora* (535 individual strains), *Nocardia* (375), *Actinomadura* (275), *Actinoplanes* (268), *Streptoverticillium* (260), *Amycolatopsis* (140), *Saccharothrix* (108), *Nocardiopsis* (95) and *Streptosporangium* (88). Presently, over 50 rare actinobacterial species are known as antibiotic producers.

The production of other eubacterial products, from organisms including cyanobacteria and myxobacteria, represents over 5000 metabolites (12%). Almost half of them are derived from two main genera, *Bacillus* (1360 species) and *Pseudomonas* (912 species). The next most prolific producers are the *Lactobacillus* and *Streptococcus* species representing 200 and 136 species, respectively. Among cyanobacteria (producing over 1500 metabolites), the most prolific include *Lyngbya Moorea* (330), *Mycrocystis* (300) and *Nostoc* (165) species, while in the myxobacter group (750 metabolites), *Sorangium* (220) and *Myxococcus* (140) species are the most frequent producers.

The microscopic fungi produce over 15,000 bioactive metabolites, and the basidiomycetes over 4000 metabolites, representing altogether half of the all microbial metabolites. The production of the fungal metabolites is distributed among approximately 1500 different species. Among microscopic fungi (more than a thousand different producer species are known), *Aspergillus* (2150 strains), *Penicillium* (1590), *Trichoderma* (567), *Fusarium* (476), *Alternaria* (260), *Chaetomium* (248), *Phoma* (235), *Acremonium* (204), *Stachybotrys* (170), *Paecilomyces* (161), *Xylaria* (127), *Emericella* (120) and *Talaromyces* (112) species are the frequent producers, each representing over 100 individual strains. The distribution of metabolites in the over 400 producing basidiomycete species is less characteristic. The genus *Ganoderma* (238 strains), *Lactarius* (93), *Cordyceps* (90) and *Stereum* (70) species are the most frequent producers.

The specificity of metabolite-producing ability of the microbes, their main chemical characters, their frameworks and the types of bioactivities of the secondary metabolites produced are more or less characteristic in different taxa.

Chemical diversity

The chemical structures of microbial metabolites cover the whole of organic chemistry. However, the numerous unique frameworks and structural elements of natural products are very rare among synthetic products and hardly accessible by pure chemical synthesis. There is an interesting specificity according to the taxonomy of producer organisms, namely between microbes, higher plants and animals, in the frequency of chemical skeletons of their products (metabolites). In general, the chemical diversity of microbes is greater than that of plant and animal products. The frequency of occurrence of the main chemical types is extremely variable in different microbial taxa. For example, the hit rate (in per cent and numbers) of *terpenoid* type compounds is less than 3%, i.e. only 320 compounds of all the *Streptomyces* products. This rate is only 1.5% in rare actinomycetales, but it is as high as 18% (3200 terpenoid compounds) of the fungal metabolites. It is interesting to note that among the higher plants, according to some estimates, the hit rate of terpenoids is at least 30%. Among eubacteria (4600), and cyanobacteria (1500), the hit rate of terpenoid compounds represents only 1.3% and 3.0%, respectively. The peptide type compounds (including numerous cyclopeptides) represent 18% of *Streptomyces*, 17% of rare actinomycetes, 10% of fungal metabolites but are as high as 53% in other eubacteriales and 62% among cyanobacterial products. The macrocyclic lactone type metabolites (including macrolactams), represent 10%, and 27% in *Streptomyces* and rare actinomycete products, but only 5.5%, 2.5%–3%, respectively, in fungal and bacterial products. The extremely high rate (27%) in the "rare" actinomycete-produced macrocyclic compounds is very interesting.

Frequently produced specific chemical structures also show great variability. Aminoglycosides are produced exclusively by actinobacterial and to a smaller extent by bacterial species. Tetracycline-type compounds, i.e. anthracyclines, angucyclines, and nigericin-like polyethers, nonactin type cyclopolylactones, actinomycins, echinomycin-type quinoxaline peptides, and polyene macrolides are derived exclusively from actinobacteria, mainly from *Streptomyces* species. Brefeldin and cytochalasin type condensed small macrolides, gliotoxin-like diketopiperazines, peptaibol antibiotics, heterocyclic compounds as simple pyrones are exclusively or very frequently among fungal metabolites. The chemical variability of microbial products, e.g. macrocyclic compounds such as lactones, cyclopeptides, unique polycondensed heterocycles, etc. is also much greater than that of the higher plant and animal products.

The specificity of biological effects, the various antimicrobial, antitumour/antiviral and other pharmacological and agricultural activities of secondary metabolites derived from different main types of producer species, is also interesting. 67% of the actinobacterial products and 65% of other eubacterial products exhibit definite antimicrobial (antibacterial, antifungal effects) activities. They are, in fact, the "real" antibiotics in the oldest and strictest points of view. In contrast, only 36% of known fungal metabolites have antimicrobial activity. 31% of actinobacterial and 32% of other eubacterial products, but only 18% of all fungal products, exhibit some kind of antitumour/cytotoxic and/or antiviral activity. A special case is the cyanobacterial products; 14% have antimicrobial, 33% antitumour and 68% exhibit other, mainly enzyme inhibitory, activities. The occurrence of non-antibiotic, "other" biological, pharmacological and agricultural effects is 38% among all actinobacteria (30% of "rare" actinobacteria), 43% of other eubacterial and 53% of the fungal products. The exact reasons for these significant differences, are certainly independent of the screening frequency of various species, and we are waiting for a genomic explanation.

Biodiversity

Antimicrobial activity

Among the 38,500 known bioactive antibiotic compounds, about 16,000 exhibit antibacterial (against Gram-positive, Gram-negative, and other bacteria, e.g. mycobacteria) activities, 6800 show antifungal (against microscopic fungi, yeasts, etc.) and about 700 have antiprotozoal activities. These types of activities, i.e. antibacterial, antifungal and antiprotozoal effects are occurring in 45%, 18% and 2.5% of all the microbial metabolites. More than 5000 compounds exhibit both antibacterial and antifungal activity. The percentage of all antimicrobial products in the actinobacterial area (*Streptomyces*, and rare actinomycetes) and other eubacteria (*Bacillus*, *Pseudomonas*, etc.) is 66% and 70%, while of the fungal metabolites (microscopic fungi and basidiomycetes), this rate is only 35%.

Antitumour and antiviral activities

The number of antitumour (cytotoxic, antimitotic) compounds in all microbial products is about 9700, representing 28%, while the total 1300 antiviral products represent 3.6% of the all bioactive microbial metabolites.

Other (chemotherapeutic-like and pharmacological) effects

For a long time, it has not been understood how and why microbial factors (metabolites) regulate diverse pharmacological processes, e.g.

inflammation. The great number of pharmacological effects which have been discovered, and the investigation of their mode of action in life processes, hopefully, will shortly provide the answers.

The most frequent chemotherapeutic-like biological activities, mainly toxic effects, against lower animals include insecticidal (1100), antiparasitic (580), anthelminthic (390), antiprotozoal and antimalarial (360), herbicidal (340), nematocidal (322), acaricidal (180), larvicidal (180), miticidal (160), molluscicidal effects and repellent actions. They totally number about 3500. There are agents active on lower plants (close to 1000 compounds) such as algicides (560), whereas others acts on ferns, bryophyta, liverworts, and others. The compounds active against higher plants, i.e. phytotoxic compounds (1650), other plant growth regulators (320) and chlorosis inducers, promoters, antiauxins, seed germination inhibitors, etc. number about 2200, while the number of agricultural feed additives is 220 and the food preservatives represent some 400 compounds. There are numerous microbial regulators, such as sporogenic, morphogenic, bulge-forming compounds, differentiation inducers and mating hormones, etc., all amounting to about 250 compounds.

The pharmacological effects of the microbial metabolites are far from negligible; they include the statins (the world's largest selling anticholesterolemic agents) and cyclosporin-like compounds, which revolutionized the whole organ transplantation process. The first and most important event was the discovery of the diverse enzyme-inhibitory effects of these metabolites. This research was initiated by H. Umezawa in the late 1960s. As a result, today there are more than 9000 compounds, i.e. almost a quarter of all the bioactive metabolites (including antibiotics), that have known enzyme inhibitory activities! The 250 or so enzymes investigated cover almost all the main types of enzymes playing roles in the living function of microbes, animals and humans. They include proteinases, protein kinases, amylases, phosphatases, elastases, DNA polymerases, HIV integrases, glycosidases, squalene synthetases, lipoxygenases, ACAT, ACE, MAO, HMG-CoA reductases, ATP-ases, esterases, deaminases, etc.

There are a large number of other pharmacological effects of these metabolites. In the past decades, about 500 pharmacological and similar tests, covering most areas of pharmacy, have revealed positive results of these metabolites. The following describes these detected activities: immunoactive agents including suppressives, stimulators, potentiators, modulators (1010 compounds), anti-inflammatory compounds (590), anti-oxidative-radical scavengers and similar agents (480), siderophores (460), anticholesterolemic agents (310), radical scavengers (275), apoptosis inducers (270), cell-cycle affectors (255), membrane active compounds (238), platelet aggregation inhibitors (187), chelate-forming agents (186), DNA effectors (170), CNS-active agents (160), antimetabolites (167), antidiabetes (153), biosurfactants (152), growth factors (140), haemolytic agents (136), quorum sensing, signalling compounds (125), anti-allergenics (124), hepatotoxins (122), anti-artherosclerotics (120), angiogenesis inhibitors (112), surface-actives (105), anti-ulcer compounds (103). Others include repellents, chlorosis inducers, antifeedants, ichthyotoxics, amoebicides, neurotoxic, proteosome inhibitory, antihypertension, analgesic, antidementia, anti-Alzheimer, interleukin inhibition, interferon inducers, antihypertensive, cardioactive effects, allosteric regulators, mutagens and anti-mutagens, transcription modulators, phage inducers, receptor agonists, pheromones and the most diverse toxins. Most of these effects occurring simultaneously may represent variability of additional pharmacological effects and manifold practical usefulness in human or animal therapy.

In summary, almost 20,000 metabolites, i.e. 54% of all known bioactive metabolites, exhibit at least one non-chemotherapeutic pharmacological activity. Of course there is a high degree of multiple overlapping in the various detected antimicrobial, antitumour and pharmacological activities of the compounds. One antibiotic compound frequently exhibits multiple "other" actions. Taking into account the antitumour/antiviral compounds, the total overlapping is about 60%. Schematically, the simplified situation of the different activities, their relative numbers, and the overlaps are illustrated on Fig. 3.2 (Bérdy, 2012).

Additionally, many secondary metabolites are used in other, non-pharmaceutical areas such

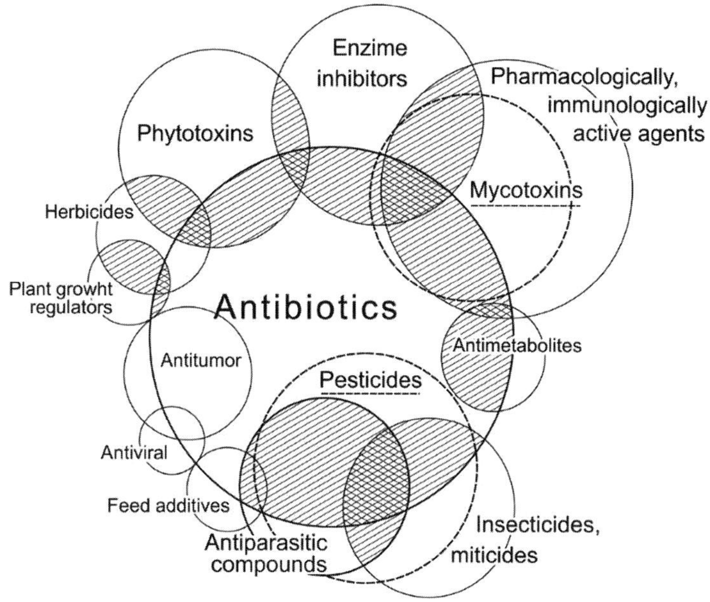

Figure 3.2 Bioactive microbial metabolites.

as biochemical reagents, feed colorants, dyes, vitamins, co-factors, odorants, sunscreens, disinfectants, preservatives and, last but not least, as starting materials for chemical synthesis, such as 6-aminopenicillanic acid.

Just to complicate the situation, there is the matter of concentration dependence in the bioactivity of numerous microbial metabolites. At high concentrations, they may act as poisons but at low (sub-inhibitory) concentrations, they may act quite differently (hormesis). Moreover, some medically useful antibiotics in low concentration may act as signal molecules, in fact, be beneficial to pathogenic bacteria.

The way ahead

For sure, plenty remains to be done. For the future discovery of the full biodiversity of the world, two main areas are on the horizon, i.e. increasing the sources of organisms producing bioactive secondary metabolites and increasing the productivity and variability of the produced metabolites by the most diverse genetic and genomic methods (Krypides, 2009; Ongley *et al.*, 2013).

The main new sources are the oceans and the endophytic organisms of plants or similar living things. Additionally, the investigation of microbes existing in the most unique surroundings (extremophiles) or geographic areas, etc. could be significant.

In microbial metabolite research, the classical and unusual terrestrial samples, the marine environment, and the endophytic microbes are in the foreground of present efforts. Of course, the genetic methods, the biochemical, genomic, and metagenomic modifications of production by microbes, the known and discovered natural leads, and the post-evolution, are perhaps the most effective approaches for the future (Cordell and Colvard, 2007).

Oceans

One important task is continuing the investigation of the oceans, which occupy 70% of the Earth's surface. The detailed investigation of the practically untouched flora and fauna, which are the hosts of about 87% of life on earth, represents the most promising, wide- range and unique biosphere from the seashore to depth of 10,000 metres. The oceans seem to be the richest source to find new microbes, better than soil samples. One millilitre of sea water contains almost 1 million free-living cells and microbes, and they are generally of different types, including many new species previously isolated from terrestrial sources (Imhoff *et al.*, 2011). Interestingly, rare actinomycetales are relatively more frequently isolated from marine habitats.

The marine macrofauna, the invertebrates, fishes, amphibians, etc. could contain a practically innumerable wide range of species. The marine microbes and lower plants (algae) are generally living in assemblage – symbiotic interactions with various invertebrates, being usually new, as compared to those that are living in terrestrial surroundings. Table 3.1 gives some data about marine microbes.

In the 21st century, the great part, i.e. 30% to 35%, of the newly discovered antibiotics, antitumour and other metabolites are produced by marine microbes (e.g. ecteniascidin-E 743 = Yondelis). and invertebrates. The real origin of compounds isolated directly from marine invertebrates (sponges, tunicates, molluscs, mussels, bryozoans, soft corals, amphibians, etc.), especially in the past, may also be derived from these symbiotic microbes. The marine animals may be assembled with bacteria, actinobacteria, fungi, protozoa, algae, etc. The symbiosis is usually beneficial for both members. The microbial population may sometimes represent over 50% of the dry weight of the animals. Especially, sponges may play a role as microbial fermentors. Sometimes there is great uncertainty as to which is the real producer. The direct evidence is when a sponge-derived product (after the isolation of the assembled microbes) is produced directly by the microbe. The identity or close similarity of independently obtained microbial and sponge products increases the possibility of this phenomenon, i.e. the fermentative origin of the compounds isolated from the native animal. It is most probable that several secondary metabolites derived from some specific sponges, e.g. from *Theonella* species, are actually produced by microbial (fungal/bacterial) symbionts or have been cloned from metagenomes obtained from the associated microbes. Recently, the introduction of a new microbial taxon "*Entotheonella*" was proposed to be the producer of metabolites by these sponges (Wilson *et al.*, 2014). Numerous marine derived compounds have identity or very close structural similarity to metabolites produced by different terrestrial microbes, such as microsclerodermins, polytheonamides, nodularins, swinholides, etc. (Radjasa *et al.*, 2011).

Endophytes

The other important types of new promising species are the endophytic microorganisms. Until the last decade, these microbes were practically unstudied as possible sources of novel natural products. However, in the past, some fungal endophytes and their hosts were reported to produce the same compound. Their products were mainly polyketides and terpenoids. They can be found in practically all plants on earth, residing in the living tissues of the host, in a variety of relationships from opportunistic and mutualistic relations to pathogenic effects. They contribute to host protection and they increase the survival value of the host. Each of the higher plants, from mosses and mangroves to green trees, are the hosts of possible producers of new microbial metabolites (Kusari *et al.*, 2012). Their productivity is similar to that of other terrestrial species, including all known chemical types and bioactivities (Brader *et al.*, 2014). In fact, taxol (paclitaxel) and other similar antitumour compounds (i.e. camptothecin, podophyllotoxin, isoquinoline alkaloids, etc.), isolated earlier from higher plants are actually derived from endophytic microbes (Sato and Kumagi, 2013).

In summary, many old secondary metabolites obtained from plant, marine and terrestrial animal sources may turn out to be partly (or fully?) derived from endophytic, symbiotic or associated microbes. We do not know how widespread this phenomenon is in nature, but certainly, it is very common in the Plant Kingdom and marine fauna (Crawford and Clardy, 2011).

Other approaches

Additionally, two promising general ways to access new types of producing species as new sources for new compounds are mentioned below.

The investigations of unique organisms, such as the extremophiles, and untapped geographical surroundings (mine waste disposals, deserts, deep seas, polar areas, etc.) are relatively old approaches. The investigation of some 40 million insect species and their products, e.g. from their gut flora, is a relatively new trend which may result in the discovery of new chemical types (Ramadhar *et al.*, 2014). The most important approaches, which will be discussed in detail in the following chapters, are the genomic approaches.

Presently, we entering into the genomic era. The best possibilities to discover the hidden biosynthetic capacity of microbes unambiguously are the biochemical genetic approaches (Nguyen *et al.*, 2012; Wilson *et al.*, 2014). The oldest and simplest method in this field, i.e. precursor-directed biosynthesis used with different microbes already has yielded over 1500 new, modified microbial metabolites. The number of metabolites derived from various mutant strains, using the most diverse genetic techniques, is over one thousand. Recently, from the mutants and recombinants of a single *Streptomyces* species, more than 350 antimycin type metabolites were identified and isolated by a combination of precursor added, multiplex genetic combinatorial biosynthesis (Yan *et al.*, 2013).

The most important methods and techniques in this area are as follows:

- Culturing the unculturable. Incorporating genes into more convenient host species, e.g. *E. coli*.
- Combinatorial biosynthesis.
- Co-cultures, interspecific cross-talk.
- Awakening the dormant, silent genes.
- Plant tissue culturing by combinatorial genetics.
- Simple genetic methods: cloning the clusters and genetic engineering.
- Genomics: DNA sequences in databases.
- Cloning of environmental DNA in the metagenomic approach, allowing identification of the unculturable species.
- Metabolomics.

Epilogue

In our days, only a small portion of the 25,000–30,000 known, clinically described, illnesses are effectively curable. We can treat most of them only symptomatically. Among the most dangerous are diseases such as tuberculosis, malaria and HIV. The biggest problem is the continuously increasing resistance and the re-emergence of old pathogens. It is problematic whether our presently available drugs, including new chemotherapeutic agents, can efficiently reduce the increasing problems. Unfortunately, it seems that, in the future, many additional new and more efficient drugs and safe agricultural agents will be required for satisfactory and continuous cures and to feed the world's increasing population. To meet this task. basic reconsiderations and cooperation are required from the scientists and the representatives of economic and political life.

References

Appelbaum, P.C. (2012). 2012 and beyond: potential for the start of a second pre-antibiotic era? J. Antimicrob. Chemother. *67*, 2062–2068.

Bérdy, J. (2005). Bioactive microbial metabolites: a personal view. J. Antibiot. *58*, 1–26.

Bérdy, J. (2012). Thoughts and facts about antibiotics: where we are now and where we are heading. J. Antibiot. *65*, 385–395.

Brader, G., Compant, S., Mitter, B., Trognitz, F., and Sessitsch. A. (2014). Metabolic potential of endophytic bacteria. Curr. Opin. Biotechnol. *27*, 30–37.

Butler, M.S., Blaskovich, M.A., and Cooper, M. (2013). Antibiotics in the clinical pipeline in 2013. J. Antibiot. *66*, 1–21.

Capon, R.J., (2012). Microbial biodiscovery: back to the future. Curr. Topics Med. Chem. *12*, 1508–1515.

Clardy, J., and Walsh, C. (2004). Lessons from natural molecules. Nature, *432*, 829–837.

Cordell, G.A., and Colvard, M.D. (2007). Natural products in a world out-of balance. Arkikov *VIII*, 97–115.

Cragg, G.M., and Newman, D.J. (2013). Natural products: a continuing story of novel drug leads. Biochim. Biophys. Acta. *1830*, 3670–3695.

Crawford, J.M., and Clardy, J. (2011). Bacterial symbionts and natural products. Chem. Commun. *47*, 7559–7566.

Davies, J. (2006). Are antibiotics naturally antibiotics? J. Ind. Microbiol. Biotechnol. 33, 496–499.

Davies, J. (2011). How to discover new antibiotics: harvesting the parvome. Curr. Opin. Chem. Biol. *15*, 5–10.

Davies, J., and Ryan, K.S. (2012). Introducing the parvome: bioactive compounds in the microbial world. ACS Chem. Biol. *7*, 252–259.

Demain, A.L., and Sanchez, S. (2009). Microbial drug discovery: 80 years of progress. J. Antibiot. *62*, 5–16.

Fernandes, P. (2006). Antibacterial discovery and development-the failure and success? Nat. Biotechnol. *24*, 1497–1503.

Greenberg, E.P. (2010). Sociomicrobiology: a personal perspective on an emerging research area. ACS Microbe Magazine.

Imhoff, J.F., Labes, A., and Wiese, J. (2011). Bio-mining the microbial treasures of the ocean: new natural products. Biotechnol. Adv. *29*, 468–482.

Krypides, N.C. (2009). Fifteen years of microbial genomics. Nat. Biotechnol. *26*, 627–632.

Kusari, S., and Spiteller, M. (2011). Are we ready for industrial production of bioactive plant secondary metabolites utilizing endophytes? Nat. Prod. Rep. *28*, 1203–1207.

Newman, D.J., and Cragg, G.M. (2012). Natural products as sources of new drugs over the 30 years from 1981 to 2010. J. Nat. Prods. *75*, 311–335.

Nguyen, Q.T., Merlo, M.E., Medema, M.H., Jankevics, A., Breitling, R., and Takano, E., (2012). Metabolomics methods for the synthetic biology of secondary metabolism. FEBS Letters *586*, 2177–2183.

Ochi, K., and Hosaka, T. (2013). New srategies for drug discovery: activation of silent or weakly expressed microbial gene clusters. Appl. Microbiol Biotehnol. *97*, 87–98.

Ongley, S.E., Bian, X,Y., Neilan, B.A., and Müller, R. (2013). Recent advances in the heterologous expression of microbial natural product biosynthetic pathways. Nat. Prod. Rep. *30*, 1121–1138.

Radjasa, O.K., Vaske, Y.M., Navarro, G., Vervoort, H.C., Tenney, K., Linington, R.G., and Crews, P. (2011). Highlights of marine invertebrate-derived biosynthetic products: their biomedical potential and possible production by microbial associations. Bioorg. Med. Chem. *19*, 6658–6674.

Ramadhar, T.R., Beemelmanns, C., Currie, C.R., and Clardy, J. (2014). Bacterial symbionts in agricultural systems provide a streategic source for antibiotic discovery. J. Antibiot. *67*, 53–58.

Sato, F., and Kumagi, H. (2013). Microbial production of isoquinoline alkaloids as plant secondary metabolites based on metabolic engineering research. Proc. Japan Acad. Sci. Ser. B *89*, 165–182.

Singh, S.B., and Palaez, F. (2008). Biodiversity, chemical diversity and drug discovery. Progr. Drug Res. *65*, 143–177.

Smith, P.M., Howitt, M.R., Panikov, N., Michaud, M., Gallini, C.A., Bohlooly-Y, M., Glickman, J., and Garrett, W.S. (2013). The microbial metabolites, short chain fatty acids, regulate Treg cell homeostasis. Science *341*, 569–573.

Walsh, C.T., and Wencewitz, T.A. (2014). Prospects for new antibiotics: A molecule-centered perspective. J. Antibiot. *67*, 7–22.

Wilson, M.C., Mori, T., Rückert, C., Uria, A.R., Helf, M.J., Takada, K., Gernert, C., Steffens, U.A., Heycke, N., Schmitt, S., *et al.* (2014). An environmental bacterial taxon with a large and distinct metabolic repertoire. Nature *506*, 58–62.

Winter, J.M., Behnken, S., and Hertweck, C. (2011). Genomic-inspired discovery of natural products. Curr. Opin. Chem. Biol. *15*, 22–31.

Wright, G.D. (2012). Antibiotics: a new hope. Chem. Biol. *19*, 3–12.

Yan, Y., Chen, J., Zhang, L., Zheng, Q., Han, Y., Zhang, H., Zhang, D.H., Awakawa, T., Abe, I., and Liu, W. (2013). Multiplexing of combinatorial chemistry in antimycin biosynthesis: expansion of molecular diversity and utility. Angew. Chem. Int. Ed. *52*, 12308–12312.

The Need for New Antibiotics

4

Arnold L. Demain and Sergio Sánchez

Abstract

The large amounts of antibiotics used in human therapy, as well as those used for farm animals and aquaculture, have resulted in the selection of pathogenic bacteria resistant to multiple drugs. New antibiotics are continuously required to combat antibiotic-resistant bacteria and pathogenic yeast. Such resistance increasingly limits the effectiveness of current antimicrobial drugs. The problem is not just antibiotic resistance, but also the appearance of an increasing number of multidrug-resistant pathogenic bacteria. Resistant bacteria were detected in hospitals and nosocomial infections have become a major problem. *Staphylococcus aureus* is responsible for half of hospital-acquired infections and causes deaths of many people around the world. Besides the problem of antibiotic resistance, new families of antibiotics are needed to enter the marketplace at regular intervals to face new diseases caused by evolving pathogens. At least 30 new diseases emerged in the 1980s and 1990s and they are growing in incidence. Also important are re-emerging diseases such as influenza and hepatitis B. Owing to the movement of the pharmaceutical industry away from natural products, especially antibiotics, the number of drug approvals in recent years has drastically dropped. However, antimicrobial pharmaceuticals are still big business and the search for new drugs must not be stopped. New screening approaches, including the search for novel targets and exploration of non-conventional places as sources of the producer microorganisms, are needed. In addition, metagenomic and genome-mining techniques have shown strong potential for discovery of new antibiotics. What is needed is a move back by the large companies to rational drug design and the use of more focused, more drug-like, compound libraries. In addition, it is desirable that small companies and academics, either in an independent manner or organized as biotechnology/university groups, increase their participation in a worldwide effort for new antibiotic discovery.

The antibiotic crisis

The period 1940 to 1970 featured the discovery and development of most of the useful antibiotics. Microbiologists knew then that technology had not yet won the war against infectious microorganisms due to resistance development in pathogenic microbes. Indeed, we know that technology will never win this war permanently and we must be satisfied to stay one step ahead of the pathogens for a long time to come. Thus, the search for new drugs must not be stopped.

Infectious disease was the leading cause of death in 1900 and is now the second most important killer in the world. It causes one-fifth of all deaths around the world and is the leading killer of children under five years of age (Hoegberg *et al.*, 2010). It is killer number three in developed nations (Kraus, 2008) and fourth in the USA (Scott, 2009). Worldwide, 17 million people die each year from bacterial infections (Butler and Buss, 2006). Despite existing antibiotic therapies and vaccines, infectious disease remains the major cause of mortality and morbidity (Vincente *et al.*, 2006). Americans are infected with bacteria at the rate of over 2.5 million per year. Fungi are also causing serious problems. For example, *Candida* infections in the USA cause 60,000–70,000 cases per year (Perlroth, 2007).

Hospital acquired infections, also known as nosocomial infections, have become a major problem. About 15 million people are admitted to US hospitals annually and about 10% of the patients acquire bacterial infections there (Overbye and Barrett, 2005; Brickner *et al.*, 2008; Carlson, 2009).

Gastrointestinal infections killed 17,000 people in the USA in 2007 compared with 7000 in 1999. Eighty-three per cent of the deaths were in people older than 65 years of age. Two-thirds of the deaths were caused by *Clostridium difficile*, mainly (94%) contracted in hospitals, nursing homes, doctors' offices and clinics. In 2004, a more virulent and drug-resistant strain evolved, producing two toxins causing problems in cells lining the intestine. Cases in US hospitals rose from 139,000 in 2000 to 336,000 in 2009 and killed 14,000 people. A 400% increase was observed in both cases and deaths from 2000 to 2013. Three million cases also occurred outside hospitals. The overall cost of *C. difficile* infections is $1 billion per year. A second cause of nosocomial infections is the norovirus where infection occurs in hospitals, prisons, dormitories, and on cruise ships. Norovirus is responsible for 800 deaths per year in USA among 20 million infections. Those stunning figures are higher than the death toll from AIDS, car accidents and prostate cancer combined.

Antibiotic resistance

The human population is ageing, and many patients undergo surgeries, or have implanted joint replacements, or are subjected to immunosuppressive therapy, making them possible victims of resistant bacteria. Bacterial sepsis has already become one of the main causes of death in the elderly. The extensive use of antibiotics has selected antibiotic-resistant strains, some of which are resistant to more than one antibiotic. The major public health problems today are (1) methicillin-resistant *S. aureus* (MRSA), (2) multidrug-resistant (MDR) and pandrug-resistant (PDR) Gram-negative bacteria, (3) MDR and extensively drug-resistant (XDR) strains of *Mycobacterium tuberculosis* (MDR-TB and XDR-TB), and (4) *Candida* species, the third leading cause of catheter-related infections (Ball and Tegos, 2012). In the USA, 70% of infections are resistant to current antibiotics (Najafi, 2011). Antibiotic resistance led to 400,000 infections in Europe in 2007 with 25,000 deaths (Bush *et al.*, 2011). Costs for these problems amounted to 1.5 billion euros. Almost 2 million US patients develop a hospital-acquired infection each year. Thus, such infections are the sixth leading cause of death in the USA. Some pathogenic bacteria in intensive care units, such as *Acinetobacter baumannii*, have been described as 'pan-resistant'. The increasing resistance of these bacteria raises fear of failure of antibiotic treatment (Spížek *et al.*, 2010).

S. aureus is responsible for half of hospital-acquired infections and kills 100,000 patients per year in the USA (Balaban and Dell'Acqua, 2005). Methicillin-resistant *S. aureus* (MRSA) alone is responsible for 19,000 deaths in the USA and 360,000 hospitalizations (Walsh and Fischbach, 2009), amounting to $3–4 billion in healthcare costs (Ball and Tegos, 2012). There was an increase in MRSA-resistance from 38% in 1992 to 65% in 2004. MRSA incidence in US intensive care units among *S. aureus* isolates was 2% in 1974, 22% in 1995, and 64% in 2004. Vancomycin had long been considered the last hope for an antibiotic against MRSA but strains of *S. aureus* resistant to vancomycin (VRSA) developed. The occurrence of antibiotic-resistant strains is apparently inevitable. A new resistant pathogen became a serious problem in 2011, i.e. carbapenem-resistant *Klebsiella* (CRK). Up to 50% of patients contracting CRK die. It has become an important hospital-acquired infection. Resistant *Escherichia coli* is also a problem. In 2003, only 4% of its infections were drug-resistant. By 2013, the figure had risen to 17% (Landro, 2013). If we continue to use antibiotics at the current rate, we may face a return to the pre-antibiotic era when a common bacterial infection could be a death sentence.

Resistant bacteria were originally detected in hospitals; however, their occurrence is now more widely distributed. The large amounts of antibiotics used in human therapy, as well as those used for farm animals and even for fish in aquaculture, have resulted in the selection of pathogenic bacteria resistant to multiple drugs (Nikaido, 2009). At present, more than 50% of produced antibiotics are used to promote animal growth (Bush *et al.*, 2011). Europe banned antibiotic use on the farm and has observed that emergence and spread of resistant bacteria has decreased (Marshall and Levy, 2012). In hospitals, resistant bacteria can survive for a prolonged time and can cause epidemics, for example, in intensive care units. The risk of infection increases with the amount of time spent in the hospital.

There are naturally antibiotic-resistant bacteria, e.g. *Pseudomonas aeruginosa* (causing fatal wound infections, burn infections and chronic and fatal infections of lungs in cystic fibrosis patients), *Stenotrophomonas maltophilia*, *Enterococcus faecium*, *Burkholderia cepacia* and *A. baumannii* (Tenover and Hughes, 1996). The opportunistic pathogen *P. aeruginosa* causes fatal wound infections, burn infections, and chronic and fatal infections of lungs in cystic fibrosis patients. Very few antibiotics inhibit this pathogen. It almost never infects uncompromised tissues; however, it can infect practically any tissue of patients compromised in some way. It also causes urinary tract infections, respiratory system infections, dermatitis, soft tissue infection, bacteraemia, bone and joint infections, gastrointestinal infections and a variety of systemic infections, particularly in patients with severe burns, and in cancer and AIDS patients who are immunosuppressed.

In addition to naturally resistant organisms, there is a continuous development of resistant pathogens, e.g. enterococci resistant to all antibiotics (Chu *et al.*, 1996); also the organisms causing tuberculosis (TB) and malaria. Resistance occurs by mutation and/or by gene acquisition. The antibiotic resistance problem has been addressed in the report of the colloquium sponsored by the American Academy of Microbiology (2009) titled 'Antibiotic resistance: An ecological perspective on an old problem' and clearly analysed by Jovetic *et al.* (2010). Environmental microorganisms are highly drug-resistant (Wright, 2010). The genes that form the environmental resistome have the potential of being transferred to pathogens and there is some evidence that clinically relevant resistance genes originated in environmental microbes.

Resistance involves inactivation by enzymes such as β-lactamases. increased efflux of the antibiotic out of cells, decreased uptake of the antibiotic, modification of the target to decrease binding of the antibiotic, amplification of the target, bypassing the essentiality of the target, sequestration of the antibiotic, protection of the target, intracellular localization, and biofilm formation (Table 4.1) (Singh and Barrett, 2006; Davies, 2007).

Bacteria and fungi forming biofilms, including staphylococci, are very resistant to antibiotics.

Table 4.1 Microorganisms that frequently develop antibiotic resistance

Microorganism	Antibiotic	Resistance mechanism[a]
Staphylococcus aureus	Methicillin	EAI, M/PTS, LAA, OTS
Mycobacterium tuberculosis	Isoniazid, rifampin	M/PTS, FAA
Klebsiella spp.	Carbapenems	EAI, M/PTS, LAA
Candida albicans	Azoles	AE, M/PTS, OTS
Aspergillus fumigatus	Echinocandin	M/PTS
Acinetobacter baumannii	Sulbactam Tobramycin	LAA, EAI M/PTS
Escherichia coli	Cephalosporins Fluoroquinolones	EAI AE
Neisseria gonorrhoeae	Tetracyclines	AE, M/PTS, LAA
Aspergillus fumigatus	Azoles	AE, M/PTS, OTS, FAA
Plasmodium falciparum	Chloroquine	AE/LAA

[a]EAI, enzymatic antibiotic inactivation; M/PTS, modification/protection of target sites; AE, active efflux; LAA, limiting antibiotic access to microbial cell; OTS, overproduction of target.

Biofilm infections are especially a problem on the surface of medical devices placed within the body of patients, and surgical fixtures. The organisms grow on wounds, scar tissue, and medical implants such as joint prostheses, spinal instruments, vascular prosthetic grafts and heart valves (Bjarnscholt *et al.*, 2013). In these situations, staphylococcal biofilm infections are not only associated with increased morbidity and mortality, but also significantly contribute to the emergence and dissemination of antibiotic resistance in the nosocomial setting (Lynch and Robertson, 2008). Biofilms are also important in native infections such as cystic fibrosis, otitis media, endocarditis, and urinary tract infections. Such chronic infections have extreme resistance to antimicrobial agents. The communities of bacteria present in a biofilm are relatively resistant to antibiotics that normally would inhibit the individual species in the free state. Some secondary metabolites actually trigger biofilm formation, e.g. sub-inhibitory levels of imipenem or tobramycin. There are relatively few antibiotics active against biofilms but new compounds are being pursued (Lynch and Abbanat, 2010).

At present, pneumonia and influenza are the greatest causes of death in the USA. Community-acquired bacterial pneumonia (CABP) caused by *Streptococcus pneumoniae* is becoming a major problem. The organism causes bacterial pneumonia resulting in 40,000 deaths in the USA each year. By 1999, 25% of US isolates of *S. pneumoniae* were penicillin resistant. Small children and the elderly are at a very high risk. This is important since the elderly are becoming a larger segment of the population as a consequence of improved living standards in developed societies (Vincente *et al.*, 2006). Resistance is also developing against macrolides such as azithromycin and clarithromycin. Macrolide resistance has reached 35% in the USA. New and more potent macrolides are sorely needed (Fernandes and Schranz, 2011). Resistance has also markedly developed to fluoroquinolones, especially in older adults. Many *S. pneumoniae* strains are not susceptible to pneumococcal vaccines.

Tuberculosis is a major health problem in the world. About two billion people in the world carry the infectious organism *M. tuberculosis*. It is the second largest infectious killer. Each year, nine to ten million new cases are diagnosed and 1.4 million people die (Phillips, 2013). Only AIDS (three million deaths/year) is more dangerous. Robert Koch found *M. tuberculosis* to be the causative agent in 1882 but the epidemic continues on the increase. Back in the 1920s, the Bacillus Calmette–Guerin (BCG) vaccine was successfully used against TB but today, it is effective only against childhood TB. Isoniazid in 1952 and rifampicin in the 1970s were very effective. However, when HIV emerged in the 1980–1990s, an infection with TB and HIV became synergistic in causing death. TB is the leading cause of death among people having HIV/AIDS. The last drug found with a new mechanism of action was rifampicin, which was discovered in 1963. No new drug was commercialized against TB (Koppal, 2004) until bedaquiline was approved in 2012. TB in the 1990s became resistant to isoniazid and rifampicin and required treatment with antibiotics such as the fluoroquinolones, amikacin, capreomycin and kanamycin (Phillips, 2013). Multidrug-resistant mycobacteria arise for a complex set of reasons, an important one being high failure rates for completion of therapeutic courses, which are often associated with a lack of resources required to comply with relatively long-term therapeutic regimens. TB chemotherapy has been going on for 50 years but much of the world's population harbours a latent or dormant form of the organism. Reactivation of the latent form is particularly common in immune-suppressed people especially those with HIV infections, those on therapy for tumour necrosis factor, and those with diabetes. It is alarming that highly resistant strains continue to evolve and we face the risk of losing control, even in the industrialized world (Russell *et al.*, 2010). Patients with drug-susceptible (DS) TB can be treated with a combination of isoniazid, rifampicin, pyrazinamide, streptomycin and/or ethambutol (for two months) followed by rifampicin (for the last 4 months), resulting in a 95% cure rate in those 6 months. However, those with multiple drug-resistant (MDR) TB are resistant to isoniazid and rifampicin (Kamholz, 2002) and cure rates are lower, i.e. 50–70%. Extremely drug-resistant (XDR) TB is characterized by resistance to isoniazid, rifampicin, fluoroquinones, amikacin, kanamycin and capreomycin with very high mortality rates.

More than 59% of the 100,000 MDR-TB cases are in India and China. At present, MDR-TB is treated with 8–10 drugs in combination lasting for 18–24 months, with 30% of the cases resulting in failure. Another problem is that of serious side-effects of these drugs, such as nephrotoxicity and ototoxicity with aminoglycosides, hepatotoxicity with ethionamide, and dysglycaemia with gatifloxacin. Also a problem is the association of TB with chronic diseases such as diabetes, which increases the risk of developing active-TB by 3-fold.

Most TB drugs do not follow Lipinski's 'rule of five' which is claimed to define optimal drug-like features. The genome sequence of *M. tuberculosis* has failed to reveal any opportunities for new anti-TB candidates. However, there is some hope. Whole-cell screening is leading to new candidates such as diarylquinolines (e.g. TMC 207), targeting ATP synthesis, and benzothiazines (e.g. BTZ 043), which target synthesis of araban, essential for cell wall formation. Redesign of old drugs is also yielding new candidates such as PNI-100480 and AZD-5847 which are new oxazolidinones which show improved activity against *M. tuberculosis.* They are in phase I clinical trials. Also, new nitroimidazoles are under study (PA-824 and OPC-67683). Knockout of an extended-spectrum Ambler class A β-lactamase has shown increased activity of carbapenems against TB. Combining meropenem with clavulanic acid is another strategy, which has shown increased *in vitro* activity on clinical isolates of *M. tuberculosis*. TB vaccines are also showing promise, with seven candidates in clinical development. A comprehensive article on TB and the challenge of new TB drug discovery is that of A. Koul (2011).

Gonorrhoea, produced by *Neisseria gonorrheae*, affects over 100 million people globally each year. Eight per cent of gonorrhoea-producing bacteria are resistant to tetracyclines. They are also resistant to penicillins, although they have been successfully treated with cephalosporins such as carbapenems. However, carbapenem-resistant enterobacteria (CRE) increased from 1.2% to 4.2% in 2011. One of these genera, *Klebsiella*, showed an increase from 1.6% to 10.4% (Landro, 2013). In 2011, the US National Institutes of Health (NIH) facilities in Bethesda, Maryland, were attacked with resistant *Klebsiella*, resulting in 11 deaths out of a total of 18 people infected. The outbreak was derived from a single patient who was discharged three weeks before the next attack occurred.

Globally, 200–500 million people are infected with malaria and 1–2 million die. The *Cinchona* plant produces the alkaloid quinine which was the first anti-malarial agent. The parasite *Plasmodium falciparum* produces the pathogenic agent and female *Anopheles* mosquitoes carry the parasite. The insecticide DDT was used for a while but its use was later outlawed. Synthetic chloroquine followed quinine in the 1930s but resistance developed. In the 1980s, sweet wormwood (*Artemisin annua*), which had been part of traditional Chinese medicine for centuries, was used against malaria. The key ingredient was artemisinin. However, artemisinin quickly decomposes in the body and has neurotoxicity, at least in animal tests. Its derivatives, artemether, artesunat and artether have been used for multidrug-resistant malaria. Another derivative, artemisone, lacks toxicity and is being clinically tested. It is reported to cure patients in two days!

Important fungal diseases are cutaneous and subcutaneous infections, mucosal invasion, and bloodstream infections. Invasive candidiasis is caused by various species of *Candida*. *Candida albicans* was the leading cause of invasive candidiasis and bloodstream infections prior to the use of fluconazole. However, in recent years, triazole-resistant species such as *Candida glabrata*, *Candida krusei* and *Candida parapsilosis* have become more dangerous. Fungal infections have become more common. From 1980 to 1997, mortality due to fungal disease increased from the 10th to the 7th most common infection. The five leading mycoses during this period were invasive candidiasis, invasive aspergillosis, histoplasmosis, cryptococcosis and coccidioidomycosis.

Candida species are among the top three or four causes of nosocomial infectious diseases and *Aspergillus fumigatus* is the most deadly and frequent mould infection of humans. The cost of treating fungal infections in the USA is $2.6 billion per year. Unfortunately, the antifungal azoles and polyenes have some safety issues.

Azoles are grouped into two classes: (a) the imidazoles (miconazole and ketoconazole), employed for topical treatment of superficial infections, and (b) triazoles such as fluconazole and itraconazole. The target of azoles is the fungal cytochrome P450-dependent lanosterol 14-α-demethylase that converts lanosterol to ergosterol. Toxic sterols form, resulting in a membrane that does not function as well as with ergosterol, leading to growth inhibition. Triazoles can induce drug–drug interactions, are only fungistatic, and their use often results in drug-resistant fungal pathogens. On the other hand, enhanced activity has been observed with the newer triazoles, i.e. posaconazole and voriconazole. However, resistance to these is developing in *Candida* species.

Polyenes (amphotericin B and nystain) are toxic due to binding to cholesterol and disruption of mammalian cell membranes. 5-Flucytosine B, an important polyene, used to be administered with amphotericin B but resistance developed. Amphotericin B, an important polyene, shows toxicity, but is fungicidal and broad-spectrum in activity. Unfortunately, its lipid-encapsulated derivatives are extremely expensive. Newer antifungals, the echinocandins, inhibit synthesis of cell wall β-1,3-glucan. They are relatively broad spectrum, have no safety issues, but resistance is developing.

Emergence of new diseases

The World Health Organization concluded that at least 30 new diseases emerged in the 1980s and 1990s (DaSilva and Iaccarino, 1999). New diseases include HIV-AIDS, Hanta virus, Ebola virus, Dengue virus, *Cryptospiridium*, Legionnaire's disease, Lyme disease, *E. coli* 0157:H7, and severe acute respiratory syndrome (SARS). Also important are re-emerging diseases such as influenza and hepatitis B. Other emerging and re-emerging diseases include West Nile Fever, hepatitis C, hantavirus pulmonary syndrome, children's diarrhoea by rotavirus, and mad cow disease. The cost of combating such diseases is more than $120 billion per year. The increases in emerging and re-emerging diseases involve the following factors: (a) increased international travel, (b) increased human population density, (c) global movement of wild, exotic animals, (d) destruction of animal habitats, (e) inadequate public health in underdeveloped countries leading to poor preventative vaccine campaigns and ineffective monitoring of water and food cleanliness, and (f) the decrease in new antibiotic discovery and development. An example of (c) is the use of exotic cats in China for meat, which has led to the host transition of the SARS coronavirus from animals to humans. With regard to (e), water pollution causes 14,000 deaths per day in third world countries. Nearly two million people die of diarrheal diseases every year, mainly due to contaminated water and poor sanitation. Other problems include diseases caused by a known pathogen but attacking new populations, and toxicity of some of the compounds in use (Strohl, 1997).

Lack of new drugs

There were 120 drug approvals by the US Food and Drug Administration (FDA) in 1996–1997 (Ryan, 2003). Owing to the movement of the pharmaceutical industry away from natural products, especially antibiotics, the number of drug approvals dropped to 36 in 2004 and then to 20 in 2005 (Mullin, 2006). Between 2003 and 2012, the figure dropped to seven (Gravitz, 2012). Although 86 new drugs were approved by the FDA in 2003, the actual number of new chemical structures hit a nine year low. In 2004, only 1.6% of the compounds in clinical development were antibiotics.

Whereas FDA drug applications peaked at 131 in 1996, the number dropped steadily to 78 in 2002 (Warner, 2003). Launches of new drugs dropped from an average of 44 per year between 1995 and 2000 to 33 per year during 2001–2006, and to 27 in 2007 (Malik, 2008).

During 1978–1980, the average number of 'new molecular entities' (NMEs), i.e. drugs with a novel molecular structure that had not been previously approved in the USA, launched by the pharmaceutical industry averaged 43 per year. In 1998–2000, the average number had dropped to 33. Evidence of the drop-off can be seen in the following data: 1996, 53; 1997, 39; 1998, 30; 1999,

35; 2000, 27; 2001, 24; 2002, 17; 2003, 21; 2005, 18 (Warner, 2004; Kostel, 2004; Wolff, 2006).

Another FDA category is 'new active substances' (NASs). In 2001, there were 37 of these, which represented a 20-year low in the number approved by the FDA (Jacobs, 2002). There has been a continuous drop since 1997.

The number of drugs classified by FDA as 'new chemical entities' (NCEs) developed by the top 20 pharmaceutical organizations continued to drop over the 15-year period between 1987 and 2002 (Handen, 2002). The number launched in 2003 was 30, the lowest in over 20 years. (Class, 2004). Furthermore, fewer than half of the NCEs approved by FDA in 2002 were really new (Willis, 2004). The rest were protein-based products, off-patent generics and derivatives of existing drugs.

Between 1983 and 1987, 16 new antibacterials were approved. Since the late 1980s, only one new antibacterial class has been discovered and commercialized, i.e. bedaquiline acting against TB. It was FDA-approved in 2012. It was the first anti-TB drug approved in over 40 years (Torrice, 2013).

Why drug discovery has dropped

Many large pharmaceutical companies have abandoned their anti-infective research programmes (Demain, 2002). The number of companies working on antimicrobial agents dropped from 35 in the 1950s to fewer than ten in 2013 (Torrice, 2013). One reason for the lack of new drug discovery is 'merger-mania' in this industry. Mergers have decreased the number of groups searching for new antibiotics. Almost 40 major mergers in the pharmaceutical industry took place between 1985 and 2005 (Daemmrich and Bowden, 2005). As recently as 2009, major companies underwent mergers, e.g. Wyeth with Pfizer and Schering-Plough with Merck. According to Drews (2001), who spent many years in the pharmaceutical industry, (i) pharmaceutical companies formed by mergers do not become more productive, and (ii) the probability of producing a blockbuster drug is not a linear function of company size. Even many drug executives now realize that mergers can actually have a negative impact on R&D productivity.

A second reason involves the nature of natural products. Among medicines used up to 1996, 80% were natural products or inspired by natural products (Harvey, 2007). Of the 868 NCEs approved between 1981 and 2002, 52% were natural or created around natural product structure. It is quite difficult to discover new natural products with antibacterial activity when those more prevalent in nature have already been discovered. As a result, there has been a misguided loss of interest by companies in natural products, especially those with antibiotic activity. The industry has opted to save funds by eliminating natural product departments or decreasing their relevance in the hunt for new drugs. Companies which have dropped or significantly reduced research on discovery of natural products/antibiotics include Merck, Wyeth (now part of Pfizer), Aventis, Eli Lilly, GlaxoSmithKline, Bristol-Myers Squibb, Abbott Labs, and Procter and Gamble (Barrett, 2005).

A third reason is the increased costs of getting a drug to the market. Since 1998, attrition rates, development times and R&D expenses have increased. The average time from patenting to commercialization in the US and in Europe increased from 9.7 years in the 1990s to 13.9 years between 2000 and 2010. Clinical development time doubled to 6 years between 1982 and 2002. This includes 1 year of phase I (involving 20–30 healthy volunteers for safety and dosage), 1.5 years for phase II (100–300 patient volunteers for efficacy and side-effects), and 3.5 years for phase III (1000–5000 patient volunteers monitoring effects of long-term use). Recently, the clinical development time has increased to 8 years. Added to this could be 2–10 years for discovery, 4 years for preclinical testing, 1 year of FDA review and approval, and 1 year of post-marketing testing. Although some estimate that the total time to get a drug on the market is 12–15 years (Burrill, 2002), the above breakdown indicates that it could be as long as 24 years. The estimated cost of getting a drug to market rose from \$500 million to \$600 million in 1999 and to \$900 million to \$1 billion in 2002 (Kettler, 1999; Di Masi, 2001; Agres, 2003). By 2003, the cost had risen to \$1.7 billion (Thayer, 2004) and by 2010, it reached \$1.8 billion (Morrow, 2010). Two-thirds of the cost was due to leads which failed in the clinic

(Wilson, 2001). Attrition rates are the proportion of failures out of the total number of projects entering a given stage of R&D. The attrition rates in phase II and phase III have increased dramatically. One half of all potential drugs failed because of adsorption, distribution, metabolism, excretion or toxicity (ADME/TOX) problems. Seventy per cent of compounds get through phase I of clinical trials; 33% of these get through phase II; and 25–30% of these get through phase III. Overall, only 8% of compounds entering trials become commercialized.

Because of the increased costs, the pharmaceutical industry's discovery efforts in the 1990s moved away from natural products to combinatorial chemistry followed by high throughput screening (HTS). This was done because it was considered that natural product extracts were not amenable to HTS. Although it was thought that combinatorial chemistry/HTS would yield many new hits and leads, the results were disappointing despite the extraordinary amount of money spent (Ausman, 2001; Horrobin, 2001). Developed in the early 1990s, speed and miniaturization were accomplished by HTS but discovery of new leads did not accelerate. HTS methods allowed 100,000–200,000 chemicals to be assayed per day (Firn and Jones, 2000; Hefti and Bolten, 2003) and combinatorial and other chemical libraries of one million compounds were available commercially. Since use of the conventional 96-microwell format for HTS could cost $1 million to screen 500,000 compounds against a single target, some companies went to 384, 1536 and even 3456 well formats to cut expenses. However, the premise that 'the more compounds screened, the more leads found' did not work. No drugs were approved resulting from HTS by 1999 (Fox *et al.*, 1999) and not a single drug derived solely by combinatorial chemistry was introduced through 2005. The USA pharmaceutical industry invested $32 billion in 2002, triple the amount invested 10 years earlier, but the number of resulting NCEs dropped. Finally, one NCE of synthetic origin produced by combinatorial chemistry was approved by FDA. It was the anti-tumour drug sorafenib (Nexavar) from Bayer, a kinase inhibitor, approved in 2006 for renal carcinoma (Newman and Cragg, 2007).

The problems were that HTS had not been applied to natural product libraries and that combinatorial chemistry had not utilized natural products as scaffolds (Demain, 1999; Kingston and Newman, 2002; Waldmann and Breinbauer, 2002). This made no sense since the role of combinatorial chemistry, like those of structure–function drug design and recombinant DNA technology two and three decades ago, was that of complementing and assisting natural product discovery and development, not replacing them. Combinatorial chemistry is great for improving leads but not for the discovery of new leads. However, when combinatorial chemistry is applied to natural products, it works. For example, workers at Vicuron (Clough *et al.*, 2003) produced 500 combinatorial analogues of the natural but poorly soluble thiazole peptide antibiotic GE2270A. Of these, eight had both good solubility and good activity. A number of other examples are cited by Ganesan (2004). The chemist Waldemann (2003) has stated that 'biological investigation of the million compound speculative combinational libraries of the first generation yielded disappointingly low hit rates'. He urged the use of biologically validated compounds as scaffolds for library generation. He further stated that, 'biologically active natural products can be regarded as chemical entities that were evolutionarily selected and validated for binding to particular protein domains'. It is encouraging that natural products are finally being used by chemists as combinatorial chemistry scaffolds for synthesis of potential drugs. It is hoped that new products will result from such efforts (Newman, 2008).

Despite the great costs of genomics research, the use of genomics has not had a major effect on drug discovery. After 10 years of bacterial genomics, there were still no promising antibacterial agents on the market or even in clinical testing resulting from genomic studies (Shlaes *et al.*, 2004; Coates and Hu, 2007). Indeed, investment in genomics and HTS have had no effect on the number of products in preclinical development or phase I clinical trials. However, instead of downgrading natural product screening, there is real opportunity in incorporating it with HTS, combinatorial chemistry, genomics, proteomics and new discoveries being made in biodiversity.

The pharmaceutical industry has been

spending about \$30 billion per year on R&D (Tralau-Stewart *et al.*, 2009). Data from 48 drug companies show that R&D spending increased by 80% from 1992 to 2002 whereas new drug launchings dropped by 35% (Thayer, 2003). Similarly, from 1998 to 2001, the annual spending on R&D of the top 20 pharmaceutical companies increased from \$26 billion in 1998, to \$31 billion in 1999, to \$35 billion in 2000, and to \$37 billion in 2001, but the number of new drug applications (NDAs) decreased from 34 to 23 to 21 and to 16 in those same years. It is obvious that the incorrect approaches mentioned above have contributed to this problem.

The drop-off in rate of discovery is not due to a decrease in total investment by the pharmaceutical industry but more likely in an over-emphasis on promotion. The amount spent by the pharmaceutical industry to market, promote, and advertise its products in 1991 was \$9.2 billion. By 2004, the amount was \$25 billion, mainly due to direct advertising to consumers, free drug samples, and salaries for drug representatives. Novartis spent 33% of sales on promotion but only 19% on R&D in 2004.

Lack of new antibiotics

The drop-off in new drug approvals is even more significant when one considers antibiotics. In 2011, less than 5% of the products in the pharmaceutical R&D pipeline were antibiotics, according to the World Health Organization. Since 1983, the rate of antibacterials commercialized has become vanishingly low. The data are as follows: 1983–1987, 16 new antibiotics; 1988–1992, 14; 1993–1997, 10; 1998–2002, 7; 2003–2007, 5; 2008–2010, 2. In 2002, there were 89 new drugs approved by FDA but none was an antibiotic. Approvals of antibiotics by FDA decreased from 17 in 1980–1984 to 2 in 2005–2007. Of 74 new therapeutic agents approved by the FDA in 2007, only two were antibiotics. Of 2700 compounds in development in 2008, only 50 were antibacterials and, of these, only ten were from large pharmaceutical companies (Carlson, 2009). In the meantime, the 'superbugs' were on the increase.

Reasons for the lack of discovery of new antibiotics were put forth by Baltz (2008a) as follows: (a) enzymes essential to the viability of pathogens are not readily 'druggable' (i.e. they may not have binding sites for inhibitors, or the compounds cannot get into the cell, or the compounds have poor pharmacological properties), (b) targets are not accessible to *in vitro* screening (e.g. if they are ribosomal or part of nascent peptidoglycan), and (c) chemical libraries lack the molecular complexity of natural antibiotics. According to Baltz (2008a), we can no longer depend on pharmaceutical companies alone to come up with new antibiotics. The effort will have to come from medical research by academia in collaboration with biotechnology and pharmaceutical companies.

Outsourcing by pharmaceutical companies for lead optimization has become a major trend in the drug industry (Clark, 2011). A major increase has occurred in the number of deals with outsourcing companies in India, China and Eastern Europe. Whether such moves will prove successful remains to be seen. There are also increasing numbers of deals between outsourcing companies and not-for-profit organizations, including universities.

Natural products versus synthetics as drugs

Between 1988 and 2008, 877 pharmaceuticals were commercialized. Of these, 60% were from natural sources or derived from them (Lefevre *et al.*, 2008). Quality appears to be more important than quantity when it comes to new drug discovery. Whereas only 0.001% of the total synthetic compounds have become drugs, 0.2–0.3% of microbial metabolites have become drugs and another 0.2–0.3% have become lead compounds for chemical modification. This is more than two orders of magnitude difference. Natural product collections have a much higher hit rate in high throughput screens than do combinatorial libraries. Breinbauer *et al.* (2002) point out that the number of compounds in a chemical library is not the important point; it is the biological relevance, design and diversity of the library and that a scaffold from nature provides viable, biologically validated starting points for the design of chemical libraries.

Products from nature are unsurpassed in their ability to provide novelty and complexity (Bull *et al.*, 2000). With respect to the number of chirality centres, rings, bridges and functional groups in the molecule, natural products are spatially more complex than synthetic compounds (Henkel *et al.* 1999). Synthetic compounds highlighted via combinatorial chemistry and *in vitro* high-throughput assays are based on small chemical changes to existing drugs, and of the thousands, perhaps millions, of chemical 'shapes' available to pharmaceutical researchers, only a few hundred are being explored. Many compounds are probably being missed (Willis 2004). Natural products differ from synthetic compounds by having more oxygen atoms and stereochemical elements such as chiral centres and polycyclic (often bridged) carbon skeletons (Ganesan, 2004). Most drugs in use today are chiral. In a survey comparing about 670,000 chemical combinatorial compounds, about 11,000 drugs, and over 3000 natural products, it was found that 82% of natural products were chiral, 55% of drugs were chiral but only 29% of combinatorial products were chiral (Feher and Schmidt, 2003).

According to Sam Danishefsky, prominent chemist at Memorial Sloan-Kettering Cancer Center in New York, it is appropriate 'to critically examine the prevailing supposition that synthesizing millions of compounds at a time is necessarily going to cut the costs of drug discovery or fill pharma pipelines with new drugs any time soon' (Borman, 2002). Danishefsky continues:

> At the risk of sounding Neanderthal, I would even put in a pitch for industry getting back to the screening of natural products. Some of the most valuable products and promising leads in oncology are naturally derived or naturally inspired. For instance, paclitaxel, a chemically established drug, came from natural product sources, as did doxorubicin, the etoposides and the latter-day camptothecins. In fact, even tamoxifen arose from natural product leads, steroid hormones. Moreover, several of today's most promising pipeline candidates in oncology-such as ecteinascidin, halichondrin, bryostatin, and of course, the epothilones- all arose from natural product screening followed by synthetic modifications. A small collection of smart compounds may be more valuable than a much larger hodgepodge collection mindlessly assembled. Thus, the decision on the part of several pharma companies to get out of the natural products business is gross foolishness. There are major teachings in these natural products that we would do well to consider. They may be reflecting eons of wisdom and refinement. The much maligned natural product collections did, after all, bring us statin, β-lactam, aminoglycoside, and macrolide blockbuster drugs. In fact, one of the most promising approaches in diversity chemistry is to produce diversity-chemistry-derived collections that benefit from or partake of the 'wisdom' of natural products.

Possible solutions to the problem

Unfortunately, from the commercial view, antibiotics are taken for only a short time by the patient, in contrast to drugs for heart disease, which are taken for life, and to drugs enhancing male sexual performance. This is one reason why large companies are abandoning the antibiotic area. Despite this dismal situation, there is hope. After all, antimicrobial pharmaceuticals are still big business. In 2002, the anti-infective market amounted to $45 billion, made up of 62% antibacterials, 13% sera, immunoglobulins and vaccines, 12% anti-HIV antivirals, 7% antifungals, and 6% non-HIV antivirals (Bush, 2004). What is needed is a move back by the large companies to rational drug design and the use of more focused, more drug-like, compound libraries.

New approaches, including new targets, are available for screening natural products (Genilloud *et al.*, 2011). For example, inhibitors of peptide deformylase and fatty acid biosynthesis, which are new targets not based on genomics, are in clinical trials. Other targets include lipid A biosynthesis and tRNA synthetases. Eighteen antibacterial drugs were in clinical trials in 2004 (Bush *et al.*, 2004). They included carbapenems, cephalosporins, glycopeptides, quinolones, a glycolipodepsipeptide, a dihydrofolate reductase inhibitor, an oxazolidinone, two peptides, and a peptide deformylase inhibitor. Furthermore, major improvements have been made in detection, characterization and purification of small molecules. Peptidoglycan synthesis is an important target of antibacterial agents (Goutam *et al.*, 2010). Thirteen enzymes are involved in the biosynthetic process in *E. coli*. Nine are cytoplasmic and four are associated with membranes.

It is possible that peptides, such as bacteriocins, could help to solve the problem (Cotter *et al.*, 2013). These small antibacterial peptides are produced by bacteria. They are potent, have low toxicity, and exhibit broad or narrow spectrum activities. Some have mechanisms not shown by current antibiotics. They contain 15 to nearly 50 amino acids, are generally positively charged, are synthesized by the ribosomes, and modified post-translationally. More than 2000 are known. Some are toxic; others, such as nisin, are broken down by the body and serve as food preservatives. A major problem is that they are not effective against Gram-negative pathogens, treatment of which costs $45 billion per year. New peptides are being pursued against biofilm infections. Novartis is examining thiopeptide LFF551 in a phase II clinical trial against *C. difficile*. Cubist Pharmaceuticals is conducting phase III trials of its lipopeptide surotomicin (CB-315) against the same bacterium. Novexatin of Novabiotics (Aberdeen, UK) is a cyclic and highly cationic (arginine-rich) peptide targeting fungal infections in toe nails. It is in phase III trials. Another group under study are the peptoids, which have a natural amino acid backbone with synthetic side chain residues conferring protease resistance and increased hydrophobicity, thus enhancing membrane permeability.

Another possible solution is the use of bacteriophage. Phage therapy has been considered for many years but never put into practice. However, phage are being considered for control of biofilm infections, ether as a preventative or a curative process (Bjarnsholt *et al.*, 2013). Phage have proven effective in animals. Phage cocktails can eradicate mucoid *P. aeruginosa* biofilms growing on the surface of a cystic fibrosis bronchial epithelial cell line. This shows that the phage can penetrate alginate, the major part of the *P. aeruginosa* biofilm matrix. Topical phage therapy is being tested in three clinical trials and results so far are encouraging with no adverse effects.

Another possible area is the combination of drugs, such as the combination of β-lactam antibiotics and β-lactamase inhibitors (Schlaes, 2013). The activity of β-lactamases has been a problem for many years (Ehmann *et al.*, 2012). Over a thousand such enzymes of different structural classes are known. The first combination treatment used was amoxicillin plus clavulanic acid. A later combination was sulbactam and clavulanic acid. The three approved β-lactamase inhibitors (clavulanic acid, tazobactam and sulbactam) have the β-lactam core. Tazobactam, a β-lactam sulfone, was the last β-lactam to be approved (1993). It was used in combination with pipericillin. Today, it is being combined with ceftolozone, a cephalosporin with activity against carbapenem-resistant and cephalosporin-resistant *P. aeruginosa*. Unfortunately, tazobactam is not active against class A carbapenemase, especially that from *K. pneumoniae*. Since 1993, there has been an increase in Gram-negative bacteria producing high levels of β-lactamase, multiple enzymes, inhibitor-resistant enzymes and Class C enzymes which are not inhibited by tazobactam and earlier β-lactamase inhibitors. No new β-lactamase inhibitors have been approved in the last 20 years. However, a new inhibitor, avibactam, is being combined with MK-7655. They are both non-β-lactam inhibitors of β-lactamase, part of a new class of inhibitors known as diazabicyclooctanes. They are more potent than tazobactam, have a broader spectrum of action including class A carbapenemases, and have a different mechanism of action. Avibactam is also being combined in second and third stage clinical trials with ceftazidine, a broad-spectrum cephalosporin active against *P. aeruginosa*, and also with ceftaroline, an approved cephalosporin active against MRSA but not against *P. aeruginosa*. MK-7655 is also being combined with imipenem plus cilastatin in a three-drug combination.

New screening technologies are also very important. For example, a recent means of discovering antifungal drugs uses an assay based on the application of the laboratory nematode *Caenorhabditis elegans* as host for *C. albicans* and other pathogenic *Candida* species (Breger *et al.*, 2007). The yeast is ingested by the worm, which causes an infection in the worm's intestinal tract and kills it. Known antifungal agents such as amphotericin B and caspofungin prolong the worm's survival. The test is done in liquid media contained in 96-well plates and appears to be a valuable discovery tool. Of 1266 compounds tested, 1.2% (15 compounds) were active. The most active were caffeic acid phenethyl ester and the fluoroquinoiline enoxacin. Another method employs *C. elegans* infected with *E. faecalis* (Moy *et al.*, 2006).

A new enzymatic technique called glycorandomization is being used to prepare glycoside libraries and to make optimized or novel glycoside antibiotics. Sugars in natural products such as antibiotics are usually members of the 6-deoxyhexose family. Over 70 different variants were found in products of bacteria, fungi and plants. Engineering the formation of new secondary metabolites in actinomycetes by glycosylation was reviewed by Salas and Mendez (2007). Novel deoxysugars can be placed on macrolide antibiotics by combinational biosynthesis (Pageni *et al.*, 2008). The presence of glycosidic residues in antibiotics is very important for their activity (Kren and Rezanta, 2008).

The marine cyanobacteria are a rich source of novel secondary metabolites (Tan, 2007). They produce more than 300 nitrogenous secondary metabolites. Most are bioactive non-ribosomal peptides (NRPs) or mixed polyketide-NRPs. This source provides a much higher discovery rate of novel secondary metabolites (>75%) than other sources of microbes.

The potential of genomics for discovery of new antibacterials has been discussed (Mills, 2006). Genome sequencing has revealed many more gene clusters for biosynthesis of secondary metabolites than the number of metabolites known at the time. These 'orphan' biosynthetic pathways (Gross, 2007) are now being activated by determination of optimum conditions for production. *Streptomyces coelicolor* was known to produce four secondary metabolites at the time that the genome sequence revealed 18 additional biosynthetic pathways. Genome sequencing of the marine organism *Salinispora tropica* revealed a circular genome of 5,183,331 base pairs (Udwary *et al.*, 2007). A large portion (9.9%) is devoted to secondary metabolism, greater than ever before seen. It contained genes encoding PKS systems of every known family, non-ribosomal peptide synthases, and hybrid clusters. The majority were novel.

Genome mining involves powerful techniques for discovery of new natural products (Zerkly and Challis, 2009; Scheffler *et al.*, 2013). In recent years, a number of additional 'silent' secondary metabolites have been found by genome mining. These include coelichelin (Lautru *et al.*, 2005) from *S. coelicolor*, geosmin (Cane *et al.*, 2006) from *Streptomyces avermitilis*, and epi-isozizaene (Lin *et al.*, 2006), germicidins (Song *et al.*, 2006) and mycothiol (Park *et al.*, 2006) from *S. coelicolor*. Coelichelin is a peptide which appears to be a siderophore involved in iron uptake and the germicidins are a group of five related compounds. Germicidin A was previously known to be an inhibitor of spore germination in *Streptomyces viridochromogenes*. Three of the germicidins were new compounds (Song *et al.*, 2006). Many new compounds have been isolated from other mined microbes (Gross, 2007). Since hundreds of microbial genomes have been sequenced, genome mining has great importance for the future of drug discovery. Multiple gene clusters encoding secondary metabolite production are common in species of *Streptomyces*, other filamentous actinomycetes and mycobacteria (Busti *et al.*, 2006). *S. coelicolor* and *S. avermitilis* contain 20–30 of these clusters. On the other hand, most other bacterial genomes lack them.

To discover new drugs from uncultivated microbes, a metagenomic approach has been recommended (Lorenz and Eck, 2005; Lefevre *et al.*, 2008). Developments in metagenomics have been reviewed by Simon and Daniel (2009).

Like the bacteria, fungi have many extra clusters of secondary metabolite biosynthesis genes (Sanchez, 2008). Genome sequencing of eight species of *Aspergillus* (*clavatus*, *flavus*, *fumigatus*, *nidulans*, *niger*, *oryzae*, *terreus* and *fischeri*) showed many more gene clusters, including PKS and NRPS sequences, than those of secondary metabolites known to be produced by these species. Sequencing of the *Aspergillus nidulans* genome revealed 27 polyketide synthases and 14 non-ribosomal peptide synthases whereas previously less than ten biosynthetic gene clusters were known (Chiang, 2008). Production of 'silent' secondary metabolites by fungi can be brought about by addition of inhibitors of DNA methyltransferase and histone deacetylase (Williams *et al.*, 2008) or by disrupting activities of such enzymes (Schwab, 2007).

It is clear that only a minor portion of microbes from nature have been cultivated in the laboratory and examined for antibiotic production. Existing bacterial species in nature are estimated to number somewhere between 10^7 and 10^9 (Schloss and Handelsman, 2004). Less than 0.3% of soil bacteria and less than 0.00001% of water-associated bacteria have been grown in common laboratory media (Amann *et al.*, 1995). One gram of soil has been estimated to contain up to 10 billion microorganisms and thousands of species (Rosello-Mora and Amann, 2001). Another estimate is that 30 g of soil contains more than 500,000 species (Doolittle, 1999). These numbers are much higher than the 5000–6000 individual prokaryotic microbes in culture collections or described in the literature (Daniel, 2004). It is interesting that in one gram of soil, 10^7 cells could be counted but only 10^4 (0.1%) could be cultivated (Kellenberger, 2001). Using a synthetic statistical approach to measure biodiversity, Hong *et al.* (2006) estimated that a 5-g sample of marine sediment contained 2000–3000 bacterial species. Estimates of the number of prokaryotic cells in soils are 4×10^7 per gram in forest soil and 2×10^9 per gram in cultivated soils and grasslands (Daniel, 2004). Of the 61 known bacterial phyla, only 30 have been cultured (Hugenholtz *et al.*, 2009). It is obvious that the vast majority of microbes in nature have never been cultured in laboratories. Some uncultured microbes were finally grown by (a) the use of low-nutrient oligotrophic growth media which prevent overgrowth by rapidly growing species; (b) using signalling molecules; (c) using inhibitors of undesirable organisms; (d) using very long periods of incubation, sometimes in the natural environment; (e) protection of cells from exogenous peroxides; (f) inclusion of humic acid; (g) use of hypoxic (1–2% oxygen) and anoxic atmospheres; (h) use of a high concentration (5%) of carbon dioxide along with high-throughput polymerase chain reaction (PCR) methodology; (i) construction of a diffusion chamber containing a simulated natural environment; and (j) encapsulating cells in gel microdroplets under low-nutrient flux conditions and detecting microcolonies by flow cytometry (Zengler *et al.* 2002, 2005; Connon and Giovannoni, 2002; Kaeberlein *et al.*, 2002; Wery *et al.*, 2003; Stevenson *et al.*, 2004; Vartoukian *et al.*, 2010).

Furthermore, of the 1.1 million fungal species estimated to exist in nature (Mueller *et al.*, 2007), only a minor portion have been cultivated. 14,000 species of microfungi are known, but there actually may be ten times that amount in nature (Hawksworth, 2001).

Donadio *et al.* (2010) proposed that in addition to retrieving microbial strains from underexplored environments, genome mining, increased sensitivity assays and high throughput screening, and even chemical derivatization of known microbial products, could be applied with the aid of 'click' chemistry. To be useful, the click reaction must be of wide scope, give consistently high yields with various starting materials, be easy to perform, be insensitive to oxygen or water, use only readily available reagents, and finally, the reaction work-up and product isolation must be simple, without the need for chromatographic purification (Kolb and Sharpless, 2003).

The global need for new antibiotics has been pointed out by Hoegberg *et al.* (2010). They suggest the following measures to correct the situation: (i) sharing of information about drug discovery among companies and students throughout the world ('open source platforms'); (ii) sharing compound libraries; (iii) providing financial incentives for early development of new antibiotics through preclinical and early clinical trials by governments, charities and international government initiatives; (iv) spreading financial risk between the financial community and new drug developers; and (v) funding internationally drug discovery and scale-up efforts of small companies in low-income countries and universities studying antibiotic resistance.

Baltz (2008b) argues that the lack of new antibiotics can be markedly changed by using high-throughput fermentations, isolating marine actinomycetes, mining genomes to find cryptic pathways, and employing combinatorial biosynthesis. The results of 25 years of combinatorial biosynthesis have been reviewed by Baltz (2006).

Aside from the application of new technologies to solve the antibiotic crisis, there are additional possible remedies. Although most of the pharmaceutical industry has discontinued screening for natural product antibiotics, the small biotechnology companies and academics have picked up the slack and have products in clinical development (Jabes, 2011). A comprehensive analysis of the sources of 252 FDA-approved drugs from 1998 to 2007 (Kneller, 2010) revealed that biotechnology companies, and universities that send their discoveries to biotechnology organizations, accounted for half of FDA approvals. Also, of those drugs responding to unmet medical needs, half were developed by biotechnology companies and their university partners. Whereas the major pharmaceutical companies had only three antibacterial products in clinical development in 2005, the biotechnology companies and small pharmaceutical organizations had 17 (Bush, 2004; Carlson, 2009). In general, about one-half to two-thirds of new products of the pharmaceutical industry are being licensed in from such companies. In 2006, small companies provided 50–60% of pharmaceutical revenues (Mullin, 2006). Of those drugs responding to unmet medical needs, half came from biotechnology/university groups. A useful move would involve more government support of small companies and academic institutions attempting to discover new antibiotics.

When one considers antibiotics in the recent pipeline, the contribution of small companies can be appreciated.

1 Amadacycline (PTK-0796, Parerek), a broad-spectrum tetracycline (an aminomethylcycline), is being tested in phase III against skin and skin structure infections (SSSI) and community-acquired pneumonia (CAP). Another tetracycline is TP-434, a fluorotetracycline (Tetraphase) in phase II.
2 Retapamulin, tiamulin and valnemulin are pleuromutilins which were approved for veterinary use in 1979 and 1999. Nabriva Therapeutics is providing the pleuromutilin BC-3781 which is in phase II for complicated SSSI in phase II. Two other pleuromutilins in phase I are BC-3205 and BC-7013.
3 Macrolides/ketolides: cethromycin (Restanga, Advanced Life Sciences) is being tested as an oral drug against CAP and anthrax, in phase III. Another is solithromycin (CEM-101), developed by Cempra for CAP and other infections.
4 ACHN-490, an aminoglycoside being developed by Achaogen, is in phase II.
5 CB-183,315 (Cubist) is a lipopeptide in phase II for oral treatment of *C. difficile*-associated diarrhoea (CDAD).
6 FabI (enoyl-ACP reductase) inhibitors include CG-400549 (Korean Crystal Genomics) in phase I. AFN-1252 (Affinium Pharmaceuticals) has completed phase I. FAB-001 (FAB Pharma) is in phase I.

Another approach involves measures by which governments might encourage large pharmaceutical companies to return to antibiotic discovery. The Infectious Diseases Society of America has suggested the following: (i) shorter approval process for new antibiotics; (ii) patent extensions; (iii) orphan drug status; (iv) tax credits; (v) limitation of liability; and (vi) advanced purchasing commitment by governments. In addition, we believe that governments throughout the world should establish institutes devoted solely to the discovery of new antibiotics. These would be devoted full-time to the use of genome mining, combinatorial biosynthesis, metagenomics, glycorandomization, and new methods of high throughput screening for inhibitors of novel targets. This would demonstrate that governments are interested in solving the terrible problems that we in the world are facing as potential victims of antibiotic-resistant pathogenic microorganisms. In the USA, the NIH has established the National Center for Advancing Translational Sciences (NCATS) (Reed *et al.*, 2012). This could result in more drugs being brought to the market, especially antibiotics which are sorely needed. It is hoped that the NCATS can bridge the gap between discovery and basic research at universities and its commercial development. NCATS has teamed up with Pfizer, AstraZeneca and Eli Lilly & Co. to search for new applications of 24 drugs that have failed in the clinic for other applications (Jarvis, 2012). NIH contributed $20 million to fund academic scientists in this effort.

The Generating Antibiotic Incentives Now (GAIN) Act, which came into effect in the US recently, will also help by offering an extra 5 years of market exclusivity for new antibiotics. 'Fast track' designation is for new drugs against MRSA, VRSA (vancomycin-resistant *S. aureus*), VRE (vancomycin-resistant *Enterococcus*), multidrug-resistant Gram-negatives (*Acinetobacter, Klebsiella, Pseudomonas, E. coli*) and tuberculosis.

The European Innovative Medicines Initiative (IMI) was started in 2012 by the European Union (Roberts, 2012). Called the 'New Drugs 4 Bad Bugs' (ND4BB) project, it is supported by government contributions of $134 million plus contributions of another $141 million by European pharmaceutical companies. The funds are supporting preclinical studies of compounds against Gram-negative bacteria, phase II and phase III clinical trials of GSK 1322322 which inhibits peptide deformylase of MRSA, AstraZeneca's MED14893, a monoclonal antibody against an *S. aureus* toxin, and AZD9773, a polyclonal antibody directed at tumour necrosis factor as a potential treatment of severe sepsis and septic shock. The programme involves pharmaceutical companies, academia and biotechnology organizations. Another $259 million programme funds studies to accelerate drug discovery for unmet medical needs. More than 500,000 small molecules are being contributed by seven drug companies, academia and small companies (Scott, 2013).

Acknowledgements

We are indebted to M.A. Ortíz and B. Ruiz for their support in the manuscript preparation. Financial support by PAPIIT, DGAPA IN201413 (UNAM), is greatly acknowledged.

References

Agres, T. (2003). Alliances eye early-stage drugs. Drug Discov. Develop. 6, 17–18.

Amann, R.I., Ludwig, W., and Schleifer, K.-H. (1995). Phylogenetic identification and in situ detection of individual microbial cells without cultivation. Microbiol. Rev. 59, 143–169.

American Academy of Microbiology (2009). Antibiotic Resistance: An Ecological Perspective On an Old Problem (ASM, Washington, DC), 32 pp.

Ausman, D.J. (2001). Screening's age of insecurity. Modern Drug Discov. *4*, 32–34.

Balaban, N., and Dell'Acqua, G. (2005). Barriers on the road to new antibiotics. The Scientist *19*, 42–43.

Ball, A.R., and Tegos, G.P. (2012). Emerging antimicrobial drug-discovery strategies: an evolving necessity. In Antimicrobial Drug Discovery: Emerging Strategies. G. Tegos, and E. Mylonakis, eds. (CAB International, Wallingford, UK), pp. 7–25.

Baltz, R. (2006). Molecular engineering approaches to peptide, polyketide and other antibiotics Nat. Biotechnol. *24*, 1533–1540.

Baltz, R. (2008a). Renaissance in antibacterial discovery from actinomycetes. Curr. Opin. Pharmacol. *8*, 557–563.

Baltz, R. (2008b). Biosynthesis and genetic engineering of lipopeptide antibiotics related to daptomycin. Curr. Top. Med. Chem. *8*, 618–638.

Barrett, J.F. (2005). Can biotech deliver new antibiotics? Curr. Opin. Microbiol. *8*, 498–503.

Bjarnsholt, T., Ciofu, O., Molin, S., Givskov, M., and Hoiby, N. (2113). Applying insights from biofilm biology to drug development – can a new approach be developed? Nat. Revs, Drug Disc. *12*, 791–808.

Borman, S. (2002). Organic lab sparks drug discovery. Chem. Eng. News *80* (40), 23–24.

Brickner, S.J., Barbachyn, M.R., Hutchinson, D.K., and Manninen, P.R. (2008). Linezolid (ZYVOXâ), the first member of a completely new class of antibacterial agents for treatment of serious Gram-positive infections. J. Med. Chem. *1*, 1981–1990.

Breger, J., Fuchs, B.B., Aperis, G., Moy, T.I., Ausubel, F.M., and Mylonakis, E. (2007). Antifungal chemical compounds identified using a *C. elegans* pathogenicity assay. PLoS Pathog. *3*, e18.

Breinbauer R., Manger, M., Scheck, M., and Waldmann, H. (2002). Natural product guided compound library development. Curr. Med. Chem. *9*, 2129–2145.

Bull, A.T., Ward, A.C., and Goodfellow, M. (2000). Search and discovery strategies for biotechnology. Microbiol. Mol. Biol. Rev. *64*, 573–606.

Burrill, G.S. (2002). Personalized medicine or blockbusterology? BioPharm *15*, 46–50.

Bush, K. (2004). Why it is important to continue antibacterial drug discovery. ASM News *70*, 282–287.

Bush, K., Macielag, M., and Weidner-Wells, M. (2004). Taking inventory: antibacterial agents currently at or beyond Phase 1. Curr. Opin. Microbiol. *7*, 466–476.

Bush, K., Courvalin, P., Dantas, G., Davies, J., Eisenstein, B., Huovinen, P., Jacoby, G.A., Kishony, R., Kreiswirth, B.N., Kutter, E., *et al.* (2011). Tackling antibiotic resistance. Nat. Rev. Microbiol. *9*, 894–896.

Busti, E., Monciardini, P., Cavaletti, L., Bamonte, R., Lazzarini, A., Sosio, M., and Donadio, S. (2006). Antibiotic-producing ability by representatives of a newly discovered lineage of actinomycetes. Microbiology *152*, 675–683.

Butler, M.S., and Buss, A.D. (2006). Natural product—The future scaffolds for novel antibiotics? Biochem. Pharmacol. *71*, 919–929.

Cane, D.E., He, X., Kobayashi, S., Omura, S., and Ikeda, H. (2006). Geosmin biosynthesis in *Streptomyces*

avermitilis. Molecular cloning, expression and mechanistic study of the germacradienol/geosmin synthase. J. Antibiot. *59*, 471–474.
Carlson, B. (2009). Biosimilar market fails to meet projections. Gen. Eng. Biotechnol. News *29* (17), 43–45.
Chiang, Y.-M., Szewczyk, E., Nayak, T., Davidson, A.D., Sanchez, J.F., Lo, H., Ho, W., Simityan, H., Kuo, E., Praseuth, A., *et al.* (2008). Molecular genetic mining of the *Aspergillus* secondary metabolome: discovery of the emericellamide biosynthetic pathway. Chem. Biol. *15*, 527–532.
Chu, D.T.W., Plattner, J.J., and Katz, L. (1996). New directions in antibacterial research. J. Med. Chem. *39*, 3853–3874.
Clark, D.E. (2011). Outsourcing lead optimization: the eye of the storm. Drug Disc. Today *16*, 147–157.
Class, S. (2004). Health care in focus. Chem. Eng. News *82* (49), 18–29.
Clough, J., Chen, S., Gordon, E.M., Hackbarth, C., Lam, S., Trias, J., White, R.J., Candiani, G., Donadio, S., Romanò, G., *et al.* (2003). Combinatorial modification of natural products: synthesis and *in vitro* analysis of derivates of thiazole peptide antibiotic GE2270 A: A ring modification. Bioorg. Med. Chem. Lett. *13*, 3409–3414.
Coates, A.R.M., and Hu, Y. (2007). Novel approaches to developing new antibiotics for bacterial infections. Br. J. Pharmacol. *152*, 1147–1154.
Connon, S.A., and Giovannoni, S.J. (2002). High-throughput methods for culturing microorganisms in very-low-nutrient media yield diverse new marine isolates. Appl. Environ. Microbiol. *68*, 3878–3885.
Cotter, P.D., Ross, P., and Hill, C. (2013). Bacteriocins – a viable alternative to antibiotics? Nat. Rev. Microbiol. *11*, 95–103.
Daemmrich, A.A., and Bowden, M.E. (2005). A rising drug industry. Chem. Eng. News *83* (25), 28–42.
Daniel, R. (2004). The soil metagenome - a rich resource for the discovery of novel natural products. Curr. Opin. Biotechnol. *15*, 199–204.
DaSilva, E., and Iaccarino, M. (1999). Emerging diseases: a global threat. Biotechnol. Adv. *17*, 363–384.
Davies, J. (2007). Resistance redux. EMBO Repts. *8*, 616–621.
Demain, A.L. (2002). Prescription for an ailing pharmaceutical industry. Nat. Biotechnol. *20*, 331.
Di Masi, J.A. (2001). Winners and losers in new drug innovation. Med. Mkting. Media, *36* (9): 99–110.
Donadio, S., Maffioli, S., Monciardini, P., Sosio, M., and Jabes, D. (2010). Antibiotic discovery in the twenty-first century: current trends and future perspectives. J. Antibiot. *63*, 423–430.
Doolittle, W. (1999). Phylogenetic classification and the universal tree. Science *284*, 2124–2128.
Drews, J. (2001). Quo vadis, biotech? (Part 2) Drug Discov. Today *6*, 21–26.
Ehmann, D.E., Jahic, H., Ross, P.L., Gu, R.-F., Hu, J., Kern, G., Walkup, G.K., and Fisher, S.L. (2012). Avibactam is a covalent, reversible, non-β-lactam β-lactamase inhibitor. Proc. Natl. Acad. Sci. U.S.A. *109*, 11663–11668.
Feher, M., and Schmidt, J. (2003). Property distributions: differences between drugs, natural products, and molecules from combinatorial chemistry. J. Chem. Inf. Comp. Sci. *43*, 218–227.
Fernandes, P., and Schranz, J. (2011). Gen. Eng. Biotechnol. News *31*(9):54–55, 2011.
Firn, R.D., and Jones, C.G. (2000). The evolution of secondary metabolism-a unifying model. Mol. Microbiol. *37*, 989–994.
Fox, J.L. (2013). Antimicrobial peptides stage a comeback. Nat. Biotechnol. *31*, 379–382.
Ganesan, A. (2004). Natural products as a hunting ground for combinatorial chemistry. Curr. Opin. Biotechnol. *15*, 584–590.
Genilloud, O., González, I., Salazar, O., Martín, J., Tormo, J.S., and Vicente, F. (2010). Current approaches to exploit actinomycetes as a source of novel natural products. J. Ind. Microbiol. Biotechnol. *38*, 375–389.
Goutam, A., Vyas, R., and Tewari, R. (2011). Peptidoglycan biosynthesis machinery: a rich source of drug targets. Crit. Rev. Biotechnol. *31*, 295–336.
Gravitz, H. (2012). Turning a new phage. Nat. Med. *18*, 1318–1320.
Gross, H. (2007). Strategies to unravel the function of orphan biosynthesis pathways: recent examples and future prospects. Appl. Microbiol. Biotechnol. *75*, 267–277.
Handen, J.S. (2002). The industrialization of drug discovery. Drug Discov. Today *7*, 83–85.
Harvey, A.L. (2007). Natural products as a screening resource. Curr. Opin. Chem. Biol. *11*, 480- 484.
Hawksworth, J. (2001). Mushrooms: the extent of the unexplored potential. Int. J. Med. Mushrooms 3:333–337.
Hefti, E., and Bolten, B.M. (2003). Advances in high-throughput screening advances- do they lead to new drugs? Gen. Eng. News *23* (21), 7–10.
Henkel, T., Brunne, R.M., Mueller, H., and Reichel, F. (1999). Statistical investigation into the structural complementarity of natural products and synthetic compounds. Angew Chem. Int. Ed. *38*, 643–647.
Hoegberg, L.D., Heddini, A., and Cars, O. (2010). The global need for effective antibiotics: challenges and recent advances. Trends Pharmacol. Sci. *31*, 509–515.
Hong, S.-H., Bunge, J., Jeon, S.O., and Epstein, S.S. (2006). Predicting microbial species richness. Proc. Natl. Acad. Sci. U.S.A. *103*, 117–122.
Horrobin, D.F. (2001). Realism in drug discovery-could Cassandra be right? Nat. Biotechnol. *19*, 1099–1100.
Hugenholtz, P., Hooper, S.D., and Kyrpides, N.C. (2009). Focus: Synergistetes. Environ. Microbiol. *11*, 1327–1329.
Jabes, D. (2011). The antibiotic R&D pipeline: an update. Curr. Opin. Microbiol. *14*, 564–566.
Jacobs, M. (2002). Pharmaceutical balancing act. Chem. Eng. News *80* (48), 5.
Jarvis, L.M. (2012). NIH initiative aims to partner academics with pharmaceutical companies to revive failed drug candidates. Chem. Eng. News *90* (22), 41–43.
Jovetic, S., Zhu, Y., Marcone, G.L., Marinelli, F., and Tramper, J. (2010). β-Lactam and glycopeptide

antibiotics: first and last line of defense? Trends Biotechnol. *28*, 596–604.

Kaeberlein, T., Lewis, K., and Epstein, S.S. (2002). Isolating 'uncultivable' microorganisms in pure culture in a simulated natural environment. Science *296*, 1127–1129.

Kamholz, S. (2002). Drug resistant tuberculosis. J. Assoc. Acad. Minor Phys. *13*, 53–56.

Kellenberger, E. (2001). Exploring the unknown: the silent revolution of microbiology. EMBO Rep. *2*, 5–7.

Kettler. H.E. (1999). Updating the cost of a new chemical entity (Office of Health Economics, London).

Kingston, D.G.I., and Newman, D.J. (2002). Mother nature's combinatorial libraries; their influence on the synthesis of drugs. Curr. Opin. Drug Discov. Develop. *5*, 304–316.

Kneller, R. (2010). Importance of new companies for drug discovery: origins of a decade of new drugs. Nat. Rev. Drug Discov. *9*, 867–882.

Kolb, H.C., and Sharpless, B.K. (2003). The growing impact of click chemistry on drug discovery. Drug Discov. Today *8*, 1128–1137.

Koppal, T. (2004). Global teams battle. Drug Discov. Devel. *4*, 28–32.

Kostel, K. (2004). Industry snapshot. The Scientist *18* (6), 45.

Koul, A., Arnoult, E., Lounis, N., Guillemont, J., and Andrieset, K. (2011). The challenge of new drug discovery for tuberculosis. Nature *469*, 483–490.

Kraus, C.N. (2008). Low hanging fruit in infectious disease drug development. Curr. Opin. Microbiol. *11*, 434–438.

Kren, V., and Rezanka, T. (2008). Sweet antibiotics-the role of glycosidic residues in antibiotic and antitumor activity and their randomization. FEMS Microbiol. Rev. *32*, 858–889.

Landro, L. (2013). Deadly bacterial infections on the rise. The Wall St. J., March 5, page A6.

Lautru, S., Deeth, R.J., Bailey, L.M., and Challis, G.L. (2005). Discovery of a new peptide natural product from *Streptomyces coelicolor* genome mining. Nat. Chem. Biol. *1*, 265–269.

Lefevre, F., Robe, P., Jarrin, C., Ginolhac, A., Zago, C., Auriol, D., Vogel, T.M., Simonet, P., and Nalin, R. (2008). Drugs from hidden bugs: their discovery via untapped resources. Res. Microbiol. *159*, 153–161.

Lin, X., Hopson, R., and Cane, D.E. (2006). Genome mining in *Streptomyces coelicolor*: molecular cloning and characterization of a new sesquiterpene synthase. J. Amer. Chem. Soc. *128*, 6022–6023.

Lorenz, P., and Eck, J. (2005). Metagenomics and industrial applications. Nat. Rev. Microbiol. *3*, 510–516.

Lynch, A.S., and Abbanat, D. (2010). New antibiotic agents and approaches to treat biofilm-associated infections. Expert Opin. Ther. Patents *20*, 1373–1387.

Lynch, A.S., and Robertson, G.T. (2008). Bacterial and fungal biofilm infections. Ann. Rev. Med. *59*, 415–428.

Malik, N.N. (2008). Drug discovery: past, present and future. Drug Discov. Today *13*: 909–912.

Marshall, B.M., and Levy, S.B. (2012). Antibiotics in the animals we eat. The Scientist 20(4): 2.

Mills, S.D. (2006). When will the genomics investment pay off for antibacterial discovery? Biochem. Pharmacol. *71*, 1096–1102.

Morrow, K.J. (2010). Using selection to advance drug discovery. Amer. Biotechnol. Lab. Mag. *28*, 24–25.

Moy, T.I., Ball, A.R., Anklesaria, G., Lewis, K., and Ausubel, F.M. (2006). Identification of novel antimicrobials using a live-animal infection model. Proc. Natl. Acad. Sci. U.S.A. *103*, 10414–10419.

Mueller, G.M., Schmit, J.P., Leacock, P.R., Buyck, B., Cifuentes, J., and Desjardin, D.E. (2007). Global diversity and distribution of macrofungi. Biodivers. Conserv. *16*, 37–48.

Mullin, R.R. (2006a). Pharma in flux. Chem. Eng. News *84* (25), 30–35.

Mullin, R. (2006b). Tufts report anticipates upturn. Cites drugmakers' actions to accelerate drug development. Chem. Eng. News *84* (4), 9.

Najafi, R. (2011). Developing new treatments for antibiotic-resistant bacteria. Drug Discov. News 7 (7), 11.

Newman, D.J. (2008). Natural products as leads to potential drugs: an old process or the new hope for drug discovery? J. Med. Chem. *51*, 2589–2599.

Newman, D.J., and Cragg, G.M. (2007). Natural products as sources of new drugs over the last 25 years. J. Nat. Prod. *70*, 461–477.

Nikaido, H. (2009). Multidrug resistance in bacteria. Annu. Rev. Biochem. *78*, 119–146.

Overbye, K.M., and Barrett, J.F. (2005). Antibiotics: where did we go wrong? Drug Discov. Today *10*, 45–52.

Pageni, B.B., Oh, T.J., Liou, K., Yoon, Y.J., and Sohng, J.K. (2008). Genetically engineered biosynthesis of macrolide derivatives including 4-amino-4,6-dideoxy-L-glucose from *Streptomyces venezuelae* YJ003-OTBP3. J. Microbiol. Biotechnol. *18*, 88–94.

Park, J.H., Cha, C.J., and Roe, J.H. (2006). Identification of genes for mycothiol biosynthesis in *Streptomyces coelicolor* A3. J. Microbiol. *44*, 121–125.

Perlroth, J., Choi, B., and Spelberg, B. (2007). Nosocomial fungal infections: epidemiology, diagnosis, and treatment. Med. Mycol. *45*, 321–346.

Phillips, L. (2013). Infectious disease: TB's revenge. Nature *493*, 14–16.

Reed, J.C., White, E.L., Aube, J., Lindsley, C., Li, M., Sklar, L., and Schreiber, S. (2012). The NIH's role in accelerating translational sciences. Nat. Biotechnol. *30*, 16–19.

Roberts, J.P. (2012). Incentives aim to boost antibiotic development. Nat. Biotechnol. *30*, 735–736.

Rosello-Mora, R., and Amann, R. (2001). The species concept for prokaryotes. FEMS Microbiol. Rev. *25*, 39–67.

Russell, D.G., Barry, C.E. III, and Flynn, J.A.L. (2010). Tuberculosis: what we do not know can, and does hurt us. Science *328*, 852–855.

Ryan, J.F. (2003). Tough times. Today's Chemist at Work *12* (1), 7.

Salas, J.A., and Mendez, C. (2007). Engineering the glycosylation of natural products in actinomycetes. Trends Microbiol. *15*, 219–232.

Sanchez, J.F., Chiang, Y.-M., and Wang, C.C.C. (2008). Bacterial hosts for natural product formation. Mol. Pharm. *5*, 226–233.

Scheffler, R.J., Colmer, S., Tynan, H., Demain, A.L., and Gullo, V.P. (2013). Antimicrobials, drug discovery, and genome mining. Appl. Microbiol. Biotechnol. *97*, 969–978.

Schloss, P.D., and Handelsman, J. (2004). Status of the microbial census. Mol. Biol. Rev. *68*, 686–691.

Schwab, E.K., Bok, J.W., Tribus, M., Galehr, J., Graessle, S., and Keller, N.P. (2007). Histone deacetylase activity regulates chemical diversity in *Aspergillus*. Eukaryot. Cell *6*, 1656–1664.

Scott, A. (2013). Europe to boost drug discovery. Chem. Eng. News *91*(7), 8.

Scott, R.W. (2009). Defensin mimetics: nature knows best. Am. Biotechnol. Lab. *27*(10), 16–19.

Shlaes, D.M., Projan, S.J., and Edwards, J.E. Jr (2004). Antibiotic discovery: state of the state. ASM News *70*, 275–281.

Simon, C., and Daniel, R. (2009). Achievements and new knowledge unraveled by metagenomic approaches. Appl. Microbiol. Biotechnol. *85*, 265–276.

Singh, S.B., and Barrett, J.F. (2006). Empirical antibacterial drug discovery – foundation in natural products. Biochem. Pharmacol. *71*, 1006–1015.

Song, L., Barona-Gomez, F., Corre, C., Xiang, L., Udwary, D.W., Austin, M.B., Noel, J.P., Moore, B.S., and Challis, G.L. (2006). Type III polyketide synthase beta-ketoacyl-ACP starter unit and ethylmalonyl-CoA extender unit selectivity discovered by *Streptomyces coelicolor* genome mining. J. Amer. Chem. Soc. *128*, 14754–14755.

Spížek, J., Novotná, J., Řezanka, T., and Demain, A.L. (2010). Do we need new antibiotics? The search for new targets and new compounds. J. Ind. Microbiol. Biotechnol. *37*, 1241–1248.

Stevenson, B.S., Eichorst, S.A., Wertz, J.T., Schmidt, T.M., and Breznak, J.T. (2004). New strategies for cultivation of previously uncultured microbes. Appl. Environ. Microbiol. *70*, 4748–4755.

Strohl, W.R. (1997). Industrial antibiotics: today and the future. In Biotechnology of Antibiotics, W.R. Strohl, ed. (Marcel Dekker, New York), pp. 1–47.

Tan, L.T. (2007). Bioactive natural products from marine cyanobacteria for drug discovery. Phytochemistry. *68*, 954–979.

Tenover, F.C., and Hughes, J.M. (1996). The challenges of emerging infectious diseases. J. Am. Med. Assoc. *275*, 300–304.

Thayer, A.M. (2003). Biomarkers emerge. Chem. Eng. News *81*(30), 33–37.

Thayer, A.M. (2004). Bluckbuster model breaking down. Mod. Drug. Disc. *7*, 23–24.

Torrice, M. (2013). Antibacterial boom and bust. Chem. Eng. News *91*(36), 34–37.

Tralau-Stewart, C.J., Wyatt, C.A., Kleyn, D.E., and Ayad, A. (2009). Drug discovery: new models for industry–academic partnerships. Drug Discov. Today *14*, 95–101.

Udwary, D.W., Zeigler, L., Asokar, R.N., Singan, V., Lapidas, A., Fenical, W., Jensen, P.R., and Moore, B.S. (2007). Genome sequencing reveals complex secondary metabolome in the marine actinomycete *Salinospora tropica*. Proc. Natl. Acad. Sci. U.S.A. *104*, 10376–10381.

Vartoukian, S.R., Palmer, R.M., and Wade, W.G. (2010). Strategies for culture of 'unculturable' bacteria. FEMS Microbiol. Lett. *309*, 1–7.

Vicente, M., Hodgson, J., Massidda, O., Tonjum, T., Henriques-Normark, B., and Ron, E.Z. (2006). The fallacies of hope: will we discover new antibiotics to combat pathogenic bacteria in time? FEMS Microbiol. Rev. *30*, 841–852.

Waldemann, H. (2003). At the crossroads of chemistry and biology. Bioorg. Med. Chem. *11*, 3045–3051.

Waldmann, H., and Breinbauer, R. (2002). Nature provides the answer. Screening *6*, 46–48.

Walsh, C.T., and Fischbach, M.A. (2009). Squashing superbugs – the race for new antibiotics. Sci. Amer. *301*(1), 44–51.

Warner, S. (2003). Pipeline anxiety: scientists pumped into new roles. The Scientist *17*(10), 46.

Warner, S. (2004). High-priced biotech drugs: are they worth it? The Scientist *18*(23), 20–24.

Wery, N., Gerike, U., Sharman, A., Chaudhuri, J.B., Hough, D.W., and Danson, M.J. (2003). Use of a packed-cell bioreactor for isolation of diverse protease-producing bacteria from Antarctic soil. Appl. Environ. Microbiol. *69*, 1457–1464.

Williams, R.B., Henrikson, J.C., Hoover, A.R., Lee, A.E., and Cichewicz, H. (2008). Epigenetic remodeling of fungal secondary metabolome. Org. Biomol. Chem. *6*, 1895–1897.

Willis, R.C. (2004). The discovery doldrums: NCEs are coming slowly; Juergen Drews explains why. Mod. Drug Discov. *4*, 23–24.

Wilson, E.K. (2002). Picking the winners. Chem. Eng. News *80* (17), 35–39.

Wolff, M.E. (2006). Dispelling discovery myths. Drug. Discov. News, *12* (7), 13.

Wright, G.D. (2010). Antibiotic resistance in the environment: a link to the clinic? Curr. Opin. Microbiol. *13*, 589–594.

Zengler, K., Toledo, G., Rappe, M., Elkins, J., Mathur, E.J., Short, J.M., and Keller, M. (2002). Cultivating the uncultured. Proc. Natl. Acad. Sci. U.S.A. *99*, 15681–15686.

Zengler, K., Paradkar, A., and Keller, M. (2005). New methods to access microbial diversity for small molecule discovery. In Natural Products: Drug Discovery and Therapeutic Medicine, L. Zhang, and A.L. Demain, eds. (Humana Press, Totowa, New Jersey), pp. 275–293.

Zerikly, M., and Challis, G.L. (2009). Strategies for the discovery of new natural products by genome mining. Chem. Bio. Chem. *10*, 625–633.

Tackling Antibiotic Resistance

5

Jaroslav Spizek and Vladimir Havlicek

Abstract

Resistance to antibiotics and other antimicrobial compounds permanently increases and it appears that the struggle against antibiotic resistance is a war we can never win. The strength of the immense numbers of microorganisms appears to overpower our drugs. However, there are hopes and chances to at least improve the situation. In the present review we want to discuss certain steps to be used to tackle antibiotic resistance including (1) preparation of new vaccines against resistant bacterial strains; (2) search for new antibiotics in both traditional and non-traditional sources; (3) search for genes specifying biosynthesis of antibiotics; (4) use of forgotten natural compounds and their transformation; and (5) search for new antibiotic targets. We want to devote a special attention to the search for new compounds that would inhibit the currently antibiotic resistant pathogenic bacteria.

Introduction

Infectious diseases account for about one-quarter of all deaths worldwide. Since the introduction of penicillin in the 1940s, antibiotics have occupied a central place in the treatment of bacterial infections facilitating many of the procedures of modern medicine such as cancer chemotherapy, organ transplantation and the care of premature babies. In spite of a considerable progress in these areas, we witness an increase in the number and distribution worldwide of pathogens resistant to antimicrobial (antibacterial, antiviral, antiparasitic and antifungal) drugs. In a recent report by the UK Chief Medical Officer (Davies *et al.*, 2013) it was stated that 'antimicrobial resistance poses catastrophic threat' and that 'The continuing increase in organisms that are resistant to one or more antimicrobial drugs is one of the greatest threats that we face today. Some forms of resistance can be transferred between different species of bacteria and parasites, which can result in organisms that are resistant to many, most or even all drugs that are currently available'.

The latest G8 Science Ministers Statement (Willetts, 2013) focused on the global challenge of antimicrobial resistance and the World Health Organization expressed concern that this rapidly growing problem may impede progress towards the Millennium Development Goals 2015. They focused on antimicrobial (i.e. anti-viral, anti-bacterial, anti-fungal and anti-parasite) drug resistance as a major health security challenge of the twenty first century and decided to act concertedly on developing the scientific input necessary to reduce antimicrobial resistance working with existing agencies such as the World Health Organization. In the meeting it was concluded that 'the global pandemic of antibiotic resistance in both community care and hospital-associated infections represents a major health and economic burden and this crisis is being exacerbated by a relative lack of innovation in generating new antibiotics: we are in danger of returning to a pre-antibiotic era'.

In the article by a group of renowned scientists the authors (Bush *et al.*, 2011) divided the questions to be answered into research priorities to control resistance, urgent actions to be taken including public education, public health, sanitation and quality of life, new antibiotics, old antibiotics, control of antibiotic use, alternatives to antibiotics and a collaborative approach. It is evident that infectious diseases belong to most important death causes in humans and apparently the most significant death cause of children and elderly. Bacteria causing serious health problems include especially the 'ESKAPE' pathogens *Enterococcus faecium, Staphylococcus aureus, Klebsiella pneumoniae, Acinetobacter baumannii, Pseudomonas aeruginosa* and *Enterobacter* species (Boucher *et al.*, 2013). Methicillin resistant strains of *Staphylococcus aureus* are particularly dangerous. The list of bacteria causing serious health problems also includes *Streptococcus pneumoniae, Mycobacterium tuberculosis*, group A streptococci, *Salmonella typhimurium, Escherichia coli* 0157-H7, *Legionella pneumophila, Borrelia burgdorferi, Helicobacter pylori*, etc.

For a long time it appeared that antibiotics are omnipotent and that the treatment of pathogenic bacteria with them would wipe out infectious diseases. However, the bacteria rapidly developed antibiotic resistance, quite often immediately after the antibiotic had been introduced to the clinical practice. Bacteria have lived on Earth for several billion years. During this time, they encountered in nature a wide range of naturally occurring antibiotics. To survive, they developed antibiotic resistance mechanisms. Therefore, it is not surprising that they have become resistant to most of the natural antimicrobial agents developed including antibiotics (Hancock, 2007). It is thus evident that we should develop activities protecting us from the danger of antibiotic resistant microorganisms. As concluded by Walsh and Wencewitz (2014), 'There is a continuous need for iterative cycles of antibiotic discovery and development to deal with the selection of resistant pathogens that emerge as therapeutic application of an antibiotic becomes widespread'. Some of the problems were also discussed in our previous reviews (Demain and Spizek, 2012; Rezanka *et al.*, 2007; Spizek *et al.*, 2010).

Preparation of new vaccines against resistant bacterial strains

Antibiotics and vaccines are essential for improvement of human health. It is important about the use of vaccines that no vaccine-resistant bacteria have been described so far. Development of new vaccines is undoubtedly a promising strategy against antibiotic resistance. In a review article by Mishra and coworkers, the authors showed that quite often antibiotics are not sufficient to control the morbidity and mortality associated with microbial infections due to the rapid and widespread occurrence of multidrug-resistant bacteria (Mishra *et al.*, 2012). They stated that vaccines and antibiotics have significantly contributed to improve health and also to increase the longevity of human beings. The fast-acting effect of antibiotics makes them indispensable to treat infected patients. Likewise, when the causative agent of the infection is unknown and in cases of superinfections with different species of bacteria, antibiotics appear to be the only therapeutic option. On the contrary, vaccines are usually not efficacious in people already infected and their action is generally limited to a much narrowed range of pathogens.

However, vaccines have contributed to the eradication of some of the most deadly infectious agents worldwide, can generate immunity to infections lasting for several years or life-long, and are able to induce herd immunity. Nonetheless, infectious diseases are still among the leading causes of morbidity and mortality worldwide. This is mainly owing to the emergence of bacterial resistance to antibiotics and the lack of efficacious medications to treat several other infectious diseases. Development of new vaccines appears to be a promising solution to these issues. Indeed, with the advent of new discovery approaches and adjuvants, it is possible today to produce vaccines virtually against every pathogen.

In addition, while vaccine-resistant bacteria have never been reported, accumulating literature is providing evidence that vaccination can reduce the raise of antibiotic resistant strains by decreasing their use. Vaccines have completely, or nearly, eradicated both viral and bacterial infections: examples of these are the eradication of smallpox, almost complete elimination of poliomyelitis, and a decrease of more than 95% in the incidence of diseases such as diphtheria, tetanus, pertussis, measles, mumps and rubella (Octavia *et al.*, 2012). Today, with the introduction of reverse vaccinology and other methodologies that allow for the identification of antigens in a high-throughput and comprehensive manner, vaccines can be developed with very high scientific and quality standards (Rappuoli *et al.*, 2011). Mishra and coworkers concluded that undoubtedly, both vaccines and antibiotics are and will continue to be important for global public health (Mishra *et al.*, 2012). Unfortunately, antibiotic resistance in bacteria is exponentially growing and the rate of discovery and development of new antibacterials is concomitantly declining (Hancock *et al.*, 2012).

Vaccines can be a promising tool to improve this situation. It is important to note that no vaccines have so far been prepared against the most important multidrug-resistant strains. Today, thanks to reverse vaccinology, novel adjuvants, and omics science, it is possible to make vaccines virtually against every pathogen (Rappuoli *et al.*, 2011). Omics biotechnologies and integrative biology will apparently accelerate the identification of vaccine candidates against difficult pathogens for which traditional vaccine development has thus far been unsuccessful (Bagnoli *et al.*, 2011). It is expected that, together, novel vaccines and antibiotics will be able to reduce the burden of infective diseases worldwide.

Search for new antibiotics

In our previous review articles (Rezanka *et al.*, 2007; Spizek *et al.*, 2010) we briefly discussed several approaches to be used for the search of new antibiotics including search for new antibiotics from non-traditional or less traditional sources, microbial genome sequences – search for genes specifying biosynthesis of antibiotics, metagenomics – analysis of DNA from the environment, forgotten natural compounds and their transformation, and investigation of new antibiotic targets in pathogenic bacteria. In this updated review we should also like to mention briefly newly approved antibiotics and antibiotics in the pipeline.

Newly approved antibiotics and antibiotics in the pipeline

Butler and Cooper (2011) summarized antibiotics in the pipeline. Those in phase III clinical trials or under NDA/MAA evaluation included fidaxomicin, amacycline, torezolic phosphate, oritavancin, dalbavancin and cethromycin, all of them derived from natural products. In their most recent review, Butler and coworkers concluded that 'The continued emergence of multidrug-resistant bacteria is a major public health concern. The identification and development of new antibiotics, especially those with new modes of action, is imperative to help treat these infections' (Butler *et al.*, 2013). In the review they listed the 22 new antibiotics launched since 2000 and provided details of two promising antibiotics, fidaxomicin and bedaquiline that can serve as examples.

Fidaxomycin is a narrow spectrum macrocyclic drug produced by the actinomycete *Dactylosporangium aurantiacum.* It is bactericidal and active against pathogenic *Clostridium difficile.* Currently only metronidazole and vancomycin are used to treat infections caused by *C. difficile.* Fidaxomicin was found to exhibit a higher global cure rate than vancomycin and a lower recurrence rate. Bedaquiline is a diarylquinoline anti-tuberculosis drug. It is the first new medicine to fight tuberculosis (TB) in more than 40 years and is specifically approved to treat multidrug-resistant TB. The last drug introduced specifically for the treatment of tuberculosis was rifampicin in 1960s. Since then, resistance to rifampicin has increased significantly. Bedaquiline has been released for the treatment of TB patients infected with rifampicin and isoniazide resistant bacteria.

Boucher and coworkers summarized the status of development and approval of systemic antibiotics in the United States as of early 2013 (Boucher *et al.*, 2013). Only two new antibiotics have been approved since IDSA's 2009 pipeline status report, and the number of new antibiotics annually approved for marketing in the United States continues to decline. The authors identified seven drugs in clinical development for treatment of infections caused by resistant Gram-negative bacteria. None of these agents was previously included in their 2009 list of antibacterial compounds in phase 2 or later development, but unfortunately none was found to address the entire spectrum of clinically relevant GNB resistance. The authors concluded that some progress in development of new antibacterial drugs that target infections caused by resistant

Gram-negative bacteria was reached but that it still remains alarmingly elusive.

Gould and Bal in their review summarized compounds in development including new generations aminoglycosides, quinolones, new β-lactams and monobactams, ketolides, tetracyclines, newer glycopeptides and efflux pump inhibitors (Gould and Bal, 2013). Of the neoglycosides plazomicin seems to be a very good candidate inhibiting bacterial protein synthesis and exhibiting dose-dependent bactericidal activity. According to the authors it is expected that data from a phase II randomized trial of this compound would become available later this year.

A number of broad spectrum quinolones that retain activity against strains resistant to the current quinolone compounds are under development including for instance delafloxacin and nemonoxacin with activity against *S. aureus*, methicillin resistant strains (MRSA) in particular. Another related compound, ACH 702, which belongs to a related class called the isothiazoloquinolones, is highly active against MRSA and also against *Mycobacterium tuberculosis* which is particularly important in the era of complicated drug-resistant tuberculosis. In the group of new oxazolinedione derivatives the authors described tedizolid and radezolid exhibiting a better activity against staphylococci, streptococci and enterococci than linezolid. In addition to the above compounds, some compounds of this class are being developed as antimycobacterial agents, viz. for instance sutezolid.

Within the group of new β-lactams and monobactams ceftaroline is a fifth-generation cephalosporin exhibiting activity against MRSA and vancomycin resistant *S. aureus*, the latter property apparently unique. New compounds inhibiting β-lactamases are also under development. Among them avibactam has a broad spectrum of activity including the KPC family. New ketolides include cethromycin and solithromycin. Cethromycin has been approved by FDA for the treatment of possible bioterrorist threats such as anthrax and plague. Based on *in vitro* studies with solithromycin it is hoped that its main application would be in the treatment of skin and soft tissue infection and community acquired pneumonia.

Among tetracyclines tigecycline is a broad-spectrum glycylglycine that has already been approved for clinical use. It is efficient in the treatment of various mixed infections caused by pathogens resistant to common tetracyclines. It is its advantage that it is not a substrate for drug efflux or the ribosomal protection proteins, mechanisms responsible for tetracycline resistance. The authors also described newer glycopeptides telavancin and oritavancin. More trials on oritavancin are expected in the near future. Efflux pump inhibitors are new compounds affecting the efflux pump in bacterial cells that drive out the intracellular antibiotic back to the exterior. According to the authors the resistance-nodulation-division family of efflux pumps seem to be the most promising.

Search for new antibiotics from non-traditional or less traditional sources

Antibiotic producers have been looked for in extreme locations such as hot springs, deep-sea bottom vents or water reservoirs with high salt content. The marine actinomycete *Salinispora* was found to be a particularly rich source of new chemical structures with promising biological activities (Daniel, 2004). The importance of the results was emphasized by Hopwood (Hopwood, 2007). In an excellent review article by Bull and Stach the authors showed that new drugs, especially antibiotics, are urgently required, and that the most propitious source remains natural products (Bull and Stach, 2007). They argued that in exploring new sources of bioactive natural products the marine environment deserves a particular attention, in view of the remarkable diversity of microorganisms and metabolic products. Recent reports of new chemical entities and first-in-class drug candidates, and confirmation of indigenous marine actinobacteria, make exciting discoveries even more likely given the unrivalled capacity of this class of bacteria to produce exploitable natural products. The authors also stressed that marine microbiology develops in many countries of different geographical origin. According to them more than 67% of marine natural products were sourced from Australia, the Caribbean,

Indian Ocean, Japan, the Mediterranean and Western Pacific Ocean sites. Marine environment has yielded many new natural products with biological activity (Gulder and Moore, 2009). According to the authors marine natural products are of extreme importance for drug discovery. The cytotoxic properties of many of these secondary metabolites are of particular interest for the development of new anti-cancer compounds.

Marine cyanobacteria continue to be an important source of bioactive compounds. Marine-derived actinomycetes, such as species of the genus *Salinispora* are highly interesting in this respect. According to Doshi and coworkers it has become essential to develop novel and more effective antibiotics (Doshi *et al.*, 2011). Current antibiotics are obtained from terrestrial life or made synthetically from intermediates. The ocean represents virtually untapped resource from which novel antibiotic compounds can be discovered. According to the authors it is the marine world that will provide the pharmaceutical industry with the next generation of antibiotics. Marine antibiotics were reported in marine bacteria (aplasmomycin, himalomycins, and pelagiomycins), sponges (Ara C, variabillin, strobilin, ircinin-1, aeroplysin, 3,5-dibromo-4-hydroxyphenylacetamide), coelenterates (asperidol and eunicin), molluscs (laurinterol and pachydictyol), tunicates (geranylhydroquinone and cystadytins), algae (cycloeudesmol, aeroplysinin-1(+), prepacifenol and tetrabromoheptanone), worms (tholepin and 3,5-dibromo-4-hydroxybezaldehyde), and actinomycetes (marinomycins C and D). This indicates that the marine environment, representing approximately half of the global diversity, is an enormous resource for new antibiotics and this source needs to be explored for the discovery of new generation antibiotics.

Marine organisms must have some characteristics that differentiate them from terrestrial organisms in many aspects, such as metabolism, behaviour, information transfer, and adaptation strategy. The data obtained so far were treated statistically by Hu and coworkers (Hu *et al.*, 2011). The authors concluded that 'Unlike terrestrial organisms, marine organisms have to adapt to extreme marine environmental conditions such as high pressure, high salt concentration, low nutrient concentration, low but steady temperature' and further that 'These differences are responsible for the diversity in the secondary metabolism of marine organisms'. It is interesting that there are more multihalogenated compounds isolated from marine algae most displaying strong bioactivity. As the industrial culturing and the application of gene technology are relatively easy, marine microorganisms are the most promising field in which to develop new drugs. The authors conclude that 'the ocean, which hosts approximately 87% of the Earth's life, offers immense potential for the discovery of pharmaceuticals'.

Microbial genome sequences – search for genes specifying biosynthesis of antibiotics

Microbial genome mining is a rapidly developing approach to discover new secondary metabolites for drug discovery (Bachmann *et al.*, 2014). Bachman and coworkers defined genome mining 'as the process of translating secondary metabolite-encoding gene sequence data into purified molecules in tubes'. For some unknown reason, the actinomycetes are amazingly prolific in the number of antibiotics they produce (Baltz, 2007). About 75% of the antibiotics are produced by actinomycetes, and about 75% of these are made by a single genus, *Streptomyces*. Strains of *Streptomyces hygroscopicus* make almost 200 antibiotics. One *Micromonospora* strain can produce 48 aminoglycosides, while *Streptomyces griseus* strains produce over 40 different antibiotics (Demain, 2014). It was proposed that the numbers of type I polyketide synthase (PKS) and non-ribosomal peptide synthetase (NRPS) or mixed PKS-NRPS genes per the producing microorganism may be used to measure the potential to produce secondary metabolites. It follows clearly that actinomycetes are most productive among microorganisms with large genomes (Baltz, 2011b). It was shown that the number of functional NRPS pathways can be estimated by counting the number of *mbtH* homologues in a microbial genome (Baltz, 2011a). Thanks to this interesting and clever approach it was possible to classify especially 'gifted' actinomycetes producing secondary metabolites (Baltz, 2014). According to

Baltz it is evident that of the ~10^{26} actinomycete colony-forming units presumably existing in the top 10 cm of soil only ~10^{7} actinomycetes have so far been screened for secondary metabolite production during the past 50 years and that thus the great majority of antibiotic producing actinomycetes still remains to be utilized (Baltz, 2005).

Streptomycetes with fully sequenced genomes include:

1 *Streptomyces coelicolor* A3(2) (Bentley *et al.*, 2002). The complete genome sequence of *S. coelicolor* greatly accelerated progress in understanding the biosynthesis of metabolites known or suspected to be produced by *S. coelicolor* and other streptomycetes and revealed that streptomycetes have far greater potential to produce bioactive natural products than previously suggested. *Streptomyces* and related genera of filamentous Gram-positive actinobacteria are an excellent source of natural compounds including antibacterials, antifungals, immunosuppressants, anticancer agents, anthelminthics and insecticides. The availability of a complete genome sequence has greatly facilitated the discovery and characterization of pathways for specialized metabolite biosynthesis in *S. coelicolor*. Thus, *S. coelicolor* specialized metabolism is likely to remain an important topic of research for the foreseeable future. 'The cost of generating near-complete bacterial genome sequences continues to drop, the methodology developed to mine the *S. coelicolor* genome for new natural products and biosynthetic pathways is increasingly being applied to other actinobacteria. Ultimately, such approaches will uncover the full metabolic potential of this remarkable phylum' (Challis, 2014).

2 *Streptomyces avermitilis* (Ikeda *et al.*, 2003). Genome mining of *Streptomyces avermitilis* was developed even further, as summarized in an excellent review by Ikeda and coworkers. The authors showed that the genome of *Streptomyces avermitilis* harbours at least 38 secondary metabolic gene clusters and 46 insertion line sequences on the genome, which have not been searched so far (Ikeda *et al.*, 2014). By heterologous expression of more than 20 exogenous biosynthetic gene clusters they confirmed the feasibility of large-deletion mutants of *S. avermitilis*. They found that biosynthetic gene clusters for polyketide compounds, for peptide compounds, terpene compounds and biosynthetic gene clusters for other compounds indicating that *S. avermitilis* can be used as a rich source of antibiotics and other biologically active compounds.

3 *Streptomyces griseus* (Ohnishi *et al.*, 2008). The complete genome sequence of *Streptomyces griseus* IFO 13350, a soil bacterium producing an antituberculosis antibiotic, streptomycin, was determined. Streptomycin is the first aminoglykoside antibiotic discovered more than 60 years ago. Comparisons with the genomes of two related species, *Streptomyces coelicolor* A3(2) and *Streptomyces avermitilis*, clarified not only the characteristics of the *S. griseus* genome but also the existence of 24 *Streptomyces*-specific proteins. The *S. griseus* genome contains 34 gene clusters or genes for the biosynthesis of known or unknown secondary metabolites. Transcriptome analysis using a DNA microarray showed that at least four of these clusters, in addition to the streptomycin biosynthesis gene cluster, were activated directly or indirectly by AdpA, which is a central transcriptional activator for secondary metabolism and morphogenesis in the A-factor (a γ-butyrolactone signalling molecule) regulatory cascade in *S. griseus*.

4 *Streptomyces scabies* 87–22 (Bignell *et al.*, 2010). *Streptomyces scabies* is a plant pathogen causing scab disease on economically important root and tuber crops. The genome sequence of *Streptomyces scabies* 87–22 was completed, and a biosynthetic cluster was identified that is predicted to synthesize a novel compound similar to coronafacic acid (CFA), a component of the virulence-associated coronatine phytotoxin produced by the plant-pathogenic bacterium *Pseudomonas syringae*. Southern analysis indicated that the *cfa*-like cluster in *S. scabies* 87–22 is likely conserved in other strains of *S. scabies* but is absent from two other pathogenic streptomycetes, *S. turgidiscabies* and *S. acidiscabies*.

Transcriptional analyses demonstrated that the cluster is expressed during plant–microbe interactions and that expression requires a transcriptional regulator embedded in the cluster as well as the *bldA* tRNA. A knockout strain of the biosynthetic cluster displayed a reduced virulence phenotype on tobacco seedlings compared with the wild-type strain. Thus, the *cfa*-like biosynthetic cluster is a newly discovered locus in *S. scabies* that contributes to host–pathogen interactions.

5 *Streptomyces cattleya* NRRL 8057 (Barbe *et al.*, 2011). *Streptomyces cattleya*, a producer of the antibitiotics thienamycin and cephamycin, is one of the rare bacteria known to produce fluorinated metabolites. The complete genome of this *Streptomyces* was analysed and found to contain two linear replicons. The genes involved in fluorine metabolism and in the biosynthesis of the antibiotic thienamycin were mapped on both replicons. *S. cattleya* has been used as a convenient model for characterizing both the enzyme responsible for the fluorination step and its substrate, as well as the metabolic pathway leading to 4-fluorothreonine.

6 *Streptomyces bingchenggensis* (Wang *et al.*, 2013). *Streptomyces bingchenggensis* is a soil bacterium that produces milbemycins, a family of macrolide antibiotics that are commercially important in crop protection and veterinary medicine. In addition, *S. bingchenggensis* produces many natural products including the polyether nanchangmycin and novel cyclic pentapeptides. To identify the gene clusters involved in the biosynthesis of these compounds and better clarify the biochemical pathways of these gene clusters, the whole genome of *S. bingchenggensis* was sequenced and the transcriptome profile was subsequently investigated by microarray. Transcriptional analysis demonstrated an extremely high expression level of the nachmangmycin gene cluster during the initial hours that subsequently decreased. However, other gene clusters appear to be silent. The genome-wide analysis of the secondary metabolite gene clusters in *S. bingchenggensis*, coupled with transcriptional analysis, will facilitate the rational development of high milbemycin-producing strains as well as the discovery of new natural products.

We propose that analysis of genes specifying biosynthesis of antibiotics should be used for the development of new derivatives of antibiotics that would be efficient against currently resistant microbial strains. According to the Microbial Genome Database for Comparative Analysis a total of 18 *Streptomyces* genomes have been sequenced also including *Streptomyces albus*, *S. collinus*, *S. davawensis*, *S. flavogriseus*, *S. fulvissimus*, different species of *S. hygroscopicus*, *S. venezuelae* and *S. violaceusniger*. This extensive knowledge can certainly be utilized for the search for new antibiotics.

Forgotten antibiotics to be revitalized

During 'the golden age' of antibiotics, thousands of biologically active compounds were discovered. However, many of them were not further developed either because better compounds were available at that time or because they could not be developed into drugs, due to their unsuitable pharmacological properties. Pulcini and coworkers in their review proposed that with respect to the spread of antimicrobial resistance in the absence of new antibiotics, potentially useful older antibiotics should be looked for (Pulcini *et al.*, 2012). The authors concluded that many potentially useful antibiotics, usually low cost and marketed as generics, and with a long history of tolerability and safety data, are not available in many countries. They proposed that temocillin and mecillinam could be useful for extended-spectrum β-lactamase–producing organisms. Fosfomycin and colistin may also be used to treat infections caused by carbapenem-resistant Gram-negative bacilli. They concluded that 31 particularly useful antibiotics are available in a limited number of countries surveyed. The authors stated that 'Urgent measures are now needed to ensure a better availability of these antibiotics on a global scale and to conduct additional research regarding the benefit of these antibiotics in current clinical infections'.

Freire-Morana and coworkers described agents in clinical development that might reach the market in the next 5–10 years (Freire-Moran *et al.*, 2011). Their review included 66 new active compounds of which 27 had a new mechanism of action or a new target. The remaining 39 agents were classified to belong to known antibacterial classes or groups, i.e. quinolone (15 agents), β-lactam (6), oxazolidinone (3), diaminopyrimidine (2), macrolide (2), pleuromutilin (2), aminoglycoside (1), ansamycin (1), FabI inhibitor (1), glycopeptide (1), metallic ion (1), streptogramin (1), tetracycline (1) and hybrid compounds (oxazolidinone/quinolone and rifamycin/fluoroquinolone) (2). They were thus considered as acting on the same target as of at least one previously licensed antibacterial agent. Therefore, they were not used in the remaining analysis. Of the 27 new active substances considered as having a new mechanism of action or a new target, there were 15 agents which could be systemically administered and thus considered potentially useful for the treatment of serious invasive infections. Of these 15 systemically administered agents, 13 were judged to have activity against at least one of the selected antibiotic resistant Gram-positive bacteria and eight against at least one of the selected antibiotic-resistant Gram-negative bacteria. Among antibiotic-resistant Gram-positive bacteria, MRSA was the most often covered by these agents (13 out of 15) and VRE the least covered (5 out of 15). Of the eight agents with activity against antibiotic-resistant Gram-negative bacteria, four had an activity based on actual data and four had assumed activity based on known class properties or mechanisms of action. Of the four agents with activity against antibiotic-resistant Gram-negative bacteria based on actual data, two acted on new or possibly new targets and new mechanisms of action.

New antibiotic targets

The information on genomes of important pathogenic bacteria such as *Bacillus anthracis, Listeria monocytogenes* and *Staphylococcus aureus* should certainly help in the search for new antibiotic targets in those pathogens. Many hundreds of genomes of pathogenic microorganisms have been completely sequenced and many additional projects are in progress. Boucher and coworkers summarized the prospects in the development of new compounds against pathogenic microorganisms including the search for new targets (Boucher *et al.*, 2013). According to the authors infections caused by antibiotic-resistant bacteria, especially the 'ESKAPE' pathogens, continue to increase in frequency and cause significant morbidity and mortality. New antimicrobial agents are hence greatly needed to treat infections caused by Gram-negative bacilli (GNB) resistant to currently available agents. The Infectious Diseases Society of America (IDSA) continues to propose legislative, regulatory, and funding solutions to this continuing crisis. In the current report the authors updated the status of development and approval of systemic antibiotics in the United States as of early 2013. Only two new antibiotics have been approved since IDSA's 2009 pipeline status report, and the number of new antibiotics annually approved for marketing in the United States continues to decline. In the study seven drugs were in clinical development for treatment of infections caused by resistant GNB. None of these agents was included in the 2009 list of antibacterial compounds in phase 2 or later development, but unfortunately none addressed the entire spectrum of clinically relevant GNB resistance. In the survey it was demonstrated some progress in development of new antibacterial drugs that target infections caused by resistant GNB has been reached but that it remains alarmingly elusive. IDSA stresses the author's conviction that the antibiotic pipeline problem can be solved by the collaboration of global leaders to develop creative incentives that will stimulate new antibacterial research and development. According to the authors a sustainable global antibacterial drug research and development enterprise should be created with the power in the short term to develop 10 new, safe, and efficacious systemically administered antibiotics by 2020 as called for in IDSA's '10 × 20 Initiative'.

Bakheet and Doig (2010), in a highly interesting review, analysed 48 non-redundant antibiotic target proteins from all bacteria, 22 antibiotic target proteins from *E. coli* only and 4243 non-drug targets from *E. coli* to identify differences in their properties and to predict new potential drug targets. They found that as compared with non-targets, bacterial antibiotic targets tend to be long, have high β-sheet and low α-helix contents, are polar, are found in the cytoplasm rather than in membranes, and are usually enzymes, with ligases particularly favoured. The authors used the sequence features to build a support vector machine model for *E. coli* proteins, allowing the assignment of any sequence to the drug target or non-target classes, with the accuracy in the training set of 94%. They identified 319 proteins (7%) in the non-target set that have target-like properties, many of which have unknown function. Sixty-three of these proteins have significant and undesirable similarity to a human protein, leaving 256 target-like proteins that are not present in humans. The authors suggested that antibiotic discovery programmes would be more likely to succeed if new targets are chosen from this set of target like proteins or their homologues. In particular, 64 are essential genes where the cell is not able to recover from a random insertion disruption. It appears to us that such bioinformatic research will be required for a successful search for new antibiotics.

Discussion and conclusions

Resistance of microorganisms to antibiotics permanently increases and represents a global threat to human health. According to Amábile-Cuevas the emergence and spread of antibiotic resistant bacteria reflects a completely Darwinian evolution, which mostly yields slight decrease in antibiotic susceptibility, along with phenotypes that are not precisely characterized as resistance and sudden changes from full susceptibility to full resistance, which are driven by a vast array of horizontal gene transfer mechanisms (Amabile-Cuevas, 2013). According to the author, antibiotics select for more than just antibiotic resistance (i.e. increased virulence and enhanced gene exchange abilities) and many non-antibiotic agents or conditions select for or maintain antibiotic resistance traits as a result of a complex network of underlying and often overlapping mechanisms.

Tackling antibiotic resistance is thus a highly complicated and never-ending prospect. It appears that the struggle against antibiotic resistance is a war we can never win. The strength of immense numbers of microorganisms, combined with the ancient force of evolution seem to overpower our drugs. However, when facing this threat we cannot remain helpless. It is estimated that about 150,000 antibiotics have been described of which only about 150 are in clinical use. This is mainly because some antibiotics are toxic for humans or poorly water soluble, have other pharmacologically unsuitable properties or have similar effects as the antibiotics already used in the clinical practice. Does it thus make sense to look for new antibiotics that would help us against increasing antibiotic resistance? According to Davies and Davies (2010), 'The importance and value of antibiotics cannot be overestimated; we are totally dependent on them for the treatment of infectious diseases, and they should never be considered mere commodities. In addition to their use in the treatment of infectious diseases, antibiotics are critical to the success of advanced surgical procedures, including organ and prosthetic transplants'.

For years antibiotics have provided an effective treatment for a number of diseases caused by microorganisms. However, as mentioned several times above, the number of microorganisms that have adapted to the current antibiotics increased dramatically. Many challenges face the discovery and development of new antibiotics, making it difficult to reach the market, especially since many pharmaceutical companies have stopped searching for antibiotics. With the advent of genome sequencing, new antibiotics are being found by the techniques of genome mining offering hope for the future (Scheffler *et al.*, 2013).

New antibiotics or their derivatives are very much needed. It appears to us that microbial natural products still represent the most promising source of new antibiotics. According to the old but still interesting mathematical model of Watve and coworkers it should be possible to identify 294,300 compounds that have not yet been described, and even according to much more conservative estimates, 150,000 new compounds could still be isolated (Watve *et al.*, 2001). Even if the mathematical model is doubted by mathematicians, it may be still reasonably estimated that only 3% of the antibiotics produced by streptomycetes have so far been isolated. Quite clearly, new groups of chemical compounds can still be discovered with the aid of modern, highly sophisticated approaches.

We are convinced that the search for new antimicrobial compounds is an absolute must, in spite of the fact that big pharma does not seem to be much interested in this process. It is apparent that many drug targets exist in pathogenic microorganisms that have not yet been investigated. With the increasing application of genome knowledge and mining it can be expected that new targets for antibiotics will be found. We propose that it is necessary to combine the efforts of both the academia and industry in the fight against antibiotic resistance. As concluded by Davies and Davies (2010), we should renew a concerted attack on antibiotic resistance, otherwise 'the preantibiotic era awaits our descendants'.

References

Amabile-Cuevas, C.F. (2013). Antibiotic resistance: from Darwin to Lederberg to Keynes. Microb. Drug Resist. *19*, 73–87.

Bachmann, B.O., Van Lanen, S.G., and Baltz, R.H. (2014). Microbial genome mining for accelerated natural products discovery: is a renaissance in the making? J. Ind. Microbiol. Biotechnol. *41*, 175–184.

Bagnoli, F., Baudner, B., Mishra, R.P.N., Bartolini, E., Fiaschi, L., Mariotti, P., Nardi-Dei, V., Boucher, P., and Rappuoli, R. (2011). Designing the next generation of vaccines for global public health. Omics J. Integrat. Biol. *15*, 545–566.

Bakheet, T.M., and Doig, A.J. (2010). Properties and identification of antibiotic drug targets. BMC Bioinformatics *11*.

Baltz, R.H. (2005). Antibiotic discovery from actinomycetes: will a renaissance follow the decline and fall? SIM News *55*, 186–196.

Baltz, R.H. (2007). Antimicrobials from actinomycetes: back to the future. Microbe *2*, 125–131.

Baltz, R.H. (2011a). Function of MbtH homologs in nonribosomal peptide biosynthesis and applications in secondary metabolite discovery. J. Ind. Microbiol. Biotechnol. *38*, 1747–1760.

Baltz, R.H. (2011b). Strain improvement in actinomycetes in the postgenomic era. J. Ind. Microbiol. Biotechnol. *38*, 657–666.

Baltz, R.H. (2014). MbtH homology codes to identify gifted microbes for genome mining. J. Ind. Microbiol. Biotechnol. *41*, 357–369.

Barbe, V., Bouzon, M., Mangenot, S., Badet, B., Poulain, J., Segurens, B., Vallenet, D., Marliere, P., and Weissenbach, J. (2011). Complete genome sequence of *Streptomyces cattleya* NRRL 8057, a producer of antibiotics and fluorometabolites. J. Bacteriol. *193*, 5055–5056.

Bentley, S.D., Chater, K.F., Cerdeno-Tarraga, A.M., Challis, G.L., Thomson, N.R., James, K.D., Harris, D.E., Quail, M.A., Kieser, H., Harper, D., *et al.* (2002). Complete genome sequence of the model actinomycete Streptomyces coelicolor A3(2). Nature *417*, 141–147.

Bignell, D.R.D., Seipke, R.F., Huguet-Tapia, J.C., Chambers, A.H., Parry, R.J., and Loria, R. (2010). *Streptomyces scabies* 87–22 contains a coronafacic acid-like biosynthetic cluster that contributes to plant–microbe interactions. Mol. Plant-Microbe Int. *23*, 161–175.

Boucher, H.W., Talbot, G.H., Benjamin, D.K., Jr., Bradley, J., Guidos, R.J., Jones, R.N., Murray, B.E., Bonomo, R.A., Gilbert, D., and Infect. Dis. Soc., A. (2013). 10 x '20 Progress-development of new drugs active against Gram-negative bacilli: an update from the infectious diseases society of America. Clin. Infect. Dis. *56*, 1685–1694.

Bull, A.T., and Stach, J.E.M. (2007). Marine actinobacteria: new opportunities for natural product search and discovery. Trends Microbiol. *15*, 491–499.

Bush, K., Courvalin, P., Dantas, G., Davies, J., Eisenstein, B., Huovinen, P., Jacoby, G.A., Kishony, R., Kreiswirth, B.N., Kutter, E., *et al.* (2011). Tackling antibiotic resistance. Nature Rev. Microbiol. *9*, 894–896.

Butler, M.S., and Cooper, M.A. (2011). Antibiotics in the clinical pipeline in 2011. J. Antibiot. *64*, 413–425.

Butler, M.S., Blaskovich, M.A., and Cooper, M.A. (2013). Antibiotics in the clinical pipeline in 2013. J. Antibiot. *66*, 571–591.

Challis, G.L. (2014). Exploitation of the *Streptomyces coelicolor* A3(2) genome sequence for discovery of new natural products and biosynthetic pathways. J. Ind. Microbiol. Biotechnol. *41*, 219–232.

Daniel, R. (2004). The soil metagenome – a rich resource for the discovery of novel natural products. Cur. Opin. Biotechnol. *15*, 199–204.

Davies, J., and Davies, D. (2010). Origins and evolution of antibiotic resistance. Microbiol. Mol. Biol. Rev. *74*, 417–433.

Davies, S.C., Fowler, T., Watson, J., Livermore, D.M., and Walker, D. (2013). Annual report of the Chief medical officer: infection and the rise of antimicrobial resistance. Lancet *381*, 1606–1609.

Demain, A.L. (2014). Importance of microbial natural products and the need to revitalize their discovery. J. Ind. Microbiol. Biotechnol. *41*, 185–201.

Demain, A.L., and Spizek, J. (2012). The antibiotic crisis. In Antimicrobial Drug Discovery: Emerging Strategies, A. Tegos, and E. Mylonakis, eds. (CABI, Wallingford, UK), pp. 26–43.

Doshi, G.V., Aggarwal, G.V., Martis, E.A., and Shanbhag, P.P. (2011). Novel antibiotics from marine sources. Int. J. Pharmaceut. Sci. Nanotechnol. *4*, 1446–1461.

Freire-Moran, L., Aronsson, B., Manz, C., Gyssens, I.C., So, A.D., Monnet, D.L., Cars, O., and Grp, E.-E.W. (2011). Critical shortage of new antibiotics in development against multidrug-resistant bacteria-time to react is now. Drug Res. Updates *14*, 118–124.

Gould, I.M., and Bal, A.M. (2013). New antibiotic agents in the pipeline and how they can help overcome microbial resistance. Virulence *4*, 185–191.

Gulder, T.A.M., and Moore, B.S. (2009). Chasing the treasures of the sea – bacterial marine natural products. Curr. Opin. Microbiol. *12*, 252–260.

Hancock, R.E.W. (2007). The end of an era? Nat. Rev. Drug Discov. *6*, 28–28.

Hancock, R.E.W., Nijnik, A., and Philpott, D.J. (2012). Modulating immunity as a therapy for bacterial infections. Nat. Rev. Microbiol. *10*, 243–254.

Hopwood, D.A. (2007). Therapeutic treasures from the deep. Nat. Chem. Biol. 3, 457–458.

Hu, G.-P., Yuan, J., Sun, L., She, Z.-G., Wu, J.-H., Lan, X.-J., Zhu, X., Lin, Y.-C., and Chen, S.-P. (2011). Statistical research on marine natural products based on data obtained between 1985 and 2008. Marine Drugs 9, 514–525.

Ikeda, H., Ishikawa, J., Hanamoto, A., Shinose, M., Kikuchi, H., Shiba, T., Sakaki, Y., Hattori, M., and Omura, S. (2003). Complete genome sequence and comparative analysis of the industrial microorganism *Streptomyces avermitilis*. Nat. Biotechnol. *21*, 526–531.

Ikeda, H., Shin-ya, K., and Omura, S. (2014). Microbial genome mining of the *Streptomyces avermitilis* genome and development of genome-minimized host for heterologous expression of biosynthetic gene clusters. J. Ind. Microbiol. Biotechnol. *41*.

Mishra, R.P.N., Oviedo-Orta, E., Prachi, P., Rappuoli, R., and Bagnoli, F. (2012). Vaccines and antibiotic resistance. Curr. Opin. Mircobiol. *15*, 596–602.

Octavia, S., Sintchenko, V., Gilbert, G.L., Lawrence, A., Keil, A.D., Hogg, G., and Lan, R. (2012). Newly emerging clones of *Bordetella pertussis* carrying *prn2* and *ptxP3* alleles implicated in Australian pertussis epidemic in 2008–2010. J. Infect. Dis. *205*, 1220–1224.

Ohnishi, Y., Ishikawa, J., Hara, H., Suzuki, H., Ikenoya, M., Ikeda, H., Yamashita, A., Hattori, M., and Horinouchi, S. (2008). Genome sequence of the streptomycin-producing microorganism *Streptomyces griseus* IFO 13350. J. Bacteriol. *190*, 4050–4060.

Pulcini, C., Bush, K., Craig, W.A., Frimodt-Moller, N., Grayson, M.L., Mouton, J.W., Turnidge, J., Harbarth, S., and Gyssens, I.C.; the ESCMID Study Group for Antibiotic Policies (2012). Forgotten antibiotics: an inventory in Europe, the United States, Canada, and Australia. Clin. Infect. Dis. *54*, 268–274.

Rappuoli, R., Mandl, C.W., Black, S., and De Gregorio, E. (2011). Vaccines for the twenty-first century society. Nat. Rev. Immunol. *11*, 865–872.

Rezanka, T., Spizek, J., and Sigler, K. (2007). Medicinal use of lincosamides and microbial resistance to them. Anti-Infect. Agents Med. Chem. *6*, 133–144.

Scheffler, R.J., Colmer, S., Tynan, H., Demain, A.L., and Gullo, V.P. (2013). Antimicrobials, drug discovery, and genome mining. Appl. Microbiol. Biotechnol. *97*, 969–978.

Spizek, J., Novotna, J., Rezanka, T., and Demain, A.L. (2010). Do we need new antibiotics? The search for new targets and new compounds. J. Ind. Microbiol. Biotechnol. 37, 1241–1248.

Walsh, C.T., and Wencewicz, T.A. (2014). Prospects for new antibiotics: a molecule-centered perspective. J. Antibiot. *67*, 7–22.

Wang, X.-J., Zhang, B., Yan, Y.-J., An, J., Zhang, J., Liu, C.-X., and Xiang, W.-S. (2013). Characterization and analysis of an industrial strain of *Streptomyces bingchenggensis* by genome sequencing and gene microarray. Genome *56*, 677–689.

Watve, M.G., Tickoo, R., Jog, M.M., and Bhole, B.D. (2001). How many antibiotics are produced by the genus Streptomyces? Arch. Microbiol. *176*, 386–390.

Willetts, D. (2013). Policy paper: G8 science ministers statement: London, 12 June 2013. Department for Business, Innovation & Skills. Available at: https://www.gov.uk/government/publications/g8-science-ministers-statement-london-12-june-2013. Retrieved 4 April 2014.

Eradication of Dormant Pathogens

Kim Lewis, Brian Conlon and Michael LaFleur

6

Abstract

Persisters are specialized survivor cells that protect bacterial and fungal populations from killing by antibiotics. Persisters are dormant phenotypic variants of regular cells rather than mutants. Most microbicidal antibiotics kill by corrupting their targets into producing toxic products; tolerance to antibiotics follows when targets are inactive. Mechanisms of persister formation are redundant, making it difficult to eradicate these cells. In *Escherichia coli*, toxins RelE and MazF cause dormancy by degrading mRNA; HipA inhibits translation by phosphorylating glutamyl tRNA synthase; and TisB forms an anion channel in the membrane, leading to a decrease in proton motive force (pmf) and ATP levels. Prolonged treatment of chronic infections with antibiotics selects for *hip* mutants that produce more persister cells. Eradication of tolerant persisters is a serious challenge. Existing antibiotics were developed to kill rapidly growing cells and have limited activity against dormant persisters. A number of compounds were recently described that have a capability of killing persisters. A potentiator of azoles, AC17, sterilizes a biofilm of *Candida albicans*. Prodrugs are converted into generally reactive compounds with bacteria-specific enzymes and kill growing and dormant cells. Lassomycin activates the ATPase of the ClpP1P2C1 protease of *Mycobacterium tuberculosis*, and kills persisters. Acyldepsipeptide (ADEP4) activates the ClpP protease, forcing the cell to self-digest, and kills both growing cells and persisters. These findings suggest a path towards developing sterilizing antibiotics.

Persisters and infectious disease

Persisters represent a small subpopulation of cells that spontaneously enter into a dormant, non-dividing state. When a population is treated with a bactericidal antibiotic, regular cells die, while persisters survive. Taking samples and plating them for colony counts over time from a culture treated with antibiotic produces a biphasic pattern, with a distinct plateau of surviving persisters. In order to kill, antibiotics require active targets, which explains tolerance of dormant persisters. By contrast, resistance mechanisms prevent antibiotics from binding to their targets.

Infectious disease is often untreatable, even when caused by a pathogen that is not resistant to antibiotics. This is the essential paradox of chronic infections. In most cases, chronic infections are accompanied by the formation of biofilms, which seems to point to the source of the problem (Costerton *et al.*, 1999; Del Pozo and Patel, 2007). Biofilms have been linked to dental disease, endocarditis, cystitis, urinary tract infection (UTI), deep-seated infections, indwelling device and catheter infections, and the incurable disease of cystic fibrosis. It appears that biofilms are largely a side-effect of medical intervention, associated with the use of indwelling devices and the increase in the number of elderly and immunosuppressed individuals. In the case of indwelling devices such as prostheses and heart valves, reoperation is the method of choice for treating the infection. Biofilms do not generally restrict penetration of antibiotics (Walters *et al.*, 2003), but do form a barrier for the larger components of the immune system (Jesaitis *et al.*, 2003; Leid *et al.*, 2002; Vuong *et al.*, 2004). The presence of biofilm-specific resistance mechanisms was suggested to account for the recalcitrance of infectious diseases (Stewart and Costerton, 2001). However, the majority of cells in the biofilm are actually highly susceptible to killing by antibiotics; only a small fraction of persisters remains alive (Spoering and Lewis, 2001). Based on these findings, we proposed a simple model of a relapsing chronic infection – antibiotics kill the majority of cells, and the immune system eliminates both regular cells and persisters from the bloodstream (Lewis, 2001). The only remaining live cells are then persisters in the biofilm (Fig. 6.1).

Once the level of antibiotic drops, persisters repopulate the biofilm, and the infection relapses. While this is a plausible model, it is not the only one. A simpler possibility is that antibiotics fail to effectively reach at least some cells *in vivo*, resulting in a relapsing infection.

Establishing potential causality between persisters and therapy failure is not trivial, since these cells form a small subpopulation with a temporary phenotype, which precludes introducing them into an animal model of infection. We reasoned that causality could be tested based on what we know about selection for high persister (*hip*) mutants *in vitro*. Periodic application of high doses of bactericidal antibiotics leads to the selection of strains that produce increased levels of persisters (Moyed and Bertrand, 1983; Wolfson *et al.*, 1990). This is precisely what happens in the course of treating chronic infections – the patient is periodically exposed to high doses of antibiotics, which may select for *hip* mutants. But *hip* mutants would only gain advantage if the drugs effectively reach, and kill the regular cells of the pathogen.

Patients with cystic fibrosis (CF) largely succumb to an incurable chronic infection of the lungs with *Pseudomonas aeruginosa* (Gibson *et al.*, 2003). The periodic application of high doses of antibiotics provides some relief by decreasing the pathogen burden, but does not clear the infection. If *hip* strains of pathogens were selected *in vivo*, they would most likely be present in a CF patient. We took advantage of a set of longitudinal *P. aeruginosa* isolates from a single patient, collected over the course of many years (Smith *et al.*, 2006). Testing persister levels by monitoring survival after challenge with a high dose of ofloxacin showed a dramatic, 100-fold increase in surviving persister cells in the last four isolates (Mulcahy *et al.*, 2010). Testing paired strains from additional patients showed that in most cases, there was a considerable increase in persister levels in the late isolate from a patient. Interestingly, most of the *hip* isolates had no increase in MIC compared to their clonal parent strain to ofloxacin, carbenicillin, and tobramycin, suggesting that classical acquired resistance plays little to no role in the recalcitrance of CF infection. These experiments directly link persisters to the clinical manifestation of the disease and suggest that persisters are responsible for the therapy failure of chronic CF infection. But why have the *hip* mutants with their striking survival phenotype evaded detection for such a long

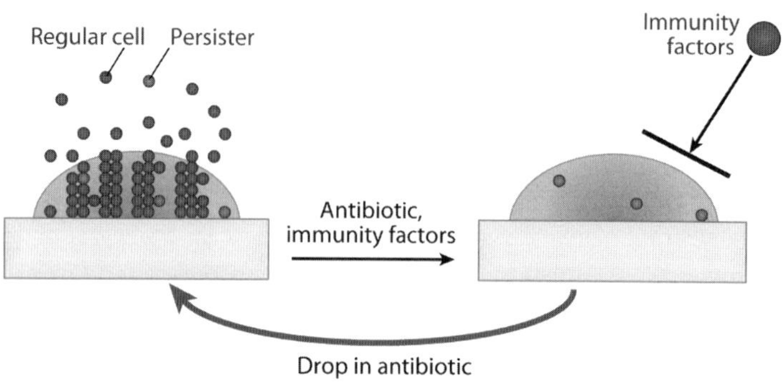

Figure 6.1 A model of relapsing biofilm infection (Lewis, 2001). A colour version of this figure is available in the plate section at the back of the book.

time? The main focus of research in antimicrobials has been on drug resistance, and the basic starting experiment is to test a clinical isolate for its ability to grow in the presence of elevated levels of different antibiotics, and record any increases in the MIC. This is also the standard test employed by clinical microbiology laboratories. *hip* mutants are of course missed by this test, which explains why they had remained undetected, in spite of a major effort aimed at understanding pathogen survival of antimicrobial chemotherapy. Given that *hip* mutants are the likely main culprit responsible for morbidity and mortality of the CF infection, it makes sense to test for their presence.

Another interesting link between persisters and disease was recently reported in the study of *Salmonella typhimurium* (Helaine *et al.*, 2014). Entrance of bacteria into macrophages sharply increased the level of drug-tolerant persisters, which was dependent on the presence of several toxin–antitoxin (TA) modules. This observation suggests that pathogens increase their tolerant population as a strategy to survive the hostile environment of the host, and at the same time benefit from acquiring multidrug tolerance.

Biofilms apparently serve as a protective habitat for persisters (Harrison *et al.*, 2005a,b, 2009; LaFleur *et al.*, 2006; Spoering and Lewis, 2001), allowing them to evade the immune system. However, a more general paradigm is that persisters will be critical for pathogens to survive antimicrobial chemotherapy whenever the immune response is limited. Such cases would include disseminating infections in immunocompromised patients undergoing cancer chemotherapy or infected with HIV. Persisters are also likely to play an important role in immunocompetent individuals in cases where the pathogen is located at sites poorly accessible by components of the immune system. These include the central nervous system, where pathogens cause debilitating meningitis and brain abscesses (Honda and Warren, 2009), and the gastrointestinal tract, where a hard-to-eradicate *Helicobacter pylori* strain causes gastroduodenal ulcers and gastric carcinoma (Peterson *et al.*, 2000). Tuberculosis infections are chronic infection due to the ability of the pathogen to evade the immune system. An acute infection may resolve as a result of antimicrobial therapy, but the pathogen often remains in a 'latent' asymptomatic form (Barry *et al.*, 2009). It is estimated that one in every three people carry latent *M. tuberculosis*, and 10% of carriers develop an acute infection at some stage in their lives. Virtually nothing is known about this latent form that serves as the main reservoir of tuberculosis. One simple possibility is that persisters are equivalent to the latent form of the pathogen. The above analysis underscores the significance of drug tolerance as a barrier to effective antimicrobial chemotherapy. Given its significance – roughly half of all cases of infection, the number of studies dedicated to tolerance is minuscule compared twith publications on resistance. The formidable barriers to study persister cells account for the lack of parity between these two comparably significant fields. Hopefully a better balance will be achieved, and the following discussion summarizes recent advances in understanding the mechanism of tolerance.

Mechanisms of persister formation

Persisters were initially discovered in 1944 (Bigger, 1944), but the mechanism of their formation eluded us for a very long time. Only recently have the molecular mechanisms of dormancy begun to emerge. The most straightforward approach to finding an underlying mechanism of a complex function is by screening a library of transposon (Tn) insertion mutants. This produces a set of candidate genes, and subsequent analysis leads to a pathway and a mechanism. This is indeed how the basic mechanisms of sporulation, flagellation, chemotaxis, virulence and many other functions have been established. However, screening a Tn insertion library of *E. coli* for an ability to tolerate high doses of antibiotics produced no mutants completely lacking persisters (Hu and Coates, 2005; Spoering *et al.*, 2006). With the development of the complete, ordered *E. coli* gene knockout library by the Mori group (Baba *et al.*, 2006), the Keio collection), it seemed reasonable to revisit the screening approach. Indeed, there always remains a possibility that transposons missed a critical gene, or the library was not large enough. The use of the Keio collection largely resolves this uncertainty.

This advanced screen also failed to produce a single mutant that failed to form persisters, suggesting a high degree of redundancy (Hansen *et al.*, 2008). However, a number of genes did affect the ability of cells to form persisters, with some knockouts resulting in a 10-fold decrease in persister formation. The majority of hits were in global regulators, DksA, DnaKJ, HupAB, and IhfAB. This is an independent indication of redundancy – a global regulator can affect expression of several persister genes simultaneously, resulting in a phenotype. The screen also produced two interesting candidate genes that may be more directly involved in persister formation – YgfA an inhibitor of nucleotide synthesis, and YigB, which may block metabolism by depleting the pool of FMN. A similar screen of a *P. aeruginosa* mutant library was reported and confirmed that the genetic mechanisms of persister formation are redundant (De Groote *et al.*, 2009).

The main conclusion from the screens is that persister formation does not follow the usual design theme of complex cellular functions – a single linear regulatory pathway controlling an execution mechanism. By contrast, persisters are apparently formed through a number of independent parallel mechanisms. There is a considerable adaptive advantage in this redundant design – no single compound will disable persister formation.

Screens for persister genes were useful in finding some possible candidate genes and pointing to redundancy of function. It seemed that a method better suited to uncover redundant elements would be transcriptome analysis. For this, persisters had to be isolated.

Persisters form a small and temporary population, making isolation challenging. The simplest approach is to lyse a population of growing cells with a β-lactam antibiotic, and collect surviving persisters. This approach allowed us to perform the first persister transcriptome (Keren *et al.*, 2004). A more advanced method aimed at isolating native persisters was developed, based on a hypothesis that persisters are dormant and will thus have diminished protein synthesis (Shah *et al.*, 2006). The strains were engineered to express unstable green fluorescent protein (GFP) under control of a ribosomal promoter. Growing cells remained GFP positive, while cells that entered dormancy rapidly dimmed due to degradation of GFP. The difference in fluorescence allowed for the sorting of the two subpopulations. The dim cells were tolerant to ofloxacin, confirming that they are persisters.

Transcriptomes obtained by both methods pointed to down-regulation of biosynthesis genes, and indicated increased expression of several toxin/antitoxin modules (RelBE, MazEF, DinJYafQ, YgiU). TA modules are found on plasmids where they constitute a maintenance mechanism (Gerdes *et al.*, 1986b; Hayes, 2003). Typically, the toxin is a protein that inhibits an important cellular function such as translation or replication, and forms an inactive complex with the antitoxin. The toxin is stable, while the antitoxin is degradable. If a daughter cell does not receive a plasmid after segregation, the antitoxin level decreases due to proteolysis, leaving a toxin that either kills the cell or inhibits propagation. TA modules are also commonly found on bacterial chromosomes, but their role had been unknown. In *E. coli*, MazF and an unrelated toxin RelE induce stasis by cleaving mRNA, which of course inhibits translation, a condition that can be reversed by expression of corresponding antitoxins (Christensen and Gerdes, 2003; Pedersen *et al.*, 2003). The ability of toxins to induce reversible stasis makes them excellent candidates for persister genes.

Ectopic expression of RelE (Keren *et al.*, 2004) or MazF (Vazquez-Laslop *et al.*, 2006) strongly increased tolerance to antibiotics. The first gene linked to persisters, *hipA* (Moyed and Bertrand, 1983), is also a toxin, and its ectopic expression causes multidrug tolerance as well (Correia *et al.*, 2006; Falla and Chopra, 1998; Korch and Hill, 2006). Interestingly, a bioinformatics analysis indicates that HipA is a member of the Tor family of kinases, which have been extensively studied in eukaryotes (Schmelzle and Hall, 2000), but have not been previously identified in bacteria. HipA is indeed a kinase; it autophosphorylates on serine 150, and site-directed mutagenesis replacing it, or other conserved amino acids in the catalytic and Mg^{2+}-binding sites abolishes its ability to stop cell growth and confer drug tolerance (Correia *et al.*, 2006). The crystal structure of HipA in complex with its antitoxin HipB was recently resolved (Schumacher *et al.*, 2009), and it has been established that the kinase inhibits protein synthesis by phosphorylating glutamyl-tRNA synthase (Germain *et al.*, 2013; Kaspy *et al.*, 2013).

Deletion of potential candidates of persister genes noted above does not produce a discernible phenotype affecting persister production, possibly due to the high degree of redundancy of these elements. In *E. coli*, there are at least 15 TA modules (Alix and Blanc-Potard, 2009; Pandey and Gerdes, 2005; Pedersen and Gerdes, 1999), and more than 80 in *M. tuberculosis* (Ramage *et al.*, 2009). Recently, a strain with 10 TA module deletions was constructed and it forms much less persisters (Maisonneuve *et al.*, 2011).

It is possible that a particular TA module is expressed under given conditions, and then becomes responsible for producing the majority of persisters. Following this logic, we considered TA modules that are controlled by known transcriptional regulators.

Several TA modules contain the Lex box and are induced by the SOS response. These are *symER, hokE, yafN/yafO*, and *tisAB/istr1* (Courcelle *et al.*, 2001; Fernandez De Henestrosa *et al.*, 2000; Kawano *et al.*, 2007; McKenzie *et al.*, 2003; Motiejunaite *et al.*, 2007; Pedersen and Gerdes, 1999; Singletary *et al.*, 2009; Vogel *et al.*, 2004). Fluoroquinolones induce the SOS response (Phillips *et al.*, 1987), and we found that ciprofloxacin can induce persister formation (Dorr *et al.*, 2009). Examination of TA deletion strains showed that the level of persisters dropped dramatically, 10- to 100-fold, in a Δ*tisAB* mutant (Dorr *et al.*, 2010). This suggests that TisB is responsible for the formation of the majority of persisters under conditions of SOS induction. The level of persisters was unaffected in strains deleted in the other Lex box containing TA modules. Persister levels observed in time-dependent killing experiments with ampicillin or streptomycin that do not cause DNA damage were unchanged in the Δ*tisAB* strain. TisB only had a phenotype in the presence of a functional RecA protein, confirming the dependence on the SOS pathway. Ectopic overexpression of *tisB* sharply increased the level of persisters. Drop in persisters in a deletion strain and increase upon overexpression gives reasonable confidence in the functionality of a candidate persister gene. The dependence of TisB-induced persisters on a particular regulatory pathway, the SOS response, further strengthens the case for TisB as a specialized persister protein. Incidentally, a *tisB* mutant is not present in the otherwise fairly complete Keio knockout library, and the small ORF might have been easily missed by Tn mutagenesis as well, evading detection by the generalized screens for persister genes.

The role of TisB in persister formation is unexpected based on what we know about this type of proteins. TisB is a small, 29 amino acid hydrophobic peptide that binds to the membrane and leads to a drop in ATP levels (Unoson and Wagner, 2008). Bacteria, plants and animals all produce antimicrobial membrane-acting peptides (Garcia-Olmedo *et al.*, 1998; Sahl and Bierbaum, 1998; Zasloff, 2002). Toxins of many TA loci found on plasmids belong to this type as well. If a daughter cell does not inherit a plasmid, the concentration of a labile antitoxin decreases, and the toxin such as the membrane-acting *hok* kills the cell (Gerdes *et al.*, 1986a). High-level artificial overexpression of TisB also causes cell death by disrupting proton motive force (pmf) (Unoson and Wagner, 2008). We recently found that TisB forms an anion channel in artificial membranes, confirming its function as a typical toxic antimicrobial peptide (Gurnev *et al.*, 2012). It is remarkable from this perspective that the membrane-acting TisB under conditions of natural (mild) expression has the exact opposite effect of protecting the cell from antibiotics. Disruption of the pmf leading to a decrease in ATP is indeed a perfect tool for rapidly creating a dormant state – antibiotic targets require energy, and a full systems shutdown will prevent them from killing persisters.

While resistance and tolerance are mechanistically distinct, there is sufficient reason to believe that tolerance may be a major cause for developing resistance. Indeed, the probability of resistance development is proportional to the size of the pathogen population, and a lingering chronic infection that cannot be eradicated due to tolerance will go on to produce resistant mutants and strains acquiring resistant determinants by transmission from other bacteria (Levin and Rozen, 2006). Combating tolerance then becomes a major component in preventing resistance.

Persister eradication

Existing antibiotics were developed to stop propagation of microbial cells or to kill growing pathogens. Killing is not a requirement for drug development, and many antibiotics are bacteriostatic. FDA does not require antibiotics to be tested against stationary or biofilm populations of pathogens. Not surprisingly, antibiotics in our armamentarium have a limited ability to kill persister cells.

The killing mechanism for the main classes of antibiotics is fairly well understood – β-lactams induce autolysins (Groicher *et al.*, 2000); aminoglycosides cause mistranslation, which produces toxic misfolded peptides (Davis *et al.*, 1986); and fluoroquinolones convert DNA gyrase/topoisomerase into an endonuclease that degrades DNA (Hooper, 2001). All of these killing mechanisms require active targets. It was recently reported that all bactericidal antibiotics kill by inducing formation of reactive oxygen species (ROS) (Kohanski *et al.*, 2007). However, this claim has been challenged by several groups (Ezraty *et al.*, 2013; Keren *et al.*, 2013; Liu and Imlay, 2013; Mahoney and Silhavy, 2013; Renggli *et al.*, 2013). It is unlikely that ROS play a significant role in killing, since there is no difference between the minimal bactericidal concentration (MBC) of the main classes of bactericidal antibiotics under aerobic vs. anaerobic conditions.

While we now appreciate the unmet need for sterilizing antibiotics, obtaining such compounds is not trivial, not only because of the intrinsic difficulty of killing dormant cells with inactive targets, but also due to the generally poor state of affairs in antibiotic discovery in general. Our ability to develop even an 'easy' novel compound, such as a bacteriostatic narrow-spectrum compound, is severely limited. Most classes of antibiotics are natural products discovered in a relatively short, 20-year period, based on screening soil Actinomycetes. Developed by Selman Waksman (Waksman and Woodruff, 1940), this 'Waksman platform' was also the only successful approach to antibiotic discovery we ever had. Overmining of the actinomycetes led to rediscovery of known compounds by the 1960s and an essential collapse of the platform. Efforts to develop a high-tech platform based on high-throughput screens and target-based rational design of synthetics have not been successful (Payne *et al.*, 2007). Almost invariably, synthetic leads fail to penetrate well into bacterial cells. This problem is especially formidable in the case of Gram-negative pathogens, equipped with powerful multidrug resistance pumps. As a result, the last several compounds to be introduced are narrow-spectrum antimicrobials that were actually discovered decades ago and were considered insufficiently attractive at the time (Lewis, 2012, 2013).

Obtaining sterilizing compounds requires rethinking the standard paradigm of finding single inhibitors of essential targets. In the following sections, we will describe recent advances in discovering anti-persister compounds.

Potentiators of antifungals

The efficacy of an antifungal agent is typically determined by MIC or MBC testing of a rapidly growing planktonic population. However, *in vivo* infections are considerably more complex, and cannot be approximated by an exponentially growing population of the pathogen (Coates and Hu, 2006). Non-growing, slow growing, or surface-attached populations of cells are considerably more tolerant to antimicrobials compared to growing cells (Coates and Hu, 2006; Lewis, 2007). Upon attachment, *Candida albicans* forms a small subpopulation (~1%) of persister cells that are tolerant to antifungals (LaFleur *et al.*, 2006), and resemble well-characterized dormant persisters formed by pathogenic bacteria. We also found clinical isolates of *C. albicans* from immunocompromised patients with oral thrush that respond poorly to therapy produced higher levels of persister cells (LaFleur *et al.*, 2010). By contrast, isolates from patients whose disease resolved within a few weeks of treatment did not show a high-persistence phenotype. Importantly, all of the isolates had the same MIC to antifungals, indicating that these are not resistant mutants. Rather, these appear to be high-persistence (*hip*) mutants. Previously, *hip* mutants were selected *in vitro* in *E. coli* by repeated application of high doses of bactericidal antibiotics and regrowing the surviving cells (Correia *et al.*, 2006; Keren *et al.*, 2004; Moyed and Bertrand, 1983). Periodic application of high doses of antifungals apparently selects for *hip* mutants of *C. albicans*. This finding provides a strong link between quiescent cells of the pathogen and the clinical manifestation of the disease.

For *Candida*, attachment to a surface is an important step in the pathogenesis of vaginitis, oral thrush, and catheter biofilm infections (De Bernardis *et al.*, 2007; Ghannoum and Radwan, 1990; Kumamoto and Vinces, 2005), and attachment has been shown to produce persister cells (LaFleur *et al.*, 2006). A biofilm formed by surface-attached cells restricts penetration of immune system components, further complicating treatment (Hoyle *et al.*, 1990; von Eiff *et al.*, 1999). Currently, high doses of topically applied miconazole (Monistat®) are used as an alternative to systemic fluconazole treatment for vaginitis and oral thrush. Fluconazole has no microbicidal activity against *C. albicans*, while miconazole is able to kill growing planktonic cells. However, even the most effective antifungals such as miconazole, amphotericin B and caspofungin are completely inactive against *C. albicans* persisters (LaFleur *et al.*, 2006), suggesting that it is unlikely to discover a single 'sterilizing' antifungal compound.

We reasoned that a molecule that disables the mechanism of formation or maintenance of quiescent cells could synergize with a conventional antimicrobial and lead to biofilm eradication. We developed a screen for potentiators of the azole class of antifungals, which is based on testing biofilms growing in a microtitre plate with the viability dye, resazurin (LaFleur *et al.*, 2011). The compound AC17, was identified as an attractive lead using this approach and while AC17 had no effect on the viability of *C. albicans*, it eradicated biofilms containing persisters when combined with miconazole (LaFleur *et al.*, 2013). To our knowledge, this is the first example of biofilm eradication by compounds that are not antiseptics. AC17 is 2-adamantanamine, and a close analogue of amantadine and rimantadine, antiviral therapeutics that inhibit the H^+ channel of influenza virus (Pinto and Lamb, 2007). AC17 did not exhibit direct activity against *C. albicans* or human cells suggesting that AC17 is not a generally toxic molecule, but rather a compound with a particular target in yeast which is absent in human cells. Indeed, our data indicate that AC17 targets the ergosterol pathway, causing fungicidal synergy with the azoles and impairing the cell membranes of persisters (LaFleur *et al.*, 2013).

Prodrugs

We considered prodrugs as a type of compounds that could kill persister cells and eradicate a chronic infection. Nitroaromatic prodrugs such as metronidazole or nitrofurantoin are benign compounds that enter into the cell and are converted into a reactive drug by nitroreductases specific to microorganisms. Since the highly activated species produced hit multiple targets, this could in principle kill both growing and dormant cells. Given that redox activated prodrugs bind covalently to their targets, this creates an irreversible sink, ensuring accumulation over time. The sink is likely to counter efflux by multidrug resistance (MDR) pumps. The dual barrier of the outer membrane and MDRs prevents most compounds from entering cells of Gram-negative bacteria (Li and Nikaido, 2004; Lomovskaya and Lewis, 1992), and is largely responsible for the paucity of broad-spectrum antibiotics. The last class of broad-spectrum compounds, the fluoroquinolones, was discovered over 50 years ago (Lewis, 2013).

It is interesting to note that all prodrug antibiotics were discovered in the 1950s (Lewis, 2013). It seems that subsequently developed validation tests based on determining specificity of action of hits precluded prodrug discovery. In an effort to eliminate generally toxic compounds, the specificity test was introduced, where the ability of a test compound to inhibit label incorporation into major biopolymers is measured (Hilliard *et al.*, 1999). Compounds that inhibit all biosynthesis are non-specific and are eliminated. By this test, metronidazole is a nuisance compound. Interestingly, metronidazole is a broad-spectrum antibiotic, but its use is limited, since nitroreductases are expressed primarily under anaerobic/microaerophilic conditions. Given the potential of prodrugs for both broad-spectrum and sterilizing activity, this class may present an attractive antibiotic discovery platform.

We decided to revisit prodrugs, and developed a screen based on the lack of specificity of mechanism of action, essentially looking for compounds that are discarded in conventional high-throughput screening (HTS) campaigns. Using the vital dye Alamar Blue, we identified hits acting against *E. coli* that inhibited reduction of this reporter of general metabolism, and then tested the compounds for cytotoxicity (Fleck *et al.*, 2013). Testing for cytotoxicity in mammalian cells eliminates generally toxic compounds. The screen successfully differentiates target-specific compounds from known prodrugs such as nitrofurantoin. A pilot HTS with 55,000 compounds produced 50 hits, and three of these with low cytotoxicity were examined. Importantly, one of the compounds, ADC111, is a nitrofurazone analogue of nitrofurantoin which is used to treat UTI (Guay, 2001; Shah and Wade, 1989). The activity of ADC111 depended on the presence of nitroreductases in *E. coli*. This discovery of a prodrug validates the screen. ADC112 is an analogue of another antimicrobial that has been used in the clinic, Tilbroquinol. The mode of action of tilbroquinol is unknown. We also identified compound ADC113 that has been previously reported to be a putative dehydratase inhibitor in mycobacteria (Bhowruth *et al.*, 2008); however, in our hands we find it to be acting as a non-specific compound, possibly as a prodrug. Taken together, the results of the pilot screen suggest that a larger HTS is likely to produce a number of novel prodrug leads. We also examined the main predictions of the prodrug model – the ability of compounds to have a broad spectrum of action, and kill dormant cells. There was little effect of MDRs on the MIC of prodrugs, in contrast to erythromycin, a model substrate of MDRs. This suggests that irreversible binding to targets may indeed impart prodrugs with good penetration properties, enabling a broad spectrum of action. Importantly, prodrugs were able to effectively kill biofilms of *E. coli* and stationary cells that produce large amounts of dormant persisters. Biofilms of *E. coli* are often associated with recurring urinary tract infection (UTI). *E. coli* forms intracellular biofilms in the bladder epithelial cells (Anderson *et al.*, 2003), and these are likely to contain persister cells. In some patients with recurring UTI, *E. coli* forms 'quiescent intracellular reservoirs' (QIRs) residing in Lamp1$^+$ endosomes of bladder epithelial cells (Mysorekar and Hultgren, 2006). QIRs are protected from the immune system and are not killed by antibiotics (Schilling and Hultgren, 2002). When bladder epithelial turnover occurs, and antibiotic treatment has ceased, viable *E. coli* re-emerge and cause a recurrence of the UTI (Silverman *et al.*, 2013). QIRs may be equivalent to persisters. Nitrofurantoin killed stationary cells much better than ciprofloxacin, and ADC111 almost completely eradicated a stationary population. An irreversibly binding compound will actually be more effective against non-growing dormant cells; in rapidly propagating bacteria, a prodrug will be diluted, diminishing the sink effect. The results of this study suggest that prodrugs are a promising type of antimicrobials capable of sterilizing and possessing broad-spectrum activity.

Protease activation by acyldepsipeptides

Another approach to killing persisters involves the activation and corruption of a target protein in a dormant persister cell. The acyldepsipeptides activate the ClpP protease, relieving it both from regulatory control and from the requirement to use ATP. Acyldepsipeptide is produced by a soil streptomycete and was discovered 29 years ago by scientists from Eli Lilly (Michel and Kastner, 1985). They evaluated the compound, found that it had modest activity against Gram-positive but not Gram-negative bacteria, and dropped it. Twenty years later, there emerged a serious need to treat infections caused by drug resistant Gram-positive pathogens. By that time, Eli Lilly was no longer in the business of antiinfectives, and Bayer picked up the acyldepsipeptides for evaluation. They produced a more active derivative of the natural product, ADEP 4 (Brotz-Oesterhelt *et al.*, 2005), determined its mechanism of action and, importantly, found that it is safe and effective in several animal models of uncomplicated disseminating infection caused by *Staphylococcus aureus* and *Enterococcus faecalis* (Brotz-Oesterhelt *et al.*, 2005). However, they found that resistance development to ADEP was very high due to null mutations in non-essential *clpP*, and the compound was dropped again. To this point, ADEP4 had been characterized solely against exponentially growing cultures. Given its mechanism of action (the energy independent activation of a protease), it was hypothesized that ADEP4 could be active against persister cells. Stationary phase *S. aureus* behave as a persister population and are completely tolerant to most conventional antibiotics. Incubation of a stationary population with 5 μg/ml ADEP 4 for 24 hours resulted in the degradation of over 400 intracellular proteins and a 4-log reduction in viable bacteria (Conlon *et al.*, 2013). Although resistant mutants occur at a frequency of 1 in 10^6 cells due to null mutation in *clpP*, it was discovered that these resistant cells are highly susceptible to various conventional antibiotics. A combination of ADEP4 with either rifampicin or linezolid led to complete eradication of a stationary phase culture in 48–72 hours (Fig. 6.2) (Conlon *et al.*, 2013). ADEP4 can also eradicate persisters formed in exponential phase that survive treatment with other antibiotics. Importantly ADEP4/rifampicin in combination can eradicate *S. aureus* biofilms *in vitro* and a deep-seated *S. aureus* biofilm infection in mice (Conlon *et al.*, 2013). Protease activation may prove to be a general mechanism of persister eradication and, apart from ADEP4, other activators of ClpP or the major intracellular bacterial protease, Lon, may be developed into therapeutics for the killing of persister cells and eradication of infection.

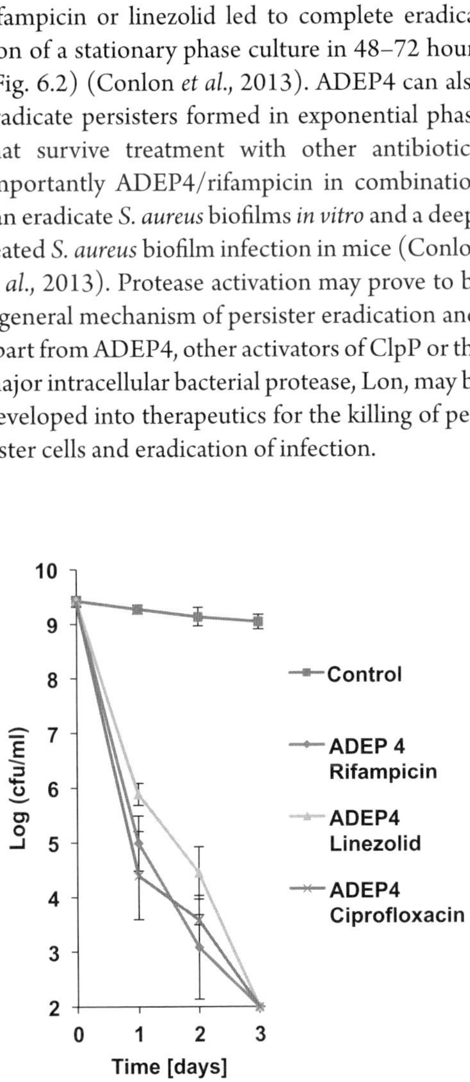

Figure 6.2 ADEP combinations eradicate stationary phase *S. aureus* (Conlon *et al.*, 2013).

Uncultured bacteria and lassomycin

Currently available antibiotics were developed to inhibit growth of rapidly growing cells. This is indeed the sufficient property that the FDA requires; an ability to kill, to say nothing of killing dormant forms of pathogens, is not a requirement. The need to kill dormant pathogens has been considered fairly recently, and we do not have a long tradition of searching for killer compounds. At the same time, antibiotic producers such as streptomycetes encounter a very similar problem we have with our pathogens: antibiotic-tolerant persisters. Competitors of streptomycetes produce persisters, and ADEP provides the first example of a natural anti-persister compound. It is reasonable to expect that a variety of compounds capable of killing persisters have evolved. However, natural product discovery in general has been in decline, due to rediscovery of known compounds from this limited resource – cultivable streptomycetes (Lewis, 2013). Using new microbial sources of compounds could reduce the problem of rediscovery and lead to novel antimicrobials. Uncultured species of bacteria account for 99% of all microbial diversity and by definition represent an unexploited source of secondary metabolites. General methods to grow previously uncultured bacteria, based on cultivation in diffusion chambers in their natural environments (Kaeberlein *et al.*, 2002; Lewis *et al.*, 2010) and on prolonged incubation *in vitro* (Buerger *et al.*, 2012a,b) have been developed. However, even with this new source, most of the effort is expended on rediscovery of known compounds or generally toxic ones. We reasoned that this problem could be addressed by a species-selective approach, whereby compounds with broad antibiotic spectra are eliminated, and only compounds active against a specific species are considered. We chose *M. tuberculosis* as a target organism for this approach, since few natural products are known to act specifically against this pathogen, and therefore most of the specific hits obtained should be new agents. There is also a considerable medical need for novel anti-TB compounds (Lechartier *et al.*, 2014) to stem the spread of extremely and totally drug-resistant strains of the pathogen. We screened extracts from a collection of soil bacteria, obtained by *in situ* cultivation and by prolonged incubation, against

M. tuberculosis and counter-screened against *S. aureus*. A novel antimicrobial, lassomycin, was purified from an extract of *Lentzea kentuckyensis* sp. Lassomycin was found to specifically target mycobacteria, and resistant mutants were affected in the C1 ATPase of the ClpP1P2C1 protease that is unique to mycobacteria, and essential (Akopian *et al.*, 2012). This agent is a highly basic lasso peptide antibiotic that is encoded in the genome. Lassomycin inhibited the protease while increasing its ATPase activity (Gavrish *et al.*, 2014). Lassomycin is highly bactericidial and, unlike conventional antibiotics, eliminated *M. tuberculosis* persisters over time. The experience with ADEP and lassomycin strongly suggest that additional natural product antibiotics capable of killing persisters are yet to be discovered.

References

Akopian, T., Kandror, O., Raju, R.M., Unnikrishnan, M., Rubin, E.J., and Goldberg, A.L. (2012). The active ClpP protease from *M. tuberculosis* is a complex composed of a heptameric ClpP1 and a ClpP2 ring. EMBO J. *31*, 1529–1541.

Alix, E., and Blanc-Potard, A. (2009). Hydrophobic peptides: novel regulators within bacterial membranes. Mol. Microbiol. *72*, 5–11.

Anderson, G.G., Palermo, J.J., Schilling, J.D., Roth, R., Heuser, J., and Hultgren, S.J. (2003). Intracellular bacterial biofilm-like pods in urinary tract infections. Science *301*, 105-107.

Baba, T., Ara, T., Hasegawa, M., Takai, Y., Okumura, Y., Baba, M., Datsenko, K.A., Tomita, M., Wanner, B.L., and Mori, H. (2006). Construction of *Escherichia coli* K-12 in-frame, single-gene knockout mutants: the Keio collection. Mol. Syst. Biol. *2*, 2006.0008.

Barry, C.E., 3rd, Boshoff, H.I., Dartois, V., Dick, T., Ehrt, S., Flynn, J., Schnappinger, D., Wilkinson, R.J., and Young, D. (2009). The spectrum of latent tuberculosis: rethinking the biology and intervention strategies. Nat. Rev. Microbiol. *7*, 845–855.

Bhowruth, V., Brown, A.K., and Besra, G.S. (2008). Synthesis and biological evaluation of NAS-21 and NAS-91 analogues as potential inhibitors of the mycobacterial FAS-II dehydratase enzyme Rv0636. Microbiology *154*, 1866–1875.

Bigger, J.W. (1944). Treatment of staphylococcal infections with penicillin. Lancet *244*, 497–500.

Brotz-Oesterhelt, H., Beyer, D., Kroll, H.P., Endermann, R., Ladel, C., Schroeder, W., Hinzen, B., Raddatz, S., Paulsen, H., Henninger, K., *et al.* (2005). Dysregulation of bacterial proteolytic machinery by a new class of antibiotics. Nat. Med. *11*, 1082–1087.

Buerger, S., Spoering, A., Gavrish, E., Leslin, C., Ling, L., and Epstein, S.S. (2012a). Microbial scout hypothesis and microbial discovery. Appl. Environm. Microbiol. *78*, 3229–3233.

Buerger, S., Spoering, A., Gavrish, E., Leslin, C., Ling, L., and Epstein, S.S. (2012b). Microbial scout hypothesis, stochastic exit from dormancy, and the nature of slow growers. Appl. Environm. Microbiol. *78*, 3221–3228.

Christensen, S.K., and Gerdes, K. (2003). RelE toxins from bacteria and Archaea cleave mRNAs on translating ribosomes, which are rescued by tmRNA. Mol. Microbiol. *48*, 1389–1400.

Coates, A.R., and Hu, Y. (2006). New strategies for antibacterial drug design: targeting non-multiplying latent bacteria. Drugs Res. Devel. *7*, 133–151.

Conlon, B.P., Nakayasu, E.S., Fleck, L.E., LaFleur, M.D., Isabella, V.M., Coleman, K., Leonard, S.N., Smith, R.D., Adkins, J.N., and Lewis, K. (2013). Activated ClpP kills persisters and eradicates a chronic biofilm infection. Nature *503*, 365–370.

Correia, F.F., D'Onofrio, A., Rejtar, T., Li, L., Karger, B.L., Makarova, K., Koonin, E.V., and Lewis, K. (2006). Kinase activity of overexpressed HipA is required for growth arrest and multidrug tolerance in *Escherichia coli*. J. Bacteriol. *188*, 8360–8367.

Costerton, J.W., Stewart, P.S., and Greenberg, E.P. (1999). Bacterial biofilms: a common cause of persistent infections. Science *284*, 1318–1322.

Courcelle, J., Khodursky, A., Peter, B., Brown, P., and Hanawalt, P. (2001). Comparative gene expression profiles following UV exposure in wild type and SOS-deficient *Escherichia coli*. Genetics *158*, 41–64.

Davis, B.D., Chen, L.L., and Tai, P.C. (1986). Misread protein creates membrane channels: an essential step in the bactericidal action of aminoglycosides. Proc. Natl. Acad. Sci. U.S.A. *83*, 6164–6168.

De Bernardis, F., Liu, H., O'Mahony, R., La Valle, R., Bartollino, S., Sandini, S., Grant, S., Brewis, N., Tomlinson, I., Basset, R.C., *et al.* (2007). Human domain antibodies against virulence traits of *Candida albicans* inhibit fungus adherence to vaginal epithelium and protect against experimental vaginal candidiasis. J. Infect. Dis. *195*, 149–157.

De Groote, V., Verstraeten, N., Fauvart, M., Kint, C., Verbeeck, A., Beullens, S., Cornelis, P., and Michiels, J. (2009). Novel persistence genes in *Pseudomonas aeruginosa* identified by high-throughput screening. FEMS Microbiol. Lett. *297*, 73–79.

Del Pozo, J., and Patel, R. (2007). The challenge of treating biofilm-associated bacterial infections. Clin. Pharmacol. Ther. *82*, 204–209.

Dorr, T., Lewis, K., and Vulic, M. (2009). SOS response induces persistence to fluoroquinolones in *Escherichia coli*. PLoS Genet. *5*, e1000760.

Dorr, T., Vulic, M., and Lewis, K. (2010). Ciprofloxacin causes persister formation by inducing the TisB toxin in *Escherichia coli*. PLoS Biol. *8*, e1000317.

von Eiff, C., Heilmann, C., and Peters, G. (1999). New aspects in the molecular basis of polymer-associated infections due to staphylococci. Eur. J. Clin. Microbiol. Infect. Dis. *18*, 843–846.

Ezraty, B., Vergnes, A., Banzhaf, M., Duverger, Y., Huguenot, A., Brochado, A.R., Su, S.Y., Espinosa, L., Loiseau, L., Py, B., *et al.* (2013). Fe–S cluster biosynthesis controls uptake of aminoglycosides in a ROS-less death pathway. Science *340*, 1583–1587.

Falla, T.J., and Chopra, I. (1998). Joint tolerance to β-lactam and fluoroquinolone antibiotics in *Escherichia coli* results from overexpression of *hipA*. Antimicrob. Agents Chemother. *42*, 3282–3284.

Fernandez De Henestrosa, A.R., Ogi, T., Aoyagi, S., Chafin, D., Hayes, J.J., Ohmori, H., and Woodgate, R. (2000). Identification of additional genes belonging to the LexA regulon in *Escherichia coli*. Mol. Microbiol. *35*, 1560–1572.

Fleck, L.E., North, E.J., Lee, R.E., Mulcahy, L.R., Casadei, G., and Lewis, K. (2013). A screen and validation of prodrug antimicrobials. Antimicrob. Agents Chemother. *58*, 1410–1419.

Garcia-Olmedo, F., Molina, A., Alamillo, J.M., and Rodriguez-Palenzuela, P. (1998). Plant defense peptides. Biopolymers *47*, 479–491.

Gavrish, E., Sit, S., Cao, S., Kandror, O., Spoering, A., Peoples, A., Ling, L., Fetterman, A., Hughes, D., Bissell, A., *et al.* (2014). Lassomycin, a ribosomally synthesized cyclic peptide, kills *Mycobacterium tuberculosis* by targeting the ATP-dependent protease ClpC1P1P2. ACS Chem. Biol. *21*, 509–518.

Gerdes, K., Bech, F.W., Jorgensen, S.T., Lobner-Olesen, A., Rasmussen, P.B., Atlung, T., Boe, L., Karlstrom, O., Molin, S., and von Meyenburg, K. (1986a). Mechanism of postsegregational killing by the *hok* gene product of the *parB* system of plasmid R1 and its homology with the *relF* gene product of the *E. coli relB* operon. EMBO J. *5*, 2023–2029.

Gerdes, K., Rasmussen, P.B., and Molin, S. (1986b). Unique type of plasmid maintenance function: postsegregational killing of plasmid-free cells. Proc. Natl. Acad. Sci. U.S.A. *83*, 3116–3120.

Germain, E., Castro-Roa, D., Zenkin, N., and Gerdes, K. (2013). Molecular mechanism of bacterial persistence by HipA. Mol. Cell *52*, 248–254.

Ghannoum, M.A., and Radwan, S.S. (1990). *Candida* Adherence to Epithelial Cells (CRC Press, Boca Raton, FL).

Gibson, R.L., Burns, J.L., and Ramsey, B.W. (2003). Pathophysiology and management of pulmonary infections in cystic fibrosis. Am. J. Respir. Crit. Care Med. *168*, 918–951.

Groicher, K.H., Firek, B.A., Fujimoto, D.F., and Bayles, K.W. (2000). The *Staphylococcus aureus lrgAB* operon modulates murein hydrolase activity and penicillin tolerance. J. Bacteriol. *182*, 1794–1801.

Guay, D.R. (2001). An update on the role of nitrofurans in the management of urinary tract infections. Drugs *61*, 353–364.

Gurnev, P.A., Ortenberg, R., Dorr, T., Lewis, K., and Bezrukov, S.M. (2012). Persister-promoting bacterial toxin TisB produces anion-selective pores in planar lipid bilayers. FEBS Lett. *586*, 2529–2534.

Hansen, S., Lewis, K., and Vulic, M. (2008). Role of global regulators and nucleotide metabolism in antibiotic tolerance in *Escherichia coli*. Antimicrob. Agents Chemother. *52*, 2718–2726.

Harrison, J.J., Ceri, H., Roper, N.J., Badry, E.A., Sproule, K.M., and Turner, R.J. (2005a). Persister cells mediate tolerance to metal oxyanions in *Escherichia coli*. Microbiology *151*, 3181–3195.

Harrison, J.J., Turner, R.J., and Ceri, H. (2005b). Persister cells, the biofilm matrix and tolerance to metal cations in biofilm and planktonic *Pseudomonas aeruginosa*. Environ. Microbiol. *7*, 981–994.

Harrison, J.J., Wade, W.D., Akierman, S., Vacchi-Suzzi, C., Stremick, C.A., Turner, R.J., and Ceri, H. (2009). The chromosomal toxin gene yafQ is a determinant of multidrug tolerance for *Escherichia coli* growing in a biofilm. Antimicrob. Agents Chemother. *53*, 2253–2258.

Hayes, F. (2003). Toxins-antitoxins: plasmid maintenance, programmed cell death, and cell cycle arrest. Science *301*, 1496–1499.

Helaine, S., Cheverton, A.M., Watson, K.G., Faure, L.M., Matthews, S.A., and Holden, D.W. (2014). Internalization of *Salmonella* by macrophages induces formation of nonreplicating persisters. Science *343*, 204–208.

Hilliard, J.J., Goldschmidt, R.M., Licata, L., Baum, E.Z., and Bush, K. (1999). Multiple mechanisms of action for inhibitors of histidine protein kinases from bacterial two-component systems. Antimicrob. Agents Chemother. *43*, 1693–1699.

Honda, H., and Warren, D.K. (2009). Central nervous system infections: meningitis and brain abscess. Infect. Dis. Clin. North Am. *23*, 609–623.

Hooper, D.C. (2001). Mechanisms of action of antimicrobials: focus on fluoroquinolones. Clin. Infect. Dis. *15*, S9–S15.

Hoyle, B.D., Jass, J., and Costerton, J.W. (1990). The biofilm glycocalyx as a resistance factor. J. Antimicrob. Chemother. *26*, 1–5.

Hu, Y., and Coates, A.R. (2005). Transposon mutagenesis identifies genes which control antimicrobial drug tolerance in stationary-phase *Escherichia coli*. FEMS Microbiol. Lett. *243*, 117–124.

Jesaitis, A.J., Franklin, M.J., Berglund, D., Sasaki, M., Lord, C.I., Bleazard, J.B., Duffy, J.E., Beyenal, H., and Lewandowski, Z. (2003). Compromised host defense on *Pseudomonas aeruginosa* biofilms: characterization of neutrophil and biofilm interactions. J. Immunol. *171*, 4329–4339.

Kaeberlein, T., Lewis, K., and Epstein, S.S. (2002). Isolating 'uncultivable' microorganisms in pure culture in a simulated natural environment. Science *296*, 1127–1129.

Kaspy, I., Rotem, E., Weiss, N., Ronin, I., Balaban, N.Q., and Glaser, G. (2013). HipA-mediated antibiotic persistence via phosphorylation of the glutamyl-tRNA-synthetase. Nat. Commun. *4*, 3001.

Kawano, M., Aravind, L., and Storz, G. (2007). An antisense RNA controls synthesis of an SOS-induced toxin evolved from an antitoxin. Mol. Microbiol. *64*, 738–754.

Keren, I., Shah, D., Spoering, A., Kaldalu, N., and Lewis, K. (2004). Specialized persister cells and the mechanism of multidrug tolerance in *Escherichia coli*. J. Bacteriol. *186*, 8172–8180.

Keren, I., Wu, Y., Inocencio, J., Mulcahy, L.R., and Lewis, K. (2013). Killing by bactericidal antibiotics does not depend on reactive oxygen species. Science 339, 1213–1216.

Kohanski, M.A., Dwyer, D.J., Hayete, B., Lawrence, C.A., and Collins, J.J. (2007). A common mechanism of cellular death induced by bactericidal antibiotics. Cell *130*, 797–810.

Korch, S.B., and Hill, T.M. (2006). Ectopic overexpression of wild-type and mutant *hipA* genes in *Escherichia coli*: effects on macromolecular synthesis and persister formation. J. Bacteriol. *188*, 3826–3836.

Kumamoto, C.A., and Vinces, M.D. (2005). Alternative *Candida albicans* lifestyles: growth on surfaces. Annu. Rev. Microbiol. *59*, 113–133.

LaFleur, M.D., Kumamoto, C.A., and Lewis, K. (2006). *Candida albicans* biofilms produce antifungal-tolerant persister cells. Antimicrob. Agents Chemother. *50*, 3839–3846.

LaFleur, M.D., Qi, Q., and Lewis, K. (2010). Patients with long-term oral carriage harbor high-persister mutants of *Candida albicans*. Antimicrob. Agents Chemother. *54*, 39–44.

LaFleur, M.D., Lucumi, E., Napper, A.D., Diamond, S.L., and Lewis, K. (2011). Novel high-throughput screen against *Candida albicans* identifies antifungal potentiators and agents effective against biofilms. J. Antimicrob. Chemother. *66*, 820–826.

LaFleur, M.D., Sun, L., Lister, I., Keating, J., Nantel, A., Long, L., Ghannoum, M., North, J., Lee, R.E., Coleman, K., *et al.* (2013). Potentiation of azole antifungals by 2-adamantanamine. Antimicrob. Agents Chemother. *57*, 3585–3592.

Lechartier, B., Rybniker, J., Zumla, A., and Cole, S.T. (2014). Tuberculosis drug discovery in the post-post-genomic era. EMBO Mol. Med. 6(8), 158–168.

Leid, J.G., Shirtliff, M.E., Costerton, J.W., and Stoodley, A.P. (2002). Human leukocytes adhere to, penetrate, and respond to *Staphylococcus aureus* biofilms. Infect Immun *70*, 6339–6345.

Levin, B.R., and Rozen, D.E. (2006). Non-inherited antibiotic resistance. Nat. Rev. Microbiol. *4*, 556–562.

Lewis, K. (2001). Riddle of biofilm resistance. Antimicrob. Agents Chemother. *45*, 999–1007.

Lewis, K. (2007). Persister cells, dormancy and infectious disease. Nat. Rev. Microbiol. *5*, 48–56.

Lewis, K. (2012). Antibiotics: Recover the lost art of drug discovery. Nature *485*, 439–440.

Lewis, K. (2013). Platforms for antibiotic discovery. Nat. Rev. Drug Discov. *12*, 371–387.

Lewis, K., Epstein, S., D'Onofrio, A., and Ling, L.L. (2010). Uncultured microorganisms as a source of secondary metabolites. J. Antibiot. (Tokyo) *63*, 468–476.

Li, X.Z., and Nikaido, H. (2004). Efflux-mediated drug resistance in bacteria. Drugs *64*, 159–204.

Liu, Y., and Imlay, J.A. (2013). Cell death from antibiotics without the involvement of reactive oxygen species. Science *339*, 1210–1213.

Lomovskaya, O., and Lewis, K. (1992). Emr, an *Escherichia coli* locus for multidrug resistance. Proc. Natl. Acad. Sci. U.S.A. *89*, 8938–8942.

McKenzie, G.J., Magner, D.B., Lee, P.L., and Rosenberg, S.M. (2003). The dinB operon and spontaneous mutation in *Escherichia coli*. J. Bacteriol. *185*, 3972–3977.

Mahoney, T.F., and Silhavy, T.J. (2013). The Cpx stress response confers resistance to some, but not all, bactericidal antibiotics. J. Bacteriol. *195*, 1869–1874.

Maisonneuve, E., Shakespeare, L.J., Jorgensen, M.G., and Gerdes, K. (2011). Bacterial persistence by RNA endonucleases. Proc. Natl. Acad. Sci. U.S.A. *108*, 13206–13211.

Michel, K.H., and Kastner, R.E. (1985). A54556 antibiotics and process for production thereof. US patent 4492650.

Motiejunaite, R., Armalyte, J., Markuckas, A., and Suziedeliene, E. (2007). *Escherichia coli dinJ-yafQ* genes act as a toxin–antitoxin module. FEMS Microbiol. Lett. *268*, 112–119.

Moyed, H.S., and Bertrand, K.P. (1983). *hipA*, a newly recognized gene of *Escherichia coli* K-12 that affects frequency of persistence after inhibition of murein synthesis. J. Bacteriol. *155*, 768–775.

Mulcahy, L.R., Burns, J.L., Lory, S., and Lewis, K. (2010). Emergence of *Pseudomonas aeruginosa* strains producing high levels of persister cells in patients with cystic fibrosis. J. Bacteriol. *192*, 6191–6199.

Mysorekar, I.U., and Hultgren, S.J. (2006). Mechanisms of uropathogenic *Escherichia coli* persistence and eradication from the urinary tract. Proc. Natl. Acad. Sci. U.S.A. *103*, 14170–14175.

Pandey, D.P., and Gerdes, K. (2005). Toxin–antitoxin loci are highly abundant in free-living but lost from host-associated prokaryotes. Nucleic Acids Res. *33*, 966–976.

Payne, D.J., Gwynn, M.N., Holmes, D.J., and Pompliano, D.L. (2007). Drugs for bad bugs: confronting the challenges of antibacterial discovery. Nature reviews *6*, 29–40.

Pedersen, K., and Gerdes, K. (1999). Multiple *hok* genes on the chromosome of *Escherichia coli*. Mol. Microbiol. *32*, 1090–1102.

Pedersen, K., Zavialov, A.V., Pavlov, M.Y., Elf, J., Gerdes, K., and Ehrenberg, M. (2003). The bacterial toxin RelE displays codon-specific cleavage of mRNAs in the ribosomal A site. Cell *112*, 131–140.

Peterson, W.L., Fendrick, A.M., Cave, D.R., Peura, D.A., Garabedian-Ruffalo, S.M., and Laine, L. (2000). Helicobacter pylori-related disease: guidelines for testing and treatment. Arch. Intern. Med. *160*, 1285–1291.

Phillips, I., Culebras, E., Moreno, F., and Baquero, F. (1987). Induction of the SOS response by new 4-quinolones. J. Antimicrob. Chemother. *20*, 631–638.

Pinto, L.H., and Lamb, R.A. (2007). Controlling influenza virus replication by inhibiting its proton channel. Mole. BioSyst. 3, 18–23.

Ramage, H.R., Connolly, L.E., and Cox, J.S. (2009). Comprehensive functional analysis of Mycobacterium tuberculosis toxin–antitoxin systems: implications for pathogenesis, stress responses, and evolution. PLoS Genet. *5*, e1000767.

Renggli, S., Keck, W., Jenal, U., and Ritz, D. (2013). Role of autofluorescence in flow cytometric analysis of *Escherichia coli* treated with bactericidal antibiotics. J. Bacteriol. *195*, 4067–4073.

Sahl, H.G., and Bierbaum, G. (1998). Lantibiotics: biosynthesis and biological activities of uniquely modified peptides from Gram-positive bacteria. Annu. Rev. Microbiol.

Schilling, J.D., and Hultgren, S.J. (2002). Recent advances into the pathogenesis of recurrent urinary tract infections: the bladder as a reservoir for uropathogenic *Escherichia coli*. Int. J. Antimicrob. Agents *19*, 457–460.

Schmelzle, T., and Hall, M.N. (2000). TOR, a central controller of cell growth. Cell *103*, 253–262.

Schumacher, M.A., Piro, K.M., Xu, W., Hansen, S., Lewis, K., and Brennan, R.G. (2009). Molecular mechanisms of HipA-mediated multidrug tolerance and its neutralization by HipB. Science *323*, 396–401.

Shah, D., Zhang, Z., Khodursky, A., Kaldalu, N., Kurg, K., and Lewis, K. (2006). Persisters: A distinct physiological state of *E. coli*. BMC Microbiol. *6*, 53–61.

Shah, R.R., and Wade, G. (1989). Reappraisal of the risk/benefit of nitrofurantoin: review of toxicity and efficacy. Adverse Drug React. Acute Poisoning Rev. *8*, 183–201.

Silverman, J.A., Schreiber, H.L.t., Hooton, T.M., and Hultgren, S.J. (2013). From physiology to pharmacy: developments in the pathogenesis and treatment of recurrent urinary tract infections. Curr. Urol. Rep. *14*, 448–456.

Singletary, L.A., Gibson, J.L., Tanner, E.J., McKenzie, G.J., Lee, P.L., Gonzalez, C., and Rosenberg, S.M. (2009). An SOS-regulated type 2 toxin–antitoxin system. J. Bacteriol. *191*, 7456–7465.

Smith, E.E., Buckley, D.G., Wu, Z., Saenphimmachak, C., Hoffman, L.R., D'Argenio, D.A., Miller, S.I., Ramsey, B.W., Speert, D.P., Moskowitz, S.M., *et al.* (2006). Genetic adaptation by *Pseudomonas aeruginosa* to the airways of cystic fibrosis patients. Proc. Natl. Acad. Sci. U.S.A. *103*, 8487–8492.

Spoering, A.L., and Lewis, K. (2001). Biofilms and planktonic cells of *Pseudomonas aeruginosa* have similar resistance to killing by antimicrobials. J. Bacteriol. *183*, 6746–6751.

Spoering, A.L., Vulic, M., and Lewis, K. (2006). GlpD and PlsB participate in persister cell formation in *Escherichia coli*. J. Bacteriol. *188*, 5136–5144.

Stewart, P.S., and Costerton, J.W. (2001). Antibiotic resistance of bacteria in biofilms. Lancet *358*, 135–138.

Unoson, C., and Wagner, E. (2008). A small SOS-induced toxin is targeted against the inner membrane in *Escherichia coli*. Mol. Microbiol. *70*, 258–270.

Vazquez-Laslop, N., Lee, H., and Neyfakh, A.A. (2006). Increased persistence in *Escherichia coli* caused by controlled expression of toxins or other unrelated proteins. J. Bacteriol. *188*, 3494–3497.

Vogel, J., Argaman, L., Wagner, E.G., and Altuvia, S. (2004). The small RNA IstR inhibits synthesis of an SOS-induced toxic peptide. Curr. Biol. *14*, 2271–2276.

Vuong, C., Voyich, J.M., Fischer, E.R., Braughton, K.R., Whitney, A.R., DeLeo, F.R., and Otto, M. (2004). Polysaccharide intercellular adhesin (PIA) protects *Staphylococcus epidermidis* against major components of the human innate immune system. Cell Microbiol. *6*, 269–275.

Waksman, S.A., and Woodruff, H.B. (1940). The soil as a source of microorganisms antagonistic to disease-producing bacteria. J. Bacteriol. *40*, 581–600.

Walters, M.C., 3rd, Roe, F., Bugnicourt, A., Franklin, M.J., and Stewart, P.S. (2003). Contributions of antibiotic penetration, oxygen limitation, and low metabolic activity to tolerance of *Pseudomonas aeruginosa* biofilms to ciprofloxacin and tobramycin. Antimicrob. Agents Chemother. *47*, 317–323.

Wolfson, J.S., Hooper, D.C., McHugh, G.L., Bozza, M.A., and Swartz, M.N. (1990). Mutants of *Escherichia coli* K-12 exhibiting reduced killing by both quinolone and β-lactam antimicrobial agents. Antimicrob. Agents Chemother. *34*, 1938–1943.

Zasloff, M. (2002). Antimicrobial peptides in health and disease. N. Eng. J. Med. *347*, 1199–1200.

Toxicity of Antibacterial Drugs

7

Steven J. Projan

Abstract

Therapeutic antibacterial drugs are considered among the safest of pharmaceuticals but this was not always the case. Indeed prior to the discovery of penicillin and, subsequently, other antibiotics, the safety profile of antibacterial drugs more closely resembled that of today's cytotoxic, chemotherapeutic agents used in oncology with narrow therapeutic windows and considerable side-effects. Today's antibiotics are, in fact, safe by design. Where agents have defined toxicities (e.g. photosensitivity induced by fluoroquinolones) they are usually class effects and, rather than 'idiosyncratic' are more frequently predictable based on pharmacokinetics and tissue distribution. As newer antibacterial drugs are being designed to be more 'pathogen specific' the expectation is that these future drugs will have even better safety profiles than today's therapeutics.

The common perception of antibacterial drugs is that they are very safe, very useful and very important therapeutic agents, indeed the safety and tolerability of antibacterials are, in general are among the best of all prescribed drugs. This is not a matter of chance and it was not always the case. In 1908 Paul Ehrlich shared the Novel Prize for Medicine or Physiology (with Ilya Metchnikoff) for his work on immunotherapeutics (indeed the first Nobel Prize in that category went to Emil von Behring, also for immunotherapy) but Ehrlich was unhappy with the efficacy of immunotherapeutics as well as their safety (e.g. 'serum sickness'). He embarked on a bona fide medicinal chemistry/drug discovery campaign to find novel organic molecules based on arsenical compounds that were active against trypanosomes and spirochaetes. It should be noted that at the turn of the twentieth century mercury was still being used as treatment for syphilis, a bacterial infection caused by *Treponema pallidum*. Even today, mercury toxicity remains an important issue. The compound that Ehrlich's group eventually discovered, arsphenamine ('606') became known as Salvarsan and was both more effective and far safer than mercury treatment (not to mention less expensive) but was not without its own safety and tolerability issues (which are a subject of controversy even today) (Baumler,1984). Eventually arsphenamine was supplanted by penicillin, where even a single, intramuscular dose was found to be effective in the early stages of the disease with a significantly better safety profile. It should be realized that today's antibacterial drugs are safe not as a matter of chance but by design; the result of decades of microbial research, the development of *in vitro* and animal models of infection, brute force science and inventive medicinal chemistry, and careful (as well as some careless) clinical research resulting in the successful treatment of hundreds of millions (if not billions) of people.

Class effects

Our current antibiotics have a relatively small number of molecular targets in bacteria. For example, the fluoroquinolones target a pair of related enzymes (gyrase and topoisomerase IV which bacteria require to maintain DNA superhelicity and resolve daughter chromosomes post DNA replication). Many compounds in this family are, to various degrees, phototoxic. Unfortunately, this side-effect appeared to track with the relative potency of these agents (e.g. sparfloxacin, clinafloxacin). The most commonly used of the fluoroquinolones has been ciprofloxacin and during the prolonged therapy in potentially exposed people following the mailing of letters with anthrax spores in 2001 many patients experienced tendinopathies including ruptured Achilles tendons. In neither the case of phototoxicity nor tendinopathies is it all clear that the toxicity is direct function of the antibiotics' mechanism of action, rather these may well represent phenomenon related to pharmacokinetics and tissue distribution (Lipsky and Baker, 1999). In contrast, the aminoglycosides (e.g. gentamycin, tobramycin) have toxicity that appears to be directly related to their mechanism of action (targeting the bacterial ribosome and inhibiting protein synthesis/mis-incorporation of amino acids into polypeptides). Indeed there are those who appear to be hypersusceptible to aminoglycoside ototoxicity based on maternally inherited genetic factors (implicating the mitochondrion as intracellular target organelle). Since mitochondria have their own protein synthesis machinery, including ribosomes which more closely resemble those of bacteria than mammals this logic would appear to hold. The case of nephrotoxicity, on the other hand, appears to be a function of pharmacokinetics and tissue distribution. Aminoglycosides bind to anion phospholipids found in the plasma membrane of proximal tubule cells (PTC) and those aminoglycosides with highest binding affinity for PTCs have the greatest degree of nephrotoxicity (Lopez-Novoa *et al.*, 2011). It is remarkable that some 70 years after their discovery we are still developing insights as to the mechanism of toxicity. This could lead one to hypothesize that all protein synthesis inhibitors that function by binding to the bacterial ribosome (which represent the large majority of protein synthesis inhibitors) would share similar toxicities. However, this is not the case. Most antibacterial drugs, such as the macrolides (erythromycin, azithromycin) or the tetracyclines (doxycycline, minocycline), have been shown to inhibit mitochondrial protein synthesis at concentrations several orders of magnitude higher than bacterial protein synthesis.

Perhaps the most important toxicity/tolerability issue for a class of antibacterial agents is allergy/anaphylaxis in response to -lactam antibiotics exemplified by the progenitor of the family, penicillin. It has been estimated that the rate of anaphylactoid reactions is about 0.01% but given that the β-lactams are the most used of antibiotics this still represents a considerable number of patients. Rates of cross-reactivity between different β-lactam drugs vary but are in the 5–25% range with the exception of the monobactam aztreonam where no cross reactivity has been observed. Indeed virtually all antibiotics can give rise to allergic/hypersensitivity reactions, they all can function as haptens. The β-lactams, which inhibit bacterial cell wall biosynthesis and can induce bacterial autolysis through binding to well-conserved cell wall biosynthetic enzymes, are something of a special case. By virtue of their structure the β-lactams would appear to as a peptide mimetic (D-ala-D-ala) and their peptidic nature may well increase their ability to elicit an immune response in people. It is important to point out that not all members of antibiotic family will necessarily share the same toxicity/tolerability issues a good example of this are the tetracyclines minocycline and doxycycline. Minocycline has been shown to have some vestibular effects leading to dizziness while this has not been observed with doxycycline, common to all members of the family, though is binding to bone which can lead to discoloration of the teeth if administered to young children (which is why tetracyclines tend not to be used in children below the age of eight. Given that tetracyclines are avid chelators of divalent cations (such as calcium) this is not surprising and demonstrates once again the role of pharmacokinetics and tissue distribution in toxicities to antibiotics.

Has genomics been misleading?

One of the dogmas of discovering and developing safer antibacterial drugs is that if there is any similarity at between mammalian and bacterial targets at the DNA, RNA and especially protein level then there will be safety issues with drugs directed towards those similar targets (Payne *et al.*, 2007). But as Sportin' Life said in Gershwin's opera *Porgy and Bess* 'It ain't necessarily so'. Indeed the protein sequences for DHPS and DHFR, the targets for the sulfa drugs and trimethoprim are quite similar. Yet targeting the folic acid biosynthetic pathway in bacteria has been very successful therapeutically (and safety-wise as well) (Projan, 2002).

Collateral damage

Perhaps the most common adverse events associated with antibacterial drugs are gastrointestinal in nature. Anything goes from mild nausea to severe vomiting, from flatulence to diarrhoea. Given the broad spectrum (and high potency) of the large majority of the antibiotics used in today's medicine it comes as little surprise that these drugs can and do disrupt the balance of our own microflora, especially, but not limited to those that reside within our gastrointestinal tracts. The emergence of *Clostridium difficile* as an important nosocomial pathogen which gives rise to severe, toxin-mediated diarrhoea (and pseudomembranous colitis), is almost always directly associate with antibiotic use. Indeed one might term *C. difficile* infection and associate diarrhoea an 'iatrogenic infection'. As we learn more about our microbiomes and their importance in the maintenance of health we may (must?) more closely examine and understand the role that antibacterial treatment has on our fellow travellers. And should we consider disruption of the microbiome a safety/tolerability issue?

What is 'safe enough'?

As we have seen from the history of the treatment of syphilis (and with everything else in drug discovery and development) safety is a relative, not an absolute concept. There is not a single antibacterial drug that does not have a long list of potential side-effects, toxicities and tolerability issues and, stated above these among the safest of the drugs that we use and there are used routinely in every health care setting. When telithromycin was being developed in order to achieve regulatory approval in the United States the sponsor conducted the largest safety trial ever performed for an antibacterial drug. Indeed the safety profile that emerged appeared better than for the comparator in the study: amoxicillin-clavulanate and this continued to be the case after over 10 million patients received the antibiotic. There were cases of severe liver toxicity observed but not necessarily at an alarming rate and perhaps in a predictable manner however the drug had attracted some very vocal detractors (none of whom were expert in antibacterial agents) and telithromycin ceased to be a therapeutic option. An analysis by two experts, David Shlaes and Robert Moellering (2008), showed that telithromycin, while not having the best safety profile of commonly used antibiotics it was certainly far from the worst (and frankly according to Shlaes and Moellering's analysis safer than penicillin).

Into the future: prophylaxis and precision medicine

The advent of the 'age of antibiotics' in the 1940s had a major impact on extending lives but it also brought with it a false sense of security and led to an overenthusiastic use of antibiotics, sometimes as a substitute for good sanitation and hygiene. We now know that resistance to these important agents can and does arise and this may have dire consequences for patients. Finding new ways to prevent and treat infections caused by resistant organisms will be a major challenge in the twenty-first century. Identifying 'at risk' patients (e.g. those colonized with *Staphylococcus aureus*) and taking pre-emptive measures (e.g. decolonization) or perhaps developing a protective vaccine may well obviate the use of the antibiotics that have made the medicine of the late twentieth and early twenty-first century possible. A new term of art has been coined 'Precision Medicine' and we now have available (although not routinely employed) the ability to identifying an infecting pathogen within two hours (if not less). This should allow for the use of narrow spectrum agents which, in and of themselves cannot obviate either toxicity or the development of resistance but may well prevent cross resistance to current antibiotics as well ameliorate, in for not prevent, horizontal gene transfer. In addition to the continued hunt for effective vaccines, currently there are multiple monoclonal antibodies in clinical development to prevent and/or treat infections caused by *Pseudomonas aeruginosa*, *Staphylococcus aureus*, *Clostridium difficile* and others. To date these monoclonal antibodies, mainly of human origin or design, have had an enviable safety record (as may well be predicted) and these may well represent the future of antibacterial treatments.

Conclusions

It is hard to discuss the safety of antibacterial agents without using the word 'idiosyncratic' but in case after case of observed toxicity there is almost always a scientific rationale for what has been observed and often enough it is clear that a better understanding of the pharmacokinetics and tissue distribution of the drugs would allow for more accurate prediction of toxicity and tolerability issues. And, indeed, most of these issues were 'discoverable' preclinically. In the twenty-first century it is no longer acceptable to 'shoot first and ask questions later' even in stand your own ground states.

References

Baumler, E. (1984). Paul Ehrlich Scientist for Life (Holmes & Meier, Teaneck, NJ)

Lipsky, B.A., and Baker, C.A. (1999). Fluoroquinolone toxicity profiles: a review focusing on newer agents. Clin. Infect. Dis. *28*, 352–361.

Lopez-Novoa, J.M., Quiros, Y., Vicente, L., Morales, A.I., and Lopez-Hernandez, F.J. (2011). New insights into the mechanism of aminoglycoside nephrotoxicity: an integrative point of view. Kidney Int. *79*, 33–45.

Payne, D., Gwynn, M.N., Holmes, J., and Pompliano, D.L. (2007). Drugs for bad bugs: confronting the challenges of antibacterial discovery. Nat. Rev. Drug Discov. *6*, 29–40.

Projan, S.J. (2002). New (and not so new) antibacterial targets – from where and when will the novel drugs come? Curr. Opin. Pharmacol. *2*, 513–522.

Shales, D.M., and Moellering, R.C. (2008). Telithromycin and the FDA: implcations for the future. Lancet Inf. Dis. *8*, 83–85.

Overuse of Antibiotics: Non-medical Applications

8

Nelson Kardos

Abstract

The decreasing effectiveness of antibiotics in treating common infections results from the spread of antimicrobial resistance (AMR), and is building up to become an epic global public health crisis. Extended periods of antibiotic overuse and misuse since their introduction have applied strong selective pressure towards high-level AMR and multiple drug resistance (MDR), rendering entire classes of antibiotics ineffective. The primary driving force for this global AMR pandemic is the widespread misuse and overuse of antibiotics, in both medical and non-medical applications. The introduction of every antibiotic product has been closely followed by emerging resistance to that antibiotic. Levels of antibiotic consumption correlate with levels of AMR. Antibiotics have been misused in all of their applications, including:

- hospital and outpatient use by physicians through unnecessary, indiscriminate or incorrect prescribing;
- by patients, through incorrect dosing and course durations;
- large-scale use in agriculture for disease treatment, prophylaxis and growth promotion in animal husbandry and food production.

These actions not only have provoked the emergence of resistant microbes, but also have provided optimal environments for the spread of and selection of resistance determinants. It has been established in many countries that the levels of antibiotic consumption consistently correlate with levels of antibiotic resistance (i.e. the more antibiotics are being used in a population, the more resistance to antibiotics there will be in bacteria responsible for infections in that population). The increase in resistance from overuse of antibiotics in turn leads to cross transmission of AMR microbes between humans, between animals, and between humans and animals and the environment. Almost two million Americans per year develop hospital-acquired infections (HAIs), resulting in 99,000 deaths per year. The vast majority of these HAI related deaths are due to AMR infections. Based on studies of the costs of infections caused by antibiotic resistant pathogens vs. antibiotic susceptible pathogens, the annual cost to the US health system of antibiotic resistant infections is $21 to $34 billion, and eight million additional hospital days. This chapter will focus on the overuse of antibiotics through non-medical applications; by their extensive use in animal agriculture and aquaculture and their effect as environmental contaminants on land and water systems. These non-human medical applications are a primary cause of AMR and ultimately impact negatively on human health.

Introduction

The global impact of antibiotics

Prior to the discovery of penicillin as the first available antibiotic, infectious disease had been the leading cause of death throughout history. The development and production of penicillin during the Second World War had a great impact as a war-winning medicine in the field of military medicine. Prior to the twentieth century, more combatants would die from infectious disease than from enemy fire. While this deadly ratio was improved at the start of the twentieth century due to advances in public health and sanitation, the toll of combat fatalities due to infection was still enormous. However, it was the availability of large volumes of low-cost penicillin to the Allied troops in the Second World War that greatly reduced mortality from infected wounds. Before penicillin, the therapeutic tools available for treating infections were few and of limited use. Penicillin was more active against a broad spectrum of infections, including those commonly found in military medicine. The widespread use of penicillin also offered a quick and effective cure for venereal disease infections (i.e. gonorrhoea and syphilis).

Penicillin was the first successful chemotherapeutic agent produced by microbes and it initiated the age of antibiotics. It represents the first therapeutic agent that destroyed bacteria *in vivo*, was not destroyed in the body, and was non-toxic to humans. Penicillin belongs to the β-lactam class of antibiotics, which are the most successful natural product group used in chemotherapy. Later developed members of this antibiotic class which are widely used in medicine today include the orally active semisynthetic penicillins (ampicillin, amoxicillin) and the cephalosporins. It has been estimated that at the end of the nineteenth century, nearly one-third of all deaths were due to infectious disease. By the end of the twentieth century, the death rate from all sources of infection drooped to levels well below 10% (Kardos and Demain, 2011). The use of antibiotics has been responsible for ending what had been since the dawn of history, the leading cause of childhood deaths (i.e. infectious diseases). No other class of medicine has had a comparable cumulative impact on reducing death rates and increasing lifespans (Kardos and Demain, 2011).

The use of penicillin was followed by newer antibiotics, whose discovery and production were derived via the same medical science and production methods pioneered in penicillin's development. The development of antibiotics, starting from penicillin, had a tremendous impact on both government policies and clinical practices for dealing with infection and disease treatment. For example, use of antibiotics enabled a fast throughput in general medical practice because they could be quickly administered or prescribed after a short visit with the general practitioner. This innovation enabled a wider use of health services by the general public and a greater demand for medical services. People with infections could now be effectively treated. This was a great change from previous social attitudes that dictated that people suffering with infectious disease were to be avoided and not treated. In addition, surgeons could undertake more complex operations on patients who were now protected from infections. It is not an exaggeration to state that the great development in modern surgery in the last 60 years could not have taken place without the availability of antibiotics as protective agents (Kardos and Demain, 2011).

The spread of antimicrobial resistance (AMR)

Until recently, the world has been living in the golden age of antibiotics. However, the world is now entering the post antibiotic age because of the growing problem of antimicrobial resistance (AMR). Infections from resistant bacteria have become more common and some pathogens have become resistant to multiple classes of antibiotics. The spread of AMR threatens to compromise the treatment of all infectious disease and is one of the most serious problems confronting both contemporary and future global public health. The World Health Organization (WHO) had proclaimed antimicrobial resistance the focus of World Health Day in 2011 in recognition of this serious and growing problem (WHO, 2011). Based on studies of the costs of infections caused by antibiotic-resistant pathogens versus antibiotic-susceptible pathogens, the annual cost to the US healthcare system if antibiotic-resistant infections is \$21 to \$36 billion, and more than eight million additional hospital days (Frieden, 2013; Scott and Douglas, 2009).

The global struggle against AMR

The loss of effective antibiotic treatment will not only compromise the ability to control routine infectious disease, but will also prevent the treatment of infectious complications in patients with other disease states. The following advanced medical treatments are dependent on antibiotics to fight infections (Fishman, 2006; WHO, 2011):

1 *Cancer chemotherapy*. Cancer patients are at risk to develop serious infections when their white blood cell count is low. Such infections can be serious, and effective antibiotics are needed to protect cancer patients from complications and death.
2 *Surgery*. Patients are at risk for infection from many surgeries including joint replacements, etc. Antibiotics are routinely given before surgery to prevent infection.
3 *Rheumatoid arthritis (RA)*. This disease reduces the patient's immune system and increases the risk of infection. Since many medicines used to treat RA can weaken the immune system, effective antibiotics are needed to ensure that arthritis patients can continue receiving treatments.
4 D*ialysis for end stage renal disease*. Such patients have a weakened immune system and a higher risk for blood stream infections. Such infections are the leading cause of death in dialysis patients. Effective antibiotics ensure that dialysis patients will continue to receive life-long treatment.
5 *Organ and bone marrow transplant*. These patients receive complex surgery and have weakened immune systems. They are at high risk for infections and the use of effective antibiotics is essential.

Without effective antibiotics, all of these procedures would have to be reduced. The reduction in antibiotic effectiveness from resistant pathogens leads to more difficult and costly treatments as well as greater morbidity and mortality and death. The Centers for Disease Control (CDC) issued a landmark report in September 2013 (Frieden, 2013) advising that AMR has become one of the most serious healthcare threats both in America and globally. The CDC reports that some two million people per year in the USA are contracting infections that are resistant to antibiotic treatment. And among such infections, there are more than 23,000 reported deaths per year as a result of AMR. These estimates are regarded as both conservative and on the low end. Such prospects present a preview of a grim post-antibiotic era.

Antibiotic overuse and misuse as primary drivers for AMR

Unlike other classes of medications, antibiotics have a unique attribute; their effectiveness is finite. Unlike other drugs, antibiotics lose their ability to treat disease because of the problem of AMR development. Therefore, in an effort to prolong the effectiveness of antibiotics for as long as possible, these drugs must be used appropriately and sparingly, and their unnecessary uses must be limited. WHO advises that six major factors together drive AMR, and interventions are needed to address all of these factors (WHO, 2011):

- lack of comprehensive national actions with civil society engagement to address AMR;
- lack of laboratory capacity and surveillance lead to inadequate information for choosing treatment for individual patients and for making policy decisions;
- overuse of antibiotics for diseases that do not require them, and underuse due to insufficient doses, duration or use of antibiotics of substandard potency;
- unnecessary use of antibiotics in food-producing animals, especially for supposed benefits in disease prevention and growth promotion;
- insufficient measures to prevent the spread of resistant bacteria in hospitals and the community and their presence as environmental contaminants.;
- inadequate momentum in research and development in the essential technologies to combat AMR (e.g. new antibiotics, diagnostics and vaccines).

The most important factors for the global spread of AMR are the overuse and inappropriate use of antibiotics in both medical and non-medical applications. For the latter category, antibiotics are overused and misused in agriculture and are a major source of environmental contamination. Table 8.1 shows how AMR emerges soon after deployment of new antibiotics, reflecting their overuse in both medical and non-medical applications. AMR has generally been observed to

of effective diagnostics; and lack of current treatment guidelines. Overuse and misuse of antibiotics occur in both the hospital setting as well as in community/primary care and long-term care settings (Scott and Douglas, 2009; WHO, 2011; Acar and Davies, 2008; Laxminarayan *et al.*, 2013; Rogers and Haines, 2005; Bud, 2007). Studies indicate that nearly 50% of antibiotic use in hospitals is unnecessary or inappropriate (IMS, 2012; Fishman, 2006). This situation is not confined to the USA, as a study of antibiotic misuse in medium-sized Swiss hospitals showed that a high proportion of hospitalized patients received antibiotics (25%), of which 47% were categorized as antibiotic misuse (defined in the study as an absence of an indication for such antibiotics) (IMS, 2012; Bugnon-Reber *et al.*, 2004). Misuse of broad-spectrum antibiotics may be the largest single factor for the spread of AMR. Studies have shown that broad spectrum antibiotics make up 50% of all antibiotic prescriptions for adults and 40% of prescriptions for children (IMS, 2012; Steinman *et al.*, 2003). Accordingly, it has been shown from data in Korea, Japan and Turkey that reduction in the use of broad-spectrum antibiotics can be a step towards a credible solution to the control of AMR (IMS, 2012; Livermore, 1990)

Inappropriate use also includes the use of sub inhibitory concentrations of antibiotics. Low concentrations of antibiotics can enrich for resistance genes in a population while having little effect on overall bacterial mortality. The tendency to mutate also increases upon exposure to subinhibitory concentrations of antibiotics. Low concentrations of antibiotics can also select for strains that increase expression of their existing resistance genes, further enhancing their resistance (Acar and Davies, 2009). Although antibiotic resistance is mainly considered to be a clinical problem, antibiotic use and overuse is not restricted to clinical settings. The majority of antibiotics consumed in the world are used in farming and animal agriculture and related settings (aquaculture). Overuse of antibiotics in agriculture leads to the spread and cross-transmission of antimicrobial-resistant microbes between humans, between animals, and between humans and animals and the environment.

AMR is most often portrayed as an undesirable consequence of antibiotic abuse or misuse. But this explanation is not a complete picture. The rate of AMR emergence is related to all uses of these drugs, and not just to their misuse. The total quantity of antibiotics put out into the environment also plays a role in AMR. Selection for AMR is not confined to the human body, and is not limited to hospitals, clinics and farms. Selection takes place anywhere an antibiotic is present in natural environments, and especially in sewage and surface water sediments. Antibiotics are often found in the latter places coupled with high densities of large microbe populations. Large amounts of antibiotics and biocides end up in sewage sludge, which make them primary sources for development of AMR. Increasing amounts of antibiotics and biocides found in waste water, sediment and sludge discharges originate from agricultural applications. The stability of an antibiotic is one key as to how it will impact development of AMR in the environment. Stable antibiotics are more likely to persist long enough to select for resistance (Acar and Davies, 2009; Rogers and Haines, 2005).

In addition to antibiotics and biocides, there are many chemical stressors that can contribute to the selective pressure to develop AMR, although these agents were not specifically designed or used to target pathogens. Many of these stressors include heavy metals, including mercury and silver in dental fillings. These metals and also micronutrients such as zinc, arsenic and copper used in animal feed may stimulate survival mechanisms in microbes that result in mutagenesis; a quick route to resistance. It has also been found that overuse of many types of non-antibiotic drugs such as quinine and antiviral drugs can also lead to AMR (Acar and Davies, 2009). Having explained the key processes for the development of evolutionary AMR, the major drivers at work for the spread of AMR can be listed as follows:

- *Travel and trade*. The growth of international travel and global trade serves to accelerate the movement of antibiotic resistance genes and strains around the world. Antibiotic-resistant pathogens move around the world as quickly as people do. Travellers can contract and bring home resistant strains, both as active infections or as unaffected carriers. International trade of food and livestock also contributes to international movement of AMR.

- *Transmission through animals and insects.* Antibiotic-resistant pathogens have been found in plants, insects and wild animals. Moreover, migratory birds can move across borders and serve as vector sources to transmit AMR. The cycle of transmission between humans and animals as well as plants operates in all directions. While non-human species can bear and transmit AMR pathogens, the burden on public health is not well understood (Acar and Davies, 2009; Rogers and Haines, 2005).
- *Antibiotics as environmental contaminants.* Selection for AMR can take place anywhere antibiotics are present. Large amounts of antibiotics and biocides are found in urban waste water as well as sediment and sewage sludge originating from agricultural applications. Until recently, AMR as an environmental problem had been largely overlooked, due to the fact that antibiotics in non-clinical settings are usually found in concentrations below those used therapeutically. However, even low levels of antibiotics can both favour and sustain development and spread of AMR in microbial communities (Acar and Davies, 2009; Laxminarayan *et al.*, 2013; Bud, 2007). Even the introduction of a low concentration of antibiotics in an environment can have significant effects on an ecosystem's stability and the selection for AMR.
- *Overuse of antibiotics for medical applications.* In healthcare settings, the overuse and inappropriate use of antibiotic are the primary drivers for development of AMR. Factors that contribute to poorly managed use of antibiotics in all healthcare settings include: poor surveillance of resistant pathogens, insufficient antibiotic susceptibility testing and use of diagnostics, and insufficient control of nosocomial infections and poor infection control, and inappropriate prescribing of antibiotics for outpatient use (WHO, 2011; Acar and Davies, 2009).
- *Overuse of antibiotics in agricultural and veterinary applications.* Antibiotics are widely used in agricultural settings, including in animals for both infection control and as growth promoters, and also in veterinary hospitals. Antibiotics are also used in horticulture and in aquaculture applications. All of these applications provide an opportunity for gene transfer and development of AMR (Acar and Davies, 2009; Rogers and Haines, 2005). It should be mentioned that all of these paths for development of AMR are not sufficiently understood, and further research is needed in order to develop effective strategies for containment. In hospital settings, resistance management often takes the form of combining the prudent use of antibiotics with hand hygiene and other infection control measures. While resistance management will need additional strategies and tools, managing antibiotic usage is a necessary part of any programme to contain AMR. There are many knowledge gaps about the prevalence of resistant strains and resistant infections. Such gaps must be filled with greater surveillance efforts and the development and use of rapid diagnostic tools. The follow on sections in this chapter explain in detail the problems associated with these listed drivers, as well as solutions for limiting these same problems. Fig. 8.1 illustrates how antibiotic resistance happens, and provides examples of how antibiotic resistance spreads.

The struggle to contain AMR, a global strategy of microbial containment

During the twentieth century Cold War struggle between the USSR and the West, an apocalypse was contained through an uneasy détente between the global adversaries. Similarly, a new global cooperation is required to contain the danger of AMR to public health. This form of conflict management requires better control of the use of presently available antibiotics, and also developing new ones. While there are no known methods to prevent development of AMR, it is possible to limits its spread and contain the threat. A successful strategy for microbial détente requires several elements: careful management of antibiotic use; effective methods for containing AMR, and reductions in the overall burden of hospital acquired

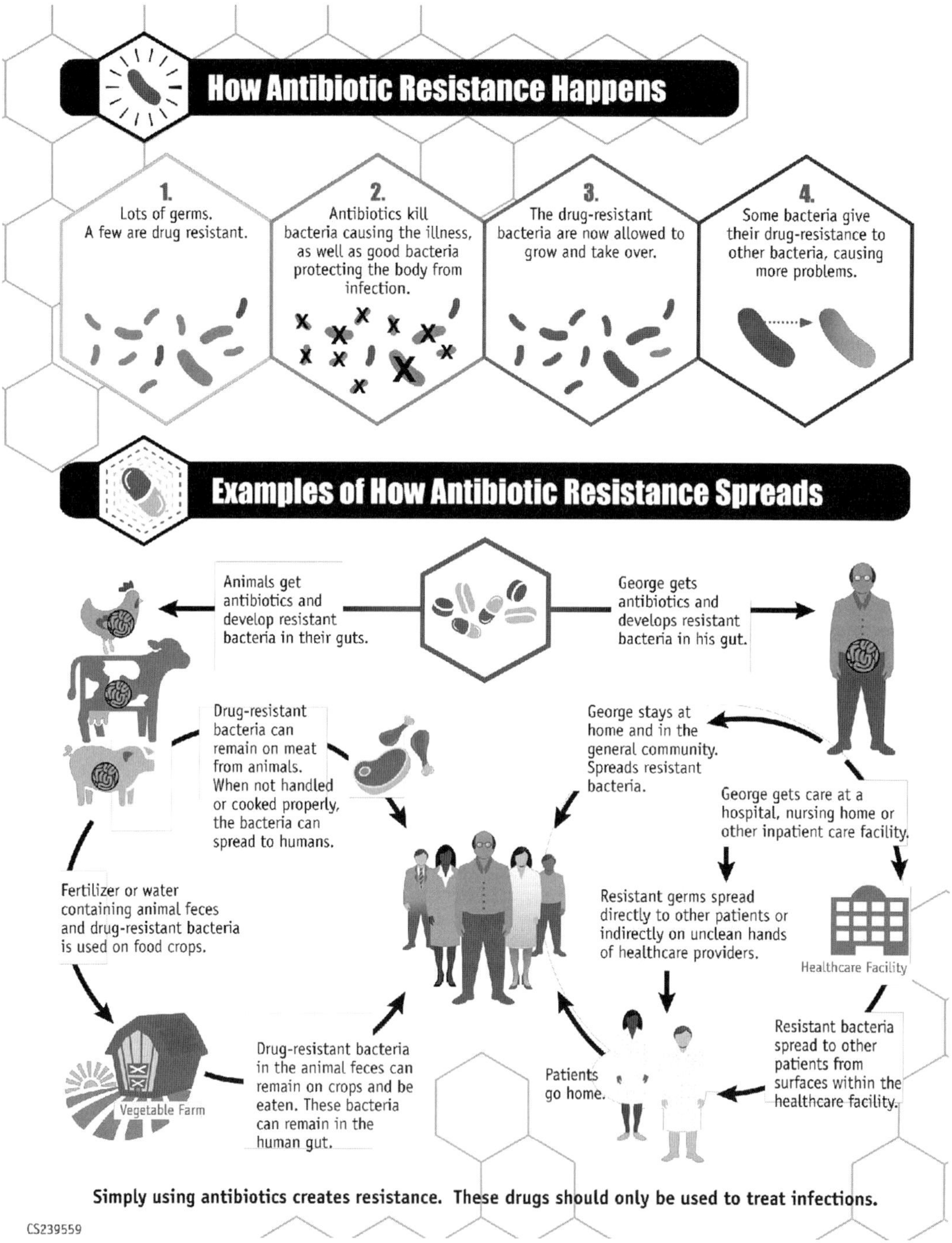

Figure 8.1 How antibiotic resistance happens, and examples of how antibiotic resistance spreads. Frieden, T. (2013). Antibiotic resistance threat in the United States, 2013. CDC (Washington, DC, Centers for Disease Control and Prevention). A colour version of this figure is available in the plate section at the back of the book.

infections; a reduction of non-medical uses for antibiotics. Adapting these strategies can enable a more stable coexistence with our microbial adversaries and preserve the effectiveness of antibiotics and also prevent the development of new forms of AMR. Antibiotics are a limited resource.

Since there are few new antibiotics in the development pipeline, implementation of a prudent antibiotic use strategy can help preserve their effectiveness into the future. Such a strategy for prudent use of antibiotics will be greatly assisted with a parallel public information programme about antibiotic resistance directed to all stakeholders: the public, policy makers and clinicians (Frieden, 2013; WHO, 2012).

Problem: overuse of antibiotics in agriculture

The problem

There is much controversy about the use of antibiotics for use in food-producing animals. In the USA, Europe, Australia and other parts of the world, antibiotics are primarily used to treat beef cattle, pigs and poultry stocks. The quantity of antibiotics sold around the world for food producing animals exceeds that for human use (WHO, 2012; Laxminarayan, 2013; Rogers and Haines, 2005). This use in animals leads to the emergence of AMR in food producing animals. Such animals become a source of resistant bacteria transferred to humans. Many antibiotics used in food producing animals are members of the same classes of antibiotics used in humans, thus cross resistance is a matter of concern with respect to foodborne pathogens. However, some classes of antibiotics are used only in animals, such as the polyether ionophores, bambermycins, and orthosomycins (Rogers and Haines, 2005). A direct effect of the emergence and spread of antimicrobial resistance in bacteria of animal origin is the loss of effectiveness of the antibiotics used for treatment of animals. This situation can greatly increase morbidity and mortality of food animals and lead to economic losses on the farm and ultimately present a risk to food security (Laxminarayan, 2013).

Farms with high rates of antibiotic use are evolutionary incubators where high antibiotic resistant bacteria and multidrug resistant bacteria thrive. In such environments, strong selection also favours the evolution of genetic mechanisms that increase the mobility of genes. Antibiotics used in agriculture may introduce new antibiotic resistant strains into the human population. Public health can be threatened when important evolutionary events occur first in bacteria populations associated with animals which then move into bacteria populations associated with humans. This movement between species can have a serious public health impact (WHO, 2011; Acar and Davies, 2013; Laxminarayan, 2013). The resistance of both commensal and pathogenic bacteria in livestock animals to antibiotics of clinical importance is now commonplace, and the situation is related to the increased use of antibiotics on the farm as growth promoters and for prophylactic applications over the past sixty years.

Scientists have established that antibiotics used in food-producing animals can harm public health through the following sequence of events (WHO, 2011; Acar and Davies, 2013; Laxminarayan, 2013; Rogers and Haines, 2005; Smith *et al.*, 2002):

- Use of antibiotics in food-producing animals allows antibiotic-resistant bacteria to thrive while susceptible bacteria are suppressed or die.
- Resistant bacteria can be transmitted from food-producing animals to humans through the food supply.
- Resistant bacteria from animal reservoirs can cause infections in humans.
- Humans can contract infections from resistant bacteria.

Animal use of antibiotics may cause antibiotic-resistant infections in humans by two different processes. First, use of antibiotics in agriculture increases the frequency of antibiotic resistance in zoonotic pathogens such as *Campylobacter* or *Salmonella* (Vickers *et al.*, 1959). These pathogens are acquired in humans through exposure to contaminated animal food products. The incidence of antibiotic resistance in zoonotic infections of humans is directly related to the prevalence of antibiotic resistant bacteria in food animals. Second, antibiotic resistant bacteria from food animals may facilitate the development of antibiotic resistance in human commensal bacteria which ordinarily colonize humans without causing infection. Commensal bacteria typically have long persistence times, frequent human to human transmission, and high bacterial loads. The risk that a human is infected with commensal antibiotic resistant bacteria increases with the prevalence of antibiotic resistance in the human population (Rogers and Haines, 2005; Bud, 2007). Several studies have shown the presence of antibiotic resistant bacteria on meat and poultry products purchased in grocery stores. One study showed that meat and poultry from five American cities had *Staphylococcus aureus* on 47% of the samples. Moreover, 96% of those samples were resistant to at least one antibiotic, and 52% of the samples were multidrug resistant (North and Christie, 1945).

The origins for use of antibiotics in agriculture

The use of antibiotics have become a high-tech solution to the needs of intensive farming for food animals after the Second World War. Between 1940 and 1980 in the USA, the number of farms raising livestock fell by 80% while the number of animals raised had doubled, and covering cattle, chickens and pigs. The use of antibiotics provided an early solution for farming of all three animals (Bud, 2007). The production of beef to meet growing demand for meet was transformed by the rise of cattle feed lots for final fattening. The use of feed lots holding more than 2000 head of cattle made rapid weight gain possible at a low cost. Historically, infectious disease had been a severe restriction to the mass raising of animals. Crowded conditions led to epidemics in intensive breeding. Epidemics in the early period intensive breeding centres made them unviable. The use of antibiotics in these operations controlled infections (Rogers and Haines, 2005; Bud, 2007; WHO, 1963). In Europe, there is less use of feed lots for raising of animals. The transport of cattle was the main beneficiary of antibiotic use in European agriculture. In Europe, the raising of cattle was mainly a by-product of milk production. Male calves were separated from mothers and raised on separate farms for beef production. This led to the transport of animals around Europe, and resulting in the mixing of healthy animals with infected ones. Infection risk in these circumstances could only be controlled with use of antibiotics (Rogers and Haines, 2005; Bud, 2007; WHO, 1963).

In these applications with all food animals (cattle, pigs, chickens) antibiotics were found to cure most infections. It was also observed that use of antibiotics helped the animals to grow faster. It was discovered in 1946 that adding antibiotics to feed caused pigs to grow faster, enabling farmers to produce more meat with less feed and less cost (Bud, 2007). This practice started in the USA and spread to Europe. The doses added to animal feed to for growth promotion were much smaller than doses used to treat infection. The larger doses given to animals to treat infection resulted in the total destruction of pathogens. However, the smaller doses used for growth promotion (i.e. 4- to 20-fold difference) only challenges disease-causing pathogens but do not destroy them (Rogers and Haines, 2005; Bud, 2007).

The scientific basis as to how antibiotics act as growth promoters in animals is not well understood. However, several studies suggest the likely mechanism of action deals with the action of antibiotics on the microflora inhibiting the animal's gastrointestinal tract (GIT) (Dibner, 2005). The GIT of vertebrate animals contains a diverse group of microflora. As many as 500 bacterial species exist in in the GI microflora, and are estimated to outnumber host cells by 10:1 (Gaskins, 2001). These GI microflora influence a variety of immunological, physiological, nutritional and protective processes of the GIT, and exert major effects on the overall health, development and performance of monogastric animals (Dibner, 2005). The microflora provide real benefits to the host animal, including both nutrition (i.e. fermentation products) and protection (prevention of colonization by pathogens). For nutrition benefits, the microflora secrete nutrients that become available to the host. These nutrients include short chain fatty acids, amino acids and vitamins B and K. The fatty acids make a significant contribution to the host's energy supply (Snel, 2002). For protection, the microflora provide resistance to colonization by pathogenic and other non-indigenous microbes, a phenomenon known as competitive exclusion (Kelly and King, 2001). Studies have shown that germ-free animals are more susceptible to colonization by pathogens than are conventionally grown animals (Koopman *et al.*, 1984). The normal microflora confer a second benefit by stimulating the development of intestinal host defences including production of the key immune cells (i.e. lymphocytes, IgA-producing cells, antibodies and phagocytic cells (Kelly and King, 2001). The experimental evidence suggests that the intestinal immune system develops in parallel with the development of the normal microflora (McCracken, 1999).

However, these benefits imparted to animal development by normal microflora come at a significant cost to the host animal. On the negative side of the ledger, the following effects come at the expense of animal growth performance (Dibner, 2005):

- The commensal bacterial (i.e. microflora) compete with the host for nutrients and lead to reduced fat digestibility.
- They secrete toxic compounds (amino acid catabolites).
- They stimulate rapid turnover of absorptive epithelial cells.
- They stimulate immune system development and inflammatory response in the GIT.

For example, it is thought that 6% of the net energy in the pig diet is lost to the microflora (Vervaeke *et al.*, 1979). The microflora bacteria also compete with the host for uptake of amino acids and incorporation into bacterial protein, thereby reducing nitrogen utilization in the host (Furuse and Yohota, 1985). Certain bacteria ferment amino acids to produce toxic catabolites which have a negative impact on intestinal cell turnover and growth performance (Dibner, 2005). These catabolite products also reduce lipid absorption and produce toxic degradation products that inhibit growth performance in the host (Eyssen, 1973). All of these factors arising from the microflora result in a major hindrance to growth performance due to loss of protein in endogenous secretions and a high expenditure of metabolic energy (Dibner, 2005).

It is thought that antibiotics interact with the microflora to decrease competition with the host for nutrients, and reduce the population of microbial metabolites that depress growth (Anderson, 1999). Antibiotics are also thought to reduce the number of opportunistic pathogens and subclinical infections (Dibner, 2005). It is presumed that the antibiotic's effects are due to the overall reduction in the numbers and species of gut bacteria. This view is supported by the observation that antibiotics do not promote the growth of germ-free animals (Coates, 1968). These listed effects of antibiotics on the reduction in GIT microflora can have a beneficial effect on the host animal, and applies for all major livestock species. A meta-analysis of more than 1000 growth experiments performed in swine over a 25-year period demonstrated that antibiotics improved growth rate in starter pigs (7–25 kg) by an average of 16.4% and feed efficiency by 6.9% (Cromwell, 2002).

The great increase in meat consumption in the developed world was made possible by intensive industrialized rearing of animals sustained by the use of antibiotics to prevent disease and accelerate growth. Antibiotics were also used on dairy farms to treat mastitis and avoid loss of milk production. A problem arose from farmers' application of antibiotics without veterinary supervision. Bad habits developed when farmers found that illicit addition of penicillin in milk could be used to prevent spoilage for the times when a tanker arrived late on the farm to pick up their milk (Bud, 2007). However, the resultant low-level contamination of milk with penicillin had effects on humans. Consumption of contaminated milk sensitized some of the consumers to penicillin, causing them to experience allergies when they took penicillin as a prescription medicine (Bud, 2007; Dibner, 2005). It was also observed in the USA in the mid-1950s that many sources of milk were not curdling into cheese because of the penicillin added by farmers. Surveys carried out in the USA and the UK during that period found that 10% of milk samples were contaminated with penicillin (Rogers and Hanes, 2005; WHO, 1963).

The presence of AMR bacteria may occur rapidly following the introduction of antibiotics as growth promoters in feed animals. A study (Frieden, 2013; Acar and Davies, 2009) has shown that tetracycline-resistance in faecal isolates from chickens increased rapidly from 10% of animals excreting less than 0.1% of organisms resistant to tetracycline (baseline), to 90% of animals excreting 100% of organisms resistant to tetracycline within 2 weeks of introducing the antibiotic to chicken feed. In contrast, there was no increase observed in the control group. Further, multidrug resistance developed even though only tetracycline was being supplemented in the feed. By 12 weeks, more than 60% of the animals from the experimental group excreted bacteria resistant to tetracycline plus one or more other antibiotics. After 4 months, AMR had spread from the experimental group to the control group, where one-third of the chickens excreted bacteria of which more than 50% of isolates were tetracycline resistant. Within 6 months, AMR had also spread to the farm workers and their immediate families. More than 30% of faecal samples from farm workers and their families contained more than 8% tetracycline-resistant organisms, versus 6.7% resistant organisms from their neighbours. Stopping the feed additives eventually reduced the incidence of tetracycline-resistant bacteria in farm dwellers.

The Infectious Disease Society of America (IDSA) has presented testimony to the US Congress several times in efforts to promote the judicious use of antibiotics in animal agriculture. At a hearing before the House of Representatives on 14 July 2010 (IDSA, 2010), IDSA presented expert testimony on the dangers to human health from use of antibiotics on the farm for growth promotion and prophylactic purposes. The ISDA delegation cited evidence from several epidemiological studies to demonstrate the direct connection between the use of antibiotics in animal agriculture and the development of antibiotic resistance in humans. One example showed that poultry treated with fluoroquinone antibiotics were a major source of fluoroquinone-resistant *Campylobacter* infections in humans, leading to treatment failures in patients. Another example cited a livestock-associated strain of MRSA that was first encountered in the Netherlands in 2003. Within a few years, that strain accounted for 40% of human MRSA infections in the Netherlands (IDSA, 2010). Similarly, some MRSA strains that originated with swine on farms in Iowa and Illinois in the USA were found to be the cause of human infections (IDSA, 2010). Moreover, many antibiotic-resistant *E. coli* strains identified in urinary tract infections and blood stream infections in humans have their origins with antibiotic-resistant strains derived from food animals on the farm (AASV, 2001) The evidence was both abundant and clear that overuse of antibiotics in food animal production had become a serious dual threat to public health. One threat concerned the acquisition of antibiotic-resistant pathogens by humans through several sources: contaminated food supply; direct contact with animals, and through environmental contamination. The second and parallel threat concerns the transfer of antibiotic resistant genes from animal to human bacterial populations.

The first country to control the use of antibiotics in food animals was the UK. In 1962, government committees in the UK reviewed the use of antibiotics in agriculture. At that time, it was considered safe to continue to use antibiotics in animal feed. It was mistakenly assumed that bacterial resistance in animals would have no direct consequences for human health. However, by 1965 reports made it clear that multiple health infections were rising quickly and there was direct evidence for its association with the use of antibiotics in animal husbandry. The Swann Commission was convened by Parliament in 1968 to investigate this situation and make recommendations (Bud, 2007). That group distinguished for agricultural use the difference between high dose therapeutic use and low dose use for growth promotion. The latter use was considered a threat to antibiotic effectiveness. In 1970, the use of penicillin and tetracycline as growth promoters in animals was banned in the UK. Some antibiotics were allowed for veterinary use in the UK when used in large doses to treat infections. With the advent of the Common Market in the 1970s, British rules on agricultural use of antibiotics were gradually adopted by all member nations (Bud, 2007). Use of antibiotics as growth promoters in animals has been phased out in Europe's EU members, starting in the UK in the 1970s. All such use was stopped throughout the EU in 2006 (Bud, 2007).

In the USA, however, the use of antibiotics as growth promoters is still extensive. It should be recognized that major animal husbandry trade associations and state farm bureaus do promulgate prudent use of antibiotics to treat food animals. These organizations emphasize to farmers that prudent use of antibiotics helps prevent antibiotic residues in marketed livestock and animal products. Failure to follow the directions of a veterinarian or drug label can result in the animal or its products not being withheld from the market long enough (withdrawal period) for drug concentrations to disappear. Improper use resulting in antibiotic residues may result in extended withdrawal times, and especially if the antibiotic use was not warranted. Such extended withdrawal times can lead to loss of market options and financial loss for the farmer. Common errors in antibiotic usage on farms include the following (FDA, 2001, 2002, 2012; Irwin, 2003; AVMA, 2001; AASV, 2001):

- incorrect diagnosis of the sick animal;
- ineffective levels of the appropriate product are administered;
- a product that has no established specific effectiveness is prescribed;
- treatment of uncomplicated viral infections with antibiotics;
- antibiotic therapy is changed too rapidly and incorrectly. In these situations, the user assumes therapeutic failure prior to correcting all contributing factors;
- non-compliance with product label or written instructions from veterinarian.

Extent of antibiotic use in agriculture in the USA

The amount of antibiotics used in the USA for food animals far exceeds the amounts used in medicine. It was not until 2009 that the FDA released full data concerning the annual amount of antibiotics sold and distributed for use in food animals. The amounts are staggering. In 2009, 13.1 million kilograms was sold for food animal use. In contrast, for 2009 the FDA reported that 3.3 million kilograms of antibiotics was sold for human medical use. Table 8.2 lists a breakdown of antibiotics by class sold in 2009 for agricultural animal use.

The numbers for antibiotic use in food animals is incredibly large. More than 3.7 million kilograms of tetracyclines were sold for animal use in the USA in 2009. That volume exceeds all of the antibiotics of all classes consumed for human use in the same year. It is estimated that about 70–80% of the antibiotics consumed in agriculture are for non-therapeutic use.

Table 8.2 Antimicrobial drugs approved for use in food-producing animals: 2009 sales and distribution data reported by drug class (Loglisci, 2010)

Drug class	Kilograms sold	% of total sold
Food animal use		
Aminoglycosides	339,678	2
Cephalosporins	41,328	0
Ionophores	3,740,627	23
	115,837	1
Macrolides	861,985	5
Penicillins	610,514	4
Sulfas	517,873	3
Tetracyclines	4,611,892	28
NIR	2,227,366	14
Subtotal	13,067,100	79.8
Human medical use	3,300,000	20.2
Total	**16,367,100**	**100.00**

Table 8.3 lists all of the antibiotics approved for use in agriculture in the USA, including animal species treated and indications for therapeutic and non-therapeutic use.

Other uses of antibiotics in agriculture

In addition for use in food animals, antibiotics are used in large quantities for other agricultural applications. These include:

- In crop production, antibiotics are sprayed onto high-value fruit and vegetable crops and ornamental plants, both in the USA and in other countries. This application includes streptomycin, gentamycin, oxytetracycline and oxolinic acid. Plant grade antibiotics are formulated as powders consisting of 17–20% active ingredient. The powders are dissolved or suspended in water to concentrations of 50–300 ppm and then applied as a fine mist. The total amount of antibiotics used on plants in the USA is estimated in the range of 20,000 to 65,000 kg/year (Acar and Davies, 2009;

Table 8.3 Selected antimicrobial agents approved for use in animal agriculture (Roger and Haines, 2005)

Antimicrobial class and drug	Use in animal agriculture: Animal species	Therapeutic	Non-therapeutic[†]	Analogues used for human therapy[‡]
Aminoglycosides				
Gentamicin	Cattle (beef and dairy), horses, swine, poultry[§]		7	Amikacin, gentamicin, neomycin, streptomycin
Neomycin	Cattle (beef and dairy), sheep, swine, poultry	7	7	
Spectinomycin	Beef cattle, swine, poultry	7	7	
Streptomycin	Cattle (beef and dairy), swine, poultry	7	7	
Aminopenicillins				
Ampicillin	Cattle (beef and dairy), horses, swine	X		Amoxicillin, Ampicillin, amoxicillin-clavulanic acid, pivampicillin
Amoxicillin	Cattle (beef and dairy), swine	7		
Cephalosporins (third generation)				
Ceftiofur	Cattle (beef and dairy), horses, swine, poultry, sheep	7	7	Ceftriaxone, cefixime, cefotaxime, ceftazidime, ceftizoxime
Fluoroquinilones				
Enrofloxacin	Beef cattle, poultry	7		Ciprofloxacin, difloxacin, gatifloxacin, levofloxacin, moxifloxacin, norfloxacin, ofloxacin, trovafloacin Nalidixic acid
Lincosamides				
Lincomycin hydrochloride	Swine, poultry	7	7	Clindamycin, lincomycin hydrochloride
Macrolides				
Erythromycin	Cattle (beef and dairy), swine, poultry, layers	7	7	Erythromycin, azithromycin
Tylosin	Cattle (beef and dairy), swine, poultry	7	7	
Tilmicosin	Cattle (beef and dairy), sheep, swine	7	7	
Penicillins				
Cloxacillin sodium	Dairy cattle	7	7	Ampicillin sublactam, cloxacillin sodium, penicillin G benzathine, penicillin G potassium, piperacillin, ticarcillin
Penicillin G procaine	Cattle (beef and dairy), horses, sheep, swine, poultry, fowl	7		
Penicillin G benzathine	Beef cattle, horses	7		
Peptides				
Bacitracin	Cattle (beef and dairy), sheep, swine, poultry, layers	7	7	Bacitracin
Sulfonamides				
Sulfadiazine	Horse	7		Sulfamethoxazole
Sulfadimethoxine	Cattle (beef and dairy), horse, poultry, fowl, fish	7	7	
Sulfamethazine	Cattle (beef and dairy), swine, poultry	7	7	

Table 8.3 (Continued)

	Use in animal agriculture			
Antimicrobial class and drug	Animal species	Therapeutic	Non-therapeutic†	Analogues used for human therapy‡
Sulfanitran	Poultry		7	
Sulfaquinoxaline	Cattle (beef and dairy), poultry	7		
Sulfathiozole	Swine		7	
Streptogramins				
Virginiamycin	Beef cattle, swine, poultry		7	Quinipristin, dalfopristin
Tetracyclines				
Chlortetracycline	Cattle (beef and dairy), sheep, swine, poultry	7	7	Tetracycline hydrochloride, doxycycline
Oxytetracycline	Cattle (beef and dairy), sheep, swine, poultry, fish, honey bees		7	
Tetracycline hydrochloride	Cattle (beef and dairy), horses, sheep, swine, poultry	7	7	

*US Congress OTA, 1995; Khachatourians, 1998; US GAO, 1999; NRC, 1999; Mellon *et al.*, 2001; Shea, 2003; Sayah *et al.*, 2005; USFDA, 2005.
†Non-therapeutic uses include prophylaxis and/or growth promotion.
‡Antimicrobials used in human medicine that are similar to or the same as antimicrobials used in animal agriculture.
§Poultry = Broilers and/or turkeys; Fowl = Quail, pheasant, duck, and/or geese; Sheep = sheep and/or goats.

McManus and Stockwell, 2001, McManus *et al.,* 2002). Difference in estimates are performed by different govt. agencies using different methods and different geographic areas surveyed for data. Compared to the use of antibiotics in animal use, the application of antibiotics for plant use is modest, estimated at less than 0.5% of the total antibiotics consumed in the USA. For example, streptomycin is applied to apple and pear crops to counter 'fire blight' infections (McManus *et al.,* 2002; McManus and Stockwell, 2001). The impact of this use on humans is not known.

- Large quantities of antibiotics such as tetracycline, florphenicol, and flumequie are used in aquaculture (Acar and Davies, 2009). There is a large opportunity for widespread lateral gene transfer in aquatic environments, affecting farmed waters, recreational waters, beaches and drinking water sources – all of which can impact human health. Large volumes of antibiotics are used in aquaculture in developing countries, such as for shrimp farming in Vietnam (Thuy, 2011). However, there has been little research done on the impact of antibiotic use in horticulture and aquaculture on human health. This is an area which merits investigation.

Prudent use of antibiotics in agriculture

Recognizing the peril of improper use of antibiotics in food animals, many international and American organizations have developed a common set of principles for 'Judicious' or 'Prudent Use of Antibiotics'. These organizations include: American Veterinary Medicine Association; Federation of Veterinarians in Europe, WHO, Australian Veterinary Association, European Platform (Rogers and Haines, 2005; Page, 2011; FVE, 2005). The guidelines outline the appropriate use of antibiotics, which include these major points:

1 Avoid using antibiotics that are important in human medicine.
2 Use narrow-spectrum antimicrobials whenever possible.
3 Antibiotic use should be limited to prevent or control disease and should not be used if the principlal intent is to improve performance.
4 Treat the fewest number of animals possible.

Everyone who administers antibiotics to animals should understand and be willing to apply these general concepts of prudent use. These principles are listed in the following 15 underlined principles, and explanatory notes are added (FDA, 2001, 2002, 2012; Irwin, 2003). Application of the judicious use of antibiotics will reduce the development of resistant zoonotic pathogens and commensals in animals and will lessen the risk of a human health impact related to the therapeutic use of antimicrobials in animals. Implementation of these principles requires the diligence and cooperation between farmers and veterinarians.

1 Preventative strategies, such as appropriate husbandry and hygiene, routine health monitoring and immunizations should be emphasized. Prevent disease through better management, emphasizing animal husbandry, biosecurity and health maintenance. Antibiotic use cannot replace sound management practices. Appropriate and timely management practices include vaccinations; parasite control; stress reduction; nutritional management; and environmental management.
2 Other therapeutic options should be considered prior to antimicrobial therapy. When animals do get sick, have an accurate diagnosis: Diagnose problems early and accurately; Many diseases have similar symptoms and signs; Not all diseases are treatable with antimicrobials; Use your veterinarian and a laboratory to help diagnose the problem.
3 Judicious (Prudent) use of antimicrobials, when under the direction of a veterinarian should me*et al*l the requirements of a valid veterinarian–client–patient relationship (VCPR). The use of prescription antimicrobials or any antimicrobial used in an extra label manner requires a valid VCPR (Rogers and Haines, 2005; Peck, 2009a,b).
4 Prescription, Veterinary Feed Directive, and extra label use of antimicrobials must me*et al*l the requirements of a valid VCPR. Federal regulations mandate a valid VCPR for dispensing and use of prescription and VFD drugs and for extra label use of drugs.
5 Extralabel antimicrobial therapy must be prescribed only in accordance with the Animal Medicinal Drug Use Clarification Act of 1996 (AMDUCA) amendments to the Food, Drug, and Cosmetic Act and its regulations. The law prohibits the extra label drug (ELDU) use of antimicrobial in animal feeds.
6 Veterinarians should work with those responsible for the care of animals to use antimicrobials judiciously regardless of the distribution system through which the antimicrobial was obtained: Since 1988, FDA has approved new therapeutic antimicrobials for us in animals as prescription-only products. However, many of the older antimicrobials are available for over the counter sale to producers.
7 Regimens for therapeutic antimicrobial use should be optimized using current pharmacological information and principles.
8 Antimicrobials considered important in treating refractory infections in human or veterinary medicine should be used in animals only after careful review and reasonable justification. Consider using other antimicrobials for initial therapy. This principle takes into account development of resistance to important antibiotics and anticipates banning agricultural use for selected antibiotics of medical importance.
9 Use narrow-spectrum antimicrobials whenever appropriate. It is found that antimicrobials with a broad spectrum of activity lead to development of resistance in non-target microorganisms more rapidly than those with a narrow spectrum because they exert a selection pressure on a greater number of microbes. Therefore, to minimize the likelihood of broad antimicrobial resistance development, where an appropriate narrow spectrum agent is available, it should be selected in preference to a broad-spectrum agent.
10 Utilize culture and susceptibility results to aid in the selection of antimicrobials when clinically relevant. Ideally, the susceptibility profile of the causal organism should be determined before therapy is started. In disease outbreaks involving high morbidity

and/or mortality or where there signs of rapid spread of disease, treatment may be started on the basis of a clinical diagnosis.

11 Therapeutic antimicrobial use should be confined to appropriate clinical indications. Inappropriate uses such as for uncomplicated viral infections should be avoided.

12 Therapeutic exposure to antimicrobials should be minimized by treating only for as long as needed for the desired clinical response. Therapeutic exposure involves both the dose and treatment duration. Treatment for too short a period can be problematic because it can lead to recrudescence of the infection. It is then likely that a higher percentage of the pathogens involved in the new episode have reduced susceptibility to the antibiotic. The continued use of antibiotics in chronic non-responsive clinical cases should be discouraged.

13 Limit therapeutic antimicrobial treatment to ill or at risk animals, treating the fewest animals indicated. In some classes of livestock, if a number of animals in a group have overt signs of disease, both sick and healthy animals will usually need to be treated with therapeutic levels of an antibiotic. This is intended to cure the clinically affected animals, reduce the spread of the disease, and arrest disease development in animals not yet showing clinical signs. Long term administration of antibiotics to prevent disease should not be practised without a clear medical justification.

14 Minimize environmental contamination with antimicrobials whenever possible. Unused antimicrobials should be properly disposed. Disposal methods should be used that will not recycle antimicrobials or resistant genetic materials or organisms to humans or animals

15 Accurate records of treatment and outcome should be used to evaluate therapeutic regimens.

Veterinary oversight of farm use of antibiotics

One of the basic principles of proper antibiotic usage is having a working relationship with a veterinarian, commonly called a VCPR (valid veterinarian–client–patient relationship) (FDA, 2002; Irwin, 2003).

The FDA considers a valid VCR to exist when:

- the licensed veterinarian has assumed clinical responsibility for the animals;
- the owner of the animals has agreed to follow the veterinarian's instructions;
- the veterinarian has sufficient direct knowledge of the animal's condition and their care;
- the veterinarian is available for follow-up evaluation in case of adverse reactions or treatment failure.

The VCPR is required if an antibiotic is administered in any manner other than expressly stated on the label, or if the antibiotic is a prescription product. Any use that deviates from label instructions (including changes in dosing levels) constitutes an extra-label drug use (ELDU) and requires a valid VCPR. An important aspect of a VCPR is the extra-label drug use (ELDU). The use of non-prescription drugs on the farm requires a valid VCPR and appropriate labelling for ELDU, including withdrawal times to avoid having residues in the animal or its food products (milk, meat, eggs). Some key restrictions for use of ELDU include (Rogers and Haines, 2005; Peck, 2009):

- ELDU is permitted only by or under the supervision of a veterinarian.
- ELDU is only allowed for FDA approved animal and human drugs.
- A valid VCPR is a prerequisite for all ELDU.
- ELDU must not lead to a volatile drug residue, which includes applying a significantly extended drug withdrawal to the treated animal.
- ELDU is for therapeutic purposes only (animal's health is suffering or threatened) and not permitted if the purpose is for growth promotion, reproductive performance, or alteration of cost of therapy.
- The user must keep detailed ELDU records for 2 years and make them available to FDA inspectors.

Restriction of antibiotics use in agriculture versus regulation

The regulations on use of animal growth promoters are undergoing serious revision in many countries. In Europe, for example, before the EU regulated antibiotic use as growth promoters, many antibiotics that were used as animal growth promoters selected for resistance to drugs that were used or were eventually developed for use in humans. The European situation has similarities in the USA. In addition, many front-line antibiotics are heavily used in animals therapeutically. The debate over strategies to limit use of antibiotics in American agriculture has been long, controversial and bitter. The FDA has consistently advocated a policy of judicious use of antibiotics in agricultural applications and to ban the use of certain therapeutic antibiotics as growth promoters since 1977 (Rogers and Haines, 2005). However, the FDA efforts to regulate restrictions on antibiotic use in agriculture have been persistently opposed by the farming lobby, the agricultural industry and their supporters in Congress. As a result, all of the FDA recommendations in this sphere depend on voluntary compliance. Voluntary guidelines for reduction of non-judicious use of antibiotics in agriculture are not likely to be as effective as mandatory regulations supported by Congressional legislation and with enforcement measures.

In this regard, a bill was introduced to Congress to restrict the use of antibiotics in agriculture. This bill is titled 'Preservation of Antibiotics for Medical Treatment' (PAMTA) (H.R. 1549, 2009; APUA, 2011). The bill was introduced in 2009 by the late Senator Ted Kennedy, where it died in committee. The bill was reintroduced to Congress in 2011 by Representative Louise Slaughter of New York as H.R. 965, and again by her in March 2013 as H.R. 1150. If ever enacted, this bill would remove from food production the non-therapeutic use of seven classes of antibiotics that are important to human health, and not used unless animals are diseased, or if drug companies can prove their use does not harm human health. The PAMTA bill is vigorously opposed by many agricultural and farming groups and their Congressional supporters. The bill is presently in committee, and outside analysts suggest that it has a very small chance of getting past committee, and then an even smaller chance of being enacted. The PAMTA bill's fortunes were further reduced by the FDA's most recently announced voluntary guidelines to reduce the use of antibiotics in agriculture (Burton, 2013; Zuraw, 2013; Beef Magazine, 2013; Tavernise, 2013; Pierson, 2013; Gee, 2013).

On 12 December 2013, the FDA issued a voluntary plan with the intention of phasing out the use of certain antibiotics used for enhanced growth production on the farm. The FDA guidance directive is entitled 'FDA Guidance 213' or 'GFI #213' (Burton, 2013; FDA, 2013; Pierson, 2013; Gee, 2013). The FDA recommendations have two principles: (i) Any use of antibiotics in animal feed and water shall require a veterinarian's oversight. This action is also termed as a veterinary feed directive (VFD), and (ii) All antibiotics used in feed and water to promote the growth of animals should be phased out. The FDA is now eliminating the present status of these antibiotics as over-the-counter (OTC) products, and changing it to VFD status. Under these new conditions, antibiotics can only be used in food producing animals to treat, prevent or control disease under the order or prescription from a veterinarian. The strategy behind GFI#213 is for the animal drug producers and suppliers to voluntarily work with the FDA to revise the presently approved product use conditions for antibiotics on the farm. The FDA's objectives are to stop the use of antibiotics for growth enhancement and feed efficiency, and also to bring their therapeutic use under veterinary oversight. Once the manufacturers change their product labelling, antibiotics can no longer be used for production purposes on the farm. Based on these guidelines, within 3 years all uses of medically important antibiotics can only be used for therapeutic and prevention applications and under the supervision of a veterinarian (FDA, 2013; Pauly, 2009). That is the FDA plan.

Following the FDA's announcement of GFI#13, leading suppliers of antibiotics for agricultural use, such as Elanco and Zoetis companies, as well as leading meat producers and trade groups advised that they would not fight the new FDA guidelines (Charles, 2013; Gee, 2013). However, while these new FDA guidelines are a positive development, critics complain that this voluntary policy to restrict antibiotic use is not likely to reduce the amount of antibiotics used in food production and will not solve the problem of overuse of antibiotics in agriculture (Hoffman, 2013). The new FDA guidelines will still let farmers and ranchers use antibiotics in feed and water as a 'preventative' step. Some public advocate groups such as 'Keep Antibiotics Working' (KAW) (Pauly, 2009) express concern that there will be no reduction in antibiotic use as farmers will either ignore the plan altogether or simply switch from using antibiotics for routine growth promotion to using the same antibiotics for routine disease prevention. KAW contends that since the same antibiotics are approved for both growth promotion and disease prevention, the producers can continue to sell large quantities to farmers for the latter application. Concerns have been expressed that the new guidelines may result in farmers using an even greater quantity of antibiotics being used in higher therapeutic doses for disease prevention Hoffman, 2013). These critics suggest that a more effective policy would be to ban all uses of antibiotics on the farm (including prevention) except when treating sick animals under the supervision of a veterinarian (Tavernise, 2013). Representative Louise M. Slaughter, the PAMTA bill sponsor, has criticized the GFI#213 guidelines as an inadequate response with no enforcement mechanism and no metrics for success. She contends that the new FDA guidelines do not change how antibiotics are used on the farm, but only how they are labelled (FDA, 2013; Tavernise, 2013). The FDA has responded to critics by stating that adherence to these voluntary guidelines will be monitored for a 3-year period. If it is found that antibiotic use in agriculture has not been reduced after 3 years, the FDA suggests they would move to binding regulations for antibiotic use in agriculture (FDA, 2013).

Eliminating non-judicious uses of antibiotics will not harm US farmers: review of the European experience

There has been great opposition from the agricultural lobby in the USA to reduce use of antibiotics for animal production and in aquaculture. Their complaint is that reduced use of antibiotics will raise the cost of food, reduce the level of production, and harm farmers financially. However, studies have shown that food animal producers in Europe have made these policy shifts without disruption to farm productivity or a negative impact on animal health.

Consider the following examples:

1. Elimination of antibiotics for farmed salmon and trout in Norway. Previously, Norwegian fish farms had used large volumes of antibiotics to control infection, peaking at 50 tons in 1987. The practice of immunizing of fish with effective vaccines replaced the use of antibiotics. While antibiotic use in farmed fish was reduced from 1987 to 2004 by 98%, the volume of farmed fish increased from 50,000 tons to 600,000 tons per year (WHO, 2012).
2. Denmark banned the prophylactic use of antibiotics in animal agriculture in phases and completely by 2000. In Denmark, all uses of antibiotics in animal agriculture require a prescription within a valid VCPR. Also, farmers, vets and pharmacies must report the use and sale of antibiotics, and farm inspections are made regularly. This ban did not lead to increased pathogen loads in animals, nor to more foodborne illnesses in humans. Moreover, this ban did not lead to greater antibiotic use in animals, nor to reduced farm productivity.
3. From 1992 to 2008, pig production in Denmark saw a 50% reduction in antibiotic consumption per kilogram of pig produced. During this same period Danish pork production has increased by 40% since the ban (WHO, 2005; Rogers and Haines, 2005; IDSA, 2010; FDA, 2012). It should also be noted that Denmark is the world's largest exporter of pork products.
4. At the same time, the ban on growth promoters has reduced human infections. Prior to the ban, Danish farmers used avoparcin as a growth promoter, which led to a strain of VRE (Vancomycin-resistant *Enterococcus*) in food animals (WHO, 2012; Rogers and Haines, 2005). Vancomycin and avoparcin are related drugs. In Europe, this strain of VRE spread to humans through the food supply, especially via contaminated meat and poultry. Before the European ban on

avoparcin use, the European population commonly carried VRE in their gastrointestinal tract. After the ban, studies showed a drop in related resistance patterns in animals as well reductions in humans (both colonization and disease) (WHO, 2012; IDSA, 2010).

Alternatives to use of antibiotics in agriculture

The primary disease prevention strategy for farm animals should be through management that emphasizes valid animal husbandry, biosecurity and health maintenance practices including proper sanitization (WHO, 2012, FVE, 2005). Such appropriate practices include: vaccinations; parasite control; stress reduction; nutrition management; and environmental management and proper animal hygiene. This is a much better strategy than relying on the prophylactic use of antibiotics.

Recommendations for action

The overuse and inappropriate use of antibiotics in agriculture have been a major drive for development and spread of AMR. What is to be done? Leading experts both in the USA and from major international health organizations such as WHO have recommended that the following actions be taken to control this situation (WHO, 2012; Rogers and Haines, 2005; IDSA, 2010; FDA, 2012; FVE, 2005):

- Data must be standardized, organized and collected regarding the scope of antibiotic consumption in animals, including both prescription data and manufacturers' distribution and sales data. The lack of adequate US antibiotic consumption data impedes a proper understanding of geographic trends in AMR. A greater understanding will enable more effective and targeted interventions to reduce unnecessary antibiotic use in agriculture. In many other countries, lack of adequate data on the use of antibiotics in agriculture prevent a proper measure of the problem as well as formulating corrective actions. This is a workable objective and has undergone successful operation in the European Union.
- Development of new vaccines for animals to prevent infections and development of rapid diagnostics to more quickly identify pathogens on the farm. These measures will reduce unnecessary use of antibiotics.
- Following WHO guidelines, the use of antibiotics that are deemed critically important for human health should be restricted for use in food production. WHO gives priority to restricting the use of third-generation cephalosporins and fluoroquinolones on the farm. New regulations may restrict or eliminate the use of these medically important antibiotics from agricultural use.
- Also include provisions for prohibiting for animal use any new drug class developed for human medicine, and of those that are used only in human medicine (e.g. linezolid, daptomycin, carbapenems, glycopeptides).
- Legislation and enforcement is required to eliminate the non-judicious use of antibiotics in plant agriculture, aquaculture and for non-therapeutic use in raising livestock. An example is the PAMTA bill that is presently buried in a Congressional committee (H.R. 1549, 2009). Such legislation will mean the end of antibiotic use for purposes of growth promotion, feed efficiency, and routine disease prevention on the farm. All remaining uses of antibiotics should be carried out under the supervision of a veterinarian and within the boundaries of a valid VCPR. The objectives are to shorten animal exposure to antibiotics and to decrease the quantity used on the farm. Such actions are endorsed by leading science and healthcare groups including the American Medical Association (AMA), the Infectious Disease Society of America (IDSA), the Alliance for the Prudent Use of Antibiotics (APDU) and the National Academy of Sciences.

Overuse of antibiotics leads to environmental contamination

Overview and exposure issues

The overuse of large volumes of antibiotics used in both medical and especially in agricultural applications result in widespread environmental contamination. It has been estimated that up to

90% of the antibiotics administered to humans and animals are excreted back into the environment without undergoing metabolic degradation (Rogers and Haines, 2005; Bud, 2007; Campagnolo, 2002). Since antibiotics are designed to be biologically active even at low concentrations, their impacts and environmental interactions can be much subtler and complex than for any other group of contaminants. Antibiotic caused environmental disturbances can impact and damage microbial ecology, increase the proliferation of antimicrobial resistance, and present threats to human health. While the extent to which antibiotics disturb ecological systems is a major concern, the issue has not been well studied, this is an important issue and further research is merited. Antibiotic contamination should be a major concern to the water industry on issues of water reuse and water reserve planning. The key concerns regarding the release of antibiotics as environmental contaminants involve (i) selection for resistance among both commensal and pathogenic bacteria; and (ii) horizontal transfer of intrinsic and extrinsic resistance genes among pathogens.

The recognition that concentrations of antibiotics at sewage treatment works and in the environment are much below therapeutic levels has led to complacent views that such lower levels are insufficient from these discharge sources to allow for selection of AMR. However this assumption may not be correct, as it should be understood that at the origins of a water conveyance line the antibiotic concentrations are higher because of less dilution. Such critical niches may serve as microbial reactors and provide sufficient reactor time and maximum contact for higher antibiotic concentrations between a wide spectrum of microorganisms (including pathogens) to enable resistant gene selection and horizontal transfer (Doughton, 2002). It is thought that the lower antibiotic concentrations in environmental water are a less important factor for resistance selection than the higher undiluted antibiotic concentrations that exist in commercial septic systems and in sewage trunk lines serving many people. More research is required in this area to better understand how AMR develops in sewage treatment plants.

Environmental degradation of antibiotics

Different classes of antibiotics have different degradation pathways. Studies on the environmental fate of antibiotics are limited and have been carried out only on the six most common classes used in medicine and agriculture: aminoglycosides, β-lactams, macrolides, quinolones, sulfonamides and tetracyclines. The major 'abiotic degradation' pathways for these compounds are (Rogers and Hains, 2005; Pauly, 2009; Huang, 2001):

- *Sorption.* The amino groups of aminoglycosides can be positively charged by protonation under acidic conditions, and which may facilitate adsorption to soil clay minerals that typically possess s negative charge. As beta lactam Antibiotics are highly polar compounds with a carboxylic acid functional group, they have weak sorption to soil. With regard to sorption to soil and sediments, tetracyclines adsorb most strongly, followed by quinolones and macrolides. Sulfonamides exhibit weak to moderate adsorption to soil.
- *Hydrolysis.* Hydrolysis is an important degradation pathway for organic pollutants in the aquatic environment. Of the six major antibiotic groups listed, β-lactams, macrolides and sulfonamides are susceptible to hydrolysis. However, hydrolysis of macrolides and sulfonamides at neutral pH is very slow and can be considered negligible.
- *Photodegradation.* This is another transformation pathway that can affect organic pollutants persistent in the surface layers of water bodies that receive appreciable amounts of sunlight. Quinolones and tetracyclines are susceptible to photodegradation
- *Biotic transformation (biodegradation):* Based on observation of the stability of various classes of antibiotics in the soil, it appears that aminoglycosides, β-lactams and some macrolides are more likely to biodegrade to a greater extent than will quinolones, sulfonamides and tetracyclines (Rogers and Haines, 2005; Huang, 2001). In general, antibiotics that degrade slowly and adsorb less strongly to soil and sediments are more likely to be emitted into surface and ground water flows and be

transported to a greater distance. Compounds with such properties are also more likely to pass through wastewater treatment facilities. It is also possible that persistent antibiotics with strong affinity towards particulate matters can be transported in the aquatic environment with particles and colloids. Most antibiotic residues are transported from terrestrial domains to aquatic domains

With regard to an antibiotic's environmental stability, sulfonamide antibiotics appear to be less susceptible to any transformation. Degradation of quinolone and tetracycline antibiotics is slow when exposure to sunlight is limited. Some macrolide antibiotics may be degraded to some extent. Aminoglycoside and β-lactam antibiotics are readily degraded in the environment and are unlikely to be persistent. Sulfonamides and fluorquinolones followed by macrolides are more likely to persist and transport in aquatic environments. Tetracyclines may persist for a significant period of time but are less mobile (Huang, 2001).

Sources of antibiotics as environmental contaminants

The US Environmental Protection Agency (EPA) has made an extensive review of the environmental fate of antibiotic residues and their originating sources. They list eight major sources of antibiotics entering the environment as follows (Rogers and Haines, 2005):

1 *Usage by individual patients and pets.* These include metabolic excreta including the unmodified parent drug and transformed metabolites. This category includes the disposal of unused and out-dated medicines to sewage systems. Also included are underground leakage from sewage system infrastructure.
2 *Release of untreated and treated hospital wastes and nursing home wastes into the domestic sewage system.* Also included are disposal of medicines by pharmacies and physicians.
3 *Release to private septics/leach fields.* Treated effluents from domestic sewage treatment plants are discharged to surface waters or recycled. There is also the overflow of untreated sewage directly into surface waters because of storm events and system failure.
4 *Transfer of sewage sludge (biosolids) to land (soil amendment/fertilization).* Included are (i) straight piping from houses where untreated sewage is discharged directly to surface waters; (ii) release from agricultural spraying of crops and trees; (iii) disposal from medicated domestic animal feed lots (CAFO).
5 *Direct release into open water.* These include activities such as bathing and swimming
6 *Discharge of regulated/controlled industrial/manufacturing waste streams.* This category also includes products disposed from clandestine drug laboratories and illicit drug use.
7 *Disposal of medical waste into domestic refuse and landfills.* This category includes leakage from defective and poorly engineered landfills.
8 Release to open water of soil and excreta from agricultural medicated feed lots.

These above-mentioned sources for antibiotic contaminants are illustrated in Fig. 8.2.

Ground and water contamination from antibiotic use in animal agriculture

Intensive animal agriculture are carried out in confined animal feeding operations (CAFO). The EPA defines a CAFO as an animal feeding facility that houses more than 1000 animal units (AU). The number of AU are based on an equivalent number of beef cattle. For example, 1000 beef cattle AU are equivalent to 2500 swine or 10,000 sheep, and a range of 36,000 to 100,000 laying hens or broilers. The USDA counted more than 11,400 CAFO operations in the USA in 1997 (Rogers and Haines, 2005). At the same time, in the last 60 years, the quantities of antibiotics used in animal agriculture have increased more than 10 times. The movement of pathogens and antimicrobial resistance in a farm environment is a complex environmental issue owing to the continuous exchange of microbes between human and animal hosts and environmental reservoirs.

These CAFO units result in the production of large volumes of manure, which is used as

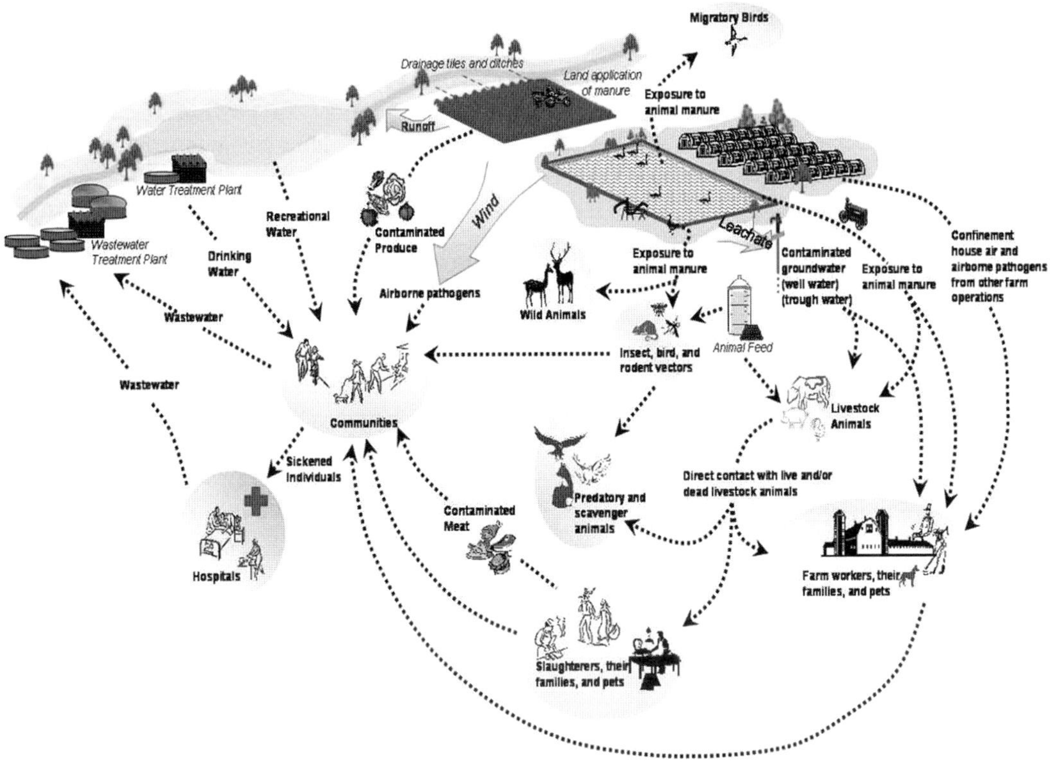

Figure 8.2 Movement of pathogens: an ecological perspective (Rogers *et al.*, 2005). A colour version of this figure is available in the plate section at the back of the book.

fertilizer for crops or spread onto land. On a per weight basis, livestock animals produce from 13 to 25 times more manure than humans. A CAFO unit with 2500 head of dairy cattle can produce as much manure as a city of 61,000 people. When farms were smaller and had small numbers of livestock present, the manure produced was largely used as a soil amendment and fertilizer, with little environmental impact. However, the advent of CAFO units has greatly changed the quantity of manure generated within small land areas. The result has been too much manure applied to too little land. The large quantities of animal waste along with large consumption of antibiotics on the farm have resulted in antibiotics and their metabolites causing long term contamination near animal confinement operations (Rogers and Haines, 2005). As much as 75–85% of antibiotics pass undigested through an animal, resulting in its waste containing high levels of AMR bacteria, plus resistance genes and undigested antibiotics. Animal manures and manure slurries can remain as significant reservoirs for environmental contamination by zoonotic pathogens and AMR bacteria for many months (Rogers and Haines, 2005; Bud, 2007; Maynard, 2003).

This waste is often stored in open air lagoons and/or spread on fields where antibiotic residues, resistant microorganisms and AMR gene reservoirs can move into the environment via aerosolization, infiltration into the groundwater, or runoff into surface water resources (Rogers and Haines, 2005). Movement of pathogens from CAFO operations can be increased by rainfall, which can stimulate release of pathogens from otherwise stable manure-treated fields. This situation leads to increased transport of pathogens and increased discharge to surface waters and increased infiltration into groundwater (Rogers and Haines, 2005). There are also important differences between human and livestock waste. Livestock animal waste from CAFO units is about 100 times more concentrated than human sewage, as human wastes are diluted with other domestic wastewaters prior to treatment (Rogers and Haines, 2005). While human waste must be treated before discharge into the environment as a matter of law, CAFO waste is not treated. Animal manures applied to land as fertilizer are not regulated in terms of pathogen reduction. Untreated animal waste contains many zoonotic pathogens that can cause disease in humans, including virus, bacteria and parasites. It is believed that most AMR bacteria found in surface waters originate from wastewater or manure runoff from antibiotic rich settings such as animal feeding operations, wastewater treatment plants, or subsequently by contaminated animal vectors (Rogers and Haines, 2005).

Aquaculture as a source of antibiotic contamination

There is considerable overuse of antibiotics in coastal aquaculture in developing countries, and which represent a serious threat to environmental contamination. While there are few published data on this topic, a detailed study was made on the use in Vietnam of antibiotics in coastal wetlands for shrimp farming (Thuy, 2011). Since 1986, shrimp production in Vietnamese coastal wetlands has greatly increased. In line with increased shrimp production is an increase in the use of antibiotics. The practice of using antibiotics for shrimp farming is not confined to Vietnam and is widely used in other Southeast Asian countries such as Thailand and the Philippines (Thuy, 2011).

Antibiotics used in Vietnamese shrimp farming have been detected in wastewater and pond sediments as well as in the coastal wetlands and resulting in significant antibiotic resistant bacteria. Most farmers use antibiotics in shrimp pond management prophylactically. Oxytetracycline was the most widely used antibiotic for a long time. However in the last decade quinolones and the combination of sulfadiazine and trimethoprim are now commonly and intensively used in Asian shrimp farming. The most common antibiotics used in Vietnam shrimp farming can be divided into five groups: (1) fluoroquinolones (enrofloxacin, noffloxacin, ciprofloxacin and oxolinic acid); (2) sulfonamides (sulfamethoxazole, sulfadiazine); (3) tetracyclines (oxytetracycline); (4) diaminopyrimidines trimethoprim, ormethoprim); (5) unclassified (griseofulvin and rifampicin) (Thuy, 2011). However, many farmers are not well informed about efficient and safe application practices. The antibiotics are used to treat both adult shrimp and shrimp larvae. Antibiotics are administered as a medicated feed, by injection and as bath formulations for topical application. While there is some government checking in Vietnam of adult shrimp for antibiotic residues, there is no supervision for shrimp larvae (Thuy, 2011).

It is found that deposition of drugs from uneaten feed or from shrimp faeces on or in under cage sediment can be a major route of environmental contamination for the antibiotics used in aquaculture. From shrimp ponds, antibiotics are released into surrounding bodies of water, enabling antibiotics to be taken up by wild fish, shellfish and crustaceans in the shrimp ponds. The antibiotics also pass on to the surrounding environment and which are consumed by humans as food. Aquaculture in general and shrimp culture in particular are the major sources of antibiotic residues in adjacent aquatic systems and accumulate in surrounding coastal wetlands in this region. While the accumulation and degradation processes of used antibiotics are complicated and dependent on multifactors, it is recognized that their accumulation leads to the selection of antibiotic-resistant bacteria. It is expected that the increased usage of antibiotics in Vietnamese shrimp farms with a lack of proper regulation will lead to a greater incidence of antibiotic-resistant bacteria. Indeed, other studies have shown that tetracycline resistance can be commonly detected in fresh and marine aquacultures at various international locations (SERC, 2003; Thuy, 2011).

Animal vectors as a source of antibiotic contamination

Studies suggest that there may be a broader role for animal vectors for the spread of antimicrobial resistance. Large use of antibiotics in animal agriculture results in antibiotic resistant bacteria being shed in animal wastes and then stored in lagoons or spread onto land and eventual travelling to aquatic environments. The resistance genes from AMR bacteria in these contaminated discharges can be transferred to non-contaminated environments through animal vectors. Such vectors include insects, rodents and birds. One study (Cole, 2005) suggests that Canadian geese may act as vectors for AMR and resistance genes in agricultural-animal providing environments. A study compared free-living Canadian geese in Craven County, Georgia versus free-living Canadian geese in Griffin Georgia. The Craven County geese used swine waste lagoons and surface waters adjacent to farm fields. The Griffin geese used crop fields that were not adjacent to animal production facilities. The Craven County geese showed a greater proportion (72%) *E. coli* isolates resistant to antibiotics vs. the Griffin geese (19%). The greater incidence of AMR in the Craven County geese was due to their interaction with CAFO origin waste containing antibiotic residues. The findings suggest the spread of pathogens and AMR bacteria from livestock operations may be significant considering the potential migration distances of Canadian geese and other migratory over long distances (Cole, 2005).

Antibiotics and water contamination: urban sources (general considerations)

Water discharged into lakes and rivers from municipal sewage treatment plants (SWT) and wastewater treatment plants (WWTP) may contain significant quantities of antibiotic residues and antibiotic resistant genes. However, the distribution of antibiotics in waste water systems and treatment plants and their effects on the environment have not been well studied and are poorly understood. It is commonly found that when antibiotic concentrations are measured in WWTP streams, their concentrations are at sub-therapeutic levels. These lower measured antibiotic levels have led to the assumption that they may not contribute to AMR. However, this complacent view does not take into account the full antibiotic load and exposure in these microbial environments. Since antibiotics have high water solubility, their measure in aqueous streams may be biased low because their concentrations in hidden reserves and niches in the waste treatment process (i.e. activated sludge, sedimentation) are not taken into account. These assumptions can lead to an underestimation of the total environmental load of the antibiotics, and may not represent an actual higher antibiotic exposure concentration to microbes in the WWTP streams (Doughton, 2002).

Such microenvironments and niches such as interfaces and biofilms will have a higher concentration of antibiotics. These higher antibiotic concentrations may enable sufficient exposure to resistant gene populations to enable horizontal gene transfer to occur. Biofilms have features that favour horizontal gene transfer, and their presence increases the transfer of antibiotic-resistant genes. While the measured concentrations of antibiotics in WWTP streams are below therapeutic levels, it should not be forgotten that the antibiotic concentrations at the origin of waste convergence lines are less diluted. At these higher concentrations, niche conditions may give sufficient reactor time to maximize contact with a cluster of both microorganisms and antibiotics originating from different sources and with different mechanisms of action, Such conditions would favour horizontal gene transfer and the proliferation and spread of AMR (Acar and Davies, 2008; Huang, 2001; Manaia, 2011). Such environments are also referred to as hot spots or cauldrons for resistance selection, and can be found in communal septic systems and sewage trunk lines served by multiorigins, and sewage sludge. As sludge products from urban WWTP are often used to fertilize agricultural crops, they can become a source of spreading of antibiotic residues and resistance genes that withstand sewage treatment processing.

The fundamentals of WWTP design are to remove solids, organic matters and nutrients, but not residual antibiotics. These treatment systems promote high growth rates and high microbial densities. Such conditions along with residual antibiotics favour the dynamics of an effective bioreactor or cauldron to promote horizontal transfer of antibiotic resistance genes away with indigenous bacteria. This situation is most pronounced with wastewater streams originating from hospitals and other sources of pharmaceuticals (i.e. nursing homes, pharmacies, manufacturers). In addition to aqueous effluent, WWTPs also produce biosolids, and these are often applied to land as a fertilizer. However, biosolids amendment to soil has been observed to elevate antibiotic resistance genes in soil. When one considers that the mass loading of antibiotic resistance genes to the environment via biosolids is about 1000 times higher than via aqueous WWTP effluent, biosolids may be a more important source of AMR into the environment and into the food supply (Doughteon, 2002; Manaia, 2011).

The dynamics of the aquatic environment as a pathway for the spread of AMR are not well understood and merits further research. It is important to identify sources of antibiotic resistance within WWTP treatment stages and their inlet and outlet discharges in order to determine which water treatment system designs will minimize the spread of antibiotic resistance genes. These points are important to evaluate when wastewater streams are considered for reuse in water conservation purposes. The dynamics of this issue are illustrated in Fig. 8.3.

A better understanding as to how antibiotic resistance spreads in the environment is needed in order to develop effective mitigation strategies. Since recycled water most likely contains both pathogens, antibiotics and antibiotic resistance genes in some form, their presence must be considered. However, for development and utilization of water conservation projects that recycle waste streams, few studies have been done to examine antibiotic resistant genes in recycled water and their impact on the environment. Moreover, little is known about the processes for removing levels of antibiotics, antibiotic-resistant bacteria, and antibiotic resistant genes from contaminated effluents (Pruden, 2014; Doughteon, 2002). In this regard, the overuse of antibiotics in both medical and non-medical applications is the primary driver for the spread of AMR through wastewater systems. It appears that both direct inputs of antibiotic resistance genes as well as antibiotics are the primary factors for elevated antibiotic resistance genes in the impacted environments. The release of antibiotics into the environment from WWTP discharges are believed to exert selection pressure along transport pathways. Antibiotics can also continue to stimulate selection and horizontal gene transfer even at very low levels, and which can be optimized under conditions within WWTP processing streams that simulate a bioreactor (Rogers and Haines, 2005; Doughton, 2002).

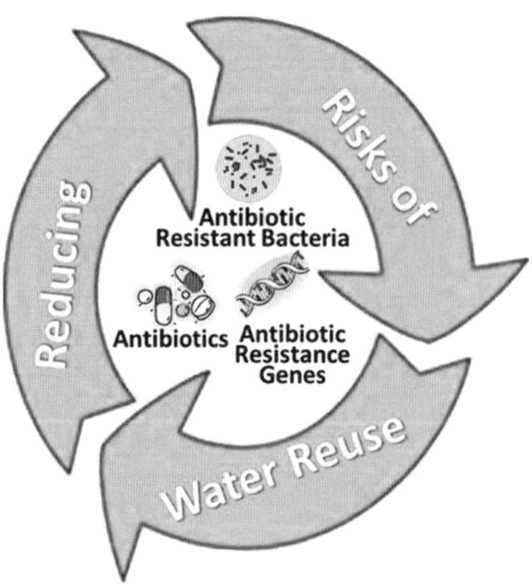

Figure 8.3 Water sustainability and reuse challenged by concerns about antibiotic resistance (Pruden, 2014).

Antibiotics and urban water wastewater treatment: examples of antibiotic contamination

A number of situations in different countries where WWTP have led to significant sources of antibiotic contamination have been documented. Some of these cases are briefly described in this section.

CHINA (Mao, 2014; Lockwood, 2013)

An international team including researchers from Rice University in the USA had previously investigated the presence of antibiotic resistance genes in North China's Hai He River basin. Further investigations examined several WWTP units in that region. The units treated both urban sewage and industrial waste water. The treated waste streams were subject to a final treatment with chlorine before discharge into local waterways. Farmers use the treated sludge – solid by-product from the treatment process as fertilizers in their fields. Bacterial

DNA was extracted from the WWTP streams at various treatment stages and also from treated sludge. Using polymerase chain reaction (PCR) methods, an antibiotic resistant gene (New Delhi metallo-β-lactamase or NDM-1) was measured. This resistant gene was first identified in India in 2010, and has since spread around the world. This resistant gene is a major concern because infected pathogens are resistant to all available antibiotic treatments, including carbapenems. Analysis of WWTP stream samples showed that there were significant levels of NDM-1 in the discharge effluents, and even higher levels in the deactivated sludge from the plant that was applied to soils. Moreover, it was found that the concentration of NDM-1 leaving the plant was greater than that coming into the plant (Mao, 2014; Lockwood, 2013). These unexpected data indicated that the NDM-1 resistance gene was proliferating inside the WWTP. It is believed that the resistance gene may be passed on to other pathogenic bacteria present in the WWTP environment such as *E. coli, Salmonella* and *Klebsiella.* It was found that the North China River basin drains an area of intensive antibiotic use for varied applications. Antibiotic plasmids were found in the river basin persisting for weeks after discharge from WWTP units, and were infecting indigenous microorganisms in that area. The essential findings were that AMR bacteria propagate and proliferate within WWTP units and then upon discharge from the plant are propagated into the environment (Mao, 2014; Lockwood, 2013).

Sweden (Börjesson, 2009)

It was found that hospital sewage discharge had a higher concentration of antibiotics than other discharge sources. The hospital pipeline also had a higher concentration of antibiotic resistance genes (including MRSA – methicillin-resistant *S. aureus*) than other environments. The researchers concluded that the higher antibiotic concentrations selected for and sustained antibiotic resistance genes. Analysis of a municipal WWTP also showed the presence of MRSA bacteria and resistance genes following treatment stages. The researchers indicate that the treatment process selected for strains with more sensitive AMR and a higher virulence. While wastewater treatment reduces the concentration of genes encoding for AMR, the concentrations at the outlet are higher than in the receiving environment (Börjesson, 2009). The correlation of incoming antibiotics to the WWTP and concentrations of ABR genes requires further investigation.

USA (O'Neill, 2011; La Para, 2011)

A study was carried out on a municipal waste water plant in Duluth, Minnesota (La Para, 2011). This plant discharges 40 million gallons per day of treated water into the St. Louis River. This plant uses advanced treatment procedures and equipment to remove large quantities of bacteria from wastewater that many other WWTP units do not use. When the plant's treated wastewater was analysed, it was found there were high levels of tetracycline resistance genes in the plant discharge. The study demonstrates that discharge of exceptionally treated municipal wastewater can have a statistically significant effect on the quantities of ABR genes in otherwise pristine surface waters (O'Neill, 2011; La Para, 2011).

The common element that these country-specific studies demonstrate is that wastewater treatment plants are interfaces between different environments, and so provide an opportunity for mobile elements (including resistance elements) to mix between pathogens, opportunistic pathogens and environmental bacteria. It also appears that the presence of antibiotics in sewage effluents selects for resistance markers that are able to spread through the microbial community, and so facilitate the spread of antimicrobial resistance. The sewage sludge products of urban and rural wastewater treatment plants are increasingly used to fertilize agricultural crops, and this results in the dispersal of unknown quantities of resistance genes and antibiotics that withstand standard sewage treatment. The level of antibiotic resistance in activated sludge is not known, and more research in this area is required. However, field evidence suggests that the activated sludge process may promote cellular interactions among all the diverse microorganisms in the swage and become an important source of AMR production and distribution (Schneider, 2012).

Antibiotic overuse in international travel and trade

General considerations

The development of widespread international travel and global trade have accelerated the movement of antibiotic resistance genes and strains around the world, and with serious consequences for public health on a global basis. The CDC has stated that in an age of expanding air travel and international trade, infectious microbes can be transmitted across borders daily, carried by infected people, animals and insects as well as contained in contaminated shipments of contained food. New forms of old diseases, such as multidrug-resistant (MDR) tuberculosis, can emerge in one region and spread throughout the world (Frieden, 2013).

With expanded international travel, civilians and military personnel alike can contract resistance strains abroad, whether as active infections or as unaffected carriers. Many possible scenarios can result in an international traveller entering a hospital to bring home a resistant pathogen (Acar and Davies, 2008). For example, consider the spread of one carbapenemase-producing Enterobacteriaceae (CRE); New Delhi metallo-β-lactamase (NDM) from India to all global regions. This CRE was first detected in a *Klebsiella pneumoniae* isolate from a patient repatriated to Sweden after being treated in a hospital in New Delhi, India (ECDC, 2011). Since then, other reports have been made of importation of NDM-producing Enterobacteriaceae in Sweden and other European countries due to cross-border transfer of patients (ECDC, 2011; Watkins, 2013). An epidemiological study in India examined the prevalence of CRE organisms in the healthcare system from 2006 to 20007, 2 years before the first case was reported in Europe. The study found that 38.5% of all carbapenem-resistant Enterobacteriaceae isolates from 14 hospitals across India produced NDM (ECDC, 2011; Kumarasamy, 2010; Castanheira, 2011). Another study looking at prevalence of NDM in seepage and tap water samples from New Delhi, India, found that NDM was generally present both in key species of Enterobacteriaceae and also in non-fermentative Gram-negative bacteria, demonstrating a wide dissemination of the NDM gene coding (ECDC, 2011; Kumarasamy, 2010; Castanheira, 2011). Cases of colonization or infection with NDM-producing Enterobacteriaceae in Europe and worldwide are reported either as imported cases in patients with a travel history to, or contact with healthcare systems on the Indian subcontinent. Such cases have been reported from all over the world (e.g. Australia, Canada, China, Japan, Kenya, Oman, and the USA (ECDC, 2011; Chen, 2011; CDC, 2010). In Europe, NDN-producing Enterobacteriaceae have been detected and reported with incremental spread initially from Sweden and the UK, as well as from Austria, Belgium, France, Germany, Switzerland, the Netherlands, Norway and the Balkan region (ECDC, 2011).

International trade of food and livestock is also a problem in the spread of AMR. Food is distributed on a global basis and many food types including produce, meat, processed foods, seafood can all carry ABR microbes and genes. As described in the subsection 'Antibiotics and urban water wastewater treatment: examples of antibiotic contamination' of this chapter, large quantities of antibiotics are used in shrimp farming in Southeast Asia, and much of that production is exported to the developed world. We have also seen in the subsection 'Aquaculture as a source of antibiotic contamination' of this chapter that sewage sludge containing antibiotic resistant genes is used as fertilizer for production of fruits and vegetables in China, and some products are exported to regional markets. Antibiotic-resistant organisms have been detected in supermarket meat in both Europe and North America in many investigations. Surveillance of foods crossing international borders is necessary. For livestock, a certification programme that qualifies only animals not colonized by certain resistant strains for export would be an effective measure (Acar and Davies, 2008). However, while such a programme may be desirable, the economic, logistic, political and regulatory difficulties to implement the same make it unlikely to happen.

Challenges form the developing world: social and economic conditions

The effect of AMR on both health and economics in developing world is largely underestimated and not well understood. These countries face unique challenges that require unique management techniques (Acar and Davies, 2008). The major drivers of AMR in developing countries relate to misuse of antibiotics, and are based on user-related factors such as: self-medication, non-compliance, misinformation, cultural and economic factors – lack of education and poor sanitation (Laxminarayan, 2013; Morga, 2011; IMS, 2012). These factors also extend to local healthcare professionals with insufficient training, leading to incorrect selection of antibiotics for treatment (Laxminarayan, 2013; Chen, 2011). The most important difference between antibiotic uses in developed and developing countries is that, for the latter group, most antibiotics are available without a medical prescription. In such situations, the consumer does not have the guidance of a clinician. Therefore, antibiotics are used indiscriminately and without regard to specific symptoms and without any information about the microorganism causing the medical problem (Laxminarayan, 2013; Morgan, 2011; IMS, 2012).

One major study concerning the development of AMR in healthy children was made in Bolivia (Spatuzza, 2002). Epidemiology studies there indicated high levels of antibiotic resistance to common E. coli infections. Investigations were carried out regarding pharmacy practices in the affected communities. It was found that more than two-thirds of those pharmacies dispensed antibiotics without a medical prescription (Spatuzza, 2002). Moreover, the quantities dispensed depended on the patient's ability to pay. It was found that these pharmacies dispensed antibiotics inappropriately to 92% of adults served and to 40% of the children served. Moreover, the local pharmacies often dispensed enough antibiotics for a 2-day supply, or less – even though the recommended course of antibiotic treatment was from one to five days (Laxminarayan, 2013; Spatuzza, 2002). This suboptimal dosing of antibiotics is a major factor for development and spread of AMR. As shown in the Bolivian study, in many developing countries, antibiotics are sold at pharmacies on a per dose basis, and customers usually buy whatever they can afford and take them until their symptoms improve or their antibiotic supply runs out. This situation results in suboptimal dosing of infections and is a major factor for developing AMR. In addition, in many cases generic drugs are offered for sale with insufficient controls on quality and potency (Laxminarayan, 2013; Spatuzza, 2002; Morgan, 2011). Such poor quality control for antibiotics sold in these locations can lead to low potency and subinhibitory doses and which can also lead to development of AMR (Acar and Davies, 2008; Laxminarayan, 2013).

To illustrate the serious nature of AMR in developing countries, consider the situation of increasing resistance to fluoroquinolone antibiotics in these locations. Based on analysis of hospital isolates, rates of resistance to fluoroquinolone antibiotics have been reported in various countries as India (44%), Pakistan (58%) and Bangladesh (65–88%) (Laxminarayan, 2013).

Another factor favouring the spread of AMR in developing countries is that humans live in closer contact with animals than they do in developed countries. This factor provides a greater opportunity for passing resistant pathogens back and forth AMR (Acar and Davies, 2008; Laxminarayan, 2013). In addition, in developing countries, hospital patients are often fed and cared for by family members, which enables transfer of pathogens from home (where animals may be present) to the clinic, and vice versa. In these countries, there is often overcrowding of patients and a lack of isolation, which conditions contribute to the spread of nosocomial infections, including tuberculosis. These same conditions exacerbate the spread of AMR (Acar and Davies, 2008). In developing countries, traffic and selling of antibiotics in markets, shops and pharmacies is largely unregulated, dispensed without prescriptions and even without the supervision of a person with pharmaceutical training (Laxminayaran, 2013; Gyansa-Luttendt, 2013; Spatuzza, 2002; Morgan, 2011). In many developing countries there are large and unregulated uses of antibiotics for agriculture, aquaculture and for intensive farming. And there is little separation of the types of antibiotics used for humans and animals, with the same antibiotics used for both applications (Acar and Davies, 2008).

On the clinical side, infection control in developing countries is often poor. Hospitals are overcrowded. In addition, insufficient equipment and trained healthcare personnel plus poor infection control surveillance lead to the spread of nosocomial infections and the spread of AMR. These resistant pathogens become a commensal reservoir of resistance genes and are spread to the community through unsafe water supplies and poor sanitation (Acar and Davies, 2008; Laxminarayan, 2013; Gyansa-Luttendt, 2013). Hospitalized patients often spread resistant infections to family members.

Self-medication by consumers with antibiotics purchased without prescription is common, especially in eastern and southern Europe, Africa, South America and Asia (Laxminarayan, 2013; Spatuzza, 2002). Consumers there have positive attitudes towards antibiotics but also poor knowledge about these drugs and diseases. The availability of antibiotics without prescription is a key enabling factor, and this results from the absence of prescription-only regulations, ineffective law enforcement, and poverty-driven practices as well as local cultural norms. Other factors that promote the spread of AMR in these countries includes suboptimal compliance on use, including taking left over antibiotics from previous treatment course and the sharing of unused drugs with other people in the community (Laxminarayan, 2013; Spatuzza, 2002; Morgan, 2011). The social conditions of patients also contribute to inappropriate prescribing practices. Many doctors prefer to prescribe wide-spectrum antibiotics for patients who they assume cannot wait for a full diagnosis or are unlikely to return because of transportation time or costs (Laxminarayan, 2013; Spatuzza, 2002). And many health workers in these countries know that if they do not prescribe a drug, the patient will likely resort to self-medication, buying cheaper over the counter drugs that are often sub-standard or insufficient for a complete treatment. A larger problem is that infectious disease spread more easily in low income communities and in developing countries in general because of their poorer hygienic conditions associated with poverty.

For developing countries, local and national authorities over there should ensure that there are tighter restrictions on the over-the-counter distribution of antibiotics. In particular, second- or third-generation drugs that are used only in hospitals in developed countries should not be available to the local community without a prescription to reduce inappropriate use. Controls on antibiotic use need to be accompanied by public and professional education to better ensure compliance. Expert boards should be established in each country to oversee antibiotic usage, prevent the distribution of counterfeit drugs, and assure the quality of generic drugs (Laxminarayan, 2013; Pyrek, 2013; Hawkings, 2008). Antibiotic use outside the health sector in developing countries, and use in animal agriculture and aquaculture needs to be regulated. This factor is especially important as it is not only a major channel for development and spread of AMR, but the export of foods from animals treated with antibiotics – including aquaculture products – can lead to transfer of AMR in the countries that import these affected food products.

Perverse economic incentives for antibiotic overuse

Another problem in developing countries for overuse of antibiotics are perverse economic incentives. From a prescription perspective, prescriber and pharmacy remuneration policies may result in the overuse and misuse of antibiotics. In many developing counties, the physician is both a prescriber and dispenser of medications. This conflict of interest leads to overprescribing of antibiotics, and often not the most effective antibiotic for the patient. In addition, many antibiotics are routinely sold in pharmacies without prescriptions on an over the counter basis to patients who request medicines. This situation is especially prevalent in China, which has the world's most rapid growth of resistance (22% average growth in one study from 1994 to 2000) (IMS, 2012; Sun, 2011). A recent study on the irrational use of medicines in rural China showed that prescription practice of rural healthcare providers and economic incentives from selling medicines are the main reasons for the high antibiotic utilization. It was shown that in Shanghai, inadequate compensation mechanisms cause hospitals to rely heavily on drug price mark-ups for the products they dispense. While this refers to all drugs, antibiotics account for almost 50% of these dispensed drugs, resulting in the overuse and irrational use of medicines in Chinese hospitals (IMS, 2012; Hu, 2001). As a common occurrence, antibiotics in China are often misused. They are used to treat viral disease or inappropriately overused with routine and overuse of third-generation antibiotics (IMS, 2012; Sun, 2011; Hu, 2001).

Conclusion

The spread of AMR threatens to compromise the treatment of all infectious disease and is one of the most serious problems confronting both contemporary and future global public health. The most important factors for the global spread of AMR are the overuse and inappropriate use of antibiotics in both medical and non-medical applications. For the latter category, antibiotics are overused and misused in agricultural animals and are a major source of environmental contamination. The overuse of large volumes of antibiotics used in both medical and especially in agricultural applications for food animal production result in widespread environmental contamination going back into the environment. Since antibiotics are designed to be biologically active even at low concentrations, their impacts and environmental interactions can be much subtler and complex than with any other group of contaminants. It is believed that most AMR bacteria found in surface waters originate from wastewater or manure runoff from antibiotic rich settings such as animal feeding operations on farms, wastewater treatment plants, or subsequently contaminated animal vectors. Water discharged into lakes and rivers from municipal sewage treatment plants (SWT) and wastewater treatment plants (WWTP) may contain significant quantities of antibiotic residues and antibiotic resistant genes. Another major factor for the spread of AMR is the development of widespread international travel and global trade. AMR can travel around the world without barriers and can create problems in countries without prior exposure to these resistant pathogens. International trade of food and livestock is also a problem in the spread of AMR. Surveillance of foods crossing international borders is likely to become necessary in the future to control the spread of AMR form overuse of antibiotics in agricultural applications. Because of the growth in international trade and travel, the overuse and irrational use of antibiotics in the developing world will have a strong impact on the spread of AMR to the developed world. Moreover, these same countries have few controls on their import, manufacture, use and disposal of antibiotics; all major factors for their overuse. Antibiotic use outside the health sector in both developed and developing countries, and use in animal agriculture and aquaculture needs to be regulated, and enforcement of the regulations. This factor is especially important as it is not only a major channel for development and spread of AMR, but the export of foods from animals treated with antibiotics can lead to transfer of AMR to the countries that import these affected food products. Controls on antibiotic use need to be accompanied by public and professional education to better ensure compliance. In summary, to preserve the potency of antibiotics, prudent use must be practised on a global basis, for both medical and non-medical applications.

References

AASV (2001). Judicious use of antimicrobials for pork producers. (American Association of Swine Veterinarians, Washington, D.C.).

Acar, J.F., and Davies, J.E. (2008). Antibiotic resistance, an ecological perspective on an old problem. American Society for Microbiology. Colloquium at Annency, France (October 12 to 14).

Anderson, D.B., McCracken, V.J., Aminov, R.I., Simpson, J.M., Mackie, R.I., Verstegen, M.W.A., and Gaskins, H.R. (1999). Gut microbiology and growth promoting antibiotics in swine. Nutr. Abstr. Rev. B Livestock Feeds and Feeding *70*, 101–188.

APUA (2011). APUA policy update. Preservation of antibiotics for medical treatment (PAMTA). Presentation to U.S. Congress. Newsletter, *29*(1) (Alliance for the Prudent Use of Antibiotics, Boston, MA).

AVMA (2001). Judicious use of antimicrobials for poultry producers (American Veterinary Medical Association, Washington, D.C.).

Beef Magazine (2013). FDA announces final guidance #213 and veterinary fed directive role for antibiotics. (Published online at: Beefmagazine.com/print/beef-quality/FDA-announces-food-guidance-213-and-veterinary-feed-directive-rule-antibiotics).

Blok, M.C., Vahl, H.A., de Lange, L., van de Braak, A.E., Hemke, G., and Hessing, M. (2002). Nutrition and Health of the Gastrointestinal Tract (Wageningen Academic Press, Wageningen, Netherlands).

Börjesson, S. (2009). Antibiotic resistant bacteria in wastewater (methicillin resistant *Staphylococcus aureus*

– MRSA- and antibiotic resistant genes). Linköping University Medical Dissertation No 1128, Sweden.
Bud, R. (2007). Penicillin, Triumph and Tragedy (Oxford University Press, Oxford, UK).
Bugnon-Reber, A.V., de Torrenté, A., Genré, D., and Troillet, N. (2004). Antibiotic misuse in medium-sized Swiss hospitals. Swiss Medical Weekly, *134*, 481–485.
Burton, T.M. (2012). FDA moves to cut use of antibiotics in food producing livestock animals. Wall Street Journal, December 11.
Campagnolo, E.R., Currier, B.W., Esteban, J.E., Johnson, K.R., Karpati, A., Kolpin, D.W., McGeehin, M., Meyer, M.T., Smith, K., and Mellon, M.C. (2002). Antimicrobial residues in animal wastes and water resources proximal to large-scale swine and poultry feeding operations. Sci. Total Environ. *299*, 89–95.
Castaheira, M., Deshpade, L.M., Mathai, D., Bell, J.M., Jones, R.N., and Mendes, R.E. (2011). Early dissemination of NDM-1 and OXA-181 producing Enterobacteriaceae in India hospitals: Report from the SENTRY antimicrobial surveillance program, 2006–2007. Antimicrob. Agents. Chemother. *55*, 1275–1278.
CDC (2010). Detection of Enterobacteriacea isolates carrying mettalo-β-lactamase – United States. MMWR. Mob Mortal Wkly Rep. *59*(24), 750. (Centers for Disease Control and Prevention, Atlanta, GA).
Charles, D. (2013). Drug companies accept FDA plans to phase out some animal antibiotic use. National Public Radio (NPR) broadcast, December 11.
Chen,Y., Zhao, Z., Jiang, Y., and Yu, Y. (2011). Emergence of BDN-1-producing *Acinetobacter baumannil* in China. J. Antimicrob. Chemother. *66*, 1255–1259.
Coates, M.E., Fuller, R., Harrison, G.F., and Levi, M. (1968). Comparison of the pregrowth of chicks in the Gusafsson germ free apparatus and in a conventional environment and without dietary supplements of penicillin. Br. J. Nutr. *17*, 141–151.
Cole, D., Drum, D.D.V., Stahlknecht, D.E., White, D.G., Lee, M.D., Ayers, S., Sobsey., M., and Maurer, J.J. (2005). Free-living Canada geese and antimicrobial resistance. Emerging Infectious Diseases. www.CDC.gov.ed. *11*(6) June pp. 935–937.
Cromwell, G.L. (2002). Why and how antibiotics are used in swine production. Anim. Biotechnol. *13*, 7–27.
Dibner, T.T., and Richards, J.D. (2005). Antimicrobial growth promoters in agriculture: history and mode of action. Poultry Science *84*, 634–643.
Doughton, C.G. (2002). Antimicrobials in the environment, less recognized issues. National Academy of Sciences Round Table on Environmental Health Sciences Research. Research section, medicine topics: human health effects of using antimicrobial agents in agriculture. Washington D.C. February 9.
ECDC (2011). Updated risk assessment on the spread of NDM and its variants in Europe (European Centre for Disease Prevention and Control, Brussels, Belgium).
Eyssen, H. (1973). Role of gut microflora on metabolism of lipids and sterols. Proc. Nutr. Soc. *32*, 59–63.
FDA Center of Veterinary Medicine (2001). Judicious use of antimicrobials for beef cattle veterinarians. (Food & Drug Administration, Washington, D.C.).
FDA Center of Veterinary Medicine (2002). The prudent use of antimicrobials for dairy producers. (Food & Drug Administration, Washington, D.C.).
FDA Center of Veterinary Medicine (2012). Guidelines for industry: the judicious use of antimicrobial drugs in food producing animals. FDA guidance #209. (Food & Drug Administration, Washington, D.C.).
FDA (2013). Phasing out certain antibiotics use in farm animals. FDA consumer update page. (December 11). (Food & Drug Administration, Washington, D.C.).
Fishman, V. (2006). Antimicrobial stewardship. Am. J. Med. *119*, 53–61.
Frieden, T. (2013). Antibiotic resistance threat in the United States, 2013. CDC (Centers for Disease Control and Prevention, Washington, D.C.).
Furose, M., and Yokota, H. (1985). Effect of the gut microflora on chick growth and utilization of protein and energy at different concentrations of dietary protein. Br. Poult. Sci. *26*, 97–104.
FVE (2005). Antibiotic resistance and prudent use of antibiotics in veterinary medicine. (Federation of Veterinarians of Europe, Brussels, Belgium).
Gaskins, H.R. (2001). Intestinal bacteria and their influence on swine growth. In: Swine Nutrition, Lewis, A.J., and Southern, L.L. eds. (CRC Press, Boca Raton, FL), pp. 585–605.
Gee, K. (2013). Food industry won't fight antibiotic rule. Wall Street Journal, December 13.
Gyansa-Luttendt, M. (2013). Antibiotic resistance in Ghana. Lancet Infect. Dis. *13*, 1006–1007.
H.R. 1549 (111th) (2009). Preservation of Antibiotics for Medical Treatment Act (PAMTA) of 2009. (U.S. House of Representatives, 11th Congress, 2009–2010, Washington, D.C.).
Hawkings, N.J., Butler, C.L., and Ward, F. (2008). Antibiotics in the community; a typology of user behavior. Patient. Educ. Cours. 73, 146–152.
Hoffman, B. (2013). New FDA rules not likely to reduce antibiotic use on the farm. In Forbes (New York, NY, USA; December 13).
Hu, S., Chen, W., Chen, K., Zhou, H., and Wang, L. (2001). Pharmaceutical cost-constraint policy: experiences in Shanghai, China. Health policy Plan. *16* (Suppl. 2), 4–9.
Huang, C.H., Renew, J.E., Smeby, K.L., Pinkston, K., and Sedlak, D.L. (2001). Assessment of potential antibiotic contaminants in water and preliminary occurrence analysis. Journal of Contemporary Water Research (JCWRE) *120*(1), 30–40.
IDSA (2010). Antibiotic resistance: promotion of judicious use of medically important antibiotics in animal agriculture. Testimony at U.S. House of Representatives, Committee on Energy and Commerce and Subcommittee on Health. (Infectious Disease Society of America, Arlington, VA).
IMS (2012). Advancing the responsible use of medicines. Applying levers for change: six primary levers of opportunity. (3) Right medicine for the right patient

– optimize antibiotic use. (Institute for Healthcare Informatics, Danbury, CT).

Irwin, K., Smith, D.R., Ebako, G.M., Ensley, S., Griffin, D.D., Wohlers, A., McManus, P.S., and Stockwell, V.O. (2003). Guidelines for the prudent use of antibiotics in food animals. NebGuide #G1485. Institute of Agriculture and Natural Resources, University of Nebraska, Lincoln Extension (January).

Kardos, N., and Demain, A. (2011). Penicillin: the medicine with the greatest impact on therapeutic outcomes. Appl. Microbiol. Biotechnol. *92*, 677–687.

Kelly, D., and King, T.P. (2001). Luminal bacteria and regulation of gut function and immunity. In Gut Environment of Pigs, Piva, A., Bach, K.E., and Lindberg, J.E., eds. (Nottingham University Press, Nottingham, UK) pp. 113–131.

Koopman, J.P., Kennis, H.M., Mulliak, J.W., Prins, R.A., Stadhouders, A.M., De Boer, H., and Heetons, M.P. (1984). Normalization of germ free mice with anaerobically cultured caecal flora of normal mice. Lab Anim. *18*, 188–194.

Kumarasamy, K.K. (2010). Emergence of a new antibiotic resistant mechanism in India, Pakistan and the UK: a molecular biology and epidemiological study. Lancet Infect. Dis. *10*, 597–602.

La Para, T.M., Tucher, R., Burch, R., McNamara, P.J., Tan, D.T., Yan, M., and Eichmiller, J.J. (2011). Tertiary treated municipal wastewater is a significant potential source of antibiotic resistance genes into Duluth-Superior harbor. Environ. Sci. Technol. *45*, 9543–9549.

Laxminarayan, R., Duse, A., Wattal, C., Zaid, A.K.M., Wertheim, H.F.L., Sumpradit, N., Vlieghe, E., Levy-Hura, G., Gould, I.M., and Goosens, H. (2013). Antibiotic resistance, the need for global solutions. Lancet Infect. Dis. *13*, 1057–1098.

Livermore, D.M. (1990). Multiple mechanisms of antimicrobial resistance in *Pseudomonas aeruginosa*: our worst nightmare? Clin. Infect. Dis. *34*, 621–640.

Lockwood, D. (2013). Multidrug resistance gene released by Chinese wastewater plants. Chemical & Engineering News *91*, 8.

Loglisci, R.F. (2010). Animals consume the lion's share of antibiotics. Food Safety News (Dec. 27). Published online: www.foodsafetynews.com/2010/12/animals-consume-lions-share-of-antibiotics

McCracken, V.J., and Gaskins, H.R. (1999). Probiotics and the immune system. In: Probiotics: A Critical Review, Tannock, G.W. ed (Horizon Scientific Press, Norfolk, UK), pp. 85–111.

McManus, P.S., and Stockwell, V.O. (2001). Antibiotic use for plant disease management in the United States. Plant Health Progress. doi:130. 1094PHP-2001-U327-01-RU.

McManus P.S., Stockwell, V.O., Sundin, G.W., and Jones, A.L. (2002). Antibiotics in plant agriculture. Annu. Rev. Phytopathol. *40*, 443–465.

Manaia, C.M., Vaz-Moreira, I., and Nunes, O.C. (2011). Antibiotic resistance in waste water and surface water and human health implications. In: Emerging Organic Contaminants and Human Health, Barceló, D. ed. (Springer Verlag, Berlin and Heidelberg, Germany), pp. 71–78.

Mao, D.,Q., Luo, Y., Mathieu, J., Wang, Q., Feng, L., Mu, Q.H., Feng, C.Y., and Alvarez, P.J.J. (2014). Persistence of extracellular DNA in river sediment facilitates antibiotic resistant gene propagation. Environ. Sci. Technol. *48*, 71–78.

Maynard, C., Fairbrother, J.M., Bekal, S., Sanschagrin, F., Levesque, R.C., Brousseau, R., Masson, L., Larivière, S., and Harel, J. (2003) Antibiotic resistance genes in enterotoxigenic *Escherichia coli* O149:K91 isolates obtained over a 23 year period from pigs. Antimicrob. Agents Chemother. *47*, 3214–3221.

Mellon, M.C., Berbrook, C., and Benbrook, K.L. (2001). Hogging it, estimates of antimicrobial abuse in livestock. Union of Concerned Scientists Pub. Cambridge. MA.

Morgan, D. (2011). Non-prescription antimicrobial use worldwide: a systematic review. Lancet Infect. Dis. *11*, 697–701.

North, E.A., and Christie, R. (1945). Observations on the sensitivity of Staphylococci to penicillin. Med. J. Aust. *2*, 424–426.

O'Neill, K. (2011). Municipal wastewater spreads antibiotic resistance. Chemical & Engineering News (October 14). Published online: www.sott.net/article/2236419-municipal-wastewater- spreads-antibiotic-resistance

Page, S. (2011). Judicious use of antimicrobials in Agriculture- principles of appropriate use. (Australian Chicken Meat Federation, North Sydney, Australia).

Paterson, D.C., and Bonoma, R.A. (2005). Extended spectrum β-lactamases: a clinical update. Clin. Microbiol. Rev. *18*, 146–152.

Pauly, A. (2009). Keep Antibiotics Working (KAW) statement on hearing regarding preservation of antibiotics for medical treatment act (PAMTA). (July 15). (Keep Antibiotics Working, Chicago, IL).

Peck, C. (2009a). Antibiotics in animals. BeefMagazine.com (November 1). Published online: www.beefmagazine.com/print/beef-quality/beef-quality-assurance/1101-antibitoics-animals-prudent

Peck, C. (2009b). Prudent use of antibiotics in beef production. Montana Farm Federation. Published online: mfbr.org/ag-newswire/2000/09/prudent-use-of-antibiotis-in-beef-production

Pierson, R., and Abbott, C. (2013). New U.S. FDA rules aim to cut antibiotic use in farm animals. Reuters (December 11). Published online: www.reuters.com/assets/print?aid=USBRE9BA0RJ20131211

Pruden, A. (2014). Balancing water sustainability and public health goals in the face of growing concern about antibiotic resistance. Environ. Sci. Technol. *48*, 5–14.

Pyrek, K. (2013). Bugs without borders: The global challenge of MDROs. Infection Control Today (February 8). Published online: www.infectioncontroltoday.com/articles/2013/02/bugs-without-borders-the-global-challenge

Rogers, S., Haines. J., and EPA (2005). Detecting and mitigating the environmental impact of fecal pathogens originating from confined animal feeding operations: Review. (Environmental Protection Agency, Office of Research & Development, Cincinnati, Ohio).

SERC (2003). Antibiotics in agriculture: policy issues package. Published online: www.secronline.org/antibiotics/fact.html

Schneider, R., and Edwards. R. (2012). Insights into antibiotic resistance through metagenomic approaches. Future Microbiol. 7, 73–89.

Scott, H., and Douglas, R. (2009). The direct medical costs of healthcare – associated infections in U.S. hospitals and the benefits of prevention. (Division of Healthcare Quality Promotion, National Center for Preparedness, Detection and Control of Infections, Center of Disease Control and Prevention, Atlanta, GA).

Smith, D.L., Harris, A.D., Johnson, J.A., Silbergeld, E.K., and Morris, J.J.O. (2002). Animal antibiotic use has an early but important impact on the emergence of antibiotic resistance in human commensal bacteria. PNAS *99*, 6434–6439.

Snel, J., Harmssen, H.J.M., Van der Wielen, P.W.J.J., and Williams, A. (2002). Dietary strategies to influence the gastrointestinal microflora of young animals and its potential to improve intestinal health. In: Nutrition and Health of the Gastrointestinal Tract, M.C. Blok, H.A. Vahl, L. de Lange, A.E. van de Braak, G. Hemke, and M. Hessing, eds. (Wageningon Academic Publishers, Wageningon, Netherlands) pp. 37–70.

Spatuzza, A. (2002). Antibiotic resistance: are we killing the cure? In WHO Perspectives in Health Magazine, the Magazine of the Pan American Health Organization Vol. 7, No. 1.

Steinman, M.A., Landefeld, C.S., and Gonzales, R. (2003). Predictions of broad spectrum antibiotic prescribing for acute respiratory tract infections in adult primary care. JAMA *289*, 719–725.

Sun, Q., Yin, J., Zuo, G.Y., and Meng, Q.Y. (2011). The irrational use of drugs in rural China: evidence from two provinces. Third International Conference for Improving Use of Medicines. November 14–18, Antalya, Turkey.

Tavernise, S. (2013). FDA restricts antibiotics use for livestock. New York Times (December 11). International Conference for improving use of medicines (#345).

Thuy, H.T.T., Nga, L.P., and Loan, T.T.C. (2011). Antibiotic contaminants in coastal wetlands from Vietnamese shrimp farming. Environ. Sci. Pollu. Res. *18*, 835–849.

Vervaeke, I.J., Decuypere, J.A., Dierick, N.A., and Hendericks, H.K. (1979). Quantitative multi evaluation of the energy metabolism influenced by virginiamycin and spiramycin used as growth promoters in pig nutrition. J. Anim. Sci. *79*, 846–856.

Vickers, H.R., Bagratuni, L., and Alexander, S. (1959). Dermatitis caused by penicillin in milk. Lancet *1*, 351–352.

Watkins, R.R., and Bonomo, R.A. (2013). Increasing prevalence of carbapenem-resistant Enterobacteriaceae and strategies to avert a coming crisis. Expert. Rev. Ann. Infect. Ther. *11*, 543–545.

WHO (1963). WHO expert committee on the public health aspect of the use of antibiotics in food and feedstuffs. (World Health Organization, Geneva, Switzerland).

WHO (2011).Website at www.who.int/world-health-day/2011. (World Health Organization, Geneva, Switzerland).

WHO (2012). The evolving threat of antimicrobial resistance. Options for action. (World Health Organization, Geneva, Switzerland). 120 pp.

Zuraw, L. (2013). Will FDA's voluntary plan actually reduce antibiotics in animal feed? Food Safety News (December 12). Published online: http://www.foodsafetynews.com/2013/12/fda-finalizes-guidance-for-phasing-out-antibiotics-in-food-animals/#.VA29fUvXYZY

Antibiotics for Emerging and Re-emerging Diseases

9

Kazuro Shiomi and Satoshi Ōmura

Abstract

Emerging and re-emerging infectious diseases are global problems, and a constant supply of new antibiotics is essential if we are to combat these diseases successfully. Among antibiotics currently used against these diseases, antiparasite antibiotics are to the fore and are the major focus of this chapter. For example, tetracyclines are used against malaria, acetylspiramycin and clindamycin are used for toxoplasmosis, amphotericin B is used for leishmaniasis, and trichomycin is deployed against trichomoniasis. Nitroimidazole compounds used for various protozoiasis are analogues of azomycin. Ivermectin is a great nematocide, used for onchocerciasis, lymphatic filariasis, and strongyloidiasis. Paromomycin is used for the treatment of cestodiasis. Recent research for new antiparasite antibiotics is also described. Viral diseases represent a significant and growing threat among the major emerging and re-emerging infectious diseases and studies in this line of research, particularly on antibiotics active against HIV and influenza virus, are briefly covered. Even within the human body, our own essential microbial flora can pose a considerable and extremely complicated threat, such as diarrheagenic *Escherichia coli* (O157:H7) which is responsible for an emerging disease. Fortunately, recent research has uncovered a new type of antibiotic that only inhibits pathogenicity and does not affect the growth of pathogenic *E. coli*. The new type of antibiotic is expected to avoid some problems of conventional antibiotics: development of resistant strains, expression of various toxins, and disruption of normal microbial flora.

Introduction

Emerging and re-emerging diseases are great threats throughout the world. No country, rich or poor, is adequately protected from either the arrival of a new disease on its territory or the subsequent disruption this can cause (WHO, 2007). Unknown diseases that have appeared in the past 20–30 years (including HIV/AIDS, Ebola, hepatitis C, Lyme disease, hantavirus pulmonary syndrome and Severe Acute Respiratory Syndrome (SARS)) are called 'emerging diseases'. They remain incurable, are a significant cause of suffering and death, and impose an enormous financial burden on society. Older diseases, previously controlled but now becoming a problem again (such as malaria, tuberculosis and bacterial pneumonia) are classified as 're-emerging diseases'.

The US National Institute of Allergy and Infectious Diseases classifies the emerging and re-emerging diseases into three groups (NIAID, 2006).

Group I contains pathogens newly recognized in the past two decades: viral diseases (Australian bat lyssavirus, hendra or equine morbilli virus, hepatitis C, hepatitis E, human herpesvirus 6, human herpesvirus 8, influenza, parvovirus B19, and SARS), bacterial diseases (*Bartonella henselae*, ehrlichiosis, *Helicobacter pylori*, and lyme borreliosis), microsporidiosis (*Encephalitozoon cuniculi*, *Encephalitozoon hellem*, *Enterocytozoon bieneusi*, and other microsporidia), and parasite diseases (acanthamebiasis and atypical babesia).

Group II are re-emerging pathogens, including prions, enterovirus 71, bacterial diseases (caused by *Clostridium difficile*, group A *Streptococcus*, and *Staphylococcus aureus*), and a fungal disease (caused by *Coccidioides immitis*).

Group III contains agents with bioterrorism potential, such as smallpox and anthrax and covers various viral, bacterial, fungal, and parasitic diseases. Diseases caused by HIV and diarrhoeagenic *Escherichia coli* (O157:H7) are well known emerging diseases, while tuberculosis and malaria are historical diseases which have been re-emerging in several parts of the world.

Resistance against antibiotics is one of the major causes for diseases to re-emerge. Methicillin-resistant *Staphylococcus aureus* (MRSA) and drug-resistant tuberculosis cannot now be treated by most antibiotics. The issue of tackling antibacterial antibiotic resistance is covered in the chapter by Spizek in this book. Many parasitic diseases are re-emerging in developed countries and malaria is re-emerging in certain regions due to drug-resistant strains. However, increases in other parasitic diseases are generally due to a variety of reasons, such as increase in international travel and trade, changes of environment, and people's particular food preferences.

In this chapter, we will focus predominantly on antibiotics derived from naturally occurring organisms that are active against parasitic diseases, together with coverage of some recent studies looking for new antiparasite antibiotics.

There are many emerging and re-emerging viral diseases, and many antiviral agents have been reported. However, no antiviral compound of microbial origin is currently being used clinically. In this chapter, we describe some research on anti-HIV and anti-influenza virus antibiotics, although fuller details of antiviral antibiotics are described in the chapter by Villa.

Diarrheagenic *E. coli* is an emerging disease which has reached epidemic status in some developed countries, and we will describe a new type of antibiotic that inhibits pathogenicity but does not inhibit the growth of such pathogenic *E. coli*.

Antiparasitic antibiotics

Antiprotozoal antibiotics

Parasites are organisms ranging from unicellular eukaryotes (protozoa) to multicellular eukaryotes (helminths). Some antibiotics are used as antiprotozoals; tetracyclines, polyethers, spiramycin, acetylspiramycin, paromomycin, clindamycin, amphotericin B, trichomycin, and fumagillin (Shiomi and Ōmura, 2004; Kita *et al.*, 2007). Antibiotics used practically are described, together with recent studies of new antibiotics.

Antimalarials

Malaria is a representative re-emerging disease and is one of the world's three major infectious diseases (the others are HIV and tuberculosis). Tetracyclines are broad-spectrum antibacterial agents active against Gram-positive and Gram-negative bacteria, mycoplasmas, rickettsiae, and chlamydiae (Fig. 9.1). They are polyketide antibiotics isolated from culture broths of the actinomycetes, *Streptomyces* spp. (Duggar, 1948; Finlay *et al.*, 1950; Porter *et al.*, 1952) or their semisynthetic derivatives. Tetracyclines are also used as supplements to quinine in the treatment of malaria caused by *Plasmodium falciparum*, when resistance to quinine has been reported in patients, and are also effective for prophylaxis of multidrug-resistant *P. falciparum* (WHO, 1995). They have potent but slow action against the asexual blood stages of all plasmodial species. They are also active against the primary intrahepatic stages of *P. falciparum*. In addition, tetracyclines have been used to treat infection caused by *Entamoeba histolytica*, *Giardia lamblia*, *Leishmania major*, *Trichomonas vaginalis*, and *Toxoplasma gondii* (Chopra and Roberts, 2001). Tetracycline treatment of filarial nematode-infected animals results in reduced numbers of adult worms and microfilariae, suggesting that tetracyclines may be beneficial for treatment of humans infected with filarial nematodes (Smith and Rajan, 2000).

A chloroplast-like organelle (apicoplast) is present in many species of the phylum Apicomplexa, containing *Plasmodium*, *Toxoplasma*, *Eimeria*, and *Cryptosporidium*. The organelle is essential for blood-stage development of *Plasmodium* and formation of the parasitophorous vacuole of *Toxoplasma*. Tetracyclines are protein synthesis inhibitors that block binding of aminoacyl-tRNA to the A site of 16S rRNA on the 30S subunit of the prokaryotic ribosome (Walsh, 2003). Malarial parasites have two extrachromosomal DNAs with

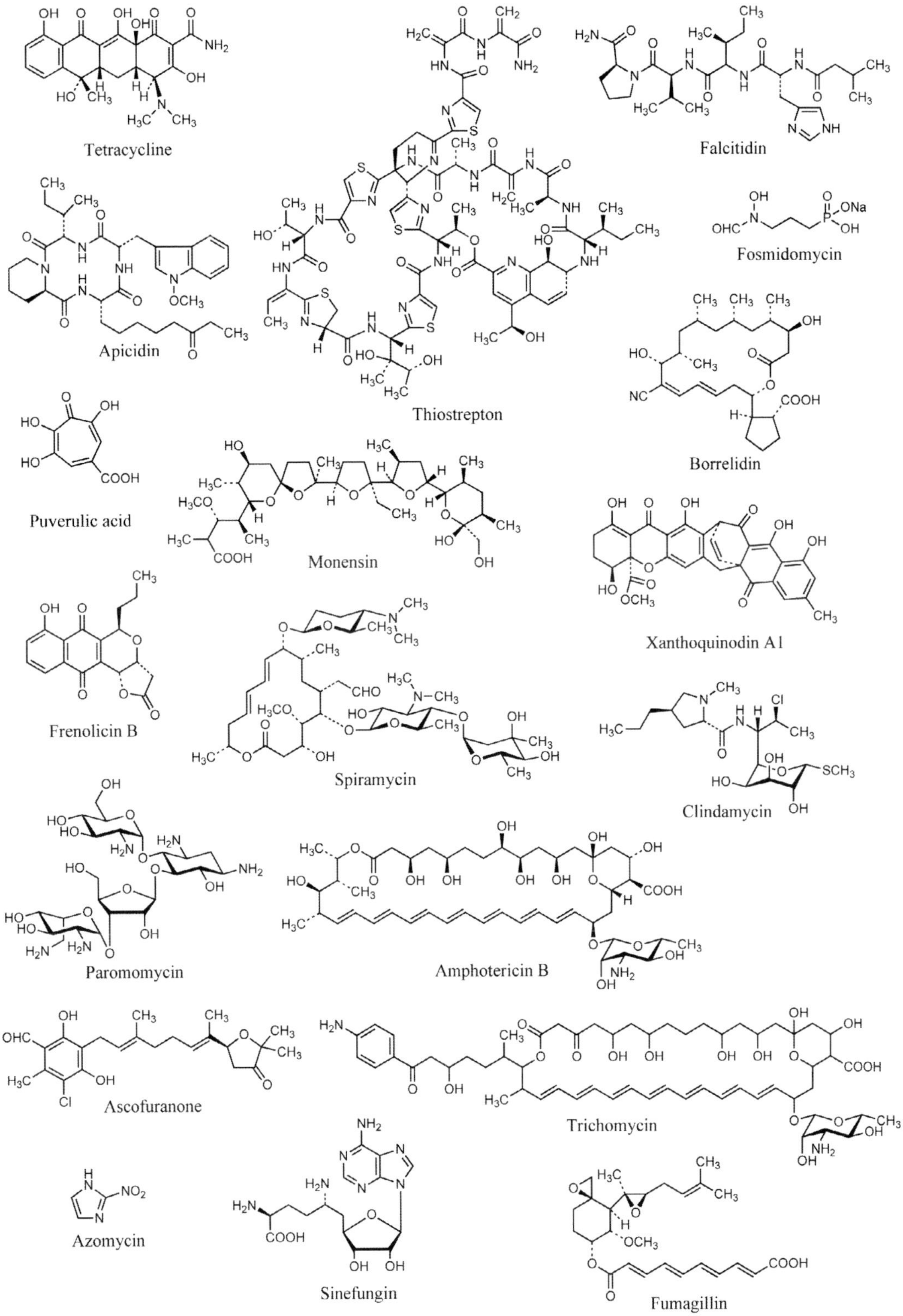

Figure 9.1 Antiprotozoal antibiotics.

prokaryotic organelle-like characteristics, apicoplast and mitochondrial DNA. Tetracyclines are suggested to inhibit apicoplast and mitochondrial protein synthesis in *Plasmodium* (Lin *et al*., 2002).

Apicidin is a cyclotetrapeptide produced by the fungus *Fusarium pallidoroseum*. It exhibits potent, broad-spectrum antiprotozoal activity *in vitro* against Apicomplexan parasites. It is orally and parenterally active against *Plasmodium berghei* malaria in mice (Darkin-Rattray *et al*, 1996). Its antiparasitic activity appears to be due to low nanomolar inhibition of Apicomplexan histone deacetylase, which induces hyperacetylation of histones in treated parasites. Antimalarial activities have also been reported for other peptide antibiotics, such as a cyclic peptide (takaokamycin) (Otoguro *et al*., 2003a), cyclic depsipeptides (enniatins) (Nilanonta *et al*., 2003), linear peptides (leucinostatin A (Otoguro *et al*., 2003a) and efrapeptins (Nilanonta *et al*., 2003)), and peptaibols (antiamebin I and zervamicins) (Nagaraj *et al*., 2001). A thiazole-containing cyclic depsipeptide, thiostrepton, produced by *Streptomyces azureus*, eliminates infection with erythrocytic forms of *P. berghei* in mice. It interacts with the GTPase binding domain of the apicoplast large subunit rRNA and inhibits protein synthesis, as it does in bacteria (Sullivan *et al*., 2000). A similar thiazole-containing cyclic peptide, micrococcin, showed much more potent antimalarial activity (IC_{50} 0.035 μM) than that of thiostrepton (IC_{50} 3.2 μM) *in vitro* (Rogers *et al*., 1998). Falcipain-2 is a cysteine protease used by *P. falciparum* to help degrade haemoglobin during the trophozoite stage of infection. An acyltetrapeptide falcitidin produced by a myxobacterium, *Chitinophaga* sp., inhibited falcipain-2 at the IC_{50} value of 6 μM (Somanadhan *et al*., 2013). Its antimalarial activity has not yet been reported.

It has recently been demonstrated that type II fatty acid biosynthesis occurs in the apicoplast, and an inhibitor of this pathway, thiolactomycin, produced by the actinomycete *Nocardia* sp., restricts the growth of *P. falciparum* and *Toxoplasma gondii* (Roberts *et al*., 2003). The type II pathway is common to plant chloroplasts and bacteria but distinct from the type I pathway of mammals. Furthermore, *Trypanosoma brucei*, which belongs to the phylum Sarcomastigophora, also uses type II fatty acid biosynthesis for myristate synthesis, and is susceptible to thiolactomycin. In some eubacteria, algae, and plants, the 1-deoxy-D-xylulose 5-phosphate (DOXP) pathway was described as an alternative non-mevalonate pathway for early steps in the biosynthesis of isoprenoids. This pathway is present in *Plasmodium*, and its DOXP reductoisomerase is inhibited by fosmidomycin, produced by *Streptomyces lavendulae* (Jomaa *et al*., 1999). Treatment with fosmidomycin for acute uncomplicated *P. falciparum* infection has been effective in clinical trials (Lell *et al*., 2003). *P. falciparum* phospholipase C hydrolyses sphingomyelin and lysocholine phospholipids, and is inhibited by a mammalian neutral sphingomyelinase inhibitor, scyphostatin, produced by the fungus *Trichopeziza mollissima* (Hanada *et al*., 2002). It also inhibits the intraerythrocytic proliferation of *P. falciparum*. Salinosporamide A (marizomib), produced by the marine actinomycete *Salinispora tropica*, is a proteasome inhibitor and is being tested in phase I trials for the treatment of multiple myeloma (Prudhomme *et al*., 2008). It was active against *Plasmodium yoelii*-infected mice and showed strong inhibitory activity against the erythrocytic stages of the parasite cycle.

Borrelidin is produced by *Streptomyces* spp. (Berger *et al*., 1949), and is a non-glycosidic 18-membered ring macrolide with a side chain of cyclopentanecarboxylic acid. It has various biological activities, such as antimicrobial, antiviral, herbicidal, insecticidal, and anti-tumour. The activities may be due to the inhibition of threonyl-tRNA synthetase and the activation of caspase-3 and caspase-8 (Kawamura *et al*., 2003). Our research group found that borrelidin shows excellent antimalarial activity against both chloroquine-sensitive and -resistant *P. falciparum in vitro* (Otoguro *et al*., 2003b). It also has antimalarial activity against chloroquine-sensitive (*P. berghei*) and -resistant (*P. yoelii*) strains in mice. The effective dose is lower than that of artemether, artesunate, and chloroquine. In the same screening, we found that polyether antibiotics have a high selective inhibitory effect against *P. falciparum in vitro*. Among these polyethers, X-206, produced by an actinomycete (Breger *et al*., 1951), shows the highest selectivity and potency (Otoguro *et al*., 2001). Although, it has potent *in vivo* antimalarial activity in mice, the therapeutic window is narrow compared with its selective toxicity *in vitro*. We also isolated five tropolone compounds from the culture broth of *Penicillium* sp. (Iwatsuki *et al.*, 2011). Among them, puberulic acid exhibited potent and selective antimalarial activity *in vitro*, suppressing 69% of malaria parasites (*P. berghei*) at 2 mg/kg (s.c.) *in vivo*.

Anticoccidials, anticryptosporidials, and antitoxoplasmals

Polyether antibiotics such as monensin (Haney *et al.*, 1968), lasalocid A, and salinomycin are produced by actinomycetes and are used as feed additives for the prevention of poultry coccidiosis. They are ionophore antibiotics that increase the influx of monovalent cations through the membrane, causing swelling and death of the parasites. The development of resistance to anticoccidial agents is now posing similar problems to those encountered with malaria resistance (Greif *et al.*, 2001). Since no active member from a new class of chemical substances had been developed recently, our group screened anticoccidial antibiotics using monensin-resistant *Eimeria tenella in vitro*. As a result, we isolated several new antibiotics, such as xanthoquinodins (Tabata *et al.*, 1993) and cytosaminomycins (Haneda *et al.*, 1994; Shiomi *et al.*, 1994). Xanthoquinodin A1, produced by the fungus *Humicola* sp., showed anticoccidial activity against a monensin-resistant strain *in vivo*. Frenolicin B was originally isolated as an antimycoplasmal antibiotic from *Streptomyces roseofulvus* by our group (Iwai *et al.*, 1978), and later was revealed to show potent anticoccidial activity *in vivo* (Ōmura *et al.*, 1985).

Spiramycin is a 16-membered macrolide produced by *Streptomyces ambofaciens* and used as an antibacterial in the veterinary field (Pinnert-Sindico *et al.*, 1955). It is also effective in treating patients with cryptosporidiosis (Moskovitz *et al.*, 1988). Its acetylated derivative, acetylspiramycin, is used as an antibacterial antibiotic for humans, and also used to treat toxoplasmosis (Maeda, 2003). Clindamycin is a semisynthetic antibiotic prepared from lincomycin, which is produced by *Streptomyces lincolnensis* (Mason and DeBoer, 1963) It is effective against toxoplasmosis and babesiosis (WHO, 1995). Spiramycins and clindamycin inhibit protein synthesis by binding to 23S rRNA on the 50S subunit of the prokaryotic ribosome. They bind adjacent to the peptidyltransferase centre of the 50S subunit, and their binding sites partially overlap (Walsh, 2003). They do not inhibit chromosomal or mitochondrial ribosomes (Pfefferkorn and Borotz, 1994). Clindamycin-resistant mutants of *Toxoplasma gondii* have a point mutation in the apicoplast large-subunit rRNA, and the position of the mutation corresponds to the position in *E. coli* predicted to bind clindamycin. The mutants show cross-resistance to macrolides. Therefore, spiramycins and clindamycin are believed to target the apicoplast ribosome (Camps *et al.*, 2002).

Paromomycin, produced by *Streptomyces rimosus*, is a member of the aminoglycoside antibiotics (Haskell *et al.*, 1959). It binds to 16S rRNA on the 30S subunit of the ribosome and increases the error rate in translation. The crystal structure of the 30S subunit complexed with paromomycin revealed that it binds in the major groove of helix 44 of the ribosome and flips out the bases A1492 and A1493 to increase the affinity of the A-site for tRNA (Carter *et al.*, 2000). Paromomycin, an anthelmintic for cestode infection, is also used for cryptosporidiosis, amoebiasis and giardiasis (Liu and Weller, 1996).

Antileishmanial and antitrypanosomals

Amphotericin B, produced by *Streptomyces nodosus*, is a widely used 38-membered ring heptaene macrolide with antifungal properties (Sternberg *et al.*, 1956). Amphotericin B or its lipid formulations (liposomal amphotericin B) are also used for visceral leishmaniasis (kala-azar) and mucocutaneous leishmaniasis that is unresponsive to pentavalent antimony compounds (WHO, 1995; Adler-Moore and Proffitt, 2002). The cell membrane of *Leishmania* parasites is composed of ergosterol-related sterols similar to those of fungi. Therefore, amphotericin B can bind to the membrane and kill the parasites. Although the growth of *Leishmania* parasites is susceptible to azoles and other ergosterol biosynthesis inhibitors, they can survive with greatly altered sterol profiles induced by continuous treatment with low concentration of the inhibitors, and they also have some ability to utilize and metabolize host sterol (Roberts *et al.*, 2003). Therefore, ergosterol biosynthesis inhibitors are not used in practice for leishmaniasis but are still in clinical studies.

Ascofuranone was originally isolated from the culture broth of the fungus *Ascochyta viciae* as an antibiotic that reduced the serum lipid levels of rats (Sasaki *et al.*, 1973). Recently, it was revealed to inhibit the trypanosome alternative oxidase, which is the terminal oxidase of the respiratory chain of long, slender bloodstream forms of *Trypanosoma brucei brucei* (Nihei *et al.*, 2002). Ascofuranone showed good therapeutic effects against *T. vivax* in mice (Yabu *et al.*, 2006).

Phenazinomycin, produced by *Streptomyces* sp., was originally isolated as an anti-tumour antibiotic by our group (Ōmura *et al.*, 1989). Recently, we found that phenazinomycin and some related phenazine antibiotics exhibit significant antitrypanosomal activities, both *in vitro* (IC_{50} 0.23 μg/ml against *T. b. brucei*) and *in vivo* (Otoguro *et al*, 2010). Sinefungin, a *S*-adenosylmethionine analogue, was isolated from the culture broth of *Streptomyces griseolus* and found to have antitrypanosomal properties (Hamill and Hoehn, 1973). It was known as a potent methyltransferase inhibitor (Pugh *et al.*, 1978). Sinefungin cured mice infected with *T. b. brucei* when administered at 5 mg/kg (i.p.) three times (Dube *et al.*, 1983). Sinefungin also showed leishmanicidal activity against *Leishmania donovani* at 0.13 μM *in vitro* (Paolantonacci *et al.*, 1985). We isolated a new analogue, sinefungin VA, which showed potent antitrypanosomal activity with an IC_{50} value of 0.0026 μg/ml *in vitro* without cytotoxicity (Niitsuma *et al.*, 2010).

Antitrichomonals

Trichomycin is used as an antifungal and antitrichomonal antibiotic. It is a mixture of polyene macrolides produced by *Streptomyces hachijoensis* (Hosoya *et al.*, 1952). The major component is a 38-membered ring heptaene macrolide, trichomycin A (Komori and Morimoto, 1990). Nitroimidazole group compounds are used as antiprotozoals. Metronidazole is used for trichomoniasis, amoebiasis and giardiasis, while benznidazole is used for American trypanosomiasis. They are derivatives of azomycin, isolated from the culture broth of *S. eurocidicus* (Maeda *et al.*, 1953). Many nitroimidazole compounds were synthesized after the antitrichomonal activity of azomycin was reported (Horie, 1956). Hitachimycin isolated by our group is a 19-membered ring lactam produced by *Streptomyces* sp. (Ōmura *et al.*, 1982), which displays antitrichomonal and antitumour activity.

Antimicrosporidial

Fumagillin was isolated from the culture broth of the fungus *Aspergillus fumigatus* as an antiphage and amoebicidal antibiotic (Eble and Hanson, 1951). It was used to control microsporidial disease due to *Nosema apis* in honeybees (Liu and Weller, 1996). Microsporidia are also responsible for opportunistic infections in patients with AIDS. Recently, fumagillin has been used topically to treat ocular microsporidiosis caused by *Encephalitozoon hellem* or *E. intestinalis* (Didier, 1997). Fumagillin is also known to possess antiangiogenic properties, and its target was revealed as methionine aminopeptidase 2 (MetAP2) (Sin *et al.*, 1997). Since microsporidia appear to lack MetAP1 and to carry only the gene for MetAP2, fumagillin should be particularly promising because the absence of MetAP1 would prevent microsporidia from compensating for the loss of MetAP2 activity when exposed to fumagillin (Didier *et al.*, 2003).

Anthelmintic antibiotics

Helminths include nematodes (roundworms), trematodes (flukes), and cestodes (tapeworms). Nematodes have a cylindrical or filamentous shape and belong to the phylum Aschelminthes. Trematodes and cestodes have a flat shape and belong to the phylum Platyhelminthes. Trematodes have a digestive tract, but cestodes do not, and are composed of many segments. Some antibiotics are used as anthelmintics, i.e. ivermectin, paromomycin, milbemycin D, destomycin A, hygromycin B, emodepside, and derquantel. Antibiotics used practically are described below, together with recent studies of new targets for antibiotics.

Avermectins and milbemycins

Avermectins and milbemycins are spiroketal 16-membered ring macrolides having good nematocidal and insecticidal activities. Avermectins have a sugar moiety at C-13, but milbemycins have no hydroxyl group at C-13 and therefore have no sugar moiety (Fig. 9.2). Avermectins were discovered by cooperative research between our group and a research group at Merck Sharp & Dohme Research Laboratories (MSD), USA (Ōmura and Crump, 2004). We isolated cultures from soil samples and undertook *in vitro* evaluation of the fermentation broths. In MSD's laboratories, promising cultures were screened in a novel model of helminth infection using mice infected with the nematode *Nematospiroides dubius*. As a result, the broth of a strain MA-4680 exhibited potent anthelmintic activity (Egerton *et al.*, 1979), the strain originating in a soil sample collected at an ocean-side site at Kawana, Shizuoka prefecture, Japan (Burg *et al.*, 1979). It was classified as a new species of actinomycetes and named *Streptomyces avermectinius* (formerly *S. avermitilis*) (Takahashi *et al.*, 2002). Eight active components were isolated from the broth and named avermectins A_{1a}-B_{2b}. Compounds of the B-series containing a 5-hydroxy group are markedly more active than the A-series, which contain a 5-methoxy group. Reduction of the 22,23-olefin of the most active avermectin B_{1a} and avermectin B_{1b} improved both the activity spectrum and safety, and the resulting 22,23-dihydro B_1 complex (as a mixture of 80% B_{1a} and 20% B_{1b}) was marketed as a veterinary drug under the generic and non-proprietary name, ivermectin, in 1981 (Chabala *et al.*, 1980).

Ivermectin, an endectocide active against a diverse variety of nematodes, insects and arachnids – is now recognized by many as one of the most significant human and animal health drugs of the last 50 years. Initially, it was launched for control of economically important gastrointestinal, lung, and kidney nematodes, as well as other nematodes, such as *Thelazia* and *Parafilaria*, in livestock. It is also used against a number of arthropod parasites, including grubs, lice, mites, and screw worms (Shoop and Soll, 2002). In companion animals, it provides good control of *Dirofilaria immitis*. Ivermectin was later found to be safe and highly effective in controlling some human nematode infections. Of greatest significance, it proved highly efficacious against microfilariae of *Onchocerca volvulus*, which causes human onchocerciasis (river blindness), an infection characterized by chronic skin and eye lesions. The unique, pioneering, multifaceted public/private partnership (PPP) established, which eventually incorporated several UN agencies and a variety of non-governmental organizations (NGOs), as well as civil society, saw the donation of ivermectin (under the brand name Mectizan®) which led to a model and highly successful mass drug administration (MDA) programme which should culminate in the elimination of onchocerciasis in the endemic areas of Africa and the Americas (Richards *et al.*, 2001). Ivermectin is a remarkably safe and effective broad-spectrum drug and side-effects when they do occur are minimal and tolerable, generally resulting from inflammatory and immune defences triggered by immobilized or dead microfilariae (Richard-Lenoble *et al.*, 2003). This is why ivermectin can be administered by non-medical individuals after a modicum of training, and why it can be used in MDA programmes run by affected communities themselves. Although ivermectin is also effective against *Loa loa* (Richard-Lenoble *et al.*, 2003), a small number of severe adverse events have occurred in individuals with high levels of *L. loa* microfilariae treated with ivermectin (Geary, 2005). More recently, ivermectin has also become an invaluable tool for a global elimination initiative against lymphatic filariasis (elephantiasis), a disease caused by infection with *Wuchereria bancrofti* and *Brugia malayi* (Richard-Lenoble *et al.*, 2003), meaning that the drug is being used, free of charge, to rid the world of two devastating diseases and improve the lives of hundreds of millions of the world's poorest people. Strongyloidiasis, distributed in the tropics and subtropics, is an intestinal disease caused by the nematode *Strongyloides stercoralis*. Ivermectin has now been found to be the most useful agent for strongyloidiasis (Zaha *et al.*, 2002). In addition to the control of nematode infections, ivermectin also shows efficacy against human scabies, caused by infection with the ectoparasitic mite, *Sarcoptes scabiei* (Heukelbach and Feldmeier, 2006).

Avermectin B_1, under the non-proprietary name abamectin, is used as a gastrointestinal nematocide through subcutaneous injection for

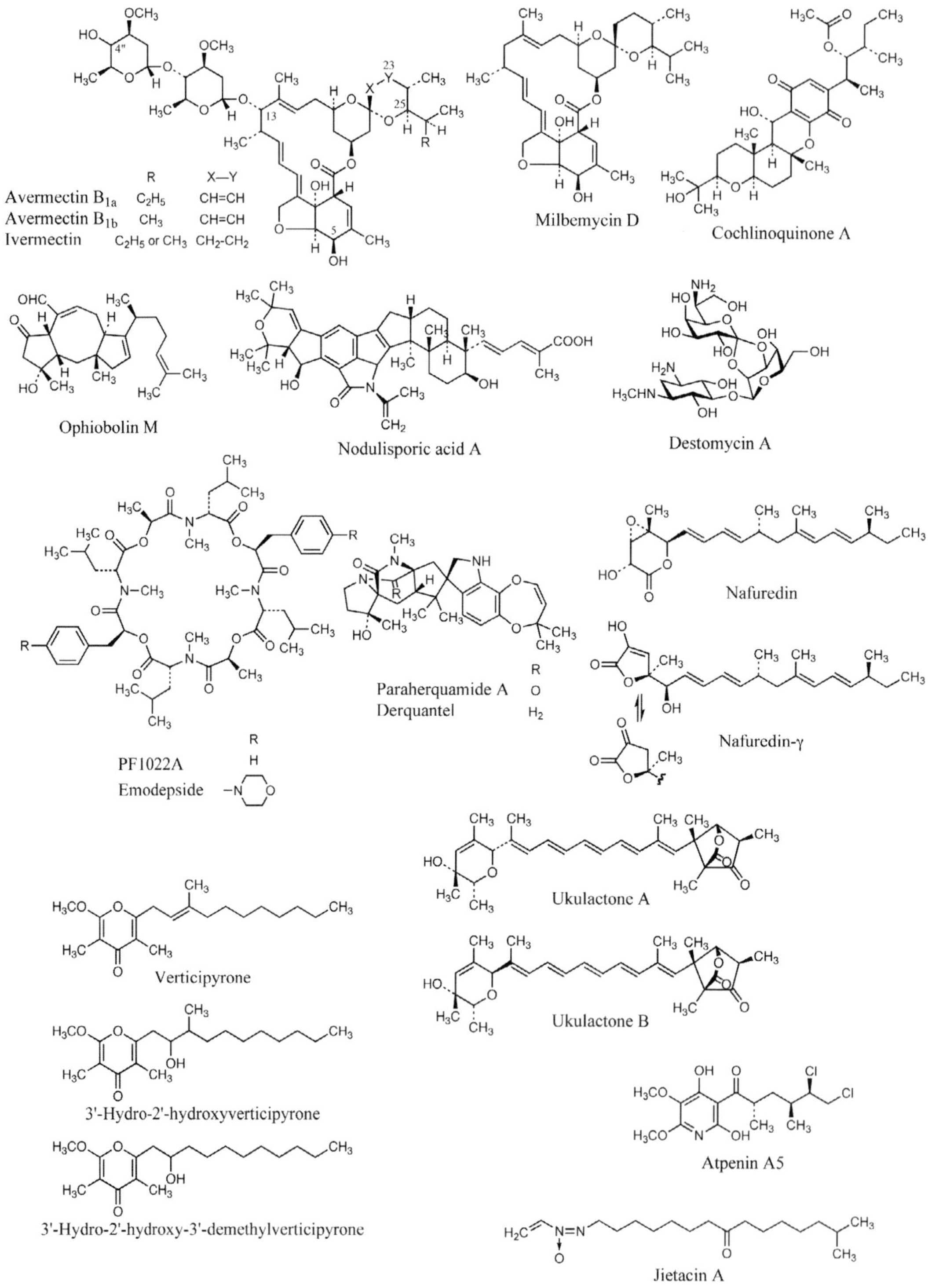

Figure 9.2 Anthelmintic antibiotics.

cattle and also as an agricultural miticide (Shoop and Soll, 2002). Doramectin is a C-25 substituted avermectin B_1 produced by mutational biosynthesis (Dutton *et al.*, 1991). Cyclohexane carboxylic acid was fed to a mutant strain of *S. avermectinius* lacking branched-chain 2-oxo acid dehydrogenase activity. Doramectin was developed for uses similar to those of ivermectin in the veterinary field. Other semisynthetic avermectins are commercially available as nematocides or insecticides (Sunazuka *et al.*, 2002). Eprinomectin, a 4″-epi-(acetylamino)-4″-deoxy derivative of avermectin B_1, is used topically for control of gastrointestinal and the other parasitic nematodes and a number of arthropods in cattle, including lactating cows. It possesses one of the lowest milk partitioning coefficients in this class of antiparasitics. Emamectin, a 4″-epi-(methylamino)-4″-deoxy derivative of avermectin B_1, is used as an agricultural insecticide, particularly against the order Lepidoptera. Selamectin, a 5-oxoimino derivative of doramectin monosaccharide, is used to control nematodes and fleas in cats and dogs.

In 2003, our group elucidated the complete genetic sequence of *S. avermectinius* (Ikeda *et al.*, 2003). The genome is the largest bacterial genome yet sequenced (9026 kb), encoding at least 7574 potential open reading frames. Analysis identified the gene cluster involved in secondary metabolite production, shedding light on the biology of microbial secondary metabolite synthesis at the genetic level and offering the prospect of creating some new metabolites.

Milbemycins were originally isolated from the culture broth of *Streptomyces hycroscopicus* (Takiguchi *et al.*, 1980). Milbemycin D is used as a nematocide in the veterinary field, and milbemectin, a mixture of milbemycins A_3 and A_4, is used as an agricultural acaricide (Sunazuka *et al.*, 2002). Milbemycin oxime, a 5-oxoimino derivative of milbemectin, has been launched as a nematocide for dogs. Nemadectin is a milbemycin group antibiotic produced by *Streptomyces cyaneogriseus* (Carter *et al.*, 1988). Moxidectin is a 23-methyloxime derivative of nemadectin, used for control of nematodes and arthropods in the veterinary field.

Avermectins target a parasite-specific glutamate-gated chloride channel (Ōmura, 2002), activating the channel, causing neurological disruption. It is an anion-selective Cys-loop receptor. The crystal structure of the homopentameric receptor of *Caenorhabditis elegans* complexed with ivermectin revealed that it binds in the transmembrane domain of the receptor and stabilizes an open-pore conformation (Hibbs and Gouaux, 2011). Although avermectins also bind to GABA- and glycine-gated chloride channels in mammals, their affinity for invertebrate receptors is approximately 100-fold higher. Knockout mice of the P-glycoprotein gene showed increased sensitivity to ivermectin (Schinkel *et al.*, 1990). This result established an important role for the P-glycoprotein in the maintenance of the blood–brain barrier, and suggested that the safety of ivermectin is due to the blood–brain barrier, dependent on P-glycoprotein, in addition to the low affinity of the receptors.

The fungal toxins cochlioquinone A and ophiobolins were isolated as nematocidal compounds. Cochlioquinone A, produced by the fungi *Cochliobolus miyabeanus* and *Helminthosporium sativum*, showed competitive inhibition against [^{3}H]ivermectin binding to membranes prepared from the free-living nematode *C. elegans* (Schaeffer *et al.*, 1990). Conversely, ophiobolin M and its analogues, produced by *Bipolaris* spp. fungi, are non-competitive inhibitors of ivermectin binding (Tsipouras *et al.*, 1996). It has been suggested that cochlioquinone A and ophiobolins may have a similar target to ivermectin. A potent insecticidal antibiotic, nodulisporic acid A, and its analogues were isolated from the culture broth of an endophytic fungus, *Nodulisporium* sp. (Ondeyka *et al.*, 1997). Nodulisporic acid A activates a glutamate-gated chloride channel, like ivermectin, but lacks activity at GABA- and glycine-gated chloride channels. However, no useful anthelmintic activity was observed *in vitro* or *in vivo* (Meinke *et al.*, 2002).

Aminoglycosides

Aminoglycoside antibiotics are protein synthesis inhibitors, and some of them are used as anthelmintics. Paromomycin is used for treatment of human and animal cestode infections (Salem and El-Allaf, 1969). It is also useful for some protozoan infections as described above. Destomycin A and hygromycin B are produced by *Streptomyces rimofaciens* and *S. hygroscopicus*, respectively (Kondo *et al.*, 1965; Mann and Bromer, 1958). They are aminoglycosides used as feed additives for treatment of nematode infections by *Ascaris, Oesophagostomum, Heterakis, Trichuris,* and *Capillaria* in the veterinary field (Ishii, 1998). Our group found that hygromycin A showed good therapeutic effect for swine dysentery caused by a spirochaete *Treponema hyodysenteriae* (Ōmura *et al.*, 1987b; Nakagawa *et al.*, 1987).

Emodepside

PF1022A is a cyclic octadepsipeptide nematocide produced by the fungus *Rosellinia* sp. (Sasaki *et al.*, 1992). Emodepside is a semisynthetic derivative of PF1022A with a morpholine ring at each of the two D-phenyllactic acids in the para position (Harder and von Samson-Himmelstjerna, 2002). *Rosellinia* sp. was genetically manipulated to express alternative substrates for PF1022A biosynthetic enzymes allowing an amino derivative (a precursor of emodepside) to be produced biologically (Yanai *et al.*, 2004). Its target was identified as a novel 110 kDa heptahelical transmembrane receptor, named HC-110R, which is similar to mammalian latrophilins (Saeger *et al.*, 2001). Latrophilins are latrotoxin (black widow spider venom) receptors and G protein-coupled receptors implicated in the regulation of exocytosis. Latrotoxin also binds to HC110-R and causes influx of external Ca^{2+}. Emodepside acts as an antagonist to latrotoxin signalling by impairing the influx of Ca^{2+}. It is highly effective against adult stages of the rat nematodes *Nippostrongylus brasiliensis* and *Strongyloides ratti,* and against the mouse nematode *Heligmosomoides polygyrus,* in an oral-dosage range of 1.0–10 mg/kg (Harder and von Samson-Himmelstjerna, 2002). It possesses resistance-breaking properties against the usual classes of anthelmintics and is used for gastrointestinal nematodes of cats and dogs.

Paraherquamide group compounds

Brevianamides, paraherquamides, and marcfortines are a group of oxindole alkaloids produced by the fungus *Penicillium* spp. (Birch and Wright, 1970; Yamazaki *et al.*, 1980; Polonsky *et al.*, 1980). Their nematocidal activity was initially detected in a gerbil and *Trichostrongylus colubriformis* model using paraherquamide A (Ostlind *et al*, 1990). They do not have insecticidal activity. A derivative of paraherquamide A, derquantel (2-deoxoparaherquamide A), has an excellent nematocidal activity and safety profile, and was launched for sheep on 2010 as a combination drug with abamectin (Epe and Kaminsky, 2013). Derquantel is an antagonist of B-subtype nicotinic acetylcholine receptors of nematodes and block cholinergic neuro-muscular transmission (Qian *et al.*, 2006). There are differences between the sensitivity of nematodes and mammals to paraherquamides due to intrinsic differences in receptor affinity between these phyla (Zinser *et al.*, 2002).

NADH-fumarate reductase inhibitors

Energy metabolism in many adult helminth parasites differs from that in larvae and hosts (Komuniecki and Tielens, 2003; Sakai *et al.*, 2012). They produce ATP in low oxygen concentration using a special respiratory system (Fig. 9.3). Phosphoenolpyruvate produced via an anaerobic glycolytic pathway is converted to oxalacetate by phosphoenolpyruvate carboxykinase, and oxalacetate is metabolized to malate and then to fumarate. Electrons from NADH are accepted by rhodoquinone through complex I (NADH-rhodoquinone reductase) and then transferred to fumarate through complex II (rhodoquinol-fumarate reductase). Therefore, the anaerobic complex II catalyses the reverse reaction of the aerobic complex II (succinate-ubiquinone reductase). The end-products of this glucose catabolism are volatile fatty acids, such as 2-methylpentanoate. This anaerobic electron transport system can provide ATP in the absence of oxygen. Rhodoquinone is used for this anaerobic respiration. Having a lower redox potential than ubiquinone enables rhodoquinone to be used for the reverse reaction of aerobic complex II. NADH-fumarate reductase (NFRD), composed of the above complex I and complex II, plays an important role in the anaerobic respiratory system. We screened for inhibitors of NADH-fumarate reductase using *Ascaris suum* (roundworm) mitochondria and discovered some helminth-specific complex I inhibitors.

Nafuredin was isolated from the culture broth of the fungus *Aspergillus niger,* obtained from a marine sponge collected in the Palau Islands, Republic of Palau (Ōmura *et al.*, 2001; Ui *et al.*, 2001). Production of nafuredin could

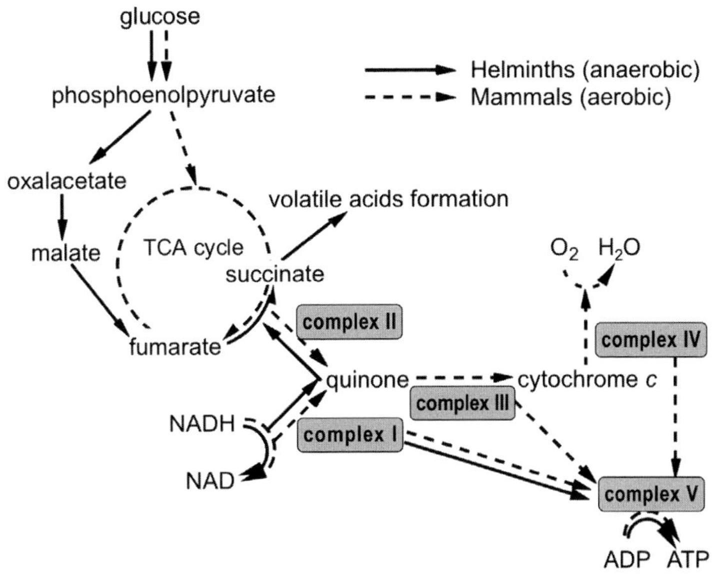

Figure 9.3 Difference of energy metabolism between aerobic mammals and anaerobic helminths.

be enhanced by the addition of natural seawater (25–100%) (Masuma *et al.*, 2001). After the isolation of nafuredin, more than five fungal strains have been found as producers of nafuredin during screening for NFRD inhibitors. It is interesting that all strains were terrestrial fungi, and moreover, all were *Trichoderma* spp. The structure of nafuredin was elucidated by NMR (Ui *et al.*, 2001), and the total synthesis revealed its absolute configuration (Takano *et al.*, 2001a,b). It has a β,γ-epoxy-δ-lactone ring with an alkenyl side chain. The biosynthesis study suggested that nafuredin is composed of nine acetates and four methionines (branched four methyl carbons).

As shown in Table 9.1, nafuredin inhibited NFRD at the IC_{50} value of 12 nM (Ōmura *et al.*, 2001). Since NFRD consists of complex I and complex II, inhibitory activity against each enzyme of *A. suum* was evaluated. Nafuredin potently inhibited complex I (IC_{50} = 24 nM) and the inhibition against complex II was very weak (IC_{50} = 80 μM). It is interesting that nafuredin inhibited not only the anaerobic adult complex I of *A. suum* (NADH-rhodoquinone reductase) but also the larval complex I (NADH-ubiquinone reductase), which has an aerobic energy metabolism like that of mammals. The IC_{50} value against larval complex I was 8.9 nM. Therefore, nafuredin is a complex I inhibitor. However, nafuredin only inhibited rat liver complex I at very high concentration (IC_{50} = 10 μM). Therefore, nafuredin seems to be a selective inhibitor of helminth complex I. From a kinetic study against NADH-rhodoquinone reductase of adult *A. suum*, nafuredin revealed competitive inhibition with rhodoquinone and non-competitive inhibition with NADH, which indicated that the site of inhibition of nafuredin is the quinone-binding domain in complex I. *Haemonchus contortus* (barberpole worm) is reported to have an NFRD system (Van Hellemond *et al.*, 1995). Therefore, *in vivo* anthelmintic activity of nafuredin was evaluated using *H. contortus*-infected sheep (Ōmura *et al.*, 2001). Nafuredin (2 mg/kg p.o.) significantly reduced the number of faecal eggs in infected sheep. This anthelmintic activity may be due to the inhibition of complex I, because nafuredin also inhibits the enzyme of *H. contortus*. Moreover, nafuredin was effective in mice infected with dwarf tapeworm, *Hymenolepis nana*. There were no signs of any side-effects and no loss of body weight during tests in either sheep (2 mg/kg p.o.) or mice (50 mg/kg p.o. and i.p.).

During total synthesis of nafuredin, it was found to be converted into a novel γ-lactone derivative, named nafuredin-γ, under mild basic conditions (Nagamitsu *et al.*, 2003; Shiomi *et

Table 9.1 Inhibitory activities (IC_{50} values) of nafuredin against electron transport enzymes

	Electron transport enzyme				
	NADH-fumarate reductase	NADH-ubiquinone reductase	NADH-rhodoquinone reductase	Rhodoquinol-fumarate reductase	Succinate-ubiquinone reductase
Complex	I+II	I	I	II	II
Ascaris suum (adult)	12	8.0	24	80,000	>100,000
Ascaris suum (L2)	NT	8.9	9	NT	NT
Haemonchus contortus (adult)	NT	86	195	NT	NT
Rat liver	1000	10,000	>100,000	NT	>100,000

NT, not tested.

al., 2005a). The epoxide of nafuredin opens and the δ-lactone recyclizes to form γ-lactone with keto–enol tautomerism. Nafuredin-γ inhibited NADH-rhodoquinone reductase of adult *A. suum* with an IC_{50} value of 2.3 nM, and did not inhibit bovine liver NADH-ubiquinone reductase at 10 μM. It showed anthelmintic activity against *H. contortus* using treatment with two oral doses each of 2 mg/kg. Thus, nafuredin-γ has similar enzyme inhibitory and anthelmintic activities as those of nafuredin. Though nafuredin was converted into nafuredin-γ in basic condition, only a part of nafuredin was converted in neutral buffer. Therefore, it is not likely that nafuredin was converted into nafuredin-γ and showed inhibitory activity. Both nafuredin and nafuredin-γ may inhibit complex I directly.

Verticipyrone was produced by the fungus *Verticillium* sp., and the structure was shown to be (*E*)-2-methoxy-3,5-dimethyl-6-(3-methyl-2-undecenyl)-4*H*-pyran-4-one by NMR studies (Ui *et al.*, 2006). MIC values of verticipyrone against *C. elegans* and *A. salina* were 20 μg/ml and 2.0 μg/ml, respectively. Verticipyrone inhibited NFRD from *A. suum*, with an IC_{50} value of 4.1 nM, and the inhibition was specific to NADH-rhodoquinone reductase (complex I). However, its inhibitory activity against NADH-ubiquinone reductase (complex I) from bovine heart was similar to that of *A. suum* complex I. The total synthesis of verticipyrone has been accomplished and some analogues of verticipyrone have been prepared (Shimamura *et al.*, 2007). The two alcohols 3′-hydro-2′-hydroxyverticipyrone and 3′-hydro-2′-hydroxy-3′-demethylverticipyrone showed much potent NFRD inhibitory activities than verticipyrone, suggesting that the newly introduced 2′-hydroxy group on the side chain may contribute to the inhibition. The analogue 3′-hydro-2′-hydroxyverticipyrone showed > 20-fold more potent inhibition against *Ascaris* complex I compared with that of verticipyrone, whereas its inhibition against bovine heart complex I was four times less potent than that of verticipyrone. Therefore 3′-hydro-2′-hydroxyverticipyrone is a selective inhibitor of *Ascaris* complex I. Alveolar echinococcosis caused by larval *Echinococcus multilocularis* is a life-threatening parasitic zoonosis. Recently, *E. multilocularis* was found to use an anaerobic NADH-fumarate reductase system for its energy metabolic pathways (Matsumoto *et al.*, 2008). Therefore, the efficacy of 3′-hydro-2′-hydroxy-3′-demethylverticipyrone (the most potent NFRD inhibitor among the analogues of verticipyrone) against larval *E. multilocularis* was evaluated. The viability of the *E. multilocularis* protoscolex was progressively reduced during *in vitro* treatment of the parasites with 3′-hydro-2′-hydroxy-3′-demethylverticipyrone, and more than 90% of the parasites were eliminated on days 5 and 18 when the compound was used at 50 μM and 5 μM, respectively.

Ukulactones A and B are produced by the fungus *Penicillium* sp. (Mori *et al.*, 2011). They have the same planar structure, with 2-oxabicyclo[2.2.1]heptane-3,5-dione and 5,6-dihydro-2*H*-pyran linked by pentaene. The configurations at C-2 of the pyran (junction of the pyran and the pentaene) are opposite between ukulactones A and B. The absolute configurations of ukulactones A and B have not yet been elucidated. Ukulactone A inhibited the growth of parasite nematode *Nippostrongylus brasiliensis* at 1 μg/ml *in vitro*. Ukulactone A inhibited *Ascaris* NFRD and is an NADH-rhodoquinone reductase (complex I)-specific inhibitor (IC_{50} = 55 nM). Its inhibition against bovine heart NADH-ubiquinone reductase (complex I) was about 500 times weaker (IC_{50} = 28 μM) than that of *Ascaris* complex I. The inhibitions against bovine complexes II and III were also weak. Therefore, ukulactone A is a selective inhibitor of *Ascaris* complex I, comparable to nafuredin. The difference between ukulactones A and B is the configuration of only one carbon. However, the inhibition by ukulactone B against *Ascaris* NFRD was about 200 times weaker than that of ukulactone A. Thus, the configuration at C-2 of the pyran may be very important for the inhibitory activity.

In the same screening, we rediscovered atpenin A5 and related compounds as NFRD inhibitors. Atpenins were originally isolated from the culture broth of *Penicillium* sp. as antifungal antibiotics by our group (Ōmura *et al.*, 1988). In contrast to the other NFRD inhibitors, atpenins inhibited complex II (succinate-ubiquinone oxidoreductase), and the inhibition was non-selective between helminths and mammals (Miyadera *et al.*, 2003). Although there had been potent inhibitors of complexes I, III, and IV, no significantly potent inhibitor of complex II had previously been identified. The IC_{50} value of atpenin A5 is 300-fold lower than that for carboxin, the most potent complex II inhibitor until then. Therefore, atpenins are very useful tools for clarifying the biochemical and structural properties of complex II. The X-ray crystallographic structure of the *E. coli* complex II was elucidated (Yankovskaya *et al.*, 2003) and the co-crystal structure of the enzyme and atpenin A5 was revealed (Horsefield *et al.*, 2006). The structures of mammalian complex II and helminth complex II have been reported (Sun *et al.*, 2005; Shimizu *et al.*, 2012). Recently, the co-crystal structure of porcine and *Ascaris* complex II with atpenin A5 was elucidated (Harada *et al.*, 2011; Shimizu *et al.*, 2013). These data may give a clue to design of helminth-specific atpenin analogues.

Other anthelmintics

Avermectin and milbemycin group macrolides are used as nematocides as described above. Some other macrolides are also reported to have nematocidal activities. We showed that the 26-membered ring macrolides, oligomycins A and G, produced by *Streptomyces* sp. have nematocidal activity *in vitro*, but they are also cytotoxic (Enomoto *et al.*, 2001). Oligomycins are complex V (H^+-transporting ATP synthase) inhibitors. A 14-membered ring macrodiolide, clonostachydiol, produced by the fungus *Clonostachys cylindrospora*, and a 16-membered ring macrodiolide, pyrenophorol (helmidiol), produced by some fungi, were active against *H. contortus in vivo* (Grabley *et al.*, 1993; Kind *et al.*, 1996). Axenomycins are 34-membered ring macrolides with sugar and naphthoquinone moieties produced by *Streptomyces lysandri*. They show *in vivo* cestocidal activity (Arcamone *et al.*, 1973; Della Bruna *et al.*, 1973). Two new macrolactams produced by the actinomycete *Nonomuraea turkmeniaca* were reported as anthelmintics (Ayers *et al.*, 2007). A 14-membered ring macrolactam, i.e. *N*-methylfluvirucin A_1 showed *in vitro* activity (EC_{90} 29 μg/ml) against *H. contortus* L3 larvae and *in vivo* activity (42% reduction in worm counts) against *Heligmosomoides polygyrus* at 50 mg/kg in mice (s.c.).

Griseulin is an aromatic nitro-containing compound with an α-pyrone moiety produced by *Streptomyces griseus* (Nair *et al.*, 1993). It gave 100% mortality after 24 hours at 1 ppm against nematodes, *Panagrellus redivivus*, *C. elegans*, and *Heterodera glycines*. MK7924 and roselipins 3A–3E have a highly methylated polyketide skeleton with mannose and showed anthelmintic activity (Kumazawa *et al.*, 2003; Ayers *et al.*, 2010). Roselipin-type compounds were originally isolated from the culture broth of the fungus *Gliocladium roseum* as diacylglycerol acyltransferase inhibitors (Ōmura *et al.*, 1999). Jietacins A and B, which were isolated from the culture broth of *Streptomyces* sp. by our group, have unique vinylazoxy moieties (Ōmura *et al.*, 1987a). They were isolated based on screening using the pine wood nematode, *Bursaphelenchus lignicolus* (Otoguro *et al.*, 1988), and they show 10-fold higher nematocidal activity than avermectin B_{1a} against *B. lignicolus in vitro*.

Antiviral antibiotics

Anti-HIV antibiotics

The human immunodeficiency virus (HIV) targets the immune system and weakens people's surveillance and defence systems against infections and some types of cancer. The most advanced stage of HIV infection results in Acquired Immunodeficiency Syndrome (AIDS), which can take 2–15 years to develop, depending on the individual (WHO, 2013a). The number of people living with HIV in 2012 was 35.3 million, people newly infected with HIV in 2012 was 2.3 million, and AIDS deaths in 2012 totalled 1.6 million (WHO 2013b). AIDS is responsible for the highest number of deaths caused by infectious diseases. Many anti-HIV compounds produced by microorganisms have been reported. Some of those compounds are described below.

Pradimicins, produced by a new species of the actinomycete *Actinomadura hibisca*, showed potent antifungal activities (Oki *et al.*, 1998). Pradimicin A (Fig. 9.4) inhibited syncytium formation in co-cultures of MOULT-4 and MOULT-4/human immunodeficiency virus $(HIV)_{HTLV\text{-}IIIB}$ cells at a concentration of more than 12.5 μg/ml (Tanabe *et al.*, 1998). Pradimicin A has a lectin-like ability to recognize D-mannopyranoside in the presence of Ca^{2+} ion. Recently, a solid-state NMR study revealed the cavity consisted of a D-alanine moiety and ABC rings of pradimicin A at the D-mannopyranoside binding site (Nakagawa *et al.*, 2011). Our group isolated a new 114-amino-acid peptide, actinohivin, using a similar syncytium formation assay (Inokoshi *et al.*, 2001; Chiba *et al.*, 2001). Actinohivin is produced by a new genus of the actinomycete *Longispora albida*, and the peptide has internal sequence triplication. Actinohivin potently inhibited T-cell line (T) and macrophage (M)-tropic syncytium formation in co-incubation of HeLa/T-env/Tat and HeLa/CD4/Lac-Z cells and of HeLa/M-env/Tat and HOS/CD4/CCR5/Lac-Z cells. It bound to gp120 from both the T- and M-tropic HIV-1 in a concentration-dependent manner (Chiba *et al.*, 2004). X-ray crystal analysis of actinohivin revealed a 3D structure containing three sugar-binding pockets. Thus, its strong specific affinity to gp120 is considered to be due to multivalent interaction of the three sugar-binding pockets with three high-mannose type glycans of gp120, via the cluster effect of lectin (Tanaka *et al*, 2009). Actinohivin exhibited nanomolar neutralizing activity against primary viruses with diverse cellular tropisms, but did not cause either the mitogenicity or cytotoxicity that are often associated with other anti-HIV lectins (Matoba *et al.*, 2010). Our group also isolated isochromophilones and chloropeptins from culture broths of *Penicillium multicolor* and *Streptomyces* sp., respectively, from the screening for gp120-CD4 binding inhibitors (Matsuzaki *et al.*, 1995; Tanaka, *et al.*, 1997). Chloropeptin I inhibited the HIV-induced cytopathic effect at the EC_{50} value of 1.6 μM. Chloropeptin II was identical to complestatin, which shows anti-complement activity (Kaneko *et al.*, 1989). Concentricolide, isolated from the fruiting bodies of the xylariaceous ascomycete *Daldinia concentrica*, inhibited syncytium formation between HIV-1 infected cells and normal cells at the EC_{50} value of 0.83 μg/ml (Qin *et al.*, 2006). Siamycins (Tsunakawa *et al.*, 1995) and FR901724 (Nakashima *et al.*, 1996) are structure-related tricyclic 21-amino-acid peptides produced by *Streptomyces* spp., which inhibit HIV–cell fusion. Feglymycin, produced by *Streptomyces* sp., is a linear 13-amino-acid peptide and inhibits HIV replication in the lower μM range. It is a gp120-CD4 binding inhibitor (Vertesy *et al.*, 1999; Ferir *et al.*, 2012).

HIV protease inhibitors, such as saquinavir and ritonavir, are important drugs for AIDS treatment. The inhibitors commercially used are all synthetic ones, but HIV protease inhibitors of microbial origin have been studied by many groups. α-MAPI, a tetrapeptide derivative previously known as a microbial alkaline protease inhibitor, was isolated from the culture broth of *Streptomyces* sp. as an HIV protease inhibitor (Stella *et al.*, 1991). It inhibited HIV protease at the IC_{50} value of 2 μM (Stefanelli *et al.*, 1995). RPI-856 A–D are hexa- and heptapeptide derivatives containing a

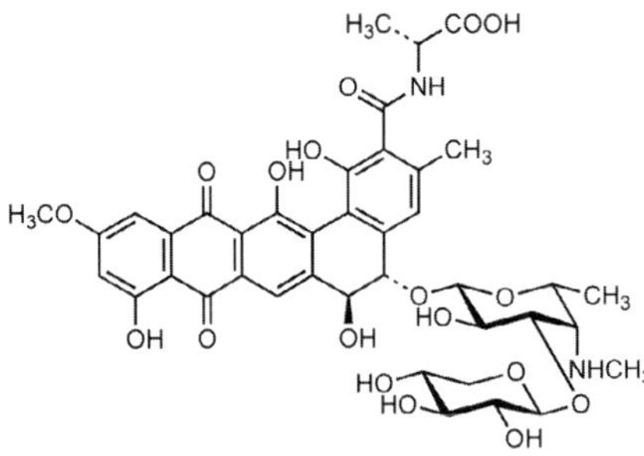

Pradimicin A

1-38 ASVTIRNAQTGRLLDSNYNGNVYTLPANGGNYQRWTGP
39-76 GDGTVRNAQTGRCLDSNYDGAVYTLPCNGGSYQKWLFY
77-114 SNGYIQNVETGRVLDSNYNGNVYTLPANGGNYQKWYTG

Actinohivin

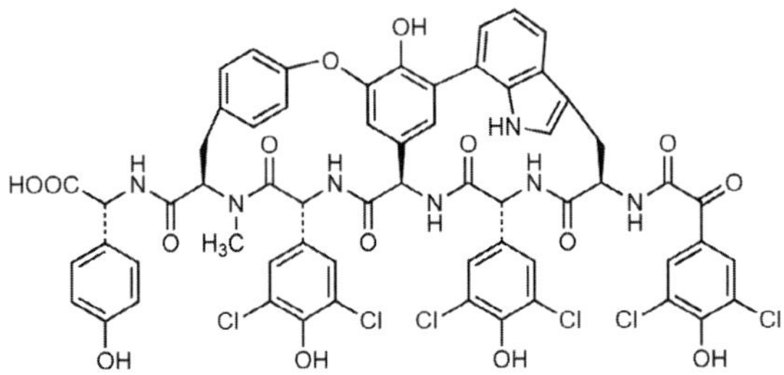

Chloropeptin I

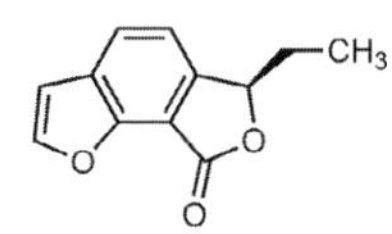

Concentricolide

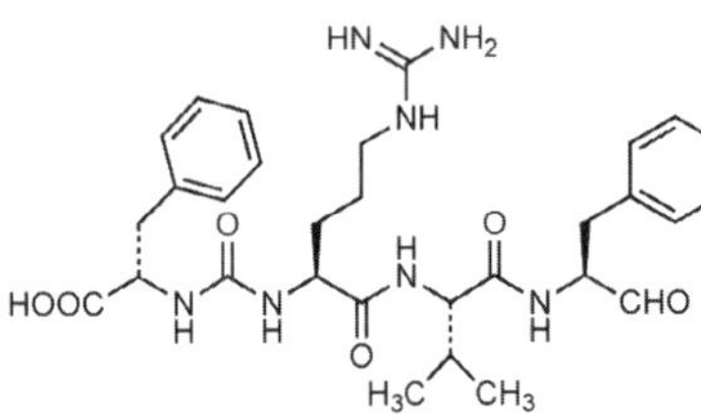

α-MAPI

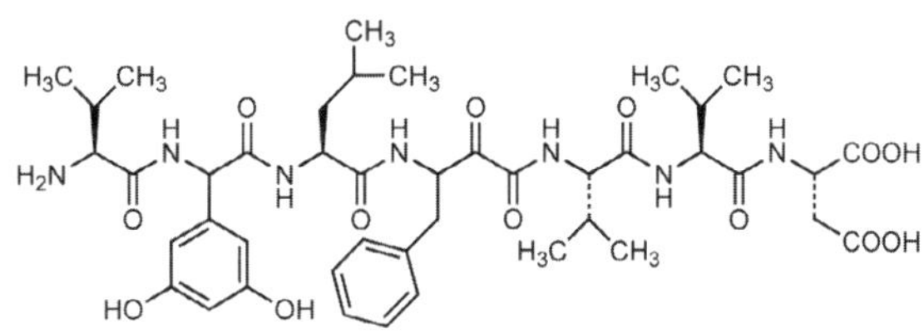

RPI-856 A

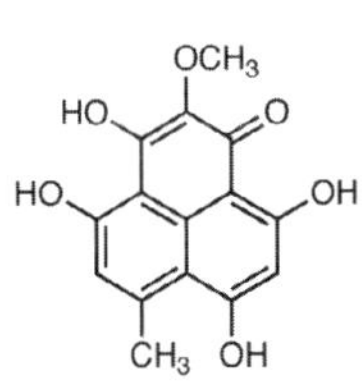

Funalenone

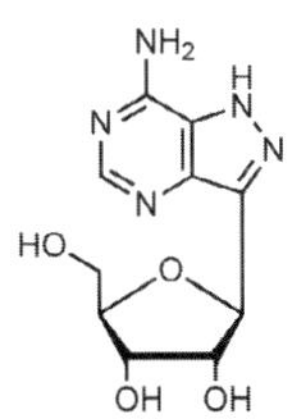

Formycin

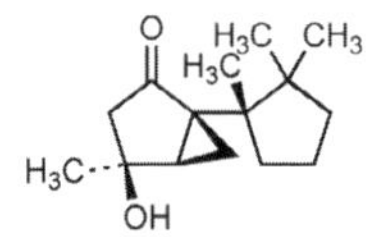

RF-2198

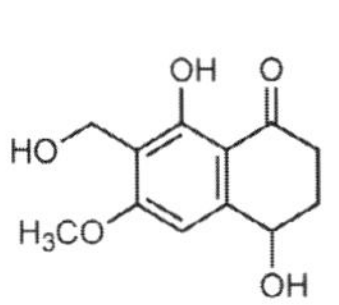

10-Norparvulenone

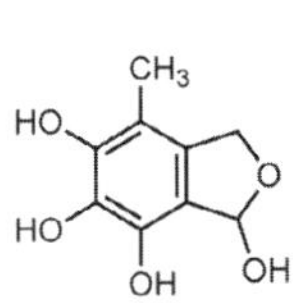

FR198248

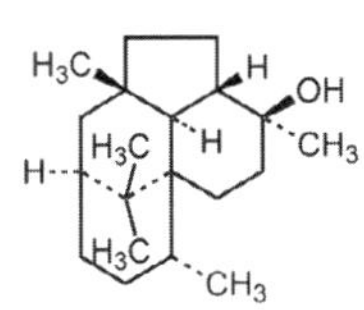

Wickerol A

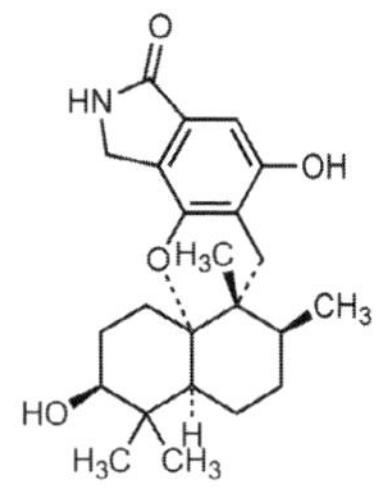

Stachyflin

Figure 9.4 Antiviral antibiotics.

dioxoethylene moiety produced by *Streptomyces* sp.. The IC_{50} value of RPI-856 A against HIV protease was 37 nM (Asano *et al.*, 1994). 4862F is a pentapeptide having *N,N,N*-trimethylated tyrosine at N-terminus. It was isolated from the culture broth of *Streptomyces albosporus* and inhibited HIV protease at the IC_{50} value of 15.3 nM (Liu *et al.*, 2013). Semicochliodinols and their analogues are bisalkylated 2,5-dihydroxybenzoquinones produced by the fungus *Chrysosporium merdarium*. They inhibit HIV protease and the IC_{50} value of semicochliodinol A was 0.37 μM (Fredenhagen *et al.*, 1997).

An HIV integrase inhibitor, raltegravir, is currently being used for AIDS treatment. HIV integrase inhibitors have also been sought from among microbial metabolites (Singh, 2011). Antibacterial equisetin, originally isolated from the culture media of *Fusarium equiseti*, inhibited 3′-end processing and strand transfer reactions at the IC_{50} value of approximately 10 μM (Burmeister *et al.*, 1974; Singh *et al.*, 1998). A new eremophilane sesquiterpenoid, integric acid, produced by the fungus *Xylaria* sp., inhibited both reactions with an IC_{50} value of 10 μM (Singh *et al.*, 1999). An inhibitor of HIV-induced cytopathic effect, chloropeptin II (complestatin), described above was reported as an HIV integrase inhibitor, and its axial-chiral isomer, isocomplestatin, showed IC_{50} values of 0.2 and 4 μM against 3′-end processing and strand transfer, respectively (Singh *et al.*, 2001). Integramycin, produced by *Actinoplanes* sp. (Singh *et al.*, 2002a), integrastatin A, produced by an unidentified fungus (Singh *et al.*, 2002b), and integracide A, produced by *Fusarium* sp. (Singh *et al.*, 2003), inhibited strand transfer reaction with IC_{50} values of 4, 1.1 and 9 μM, respectively. We found that phenalenone-type compounds inhibit HIV integrase, and funalenone, previously isolated from the culture broth of *Aspergillus niger* as a collagenase inhibitor (Inokoshi *et al.*, 1999), showed inhibition against strand transfer reaction at the IC_{50} value of 10 μM (Shiomi *et al.*, 2005b). Funalenone showed anti-HIV activity *in vitro* with an IC_{50} value of 1.7 μM.

Anti-influenza virus antibiotics

A pandemic of influenza virus infections occurs when a new type of virus emerges and transmits among humans (WHO, 2013c). Such viruses cause emerging infectious diseases because they emerge, circulate, and cause large outbreaks outside of the normal influenza season. The most notorious pandemic for which data are available was the 'Spanish flu (H1N1)' in 1918–1919 that caused an estimated 20–40 million or more deaths worldwide. In 2009, an influenza virus that had not ever been seen before, emerged, spread across the world and caused the 2009 H1N1 pandemic. The virus has been widely circulating across the globe since 2009, and is now established in human populations as a seasonal influenza virus. Humans can also be infected with influenza viruses that are routinely circulating in animals, such as avian influenza virus subtypes H5N1 and H9N2 and swine influenza virus subtypes H1N1 and H3N2. Some may occasionally infect humans and may cause disease ranging from mild conjunctivitis to severe pneumonia and even death. If such a virus acquired the capacity to spread easily among people, it could start an epidemic or a pandemic. Two types of drugs are available for influenza virus infections, neuraminidase inhibitors and M2 protein inhibitors. Neuraminidase inhibitors such as oseltamivir are used widely. M2 protein inhibitors such as amantadine are not commonly being used recently, because resistant strains have increased. Favipiravir, a viral RNA-dependent RNA polymerase inhibitor, is under development now (Furuta *et al.*, 2009). It acts as a purine or purine nucleoside in viral RNA replication. Some anti-influenza virus compounds produced by microorganisms are described below.

Several nucleosides produced by actinomycetes have been reported to inhibit influenza viruses. A C-nucleoside isomer of adenosine, formycin, was originally isolated from the culture broth of *Nocardia interforma* as an anti-tumour antibiotic (Hori *et al.*, 1964). It showed antiviral activity against the A/WSN (H1N1) strain of influenza virus at 0.3 μg/ml (Ishida *et al.*, 1967). An indole nucleoside antibiotic SF-2140 produced by *Actinomadura* sp. showed proliferation inhibition activity against the A/PR/8/34 (H1N1) strain of influenza virus at the IC_{50} value of 6.3 μg/ml (Ito *et al.*, 1984). JBIR-68 produced by *Streptomyces* sp. reduced plaque formation of influenza virus A/WSN/33 at 50 μM (Takagi *et al.*,

2010). Its structure was revealed as 5′-*O*-geranyl-5,6-dihydrouridine. Actinomycetes produce some other groups of anti-influenza virus antibiotics. A benzonaphtacenequinone antibiotic, pradimicin A, produced by *Actinomadura hibisca* has anti-HIV activity as described above. It also showed anti-influenza virus activity with an IC_{50} value of 6.8 μg/ml against the Victoria (H3N2) strain (Oki *et al.*, 1990). A 14-membered ring macrolactam antibiotic, fluvirucin B_1, was produced by *Nonomuria pusilla* and exhibited potent inhibitory activity against the Victoria strain of influenza virus at the IC_{50} value of 2.3 μg/ml (Naruse *et al.*, 1991; Tomita *et al.*, 1991). Quartromicins are macrocyclic antibiotics containing four tetronic acid moieties produced by *Amycolatopsis orientalis* (Tsunakawa *et al.*, 1992). Quartromicin D_1 showed anti-influenza virus activity against the Victoria (H3N2) strain at the IC_{50} value of 6.8 μg/ml. Fattiviracin A1 is a 22-membered ring macrodiolides consisting of four D-glucose units and two (C24 and C33) hydroxy fatty acids produced by *Streptomyces microflavus*. It has inhibitory activity against influenza A virus (H1N1) at the IC_{50} value of 2.05 μg/ml (Yokomizo *et al*, 1998).

Anti-influenza virus antibiotics are also produced by fungi. RF-2198, a cuparane-type sesquiterpene, produced by *Metarhizium* sp. inhibited the growth of influenza virus A/WSN/33 with an IC_{50} value of 0.1 μg/ml (Fujiwara *et al.*, 2001). Our group isolated a naphthalenone derivative with three hydroxyl groups, 10-norparvulenone, from the culture broth of *Microsphaeropsis* sp. It exhibited 64.8% of host cell survival at 25 μg/ml when A/PR/8/34 influenza virus was used for infection (Fukami *et al.*, 2000). It inhibited the viral neuraminidase at the IC_{50} value of 21 μg/ml. Isoaspulvinone E produced by *Aspergillus terreus* was reported recently (Gao *et al.*, 2013). It has a butenolide ring and two phenyl rings with three hydroxyl groups. It showed anti-influenza virus activity against A/PR/8/34 strain at the IC_{50} value of 32.3 μg/ml, and it also inhibited the viral neuraminidase at the IC_{50} value of 62.1 μg/ml. FR198248 is a dihydroisobenzofuran derivative with four hydroxyl groups (Nishihara *et al.*, 2001a). It was produced by another strain of *A. terreus* and inhibited the A/PR/8/34 strain and the B/Yamagata/16/88 strain at the IC_{50} values of 11.0 μM and 16.1 μM, respectively. It prolonged the survival of A/PR/8/34 virus-infected mice with the ED_{50} value of 1.98 mg/kg (10 days after infection). A study of FR198248 derivatives suggested the importance of its phenolic hydroxyl group for anti-influenza virus activity (Nishihara *et al.*, 2001b). 10-Norparvulenone and isoaspulvinone E may be related compounds having multihydroxyl groups. We recently reported on two diterpenes, wickerols A and B produced by *Trichoderma atroviride* (Yamamoto *et al.*, 2012). Their skeleton is a unique 3*H*-5a,9-methanocycloocta[*cd*]indene (fused 6-5-6-6 ring system). Wickerol A inhibited influenza virus A/PR/8/34 at the IC_{50} value of 0.07 μg/ml, while it inhibited the growth of the host MDCK cells at the IC_{50} value of 7 μg/ml. Wickerol B is 8-hydroxyl-wickerol A, and its antiviral activity (IC_{50} 5.0 μg/ml) was much weaker than wickerol A. Stachyflin isolated from solid-state fermentation of *Stachybotrys* sp. is a sesquiterpene derivative antibiotic (Minagawa *et al.*, 2002). It has a pentacyclic skeleton including a *cis*-fused decalin and a pyranoisoindolinone ring system. It showed anti-influenza virus activity against A/WSN/33 (H1N1), A/PR/8/34 (H1N1), and A/Okuda/57 (H2N2) strains at the IC_{50} value of 0.003, 0.30, and 0.64 μg/ml, respectively, and has H1 and H2 subtype-specific anti-influenza A virus activity (Yoshimoto *et al.*, 1999). It was recently revealed to inhibit replication of viruses of subtypes H5 and H6 (Motohashi *et al.*, 2013). It showed antiviral activity *in vivo*, and its target is the viral haemagglutinin.

Anti-pathogenic *E. coli* antibiotics

Pathogenic *Escherichia coli* strains such as serotype O157:H7 cause epidemics in some developed countries. Infection with *E. coli* O157:H7 was first described in the United States in 1982 (WHO, 2002; Takeda, 2011). Subsequently, it has emerged rapidly and outbreaks of infection have been reported in Australia, Canada, Japan, United States, in various European countries, and in southern Africa. Symptoms associated with this organism were quite severe with abdominal cramps and bloody diarrhoea, which was named as haemorrhagic colitis. The infection is sometimes fatal, particularly in children. The other serotypes also caused outbreaks such as O111:H8 in Japan and O104:H4 in Germany, both occurring in 2011 (Misaki *et al.*, 2012; Robert Koch-Institute, 2011). Such *E. coli* strains are called enterohaemorrhagic *E. coli* (EHEC), and they are defined by the ability to produce Vero toxin that is cytotoxic to Vero cells. The pathogenic *E. coli* are classified into EHEC, enteropathogenic *E. coli* (EPEC) and four other groups (Kaper *et al.*, 2004). They have an apparatus called type III secretion system (T3SS) that allows effector proteins to be injected directly into the host cells for virulence, attachment, and entry.

T3SS is highly conserved in many Gram-negative pathogenic bacteria, including the Enterobacteriaceae (pathogenic *E. coli*, *Shigella*, *Salmonella*, *Yersinia*, etc.), *Bordetella*, *Pseudomonas*, *Chlamydia*, and *Burkholderia* (Coburn *et al.*, 2007). The ~3.5 MDa needle complex contains a basal structure spanning the periplasm that is embedded within the inner and outer bacterial membranes followed by a needle-like structure that protrudes into the extracellular space (Chatterjee *et al.*, 2013). T3SS is involved in establishing disease processes but is not essential for bacterial growth or homeostasis. Thus, the T3SS is expected to be a potential therapeutic target, and T3SS inhibitors could avoid some common problems associated with conventional antibiotics: development of resistant strains, expression of various toxins, and disruption of normal microbial flora (Duncan *et al.*, 2012). We have screened T3SS inhibitors using T3SS-mediated haemolysis by EPEC (Kimura *et al.*, 2011). In EPEC, two translocases, EspB and EspD, are delivered into the host through the T3SS and build a physical pore on the host membrane. T3SS possesses the capacity to deliver translocases into the membrane of red blood cells, which results in haemolysis and subsequently in the release of haemoglobin into the surrounding fluid. We evaluated T3SS activity by haemolysis. We isolated new compounds, guadinomines A, B, C_1, C_2, and D produced by *Streptomyces* sp. from this screening (Iwatsuki *et al.*, 2008a; Iwatsuki *et al.*, 2008b). Guadinomines consist of a 1-carbamoyl-2-imidazolidinimine moiety and an octanoylalanylvaline moiety in common (Fig. 9.5). Guadinomine B inhibited T3SS-mediated haemolysis at the IC_{50} value of 0.07 µg/ml, while its antibacterial activity was weak. We also isolated aurodox from a culture broth of *Streptomyces* sp. as an inhibitor of T3SS-mediated haemolysis of EPEC (Kimura *et al.*, 2011). Aurodox is known as an antibacterial antibiotic that inhibits protein synthesis by binding bacterial elongation factor Tu (Vogeley *et al.*, 2001). Although, it inhibited the growth of EPEC at the IC_{50} value of 40 µg/ml, the IC_{50} value of aurodox against T3SS-mediated haemolysis was 1.5 µg/ml. Therefore, aurodox showed much

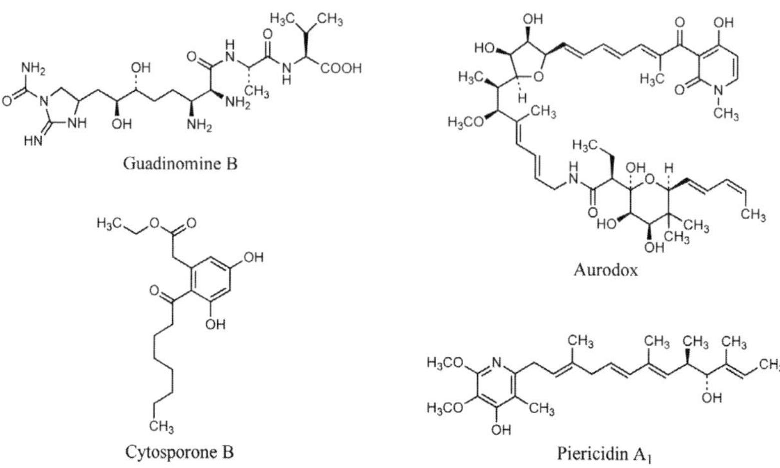

Figure 9.5 Anti-pathogenic *E. coli* antibiotics.

more potent inhibition of T3SS than antibacterial activity. Aurodox specifically inhibited the secretion of T3SS-related proteins, such as EspB, EspF and Map, without affecting the expression of the housekeeping protein GroEL. Transcription analysis using quantitative real-time PCR showed that aurodox inhibited transcription of the *ler* gene, which encodes a positive regulator of the T3SS gene cluster. Furthermore, an *in vivo* infection study using mice indicated that the oral administration of aurodox allowed the mice to survive a lethal dose of *Citrobacter rodentium* (a mouse EPEC strain). Thus, aurodox may contribute to murine survival by the elimination of T3SS-mediated disease processes. Many T3SS inhibitors have been reported and most of them are synthetic ones. There are some *in vivo* studies of the T3SS inhibitors, but they have employed the drugs as prophylactics, with only aurodox resolving an ongoing infection (Duncan *et al.*, 2012).

Recently, a few microbial products were reported as T3SS inhibitors. Cytosporone B produced by some fungi (Brady *et al.*, 2000) was selected from chemical library compounds. It inhibited the secretion of *Salmonella* pathogenicity island 1-associated effector proteins 59% at 6.25 μM (Li *et al.*, 2013). Although piericidins are known to inhibit electron transport complex I (Okun *et al.*, 1999), piericidin A_1 and its derivative Mer-A 2026B, isolated from a marine actinomycetes extracts library, inhibited *Yersinia pseudotuberculosis* from triggering T3SS-dependent activation of the host transcription factor NF-κB in HEK293T cells (Duncan *et al.*, 2013). Piericidin A_1 at a concentration of 71 μM reduced secretion of the T3SS effector YopE by 45%. Piericidins are suggested to inhibit an earlier stage of T3SS such as needle assembly.

Conclusions

Emerging and re-emerging diseases are a major and continuing threat to global public health. They will remain so because relatively few drugs are generally available to combat these diseases and many that are available quickly lose their effectiveness. However, several antibiotics do remain effective against some diseases, such as tigecycline and colistin for New Delhi metallo-β-lactamase 1-positive Enterobacteriaceae (Kumarasamy *et al.*, 2010). For others, the picture is not particularly clear. For example, although ivermectin resistance develops in helminth parasites in animals, despite 30 years of ivermectin monotherapy in humans, no resistance in human parasites has yet been detected (Crump and Ōmura, 2011). Nevertheless, it is abundantly clear that a continuing supply of new antibiotics, especially those with novel modes of action, will be essential to circumvent or overcome emerging and re-emerging diseases. Living microorganisms represent a potentially limitless source of such compounds.

Natural compounds are highly diverse and often produce highly specific biological activities. They have drug-like structures in comparison to synthetic compounds (Feher and Schmidt, 2003). Among various diseases, natural compounds are particularly suitable for infectious diseases. It was reported that 47.1% of small-molecule antiinfective drugs (antibacterials, antifungals, antiparasitics, and antivirals) were natural compounds or their derivatives (Newman and Cragg, 2012). The major natural compounds among antiinfective drugs are microbial products. Their producers, microorganisms, exhibit a vast diversity. Comparing the estimated numbers and predicted numbers of microorganism species, only a mere 12% of bacteria and 5% of fungi have so far been isolated (Bull *et al.*, 1992). Thus, many efforts are being undertaken to culture hitherto uncultured microorganisms (Lewis *et al.*, 2010). Recently, genome mining of microorganisms is rapidly developing in an effort to discover new secondary metabolites as lead compounds for drug discovery (Bachmann *et al.*, 2013). Besides such up-to-date studies, traditional screenings for new antibiotics are also being accelerated and expanded, especially through the use of advanced technology and instrumentation, particularly with regard to purification and structure elucidation.

References

Adler-Moore, J., and Proffitt, R.T. (2002). AmBisome: liposomal formulation, structure, mechanism of action and preclinical experience. J. Antimicrob. Chemother. *49* (Suppl. S1), 21–30.

Arcamone, F., Barbieri, W., Franceschi, G., Penco, S., and Vigevani, A. (1973). Axenomycins. I. Structure of chromophore and sugar moieties. J. Am. Chem. Soc. 95, 2008–2009.

Asano, T., Matsuoka, K., Hida, T., Kobayashi, M., Kitamura, Y., Hayakawa, T., Iinuma, S., Kakinuma, A., and Kato, K. (1994). Novel retrovirus protease inhibitors, RPI-856 A, B, C, and D, produced by *Streptomyces* sp. AL-322. J. Antibiot. *47*, 557–565.

Ayers, S., Zink, D.L., Mohn, K., Powell, J.S., Brown, C.M., Murphy, T., Grund, A., Genilloud, O., Salazar, O., Thompson, D., and Singh, S.B. (2007). Anthelmintic macrolactams from *Nonomuraea turkmeniaca* MA7364. J. Nat. Prod. *70*, 1371–1373.

Ayers, S., Zink, D.L., Mohn, K., Powell, J.S., Brown, C.M., Bills, G., Grund, A., Thompson, D., and Singh, S.B. (2010). Anthelmintic constituents of *Clonostachys candelabrum*. J. Antibiot. *63*, 119–122.

Bachmann, B.O., Van Lanen, S.G., and Baltz, R.H. (2013). Microbial genome mining for accelerated natural products discovery: is a renaissance in the making? J. Ind. Microbiol. Biotechnol. *41*, 175–184.

Berger, J., Jampolsky, L.M., and Goldberg, M.W. (1949). Borrelidin, a new antibiotic with antiborrelia activity and penicillin-enhancement properties. Arch. Biochem. *22*, 476–478.

Birch, A.J., and Wright, J.J. (1970). Studies in relation to biosynthesis – XLII. The structural elucidation and some aspects of the biosynthesis of the brevianamides-A and -E. Tetrahedron *26*, 2329–2344.

Brady, S.F., Wagenaar, M.M., Singh, M.P., Janso, J.E., and Clardy, J. (2000). The cytosporones, new octaketide antibiotics isolated from an endophytic fungus. Org. Lett. *2*, 4043–4046.

Breger, J., Rachlin, A.I., Scott, W.E., Sternbach, L.H., and Goldberg, M.W. (1951). The isolation of three new crystalline antibiotics from *Streptomyces*. J. Am. Chem. Soc. *73*, 5295–5298.

Bull, A.T., Goodfellow, M., and Slater, J.H. (1992). Biodiversity as a source of innovation in biotechnology. Ann. Rev. Microbiol. *46*, 219–246.

Burg, R.W., Miller, B.M., Baker, E.E., Birnbaum, J., Currie, S.A., Hartman, R., Kong, Y.-L., Monaghan, R.L., Olson, G., Putter, I., *et al.* (1979). Avermectins, new family of potent anthelmintic agents: producing organism and fermentation. Antimicrob. Agents Chemother. *15*, 361–367.

Burmeister, H.R., Bennett, G.A., Vesonder, R.F., and Hesseltine, C.W. (1974). Antibiotic produced by *Fusarium equiseti* NRRL 5537. Antimicrob. Agents Chemother. *5*, 634–639.

Camps, M., Arrizabalaga, G., and Boothroyd, J. (2002). An rRNA mutation identifies the apicoplast as the target for clindamycin in *Toxoplasma gondii*. Mol. Microbiol. *43*, 1309–1318.

Carter, A.P., Clemons, W.M., Brodersen, D.E., Morgan-Warren, R.J., Wimberly, B.T., and Ramakrishnan, V. (2000). Functional insights from the structure of the 30S ribosomal subunit and its interactions with antibiotics. Nature *407*, 340–348.

Carter, G.T., Nietsche, J.A., Hertz, M.R., Williams, D.R., Siegel, M.M., Morton, G.O., James, J.C., and Borders, D.B. (1988). LL-F28249 antibiotic complex: a new family of antiparasitic macrocyclic lactones. Isolation, characterization and structures of LL-F28249 α, β, γ, λ. J. Antibiot. *41*, 519–529.

Chabala, J.C., Mrozik, H., Tolman, R.L., Eskola, P., Lusi, A., Peterson, L.H., Woods, M.F., Fisher, M.H., Campbell, W.C., Egerton, J.R., and Ostlind, D.A. (1980). Ivermectin, a new broad-spectrum antiparasitic agent. J. Med. Chem. *23*, 1134–1136.

Chatterjee, S., Chaudhury, S., McShan, A.C., Kaur, K., and De Guzman, R.N. (2013). Structure and biophysics of type III secretion in bacteria. Biochemistry, *52*, 2508–2517.

Chiba, H., Inokoshi, J., Okamoto, M., Asanuma, S., Matsuzaki, K., Iwama, M., Mizumoto, K., Tanaka, H., Oheda, M., Fujita, K., *et al.* (2001). Actinohivin, a novel anti-HIV protein from an actinomycete that inhibits syncytium formation: isolation, characterization, and biological activities. Biochem. Biophys. Res. Commun. *282*, 595–601.

Chiba, H., Inokoshi, J., Nakashima, H., Ōmura, S., and Tanaka, H. (2004). Actinohivin, a novel anti-human immunodeficiency virus protein from an actinomycete, inhibits viral entry to cells by binding high-mannose type sugar chains of gp120. Biochem. Biophys. Res. Commun. *316*, 203–210.

Chopra, I., and Roberts, M. (2001). Tetracycline antibiotics: mode of action, applications, molecular biology, and epidemiology of bacterial resistance. Microbiol. Mol. Biol. Rev. *65*, 232–260.

Coburn, B., Sekirov, I., and Finlay, B.B. (2007). Type III secretion systems and disease. Clin. Microbiol. Rev. *20*, 535–549.

Crump, A., and Ōmura, S. (2011). Ivermectin, 'wonder drug' from Japan: the human use perspective. Proc. Jpn. Acad., Ser. B *87*,13–28.

Darkin-Rattray, S.J., Gurnett, A.M., Myers, R.W., Dulski, P.M., Crumley, T.M., Allocco, J.J., Cannova, C., Meinke, P.T., Colletti, S.L., Bednarek, M.A., *et al.* (1996). Apicidin: a novel antiprotozoal agent that inhibits parasite histone deacetylase. Proc. Natl. Acad. Sci. U.S.A. *93*, 13143–13147.

Della Bruna, C., Ricciardi, M.L., and Sanfilippo, A. (1973). Axenomycins, new cestocidal antibiotics. Antimicrob. Agents Chemother. *3*, 708–710.

Didier, E.S. (1997). Effects of albendazole, fumagillin, and TNP-470 on microsporidial replication *in vitro*. Antimicrob. Agents Chemother. *41*, 1541–1546.

Didier, E.S., Martin, A.D., Stovall, M.E., Alvarez, X., Mittleider, D., Green, L.C., Bowers, L.C., Plauche, A.K., Didier, P.J., and Brindley, P.J. (2003). Methionine aminopeptidase 2 expression in microsporidia. J. Eukaryot. Microbiol. *50*, 569–571.

Dube, D.K., Mpimbaza, G., Allison, A.C., Lederer, E., and Rovis, L. (1983). Antitrypanosomal activity of sinefungin. Am. J. Trop. Med. Hyg. *32*, 31–33.

Duggar, B.M. (1948). Aureomycin: a product of the continuing search for new antibiotics. Ann. N. Y. Acad. Sci. *51*, 177–181.

Duncan, M.C., Linington, R.G., and Auerbucha, V. (2012). Chemical inhibitors of the type three secretion system: disarming bacterial pathogens. Antimicrob. Agents Chemother. *56*, 5433–5441.

Duncan, M.C., Wong, W.R., Dupzyk, A.J., Bray, W.M., Linington, R.G., and Auerbuch, V. (2013). An NF-κB-based high throughput screen identifies piericidins as inhibitors of the *Yersinia pseudotuberculosis* type III secretion system. Antimicrob. Agents Chemother. *58*, 1118–1126.

Dutton, C.J., Gibson, S.P., Goudie, A.C., Holdom, K.S., Pacey, M.S., Ruddock, J.C., Bu'Lock, J.D., and Richards, M.K. (1991). Novel avermectins produced by mutational biosynthesis. J. Antibiot. *44*, 357–365.

Eble, T.E., and Hanson, F.R. (1951). Fumagillin, an antibiotic from *Aspergillus fumigatus* H-3. Antibiot. Chemother. *1*, 54–58.

Egerton, J.R., Ostlind, D.A., Blair, L.S., Eary, C.H., Suhayda, D., Cifelli, S., Riek, R.F., and Campbell, W.C. (1979). Avermectins, new family of potent anthelmintic agents: efficacy of the B1a component. Antimicrob. Agents Chemother. *15*, 372–378.

Enomoto, Y., Shiomi, K., Matsumoto, A., Takahashi, Y., Iwai, Y., Harder, A., Kölbl, H., Woodruff, H.B., and Ōmura, S. (2001). Isolation of a new antibiotic oligomycin G produced by *Streptomyces* sp. WK-6150. J. Antibiot. *54*, 308–313.

Feher, M., and Schmidt, J.M. (2003). Property distributions: differences between drugs, natural products, and molecules from combinatorial chemistry. J. Chem. Inf. Comput. Sci. *43*, 218–227.

Ferir, G., Hanchen, A., Francois, K.O., Hoorelbeke, B., Huskens, D., Dettner, F., Sussmuth, R.D., and Schols, D. (2012). Feglymycin, a unique natural bacterial antibiotic peptide, inhibits HIV entry by targeting the viral envelope protein gp120. Virology *433*, 308–319.

Finlay, A.C., Hobby, G.L., P'an, S.Y., Regna, P.P., Routien, J.B., Seeley, D.B., Shull, G.M., Sobin, B.A., Solomons, I.A., Vinson, J.W., and Kane, J.H. (1950). Terramycin, a new antibiotic. Science *111*, 85.

Fredenhagen, A., Petersen, F., Tintelnot-Blomley, M., Rosel, J., Mett, H., and Hug, P. (1997). Semicochliodinol A and B: inhibitors of HIV-1 protease and EGF-R protein tyrosine kinase related to asterriquinones produced by the fungus *Chrysosporium merdarium*. J. Antibiot. *50*, 395–401.

Fujiwara, T., Hatsutori, N., Kawamura, Y., Matsumoto, K., Kamigauchi, T., and Itazaki, H. (2001). Cuparane-type sesquiterpene compound which has anti-influenza A viral activity. Jpn. Patent 3152532.

Fukami, A., Nakamura, T., Kim, Y.-P., Shiomi, K., Hayashi, M., Nagai, T., Yamada, H., Komiyama, K., and Ōmura, S. (2000). A new anti-influenza virus antibiotic, 10-norparvulenone from *Microsphaeropsis* sp. FO-5050. J. Antibiot. 53, 1215–1218.

Furuta, Y., Takahashi, K., Shiraki, K., Sakamoto, K., Smee, D.F., Barnard, D.L., Gowen, B.B., Julander, J.G., and Morrey, J.D. (2009). T-705 (favipiravir) and related compounds: Novel broad-spectrum inhibitors of RNA viral infections. Antiviral Res. *82*, 95–102.

Gao, H., Guo, W., Wang, Q., Zhang, L., Zhu, M., Zhu, T., Gu, Q., Wang, W., and Li, D. (2013). Aspulvinones from a mangrove rhizosphere soil-derived fungus *Aspergillus terreus* Gwq-48 with anti-influenza A viral (H1N1) activity. Bioorg. Med. Chem. Lett. *23*, 1776–1778.

Geary, T.G. (2005). Ivermectin 20 years on: maturation of a wonder drug. Trends Parasitol. *21*, 530–532.

Grabley, S., Hammann, P., Thiericke, R., Wink, J., Philipps, S., and Zeeck, A. (1993). Secondary metabolites by chemical screening. 21. Clonostachydiol, a novel anthelmintic macrodiolide from the fungus *Clonostachys cylindrospora* (strain FH-A 6607). J. Antibiot. *46*, 343–345.

Greif, G., Harder, A., and Haberkorn, A. (2001). Chemotherapeutic approaches to protozoa: Coccidiae—current level of knowledge and outlook. Parasitol. Res. *87*, 973–975.

Hamill, R.L., and Hoehn, M.M. (1973). A 9145, a new adenine-containing antifungal antibiotic. I. Discovery and isolation. J. Antibiot. *26*, 463–465.

Hanada, K., Palacpac, N.M.Q., Magistrado, P.A., Kurokawa, K., Rai, G., Sakata, D., Hara, T., Horii, T., Nishijima, M., and Mitamura, T. (2002). *Plasmodium falciparum* phospholipase C hydrolyzing sphingomyelin and lysocholinephospholipids is a possible target for malaria chemotherapy. J. Exp. Med. *195*, 23–34.

Haneda, K., Shinose, M., Seino, A., Tabata, N., Tomoda, H., Iwai, Y., and Ōmura, S. (1994). Cytosaminomycins, new anticoccidial agents produced by *Streptomyces* sp. KO-8119. I. Taxonomy, production, isolation and physico-chemical and biological properties. J. Antibiot. *47*, 774–781.

Haney, M.E., Jr., and Hoehn, M.M. (1968). Monensin, a new biologically active compound. I. Discovery and isolation. Antimicrob. Agents Chemother. *1967*, 349–352.

Harada, S., Sasaki, T., Shindo, M., Kido, Y., Inaoka, D.K., Omori, J., Osanai, A., Sakamoto, K., Mao, J., Matsuoka, S., *et al.* (2011). Crystal structure of porcine heart mitochondrial complex II bound with atpenin A5. Protein Data Bank ID 3AEE.

Harder, A., and von Samson-Himmelstjerna, G. (2002). Cyclooctadepsipeptides – a new class of anthelmintically active compounds. Parasitol. Res. *88*, 481–488.

Haskell, T.H., French, J.C., and Bartz, Q.R. (1959). Paromomycin. I. Paromamine, a glycoside of D-glucosamine. J. Am. Chem. Soc. *81*, 3480–3481.

Heukelbach, J., and Feldmeier, H. (2006). Scabies. Lancet *367*, 1767–1774.

Hibbs, R.E., and Gouaux, E. (2011). Principles of activation and permeation in an anion-selective Cys-loop receptor. Nature *474*, 54–60.

Hori, M., Ito, E., Takita, T., Koyama, G., Takeuchi, T., and Umezawa, H. (1964). A new antibiotic, formycin. J. Antibiot., Ser. A *17*, 96–99.

Horie, H. (1956). Anti-trichomonas effect of azomycin. J. Antibiot. *9*, 168.

Horsefield, R., Yankovskaya, V., Sexton, G., Whittingham, W., Shiomi, K., Ōmura, S., Byrne, B., Cecchini, G., and Iwata, S. (2006). Structural and computational analysis of the quinone-binding site of complex II (succinate-ubiquinone oxidoreductase). A mechanism of electron transfer and proton conduction during ubiquinone reduction. J. Biol. Chem. *281*, 7309–7316.

Hosoya, S., Komatsu, N., Soeda, M., and Sonoda, Y. (1952). Trichomycin, a new antibiotic produced by *Streptomyces hachijoensis* with trichomonadicidal and antifungal activity. Jpn. J. Exp. Med. *22*, 505–509.

Ikeda, H., Ishikawa, J., Hanamoto, A., Shinose, M., Kikuchi, H., Shiba, T., Sakaki, Y., Hattori, M., and Ōmura, S. (2003). Complete genome sequence and comparative analysis of the industrial microorganism *Streptomyces avermitilis*. Nat. Biotechnol. *21*, 526–531.

Inokoshi, J., Shiomi, K., Masuma, R., Tanaka, H., Yamada, H., and Ōmura, S. (1999). Funalenone, a novel collagenase inhibitor produced by *Aspergillus niger*. J. Antibiot. *52*, 1095–1100.

Inokoshi, J., Chiba, H., Asanuma, S., Takahashi, A., Ōmura, S., and Tanaka, H. (2001). Molecular cloning of actinohivin, a novel anti-HIV protein from an Actinomycete, and its expression in *Escherichia coli*. Biochem. Biophys. Res. Commun. *281*, 1261–1265.

Ishida, N., Homma, M., Kumagai, K., Shimizu, Y., Matsumoto, S., and Izawa, A. (1967). Antiviral activity of formycin. J. Antibiot., Ser. A *20*, 49–52.

Ishii, T. (1998). Veterinary Parasitology 1 (Kodansha Scientific, Tokyo, Japan).

Ito, T., Ohba, K., Koyama, M., Sezaki, M., Tohyama, H., Shomura, T., Fukuyasu, H., Kazuno, Y., Niwa, T., Kojima, M., and Niida, T. (1984). A new antiviral antibiotic SF-2140 produced by *Actinomadura*. J. Antibiot. *37*, 931–934.

Iwai, Y., Kora, A., Takahashi, Y., Hayashi, T., Awaya, J., Masuma, R., Oiwa, R., and Ōmura, S. (1978). Production of deoxyfrenolicin and a new antibiotic, frenolicin B by *Streptomyces roseofulvus* strain AM-3867. J. Antibiot. *31*, 959–965.

Iwatsuki, M., Uchida, R., Yoshijima, H., Ui, H., Shiomi, K., Matsumoto, A., Takahashi, Y., Abe, A., Tomoda, H., and Ōmura, S. (2008a). Guadinomines, type III secretion system inhibitors, produced by *Streptomyces* sp. K01–0509. I. Taxonomy, fermentation, isolation and biological properties. J. Antibiot. *61*, 222–229.

Iwatsuki, M., Uchida, R., Yoshijima, H., Ui, H., Shiomi, K., Kim, Y.-P., Hirose, T., Sunazuka, T., Abe, A., Tomoda, H., and Ōmura, S. (2008b). Guadinomines, type III secretion system inhibitors, produced by *Streptomyces* sp. K01-0509. II. Physico-chemical properties and structure elucidation. J. Antibiot. *61*, 230–236.

Iwatsuki, M., Takada, S., Mori, M., Ishiyama, A., Namatame, M., Nishihara-Tsukashima, A., Nonaka, K., Masuma, R., Otoguro, K., Shiomi, K., and Ōmura, S. (2011). *In vitro* and *in vivo* antimalarial activity of puberulic acid and its new analogs, viticolins A-C, produced by *Penicillium* sp. FKI-4410. J. Antibiot. *64*, 183–188.

Jomaa, H., Wiesner, J., Sanderbrand, S., Altincicek, B., Weidemeyer, C., Hintz, M., Turbachova, I., Eberl, M., Zeidler, J., Lichtenthaler, H.K., *et al.* (1999). Inhibitors of the nonmevalonate pathway of isoprenoid biosynthesis as antimalarial drugs. Science *285*, 1573–1576.

Kaneko, I., Kamoshida, K., and Takahashi, S. (1989). Complestatin, a potent anti-complement substance produced by *Streptomyces lavendulae*. I. Fermentation, isolation and biological characterization. J. Antibiot. *42*, 236–241.

Kaper, J.B., Nataro, J.P., and Mobley, H.L. (2004). Pathogenic *Escherichia coli*. Nat, Rev, Microbiol. *2*, 123–140.

Kawamura, T., Liu, D., Towle, M.J., Kageyama, R., Tsukahara, N., Wakabayashi, T., and Littlefield, B.A. (2003). Anti-angiogenesis effects of borrelidin are mediated through distinct pathways: threonyl-tRNA synthetase and caspases are independently involved in suppression of proliferation and induction of apoptosis in endothelial cells. J. Antibiot. *56*, 709–715.

Kimura, K., Iwatsuki, M., Nagai, T., Matsumoto, A., Takahashi, Y., Shiomi, K., Ōmura, S., and Abe, A. (2011). A small-molecule inhibitor of the bacterial type III secretion system protects against *in vivo* infection with *Citrobacter rodentium*. J. Antibiot. *64*, 197–203.

Kind, R., Zeeck, A., Grabley, S., Thiericke, R., and Zerlin, M. (1996). Secondary metabolites by chemical screening. 30. Helmidiol, a new macrodiolide from *Alternaria alternata*. J. Nat. Prod. *59*, 539–540.

Kita, K., Shiomi, K., and Ōmura, S. (2007). Advances in drug discovery and biochemical studies. Trends Parasitol. *23*, 223–229.

Komori, T., and Morimoto, Y. (1990). Trichomycin A, a heptaene macrolide: isolation and characterization. J. Antibiot. *43*, 904–906.

Komuniecki, R., and Tielens, A.G.M. (2003). Carbohydrate and energy metabolism in parasitic helminths. In Molecular Medical Parasitology, J.J. Marr, T.W. Nielsen, and R.W. Komuniecki, eds. (Academic Press, London, UK), pp. 339–358.

Kondo, S., Sezaki, M., Koike, M., Shimura, M., Akita, E., Satoh, K., and Hara, T. (1965). Destomycins A and B, two new antibiotics produced by *Streptomyces*. J. Antibiot., Ser. A *18*, 38–42.

Kumarasamy, K.K., Toleman, M.A., Walsh, T.R., Bagaria, J., Butt, F., Balakrishnan, R., Chaudhary, U., Doumith, M., Giske, C.G., Irfan, S., *et al.* (2010). Emergence of a new antibiotic resistance mechanism in India, Pakistan, and the UK: a molecular, biological, and epidemiological study. Lancet Infect. Dis. *10*, 597–602.

Kumazawa, S., Kanda, M., Utagawa, M., Chiba, N., Ohtani, H., and Mikawa, T. (2003). MK7924, a novel metabolite with nematocidal activity from *Coronophora gregaria*. J. Antibiot. *56*, 652–654.

Lell, B., Ruangweerayut, R., Wiesner, J., Missinou, M.A., Schindler, A., Baranek, T., Hintz, M., Hutchinson, D., Jomaa, H., and Kremsner, P.G. (2003). Fosmidomycin, a novel chemotherapeutic agent for malaria. Antimicrob. Agents Chemother. *47*, 735–738.

Lewis, K., Epstein, S., D'Onofrio, A., and Ling, L.L. (2010). Uncultured microorganisms as a source of secondary metabolites. J. Antibiot. *63*, 468–476.

Li, J., Lv, C., Sun, W., Li, Z., Han, X., Li, Y., and Shen, Y. (2013). Cytosporone B, an inhibitor of the type III secretion system of *Salmonella enterica* serovar Typhimurium. Antimicrob. Agents Chemother. *57*, 2191–2198.

Lin, Q., Katakura, K., and Suzuki, M. (2002). Inhibition of mitochondrial and plastid activity of *Plasmodium falciparum* by minocycline. FEBS Lett. *515*, 71–74.

Liu, L.X., and Weller, P.F. (1996). Antiparasitic drugs. N. Engl. J. Med. *334*, 1178–1184.

Liu, X., Gan, M., Dong, B., Zhang, T., Li, Y., Zhang, Y., Fan, X., Wu, Y., Bai, S., Chen, M., *et al.* (2013). 4862F, a new inhibitor of HIV-1 protease, from the culture of *Streptomyces* I03A-04862. Molecules *18*, 236–243.

Maeda, K., Osato, T., and Umezawa, H. (1953). A new antibiotic, azomycin. J. Antibiot., Ser. A *6*, 182.

Maeda, T. (2003). Toxoplasmosis. Nippon Rinsho (Jpn. J. Clin. Med.) *61* (Suppl. 2), 603–607.

Mann, R.L., and Bromer, W.W. (1958). Isolation of a second antibiotic from *Streptomyces hygroscopicus*. J. Am. Chem. Soc. *80*, 2714–2716.

Mason, D.J., Dietz, A., and DeBoer, C. (1963). Lincomycin, a new antibiotic. I. Discovery and biological properties. Antimicrob. Agents Chemother. *1962*, 554–559.

Masuma, R., Yamaguchi, Y., Noumi, M., Ōmura, S., and Namikoshi, M. (2001). Effect of sea water on hyphal growth and antimicrobial metabolite production in a marine fungi. Mycoscience *42*, 455–459.

Matoba, N., Husk, A.S., Barnett, B.W., Pickel, M.M., Arntzen, C.J., Montefiori, D.C., Takahashi, A., Tanno, K., Ōmura, S., Cao, H., *et al.* (2010). HIV-1 neutralization profile and plant-based recombinant expression of actinohivin, an Env glycan-specific lectin devoid of T-cell mitogenic activity. PLoS One, *5*, e11143.

Matsumoto, J., Sakamoto, K., Shinjyo, N., Kido, Y., Yamamoto, N., Yagi, K., Miyoshi, H., Nonaka, N., Katakura, K., Kita, K., and Oku, Y. (2008). Anaerobic NADH-fumarate reductase system is predominant in the respiratory chain of *Echinococcus multilocularis*, providing a novel target for the chemotherapy of alveolar echinococcosis. Antimicrob. Agents. Chemother. *52*, 164–170.

Matsuzaki, K., Ikeda, H., Masuma, R., Tanaka, H., and Ōmura, S. (1995). Isochromophilones I and II, novel inhibitors against gp120-CD4 binding produced by *Penicillium multicolor* FO-2338. I. Screening, taxonomy, fermentation, isolation and biological activity. J. Antibiot. *48*, 703–707.

Meinke, P.T., Smith, M.M., and Shoop, W.L. (2002). Nodulisporic acid: its chemistry and biology. Curr. Top. Med. Chem. *2*, 655–674.

Minagawa, K., Kouzuki, S., Yoshimoto, J., Kawamura, Y., Tani, H., Iwata, T., Terui, Y., Nakai, H., Yagi, S., Hattori, N., *et al.* (2002). Stachyflin and acetylstachyflin, novel anti-influenza A virus substances, produced by *Stachybotrys* sp. RF-7260. I. Isolation, structure elucidation and biological activities. J. Antibiot. *55*, 155–164.

Misaki, T., Nagira, M., Yahata, Y., Tada, Y., Taniguchi, K., Okabe, N., Demura, N., and Sata, T. (2012). Epidemiological investigation of EHEC O111 food poisoning outbreak linked to chain restaurants, April 2011—Toyama. Infect. Agents Surv. Rep. *33*, 118.

Miyadera, H., Shiomi, K., Ui, H., Yamaguchi, Y., Masuma, R., Tomoda, H., Miyoshi, H., Osanai, A., Kita, K., and Ōmura, S. (2003). Atpenins, potent and specific inhibitors of mitochondrial complex II (succinate-ubiquinone oxidoreductase). Proc. Natl. Acad. Sci. U.S.A. *100*, 473–477.

Mori, M., Morimoto, H., Kim, Y.-P., Ui, H., Nonaka, K., Masuma, R., Sakamoto, K., Kita, K., Tomoda, H., Shiomi, K., and Ōmura, S. (2011). Ukulactones A and B, new NADH-fumarate reductase inhibitors produced by *Penicillium* sp. FKI-3389. Tetrahedron *67*, 6582–6586.

Moskovitz, B.L., Stanton, T.L., and Kusmierek, J.J. (1988). Spiramycin therapy for cryptosporidial diarrhoea in immunocompromised patients. J. Antimicrob. Chemother. *22* (Suppl. B), 189–191.

Motohashi, Y., Igarashi, M., Okamatsu, M., Noshi, T., Sakoda, Y., Yamamoto, N., Ito, K., Yoshida, R., and Kida, H. (2013). Antiviral activity of stachyflin on influenza A viruses of different hemagglutinin subtypes. Virol. J. *10*, 118.

Nagamitsu, T., Takano, D., Shiomi, K., Ui, H., Yamaguchi, Y., Masuma, R., Harigaya, Y., Kuwajima, I., and Ōmura, S. (2003). Total synthesis of nafuredin-γ, a γ-lactone related to nafuredin with selective inhibitory activity against NADH-fumarate reductase. Tetrahedron Lett. *44*, 6441–6444.

Nagaraj, G., Uma, M.V., Shivayogi, M.S., and Balaram, H. (2001). Antimalarial activities of peptide antibiotics isolated from fungi. Antimicrob. Agents Chemother. *45*, 145–149.

Nair, M.G., Chandra, A., and Thorogood, D.L. (1993). Griseulin, a new nitro-containing bioactive metabolite produced by *Streptomyces* spp. J. Antibiot. *46*, 1762–1763.

Nakagawa, A., Fujimoto, T., Ōmura, S., Walsh, J.C., Stotish, R.L., and George, B. (1987). Hygromycin A, an antitreponemal substance. II. Therapeutic effect for swine dysentery. J. Antibiot. *40*, 1627–1635.

Nakagawa, Y., Doi, T., Masuda, Y., Takegoshi, K.; Igarashi, Y., and Ito, Y. (2011). Mapping of the primary mannose binding site of pradimicin A. J. Am. Chem. Soc. *133*, 17485–17493.

Nakashima, H., Ichiyama, K., Inazawa, K., Ito, M., Hayashi, H., Nishihara, Y., Tsujii, E., and Kino, T. (1996). FR901724, a novel anti-human immunodeficiency virus (HIV) peptide produced by Streptomyces, shows synergistic antiviral activities with HIV protease inhibitor and 2′,3′-dideoxynucleosides. Biol. Pharm. Bull. *19*, 405–412.

Naruse, N., Tenmyo, O., Kawano, K., Tomita, K., Ohgusa, N., Miyaki, T., Konishi, M., and Oki, T. (1991). Fluvirucins A_1, A_2, B_1, B_2, B_3, B_4 and B_5, new antibiotics active against influenza A virus. I. Production, isolation, chemical properties and biological activities. J. Antibiot. *44*, 733–740.

Newman, D.J., and Cragg, G.M. (2012). Natural products as sources of new drugs over the 30 years from 1981 to 2010. J. Nat. Prod. *75*, 311–335.

NIAID (2006). Emerging and re-emerging infectious diseases. In NIAID Profile Fiscal Year 2005 (National Institute of Allergy and Infectious Diseases, Bethesda, MA), pp. 84–93.

Nihei, C., Fukai, Y., and Kita, K. (2002). Trypanosome alternative oxidase as a target of chemotherapy. Biochim. Biophys. Acta *1587*, 234–239.

Niitsuma, M., Hashida, J., Iwatsuki, M., Mori, M., Ishiyama, A., Namatame, M., Nishihara-Tsukashima, A., Matsumoto, A., Takahashi, Y., Yamada, H., *et al.* (2010). Sinefungin VA and dehydrosinefungin V, new antitrypanosomal antibiotics produced by *Streptomyces* sp. K05–0178. J. Antibiot. *63*, 673–679.

Nilanonta, C., Isaka, M., Chanphen, R., Thongorn, N., Tanticharoen, M., and Thebtaranonth, Y. (2003). Unusual enniatins produced by the insect pathogenic fungus *Verticillium hemipterigenum*: isolation and studies on precursor-directed biosynthesis. Tetrahedron *59*, 1015–1020.

Nishihara, Y., Tsujii, E., Yamagishi, Y., Sakamoto, K., Tsurumi, Y., Furukawa, S., Ohtsu, R., Kino, T., Hino, M., Yamashita, M., and Hashimoto, S. (2001a). FR198248, a new anti-influenza agent isolated from *Aspergillus terreus* No. 13830. I. Taxonomy, fermentation, isolation, physico-chemical properties and biological activities. J. Antibiot. *54*, 136–143.

Nishihara, Y., Takase, S., Tsujii, E., Hatanaka, H., and Hashimoto, S. (2001b). New anti-influenza agents, FR198248 and its derivatives. II. Characterization of FR198248, its related compounds and some derivatives. J. Antibiot. *54*, 297–303.

Oki, T., Konishi, M., Tomatsu, K., Tomita, K., Saitoh, K., Tsunakawa, M., Nishio, M., Miyaki, T., and Kawaguchi, H. (1988). Pradimicin, a novel class of potent antifungal antibiotics. J. Antibiot. *41*, 1701–1704.

Oki, T., Tenmyo, O., Hirano, M., Tomatsu, K., and Kamei, H. (1990). Pradimicins A, B, and C: new antifungal antibiotics. II. *In vitro* and *in vivo* biological activities. J. Antibiot. *43*, 763–770.

Okun, J.G., Lümmen, P., and Brandt, U. (1999). Three classes of inhibitors share a common binding domain in mitochondrial complex I (NADH:ubiquinone oxidoreductase). J. Biol. Chem. *274*, 2625–2630.

Ōmura, S. (1979). Avermectins, new family of potent anthelmintic agents: producing organism and fermentation. Antimicrob. Agents Chemother. *15*, 361–367.

Ōmura, S. (2002). Mode of action of avermectin. In Macrolide Antibiotics. Chemistry, Biology, and Practice, 2nd Ed., S. Ōmura, ed. (Academic Press, San Diego, CA), pp. 571–576.

Ōmura, S., and Crump, A. (2004). The life and times of ivermectin – a success story. Nat. Rev. Microbiol. *2*, 984–989.

Ōmura, S., and Shiomi, K. (2007). Discovery, chemistry, and chemical biology of microbial products. Pure Appl. Chem. *79*, 581–591.

Ōmura, S., Nakagawa, A., and Tanaka, Y. (1982). New macrocyclic antibiotics, irumamycin and hitachimycin (stubomycin). In Trends in Antibiotic Research, Umezawa, H., Demain, A.L., and Hata, T. eds. (Japan Antibiotic Research Association, Tokyo, Japan), pp. 135–145.

Ōmura, S., Tsuzuki, K., Iwai, Y., Kishi, M., Watanabe, S., and Shimizu, H. (1985). Anticoccidial activity of frenolicin B and its derivatives. J. Antibiot. *38*, 1447–1448.

Ōmura, S., Otoguro, K., Imamura, N., Kuga, H., Takahashi, Y., Masuma, R., Tanaka, Y., Tanaka, H., Su, X., and You, E. (1987a). Jietacins A and B, new nematocidal antibiotics from a *Streptomyces* sp. Taxonomy, isolation, and physico-chemical and biological properties. J. Antibiot. *40*, 623–629.

Ōmura, S., Nakagawa, A., Fujimoto, T., Saito, K., Otoguro, K., and Walsh, J.C. (1987b). Hygromycin A, an antitreponemal substance. I. Screening method and therapeutic effect for *Treponema hyodysenteriae*-caused infection in CF-1 mice. J. Antibiot. *40*, 1619–1626.

Ōmura, S., Tomoda, H., Kimura, K., Zhen, D.-Z., Kumagai, H., Igarashi, K., Imamura, N., Takahashi, Y., Tanaka, Y., and Iwai, Y. (1988). Atpenins, new antifungal antibiotics produced by *Penicillium* sp. Production, isolation, physicochemical and biological properties. J. Antibiot. *41*, 1769–1773.

Ōmura, S., Eda, S., Funayama, S., Komiyama, K., Takahashi, Y., and Woodruff, H.B. (1989). Studies on a novel antitumor antibiotic, phenazinomycin: taxonomy, fermentation, isolation, and physicochemical and biological characteristics. J. Antibiot. *42*, 1037–1042.

Ōmura, S., Tomoda, H., Tabata, N., Ohyama, Y., Abe, T., and Namikoshi, M. (1999). Roselipins, novel fungal metabolites having a highly methylated fatty acid modified with a mannose and an arabinitol. J. Antibiot. *52*, 586–589.

Ōmura, S., Miyadera, H., Ui, H., Shiomi, K., Yamaguchi, Y., Masuma, R., Nagamitsu, T., Takano, D., Sunazuka, T., Harder, A., *et al.* (2001). An anthelmintic compound, nafuredin, shows selective inhibition of complex I in helminth mitochondria. Proc. Natl. Acad. Sci. U.S.A. *98*, 60–62.

Ondeyka, J.G., Helms, G.L., Hensens, O.D., Goetz, M.A., Zink, D.L., Tsipouras, A., Shoop, W.L., Slayton, L., Dombrowski, A.W., Polishook, J.D., *et al.* (1997). Nodulisporic acid A, a novel and potent insecticide from a *Nodulisporium* sp. Isolation, structure determination, and chemical transformations. J. Am. Chem. Soc. *119*, 8809–8816.

Ostlind, D.A., Mickle, W.G., Ewanciw, D.V., Andriuli, F.J., Campbell, W.C., Hernandez, S., Mochales, S., and Munguira, E. (1990). Efficacy of paraherquamide against immature *Trichostrongylus colubriformis* in the gerbil (*Meriones unguiculatus*). Res. Vet. Sci. *48*, 260–261.

Otoguro, K., Liu, Z.-X., Fukuda, K., Li, Y., Iwai, Y., Tanaka, H., and Ōmura, S. (1988). Screening for new nematocidal substances of microbial origin by a new method using the pine wood nematode. J. Antibiot. *41*, 573–575.

Otoguro, K., Kohana, A., Manabe, C., Ishiyama, A., Ui, H., Shiomi, K., Yamada, H., and Ōmura, S. (2001). Potent antimalarial activities of polyether antibiotic, X-206. J. Antibiot. *54*, 658–663.

Otoguro, K., Ui, H., Ishiyama, A., Arai, N., Kobayashi, M., Takahashi, Y., Masuma, R., Shiomi, K., Yamada, H., and Ōmura, S. (2003a). *In vitro* antimalarial activities of the microbial metabolites. J. Antibiot. *56*, 322–324.

Otoguro, K., Ui, H., Ishiyama, A., Kobayashi, M., Togashi, H., Takahashi, Y., Masuma, R., Tanaka, H., Tomoda, H., Yamada, H., and Ōmura, S. (2003b). *In vitro* and *in vivo* antimalarial activities of a non-glycosidic 18-membered macrolide antibiotic, borrelidin, against drug-resistant strains of *Plasmodia*. J. Antibiot. *56*, 727–729.

Otoguro, K., Ishiyama, A., Iwatsuki, M., Namatame, M., Nishihara-Tukashima, A., Nakashima, T., Shibahara, S., Kondo, S., Yamada, H., and Ōmura, S. (2010). *In vitro* and *in vivo* anti-*Trypanosoma brucei* activities of phenazinomycin and related compounds. J. Antibiot. *63*, 579–581.

Paolantonacci, P., Lawrence, F., and Robert-Géro, M. (1985). Differential effect of sinefungin and its analogs on the multiplication of three *Leishmania* species. Antimicrob. Agents Chemother. *28*, 528–531.

Pfefferkorn, E.R., and Borotz, S.E. (1994). Comparison of mutants of *Toxoplasma gondii* selected for resistance to azithromycin, spiramycin, or clindamycin. Antimicrob. Agents Chemother. *38*, 31–37.

Pinnert-Sindico, S., Ninet, L., Preud'homme, J., and Cosar, C. (1955). A new antibiotic, spiramycin. Antibiot. Ann. *1954–1955*, 724–727.

Polonsky, J., Merrien, M.-A., Prangé, T., Pascard, C., and Moreau, S. (1980). Isolation and structure (x-ray analysis) of marcfortine A, a new alkaloid from *Penicillium roqueforti*. J. Chem. Soc., Chem. Commun. 601–602.

Porter, J.N., Hewitt, R.I., Hesseltine, C.W., Krupka, G., Lowery, J.A., Wallace, W.S., Bohonos, N., and Williams, J.H. (1952). Achromycin: a new antibiotic having trypanocidal properties. Antibiot. Chemother. *2*, 409–410.

Prudhomme, J., McDaniel, E., Ponts, N., Bertani, S., Fenical, W., Jensen, P., and Le Roch, K. (2008). Marine actinomycetes: a new source of compounds against the human malaria parasite. PLoS One 3, e2335.

Pugh, C.S.G., Borchardt, R.T., and Stone, H.O. (1978). Sinefungin, a potent inhibitor of virion mRNA (guanine-7-)-methyltransferase, mRNA(nucleoside-2′-)-methyltransferase, and viral multiplication. J. Biol. Chem. *253*, 4075–4077.

Qian, H., Martin, R.J., and Robertson, A.P. (2006). Pharmacology of *N*-, *L*- and *B*-subtypes of nematode nAChR resolved at the single-channel in *Ascaris suum*. FASEB J. *20*, 2606–2608 (E2108–E2116)

Qin, X.-D., Dong, Z.-J., Liu, J.-K., Yang, L.-M., Wang, R.-R., Zheng, Y.-T., Lu, Y., Wu, Y.-S., and Zheng, Q.-T. (2006). Concentricolide, an anti-HIV agent from the ascomycete *Daldinia concentrica*. Helv. Chim. Acta *89*, 127–133.

Richard-Lenoble, D., Chandenier, J., and Gaxotte, P. (2003). Ivermectin and filariasis. Fund. Clin. Pharmacol. *17*, 199–203.

Richards, F.O., Jr, Boatin, B., Sauerbrey, M., and Sékétéli, A. (2001). Control of onchocerciasis today: status and challenges. Trends Parasitol. *17*, 558–563.

Robert Koch Institute. (2011). Final Presentation and Evaluation of Epidemiological Findings in the EHEC O104:H4 Outbreak Germany 2011. (Robert Koch Institute Berlin, Germany).

Roberts, C.W., McLeod, R., Rice, D.W., Ginger, M., Chance, M.L., and Goad, L.J. (2003). Fatty acid and sterol metabolism: potential antimicrobial targets in apicomplexan and trypanosomatid parasitic protozoa. Mol. Biochem. Parasitol. *126*, 129–142.

Rogers, M.J., Cundliffe, E., and McCutchan. T.F. (1998). The antibiotic micrococcin is a potent inhibitor of growth and protein synthesis in the malaria parasite. Antimicrob. Agents Chemother. *42*, 715–716.

Saeger, B., Schmitt-Wrede, H.-P., Dehnhardt, M., Benten, W.P.M., Krücken, J., Harder, A., von Samson-Himmelstjerna, G., Wiegand, H., and Wunderlich, F. (2001). Latrophilin-like receptor from the parasitic nematode *Haemonchus contortus* as target for the anthelmintic depsipeptide PF1022A. FASEB J. *15*, 1332–1334.

Sakai, C., Tomitsuka, E., Esumi, H., Harada, S., and Kita, K. (2012). Mitochondrial fumarate reductase as a target of chemotherapy: from parasites to cancer cells. Biochim. Biophys. Acta *1820*, 643–651.

Salem, H.H., and El-Allaf, G. (1969). Treatment of *Taenia saginata* and *Hymenolepis nana* infections with paromomycin. Trans. R. Soc. Trop. Med. Hyg. *63*, 833–836.

Sasaki, H., Hosokawa, T., Sawada, M., and Ando, K. (1973). Isolation and structure of ascofuranone and ascofranol, antibiotics with hypolipidemic activity. J. Antibiot. *26*, 676–680.

Sasaki, T., Takagi, M., Yaguchi, T., Miyadoh, S., Okada, T., and Koyama, M. (1992). A new anthelmintic cyclodepsipeptide, PF1022A. J. Antibiot. *45*, 692–697.

Schaeffer, J.M., Frazier, E.G., Bergstrom, A.R., Williamson, J.M., Liesch, J.M., and Goetz, M.A. (1990). Cochlioquinone, a nematocidal agent which competes for specific [^{3}H]ivermectin binding sites. J. Antibiot. *43*, 1179–1182.

Schinkel, A.H., Smit, J.J.M., van Tellingen, O., Beijnen, J.H., Wagenaar, E., van Deemter, L., Mol, C.A.A.M., van der Valk, M.A., Robanus-Maandag, E.C., te Riele, H.P.J., Berns, A.J.M., and Borst, P. (1994). Disruption of the mouse mdr1a P-glycoprotein gene leads to a deficiency in the blood–brain barrier and to increased sensitivity to drugs. Cell 77, 491–502.

Shimamura, H., Sunazuka, T., Izuhara, T., Hirose, T., Shiomi, K., and Ōmura, S. (2007). Total synthesis and biological evaluation of verticipyrone and analogues. Org. Lett. *9*, 65–67.

Shimizu, H., Osanai, A., Sakamoto, K., Inaoka, D.K., Shiba, T., Harada, S., and Kita, K. (2012). Crystal structure of mitochondrial quinol–fumarate reductase from the parasitic nematode *Ascaris suum*. J. Biochem. *151*, 589–592.

Shimizu, H., Shiba, T., Inaoka, D.K., Osanai, A., Kita, K., Sakamoto, K., and Harada, S. (2013). Mitochondrial rhodoquinol-fumarate reductase from the parasitic nematode *Ascaris suum* with the specific inhibitor atpenin A5. Protein Data Bank ID 3VRA.

Shiomi, K., and Ōmura, S. (2004). Antiparasitic agents produced by microorganisms. Proc. Jpn. Acad., Ser. B *80*, 245–258.

Shiomi, K., Haneda, K., Tomoda, H., Iwai, Y., and Ōmura, S. (1994). Cytosaminomycins, new anticoccidial agents produced by *Streptomyces* sp. KO-8119. II. Structure elucidation of cytosaminomycins A, B, C and D. J. Antibiot. *47*, 782–786.

Shiomi, K., Ui, H., Suzuki, H., Hatano, H., Nagamitsu, T., Takano, D., Miyadera, H., Yamashita, T., Kita, K., Miyoshi, H., *et al.* (2005a). A γ-lactone form nafuredin, nafuredin-γ, also inhibits helminth complex I. J. Antibiot. *58*, 50–55.

Shiomi, K., Matsui, R., Isozaki, M., Chiba, H., Sugai, T., Yamaguchi, Y., Masuma, R., Tomoda, H., Chiba, T., Yan, H., *et al.* (2005b). Fungal phenalenones inhibit HIV-1 integrase. J. Antibiot. *58*, 65–68.

Shoop, W., and Soll, M. (2002). Ivermectin, abamectin and eprinomectin. In Macrocyclic Lactones in Antiparasitic Therapy, J. Vercruysse, and R.S. Rew, eds. (CABI Publishing, Wallingford, UK), pp. 1–29.

Sin, N., Meng, L., Wang, M.Q.W., Wen, J.J., Bornmann, W.G., and Crews, C.M. (1997). The anti-angiogenic agent fumagillin covalently binds and inhibits the methionine aminopeptidase, MetAP-2. Proc. Natl. Acad. Sci. U.S.A. *94*, 6099–6103.

Singh, S.B. (2011), Discovery and development of natural product inhibitors of HIV-1 integrase. In HIV-1 Integrase. Mechanism and Inhibitor Design, N. Neamati, ed. (John Wiley & Sons, Hoboken, NJ) 309–323.

Singh, S.B., Zink, D.L., Goetz, M.A., Dombrowski, A.W., Polishook, J.D., and Hazuda, D.J. (1998). Equisetin and a novel opposite stereochemical homolog phomasetin, two fungal metabolites as inhibitors of HIV-1 integrase. Tetrahedron Lett. *39*, 2243–2246.

Singh, S.B., Zink, D., Polishook, J., Valentino, D., Shafiee, A., Silverman, K., Felock, P., Teran, A., Vilella, D., Hazuda, D.J., and Lingham, R.B. (1999). Structure and absolute stereochemistry of HIV-1 integrase inhibitor integric acid. A novel eremophilane sesquiterpenoid produced by a *Xylaria* sp. Tetrahedron Lett. *40*, 8775–8779.

Singh, S.B., Jayasuriya, H., Salituro, G.M., Zink, D.L., Shafiee, A., Heimbuch, B., Silverman, K.C., Lingham, R.B., Genilloud, O., Teran, A., *et al.* (2001). The complestatins as HIV-1 integrase inhibitor. Efficient isolation, structure elucidation, and inhibitory activities of isocomplestatin, chloropeptin I, new complestatins, A and B, and acid-hydrolysis products of chloropeptin I. J. Nat. Prod. *64*, 874–882.

Singh, S.B., Zink, D.L., Heimbach, B., Genilloud, O., Teran, A., Silverman, K.C., Lingham, R.B., Felock, P., and Hazuda, D.J. (2002a). Structure, stereochemistry, and biological activity of integramycin, a novel hexacyclic natural product produced by *Actinoplanes* sp. that inhibits HIV-1 integrase. Org. Lett. *4*, 1123–1126.

Singh, S.B., Zink, D.L., Quamina, D.S., Pelaez, F., Teran, A., Felock, P., and Hazuda, D.J. (2002b). Integrastatins: structure and HIV-1 integrase inhibitory activities of two novel racemic tetracyclic aromatic heterocycles produced by two fungal species. Tetrahedron Lett. *43*, 2351–2354.

Singh, S.B., Zink, D.L., Dombrowski, A.W., Polishook, J.D., Ondeyka, J.G., Hirshfield, J., Felock, P., and Hazuda, D.J. (2003). Integracides: tetracyclic triterpenoid inhibitors of HIV-1 integrase produced by *Fusarium* sp. Bioorg. Med. Chem. *11*, 1577–1582.

Smith, H.L., and Rajan, T.V. (2000). Tetracycline inhibits development of the infective-stage larvae of filarial nematodes *in vitro*. Exp. Parasitol. *95*, 265–270.

Somanadhan, B., Kotturi, S.R., Leong, C.Y., Glover, R.P., Huang, Y., Flotow, H., Buss, A.D., Lear, M.J., and Butler, M.S. (2013). Isolation and synthesis of falcitidin, a novel myxobacterial-derived acyltetrapeptide with activity against the malaria target falcipain-2. J. Antibiot. *66*, 259–264.

Stefanelli, S., Cavaletti, L., Sarubbi, E., Ragg, E., Colombo, L., and Selva, E. (1995). GE20372 factor A and B. New HIV-1 protease inhibitors, produced by *Streptomyces* sp. ATCC 55925. J. Antibiot.*48*, 332–334.

Stella, S., Saddler, G., Sarubbi, E., Colombo, L., Stefanelli, S., Denaro, M., and Selva, E. (1991). Isolation of α-MAPI from fermentation broths during a screening program for HIV-1 protease inhibitors. J. Antibiot. *44*, 1019–1022.

Sternberg, T.H., Wright, E.T., and Oura, M. (1956). A new antifungal antibiotic, amphotericin B. Antibiot. Ann. *1955–1956*, 566–573.

Sullivan, M., Li, J., Kumar, S., Rogers, M.J., and McCutchan, T.F. (2000). Effects of interruption of apicoplast function on malaria infection, development, and transmission. Mol. Biochem. Parasitol. *109*, 17–23.

Sun, F., Huo, X., Zhai, Y., Wang, A., Xu, J., Su, D., Bartlam, M., and Rao, Z. (2005). Crystal structure of mitochondrial respiratory membrane protein complex II. Cell *121*, 1043–1057.

Sunazuka, T., Omura, S., Iwasaki, S., and Ōmura, S. (2002). Chemical modification of macrolides. In Macrolide Antibiotics. Chemistry, Biology, and Practice, 2nd Ed., S. Ōmura, ed. (Academic Press, San Diego, CA), pp. 99–180.

Tabata, N., Suzumura, Y., Tomoda, H., Masuma, R., Haneda, K., Kishi, M., Iwai, Y., and Ōmura, S. (1993). Xanthoquinodins, new anticoccidial agents produced by *Humicola* sp.: production, isolation, and physico-chemical and biological properties. J. Antibiot. *46*, 749–755.

Takagi, M., Motohashi, K., Nagai, A., Izumikawa, M., Tanaka, M., Fuse, S., Doi, T., Iwase, K., Kawaguchi, A., Nagata, K., *et al.* (2010). Anti-influenza virus compound from *Streptomyces* sp. RI18. Org. Lett. *12*, 4664–4666.

Takahashi, Y., Matsumoto, A., Seino, A., Ueno, J., Iwai, Y., and Ōmura, S. (2002). *Streptomyces avermectinius* sp. nov., an avermectin-producing strain. Int. J. Syst. Evol. Microbiol. *52*, 2163–2168.

Takano, D., Nagamitsu, T., Ui, H., Shiomi, K., Yamaguchi, Y., Masuma, R., Kuwajima, I., and Ōmura, S. (2001a). Absolute configuration of nafuredin, a new specific NADH-fumarate reductase inhibitor. Tetrahedron Lett. *42*, 3017–3020.

Takano, D., Nagamitsu, T., Ui, H., Shiomi, K., Yamaguchi, Y., Masuma, R., Kuwajima, I., and Ōmura, S. (2001b). Total synthesis of nafuredin, a selective NADH-fumarate reductase inhibitor. Org. Lett. *3*, 2289–2291.

Takeda, Y. (2011). *Vibrio parahaemolyticus*, enterotoxigenic *Escherichia coli*, enterohemorrhagic *Escherichia coli* and *Vibrio cholerae*. Proc. Jpn. Acad., Ser. B *87*, 1–12.

Takiguchi, Y., Mishima, H., Okuda, M., Terao, M., Aoki, A., and Fukuda, R. (1980). Milbemycins, a new family of macrolide antibiotics: fermentation, isolation and physicochemical properties. J. Antibiot. *33*, 1120–1127.

Tanabe, A., Nakashima, H., Yoshida, O., Yamamoto, N., Tenmyo, O., and Oki, T. (1988). Inhibitory effect of new antibiotic, pradimicin A on infectivity, cytopathic effect and replication of human immunodeficiency virus *in vitro*. J. Antibiot. *41*, 1708–1710.

Tanaka, H., Matsuzaki, K., Nakashima, H., Ogino, T., Matsumoto, A., Ikeda, H., Woodruff, H.B., and Ōmura, S. (1997). Chloropeptins, new anti-HIV antibiotics inhibiting gp120-CD4 binding from *Streptomyces* sp. I. Taxonomy, fermentation, isolation, and physico-chemical properties, and biological activities. J. Antibiot. *50*, 58–65.

Tanaka, H., Chiba, H., Inokoshi, J., Kuno, A., Sugai, T., Takahashi, A., Ito, Y., Tsunoda, M., Suzuki, K., Takénaka, A., *et al.* (2009). Mechanism by which the lectin actinohivin blocks HIV infection of target cells. Proc. Nat. Acad. Sci. U.S.A. *106*, 15633–15638.

Tomita, K., Oda, N., Hoshino, Y., Ohkusa, N., and Chikazawa, H. (1991). Fluvirucins A_1, A_2, B_1, B_2, B_3, B_4 and B_5, new antibiotics active against influenza A virus. IV. Taxonomy on the producing organisms. J. Antibiot. *44*, 940–948.

Tsipouras, A., Adefarati, A.A., Tkacz, J.S., Frazier, E.G., Rohrer, S.P., Birzin, E., Rosegay, A., Zink, D.L., Goetz, M.A., Singh, S.B., and Schaeffer, J.M. (1996). Ophiobolin M and analogues, noncompetitive inhibitors of ivermectin binding with nematocidal activity. Bioorg. Med. Chem. *4*, 531–536.

Tsunakawa, M., Tenmyo, O., Tomita, K., Naruse, N., Kotake, C., Miyaki, T., Konishi, M., and Oki, T. (1992). Quartromicin, a complex of novel antiviral antibiotics. I. Production, isolation, physico-chemical properties, and antiviral activity. J. Antibiot. *45*, 180–188.

Tsunakawa, M., Hu, S.-L., Hoshino, Y., Detlefson, D.J., Hill, S.E., Furumai, T., White, R.J., Nishio, M., Kawano, K., Yamamoto, S., *et al.* (1995). Siamycins I and II, new anti-HIV peptides: I. Fermentation, isolation, biological activity and initial characterization. J. Antibiot. *48*, 433–434.

Ui, H., Shiomi, K., Yamaguchi, Y., Masuma, R., Nagamitsu, T., Takano, D., Sunazuka, T., Namikoshi, M., and Ōmura, S. (2001). Nafuredin, a novel inhibitor of NADH-fumarate reductase, produced by *Aspergillus niger* FT-0554. J. Antibiot. *54*, 234–238.

Ui, H., Shiomi, K., Suzuki, H., Hatano, H., Morimoto, H., Yamaguchi, Y., Masuma, R., Sunazuka, T., Shimamura, H., Sakamoto, K., *et al.* (2006). Verticipyrone, a new NADH-fumarate reductase inhibitor, produced by *Verticillium* sp. FKI-1083. J. Antibiot. *59*, 785–790.

Van Hellemond, J.J., Klockiewicz, M., Gaasenbeek, C.P.H., Roos, M.H., and Tielens, A.G.M. (1995). Rhodoquinone and complex II of the electron transport chain in anaerobically functioning eukaryotes. J. Biol. Chem. *270*, 31065–31070.

Vertesy, L., Aretz, W., Knauf, M., Markus, A., Vogel, M., and Wink, J. (1999). Feglymycin, a novel inhibitor of the replication of the human immunodeficiency virus. Fermentation, isolation and structure elucidation. J. Antibiot. *52*, 374–382.

Vogeley, L., Palm, G.J., Mesters, J.R., and Hilgenfeld, R. (2001). Conformational change of elongation factor Tu (EF-Tu) induced by antibiotic binding. Crystal structure of the complex between EF-Tu·GDP and aurodox. J. Biol. Chem. *276*, 17149–17155.

Walsh, C. (2003). Antibiotics: Actions, Origins, Resistance. (ASM Press, Washington, DC).

WHO (1995). WHO Model Prescribing Information: Drugs Used in Parasitic Diseases—Second Edition (World Health Organization, Geneva, Switzerland).

WHO (2002). Fact sheet No. 124 Emerging Foodborne Diseases. Revised January 2002 (World Health Organization, Geneva, Switzerland).

WHO (2007). The World Health Report 2007: A Safer Future: Global Public Health Security in the 21st Century (World Health Organization, Geneva, Switzerland).

WHO (2013a). Fact sheet No. 360 HIV/AIDS. Updated October 2013 (World Health Organization, Geneva, Switzerland).

WHO (2013b). Core epidemiological slides HIV/AIDS estimates. November 2013 (World Health Organization, Geneva, Switzerland).

WHO (2013c). Influenza virus infections in humans. July 2013 (World Health Organization, Geneva, Switzerland).

Yabu, Y, Suzuki, T., Nihei, C., Minagawa, N., Hosokawa, T., Nagai, K., Kita, K., and Ohta, N. (2006). Chemotherapeutic efficacy of ascofuranone in *Trypanosoma vivax*-infected mice without glycerol. Parasitol. Int. *55*, 39–43.

Yamamoto, T., Izumi, N., Ui, H., Sueki, A., Masuma, R., Nonaka, K., Hirose, T., Sunazuka, T., Nagai, T., Yamada, H., *et al.* (2012). Wickerols A and B: novel anti-influenza virus diterpenes produced by *Trichoderma atroviride* FKI-3849. Tetrahedron *68*, 9267–9271.

Yamazaki, M., Fujimoto, H., Okuyama, E., and Ohta, Y. (1980). On the structure of two new metabolites isolated from *Penicillium paraherquei* and *Aspergillus silvaticus*. Mycotoxins *10*, 27–28.

Yanai, K., Sumida, N., Okakura, K., Moriya, T., Watanabe, M., and Murakami, T. (2004). *Para*-position derivatives of fungal anthelmintic cyclodepsipeptides engineered with *Streptomyces venezuelae* antibiotic biosynthetic genes. Nat. Biotechnol. *22*, 848–855.

Yankovskaya, V., Horsefield, R., Törnroth, S., Luna-Chavez, C., Miyoshi, H., Léger, C., Byrne, B., Cecchini, G., and Iwata, S. (2003). Architecture of succinate dehydrogenase and reactive oxygen species generation. Science *299*, 700–704.

Yokomizo, K., Miyamoto, Y., Nagao, K., Kumagae, E., Habib, E.-S.E., Suzuki, K., Harada, S., and Uyeda, M. (1998). Fattiviracin A1, a novel antiviral agent produced by *Streptomyces microflavus* strain No. 2445. II. Biological properties. J. Antibiot. *51*, 1035–1039.

Yoshimoto, J., Kakui, M., Iwasaki, H., Fujiwara, T., Sugimoto, H., and Hattori, N. (1999). Identification of a novel HA conformational change inhibitor of human influenza virus. Arch. Virol. *144*, 865–878.

Zaha, O., Hirata, T., Kinjo, F., and Saito, A. (2002). Ivermectin in clinical practice. In Macrolide Antibiotics. Chemistry, Biology, and Practice, 2nd Ed., S. Ōmura, ed. (Academic Press, San Diego, CA), pp. 403–419.

Zinser, E.W., Wolfe, M.L., Alexander-Bowman, S.J., Thomas, E.M., Davis, J.P., Groppi, V.E., Lee, B.H., Thompson, D.P., and Geary, T.G. (2002). Anthelmintic paraherquamides are cholinergic antagonists in gastrointestinal nematodes and mammals. J. Vet. Pharmacol. Therap. *25*, 241–250.

Endophytes as a Potential Source of New Antibiotics

10

Silvia Guzmán-Trampe, Karol Rodríguez-Peña, Allan Espinosa-Gómez, Rosa E. Sánchez-Fernández, Martha L. Macías-Rubalcava, Luis B. Flores-Cotera and Sergio Sánchez

Abstract

Antibiotics are useful compounds for treatment of human, farm animal and aquaculture infections. However, due to resistance development of pathogenic microbes to most of the useful antibiotics, there is a continuous necessity for new and powerful anti-infective compounds. This situation encourages the search for new alternatives for the isolation of new compounds with antimicrobial activity. Because of their proficiency in producing secondary metabolites with therapeutic properties, plants have attracted attention since ancient times. Recently, plant endophytic microorganisms have been also shown to be an important and novel source of natural bioactive products with antimicrobial, antiviral, anti-tumour, antiparasitic, and agricultural activities. Endophytes commonly live associated to the plant vascular tissues, without causing any apparent damage to their host or symptoms of disease. The bioactive compounds they produce, usually vary depending on the plant host taxonomy and forest type. Fungi and actinomycetes have been the main source of new bioactive natural products from endophytes. Occasionally, endophytes can synthesize the same metabolites produced by the host plant. This chapter mainly reviews the progress that has been achieved on the production, by endophytic microbes, of the same or similar bioactive compounds originated from their host plants, as well as other secondary metabolites apparently not produced by the plant, including antituberculosis and antiparasitic compounds. Furthermore, the potential agricultural uses of endophytic compounds as antifungal, nematicidal, antiviral, insecticidal and phytotoxic activities, is also reviewed. Finally, we mention some examples of new compounds with antimicrobial properties chemically derived from natural products produced by endophytes.

Introduction

The search for new natural products NPs has increased in the last few years, mostly because there is a constant race between pathogenic microorganisms and degenerative diseases, against pharmaceutical companies. NPs or molecules inspired by them have been the core for development of around 80% of commercial drugs (Newman and Cragg, 2012). Despite the extraordinary advances in combinatorial chemistry for the modification or creation of new molecules, it cannot compete against nature for designing novel compounds with improved activities and reduced side-effects for human health. It has been demonstrated that NPs are better absorbed than synthetic drugs, probably due to the natural purposes they fulfil for plants, microorganisms, or animal necessities.

Natural products are molecules obtained from living organisms. The best producers include plants, microorganisms (bacteria, parasites, fungi), marine organisms (corals, algae, sponges), arthropods and animals.

According to Buttler (2008) over 100 NP derived substances are in clinical trials and a similar amount are in preclinical development. This may sound impressive, but more interesting is the fact that many of these are derived from plants and their microorganisms (Harvey, 2008).

Plants have attracted attention since ancient times because of their medicinal properties. Strong efforts have been made to characterize plants employed in traditional medicine, their pharmacological properties and isolation of new molecules (Osman *et al.*, 2012). Among the known plant species on earth (more than 300,000 different species), only a small fraction have been investigated for the presence of antimicrobial compounds and only 1–10% of plants are used by humans (Shahid *et al.*, 2009; Osman *et al.*, 2012). Unfortunately, it may take several years for a plant to reach its reproductive stage; therefore tree logging and plant collection may cause damage to populations driving them to disappear. This situation has led researchers to deeply explore these secondary metabolite factories (Strobel, 2012).

Conversely, microorganisms can be isolated from almost everywhere on planet, they have been found living in soil, water, animal tissues, and recently even on the thin atmospheric air, where nutrient and surface support are normally insignificant. Despite their small size, these organisms count as one of the most abundant forms of life on earth, mostly because they have adapted, through evolution, to harsh and extreme habitats and are well furnished with chemical arsenals to display any kind of ecological interactions like competition, stress response and defence.

Despite the vast diversity of compounds, there will always be a demand for new ecological niches for potential sources of natural bioactive agents from renewable, eco-friendly and easily obtainable sources (Liu *et al.*, 2001).

Many years ago, scientists discovered microorganisms living inside vascular tissues of plants, contrasting with the conventional thought of microorganisms living exclusively on plant surfaces in epiphytic or parasitic relationships. Therefore, the term endophyte was used for the first time in 1866 by de Bary (Hyde and Soytong, 2008) after the observation of hyphae from different wheat seeds. Endophyte definition, however, has been largely modified and discussed thereafter.

Endophyte definition

The term, endophyte, can be described based on the following criteria:

First, etymologically endophyte means 'endon': inside and 'phyte': plant. Nonetheless, the term endophyte has been discussed and changed, mainly because it was not inclusive of all its properties. Let us explain: Inicial attempts to define the term referred to fungi inhabiting monocot plants (mainly herbaceous genus, with some exceptions), such as grasses, wheat and a large variety of crops. Nevertheless, many endophytes have been isolated from dicot plants as well (also called woody plants), so the definition had to be updated to include large perennial plants (Saikkonen, 2004).

Second, initial endophyte definitions fails to include prokaryotic microorganisms such as bacteria and included only fungi, this makes sense due to the large amount of fungi obtained from initial endophytic research (Schulz and Boyle, 2006). Nowadays, multiple endophytic bacteria have been isolated from plants, including the families Pseudomonaceae and Actinomycetaceae, both with enormous bioactive potential (Brader *et al.*, 2014; Ambrose *et al.*, 2013; Bulgarelli *et al.*, 2013).

Third, association type within endophyte and host must be taken into account as well, Wilson (1995) referred to endophytes as bacteria or fungi that live inside plants without causing any apparent symptom of disease or damage.

Taking the above into account, we define *endophytes* as microorganisms whose life cycle (or a part of it) is completed strictly inside a plant (woody or herbaceous) without causing any apparent damage or symptoms of disease. These are often related to vascular plant tissue and display different types of host–endophyte interactions. The term microorganism comprehends prokaryotes and fungi which could be isolated from anywhere inside plants.

Distribution in the plant

Endophytes are commonly associated with plant vascular tissues, essentially because these are responsible for sugar and nutrient transport inside plants (Selim *et al.*, 2012). Fungi distribution has been classified into the following two types.

According to their distribution inside plants

Endophytes can be found everywhere inside the plants. Different terms had been used to relate them to the plant area they inhabit (Bulgarelli *et al.*, 2013). We will describe two brief examples:

- Rizosphere: referring to microorganisms living inside roots. Some use the term laimosphere, as well (Gaiero *et al.*, 2013).
- Phyllosphere: refers to microorganisms that inhabit leaves or any plant tissues above the soil (which means stem, bark, branch, leaf) (Lindow and Brandl, 2003).

We would rather use the term 'endosphere', which in our consideration, better describes the idea of any liveable area inside plants including roots, leaves, stems, flowers, and fruits.

According to their ecological interactions

Initial attempts classified endophytic fungi into clavicipitaceous (C) and non-clavicipitaceous (NC) groups. Recently, this classification has been modified and based on colonization patterns, mechanism of transmission, biodiversity levels and ecological function (Rodríguez *et al.*, 2009). They have been divided into:

- Class I or (C endophytes): mainly isolated from grasses, phylogenetically related, seed-transmission (vertical). Confer stress tolerance and decrease insect and animal herbivory.
- Class II, III and IV (NC endophytes): class III and IV have growth restrictions, shoot and root, respectively, whereas class II can grow in shoots, roots and rhizome. Class III and IV have horizontal transfer while class II has horizontal and vertical transmission.

Bacterial endophytes are not so easily grouped, mainly because there is no association between a particular plant environments and a specific bacterial strain. However, they are divided into obligate and facultative. Obligate bacterial endophytes are forced to complete their entire life cycle inside plant tissues. Facultative endophytic bacteria are able to survive in the soil, plant surface, and inside the plants (Ambrose *et al.*, 2013).

Biological properties

Their role in plants (endophyte–plant associations)

There is a mutual benefit obtained from the endophyte–plant relationship. The plant provides energy and supplies nutrients, as well as protection against environmental stress and competition. Conversely, the endophyte indirectly influences plant growth, survival and reproduction of the host through the production of NPs or secondary metabolites that protect plants against herbivory, parasitism and infection, as well as other molecules that are responsible for adaptation against drought, UV light and hypersalinity (Selim *et al.*, 2012). This effect is also called allelopathy.

Many endophytic bacteria have been shown to produce molecules that improve plant fitness and adaptation like phytohormones, siderophores, anti-parasitic agents, antibiotics, antifungal, and herbicidal compounds (Brader *et al.*, 2014; Rashid *et al.*, 2012).

Ecological roles

Typically, the endophyte–host interaction might fall into two different ecological classes, commensalism and mutualism. The first is defined as a relationship in which one organism benefits without affecting the other, and in the second type, both parts are equally benefited. This last one can also be considered symbiosis.

What is considered beneficial for one plant could be pathogenic for a different species. Depending on this, certain endophytes might be parasites for other hosts, depending on the balance between mechanisms of virulence and pathogenicity (Saikkonen *et al.*, 2004).

Plants provide endophytes with very unique environments, nutrient disposition and protection. Meanwhile, endophytes offer an enormous production of metabolites to protect their host from environmental stress, herbivory, or parasitism. The endophyte–host interaction depends on the balance between plant defences and virulence factors produced by the endophyte. Shulz and Boyle (2005), proposed the balanced antagonism model to explain the delicate equilibrium in this interaction. The non-pathogenic endophyte state can change depending on the host and endophyte stress. When stressed, either by disease, competition, herbivory or environmental factors, the production and excretion of endophyte virulence factors increases, leading to an unbalanced relationship. At this stage, the endophyte becomes a pathogenic parasite and there is a negative selection against it. When there is no herbivore pressure, environmental stress or microbial infection, the cost of having an endophyte is elevated and can be negatively selected.

Fig. 10.1 shows the balanced interactions between the host plant and the endophytes in different parts of the plant. When the environmental factors as chemical communication between species and the natural interactions with different herbivores, insects, etc., are balanced, the endophytes interact as we mentioned above. When the balance is broken, in some cases the endophytes could become pathogens.

Pharmaceutical importance of endophytes

Since they live in a vast variety of habitats, endophytes are a particularly attractive source of unexplored natural products. It has been estimated that there are at least 1 million species of endophytic fungi alone (Dreyfuss and Chapela, 1994). It is assumed that endophytes are a rich and reliable source of genetic diversity, capable of producing novel NP's. The pharmaceutical importance of endophytes has been demonstrated extensively, it has been reported that endophytic fungi and bacteria can produce novel bioactive metabolites with antimicrobial, antiviral, antitumour and antiparasitic activities among many

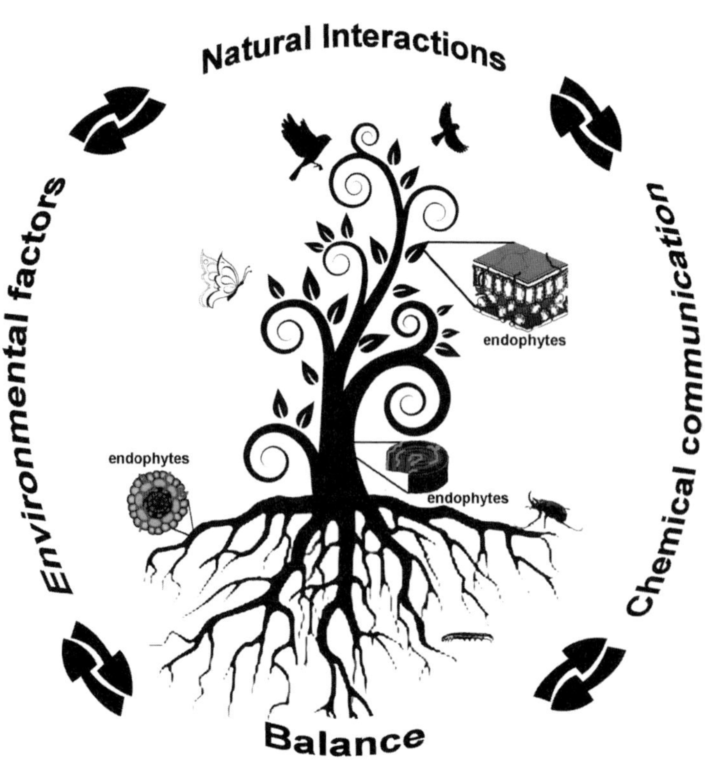

Figure 10.1 Depiction of interactions among the host plant with endophytes and diverse environmental factors. A colour version of this figure is available in the plate section at the back of the book.

others, like agricultural applications (Tan and Zou, 2001; Zhang *et al.*, 2006).

Bacteria, actinomycetes and fungi have been the main sources of new bioactive natural products. Some examples of compounds isolated from bacterial endophytes include ecomycins produced by *Pseudomonas viridiflava* (Miller *et al.*, 1998), and the peptidic antibiotics munumbicin (Castillo *et al.*, 2002) and kakadumycin (Castillo *et al.*, 2003) isolated from *Streptomyces* strains. Endophytic fungi have been shown to produce alkaloids such as chytochalasins (Wagenaar *et al.*, 2000) and camptothecin (Guo *et al.*, 2008), and antibiotics like cryptocin and brefeldin produced by the *Cladosporium* genus (Wang *et al.*, 2007). Interestingly, only a tiny fraction of plant-associated bacteria and fungi has been described so far, denoting a rising source for the discovery of novel secondary metabolites.

On some occasions, endophytes can produce the same metabolites as their host. Thus compounds formerly thought to be synthesized only by plants are actually recognized to be also produced by endophytes (Kusari *et al.*, 2013). Therefore, endophytes can be used instead of plants for production of industrial compounds (Zhou *et al.*, 2010). This is why one of the most important potential advantages in using endophytes could be the elimination of the need to harvest and extract the slow growing and in some cases endangered species of plants. As a consequence, drug production costs might decrease and therefore be available for a plethora of patients.

Endophytes as producers of the same metabolites as the host plant

Long time ago man acknowledged that plants possessed certain therapeutic and healing properties, in many cases linked with their antibiotic properties. Species, such as lemon balm (*Melissa officinalis*), tea tree (*Melaleuca alternifolia*) and garlic (*Allium sativum*), have been described as broad-spectrum antibiotic agents. Furthermore, essential oils from several plants have been used for their antibacterial and antifungal properties. More recent studies have found that numerous plant species produce a broad range of NPs, many of them with significant antimicrobial activity. Endophytes have also provided a broad variety of novel antimicrobial metabolites of many different chemical classes including alkaloids, terpenoids, lignans, quinones, peptides, steroids, flavonoids, xanthones, isocoumarins, cytochalasines and others (Zhao *et al.*, 2010; Tan and Zou, 2001; Gunatilaka, 2006; Schulz *et al.*, 2002, 2005). Nevertheless, the research on endophytic fungi has recently gained significant momentum owing to the discovery that some fungal endophytes are capable of synthesizing the same or similar compounds as their respective host plants. In the past few years, endophytes are being increasingly screened for the production of valuable compounds of plant origin like paclitaxel (Stierle *et al.*, 1993), camptothecin (Puri *et al.*, 2005; Kusari *et al.*, 2009d), podophyllotoxin (Eyberger *et al.*, 2006; Puri *et al.*, 2006), deoxypodophyllotoxin (Kusari *et al.*, 2009a), vinblastine (Kumar *et al.*, 2013), hypericin and emodin (Kusari and Spiteller, 2011; Kusari *et al.*, 2008; Kusari *et al.*, 2009b) and azadirachtin A and B (Kusari *et al.*, 2012b), which possess antibacterial and/or antifungal properties. In particular, it is of great interest that these compounds, originally obtained from plants, are among the very few NPs whose synthesis has also been reported with endophytic fungi and curiously, all are among the most potent antitumour compounds derived from plants. Further details about the current developments with respect to endophytes are given below.

Paclitaxel

Paclitaxel (Taxol®) is a highly functionalized tetracyclic diterpenoid found in each of the world's known yew (*Taxus*) species. Paclitaxel is a fungicide with notable activity against oomycetes, e.g. *Phytophthora*, *Pythium* and *Aphanomyces*, one of the main groups of fungi causing diseases in plants. In fact it has been suggested that such antifungal activity could be one of its ecological functions (Schueffler and Anke, 2011; Strobel and Daisy, 2003). Moreover, paclitaxel together with some of its derivatives represent a group of important anticancer agents, especially useful in the therapy against breast cancer (Strobel *et al.*, 2004). Paclitaxel and the taxanes baccatin III and 10-diacetylbacattin III were first discovered in the bark of *Taxus brevifolia* (Wani *et al.*, 1971). However, obtaining paclitaxel from plant material is very difficult, while the chemical synthesis, which requires about 28 reaction steps, is economically unfeasible. *Taxomyces andreanae* an endophytic fungus isolated from *T. brevifolia* was the first fungus reported to produce paclitaxel under laboratory conditions (Stierle *et al.*, 1993). These and several other scientists have since reported the isolation of paclitaxel from a number of endophytic fungi associated, not only with *Taxus* sp. trees, but also with other host plants. The production of paclitaxel by *Pestalotiopsis microspora* isolated from *Taxus wallichiana* and a second isolate of *P. microspora* from bald cypress, *Taxodium distichum*, was reported by Li *et al.* (1996). To date, a number of fungal species have been isolated from different trees reported to produce paclitaxel and/or taxane analogues (i.e. baccatin III, 10-deacetylbaccatin III). These comprise at least 19 genera of endophytic fungi: *Alternaria, Aspergillus, Botryodiplodia, Botrytis, Cladosporium, Ectostroma, Fusarium, Metarhizium, Monochaetia, Mucor, Ozonium, Papulaspora, Periconia, Pestalotia, Pestalotiopsis, Phyllosticta, Pithomyces, Taxomyces* and *Tubercularia* (Strobel and Daisy, 2003). The hosts of the reported paclitaxel-producing fungi include *Taxus* (e.g. *T. baccata, T. cuspidata, T. media and T. yunnanensis*) and other non-*Taxus* species (e.g. *Podocarpus* sp., *T. distichum, Terminalia arjuna, Torreya grandifolia and Wollemia nobilis*). Flores-Bustamante *et al.* (2010) and Zhou *et al.* (2010) have recently reviewed the published work on paclitaxel-producing endophytes. Even though paclitaxel and other taxanes have been widely reported to be fungal products, currently the reported levels of paclitaxel in fungal cultures remain too low for industrial production. Thus, 20 years after the discovery of *T. andreanae*, not one of these reported fungal endophytes has yet reached the original expectation as a source for reliable production of the large quantities of paclitaxel needed for cancer treatment. In addition, it seems that the synthesis of paclitaxel by fungi has not been corroborated in interlaboratory research, even in *T. andreanae*, the first strain described (Flores Bustamante *et al.*, 2010; Heinig *et al.*, 2013; Staniek *et al.*, 2009).

Camptothecin

The reports on the production of paclitaxel from fungal endophytes certainly spurred interest of several research groups to study the endophytes of other plants as alternative sources of NPs of plant origin. Camptothecin and 10-hydroxycamptothecin are two pentacyclic quinoline alkaloids initially isolated from *Camptotheca acuminata* (Nyssaceae) a tree native to mainland, China (Wall *et al.*, 1966). Later, camptothecin was also found in two other plants used in traditional medicine, the small tree *Nothapodytes foetida* (Icacinaceae) and *Nothapodytes nimmoniana* (Graham) of Indian origin (Schueffler and Anke, 2011). Plants of *N. nimmoniana* and *C. acuminata* are known to use camptothecin as chemical defence against a wide variety of insect pests and plant pathogenic fungi (Sirikantaramas *et al.*, 2009). Camptothecin is active against *Alternaria alternata, Epicoccum nigrum, Pestalotia guepinii, Drechslera* sp., *Fusarium avenaceum*, which are known fungal pathogens that infect trees (Schueffler and Anke, 2011). In fact, the fungal topoisomerase I–DNA complex, an enzyme required in DNA replication and transcription, is a target of camptothecin and 10-hydroxycamptothecin. A given fungus will be killed in presence of camptothecin unless it intrinsically possesses resistance to this compound (Kusari *et al.*, 2011a, 2012a). Furthermore, camptothecin and its analogues are considered within a selected group of the most expensive and relevant precursors of anticancer agents recovered from plant sources (Puri *et al.*, 2005). Although camptothecin is not used in cancer therapy, because of low solubility in aqueous media and toxicity issues, it is an important precursor for the semisynthesis of irinotecan (Hycamtin) and topotecan (Camtostar), two FDA-approved anticancer agents, which are considered to be among the most important anticancer agents derived from plant sources. The former compound is used for the treatment of colorectal carcinomas, and the latter for the treatment of ovarian cancer and non-small-cell lung cancer (Kharwar *et al.*, 2011). Although total synthesis has been accomplished by several groups, at present the main source of camptothecin is still plant material from wild trees of *C. acuminata* and *N. nimmoniana* (Schueffler and Anke, 2011). This has resulted in extensive harvest of the latter tree in India to meet the present demand of camptothecin. Consequently, other alternative sources are necessary to produce camptothecin and its analogues (Shaanker *et al.*, 2008; Zhao *et al.*, 2010).

Several endophytes isolated from plants have been reported to synthesize these compounds. An endophytic fungus from *N. foetida*, i.e. *Entrophosphora infrequens* (*phylum* Glomeromycota) was shown to produce camptothecin in culture (Puri *et al.*, 2005). Rehman *et al.* (2008) described a strain of *Neurospora* sp. isolated from the same plant species that produces camptothecin. More recently, Kusari *et al.* (2009d) reported the production of camptothecin and two analogues, 10-hydroxycamptothecin and 9-methoxycamptothecin, from *Fusarium solani* an endophytic fungus isolated from the inner bark of *C. acuminata*. However, subculturing of the fungus under axenic conditions gave very low levels of camptothecin. In this case, a host plant enzyme absent in the fungus, strictosidine synthase, was necessary to perform the key step of camptothecin production (Kusari *et al.*, 2011b,c). Another *F. solani* strain isolated from *Apodytes dimidiata* (Icacinaceae) was able to produce the same compounds (Shweta *et al.*, 2010). Furthermore, the endophytic *Xylaria* sp. was isolated from *C. acuminata*, which produced 10-hydroxycamptothecin but not camptothecin (Liu *et al.*, 2010). An endophytic fungus, which belongs to the family *Phycomycetes* was isolated from *N. foetida*. This fungus produced camptothecin when grown in a synthetic liquid medium both in shake flask and in bench-scale fermentations. The biological activity of the fungal compound was comparable with the standard when tested *in vitro* against several human cancer cell lines (Puri *et al.*, 2005).

Podophyllotoxin

Podophyllotoxin and desoxypodophyllotoxin are aryltetralin lignans which are naturally synthesized by *Podophyllum* species (Berberidaceae), perennial herbs with well-known use in traditional medicine (Liu *et al.*, 2007, and references cited therein). Podophyllotoxin was first isolated from *Podophyllum emodi*; however, in recent years podophyllotoxin and its analogues were found in other plant species such as trees belonging to *Juniperus* plants (mainly genera of *Diphylleia*, *Dysosma*, *Sabina*) and *Cupressaceae* (Zhao *et al.*, 2010; Schueffler and Anke, 2011; Kusari *et al.*, 2011d).

Podophyllotoxin and its analogues are well recognized for their antibacterial, antiviral and antioxidant activities. Deoxypodophyllotoxin has significant antimicrobial potency against the Gram-positive *Staphylococcus aureus* subsp. *aureus* and the Gram-negative *Klebsiella pneumoniae* and *Escherichia coli* (Kusari *et al.*, 2009a). In addition, podophyllotoxin and analogues are precursors for the semisynthesis of etoposide and teniposide, two compounds widely used in cancer therapy (Zhao *et al.*, 2010). Podophyllotoxin interacts with tubulin, blocking mitosis, whereas the two semisynthetic compounds interact with the DNA/topoisomerase II complex leading to double-strand breaks and cell cycle arrest.

Podophyllum hexandrum, one source of podophyllotoxin, is an endangered plant species (Nosov, 2012). The destruction of wild populations of the primary source plants, together with the intricacies involved in total synthesis of podophyllotoxin, have led many researchers to search for alternative sources of the compound.

Two fungi, *Phialocephala fortinii* (Eyberger *et al.*, 2006) and *Trametes hirsuta* (Puri *et al.*, 2006) isolated from two different *Podophyllum* species were reported to produce podophyllotoxin. Kusari *et al.* (2009a) recently isolated and studied the endophytic fungi harboured in *J. communis* sampled from Germany and India. This work lead to the discovery of the endophytic fungus *Aspergillus fumigatus* Fresenius, harboured in *Juniperus communis* L. Horstmann, which produces podophyllotoxin and deoxypodophyllotoxin. In addition, *Fusarium oxysporum*, isolated from *Juniperus recurva* (Cupressaceae), has been shown to synthesize podophyllotoxin (Kour *et al.*, 2008). Moreover, Yang *et al.* (2003) recovered six endophytic fungi from *Sinopodophyllum hexandrum*, *Diphylleia sinensis* and *Dysosma veitchii*, whereas Lu *et al.* (2006) recovered an endophytic strain of *Alternaria* sp. from *Sabina vulgaris*, all of which have been reported to produce podophyllotoxin.

Azadirachtin A

Azadirachta indica A. Juss. (synonym *Melia azadirachta*), commonly known as the Indian neem, is a versatile medicinal plant used for more than 2000 years in India and other countries. Neem products are believed to be antibacterial, anthelmintic, antifungal and antiviral among others. Numerous natural bioactive compounds have been found and characterized from different parts of the neem tree. Neem contains azadirachtin A and structural analogues (exclusive to the neem tree), substances that possess considerable toxicity towards many insects (may act as insecticides). Recently, the endophytic fungus *Eupenicillium parvum* from this plant was reported to be capable of producing azadirachtin A and B under *in vitro* axenic conditions (Kusari *et al.*, 2012b). Moreover, a significant number of endophytic actinomycetes isolated from the same plant showed significant inhibitory activity against one or more pathogenic fungi and bacteria, suggesting their antibiotic and/or antifungal potential. Isolates of *Streptomyces, Nocardia* sp. and *Streptosporangium* sp. have shown significant antagonistic activity against root pathogens, including *Pythium* and *Phytophthora* sp. (Verma *et al.*, 2009).

Hypericin

Hypericum has been an important medicinal plant throughout ancient history of Greece, India, China, Egypt and other countries of Europe. Hypericin, a naphthodianthrone derivative, is one of the main constituents of *Hypericum perforatum* (St. John's Wort) that is synthesized by the plant as a chemical defence against a number of insect pests (Kusari *et al.*, 2013). Hypericin and emodin, another NP from the plant, have been reported to possess antibacterial, antifungal, antiviral and cytotoxic activities (Radulovic *et al.*, 2007; Saddiqe *et al.*, 2010). The antimicrobial activities of these two compounds against several bacteria and fungi include: the Gram-positive *S. aureus* and the Gram-negative *K. pneumoniae* sp., *Pseudomonas aeruginosa, Salmonella enterica* ssp. *enterica* and *E. coli*, as well as against some fungal organisms, such as *Aspergillus niger* and *Candida albicans* (Kusari *et al.*, 2008). Hypericin has been found to be a potent antiviral agent against a number of viruses including HIV-1, HSV-1, BVDV, BIV and influenza A (Kusari *et al.*, 2008). In addition, the compound has been in use for a long time as an antidepressant due to its monoamine oxidase (MAO) inhibiting capacity. An endophytic fungus identified as *Thielavia subthermophila* was isolated from *H. perforatum*, collected in India (Kusari *et al.*, 2008). The fungus produced hypericin simultaneously with emodin. It was proposed that emodin anthrone is the common precursor of both hypericin and emodin in the fungal endophyte (Kusari *et al.*, 2009c).

Vinblastine

Vinblastine and vincristine are indole alkaloids originally discovered in periwinkle (*Catharanthus roseus* or *Vinca rosea*) a plant of the family *Apocyanaceae*. Both, vinblastine and vincristine are valuable anti-cancer drugs with use in the treatment of lymphoma and leukaemia, respectively. About 500 kg of periwinkle leaves are needed to produce just one gram of pure vincristine, accordingly current production of this compound using plants is limited and expensive (Kharwar *et al.*, 2011; Chandra, 2012). Several vinblastine producing endophytic fungi have been isolated during the past years, in all cases from *C. roseus*. Kumar *et al.*, 2013 discovered that *F. oxysporum* could produce vinblastine and vincristine in appreciable amounts (76 μg/l and 67 μg/l, respectively). Another strain of *Alternaria* sp. was isolated from phloem of the plant material and reported to be a vinblastine producer (Zhao *et al.*, 2010). Although these first results indicate that some endophytic fungi are a potential source to produce either vinblastine or vincristine, follow-up studies remain to be done.

Piperine and other compounds

Endophytic fungi can produce other antimicrobial compounds originally found in their hosts (Zhao *et al.*, 2010). Piperine was obtained from the endophytic fungus *Periconia* sp. isolated from *Piper longum*. Piperine has shown antifungal activity against major plant pathogens such as *Phytophthora capsici*. Besides being a potential cancer preventative agent, piperine also enhances the bioavailability of several drugs, such as sulfadiazine, tetracycline, streptomycin, rifampicin, pyrazinamide, isoniazid among others (Verma *et al.*, 2011). This is an instance of a NP that may act synergistically with antibiotics against infectious diseases. *A. niger*, isolated from *Rosa damacaena*, produces 2-phenylethanol, a constituent of rose oil, with antiseptic, disinfectant and antimicrobial properties (Wani *et al.*, 2010). The maytansinoids are NPs originally extracted from members of the higher plant families *Celastraceae*, *Rhamnaceae* and *Euphorbiaceae*. The structures of the ansamitocins produced by the actinomycete *Actinosynnema pretiosum* ssp. *auranticum* are similar to those of the maytansinoids from higher plants. This family of secondary metabolites have shown antimicrobial activity against many Gram-positive and some Gram-negative bacteria, antiviral activity towards bacteriophages and poxviruses, as well as very potent anti-tumour activity (Yu *et al.*, 2002).

The plant-synthesized antimicrobials provide protection for plants against attacks by insects, pathogenic bacteria or fungi. So, plants are a rich source of the most diverse chemical structures. It is clear that the few instances cited above are only early efforts for the discovery of endophytes able to produce antimicrobials and other bioactive molecules originally obtained from plants. However, the plant-synthesized antimicrobial compounds have enormous, but largely untapped, therapeutic potential for treating infectious diseases. The isolation and study of endophytes with the capacity to synthesize similar or identical compounds as their host plants could be a key step leading to the future development of fermentation processes required for the economically feasible mass production of a broad range of bioactive compounds.

Antibiotics from endophytic microorganisms

Throughout many years, it has been demonstrated that endophytes are important sources of novel structures with interesting biological characteristics. As normally happens when studying the biological activities of endophytes, diverse kinds of test are performed to determine the possible applications of their metabolites, such as antifungal, antibacterial, antiparasitic, antitumoral, etc.

For the purpose of this book, we are presenting only some examples where the antibiotic assays for endophyte metabolites were satisfactory. In this section, we are omitting those metabolites that are the same as those from the host plant, since these have been reviewed previously.

The list of new metabolites produced by endophytes is considerable and continuously grows, totalling around 350 characterized compounds by the year 2006. Many reviews about compounds produced by endophytes have been reported. They have focused primarily on bioactive compounds from fungi (Gunatilaka, 2006; Gutierrez *et al.*, 2012) but also from bacteria (Christina *et al.*, 2013).

Antifungal and antibacterial compounds

Compounds with antimicrobial activity have been reported from several endophytes. A total of 14 compounds were produced by the fungal endophyte *Ampelomyces* sp., isolated from the flowered plant *Urospermum picroides*. However, only two of them (3-*O*-methylalaternin and altersolanol A), presented antimicrobial activities against the Gram-positive pathogens, *S. aureus*, *Staphylococcus epidermidis* and *Enterococcus faecalis* at minimal inhibitory concentrations (MIC) of 12.5 μg/ml and 12.5–25 μg/ml, respectively. Interestingly, 3-*O*-methylalaternin was also identified as constituent of an extract derived from a healthy plant sample of the host plant *U. picroides* (Aly *et al.*, 2008).

A study searching for bacterial endophytes from surface sterilized leaf and stem segments of the agro-economic plant *Coffee arabica* L. showed that an unidentified bacterium was capable of secreting antimicrobial secondary metabolites found in the crude ethyl acetate extract. A disc diffusion assay against various bacteria gave positive inhibition of *Bacillus subtilis* (MTCC 121), *E. coli* (MTCC 7410) *Proteus mirabilis* (MTCC 245), *Shigella flexnerri* (MTCC 731), *S. aureus* (MTCC 7443) and *Xanthomonas campestris* (7908) (Baker and Shreedharmurthy, 2012).

Talaromyces wortmannii, an endophytic fungus isolated from healthy inner tissues of *Aloe vera*, produced several compounds, a new atropisomer, a new wortmannin derivative and two wortmannin derivatives that were hitherto known only as synthetic products, as well as seven known metabolites. Evaluation of these compounds for antibiotic activity against several pathogens, gave positive results against Gram-positive (methicillin-resistant *S. aureus* (MRSA), *S. epidermidis*,

Streptococcus pneumonia and *E. faecalis*) and the Gram-negative *E. coli, K. pneumoniae, P. aeruginosa, Acinetobacter baumannii, Enterobacter* sp. and *Enterococcus cloacae*. The results showed that two compounds already described, the skyrin and rugulosin A exhibited considerable antibiotic activity against the Gram-positive pathogenic bacteria with MIC values ranging between 4 and 16 μg/ml. The new biemodin also showed strong activity against Gram-positive bacteria, especially against MRSA, but was less active compared to skyrin and rugulosin A (Bara *et al.*, 2013).

The endophytic *Cryptosporiopsis* sp., was isolated from the invasive plant *Clidemia hirta* and analysed for its secondary metabolites. This lead to the isolation of 1-(2,6-dihydroxyphenyl)pentan-1-one and (Z)-1-(2-(2-butyryl-3-hydroxyphenoxy)-6-hydroxyphenyl)-3-hydroxybut-2-en-1-one. Both compounds are novel and presented activity against several bacterial pathogens. The first compound was active against *Bacillus cereus, E. coli* and *S. aureus* with IC_{50} values varying from 18 to 30 μg/ml, and the second one displayed activity against *Pseudomonas fluorescens* with an IC_{50} value of 6 μg/ml (Zilla *et al.*, 2013).

The endophyte *Phomopsis* sp. CMU-LMA, isolated from the cooking herb *Alpinia malaccensis*, contains eight polyketide compounds that can be classified into four groups: the substituted 10-membered lactones Sch-642305 and LMA-P1, the 14-membered lactone benquoine, the six-membered lactones LMA-P2 and DHTO, and the linear derivatives LMA-P3, benquinol and DHTTA. Benquoine and Sch-642305 exhibited antimicrobial activity against the Gram-positive *Micrococcus luteus* and *B. subtilis*, but not against the Gram-negative *E. coli* (Adelin *et al.*, 2011).

The research on antifungal endophyte compounds has two different goals; one is to eliminate fungal human pathogens and the other is to combat plant fungal pathogens. Even though here we have focused more on examples for human pathogenic fungi, more examples of plant pathogenic fungi will be reviewed in the agricultural application section.

The activity-guided fractionation of the ethyl acetate extract of *Botryosphaeria rhodina*, an endophyte obtained from the stems of the medicinal plant *Bidens pilosa*, showed the presence of four depsidones (botryorhodine A–D), which show antifungal activity against a range of pathogenic fungi including *Aspergillus terreus* (MIC 26.03 μM for botryorhodine A and 49.70 μM for B) and the plant pathogen *F. oxysporum* (MIC 191.60 μM for botryorhodine A and 238.80 μM for B) (Abdou *et al.*, 2010).

In addition, phomodione, usnic acid and cercosporamide were isolated from the culture broth of the fungal plant pathogen *Phoma* sp., discovered as an endophyte on a Guinea plant (*Saurauia scaberrinae*). These three compounds are well known anti-infectives, but phomodione exhibited a minimum inhibitory concentration of 1.6 μg/ml against *S. aureus* using the disk diffusion assay, and was active against the oomycete *Pythium ultimum*, the ascomycete *Sclerotinia sclerotiorum* and the basidiomycete *Rhizoctonium solani*, respectively, at concentrations ranging from 3 to 8 μg/ml. None of these compounds was effective against *E. coli* at 500 μg or lower, but minimum inhibitory concentrations (MICs) of all three compounds were approximately the same on *S. aureus*, indicating that they may be much more effective against Gram-positive bacteria. Antimycotic activity was tested using representative species from three classes of fungi. In each case, usnic acid was less effective than cercosporamide and phomodione. The phomodione MICs, were comparable to those of cercosporamide against representatives of all three fungi. Relative toxicities of these compounds have not been evaluated yet (Hoffman *et al.*, 2008).

Isolation of 138 microbes from leaves of 24 plant species native to the French Guiana were tested against *S. aureus* ATCC 29213 and *C. albicans* ATCC 10213 (Casella *et al.*, 2013). Three active fungal extracts were fractionated, resulting in the isolation of eight compounds. Seven had been previously described in the literature, including the following: acremonisol A, semicochliodinol A, cochliodinol, griseofulvin, pyrenocin A, novae zelandin A and alterperylenol. A new compound was isolated from *Lewia infectoria* SNB-GTC2402, identified by spectroscopic analysis and named pyrrocidine C (Fig. 10.2a). As in pyrrocidines A and B, this compound is a *cis*-substituted decahydrofluorene with a quaternary carbon at C-5 and

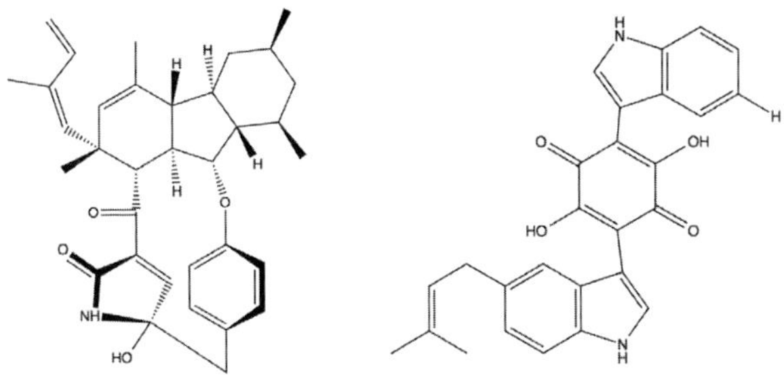

a) b)

Figure 10.2 Structures of (a) pyrrocidine C and (b) semicochliodinol A.

opposite stereochemistry at C-8 corresponding to C-6 of pyrrocidines A and B. Pyrrocidine C was active against *S. aureus* ATCC 29213 with a MIC value of 2 μg/ml, but inactive against *E. coli*. Alterperylenol predominated in the fraction and was moderately active against *S. aureus* ATCC 29213 (MIC = 32 μg/ml). Both compounds are presumably responsible for the antibacterial activity of the crude extract (MIC = 64 μg/ml). Pyrrocidines A and B and the hirsutellones are antibacterial, with approximately identical MIC values to those of Pyrrocidine C on the Gram-positive bacteria tested. The ethyl acetate extract from one of the endophytes isolated, *Chaetomium globosum* SNB-GTC2114, presented some compounds reported previously, such as acremonisol A, semicochliodinol A and cochliodinol. Acremonisol A has been previously isolated from an algal fungal endophyte, *Acremonium* sp., and in this work presented weak inhibition against *S. aureus* ATCC 29213 (64 μg/ml). Semicochliodinol A (Fig. 10.2b), the other compound isolated, shows activity against *S. aureus* ATCC 29213 (MIC = 2 μg/ml). A previous work reported that this compound, isolated from *Chrysosporium merdarium*, presented inhibition of 1′HIV-1 protease (IC_{50} = 0.37 μM), tyrosine kinase protein (EGF-R PTK) of the epidermal growing receptor (IC_{50} = 20 μM), and human cathepsin D (IC_{50} = 2.5 μM) (Fredenhagen *et al.*, 1996). Cochliodinol is a common fungal metabolite with antimicrobial and cytotoxic activity. In this study, cochliodinol was active against all targets, demonstrating antimicrobial activities with MIC values of 2 μg/ml against *C. albicans* ATCC 10213 and 4 μg/ml against *S. aureus* ATCC 29213 (Casella *et al.*, 2013).

Anti-tuberculosis compounds

3-Nitropropionic acid (3-NPA) is a well-known compound that exhibits potent antimycobacterial activity. In a study performed with different Thai medicinal plants, searching for endophyte producers of this compound (Chomcheon *et al.*, 2005), seven 3-NPA-producing endophytes were isolated; two of them, identified as *Phomopsis* sp., were isolated from a leaf of *Urobotrya siamensis* and from a twig of *Mesua ferrea*. The other five strains (unidentified fungi), were isolated from the twigs of *Grewia* sp., *Rhododendron lyi*, *Tadehagi* sp. and *Gmelina elliptica*. 3-NPA exhibited potent antimycobacterial activity against *Mycobacterium tuberculosis* H37Ra with a minimum inhibition concentration (MIC) value of 3.3 μM. The mode of action of 3-NPA against mycobacteria may be related to the inhibition of mycobacterial succinate dehydrogenase. Recent proteomic studies on *M. tuberculosis* revealed the presence of a succinate-semialdehyde dehydrogenase gene, and therefore 3-NPA could be a potential lead chemotype for this enzyme target.

From the endophytic fungus *Phomopsis* sp. PSU-D15, isolated from leaves of *Garcinia dulcis*, three metabolites termed phomoenamide, phomonitroester and deacetylphomoxanthone B, were isolated together with three known compounds, dicerandrol A, (1S,2S,4S)-*p*-menthane-1,2,4-triol, and uridine. Phomoenamide (Fig. 10.3) was the only compound that was evaluated against *M. tuberculosis* H37Ra, due to the required amount necessary for the trials. This compound presents a MIC value of 6.25 μg/ml (Rukachaisirikul *et al.*, 2008).

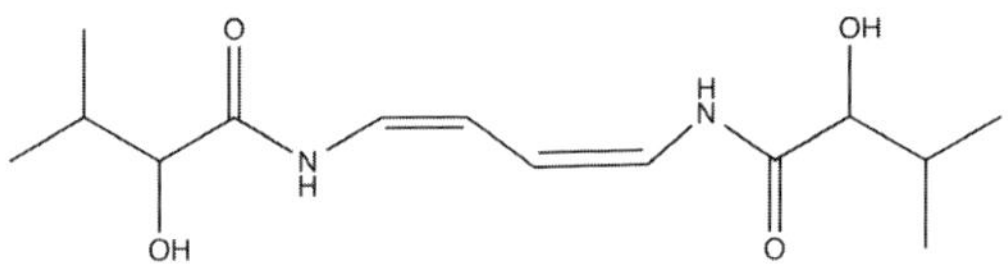

Figure 10.3 Structure of phomoenamide.

Antiviral compounds

Human cytomegalovirus (hCMV) is an ubiquitous opportunistic pathogen. It causes disease in congenitally infected immune-deficient infants and adults. A major function of the hCMV protease in infected cells is to proteolytically process the 44-kDa hCMV assembly protein precursor, to yield the 37-kDa mature assembly protein. This mature protein is essential for assembling the virus nucleocapsid. In the presence of inhibitors of this protease, cleavage is disrupted and the formation of the infectious virus is thus prevented.

In this way, two novel human cytomegalovirus protease inhibitors, cytonic acids A (Fig. 10.4a) and B (Fig. 10.4b), have been isolated from the solid-state fermentation of the endophytic fungi *Cytonaema* sp. isolated from *Quercus* sp. in the UK (Guo *et al.*, 2000).

A novel pentacyclic indolosesquiterpene, named xiamycin and its methyl ester have been obtained from *Streptomyces* sp. GT2002/1503, an endophyte from the stem of the mangrove plant *Bruguiera gymnorrhiza*. To evaluate its bioactivity profile, xiamycin was subjected to a panel of biological assays involving antimicrobial and antiviral test systems against entry by CXCR4 (X4) and CCR5 (R5) tropic HIV-1. Interestingly, xiamycin specifically blocks R5, but has no effects on X4 tropic HIV-1 infection. Albeit only being moderate, the antiviral activity is intriguing in light of the reported activity of the related fungal metabolite tubingensin against herpes simplex virus type 1. Thus, it seems that the pentacyclic carbazole system could be a suitable scaffold for antiviral agents. Xiamycin represents one of the first examples of indolosesquiterpenes isolated from prokaryotes (Ding *et al.*, 2010).

Chemical investigation of the endophytic fungus *Emericella* sp. (HK-ZJ) isolated from the mangrove plant *Aegiceras corniculatum* led to isolation of six isoindolone derivatives termed emerimidine A and B and emeriphenolicins A and D, and six previously reported compounds named aspernidine A and B, austin, austinol, dehydroaustin and acetoxydehydroaustin, respectively. The anti-influenza A viral (H1N1) activities of eight compounds were also evaluated using cytopathic effect (CPE) inhibition assay. Even though all the compounds were tested for *in vitro* activity against H1N1 replication in MDCK cells, only emerimidine A and B showed moderate inhibitory effects with IC_{50} values of 42.07 μg/ml and 62.05 μg/ml (Zhang *et al.*, 2011).

Antiparasitic compounds

An endophytic streptomycete (NRRL 30566) isolated from a fern leaved grevillea (*Grevillea pteridifolia*, Proteaceae) tree growing in the Northern Territory of Australia, produces the novel antibiotics kakadumycins, related to the quinoxaline antibiotic, echinomycin. Each contains, by virtue of their amino acid compositions, alanine, serine and an unknown amino acid. Both, kakadumycin A and echinomycin have impressive activity against the malarial parasite *Plasmodium falciparum* with LD_{50s} in the range of 7–10 ng/ml.

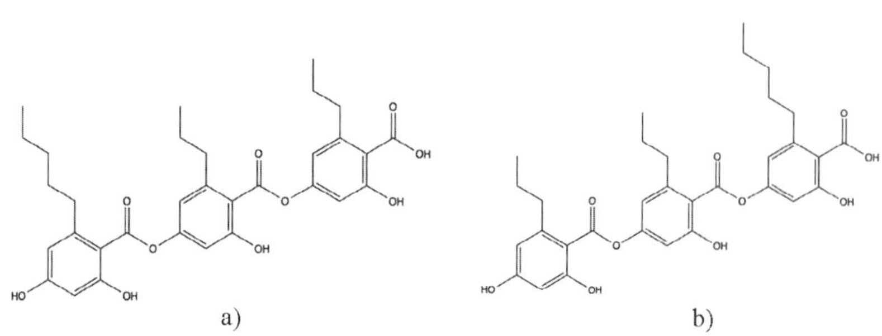

Figure 10.4 Structure of (a) cytonic acid A and (b) cytonic acid B.

Kakadumycin A also has a wide antibiotic activity against Gram-positive bacteria and generally displays better bioactivity than echinomycin. For instance, against *Bacillus anthracis* strains, kakadumycin A has minimum inhibitory concentrations of 0.2–0.3 µg/ml in contrast to echinomycin at 1.0–1.2 µg/ml. In macromolecular synthesis assays both kakadumycin A and echinomycin have similar effects on the inhibition of RNA synthesis (Castillo *et al.*, 2003).

Streptomyces NRRL 3052, an endophyte from the medicinal plant snakevine (*Kennedia nigriscans*), is native to the Northern Territory of Australia. In the screening of its bioactivity, the major components were four functionalized peptides with masses of 1269.6, 1298.5, 1312.5 and 1326.5 Da. Those peptides were characterized as munumbicins A, B, C and D, newly described antibiotics with a wide spectrum of activity against *Plasmodium* sp. and against many human as well as plant pathogenic fungi and bacteria. In this work, munumbicin B had an MIC of 2.5 µg/ml against a methicillin-resistant strain of *S. aureus*, whereas munumbicin A was not active against this organism. The most impressive biological activity of any of the munumbicins was that of munumbicin D against the malarial parasite *P. falciparum*, having an IC_{50} of 4.5 ± 0.07 ng/ml (Castillo *et al.*, 2002). The munumbicins also showed activity against the *Bacillus anthracis* and the multidrug-resistant *M. tuberculosis*.

Relevant work that led to greater understanding of the potential of secondary metabolites from endophytes is that performed by Wiyakrutta *et al.* (2004). They studied a total of 81 Thai medicinal plant species collected from forests in four geographical regions of Thailand. These were examined for the presence of endophytic fungi with biological activity. Among 582 pure isolates obtained, 360 morphologically distinct fungi were selected and tested for biological activity. Six extracts from these fungi inhibited *P. falciparum* (IC_{50} of 1.2–9.1 µg/ml), while 92 inhibited *M. tuberculosis* (MIC 0.0625–200 µg/ml). In addition, strong anti-viral activity against herpes simplex virus type 1 was observed from 40 isolates (IC_{50} of 0.28–50 µg/ml).

The mangrove has shown to be a great place to find endophytes, producers of novel antimicrobial metabolites. One of them is Cytosporone B, which shows broad activities against fungi, with a MIC against *A. niger*, *Trichoderma* sp. and *Fusarium* sp. of 0.125 mg/ml, 62.5 µg/ml and 62.5 µg/ml, respectively. This antibiotic was purified from the fermentation broth of an endophytic fungus, *Dothiorella* sp., isolated from the mangrove plant *Avicennia marina* at the estuary of the Jiulong River, Fujian Province, China. Another endophytic fungus, *Fusarium proliferatum*, isolated from the South China Sea, showed an obvious inhibitory effect on yeast and mould. The studies revealed that it is able to produce a series of antibiotics, including griseofulvin and anthracenediones as well as five anthraquinones.

A strain of the fungus *Verruculina enalia* (No 2606), an ascomycete, was isolated from the decayed wood of a Cassurina tree from a salt lake in the Bahamas. Although *V. enalia* is a very common tropical species found on mangrove wood worldwide, two new phenolic compounds, enalin A and B, were isolated together with hydroxylmethyl furfural and three cycloidpeptides from its fermentation broth. Enalin A is a coumaranone, a type of compound distributed widely from microorganisms to higher plants and having antimicrobial, antifungal, phytotoxic and antidiabetic activities (Cheng *et al.*, 2009).

Potential agricultural uses of endophytic secondary metabolites

With global population projected to increase above nine billion by 2050, food security (the availability of food and one's access to it) is increasingly important. The use of pesticides is a very important issue but its application has been controversial, it represents an important help in crop protection but has caused harm in human health and the environment (Verger and Boobis, 2013). An alternative for crop protection is the application of natural products or natural product inspired compounds isolated from endophytic microorganisms.

Endophytic bacteria and fungi have the ability to protect their host plant from harmful microbes and pests directly, by antagonism or competition for the niche, that is, the space and nutrients, or indirectly by up-regulating or inducing/priming the plant defence system to respond faster and more efficiently towards invading pathogens (Gao *et al.*, 2010; Currie *et al.*, 2014; Hardoim *et al.*, 2012). A wide variety of endophytic microorganisms with antagonistic activity against fungal, bacterial and oomycete plant pathogens have been reported. Direct antagonism towards pathogens can be attained through the production of antifungal substances or fungal growth inhibitors and by antibiotics or other antibacterial metabolites (Gao *et al.*, 2010; Kumar and Kaushik, 2012). Endophytes produce chemically diverse metabolites that exhibit potential applications in agriculture, like antifungals, antibacterials, insecticides, nematicides, among others (Kusari *et al.*, 2012b; Kumar and Kaushik, 2012).

Nematicidal metabolites

There are few reports of nematicidal metabolites isolated from endophytes, one example is the isolation of three bioactive metabolites from the fungal strain *Geotrichum* sp. AL4, cultivated from the leaves of neem tree, *A. indica* (Meliaceae). The two new chlorinated, epimeric 1,3-oxazinane derivatives (1-[(2*R*,4*S*,5*S*)-2-chloro-4-methyl-1,3-oxazinan-5-yl]ethanone) (Fig. 10.5a) and 1-[(2*R*,4*S*,5R)–2-chloro-4-methyl-1,3-oxazinan-5-yl]ethanone) (Fig. 10.5b) and the known 1-(2,4-dihydroxyphenyl)-ethanone (Fig. 10.5c) were active against the free-living nematode *Panagrellus redivivus* (Linn.) and the pine-wood parasite nematode *Bursaphelenchus xylophilus* (Steiner et Buhrer). These compounds caused the mortality of *B. xylophilus* in 62.3%, 64.2% and 54.8%, respectively and killed *P. redivivus* by 53.3%, 47.2% and 55.8%, respectively, after 48 hours of exposure at 100 µg/ml (Li *et al.*, 2007).

Another example is the isolation of 3-hydroxypropionic acid (3-HPA) (Fig. 10.5d) with nematicidal activity from several endophytic fungi. Endophytes were isolated from above ground tissues of woody plants and their extracts were tested for nematicidal activity against the plant-parasitic nematode *Meloidogyne incognita* (Kofoid and White) and the saprophytic *Caenorhabditis elegans*. Among these endophytes, the culture media extracts of five strains caused mortality, mainly of *M. incognita*. The producer strains were identified as *Diaporthe phaseolorum* (anamorph: *Phomopsis phaseoli*) and the anamorphic species of *Melanconis stilbostoma*: *Melanconium betulinium* and *Melanconium bicolor*. These fungi were isolated from an unidentified tree near Les Nourages (French Guyana) and from *Betula pendula* (Betulaceae) Roth in northern Germany, respectively.

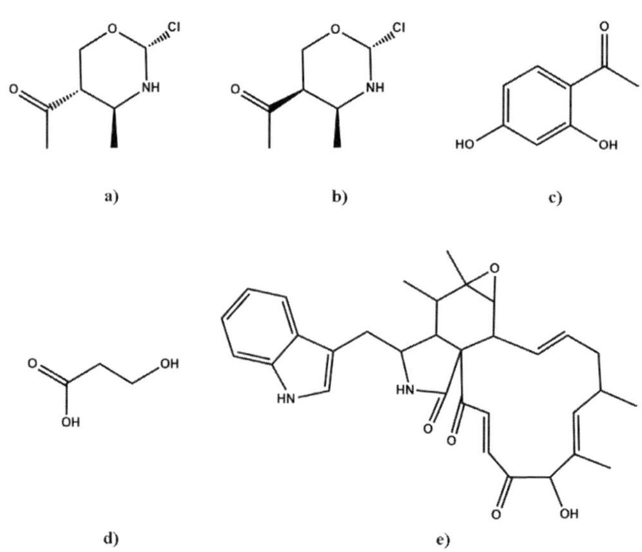

Figure 10.5 Structures of (a) (1-[(2*R*,4*S*,5*S*)–2-chloro-4-methyl-1,3-oxazinan-5-yl]ethanone, (b) 1-[(2*R*,4*S*,5*R*)–2-chloro-4-methyl-1,3-oxazinan-5-yl]ethanone, (c) 1-(2,4-dihydroxyphenyl)-ethanone, (d) 3-hydroxypropionic acid and (e) Chaetoglobosin A.

The chemical study of the active extracts led to the isolation of 3-HPA. This compound showed a LC_{50} of 12.5–15 μg/ml against *M. incognita* and 75 μg/ml for *C. elegans*. In addition, it causes the death of the *M. incognita* nematode and reduces motility in the migration test in Petri dishes, where none of the larvae reached the roots of lettuce seedlings at 50 μg 3-HPA/ml. This compound did not show cytotoxic, antifungal and antibacterial activity at 100 μg/mL, but it showed a weak phytotoxic activity at 600 μg/ml. Therefore, this compound is selective and possibly an ecologically safe nematicide (Schwarz *et al.*, 2004).

The oxazine derivates, 1-(2,4-dihydroxyphenyl)-ethanone and 3-hydroxypropionic acid are active compounds with a simple chemical structure and they are a promising alternative for a commercial nematicide.

Because *Meloidogyne* spp. is one of the most important plant-parasitic nematodes that causes great damage to a wide range of crops worldwide (Hu *et al.*, 2013), there have been efforts for the possible control of this nematode (Meyer *et al.*, 2000; Köpcke *et al.*, 2002; Nitao *et al.*, 2002, Jia *et al.*, 2013).

A recent study of the endophyte *C. globosum* NK102 from the bark of *Cupressus* sp. (Cupressaceae) showed that this strain produces Chaetoglobosin A (Fig. 10.5e) with nematicidal activity against *M. incognita*. Chaetoglobosin A inhibited egg hatching after 14 days of exposure, caused mortality of second stage juveniles after 48 hours of exposure, inhibited penetration in cucumber, reduced egg reproduction per plant and plant growth was improved. The best results for these effects were obtained with a concentration of 300 μg/ml for mortality (LC_{50} of 126.5 μg/ml) and penetration (59.4%), and a dose of 30 mg/kg for the reduction in 63% of egg reproduction in cucumber. The study confirmed that Chaetoglobosin A could be used as a nematicide due to the inhibition of one of the critical stages of the life cycle of *M. incognita*, penetration to host plant, and the high mortality of the nematode (Hu *et al.*, 2013).

Phytotoxic compounds

Endophytic fungi produce phytotoxic compounds that help them colonize their host plants (Schulz and Boyle, 2005) and thus they could be used for weed control.

Brefeldin A or Ascotoxin (Fig. 10.6a) is a lactone antibiotic known as an antiviral, antifungal, phytotoxic and anticancer compound, it has been previously isolated from *Alternaria, Penicillium, Curvularia, Eupenicillium brefeldianum* ZJB082702, *Ascochyta, Cercospora, Phyllosticta, Paecilomyces* sp., and *Aspergillus clavatus* (Vurro *et al.*, 1998; Wang *et al.*, 2002; Khan *et al.*, 2012), but recently it has also been isolated from the

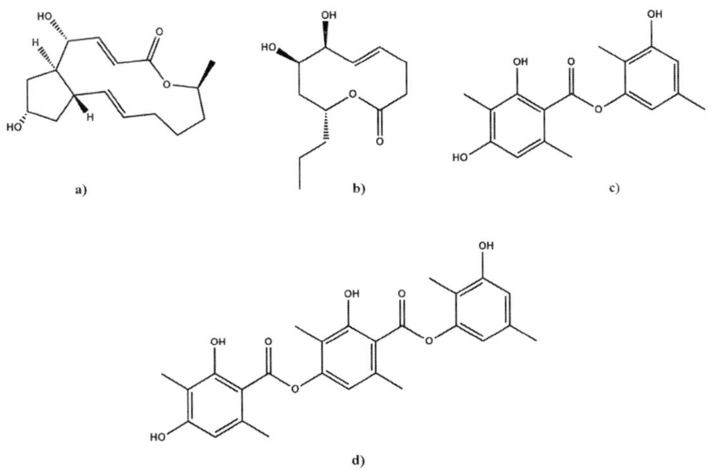

Figure 10.6 Structures of (a) Brefeldin A, (b) (6*S*,7*R*,9*R*)–6,7-dihydroxy-9-propylnon-4-eno-9-lactone, (c) 2,5-dimethylphenyl-2,4-dihydroxy-3,6-dimethylbenzoate and (d) 3-hydroxy-2,5-dimethylphenyl-4-[(2,4-dihydroxy-3,6-dimethylbenzoyl)oxy]–2-hydroxy-3,6- dimethylbenzoate.

endophyte *Paraconiothyrium* sp. This fungus inhabits the roots of *Capsicum annuum* (Solanaceae), *Cucumis sativus* (Cucurbitaceae) and *Glycine max* and was identified as a new species of *Paraconiothyrium*. Brefeldin A showed phytotoxic activity against seeds from *Lactuca sativa* and the weed *Echinochloa crus-galli*, it suppressed the germination of the seeds and caused 100% inhibitory effects on *E. crus-galli* seed germination (Khan *et al.*, 2012).

Another phytotoxic compound is a novel nonenolide (6*S*,7*R*,9*R*)-6,7-dihydroxy-9-propylnon-4-eno-9-lactone (Fig. 10.6b), isolated from the culture of the endophytic fungus *Phomopsis* sp. HCCB03520 that lives in the stems of *Achyranthes bidentata* (Amaranthaceae) in China (Ge *et al.*, 2012). Nonenolide showed phytotoxic activity on germination and radicle growth of *Medicago sativa*, *Trifolium hybridum* and *Buchloe dactyloides*. The IC_{50} for germination were 15.8, 24.2 and 31.2 μg/ml, respectively, and for radicle growth of the plants were 31.9, 63.3, 130.9 μg/ml, respectively. Although inhibition was lower than the positive control, (2-(2,4-dichlorophenoxy) acetic acid (2,4-D), this compound is an alternative for commercial herbicides (Yang *et al.*, 2012).

It is known that depsides have allelophatic activities like the inhibition of seed germination and growth of plants (Dayan and Romagni, 2001). The depsides 3-hydroxy-2,5-dimethylphenyl-2,4-dihydroxy-3,6-dimethylbenzoate (Fig. 10.6c) and 3-hydroxy-2,5-dimethylphenyl-4-[(2,4-dihydroxy-3,6-dimethylbenzoyl) oxy]-2-hydroxy-3,6- dimethylbenzoate (Fig. 10.6d) isolated from *Cladosporium uredinicola*, an endophytic fungus isolated from guava fruit (*Psidium guajava*, Myrtaceae), showed inhibitory effect in photosynthesis in spinach thylakoids. In order to elucidate the mode of action of depsides (Fig. 10.6c and 10.6d), Medeiros *et al.* (2012) investigated their effects on electron transport rate, where both inhibited the Hill reaction with IC_{50} values of 75 and 40 μmol/l, respectively, for uncoupled electron transport flow. Further investigation showed that the depsides inhibited the water-splitting enzyme and have their interaction sites localized in Photosystem II [$P_{680} - Q_A$] similar to the commercial herbicide Diuron (DCMU).

Still, more investigation is needed to determine the mode of action of the phytotoxic nonenolides and on the toxicity of all compounds to other organisms for the possible use of these metabolites for crop protection.

Antifungal metabolites

Numerous fungal endophytes produce secondary metabolites with diverse chemical groups, and some of these compounds are antifungal and antibacterial which strongly inhibit the growth of other microorganisms including plant pathogens (Kumar and Kaushik, 2012). Consequently, many endophyte species may have a great economic importance.

Miscellaneous naphthoquinone spiroketals produced by the endophytic fungus *Edenia gomezpompae* isolated from the leaves of *Callicarpa acuminata* (Verbenaceae) displayed interesting antimicrobial and phytotoxic activity. The bioactivity of the pure compounds was tested against three endophytic fungi: *Colletotrichum* sp., *Phomopsis* sp. and *Guignardia manguifera* isolated from the same host plant, and against four economically important plant pathogenic microorganisms; two fungoid oomycetes, *Phythophtora capsici* and *Phythophtora parasitica*, and the fungi *F. oxysporum* and *Alternaria solani*. Preussomerin EG_1 (Fig. 10.7a), EG_2 (Fig. 10.7b) and EG_3 (Fig. 10.7c) displayed significant growth inhibition activity against all the plant pathogens. IC_{50} values for the four plant pathogens were from 20 to 170 μg/ml. Preussomerin EG_1 showed the strongest bioactivity and was found to be more active than the commercial fungicide Captan® on the growth of *P. capsici* (Macías-Rubalcava *et al.*, 2008, 2014).

Another example is the study of *Cryptosporiopsis* sp., a fungus residing in leaves and branches of *Zanthoxylum leprieurii* (Rutaceae). *Cryptosporiopsis* sp. produces metabolites that inhibit the motility of zoospores and cause lysis of the grapevine downy mildew pathogen *Plasmopara viticola*, within 30 and 60 minutes, respectively, after treatment with 10–25 μg/ml. Metabolites were identified as the polyketides cryptosporiopsin A (Fig. 10.8a) and hydroxypropan-2',3'-diol orsellinate (Fig. 10.8b), two natural cyclic pentapeptides (Fig. 10.8c) and (–)-phyllostine (Fig. 10.8d) (Talontsi *et al.*, 2012).

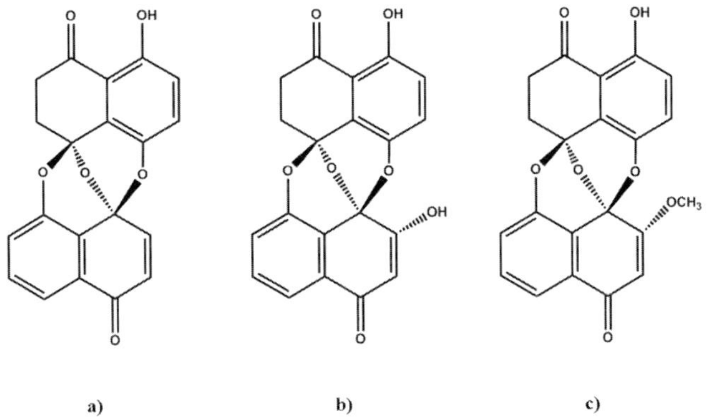

Figure 10.7 Structures of Preussomerins (a) EG_1, (b) EG_2 and (c) EG_3.

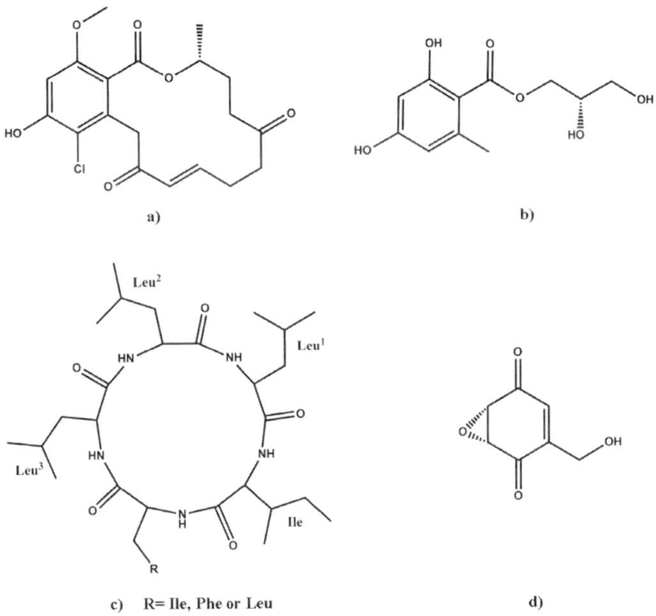

Figure 10.8 Structures of (a) cryptosporiopsin A, (b) hydroxypropan-2′,3′-diol orsellinate, (c) cyclic pentapeptides and (d)(–)-phyllostine.

The compounds cryptosporiopsin A, the two peptapeptides and (–)-phyllostine also inhibited mycelial growth of the plant pathogenic fungi and oomycetes *Rhizoctonia solani*, *P. ultimum* and *Aphanomyces cochlioides*. In order to evaluate the toxicity of the compounds, bioassays on brine shrimp were carried out. The six compounds showed weak toxicity at a concentration of 10 µg/ml, with mortality rates between 10 and 20% (Talontsi *et al.*, 2012).

A previous study showed that Protein Kinase C (PKC) maintains the motility of zoospores, which is critical for the completion of disease cycles in peronosporomycete plant pathogens, thus isolated compounds from *Cryptosporiosis* sp. surely inhibit PKC in *P. viticola* zoospores (Islam and Tiedemann, 2011).

Phomopsis sp. By254, an endophytic fungus residing in the roots of *Gossypium hirsutum* (Malvaceae), produces three cytochalasins: epoxycytochalasin H (Fig. 10.9a), cytochalasin N (Fig. 10.9b) and cytochalasin H (Fig. 10.9c). All

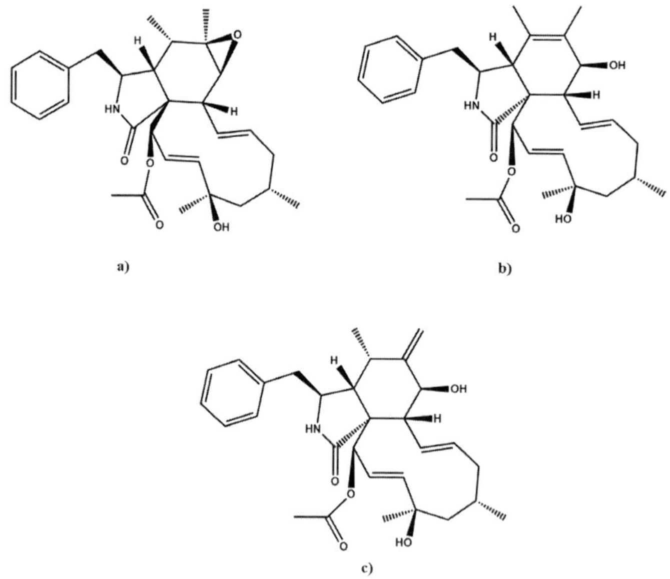

Figure 10.9 Structures of cytochalasins (a) epoxycytochalasin H, (b) cytochalasin N and (c) cytochalasin H.

the compounds significantly inhibited the growth of seven pathogenic fungi *S. sclerotiorum*, *Bipolaris maydis*, *F. oxysporum*, *Botrytis cinerea*, *Bipolaris sorokiniana*, *Gaeumannomyces Graminis* var. *tritici* and *Rhizoctonia cerealis*, with IC_{50} values ranging from 0.1 to 5 μg/ml (*Fu et al.*, 2011).

An important genus of endophytic fungi is *Trichoderma*. *Trichoderma* spp. produce secondary metabolites, enzymes, and it is known as a mycoparasite. *Trichoderma* inhibits plant pathogens through all those mechanisms and by enhancing systemic plant resistance (Lorito *et al.*, 2010; Harman, 2011). All of these properties have led to the development of biofungicides with *Trichoderma* spp., such as Sentinel™ which targets fungal rot in grapes caused by *Botrytis* and Vinevax™, a mixture of strains of *Trichoderma harzianum* formulated as a dowel that is inserted into holes drilled into the trunks of grapevines which enables internal colonization of tissue and suppression of a range of woody trunk diseases caused by fungi such as *Botryosphaeria* and *Phaeomoniella* (Glare *et al.*, 2012).

Insecticidal metabolites

Clavicipitaceous endophytes of grasses arose from insect parasitic ancestors and then progressed to forms that are epibiotic or endophytic. This evolution may in part explain why endophytes produce toxins that affect insects and other animals (Rodriguez *et al.*, 2009).

The fungal endophyte *Acremonium coenophialum* infects tall fescue plants (*Festuca arundinacea* Schreb Poaceae) and produces a mixture of loline alkaloids which deter the feeding of aphids, *Rhopalosiphum padi* (L.), *Schizaphis graminum* (Rondani), and milkweed bugs, *Oncopeltus fasciatus* Dallas. Further studies on *N*-formyl loline (Fig. 10.10a), *N*-acetyl loline (Fig. 10.10b), *N*-methyl loline (Fig. 10.10c) and synthetic loline derivatives showed that they also modify the feeding behaviour and weight of fall armyworm larvae, *Spodoptera frugiperda* Smith, and European corn borer larvae, *Ostrinia nubilalis* Hubner, besides they are toxic against greenbugs, *S. graminum* Rondani. *N*-formyl loline, *N*-acetyl loline and *N*-methyl loline have LC_{50} values of 437, 343 and 361 μg/ml, respectively (Riedell *et al.*, 1991).

Another alternative chemical for pest control is the use of insect repellents. In an attempt to develop an alternative for stinkbug control, the major rice pest in Asia, the chemical study of repellent endophytic fungus extracts was carried out. The fungus identified as a close species to *Biscogniauxia atropunctata* was isolated from a green foxtail *Setaria viridis* (L.) Beauv (Poaceae).

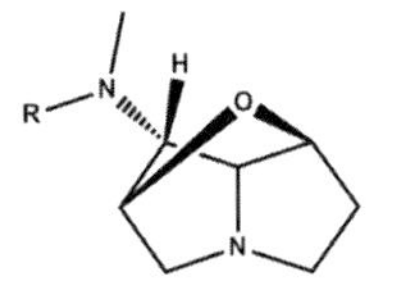

a) R = -COH
b) R = $-COCH_3$
c) R = $-CH_3$

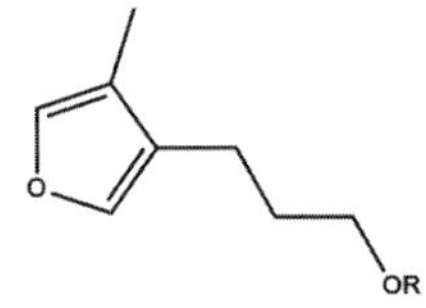

d) R=-H
e) R = $-COCH_3$
f) R = $-COCH_2CH_3$
g) R = $-CO(CH_2)_2CH_3$
h) R = $-COC_6H_5$

Figure 10.10 Structures of (a) *N*-formyl loline, (b) *N*-acetyl-loline, (c) *N*-methyl-loline, (d) 3-(4-Methylfuran-3-yl)propan-1-ol and its derivatives: (e) acetyl derivative, (f) propionyl derivative, (g) butyryl derivative and (h) benzoyl derivative.

3-(4-Methylfuran-3-yl)propan-1-ol (Fig. 10.10d) was isolated and assessed for its repellent effect on the white-spotted stinkbug *Eysarcoris ventralis* (Westwood) and four acyl derivatives (Fig 10.10e–h) were prepared from this compound.

At a concentration of 0.1 µmol/filter paper, the acetyl derivative (Fig. 10.10e) produced 90.2% repellency compared with 44.6% repellency of the positive control naphthalene. At the same concentration, the propionyl and butyryl derivatives (Fig 10.10f and g) produced repellency of 58 and 51.8%, respectively. The benzoyl derivative (Fig. 10.10h) showed no effect. At a concentration of 1 µmol/filter paper, 3-(4-methylfuran-3-yl) propan-1-ol (Fig. 10.10d) and its propionyl and butyryl derivatives (Fig. 10.10f and g) showed high repellency (89.3%, 89.3% and 96.4%, respectively), comparable with naphthalene, whereas the benzoyl derivative (Fig. 10.10h) produced moderate repellency (40.7%). These results showed that the acetyl derivative (Fig. 10.10e) was most effective and could be an effective alternative for pest control.

Conversely, if 3-(4-methylfuran-3-yl)propan-1-ol is accumulated in the foxtail plants, the plant–fungal symbiosis enhances the resistance of the foxtail plants to the stinkbugs through the repellent effect produced by the compound.

The study shows that using attractants or repellents to control pests, instead of insecticides, is a promising alternative method for protecting the rice plant (Nakajima *et al.*, 2010).

The investigation on insecticidal metabolites from endophytic fungi has led to industrial development to exploit endophytes as biopesticides. *Neotyphodium* spp. were one of the firsts fungal endophytes discovered that infect *Lolium* spp. (Poaceae) grasses. They produce insecticidal loline alkaloids with antifeedant activity against *R. padi*, *Schizaphis Graminum*, and *Oncopeltus faciatus*. *Neotyphodium* spp. is now being commercialized in many temperate grassland areas in New Zealand, Australia, USA, and South America under trade names such as AR1™ and AR37™ in ryegrass and MaxQ® in tall fescue (Schardl *et al.*, 2007; Glare *et al.*, 2013).

Volatile organic compounds are potentially important for biocontrol and control of fruit post-harvest diseases

Volatile organic compounds (VOCs) are important metabolites in chemical ecology of fungi and they can be used as mycofumigation agents -use of antimicrobial volatiles produced by fungi for the control of other organisms (Stinson *et al*, 2003a). Among fungi that produce VOCs are the genera *Muscodor*, *Nodulisporium*, *Phoma*, *Hypoxylon*, *Gliocladium*, and *Ascocoryne* (Stinson *et al.*, 2003b; Mends *et al.*, 2012).

Muscodor is a fungal genus that, so far, has only been isolated as an endophyte from certain tropical tree and vine species in certain regions of Central America, Central and Southeast Asia, Australia and Thailand. Twelve species of *Muscodor* are known: *Muscodor albus* from small limbs of *Cinnamomum zeylanicum* (Lauraceae) from Honduras (Worapong *et al.*, 2001; Sopalun *et al.*, 2003); *M. roseus* from two monsoonal rainforest tree species, *G. pteridifolia* (Proteaceae) and *Erythophleum chlorostachys* (Caesalpiniaceae) from Northern Australia (Worapong *et al.*, 2002); *M. vitigenus* from *Paullinia paullinioides* (Sapindaceae), a liana growing in the understory of the rainforests of the Peruvian Amazon (Daisy *et al.*, 2002); *Muscodor crispans* from *Ananas ananassoides* (Bromeliaceae), a wild pineapple in the Bolivian Amazon Basin (Mitchell *et al.*, 2008); *Muscodor yucatanensis* from the leaves of *Bursera simaruba* (Burceraceae) in Quintana Roo, Mexico (Macías-Rubalcava *et al.*, 2010); *Muscodor cinnamomi* from *Cinnamomum bejolghota* (Lauraceae) (Suwannarach *et al.*, 2010); *Muscodor fengyangensis* from *Actinidia chinensis* (Actinidiaceae) (Zhang *et al.*, 2010); *Muscodor sutura* from *Prestonia trifidi* (Apocynaceae) (Kudalkar *et al.*, 2012); *Muscodor musae* from *Musa acuminate* (Musaceae); *Muscodor oryzae* from *Oryza rufipogon* (Poaceae); *Muscodor suthepensis* from *C. bejolghota* (Lauraceae); and *Muscodor equiseti* from *Equisetum debile* (Equisetaceae) the last tree species are medicinal plants of northern Thailand (Suwannarach *et al.*, 2013).

A remarkable property of *Muscodor* species is their capacity to produce a mixture of toxic VOCs that are lethal to a wide variety of plant pathogenic fungi, Gram-positive and Gram-negative bacteria, human pathogens and plants. They also inhibit germination and root elongation of plants. The mixture of VOCs consists primarily of alcohols, acids, esters, ketones and lipids, with 1-butanol, 3-methyl-acetate (isoamyl acetate) being the most active compound (Strobel *et al.*, 2001; Strobel and Daisy 2003; Ezra *et al.*, 2004b,c; Strobel, 2006; Macías-Rubalcava *et al.*, 2010).

The VOCs and the genus are thus potentially important for biocontrol. Potential applications for mycofumigation by members of the *Muscodor* genus are currently being investigated, especially *M. albus* VOCs and include uses for treating plant diseases, buildings, soils, agricultural produce and many more. Mycofumigation is very important for the treatment of fruits in storage and transit. Biological fumigation with *M. albus* volatiles could thus be used to inhibit or kill plant pathogens in enclosed areas without requiring direct contact with the treated product (Mercier and Jimenez, 2004). Soil treatments have also been effectively used in both field and greenhouse situations (Mercier and Manker, 2005; Stinson *et al.*, 2003b; Jacobsen *et al.*, 2004). The soils are pretreated with a *M. albus* formulation in order to preclude the development of infected seedlings. *M. albus* is now being produced, using solid state fermentation by the ton, for its use in many practical applications (Strobel, 2011).

Furthermore, *M. albus* has proved to be an effective nematicidal treatment by causing mortality to the most economically important plant-parasitic nematodes in Washington State, *Paratrichodorus allius*, *Pratylenchus penetrans*, *Meloidogyne chitwoodi* and has nematostatic properties against *Meloidogyne hapla*. Nematode reduction was observed both in the rhizosphere soil and in the roots of potato, pepper, tobacco and bean (Riga *et al.*, 2008).

AgraQuest, an agricultural biotech company in Davis, CA, is developing *M. albus* for numerous agricultural applications (Strobel, 2011). It turns out that the VOCs of *M. albus* appear safe, effective and environmentally friendly and may serve as a replacement for soil treatments (Jacobsen *et al.*, 2004; Strobel, 2011).

Thus, the occurrence of endophytic microbes with pesticidal abilities is leading to exciting new opportunities because it overcomes delivery issues often associated with biopesticides (Glare *et al.*, 2012). The compounds described can be further adopted as greener biopesticides upon testing their toxicity on non-target organisms and its adverse impact on human health (Kumar and Kaushik, 2012). Thus, it is important to continue the study of endophytes, the isolation and characterization of their active compounds and their evaluation in target and non-target organisms.

Compounds with antibiotic activity chemically derived from natural endophytic secondary metabolites

As we know, new antibiotics are necessary to face the threat of multidrug-resistant (MDR) pathogens and some diseases that afflict humankind in diverse ways, such as cancer, a leading cause of mortality, which also diminishes the quality of life in several areas. However, from the early 1970s to 1999 the innovative antibiotic pipeline dried up. According to the World Health Organization, in 2011, less than 5% of the products in the pharmaceutical R&D pipeline were antibiotics. Of 2700 compounds in development in 2008, only 50 were antibacterials and, of these, only 10 were from large pharmaceutical companies (Carlson, 2009). In the meantime, 'superbugs' were on the increase. Furthermore, the dearth of new classes of antibiotics and the disengagement of most pharmaceutical companies from drug research because of economic and regulatory changes, have hindered the landscape in the recent years, leading us to a pre-antibiotic era (Donadio *et al.*, 2010; Butler *et al.*, 2013).

But we have a promising future thanks to the abundant source that we know as endophytes. These studies have increased radically since the Stierle *et al.* (1993) discovery, and this has led to the discovery of many more compounds. It should be noted that most of the bioactive natural products isolated from endophytes are antibiotics.

Strobel and Daisy (2003) have summarized the discovery of antibiotics from the discovery of penicillin to most of the novel antibiotics isolated from endophytes up to that year, but here we present some more, published since then (Table 10.1).

Moreover, it is known that bioactive natural compounds produced by microbes have almost limitless potential in pharmaceutical applications. In order to solve these complications, we need identification and subsequent organic synthesis of such products to generate lead compounds for research and development that will result in the creation of new, highly practical and widely used pharmaceutical products (Sunazuka *et al.*, 2008).

Although the number of studies about endophytic microorganisms has risen, it is still a little too soon to start talking about chemically derived compounds from endophytic antibiotics (Table 10.2). At the present time, there are very few publications because this is a relatively recent area of study, where some analogues to endophytic compounds are synthesized, and not all of them are biologically active. In fact, most of these compounds were found years ago through another source.

For example, derivatives of podophyllotoxin, etoposide and teniposide, are currently used as chemotherapeutic agents against various cancer diseases (Puri *et al.* 2005). The fungus, which belongs to the family Phycomycetes, isolated from the inner bark of the plant *N. foetida* (India), produces the known anticancer drug camptothecin (Fig. 10.11a) (Puri *et al.*, 2005), but in the case of the derivatives tested of altenusin (Fig. 10.11b), they did not show cytotoxic activity (Ye *et al.*, 2013).

Amongst the most common ways to obtain NP derivatives and analogues, is through biosynthetic engineering and combinatorial chemistry. Mutasynthesis (Kennedy, 2008) has also been used to create macrocyclic compounds related to the antibiotic daptomycin (Fig. 10.11c), the anti-cancer compound cryptophicin (Kopp and Marahiel, 2007) and vancomycin analogues, with the aid of oxidative modifications carried out by cytochrome P450 enzymes (Lamb, 2007).

Current situation and compounds undergoing clinical evaluation

Since the year 2000, 22 new antibiotics have been launched worldwide (two NPs, 10 NPs derived (Table 10.3) and 10 synthetic) and five first-in-class antibiotics have been introduced during this period of time: linezolid (oxazolidinone, S, 2000), daptomycin (lipopeptide, NP, 2003), retapamulin (Fig. 10.12), pleuromutilin (NP-derived, 2007), fidaxomicin (tiacumicin, NP, 2010) and bedaquiline (diarylquinoline, S, 2012) (Butler *et al.*, 2013).

The bioactive secondary metabolites produced by microorganisms associated to plants have shown a great diversity and some of these could be candidates for discovery of new drugs, but it does not end there. Noteworthy in the microbial area, the main sources of natural products to date have been fungi and terrestrial actinomycetes, but there is a growing interest in samples from marine microorganisms (actinomycetes and fungi) (Lam, 2006; Fenical and Jensen, 2006; Bull and Stach, 2007). For example, salinosporamide A (Fig. 10.13), from *Salinospora tropica*, which potently inhibits the 20S proteasome and has anti- cancer activity in experimental models (Fenical and Jensen, 2006), at present it is undergoing clinical trials.

Until 2008, over 100 new products were in clinical development, particularly as anti-cancer agents and anti-infectives (Harvey, 2008). Currently, according to Butler *et al.* (2013), there

Table 10.1 Some potential antibiotics produced by endophytic microorganisms, discovered during the last 10 years

Compound	Source	Host	Biological activities	References
Altenusin	*Alternaria* sp. UFMGCB 55	*Trixis vauthieri*	Inhibits trypanothione reductase from *T. cruzi*	Cota *et al.* (2008)
	Alternaria sp.	*Sonneratia alba*	Against *P. brasiliensis* and *S. pombe* Against resistant coliform bacteria and *C. albicans*	Johann *et al.* (2012) Kjer *et al.* (2009)
Altersetin	*Alternaria* spp.	*Vinca minor* and *Eonymus europaeus* (fruit)	Against Gram-positive cocci	Hellwig *et al.* (2002)
Asperfumoid, Asperfumine, Physcion, Helvolic acid	*Aspergillus fumigatus* CY108	*Cynodon dactylon*	Antifugal and antitumour activitites	Liu *et al.* (2004)
Camptothecin	Endophytic fungus	*Nothapodytes foetida*	Anticancer activity	Puri *et al.* (2006)
Citrinin, Emodin, Janthione	*Peniciclium janthinellum*	*Melia azedarach* (fruit)	Against *P. aeruginosa*, *B. subtilis* and *L. mexicana*	Marinho *et al.* (2005)
Phomopsins A, B, C Cytosporene B and C	*Phomopsis* sp.	*Excoecaria agallocha*	Against *S. aureus*, *E. coli*, *S. enteritidis*, *C. albicans* and *F. oxyporum*	Huang *et al.* (2008)
Cytosporene D and E	*Cytospora* sp. and *Diaporthe* sp.	*Conocarpus erecta* and *Forsteronia spicata*	Against *S. aureus*, *E. faecalis*, *E. coli*, and *C. albicans*	Brady *et al.* (2000)
Coronamycin	*Streptomyces* sp. (MSU-2110)	*Monstera* sp.	Against *P. ultimum*, *A. cochlioides*, *P. cinnamomi*, *S. pneumonia*, *C. neoformans* and *P. falciparum*	Ezra *et al.* (2004a)
Griseofulvin, Dechlorogriseofulvin	*Xylaria* sp. F0010	*Abies holophylla*	Antifungal activity	Park *et al.* (2005)
Kakadumycin A and Echinomycin	*Streptomyces* sp. NRRL 30566	*Grevillea pteridifolia*	Against Gram-positive bacteria, *P. falciparum* and cytotoxic activity (primary human cell line, human mammary epithelial cells)	Castillo *et al.* (2003)
Munumbycins (A, B, C and D)	*Streptomyces* sp. NRRL 30562	*Kennedia nigriscans*	Against *B. anthracis*, *S. pneumoniae*, *E. faecalis*, *S. aureus*, *M. tuberculosis* and *P. falciparum*	Castillo *et al.* (2002)
Munumbycins (E-4 and E-5)	*Streptomyces* sp. NRRL 3052	*Kennedia nigriscans*	Against *S. aureus*, *E. coli*, *B. subtilis*, *P. ultimum* and *P. falciparum*	Castillo *et al.* (2006)
Podophyllotoxin	*Trametes hirsuta*	*Podophyllum hexandrum*	Anticancer activity	Puri et al. (2005)
Phomol	*Phomopsis* spp.	*Erythrina crista-galli*	Antifungal, antibacterial, anti-inflammatory and weak cytotoxic activity	Weber *et al.* (2004)
Phomopsilactone	*Streptomyces* sp. GT2002/1503	*Cassia spectabilis*	Against *C. cladosporioides* y *C. sphaerospermum* and cytotoxic activity (HeLa)	Silva *et al.* (2005)
Xyamycins	*Streptomyces* sp. GT2002/1	*Bruguiera gymnorrhiza*	Selective anti-VIH activity	Ding *et al.* (2010)

Table 10.2 Compounds NP-derived initially produced by endophytic microorganisms

Compound	Lead	Biological activity	References
Xinshengin, Phialophoriol	Altenusin	Cytotoxicity tested but not found	Ye *et al.* (2013)
(*trans*) and (*cis*)-Emodin-physcion bianthrone	Physcion	Cytotoxicity against HL-60 and A-549 cell lines	Lee *et al.* (2013)
1,2-dihydrohelvolic acid	Helvolic acid	Antibacterial (*S. aureus*)	Lee *et al.* (2008)
Pennicitrinone C and penicitrinol B	Citrinin	Antioxidative activity	Lu *et al.* (2008)

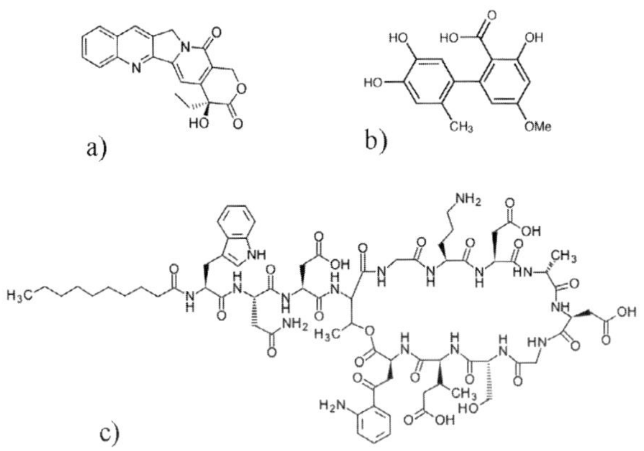

Figure 10.11 Structures of (a) camptothecin, (b) altenusin and (c) daptomycin.

Table 10.3 Antibiotics of NPs origin launched since 2000 with antibiotic class, biological activity, lead source and NP-lead source organism

Drug name	Class	Lead	Biological activity	NP-lead source organism
Telithromycin	Macrolide	Erythromycin (NP-derived)	Gram-positive/Gram-negative	Actinomycete
Biapenem	Carbapenem	Thienamycin (NP-derived)	Gram-positive/Gram-negative	Actinomycete
Ertapenem	Carbapenem	Thienamycin (NP-derived)	Gram-positive	Actinomycete
Daptomycin	Lipopeptide	Daptomycin (NP)	Gram-positive/Gram-negative	Actinomycete
Doripenem	Carbapenem	Thienamycin (NP-derived)	Gram-positive/Gram-negative	Actinomycete
Tigecycline	Tetracycline	Tetracycline (NP-derived)	Gram-positive/Gram-negative	Actinomycete
Retapamulin	Pleuromutilin	Pleuromutilin (NP-derived)	Gram-positive/Gram-negative	Fungus
Ceftobiprole medocaril	Cephalosporin	Cephalosporin (NP-derived)	Gram-positive/Gram-negative	Fungus
Tebipenem pivoxil	Carbapenem	Thienamycin (NP-derived)	Gram-positive/Gram-negative	Actinomycete
Telavancin	Glycopeptide	Vancomycin (NP-derived)	Gram-positive	Actinomycete
Ceftaroline fosamil	Cephalosporin	Cephalosporin (NP-derived)	Gram-positive/Gram-negative	Fungus
Fidaxomicin	Tiaucumicin	Tiaucumicin (NP)	Gram-positive	Actinomycete

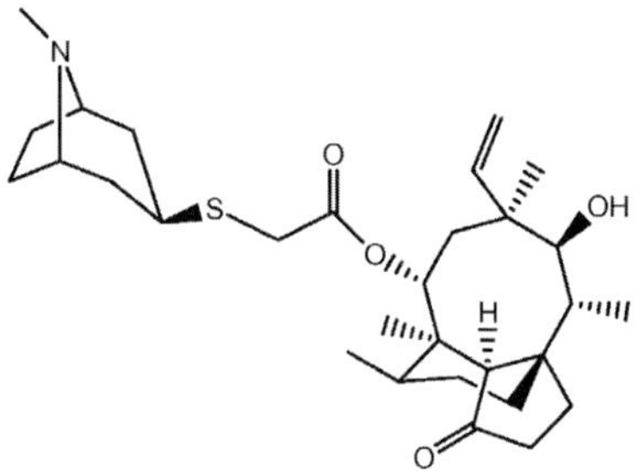

Figure 10.12 Structure of retapamulin.

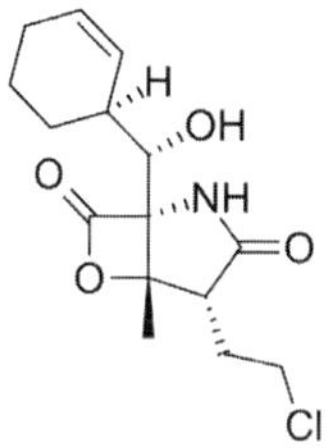

Figure 10.13 Structure of salinosporamide A.

are a total of 49 compounds and 6 β-lactam/β-lactamase inhibitor combinations currently undergoing clinical trials, of which, 16 are NP-derived. This will likely change in a few years, when the list will contain many more products obtained from endophytic organisms.

These are challenging, but also exciting times for antibiotic research, so it is imperative that we continue the search for new antibacterial drugs. Nature has given us the tools, now we need to keep looking and the endophytes with their entire full potential, are a great way to start.

Conclusions

Hundreds of plants have been studied and most of them have been proved to be rich in endophytic microorganisms. Indeed, it has been estimated that there exist at least one million species of endophytic fungi alone. These endophytes are a rich and reliable source of genetic diversity and are capable of producing novel NP's. Therefore, many novel and valuable bioactive compounds exhibiting antimicrobial, parasiticidal, and insecticidal activities have been obtained from them.

Microbial fermentation appears to be a promising alternative to establish a renewable, inexhaustible, cost-effective supply of bioactive compounds. The culture medium for the endophytic microorganisms is simple, inexpensive, with abundant supply and the production costs are relatively low. The fermentation periods are short and the process conditions can be optimized to provide the best growth and bioactive compound production conditions.

Modern genetic and metabolic strategies such as genetic engineering and metabolic engineering can be utilized to understand and manipulate these microorganisms to make them more efficient in producing the desired bioactive compounds (Kusari *et al.*, 2013).

In summary, after more than two decades of research, much progress has been achieved in plant endophytic microorganisms as novel and important microbial sources for producing bioactive compounds. However, there are still many issues related with the production and yield of bioactive compounds and the elucidation of the biosynthetic pathways and their regulation, which will need to be clarified and resolved in order to get competitive level productions.

Acknowledgements

Our thanks go to Dr Elizabeth Langley for her time and assistance in proofreading this manuscript. We are indebted to Marco A. Ortíz and Beatriz Ruiz for their support in the manuscript preparation. Financial support by PAPIIT, DGAPA IN201413 (UNAM) and CONACyT 179194, is gratefully acknowledged.

References

Abdou, R., Scherlach, K., Dahse, H.-M., Sattler, I., and Hertweck, C. (2010). Botryorhodines A-D, antifungal and cytotoxic depsidones from *Botryosphaeria rhodina*, an endophyte of the medicinal plant *Bidens pilosa*. Phytochemistry *71*, 110–116.

Adelin, E., Servy, C., Cortial, S., Lévaique, H., Martin, M.-T., Retailleau, P., Le Goff, G., Bussaban, B., Lumyong, S., and Ouazzani, J. (2011). Isolation, structure elucidation and biological activity of metabolites from Sch-642305-producing endophytic fungus *Phomopsis* sp. CMU-LMA. Phytochemistry 72, 2406–2412.

Aly, A.H., Edrada-Ebel, R., Wray, V., Müller, W.E.G., Kozytska, S., Hentschel, U., and Ebel, R. (2008). Bioactive metabolites from the endophytic fungus *Ampelomyces* sp. isolated from the medicinal plant *Urospermum picroides*. Phytochemistry, *69*, 1716–1725.

Ambrose, C., Varghese, C., and Bhore, S. (2013). Endophytic bacteria as sources of novel antibiotics: An overview. Pharmacog. Rev. *7*, 11–17.

Baker, S., and Shreedharmurthy, S. (2012). Antimicrobial activity and biosynthesis of nanoparticles by endophytic bacterium inhabiting *Coffee arabica* L. Sci. J. Biol. Sci. *1*, 107–113.

Bara, R., Aly, A.H., Pretsch, A., Wray, V., Wang, B., Proksch, P., and Debbab, A. (2013). Antibiotically active metabolites from *Talaromyces wortmannii*, an endophyte of *Aloe vera*. J. Antibiot. (Tokyo). *66*, 491–493.

Brader, G., Compant, S., Mitter, B., Trognitz, F., and Sessitsch, A. (2014). Metabolic potential of endophytic bacteria. Curr. Opin. Biotechnol. *27*, 30–37.

Brady, S.F., Wagenaar, M.M., Singh, M.P., Janso, J.E., and Clardy, J. (2000). The cytosporones, new octaketide antibiotics isolated from an endophytic fungus. Org. Lett. *14*, 4043–4046.

Bulgarelli, D., Schlaeppi, K., Spaepen, S., Themaat, E., and Schulze-Lefert, P. (2013). Structure and functions on the bacterial microbiota of plants. Ann. Rev. Plant Biol. *64*, 807–838.

Bull, A.T., and Stach, J.E. (2007). Marine actinobacteria: new opportunities for natural product search and discovery. Trends Microbiol. *15*, 491–499.

Butler, M.S. (2008). Natural products to drugs: natural product-derived compounds in clinical trials. Nat. Prod. Rep. *25*, 475–516.

Butler, M.S., Blaskovich, M.A., and Cooper, M.A. (2013). Antibiotics in the clinical pipeline in 2013. J. Antibiot. *66*, 571–591.

Carlson, B. (2009). Biosimilar market fails to meet projections. Gen. Eng. Biotechnol. News *29* (17), 43–45.

Casella, T.M., Eparvier, V., Mandavid, H., Bendelac, A., Odonne, G., Dayan, L., Duplais, C., Espindola, L.S., and Stien, D. (2013). Antimicrobial and cytotoxic secondary metabolites from tropical leaf endophytes: Isolation of antibacterial agent pyrrocidine C from *Lewia infectoria* SNB-GTC2402. Phytochemistry *96*, 370–377.

Castillo, U.F., Strobel, G., Ford, E., Hess, W., Porter, H., Jensen, J., Albert, H., Robinson, R., Condron, M., Teplow, D., *et al.* (2002). Munumbicins, wide-spectrum antibiotics produced by *Streptomyces* NRRL 30562, endophytic on *Kennedia nigriscans*. Microbiology *148*, 2675–2685.

Castillo, U.F., Harper, J., Strobel, G., Sears, J., Alesi, K., Ford, E., Lin, J., Hunter, M., Maranta, M., Ge, H., *et al.* (2003). Kakadumycins, novel antibiotics from *Streptomyces* sp. NRRL 30566, and endophyte of *G. pteridifolia*. FEMS Microbiol. Lett. *224*, 183–190.

Castillo, U.F., Strobel, G., Mullenberg, K., Condron, M., Teplow, D., Folgiano, V., Gallo, M., Ferracane, R., Mannina, L., Viel, S., *et al.* (2006). Munumbicins E-4 and E-5: novel broad spectrumantibiotics from *Streptomyces* NRRL3052. FEMS Microbiol. Lett. *255*, 296–300.

Chandra, S. (2012). Endophytic fungi: novel sources of anticancer lead molecules. Appl. Microbiol. Biotechnol. 95, 47–59.

Cheng, Z., Pan, J.-H., Tang, W., Chen, Q., and Lin, Y. (2009). Biodiversity and biotechnological potential of mangrove-associated fungi. J. For. Res. *20*, 63–72.

Chomcheon, P., Wiyakrutta, S., Sriubolmas, N., Ngamrojanavanich, N., Isarangkul, D., and Kittakoop, P. (2005). 3-Nitropropionic acid (3-NPA), a potent antimycobacterial agent from endophytic fungi: is 3-NPA in some plants produced by endophytes? J. Nat. Prod. *68*, 1103–1105.

Christina, A., Christapher, V., and Bhore, S.J. (2013). Endophytic bacteria as a source of novel antibiotics: An overview. Pharmacogn. Rev. 7, 11–16.

Cota, B., Rosa, L.H., Caligiorne, R.B., Rabello, A., Alves, T., Rosa, C.A., and Zani, C.L. (2008). Altenusin, a biphenyl isolated from the endophytic fungus *Alternaria* sp., inhibits trypanothione reductase from *Trypanosoma cruzi*. FEMS Microbiol. Lett. *285*, 77–182.

Currie, A.F., Wearn, J., Hodgson, S., Wendt, H., Broughton, S., and Jin, L. (2014). Foliar fungal endophytes in herbaceous plants: A marriage of convenience? In Advances in Endophytic Research, V.C. Verma, and A.C. Gange, eds. (Springer-Verlag, Berlin, Germany), pp. 61–81.

Daisy, B., Strobel, G., Ezra, D., Castillo, U., Baird, G., and Hess, W.M. (2002). *Muscodor vitigenus* sp.nov. an endophyte from *Paullinia paullinoides*. Mycotaxon. *84*, 39–50.

Dayan, F.E., and Romagni, J.G. (2001). Lichens as a potential source of pesticides. Pesticide Outlook. *12*, 229.

Ding, L., Münch, J., Goerls, H., Maier, A., Fiebig, H.-H., Lin, W.-H., and Hertweck, C. (2010). Xiamycin, a pentacyclic indolosesquiterpene with selective anti-HIV activity from a bacterial mangrove endophyte. Bioorg. Med. Chem. Lett. *20*, 6685–6687.

Donadio, S., Maffioli, S., Monciardini, P., Sosio, M., and Jabes, D. (2010). Antibiotic discovery in the twenty-first century: current trends and future perspectives. J. Antibiot. 63, 423–430.

Dreyfuss, M.M., and Chapela, I.H. (1994). Potential of fungi in the discovery of novel, low-molecular weight pharmaceuticals, In The Discovery of Natural Products with Therapeutic Potential, V.P. Gullo, ed. (Butterworth-Heinemann, London, UK), pp. 49–80.

Eyberger, A.L., Dondapati, R., and Porter, J.R. (2006). Endophyte fungal isolates from *Podophyllum peltatum* produce podophyllotoxin. J. Nat. Prod. *69*. 1121–1124.

Ezra, D., Castillo, U.F., Strobel, G.A., Hess, W.M., Porter, H., Jensen, J.B., Condron, M., Teplow, D.B., Sears, J., Maranta, M., *et al.* (2004a). Coronamycins, peptide antibiotics produced by a verticillate *Streptomyces* sp. (MSU-2110) endophytic on *Monstera* sp. Microbiology. *150*, 785–793.

Ezra, D., Hess, W.H., and Strobel, G.A. (2004b). New endophytic isolates of *M. albus*, a volatile antibiotic–producing fungus. Microbiology *150*, 4023–4031.

Ezra, D., Jasper, J., Rogers, T., Knighton, B., Grimsrud, E., and Strobel, G.A. (2004c). Proton-transfer reaction- mass spectroscopy as a technique to measure volatile emissions of *Muscodor albus*. Plant Sci. *166*, 1471–1477.

Fenical, W., and Jensen, P.R. (2006). Developing a new resource for drug discovery: marine actinomycete bacteria. Nat. Chem. Biol. *2*, 666–673.

Flores-Bustamante, Z.R., Rivera-Orduna, F.N., Martinez-Cardenas, A., and Flores-Cotera, L.B. (2010). Microbial paclitaxel: advances and perspectives. J. Antibiotics. *63*, 460–467.

Fredenhagen, A., Petersen, F., Tintelnot-Blomley, M., Rösel, J., Mett, H., and Hug, P. (1996). Semicochliodinol of HIV-1 protease and EGF-R protein tyrosine kinase related to asterriquinones produced by the fungus *Chrysosporium merdarium*. J. Antibiot. (Tokyo). *50*, 395–401.

Fu, J., Zhou, Y., Li, H., Ye, Y., and Guo, J. (2011). Antifungal metabolites from *Phomopsis* sp. By254, an endophytic fungus in *Gossypium hirsutum*. Afr. J. Microbiol. Res. *5*, 1231–1236.

Gaiero, J., McCall, C., Thompson, K., Day, N., Best, A., and Dunfield, K. (2013). Inside the root microbiome: Bacterial root endophytes and growth promotion. Am. J. Bot. *100*, 1738–1750.

Gao, F., Dai, C., and Liu, X. (2010). Mechanisms of fungal endophytes in plant protection against pathogens. Afr. J. Microbiol. Res. *4*, 1346–1351.

Ge, M., Yin, Y., Chen, Y., Luo, M., and Chen, D. (2012). A Novel Phytotoxic nonenolide from *Phomopsis* sp. HCCB03520. Chem. Biodivers. *9*, 403–408.

Glare, T., Caradus, J., Gelernter, W., Jackson, T., Keyhani, N., Köhl, J., Marrone, P., Morin, L., and Stewart, A. (2012). Have biopesticides come of age? Trends Biotechnol. *30*, 250–258.

Gunatilaka, A.A.L. (2006). Natural products from plant-associated microorganisms: distribution, structural diversity, bioactivity, and implications of their occurrence. J. Nat. Prod. *69*, 509–526.

Guo, B., Dai, J.-R., Ng, S., Huang, Y., Leong, C., Ong, W., and Carté, B. (2000). Cytonic acids A and B: novel tridepside inhibitors of hCMV protease from the endophytic fungus *Cytonaema* species. J. Nat. Prod. *63*, 602–604.

Guo, B., Wang, Y., Sun, X., and Tang, K. (2008). Bioactive natural products from endophytes: a review. Appl. Biochem. Microbiol. *44*, 136–142.

Gutierrez, R.M.P., Gonzalez, A.M.N., and Ramirez, A.M. (2012). Compounds derived from endophytes: a review of phytochemistry and pharmacology. Curr. Med. Chem. *19*, 2992–3030.

Hardoim, P., Nissinen, R., and van Elsas, J.D. (2012). Ecology of bacterial endophytes in sustainable agriculture. In Bacteria in Agrobiology: Plant Probiotics, D.K. Maheshwari ed. (Springer-Verlag, Berlin, Germany), pp. 97–126.

Harman, G.E. (2011). *Trichoderma*—not just for biocontrol anymore. Phytoparasitica *39*, 103–108.

Harvey, A. (2008). Natural products in drug discovery. Drug Discov. Today *13*, 894–901.

Heinig, U., Scholz, S., and Jennewein, S. (2013). Getting to the bottom of taxol biosynthesis by fungi. Fungal Divers. *60*, 161–170.

Hellwig, V., Grothe, T., Mayer-Bartschmid, A., Endermann, R., Geschke, F.U., Henkel, T., and Stadler, M. (2002). Altersetin, a new antibiotic from cultures of endophytic *Alternaria* spp. Taxonomy, fermentation, isolation, structure elucidation and biological activities. J. Antibiot. *55*, 881–892.

Hoffman, A.M., Mayer, S.G., Strobel, G.A., Hess, W.M., Sovocool, G.W., Grange, A.H., Harper, J.K., Arif, A.M., Grant, D.M., and Kelley-Swift, E.G. (2008). Purification, identification and activity of phomodione, a furandione from an endophytic *Phoma* species. Phytochemistry *69*, 1049–1056.

Hu, Y., Zhang, W., Zhang, P., Ruan, W., and Zhu, X. (2013). Nematicidal activity of chaetoglobosin A poduced by *Chaetomium globosum* NK102 against *Meloidogyne incognita*. J. Agric. Food Chem. *61*, 41–46.

Huang, Z., Cai, X., Shao, C., She, Z., Xia, X., Chen, Y., Yang, J., Zhou, S., and Lin, Y. (2008). Chemistry and weak antimicrobial activities of phomopsins produced by mangrove endophytic fungus *Phomopsis* sp. ZSU-H76. Phytochemistry *69*, 1604–1608.

Hyde, K.D., and Soytong, K. (2008). The fungal endophyte dilema. Fungal Divers. *33*, 163–173.

Islam, M.T., and Tiedemann, A.V. (2011). 2,4-Diacetylphloroglucinol suppresses zoosporogenesis and impairs motility of *Peronosporomycete zoospores*. World J. Microbiol. Biotechnol. *27*, 2071–2079.

Jacobsen, B.J., Zidack, N.K., Strobel, G.A., Ezra, D., Grimme, E., and Stinson, A.M. (2004). Mycofumigation with *Muscodor albus* for control of soilborne microorganisms. IOBC/WPRS Bull. *27*, 103–113.

Jia, C., Ruan, W.-B., Zhu, M.-J., Ren, A.-Z., and Gao, Y.-B. (2013). Potential antagnosim of cultivated and wild grass–endophyte associations towards *Meloidogyne incognita*. Biol. Control. *64*, 225–230.

Johann, S., Rosa, L.H., Rosa, C.A., Perez, P., Cisalpino, P.S., Zani, C.L., and Cota, B. (2012). Antifungal activity of altenusin isolated from the endophytic fungus *Alternaria* sp. against the pathogenic fungus *Paracoccidioides brasiliensis*. Rev. Iberoam. Micol. *29*, 205–209.

Kennedy, J. (2008). Mutasynthesis, chemobiosynthesis, and back to semi-synthesis: combining synthetic chemistry and biosynthetic engineering for diversifying natural products. Nat. Prod. Rep. *25*, 25–34.

Khan, A.L., Hamayun, M., Hussain, J., Kang, S.-M., and Lee, I.-J. (2012). The newly isolated endophytic fungus *Paraconiothyrium* sp. LK1 produces ascotoxin. Molecules. *17*, 1103–1112.

Kharwar, R.N., Mishra, A., Gond, S.K., Stierle, A., and Stierle, D. (2011). Anticancer compounds, derived from fungal endophytes: their importance and future challenges. Nat. Prod. Rep. *28*, 1208–1228.

Kjer, J., Wray, V., Edrada-Ebel, R.A., Ebel, R., Pretsch, A., Lin, W.H., and Proksch, P. (2009). Xanalteric acids I and II and related phenolic compounds from an endophytic *Alternaria* sp. isolated from the mangrove plant *Sonneratia alba*. J. Nat. Prod. *72*, 2053–2057.

Köpcke, B., Johansson, M., Sterner, O., and Anke, H. (2002). Biologically active secondary metabolites from the ascomycete A111–95.1. Production, isolation and biological activities. J. Antibiot. *55*, 36–40.

Kopp, F., and Marahiel, M.A. (2007). Where chemistry meets biology: the chemoenzymatic synthesis of nonribosomal peptides and polyketides. Curr. Opin. Biotechnol. *18*, 513–520.

Kour, A., Shawl, A.S., Rehman, S., Sultan, P., Qazi, P.H., Suden, P., Khajuria, R.K., and Verma, V. (2008). Isolation and identification of an endophytic strain of *Fusarium oxysporum* producing podophyllotoxin from *Juniperus recurva*. World J. Microbiol. Biotechnol. *24*, 1115–1121.

Kudalkar, P., Strobel, G., Hassan, S., Geary, B., and Sears, J. (2012). *Muscodor sutura*, a novel endophytic fungus with volatile antibiotic activities. Mycoscience. *53*, 319–325.

Kumar, S., and Kaushik, N. (2012). Metabolites of endophytic fungi as novel source of biofungicide: a review. Phytochem. Rev. *11*, 507–522.

Kumar, S., and Kaushik, N. (2013). Endophytic fungi isolated from oil-seed crop *Jatropha curcas* produces oil and exhibit antifungal activity. PLoS One *8*, e56202.

Kusari, S., and Spiteller, M. (2011). Are we ready for industrial production of bioactive plant secondary metabolites utilizing endophytes? Nat. Prod. Rep. *28*, 1203–1207.

Kumar, A., Patil, D., Rajamohanan, P.R., and Ahmad, A. (2013). Isolation, purification and characterization of vinblastine and vincristine from endophytic fungus *Fusarium oxysporum* isolated from *Catharanthus roseus* PLoS One *8*, e71805.

Kusari, S., Lamshöft, M., Zühlke, S., and Spiteller, M. (2008). An endophytic fungus from *Hypericum perforatum* that produces hypericin. J. Nat. Prod. *71*, 159–162.

Kusari, S., Lamshöft, M., and Spiteller, M. (2009a). *Aspergillus fumigatus* Fresenius, an endophytic fungus from *Juniperus communis* L. Horstmann, as a novel source of the anticancer pro-drug deoxypodophyllotoxin. J. Appl. Microbiol. *107*, 1019–1030.

Kusari, S., Zühlke, S., Borsch, T., and Spiteller, M. (2009b). Positive correlations between hypericin and putative precursors detected in the quantitative secondary metabolite spectrum of hypericum. Phytochemistry *70*, 1222–1232.

Kusari, S., Zühlke, S., Kosuth, J., Cellarova, E., and Spiteller, M. (2009c). Light-independent metabolomics of endophytic *Thielavia subthermophila* provides insight into microbial hypericin biosynthesis. J. Nat. Prod. *72*, 1825–1835.

Kusari, S., Zühlke, S., and Spiteller, M. (2009d). An endophytic fungus from *Camptotheca acuminata* that produces camptothecin and analogues. J. Nat. Prod. *72*, 2–7.

Kusari, S., Kosuth, J., Cellarova, E., and Spiteller, M. (2011a). Survival-strategies of endophytic *Fusarium solani* against indigenous camptothecin biosynthesis. Fungal Ecol. *4*, 219– 223.

Kusari, S., Zühlke, S., and Spiteller, M. (2011b). Effect of artificial reconstitution of the interaction between the plant *Camptotheca acuminata* and the fungal endophyte *Fusarium solani* on camptothecin biosynthesis. J. Nat. Prod. *74*, 764–775.

Kusari, S., Zühlke, S., and Spiteller, M. (2011c). Correlations between camptothecin and related metabolites in *Camptotheca acuminata* reveal similar biosynthetic principles and in planta synergistic effects. Fitoterapia. *82*, 497–507.

Kusari, S., Zühlke, S., and Spiteller, M. (2011d). Chemometric evaluation of the anti-cancer pro-drug podophyllotoxin and potential therapeutic analogues in *Juniperus* and *Podophyllum* species. Phytochem. Anal. *22*, 128–143.

Kusari, S., Hertweck, C., and Spiteller, M. (2012a). Chemical ecology of endophytic fungi: origins of secondary metabolites. Chem. Biol. *19*, 792–798.

Kusari, S., Verma, V.C., Lamshöft, M., and Spiteller, M. (2012b). An endophytic fungus from *Azadirachta indica* A. Juss. that produces azadirachtin. World J. Microbiol. Biotechnol. *28*, 1287–1294.

Kusari, S., Pandey, S.P., and Spiteller, M. (2013). Untapped mutualistic paradigms linking host plant and endophytic fungal production of similar bioactive secondary metabolites. Phytochemistry *91*, 81–87.

Lam, K.S. (2006). Discovery of novel metabolites from marine actinomycetes. Curr. Opin. Microbiol. *9*, 245–251.

Lamb, D.C., Waterman, M.R., Kelly, S.L., and Guengerich, F.P. (2007). Cytochromes P450 and drug discovery. Curr. Opin. Biotechnol. *18*, 504–512.

Lee, S.-Y., Kinoshita, H., Ihara, F., Igarashi, Y., and Nihira, T. (2008). Identification of novel derivative of helvolic acid from *Metarhizium anisopliae* grown in medium with insect component. J. Biosc. Bioeng. *105*, 476–480.

Lee, Y.M., Min, J.K., Li, H., Zhang. P., Bao, B., Lee, K.J., and Jung, J.H. (2013). Marine-derived *Aspergillus* species as a source of bioactive secondary metabolites. Mar. Biotechnol. *15*, 499–519.

Li, G.-H., Yu, Z.-F., Li, X., Wang, X.-B., Zheng, L.-J., and Zhang, K.-Q. (2007). Nematicidal metabolites produced by the endophytic fungus *Geotrichum* sp. AL4. Chem. Biodivers. *4*, 1520–1524.

Li, J.Y., Strobel, G.A., Sidhu, R., Hess, W.M., and Ford, E. (1996). Endophytic taxol producing fungi from Bald Cypress *Taxodium distichum*. Microbiology. *142*, 2223–2226.

Lindow, S., and Brandl, M. (2003). Microbiology of the phyllosphere. Appl. Environ. Microbiol. *69*, 1875–1883.

Liu, C., Wong, Z., Hong, L., and Ren, T. (2001). Antifungal activity of *Artemisa annua* endophyte cultures against phytopathogenic fungi. J. Biotechnol. *88*, 277–282.

Liu, J.Y., Song, Y.C., Zhang, Z., Wang, L., Guo, Z.J., Zou, W.X., and Tan, R.X. (2004). *Aspergillus fumigatus* CY018, an endophytic fungus in *Cynodon dactylon* as a versatile producer of new and bioactive metabolites. J. Biotechnol. *114*, 279–287.

Liu, K., Ding, X., Deng, B., and Chen, W. (2010). 10-Hydroxycamptothecin produced by a new endophytic *Xylaria* sp., M20, from *Camptotheca acuminata*. Biotechnol. Lett. *32*, 689–693.

Liu, Y.Q., Yang, L., and Tian, X. (2007). Podophyllotoxin: current perspectives. Curr. Bioact. Compd. *3*, 37–66.

Lorito, M., Woo, S.L., Harman, G.E., and Monte, E. (2010). Translational research on *Trichoderma*: from 'omics to the field. Annu. Rev. Phytopathol. *48*, 395–417.

Lu, L., He, J., Yu, X., Li, G., and Zhang, X. (2006). Studies on isolation and identification of endophytic fungi strain SC13 from pharmaceutical plant *Sabina vulgaris* Ant. and metabolites. Acta Agriculturae Boreali-Occidentalis Sinica. *15*, 85–89.

Lu, Z.-Y., Lin, Z.-J., Wang, W.-L., Du, L., Zhu, T.-J., Fang, Y.-C., Gu, Q.-Q., and Zhu, W.-M. (2008). Citrinin dimers from the halotolerant fungus *Penicillium citrinum* B-57. J. Nat Prod. *71*, 543–546.

Macías-Rubalcava, M.L., Hernández-Bautista, B.E., Jiménez-Estrada, M., Hernández-Ortega, S., Saucedo-García, A., Muria-González, M.J., González-Villaseñor, M.C., Glenn, A.E., Hanlin, R.T., and Anaya, A.L. (2008). Naphthoquinone spiroketal with allelochemical activity from the newly discoverd endophytic fungus *Edenia gomezpompae*. Phytochemistry *69*, 1185–1196.

Macías-Rubalcava, M.L., Hernández-Bautista, B.E., Oropeza, F., Duarte, G., González, M.C., Glenn, A.E., Hanlin, R.T., and Anaya, A.L. (2010). Allelochemical effects of volatile compounds and organic extracts from *Muscodor yucatanensis*, a tropical endophytic fungus from *Bursera simaruba*. J. Chem. Ecol. *36*, 1122–1131.

Macías-Rubalcava M.L., Ruiz-Velasco Sobrino M.E., Meléndez-González C., and Hernández-Ortega S. (2014). Naphthoquinone spiroketals and organic extracts from the endophytic fungus *Edenia gomezpompae* as potential herbicides. J. Agric. Food Chem. *62*, 3553–3562.

Marinho, M.R., Rodrigues-Filho, E., Moitinho, M., and Santos, L.S. (2005). Biologically active polyketides produced by *Penicillium janthinellum* isolated as an endophytic fungus from fruits of *Melia azedarach*. J. Brazilian Chem. Soc. *16*, 280–283.

Medeiros, L.S., Sampaio, O.M., da Silva, M.F.G.F., Rodrigues-Filho, E., and Veiga, T.A.M. (2012). Evaluation of herbicidal potential of depsides from *Cladosporium uredinicola*, an endophytic fungus found in Guava fruit. J. Braz. Chem. Soc. *23*, 1551–1557.

Mends, M.T., Yu, E., Strobel, G.A., Riyaz-Ul-Hassan, S., Booth, E., Geary, B., Sears, J., Taatjes, C.A., and Hadi, M.Z. (2012). An endophytic *Nodulisporium* sp. producing volatile organic compounds having bioactivity and fuel potential. J. Pet. Environ. Biotechnol. *3*, 117–123.

Mercier, J., and Jimenez, J.I. (2004). Control of fungal decay of apples and peaches by the biofumigant fungus *Muscodor albus*. Postharvest Biol. Technol. *31*, 1–8.

Mercier, J., and Manker, D. (2005). Biocontrol of soil-borne disease and plant growth enhancement in green house soilless mix by the volatile-producing fungus *Muscodor albus*. Crop Prot. *24*, 355–362.

Meyer, S.L.F., Massoud, S.I., Chitwood, D.J., and Roberts, D.P. (2000). Evaluation of *Trichoderma virens* and *Burkholderia cepacia* for antagonistic activity against root-knot nematode, *Meloidogyne incognita*. Nematology *2*, 871–879.

Miller, C.M., Miller, R., Garton-Kenny, D., Redgrave, B., Sears, J., Condron, M., Teplow, D., and Strobel, G. (1998). Ecomycins, unique antimycotics from *Pseudomonas viridiflava*. J. Appl. Microbiol. *84*, 937-44.

Mitchell, A.M., Strobel, G.A., Hess, W.M., Vargas, P.N., and Ezra, D. (2008). *Muscodor crispans*, a novel endophyte from *Ananas ananassoides* in the Bolivian Amazon. Fungal Divers. *31*, 37–43.

Nakajima, H., Ishihara, A., Sawa, Y., and Sakuno, E. (2010). 3-(4-Methylfuran-3-yl)propan-1-ol: a white-spotted Stinkbug (*Eysarcoris ventralis*) repellent produced by an endophyte isolated from green foxtail. J. Agric. Food Chem. *58*, 2882–2885.

Newman, D., and Cragg, G. (2012). Natural products as sources of new drugs over the 30 years from 1981 to 2010. J. Nat. Prod. *75*, 311–335.

Nitao, J.K., Meyer, S.L.F., Oliver, J.E., Schmidt, W.F., and Chitwood, D.J. (2002). Isolation of flavipin, a fungus compound antagonistic to plant-parasitic nematodes. Nematology *4*, 55–63.

Nosov, A.M. (2012). Application of cell technologies for production of plant-derived bioactive substances of plant origin. Appl. Biochem. Microbiol. *48*, 609–624.

Osman, K., Evangelopoulos, D., Basavannacharya, C., Gupta, A., McHugh, T.D., Bhakta, S., and Gibbons, S. (2012). An antibacterial from *Hypericum acmosepalum* inhibits ATP-dependent MurE ligase from *Mycobacterium tuberculosis*. Intern. J. Antimicrob. Agents. *39*, 124–129.

Park, J.H., Choi, G.J., Lee, H.B., Kim, K.M., Jung, H.S., Lee, S.W., Jang, K.S., Cho, K.Y., and Kim, J.C. (2005). Griseofulvin from *Xylaria* sp. strain F0010, an endophytic fungus of *Abies holophylla* and its antifungal activity against plant pathogenic fungi. J. Microbiol. Biotechnol. *15*, 112–117.

Puri, S.C., Verma, V., Amina, T., Qazi, G.N., and Spiteller, M. (2005). An endophytic fungus from *Nothapodytes foetida* that produces camptothecin. J. Nat. Prod. *68*, 1717–1719.

Puri, S.C., Nazir, A., Chawla, R., Arora, R., Riyaz-ul Hasan, S., Amna, T., Ahmed, B., Verma, V., Singh, S., Sagar, R., *et al.* (2006). The endophytic fungus *Trametes hirsuta* as a novel alternative source of podophyllotoxin and related aryl tetralin lignans. J. Biotechnol. *122*, 494–510.

Radulovic, N., Stankov-Jovanovic, V., Stojanovic, G., Smelcerovic, A., Spiteller, M., and Asakawa, Y. (2007). Screening of *in vitro* antimicrobial and antioxidant activity of nine *Hypericum* species from the Balkans. Food Chem. *103*, 15–21.

Rashid, S., Charles, T., and Glick, B. (2012). Isolation and characterization of new plant growth-promoting bacterial endophytes. Appl. Soil Ecol. *61*, 217–224.

Rehman, S., Shawl, A.S., Kour, A., Andrabi, R., Sudan, P., Sultan, P., Verma, V., and Qazi, G.N. (2008). An endophytic *Neurospora* sp. from *Nothapodytes foetida* producing camptothecin. Appl. Biochem. Microbiol. *44*, 203–209.

Riedell, W.E., Kieckhefer, R.E., Petroski, R.J., and Powell, R.G. (1991). Naturally-occurring and synthetic loline alkaloid derivatives: Insect feeding behavior modification and toxicity. J. Entomol. Sci. *26*, 122–129.

Riga, E., Lacey, L.A., and Guerra, N. (2008). *Muscodor albus*, a potential biocontrol agent against plant-parasitic nematodes of economically important vegetable crops in Washington State, USA. Biol. Control *45*, 380–385.

Rodriguez, R., White Jr. J., Arnold, A., and Redman, R. (2009). Fungal endophytes: diversity and functional roles. New Phytol. *182*, 314–330.

Rukachaisirikul, V., Sommart, U., Phongpaichit, S., Sakayaroj, J., and Kirtikara, K. (2008). Metabolites from the endophytic fungus *Phomopsis* sp. PSU-D15. Phytochemistry *69*(3), 783–787.

Saddiqe, Z., Naeem, I., and Maimoona, A. (2010). A review of the antibacterial activity of *Hypericum perforatum* L. J. Ethnopharmacol. *131*, 511–521.

Saikkonen, K., Wäli, P., Helander, M., and Faeth, S. (2004). Evolution of endophyte-plant symbioses. Trends Plant Sci. *9*, 275–280.

Schardl, C.L., Grossman, R.B., Nagabhyru, P., Faulkner, J.R., and Mallik, U.P. (2007). Loline alkaloids: currencies of mutualism. Phytochemistry *68*, 980–996.

Schueffler, A., and Anke, T. (2011). Antimicrobial compounds from tree endophytes. In Endophytes of Forest Trees: Biology and Applications, A.M. Pirttil, and A.C. Frank, eds. (Springer-Verlag, Berlin, Germany) Series Forestry Sciences Vol. 80, pp. 265–294.

Schulz. B., and Boyle, C. (2005). The endophytic continuum. Mycol. Res. *109*, 661–686.

Schulz. B., and Boyle, C. (2006). What are endophytes? In Microbial Rot Endophytes, B. Schulz, C. Boyle, and T. Sieber, eds. (Springer-Verlag, Berlin, Germany), Vol. 9, pp. 1–13.

Schulz, B., Boyle, C., Draeger, S., Anne-Katrin Rommert, A.K., and Krohn, K. (2002). Endophytic fungi: a source of novel biologically active secondary metabolites. Mycol. Res. *106*, 996–1004.

Schwarz, M., Köpcke, B., Weber, R.W.S., Sterner, O., and Anke, H. (2004). 3-Hydroxypropionic acid as a nematicidal principle in endophytic fungi. Phytochemistry *65*, 2239–2245.

Selim, K.A., El-beih, A., T AbdEl-Rahman, T., and A El-Diwany A. (2012). Biology of endophytic fungi. Curr. Res. Environ. Appl. Mycol. *2*, 31–82.

Shaanker, R.U., Ramesha, B.T., Ravikanth, G., Gunaga, R.P., Vasudeva, R., and Ganeshaiah, K.N. (2008). Chemical profiling of *Nothapodytes nimmoniana* for camptothecin, an important anticancer alkaloid: towards the development of a sustainable production system. In: Bioactive Molecules and Medicinal Plants, K.G. Ramawat, and J.M. Merillon, eds. (Springer-Verlag, Berlin, Germany). pp. 197–213.

Shahid, M., Shahzad, A., Sobia, F., Sahai, A., Tripathi, T., Singh, A., Khan, H.M., and Umesh. (2009). Plant natural products as a potential source for antibacterial agents: recent trends. Anti-Infective Agents Med. Chem. *8*, 211–225.

Shweta, S., Zuehlke, S., Ramesha, B.T., Priti, V., Kumar, P.M., Ravikanth, G., Spiteller, M., Vasudeva, R., and Shaanker, R.U. (2010). Endophytic fungal strains of *Fusarium solani*, from *Apodytes dimidiata* E. Mey. Ex Arn (Icacinaceae) produce camptothecin, 10-hydroxycamptothecin and 9-methoxy camptothecin. Phytochemistry *71*, 117–122.

Silva, G.H., Teles, H.L., Trevisan, H.C., Bolzani, V.S., Young, M., Pfenning, L.H., Eberlin, M.N., Haddad, R., Costa-Neto, C.M., and Araújo, A.R. (2005). New bioactive metabolites produced by *Phomopsis cassiae*, an endophytic fungus in *Cassia spectabilis*. J. Braz. Chem. Soc. *16*, 1463–1466.

Sirikantaramas, S., Yamazaki, M., and Saito, K. (2009). A survival strategy: The coevolution of the camptothecin biosynthetic pathway and self-resistance mechanism. Phytochemistry *70*, 1894–1898.

Sopalun, K., Strobel, G.A., Hess, W.M., and Worapong, J. (2003). A record of *Muscodor albus*, an endophyte from *Myristica fragrans* in Thailand. Mycotaxon *8*, 239–248.

Staniek, A., Woerdenbag, H.J., and Kayser, O. (2009). *Taxomyces andreanae*: a presumed paclitaxel producer demystified? Plant. Med. *75*, 1–6.

Stierle, A., Strobel, G.A., and Stierle, D. (1993). Taxol and taxane production by *Taxomyces andreanae*. Science *260*, 214–216.

Stinson, A.M., Zidack, N.K., Strobel, G.A., and Jacobsen, B.J. (2003a). Mycofumigation with *Muscodor albus* and *Muscodor roseus* for control of seedling diseases of sugar beet and verticillium wilt of eggplant. Plant Dis. *87*, 1349–1354.

Stinson, M., Ezra, D., Hess, W.M., Sears, J., and Strobel, G.A. (2003b). An endophytic *Gliocladium* sp. of *Eucryphia cordifolia* producing selective volatile antimicrobial compounds. Plant Sci. *165*, 913–922.

Strobel, G.A. (2006). *Muscodor albus* and its biological promise. J. Ind. Microbiol. Biotechnol. *33*, 514–522.

Strobel, G.A. (2011). *Muscodor* species- endophytes with biological promise. Phytochem Rev. *10*, 165–172.

Strobel, G.A. (2012). Genetic diversity of microbial endophytes and their biotechnical applications. In Genomics applications for the developing world, K. Nelson, ed. (Springer-Verlag, Berlin, Germany), pp. 249–262.

Strobel, G.A., and Daisy, B. (2003). Bioprospecting for microbial endophytes and their natural products. Microbiol. Mol. Biol. Rev. *67*, 491–502.

Strobel, G.A., Dirksie, E., Sears, J., and Markworth, C. (2001). Volatile antimicrobials from *Muscodor albus*, a novel endophytic fungus. Microbiol. *147*, 2943–2950.

Strobel, G.A., Daisy, B., Castillo, U., and Harper, J. (2004). Natural products from endophytic microorganisms. J. Nat. Prod. *67*, 257–268.

Sunazuka, T., Hirose, T., and Omura, S. (2008). Efficient total synthesis of novel bioactive microbial metabolites. Acc. Chem. Res. *41*, 302–314.

Suwannarach, N., Bussaban, B., Hyde, K.D., and Lumyong, S. (2010). *Muscodor cinnamomi*, a new endophytic species from *Cinnamomum bejolghota*. Mycotaxon *114*, 15–23.

Suwannarach, N., Bussaban, B., Hyde, K.D., and Lumyong, S. (2013). Molecular and morphological evidence support four new species in the genus *Muscodor* from northern Thailand. Ann. Microbiol. *63*, 1341–1351.

Talontsi, F.M., Facey, P., Tatong, M.D., Islam, T.M., Frauendorf, H., Draeger, S., Tiedemann, A.V., and Laatsch, H. (2012). Zoosporicidal metabolites from an endophytic fungus *Cryptosporiopsis* sp. of *Zanthoxylum leprieurii*. Phytochemistry *83*, 87–94.

Tan, R.X., and Zou, W.X. (2001). Endophytes: a rich source of functional metabolites. Nat. Prod. Rep. *18*, 448–459.

Verger, P.J.P., and Boobis, A.R. (2013). Reevaluate pesticides for food security and safety. Science *341*, 717–718.

Verma, V.C., Gond, S.K., Kumar, A., Mishra, A., Kharwar, R.N., and Gange, A.C. (2009). Endophytic actinomycetes from *Azadirachta indica* A. Juss.: Isolation, diversity, and anti-microbial activity. Microbial. Ecol. *57*, 749–756.

Verma, V.C., Lobkovsky, E., Gange, A.C., Singh, S.K., and Prakash, S. (2011). Piperine production by endophytic fungus *Periconia* sp. isolated from *Piper longum* L. J. Antibiot. *64*, 427–431.

Vurro, M., Evidente, A., Andolfi, A., Zonno, M.C., Giordano, F., and Motta, A. (1998). Brefeldin A and α,β-dehydrocurvularin, two phytotoxins from *Alternaria zinniae*, a biocontrol agent of *Xanthium occidentale*. Plant Sci. *138*, 67–79.

Wagenaar, M., Corwin, J., Strobel, G., and Clardy, J. (2000). Three new chytochalasins produced by an endophytic fungus in the genus *Rhinocladiella*. J. Nat. Prod. *63*, 1692–1695.

Wall, M.E., Wani, M.C., Cook, C.E., Palmer, K.H., Mcphail, A.T., and Sim, G.A. (1966). Plant antitumor agents. I. Isolation and structure of camptothecin a novel alkaloidal leukemia and tumor inhibitor from *Camptotheca acuminata*. J. Am. Chem. Soc. *88*, 3888–3890.

Wang, F., Jiao, R., Cheng, A., Tan, S., and Song, Y. (2007). Antimicrobial potentials of endophytic fungi residing in *Quercus variabilis* and brefeldin A obtained from *Cladosporium* sp. World J. Microbiol. Biotechnol. *23*, 79–83.

Wang, J., Huang, Y., Fang, M., Zhang, Y., Zheng, Z., Zhao, Y., and Su, W. (2002). Brefeldin A, a cytotoxin produced by *Paecilomyces* sp. and *Aspergillus clavatus* isolated from *Taxus mairei* and *Torreya grandis*. FEMS Immunol. Med. Microb. *34*, 51–57.

Wani, M.A., Sanjana, K., Kumar, D.M., and Lal, D.K. (2010). GC-MS analysis reveals production of 2-phenylethanol from *Aspergillus niger* endophytic in rose. J. Basic Microbiol. *50*, 110–114.

Wani, M.C., Taylor, H.L., Wall, M.E., Coggon, P., and Mcphail, A.T. (1971). Plant antitumor agents VI. Isolation and structure of taxol, a novel antileukemic and antitumor agent from *Taxus-brevifolia*. J. Am. Chem. Soc. *93*, 2325–2337.

Weber, D., Sterner, O., Anke, T., Gorzalczancy, S., Martin, V., and Acevedo, C. (2004). Phomol, a new antinflammatory metabolite from an endophyte of the medicinal plant *Erythrina crista-galli*. J. Antibiot. *57*, 559–563.

Wilson, D. (1995). Endophyte –the evolution of the term, a clarification of its use and definition. Oikos. *73*, 274–276.

Wiyakrutta, S., Sriubolmas, N., Panphut, W., Thongon, N., Danwiset-, K., Ruangrungsi, N., and Meevootisom, V. (2004). Endophytic fungi with anti-microbial, anti-cancer and anti-malarial activities isolated from Thai medicinal plants. World J. Microbiol. Biotechnol. *20*, 265–272.

Worapong, J., Strobel, G.A., Ford, E., Li, J.Y., and Hess, W.M. (2001). *Muscodor albus* anam. sp. nov., an endopythe from *Cinnamomum zeylanicum*. Mycotaxon. *79*, 67–79.

Worapong, J., Strobel, G.A., Daisy, B., Castillo, U., Baird, G., and Hess, W.M. (2002). *Muscodor roseus* anam sp. nov., an endophyte from *Grevillea pteridifolia*. Mycotaxon *81*, 463–475.

Yang, X., Guo, S., Zhang, L., and Shao, H. (2003). Selection of producing podophyllotoxin endophytic fungi from podophyllin plant. Nat. Prod. Res. Dev. *15*, 419–422.

Yang, Z., Ge, M., Yin, Y., Chen, Y., Luo, M., and Chen, D. (2012). A novel phytotoxic nonenolide from *Phomopsis* sp. HCCB03520. Chem. Biodiv. *9*, 403–408.

Ye, F., Chen, G.-D., He, J.-W., Li, X.-X., Sun, X., Guo, L.-D., Li, Y., and Gao, H. (2013). Xinshengin, the first altenusin with tetracyclic skeleton core from *Phialophora* spp. Tetrah. Lett. *54*, 4551–4554.

Yu, T.W., Bai, L., Clade, D., Hoffmann, D., Toelzer, S., Trinh, K.Q., Xu, J., Moss, S.J., Leistner, E., and Floss, H.G. (2002). The biosynthetic gene cluster of the maytansinoid antitumor agent ansamitocin from *Actinosynnema pretiosum*. Proc. Natl. Acad. Sci. U.S.A. *99*, 7968–7973.

Zhang, C., Wang, G., Mao, L., Komon-Zelazowska, M., Yuan, Z., Lin, F., Druzhinina, I.S., and Kubicek, C.P. (2010). *Muscodor fengyangensis* sp. nov. from southeast China: morphology, physiology and production of volatile compounds. Fungal Biol. *114*, 797–808.

Zhang, G., Sun, S., Zhu, T., Lin, Z., Gu, J., Li, D., and Gu, Q. (2011). Antiviral isoindolone derivatives from an endophytic fungus *Emericella* sp. associated with *Aegiceras corniculatum*. Phytochemistry *72*, 1436–1442.

Zhang, H., Song, Y., and Tan, X. (2006). Biology and chemistry of endophytes. Nat. Prod. Rep. *18*, 753–771.

Zhao, J., Zhou, L., Wang, J., Shan, T., Zhong, T., Liu, X., and Gao, X. (2010). Endophytic fungi for producing bioactive compounds originally from their host plants. In Current Research Technology and Educstion Topics in Applied Microbiology and Microbial Biotechnology, A. Méndez-Vilas, ed. (Formatex Research Center, Badajos, Spain), pp. 567–576.

Zhou, X., Zhu, H., Liu, L., Lin, J., and Tang, K. (2010). A review: recent advances and future propects of taxol-producing endophytic fungi. Appl. Microbiol. Biotechnol. *86*, 1707–1717.

Zilla, M.K., Qadri, M., Pathania, A.S., Strobel, G.A., Nalli, Y., Kumar, S., Guru, S.K., Bhushan, S., Singh, S.K., Vishwakarma, R.A., *et al.* (2013). Bioactive metabolites from an endophytic *Cryptosporiopsis* sp. inhabiting *Clidemia hirta*. Phytochemistry 95, 291–297.

Antibiotics from Microorganisms from Hot Springs/Geysers

11

Girish B. Mahajan

Abstract

Nature has provided mankind with abundant resources in the form of bioactive compounds which have been isolated from microbial resources. Novel natural products can have many innumerable potential uses, especially in the area of new drug discovery. Extensive research in this field has reported several terrestrial mesophilic sites. The marine environment has recently evinced more interest, and several thousands of novel bioactive compounds have been identified from various marine microorganisms. The constant need for new leads from the healthcare sector has sparked off a relentless quest for novel compounds. A difference in the habitat of the microorganisms yields different types of microorganisms, thus resulting in some variation in the types of compounds obtained from them. Physically and chemically extreme ecological units on the globe harbour special groups of microbes which are genetically well acclimatized to the stress conditions, and their adaptive ability makes them unique producers of different classes of compounds. Hot springs or geysers are such ecological niches, which have been studied mainly from their biodiversity point of view. There have been occasional reports of their harbouring a huge cache of bioactive compounds from the drug discovery perspectives. In this article, we have tried to review present some such ecological niches.

Introduction

The Vedas are renowned as the oldest literature, which influences civilization and culture, especially in India. They have been orally passed down through many generations. The Vedic thoughts have survived the ideological and political on slots for over three millennia. No one has claimed to have composed any portion of Vedas. They are the intuitive expressions of ancient sages in their moments of divine insight. After deep contemplation about contents of Vedas, several seekers compiled the wise words and sayings of the gurus (Preachers) from whom they attained spiritual knowledge and wisdom. This collection of teachings became known as the Upanishads and the seekers themselves were given the name Upanishads that literally meant '*came near to*' the gurus. The Chhandogya Upanishad is one such scripture revealed to some rishi's 1200 B.C. as proposed by Professor R.D. Ranade (http://www.thenagain.info/webchron/india/Upanishads.html).

In this Upanishad, Agni (that is the presiding deity of fire) instructs the devotee Upakosala Kämaläyana:

> IV-xi-2: 'He who knows it thus and meditates on it, destroys sinful acts, wins the region (of fire), reaches the full length of life, lives gloriously, and his descendants never perish. We protect him in this world and in the next, who knows it thus and meditates on it' (Swami Swahananda, 1959). The Sanskrit version of the Chandogya Upanishat can be read at the web page http://www.astrojyoti.com/pdfs/DevanagariFiles/ChandogyaUpanishat.pdf (page numbers 116 and 117 in this article in this link). The energy in fire gets transformed into various forms which are useful to mankind in some or other way. Part of this heat energy existing in the earth bestows

endurance to surviving microbes to tolerate it. The microbes tolerate it and survive in such areas with high temperature by adjusting their metabolic pathways, the outcome of which is secretion of certain chemicals. These heat induced microbial compounds display various bioactivity, some of which are reported to aid in the alleviation of pains afflicting humans. Hot springs/geysers are such major niches on the earth where we find such heat-tolerating microbes.

Need for new bioactive compounds and biodiversity

For researchers involved in the discovery of novel bioactive microbial products, the microbial diversity is a key factor for the novelty of the molecules. Although wide diversity is observed among the microbes with reference to their habitat, metabolism, and extremity tolerance, the microbes with an established record of synthesis of novel pharmaceutically important lead compounds are very limited. Different communities of microbes co-exist in extreme terrestrial regions and oceans, and they constitute an untapped source of bioactive compounds.

Natural product compounds (NPCs), especially those mined from microbes (bacteria and lower eukaryotes), are established resources for a variety of remedial agents (Newman and Cragg, 2009). The soil, despite teeming with billions of microbes, enables only some of them to survive the struggle for existence. In recent times, many compounds used in the treatment and management of cancer, infections due to drug-resistant microbes (bacteria, fungi and viruses), and immunosuppressive disorders, have been derived from microbial sources. Microbial metabolites are the most important chemotherapeutic agents used in oncology. This property of microbial products was identified as early as in 1940, with the discovery of actinomycin from *Streptomyces* (Demain and Sanchez, 2009). Since then, many compounds with anticancer properties have been isolated from microorganisms. Twenty-five small molecules and two biologicals have been approved by the US Food and Drug Administration (FDA)'s Center for Drug Evaluation and Research (CDER). Further, there has been a 31% decline in approvals from 2012 (Asher, 2014). The last four novel classes of antibacterial compounds to reach the clinic were all discovered (empirically) before 1987 (Silver, 2011).

Hot springs, geysers and microbes

The Encyclopaedia Britannica has defined hot springs as, 'the spring with water at temperatures substantially higher than the air temperature of the surrounding region. Most hot springs discharge groundwater that is heated by shallow intrusions of magma (molten rock) in volcanic areas. Some thermal springs, however, are not related to volcanic activity. In such cases, the water is heated by convective circulation; groundwater percolating downwards reaches depths of a kilometre or more where the temperature of rocks is high because of the normal temperature gradient of the Earth's crust—about 30 °C (54 °F) per kilometre in the first 10 km (6 miles)' (http://www.britannica.com/EBchecked/topic/272775/hot-spring, *Encyclopaedia Britannica* – hot spring). It further defines the geyser as a hot spring that intermittently spouts jets of steam and hot water. The term is derived from the Icelandic word *geysir*, meaning 'to gush' (http://www.britannica.com-/EBchecked/topic/232281/geyser, *Encyclopaedia Britannica* – geyser). Permanent exposure to high temperature ranges, high free water content, moisture content, and a typical chemical composition in such areas facilitates the proliferation of certain genera of microbes. Such surviving microbes adapt their metabolism to survive such extreme environment. Often, the collection of soil samples very close to such hot springs or geysers facilitates the isolation of native microbes from the region.

The range of colours displayed in/near hot springs is due to the growth of microbial mats that includes several types of thermophilic microbes such as eubacteria, species of archaea and algae. Various studies on the colour of such mats have been reported in the past (Hiraishi *et al.*, 1999; Doemel and Thomas, 1977). Quinones are the major reasons for the colour of the microbial mats. The quinone profiles and the physiological types of microbial mats were dramatically different depending on the temperature at which these mats were formed *in situ* (Hiraishi *et al.*, 1999).

Under high temperatures and low/high pH conditions, the microbial flora (especially chemotropic sulfur bacteria, cyanobacteria, and anoxygenic phototrophic bacteria) metabolize a unique continuum of hot spring chemicals and metals, with potential energy sources such as molecular hydrogen, dissolved sulfides, methane, iron, ammonia, and arsenic (Castenholz, 1984, 1988; Castenholtz and Pierson, 1995). Siliceous alkaline hot springs of Yellowstone Park are formed predominantly by two organisms, a unicellular blue-green alga, *Synechococcus lividus*, and a filamentous, gliding, photosynthetic bacterium, *Chloroflexus aurantiacus* (Doemel and Thomas, 1977). David *et al.* (1998) had reviewed various ecological studies of terrestrial hot spring microbial communities that provided a fresh perspective of microbial biodiversity, and the composition, structure, and function of the microbial communities.

Growth and survival of microorganisms is governed by numerous physicochemical factors in their surrounding environment. Most forms of life thrive on the earth in regions where temperatures are generally moderate, i.e. at temperatures from 4°C to 40°C, at pH between pH 5 and 8.5 and where salinity, hydrostatic pressure and ionizing radiations are low. Unlike many microorganisms that cannot survive outside of moderate conditions, extremophiles flourish optimally when one or several of these parameters are in the extreme range (Brock, 1978; Madigan *et al.*, 2009). Temperature is one of the key factors, which can be extreme either positive or subzero temperatures. High-temperature environments are generally associated with volcanic activity or sulphur depositions (Viggó *et al.*, 2013).

Hot springs, geysers, or geothermal vent microbes, encompass a group that is relatively less studied as potential sources of new bioactive compounds with different mode of actions. In such extreme and nutrient-deprived environments, their original relationships in their own communities and with other organisms in the same habitats, their contributions to survival support in such niches, as well as their evolutionary patterns and their potential to be used to produce bioactive compounds are not well understood. Hence, there is a need to review and investigate this reservoir to identify the unique microbes producing bioactive compounds of medical interest (Cheeptham *et al.*, 2013).

Hot spring sample collections and isolation of microbes

We aimed at collecting samples from hot springs by maintaining the temperature as close possible to the native area, until they were processed in the laboratory for isolation of microbes. To achieve this, we followed a simple method in our laboratory. About 5 g soil/mud/organic scrapings from the vicinity of the hot spring area were collected in a sterile plastic bottle of about 15 ml capacity using a disinfected spatula. The bottle was tightly covered and kept in thermal insulation bags for maintaining the required temperature. The thermal container could maintain temperatures for a maximum of about six hours, within which the samples were taken to our laboratories and processed for isolation of microbes.

The methods used for isolation, purification and maintenance of microorganisms were standard methods for isolation of thermopiles (Panda and Sahu, 2013; Tayyab *et al.*, 2011; Chang and Kang, 2004; Jessen and Orlygsson, 2012; Narayan *et al.*, 2008). The colonies were broadly identified by their colony morphologies and light microscopy. Some interesting actinomycetes were observed by scanning electron microscopy (SEM) for classification. The SEM image of one such streptomycetes sp. is presented in Fig. 11.1.

Antibiotic-producing bacteria

Hot spring environment differs greatly from other terrestrial habitats, and hence, the biological features of actinomycetes and their allocations are expected to be different from those of soil actinomycetes (Castenholtz and Pierson, 1995). Research on the biodiversity of hot spring actinomycetes is not only important for molecular biology studies, but also for its commercial exploitation as novel drugs.

In 2012, Chitti Thawai isolated some actinomycetes from the sediments collected from the

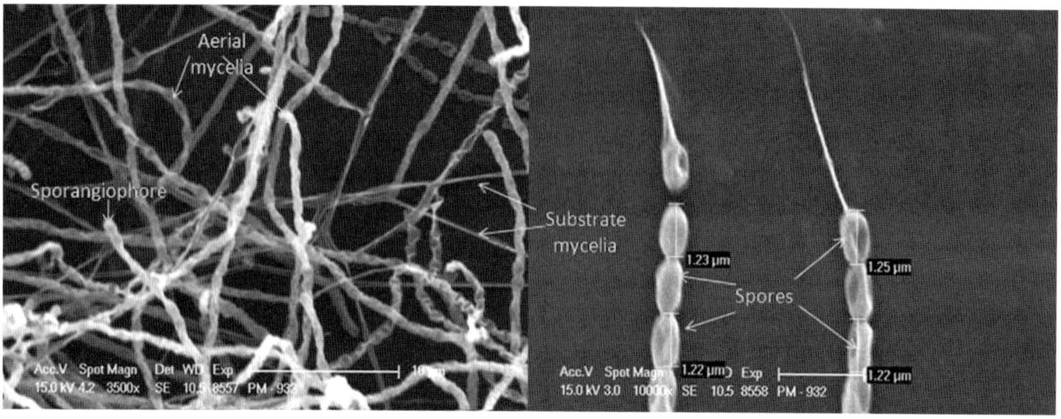

Figure 11.1 Scanning electron microscopy of *Streptomycetes* strain (PM-932) from a hot spring at Himachal Pradesh, India. A colour version of this figure is available in the plate section at the back of the book.

hot spring pond located at Krabi and Trang province, Thailand. They belonged to the genera *Streptomyces, Micromonospora, Microbiospora* and *Planosporangium*. The extracts of *Streptomyces, Micromonospora, Microbiospora* showed anti-Gram-positive bacterial activity, including anti-MRSA (methicillin-resistant *Staphylococcus aureus*) (http://semenarska.rs/UNS-PSU/radovi/2/43%20Isolation%20215-219.pdf, *Isolation and Characterization of Antibiotic producing Actinomycetes from hot spring sediment of Thailand*).

Esikova *et al.* (2002), had reported in 2002 of the presence of active metabolites from two strains of *Bacillus* sp. (VK2 and VK21). The active compounds were heat stable, butanol-extractable, and resistant to the action of trypsin and chymotrypsin. Moreover they were stable at a pH range of 2.0–9.0.

A *Bacillus licheniformis* strain, 189, isolated from a hot spring environment in Azores, Portugal, strongly inhibited the growth of Gram-positive bacteria. It produced a peptide antibiotic (Molecular mass: 3249.7 Da) at 50 °C that was highly resistant to several proteolytic enzymes and was stable at wide pH range (pH 3.5–8.0). It displayed very high thermostability retaining about 85% and 20% of its activity, after 6 hours at 50°C and 100°C, respectively (Mendo *et al.*, 2004).

The anti-infective potential of hot-spring bacteria was also reported by Pednekar *et al.* (2011). They reported the isolation of 65 bacterial strains from the hot springs of Vajreshwari and Ganeshpuri, Maharashtra- India. Three of these 65 strains were reported to possess broad-spectrum antibacterial activity, especially against strains of MRSA, Vancomycin resistant enterococci (VRE) and *Escherichia coli*.

A bacterial strain designated as *Brevibacillus laterosporus* strain BPM3, isolated from the mud samples from a natural hot water spring at Nambar Wild Life Sanctuary, Assam, India, strongly inhibited the growth of many phytopathogenic fungi (*Fusarium oxysporum, F. semitectum, Magnaporthe grisea* and *Rhizoctonia oryzae*), and Gram-positive bacterium (*Staphylococcus aureus*). Antifungal and antibacterial compounds were isolated from the bacterium which also inhibited the growth of these targeted pathogens. The compounds were purified, and a spectroscopic analysis of the purified fraction having R(f) 0.22 showed antifungal and antibacterial activity, indicating the presence of C-H, carbonyl group, dimethyl group, -CH(2), and methyl group (Saikia *et al.*, 2011). However, these compounds have not yet been fully characterized.

Fusaricidin B

At our laboratory at the research centre of Piramal Enterprises Limited, Mumbai, India, the compound fusaricidin B (Fig. 11.2) was isolated from a eubacteria from an in-house wild strain of microbe with ID PM0834793 (Mahajan *et al.*, 2013). It

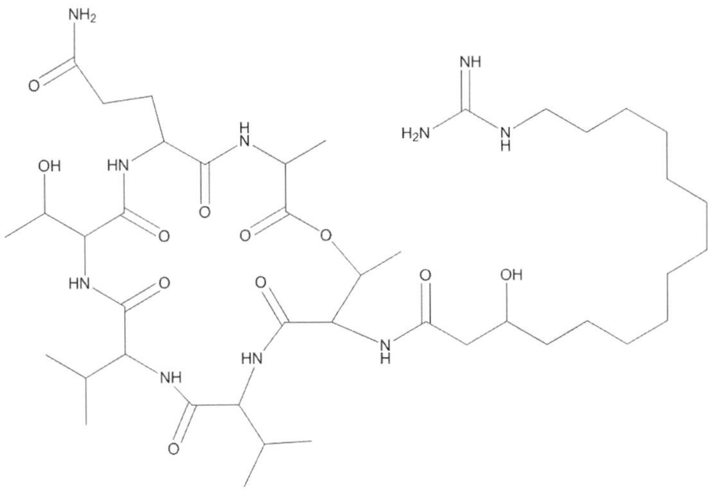

Figure 11.2 Structure of fusaricidin B.

was deposited in the Microbial Type Culture Collection and Gene Bank (MTCC), a national facility established in 1986 and funded jointly by the Department of Biotechnology (DBT), and the Council of Scientific and Industrial Research (CSIR), Government of India at Institute of Microbial Technology (IMTECH), Chandigarh. The accession number for the strain is MTCC 5620. Further, the phylogeny study of the culture using 16S-rRNA studies confirmed that it was a strain of *Paenibacillus polymyxa* (Mahajan *et al.*, 2013). The soil sample from which this bacteria was isolated had been collected from the area near some hot-springs (Figs. 11.3 and 11.4) located in Uttarkashi district of Uttaranchal state in India (north latitude 30° 27′ 18″ and 31° 27′ 42″ and east longitude 77° 48′ 26″ and 79° 24′ 00″) (http://cgwb.gov.in/district_profile/uttarakhand/uttarakashi.pdf, Ground water brochure, district Uttarkashi).

The compound was isolated during screening of biodiversity for anti-infective compounds. While examining the anti-infective bioactivity profiling of the compound fusaricidin B, we found that it showed very potent activity against *Mycobacterium smegmatis* strains in agar well diffusion assays. From these data we confirmed its activity against *M. tuberculosis*. This was confirmed at

Figure 11.3 Hot spring site at Himachal Pradesh (A). A colour version of this figure is available in the plate section at the back of the book.

Figure 11.4 Hot spring site at Himachal Pradesh (B). A colour version of this figure is available in the plate section at the back of the book.

Super Religare Laboratories Ltd (SRL), India. The compound displayed very potent activity against both multidru- resistant (MDR) *Mycobacterium tuberculosis* MDR-TB and non-MDR- *M. tuberculosis* strains.

The minimum inhibitory concentration of the compound against Non-MDR-TB strain was between 0.05–1.00 μg/ml and against MDR-TB strain it was 1–2 μg/ml (Mahajan *et al.*, 2013).

Other compounds from hot spring microbes

Ethanol production was studied by *Thermoanaerobacter* J1 isolated from a hot spring in Iceland. The strain produces 1.70 mol ethanol/mol glucose and 1.25 mol ethanol/mol xylose and shows a broad substrate spectrum, degrading various sugars and starch but not cellulose-containing substrates (Jessen and Orlygsson, 2012).

A novel moderately thermophilic, anaerobic, ethanol-producing bacterial strain, 45B(T)[*Caloramator boliviensis* sp.], was isolated from a mixed sediment water sample collected from a hot spring at Potosi, Bolivia (Crespo *et al.*, 2012). Acetate, lactate, propionate, carbon dioxide and hydrogen were also produced in minor quantities. 1,3-Propanediol was produced when glycerol-containing medium was supplemented with yeast extract.

Cyanobacteria such as *Oscillatoria limosa* and *Synechococcus lividus* were isolated from hot springs of Saudi Arabia. They were reported to produce microcystins (Mohamed, 2008). Microcystins are cyclic non-ribosomal peptides. They are cyanotoxins and can be very toxic for plants and animals including humans. Their hepatotoxicity may cause serious damage to the liver (http://www.pwrc.usgs.gov/health/rattner/rattner_blackwaternwr.cfm, *Harmful Algal Blooms and Bird Die-offs in Chesapeake Bay: A Potential Link?*).

Future scope

The hot spring microorganisms have been explored more often from the view of validating their contribution to biodiversity, modulation and specificity of their enzymes, especially enzymatic reaction requiring high temperature. The subject of bioactivity profiling and mapping needs to be well established like several other unconventional ways. Nevertheless, these niches have been explored as observed by some reports. As such thermophiles have been of great scientific curiosity for the past few decades, mainly with respect to their biotechnological potential and the thermostable enzymes they produce. Optimal cultivation techniques for these organisms are required, therefore, not only for basic research, but also for evaluation of their thermostable microbial products. Processing these metabolites in a fermentor at elevated temperatures may be advantageous in terms of increased solubility of the compounds in the vehicle medium, improved mass transfer due to decreased viscosity, and increased diffusion rates. However, the cultivation of thermophiles also has many associated problems. A high cultivation temperature can give unexpected problems affecting the choice of reactor design and construction materials, and with the heating and cooling of the fermentor (http://dx.doi.org/10.1016/S0300-9629(97)00002-9, *Thermophiles and fermentation technology*). Hence exploring hot spring microbes for new bioactive compounds needs to go hand-in-hand with advances in fermentation technology to make high temperature fermentation more conventional, as to make it more suitable for scale-up and commercialization.

Acknowledgments

The author wishes to thank the Management of Piramal Enterprises Limited (PEL) for their encouragement in the dissemination of science, which deals with the discovery of new drugs for alleviating disease on this planet. Special thanks to Dr R. K. Katti, a thinker, philosopher from Pune-India, author of the acclaimed book *Brahman, The Supreme Being in Bramhasutras,* who helped in editing the Vedic part of the article. The author is also very grateful to his colleagues – Ms Ashwini Vartak, Mr Prashant Shanbhag, Ms Rajashri Parab, Ms Nidhi Tomar and Dr Sarita Bhave – from his department at PEL for their enthusiastic support in such work. Finally, the author is highly obliged to Ms Lakshmi Balachandran for her excellent touch in editing the manuscript.

References

Asher, M. (2014). 2013 FDA drug approvals. Nat. Rev. Drug Discov. *13*, 82–89.

Brock, T.D. (1978). Thermophilic Microorganisms and Life at High Temperatures (Springer-Verlag Press, New York, USA).

Castenholz, R.W. (1984). Composition of hot spring microbial mats: a summary. In Microbial Mats: Stromatolites, Y. Cohen, R.W. Castenholtz, and H.O. Halvorson, eds. (Alan R. Liss, Inc., New York, USA), pp. 101–119.

Castenholz, R.W. (1988). The green sulphur and nonsulfur bacteria of hot springs. In Green Photosynthetic Bacteria, J.M. Olson, J.G. Ormerod, J. Amez, E. Stackebrandt, and H.G. Truper, eds. (Plenum Press, New York, USA), pp. 237–245.

Castenholtz, R.W., and Pierson, B.K. (1995). Ecology of thermophilic anoxygenic phototrophs. In Anoxygenic photos-ynthetic bacteria, Blankenship, R.E., Madigan, M.T., and Bauer, C.E. eds. (Kluwer Academic Publishers, Dordrecht, The Netherlands), pp. 87–103.

Chang, S.S., and Kang, D.H. (2004). Alicyclobacillus spp. in the fruit juice industry: history, characteristics, and current isolation/detection procedures. Crit. Rev. Microbiol. Crit. *30*(2), 55–74.

Cheeptham, N., Sadoway, T., Rule, D., Watson, K., Moote, P., Soliman, L.C., Azad, N., Donkor, K.K., and Horne, D. (2013). Cure from the cave: volcanic cave actinomycetes and their potential in drug discovery. Int. J. Spel. *42*(1), 35–47.

Crespo, C., Pozzo, T., Karlsson, E.N., Alvarez, M.T., and Mattiasson, B. (2012). *Caloramator boliviensis* sp. nov., a thermophilic, ethanol-producing bacterium isolated from a hot spring. Int. J. Syst. Evol. Microbiol. *62*(7), 1679–1686.

David, M.W., Michael, J.F., Stephen, C.N., and Mary, M.B. (1998). A natural view of microbial biodiversity within hot spring cyanobacterial mat communities. Microbiol. Molecul. Biol. Rev. *62*(4), 1353–1370.

Demain, A.L., and Sanchez, S. (2009). Microbial drug discovery: 80 years of progress. J. Antibiot. *62*, 5–16.

Doemel, W.N., and Thomas, D.B. (1977). Structure, growth, and decomposition of laminated algal-bacterial mats in alkaline hot springs. Appl. Environ. Microbiol. *34*(4), 433–452.

Esikova, T.Z., Temirov, I., Sokolov, S.L., and Alakhov, I. (2002). Secondary antimicr-obial metabolites produced by therm-ophilic Bacillus spp. strains VK2 and VK21. Prikl. Biokhim. Mikrobiol. *38*(3), 261–267.

Hiraishi, A., Umezawa, T., Yamamoto, H., Kato, K., and Maki, Y. (1999). Changes in quinone profiles of hot spring microbial mats with a thermal gradient. Appl. Environ. Microbiol. *65*(1), 198–205.

Jessen, J.E., and Orlygsson, J. (2012). Production of ethanol from sugars and lignocellulosic biomass by Thermoanaerobacter J1 isolated from a hot spring in Iceland. J. Biomed. Biotechnol. *2012*, 186982.

Madigan, M.T., Martinko, J.M., Dunlap, P.V., and Clark, D.B. (2009). Biology of Microorganisms (Pearson/Benjamin Cummings, San Francisco, USA).

Mahajan, G.B., Mishra, P.D., Vartak, A., Shanbhag, P., and Sawant, A. (2013). Compounds for the treatment of tuberculosis. Patent No WO 2013/0801–67 A1, June 6 (Piramal Enterprises Limited).

Mendo, S., Faustino, N.A., Sarmento, A.C., Amado, F., and Moir, A.J. (2004). Purification and characterization of a new peptide antibiotic produced by a thermotolerant *Bacillus licheniformis* strain. Biotechnol. Lett. *26*, 115–119.

Mohamed, Z.A. (2008). Toxic cyanobacteria and cyanotoxins in public hot springs in Saudi Arabia. Toxicon. *1*(1), 17–27.

Narayan, V.V., Hatha, M.A., Morgan, H.W., and Rao, D. (2008). Isolation and characterization of aerobic thermophilic bacteria from the savusavu hot springs in Fiji. Microbes. Environ. *23*(4), 350–352.

Newman, D.J., and Cragg, G.M. (2009). Microbial antitumor drugs: natural products of microbial origin as anticancer agents. Curr. Opin. Investig. Drugs. *10*(12), 1280–1296.

Panda, M.K., Sahu, M.K., and Tayung, K. (2013). Isolation and characterization of a thermophilic *Bacillus* sp. with protease activity isolated from hot spring of Tarabalo, Odisha, India. Iran. J. Microbiol. *5*(2), 159–165.

Pednekar, P., Jain, R., and Mahajan, G. (2011). Anti-infective potential of hot-spring bacteria. J. Glob. Infect. Dis. *3*, 241–245.

Saikia, R., Gogoi, D.K., Mazumder, S., Yadav, A., Sarma, R.K., Bora, T.C., and Gogoi, B.K. (2011). *Brevibacillus laterosporus* strain BPM3, a potential biocontrol agent isolated from a natural hot water spring of Assam, India. Microbiol. Res. *166*, 216–225.

Silver, L.L. (2011). Challenges of antibacterial discovery. Clin. Microbiol. Rev. *24*, 71–109.

Swami Swahananda (1956). The Chhandogya Upanishad (IV-xi-2), 1200-B.C. (Sri Ramakrishna Math Press, Chennai, India).

Tayyab, M., Rashid, N., and Akhtar, M. (2011). Isolation and identification of lipase producing thermophilic *Geobacillus* sp. SBS-4S: cloning and characterization of the lipase. J. Biosci. Bioeng. *111*(3), 272–278.

Viggó, M., Parag, V., Jana, K., Francesca, M., Mauro, M., Carlo, C., Angeles, A., Domenica, H., Eyjólfur, R., Sveinn, M., *et al.* (2013). A laboratory of extremophiles: Iceland Coordination Action for Research Activities on Life in Extreme Environments (CAREX) field campaign. Life *3*, 211–233.

New Sources of Antibiotics: Caves

12

Naowarat Cheeptham and Cesareo Saiz-Jimenez

Abstract

Caves are regarded as extreme habitats because their conditions are unique and harsh for all organisms including microbial life. Though demanding, these irreplaceable habitats have been demonstrated to support a great diversity of microbial communities that are believed to hold great promise as new sources of antibiotics and industrially relevant compounds. This chapter highlights recent findings that convincingly solidify the concept that caves are potentially homes to novel and rare microorganisms and tomorrow's antibiotics.

From oceans to caves

More than 70% of the planet's surface is covered by oceans. Although the diversity of life in the terrestrial environment is remarkable, the greatest biodiversity occurred in the oceans, which is assumed to contain 80% of all life on earth. Some authors reported that the marine environment is an untapped source of novel bioactive compounds (BACs) (Zhang *et al.*, 2005; Lam, 2006).

The exploration of marine organisms in search of BACs resulted in the finding of trabectedin, also known by its brand name Yondelis, initially extracted from the ascidian *Ecteinascidia turbinata*, which was found to have anticancer activity in 1969 (Cuevas and Francesch, 2009). Further, the search was extended to marine microorganisms and Feling *et al.* (2003) reported the cultivation of a group of obligate marine actinobacteria, *Salinospora*, and the extraction of BACs with antibiotic and anticancer properties. These authors stated that analogous soil-derived actinobacteria have been the single most significant source of naturally occurring antibiotics and that the ocean represented an overlooked habitat from which to isolate BAC-producing actinobacteria. In fact, unlike the discovering by chance of penicillin and subsequent research of Chain *et al.* (1940), who were able to demonstrate penicillin's ability to kill infectious bacteria, an intensive screening programme was developed by Waksman and collaborators for soil bacteria. They were looking for growth inhibition zones surrounding isolated bacteria on agar plates, which resulted in the discovery of streptomycin (Schatz *et al.*, 1944).

Soils have been considered a natural habitat for actinomycetes and this group of microorganisms has been screened since the 1940s, yielding thousands of biologically active secondary metabolites. However, the same or similar compounds have frequently been found in the screening programmes of soils and for this reason the ecological assessment of environments other than the terrestrial attracted interest in recent years. Lam (2006) believed that marine environments are extremely different from terrestrial ones, and therefore he assumed that marine actinomycetes have different characteristics and might produce different types of BACs. Currently, there is an enhanced interest for screening actinomycetes in different environments, particularly the marine ones (Manivasagan *et al.*, 2013; Subramani and Aalbersberg, 2013; Tiwari and Gupta, 2013).

Two decades ago, Williams *et al.* (1993) stated that one approach to the isolation of novel actinomycetes is to concentrate on understudied environments or substrates while using appropriate selective isolation techniques or to investigate habitats in which one or more of the environmental factors (e.g. temperature, pH, aeration, or osmotic stress) are extreme. In this context, caves seem to be a promising environment for finding novel actinomycetes as many of the extreme conditions are met.

The subterranean environment offers one of the best possible sites to search for microbial life. Caves harbour an especially interesting unique geological composition and variety of mineral-utilizing and heterotrophic microorganisms, which should be explored. Caves are one type of feature that characterizes a karst landscape, which develops in soluble rocks (e.g. limestone, dolomite, gypsum, halite, etc.) (Fig. 12.1). Karst systems constitute ~15–20% of the Earth's ice-free land surface. Caves formed in rocks other than carbonates, like basalt or sandstone, are classified as pseudokarst. Caves are currently considered as laboratories of geological, ecological and geomicrobiological interest, with diverse environments that differ in physical, chemical and microbiological composition (Lee *et al.*, 2012). In addition, caves can also be seen as reservoirs for natural products exploration because the subterranean environment contains stable physicochemical conditions, which enhance microbial growth under starvation conditions (Northup *et al.*, 2011) and these microorganisms can have new and not yet discovered secondary metabolite production. The majority of caves thus represent a nutrient limited environment which is diverse and high in minerals, where life is confined to highly specialized microorganisms able to promote a variety of enzymatic and rock-weathering processes and biomineralization (Barton, 2006; Kelly *et al.*, 2011).

On 11 April 2012, a ground-breaking study was published by a group of researchers led by Dr Gerry Wright, Professor of Biochemistry and Biomedical Sciences at McMaster University, Canada, and Dr Hazel Barton, one of the trail blazers in the field of cave microbiology, an associate professor at the Department of Biology, University of Akron, OH, USA. Their findings are pivotal in new BAC discovery research. This study set out to investigate whether antibiotic resistance exists in the cave microbial communities especially in the deepest and isolated regions of the Lechuguilla Cave, New Mexico (Bhullar *et al.*, 2012). Twenty different isolation and culture media were used and more than 500 distinctive isolates were

Figure 12.1 Yellow colonizations on the walls of a gypsum cave (Sorbas, Spain). A colour version of this figure is available in the plate section at the back of the book.

retrieved. Ninety-three strains of those isolated bacteria were subjected to antibiotic resistance screen against 26 different commercially available antibiotics. Additionally, the genetic and enzymatic studies of the antibiotic resistant mechanisms were investigated. The resulting findings were astonishingly unexpected, these subterranean bacteria were observed to be highly resistant to the antibiotics screened similar to those seen in surface bacteria. As mentioned above, caves can be used as an excellent laboratory and in this case the over four million years of isolation has signified the Lechuguilla Cave as the perfect place to examine resistance gene prevalence in environmental microorganisms and the influence of anthropogenic use of antibiotics in human history. The results of this study not only suggested that antibiotic resistome is innate, ancient, and persistent in microbial genomes and that it is not exactly a direct result of our use of anthropogenic substances, but also created new-found excitement in our quest to find new antibiotics due to the incredible genetic diversity of the antibiotic resistome. This may indicate that there are a great number of undiscovered antimicrobial bioactive compounds out there, in the caves and elsewhere, to be found.

Microbes found in caves and their potential

Though there have been a growing number of studies published on microbial communities of caves and cave-related habitats (Table 12.1) nevertheless the connection between cave microbial communities and their potential in new drug discovery has only been slightly studied compared to studies done on microorganisms isolated from other extreme habitats as sources for drug producers (Cheeptham, 2013). Yet, this trend is changing due to the urgent and dire need to discover tomorrow's antibiotics to fight against existing and emerging antibiotic resistance in many clinical and common pathogens. Recently, such studies have found their way into the cave microbiology field and the results, though preliminary, have been nothing but encouraging in demonstrating that caves may be potential sources for an untapped population of microorganisms with specific bioactive metabolites and compounds that are well suited for many areas of biotechnology in particular industrial and medical. Additionally, environments as a whole are still poorly understood in regards to microbial interactions and their roles in cave formation, cave degradation and bioprecipitation/biomineralization. Studies in caves have recently led to unravelling several new microbial taxa, genera and species. Newly discovered microorganisms have been isolated from caves and cave-like habitats around the world in which the reports on these novel organisms have been increasingly emerging in the past few years. Let us take a brief look at a number of articles published in the *International Journal of Systematic and Evolutionary Microbiology* (IJSEM) from 1999 to 2013. In addition to the 34 new species of the class *Actinobacteria* reported by Jurado *et al.* (2010) for the period 2000–2009, 15 novel species should be added up to 2013. These includes *Beutenbergia cavernae* (Groth *et al.*, 1999b), *Yaniella fodinae* (Dhanjal *et al.*, 2011a), *Streptomyces zinciresistans* (Lin *et al.*, 2011), *Streptomyces youssoufiensis* (Hamdali *et al.*, 2011), *Agrococcus carbonis* (Dhanjal *et al.*, 2011b), *Intrasporangium chromatireducens* (Liu *et al.*, 2012), *Cellulomonas carbonis* (Shi *et al.*, 2012), *Nonomuraea monospora* (Nakaew *et al.*, 2012), *Nocardioides albertanoniae* (Alias-Villegas *et al.*, 2013a), *Sphingopyxis italica* (Alias-Villegas *et al.*, 2013b), *Actinotalea ferrariae* (Li *et al.*, 2013), *Kribbella albertanoniae* (Everest *et al.*, 2013), *Amycolatopsis cihanbeyliensis* (Tatar *et al.*, 2013), *Spelaeicoccus albus* (Lee, 2013a), and *Humibacter antri* (Lee, 2013b). However, a few more species have been described in other journals: *Microlunatus cavernae* (Cheng *et al.*, 2013a) and *Saccharopolyspora cavernae* (Cheng *et al.*, 2013b).

Not only in the class *Actinobacteria*, but in *Archaea*, *Cyanobacteria*, green microalgae and fungi cave diversity remains largely unexplored and a few novel taxa have been discovered recently, such as the *Archaea Methanobacterium movilense* (Schirmack *et al.*, 2013), the cyanobacteria *Iphinoe spelaeobios*, *Loriellopsis cavernicola* (Lamprinou *et al.*, 2011), *Toxopsis calypsus* (Lamprinou *et al.*, 2012), *Chalicogloea cavernicola* (Roldán *et al.*, 2012) and the fungi *Aspergillus thesauricus*, *Aspergillus baeticus* (Nováková *et al.*, 2012),

Table 12.1 Review of microorganisms found and methods used in cave and related subsurface microbial community study

	Sample types and environments	Objectives	Techniques	Microorganisms found	Reference
1	Surfaces of the rock art paintings of Atlanterra shelter, Spain	To study bacterial present in the rock art paintings	Hygicult (dipslides) for isolation and identification of total cellular fatty acid methyl esters	Members of *Bacillus* species, predominantly *B. megaterium*	Gonzalez *et al*. (1999)
2	Surfaces of cave walls from two karstic caves, Altamira and Tito Bustillo in northern Spain	To isolate and identify microorganisms contributing to deterioration of rock art in caves	Culture methods (with contact plates, cotton swabs, and porous plastic plugs and various selected isolation media. Identification according to morphological, physiological, and chemotaxonomic characteristics	Members of *Streptomyces* (predominant), *Nocardia*, *Rhodococcus*, *Nocardioides*, *Amycolatopsis*, *Saccharothrix*, *Brevibacterium*, *Microbacterium*, and coccoid actinomycetes (family Micrococcaceae)	Groth *et al*. (1999a)
3	Dripping waters and ceiling rocks in Altamira Cave, Santillana del Mar, Spain	To study and compare microbial populations in dripping waters and ceiling rocks in a karstic cave (to better understand microbial involvement on rock art deterioration in caves)	Culture method (Petrifilm aerobic count plates and many selected agar media for isolation)	In dripping waters, Gram-negative rods and cocci (Enterobactericeae and Vibrionaceae) were found while mainly *Streptomyces* were observed in ceiling rock samples	Laiz *et al*. (1999)
4	Various cave samples: stream sediments, black mud, water drips, snottites (stalactitic forms) from Cueva de Villa Luz, Tabasco, Mexico	To study microbial contribution in the development of acidic environments and cave enlargement in a hydrogen-sulfide-rich karst environment	Combination of methods; culture-dependent and -independent approaches (various isolation media, SEM, ESD, 16S rRNA, sequencing	A number of sulfur oxidizing and reducing bacteria were found depending on sample types: *Thiobacillus* spp., *Thiothrix* spp. and *Acidimicrobium ferrooxidans*	Hose *et al*. (2000)
5	Water samples, microbial mats and biofilms from pools and spring of the Cesspool Cave, Virginia, USA	To determine acid-producing bacterial community composition and its geomicrobiological functions in cave	Combination of methods; molecular and liquid enrichment culturing techniques (16S rRNA sequencing, SEM, and sulfur-oxidizing bacteria enrichment liquid culture for 3 months)	Genus *Thiothrix* and the *Flexibacter-Cytophaga-Bacteriodes* phylum and may be *Helicobacter* or *Thiovulum* group for some remaining strains.	Engel *et al*. (2001)
6	Microbial mentles and water from Weebubie Cave, Nullarbor Plain, western Australia	To examine a cave biomineral of novel morphology	FESEM, TEM, XRD, FTIR and SAED	Bacterial filaments associated with calcite crystals	Contos *et al*. (2001)
7	Pull fingers from Hidden Cave, New Mexico, USA	To preliminary observe for possible bacterial involvement in speleothem formation	SEM, X-ray diffraction and stable isotopic analyses	Calcified filaments and micro-rods observed in samples indicating filamentous bacteria according to the size and morphology	Melim *et al*. (2001)

8	Speleothems (active stalactites, wall concretions), rock walls, ceiling, and sediments from galleries of Grotta dei Cervi, Porto Badisco, Italy	To investigate culturable heterotrophic microbial communities present in the cave	Tryptone soy agar, malt-yeast extract-agar, starch-casein agar, glycerol-asparagin-agar, and water-agar (incubation up to 8 weeks)	Members of *Agromyces*, *Arthrobacter*, *Rhodococcus*, *Streptomyces*	Groth *et al.* (2001)
9	Speleothem samples in basaltic sea caves, Kauai, Hawaii, USA	To describe kerolite and associated clay minerals	X-ray diffraction, energy dispersive X-ray microanalysis, SEM, TEM	Bacteria and cyanobacteria	Leveille *et al.* (2002)
10	Bacterial mats (slimes) in Four Windows Cave, a lava tube in El Malpais National Monument, New Mexico, USA	To characterize microbial communities of the lava wall slime using culture-independent methods	16S rRNA sequencing (with RFLP) and SEM	Members of Actinobacteria, Chloroflexi, Verrucomicrobia and Betaproteobacteria	Northup *et al.* (2004)
11	Bat guano samples in various Japanese caves (20 limestone and volcanic caves)	To examine yeasts in bat-inhabited caves	YM agar with selected antibiotics for isolation and D1/D2 26S rRNA sequencing for identification	Members of *Trichosporon* species, *Candida palmioleophila*, *C. lusitaniae*, *Debaryomyces hansenii*, *Hanseniaspora* spp., *Saccharomyces cerevisiae*, *S. kluyveri*, *Williopsis californica*, *Zysosaccharomyces floresntinus*, and *Crytococcus podzolicus*	Sugita *et al.* (2005)
12	Various geological samples from Jack Bradley Cave, Kentucky, and Carlsbad Cavern, New Mexico, USA	To describe an improved protocol to recover extremely low amounts of DNA from calcium-rich geologic samples	Improved DNA extraction protocol (use of Poly-dIdC, EGTA, and Slide-A-Lyzer MINI small-volume dialysis cassettes)	Contaminant-free, genomic DNA from a low-biomass, $CaCO_3$-rich cave samples	Barton *et al.* (2006)
13	Soil samples from Niu Cave, an earth-cave, in Guizhou Province, China	To investigate the bacterial diversity in a cave composed of soil rather than limestone or calcareous rocks)	Molecular approach (DNA extraction, PCR, 16S rRNA, cloning and phylogenetic analysis)	Members of Proteobacteria, Acidobacteria, Planctomyces, Chloroflexi (Green non-sulfur bacteria), Bacteroidetes, Gemmatimonadetes, Nitrospirae, and Actinobacteria	Zhou *et al.* (2007)
14	Water and bacterial mats (biofilms) samples from a subsurface thermal spring, Franz-Josef-Quelle, in Bad Gastein, Austria	To determine communities of *Archaea* and B*acteria* in a subsurface radioactive thermal spring in Austrian central Alps and to find ammonia-oxidizing *Crenarchaeota*	Molecular approach (DNA extraction, PCR, 16S rRNA, RFLP and phylogenetic analysis) and SEM	Crenarchaeota and Euryarchaeota, members of the Proteobacteria (Alpha-, Beta-, Gamma- and Deltaproteobacteria), Bacteroidetes and Planctomycetes	Weidler *et al.* (2007)
15	Biofabrics in a geothermal mine adit (karst environment), Colorado, USA	To study microbial involvement and influences in mineral deposition	16S rRNA (molecular phylogenetic analyses), PCR (functional gene analysis), ESEM and EDS	*Thermus* spp., Crenarchaeota, Chloroflexi, and Gammaproteobacteria	Spear *et al.* (2007)

Table 12.1 (Continued)

	Sample types and environments	Objectives	Techniques	Microorganisms found	Reference
16	Cave soil sample, Phanangkoi, Northern Thailand (karstic cave)	To isolate less-intensively studied bacteria with antimicrobial activity for potential drug discovery	Humic acid–vitamin agar with selected antibiotics for isolation. Identification using chemotaxonomic and phenotypic characteristics and 16S rRNA.	A new strain of *Spirillospora albida*, an actinomycetal bacterium	Nakaew *et al*. (2009a)
17	Cave soil samples, Phanangkoi and Phatup caves, Northern Thailand (karstic cave)	To isolate rare actinomycetes from caves using selective methods	Humic acid–vitamin, humic acid vitamin-gellan gum, and starch and casein agar with selected antibiotics for isolation. Identification using chemotaxonomic and phenotypic characteristics and 16S rRNA	Members of *Spirillospora*, *Catellatospora*, *Nonomuraea*	Nakaew *et al*. (2009b)
18	Cave popcorn from an unnamed karstic cave in Kentucky, USA	To investigate bacterial calcium carbonate precipitation in cave environments	Conventional isolation on B-4 and B-4C media and 16S rRNA sequencing (identification) and SEM	Members of Alpha-, Beta-, and Gammaproteobacteria Firmicutes and Actinobacteria	Banks *et al*. (2010)
19	Rock walls, speleotherms surfaces and soil samples from karstic caves in Turkey	To isolate actinomycetes from caves	Starch casein agar with selected antibiotics for isolation. Identification using chemotaxonomic and phenotypic characteristics	Members of *Streptomyces*	Yücel and Yamaç (2010)
20	Speleothems (stalactites and stalagmites) from Buracos Cave (lava tube), Terceira Island (Azores), Portugal	To investigate microbial contribution to iron speleothems for a better understanding on roles of microbes in speleothem construction	Microscopy (SEM, LT-SEM, SEM-EDS) and molecular approach (16S rRNA sequencing, PCR-DGGE)	From DGGE, mainly Proteobacteria and Actinobacteria were found while *Gallionella* sp. and *Leptothrix* sp. were observed from SEM but not DGGE	De los Rios *et al*. (2011)
21	Rock chips covered with either microbial mats and secondary minerals from basaltic lava caves in Azores (Portugal), Hawaii and New Mexico (USA)	To study microbial communities in microbial mats and mineral-like deposits in lava tube caves	Combination of SEM and molecular methods (DNA extraction, PCR, 16S rRNA sequencing and phylogenetic analysis)	Fourteen phyla of bacteria across three locations (Actinobacteria, Acidobacteria, Proteobacteria Alpha-, Beta-, Gamma-, Delta-and Epsilonproteobacteria), Bacteroidetes, Chloroflexi, Nitrospirae, Verrucomicrobia, Gemmatimonadetes, Planctomycetes)	Northup *et al*. (2011)

Ochroconis lascauxensis and *Ochroconis anomala* (Martin-Sanchez *et al.*, 2012).

In spite of a majority of the new species revealed in caves are members of actinobacteria, however, important notes need to be made herein in regards to the fact that different isolation media, growth condition and techniques have been employed in those studies. Those caves have different characteristics and each is unique in physical, chemical, biological and ecological factors that may contribute to individuality of microbial community specific to each environment. Although, in some studies shown in Table 12.1, it is observed that there may be some commonalities among microbial communities in caves.

Two studies on genetic diversity of rare actinomycetes (non-streptomycetes) and their potential BAC production were investigated in Pha Tup Cave Forest Park and Phanangkhoi cave in northern Thailand. Members of very rare genera of *Spirillospora, Catellatospora* and *Nonomuraea* were isolated in these studies (Nakaew *et al.*, 2009a,b). The crude extract of *Spirillospora* fermentation broth was found to show antimicrobial activity against Gram-positive bacteria. These examples are just a starting point of what yet may come in regards to discovery of new microorganisms dwelling in caves and their potential in new drug discovery. There are still more rooms to explore further for the indigenous microbial communities in this extreme subterranean habitat of caves and to understand in addition in regards to their relationships and contributions in this ecological niche. Diversity of actinomycetes in karstic caves has been studied in Turkey, a total of 290 actinomycete strains were isolated from soil, speleothem and dripping water samples collected from 19 caves when using starch casein, tryptone yeast extract (ISP-1) and yeast extract and malt extract (ISP-2) agar as isolation media (Yamaç *et al.*, 2011). Numerical classification studies of the selected 131 isolates were done according to their morphology and pigmentation properties compared to the reference *Streptomyces* strains. It was reported that 127 isolates tested were grouped in different clusters from reference *Streptomyces* strains. Again, this is very supportive of the notion that caves are home to many novel microorganisms. Additionally, in a related study of these Turkish caves, *Streptomyces* sp. 1492 was found to show antimicrobial activity against methicillin resistant *Staphylococcus aureus* (MRSA), vancomycin resistant *Enterobacter faecium* (VRE) and *Acinetobacter baumannii* (Yücel and Yamaç, 2010). With an increasing concern on antibiotic resistance in clinical and environmental pathogens, this finding shows potential. A separate study from Egypt looked into bioperspecting of actinomycetes to fight against ringworm fungal infection/mycoses (Saleh *et al.*, 2013). *Streptomyces* sp. MECO$_2$, an actinomycete strain that was isolated from ancient stone monuments in Tell Basta Tombs, Zagazig City, Egypt, was found to demonstrate promising antitrichophyton activity against the dermatophyte *Trichophyton mentagrophytes*. Though, this actinomycete isolate is not a subterranean example, however, this study is at a level associated with one of the lifestyles (rock dwellers) that are commonly found in cave environments. Two members in the Subsurface Life in Mineral Environments (SLIME) team of the Northup laboratory at the University of New Mexico, USA reported on a dynamic study regarding biotic and abiotic factors, of five New Mexican carbonated caves, that influence antimicrobial activity found in cave microorganisms (Montano and Henderson, 2013). Various samples of rock, cave soil, water, moonmilk and ferromanganese deposits of the Spider Cave, Backcountry Cave, Carlsbad Cavern: Left Hand Tunnel, Fort Stanton Cave, and Lechuguilla Cave were used for bacterial isolation. Three isolation media, namely R2A, half-strength R2A enriched with carbonate rock and Actinobacteria Isolation Agar, were used. The inoculation was done in the caves while the incubation was done at 15°C in the dark in the lab until colonies were observed. A total of 1281 isolates were obtained and 1025 isolates were subjected to antimicrobial activity screen. It was documented that these cave bacteria exhibited antimicrobial activity. This study also demonstrated the relationship between abiotic and biotic factors on antimicrobial activity pattern in these isolated cave bacteria. It was concluded that varying nutrient levels within the caves do not influence antimicrobial activity in cave bacterial isolates, or do so with other factors that still need to be identified. Importantly, it was observed and summarized that the deeper and the more oligotrophic the caves get, the more antimicrobial activity were found in cave bacterial isolates. However, the authors stated the need for more statistical analyses and further studies to be done in the aspects of ecological factors and the pattern of antimicrobial activity in cave bacterial isolates.

A few more recent studies that can certainly be used as solid proof for potential of cave microbiomes in new drug discovery were published earlier in 2013 in the *International Journal of Speleology*. The first report on potential of heterotrophic bacteria isolated from the Gallery with pre-historical drawings in Magura Cave, Bulgaria, was discussed herein. This study suggested that heterotrophic bacteria isolated held promising potential in industrially relevant psychrotolerant enzymes and bioactive metabolites (Tomova *et al.*, 2013). Of the isolates tested for hydrolytic enzyme production, antimicrobial and haemolytic activity, it was found that 87% produced protease, 64% xanthan lyase, 40% lipase, 40% β-glucosidase, 21% phytase while 75% exhibited antimicrobial and haemolytic activity.

Another great evidence were two studies on cave microbiomes and their potential in drug discovery from the Helmcken Falls Cave in Wells Gray Provincial Park, Clearwater, British Columbia, Canada (Cheeptham *et al.*, 2012; Rule and Cheeptham, 2013). This is the first report on such studies from a volcanic cave in western Canada (Fig. 12.2). Though preliminary, these were very encouraging results. Cheeptham *et al.* (2012) found that, of 400 cave bacterial isolates tested, 26.5% exhibited activity against *Klebsiella pneumoniae*, 10.25% against *Micrococcus luteus*, 9.25% against *Mycobacterium smegmatis*, 7.5% against *Candida albicans*, 6.25% *Pseudomonas aeruginosa*, 2.25% against *Acinetobacter baumannii*, 2% against MRSA, 1.75% against *Escherichia coli*, and 1% against extended spectrum β-lactamase *E. coli*. The majority of the cave bacterial isolates were found to be members of *Streptomyces* genus according to 16S rDNA sequencing results and microscopic and macroscopic morphological study. Based on the adaptations of cave microorganisms to living in the darkness, Rule and Cheeptham (2013) further reported a subsequent study on the photochemotherapeutic potential of *Streptomyces* isolated from the same cave. The objective of this study was to investigate whether there is any effect of UV exposure on antimicrobial metabolites found in the cave bacterial isolates. In the course of our screening, two plug assay experiments were designed, one in the dark and one under UV light exposure, to screen for

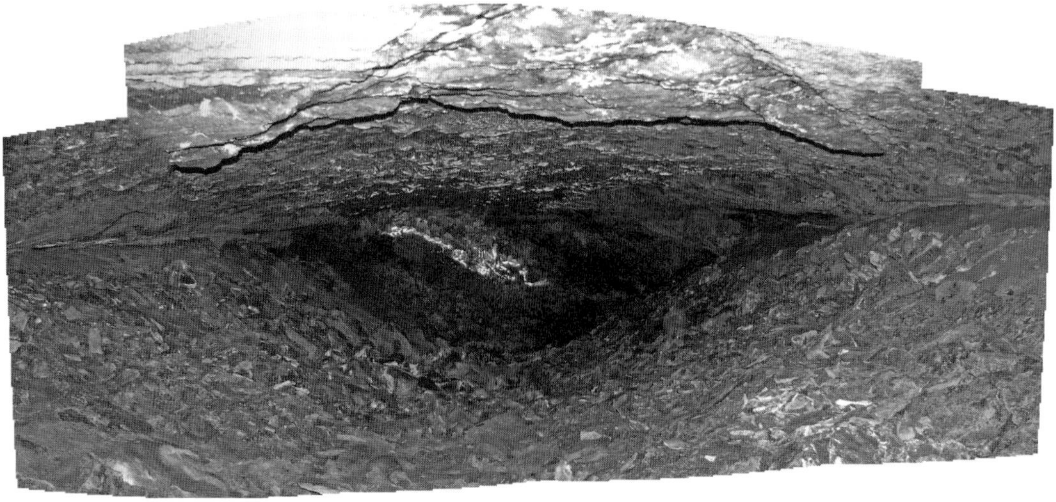

Figure 12.2 Helmcken Falls Cave, British Columbia, Canada. Volcanic caves have a great potential for biodiversity studies. Image by Kent Watson. A colour version of this figure is available in the plate section at the back of the book.

antimicrobial activity in the previously tested cave bacterial isolates. It was found that, with statistical analyses, there was not a definite pattern whether UV exposure negatively or positively changes in the cave bacteria's antimicrobial activity. Some of the cave bacteria demonstrated losing activity while others gained activity after being exposed to UV light. Further study is needed to be done to better understand the impact of UV light in regards to activation/deactivation of antimicrobial activity of the cave bacterial isolates. Nonetheless, this study confirms that cave actinomycetes are potential sources of novel antimicrobial agents.

A recent yet-to-be-published study has been discussed herein. Dr Hazel Barton and her group have led revolutionary studies in one of the most isolated caves on Earth, Lechugilla Cave in Carlsbad Caverns National Park, New Mexico, USA. This cave system is a very deep and large system with a more than 200 km complex maze of geological isolation from surface (H.A. Barton, 2013, personal communication). One of actinomycetes, isolated from a pulverized rock sample collected from a remote location (1250 feet deep from the surface) that had seen only one visit and with no signs of human access at the actual sample site, showed promising antimicrobial activity. The fermentation broth of this strain was subjected to further studies and through a series of mass spectrometric techniques, a series of not-previously reported metabolites, *C*-glycosyl angucyclines, later designated as lechacyclines A to C, were revealed. The actinomycete strain was later named *Streptomyces cunninghamii*. The to-be-published results of this dedicated study in a remote and oligotrophic location in an isolated cave system of more than four million years has not only exemplified but also solidified the harsh and arduous reality of drug discovery that tomorrow's drugs can potentially be found in cave microbiomes.

More rooms to explore in caves

Microbial processes in caves often involve redox reactions produced by aerobic (chemolithotrophic) or by anaerobic (heterotrophic) microorganisms which obtain energy from the oxidation of organic matter and reduce inorganic compounds. Biogenic minerals can be formed directly by microorganisms through enzymes or by producing substances that lead to the precipitation of minerals (e.g. by changing the pH of their surroundings). The microorganisms produce a wide variety of cave minerals: iron oxides-hydroxides, manganese oxides but other less common forms such as celestite, fibroferrite, opal, endellite, pyrolusite, and marcasite. More than 19 cave minerals are known to be formed by microorganisms (Hill and Forti, 1997). Comparatively less is know on the capability of cave microorganisms to produce bioactive substances of interest for pharmacology and medicine.

Mineral formations and deposits could also be a source for the isolation of intriguing and unknown microorganisms (Saiz-Jimenez, 2013). Black deposits are widely found in caves and usually they are found as crust-forming deposits on speleothems on cave ceilings and walls (Spilde *et al.*, 2005) often rich in manganese and iron (Miller *et al.*, 2012a). These deposits result from the activity of Fe-/Mn-oxidizing bacteria (Northup *et al.*, 2003).

For all the above studies, scanning electron microscopy (SEM) combined with ancillary chemical analyses proved to be essential tools to assess microbe–mineral interactions as well as to describe unusual mineral formations and never-seen microbial features. For example, Melim *et al.* (2008) reported intriguing filamentous structures consisting of long reticulated filaments (up to 75 μm long and 0.5 μm in diameter) in calcite speleothems from limestone caves in New Mexico, and in basalt lava tubes in Cape Verde Islands. They were characterized as hollow filaments with hexagonal and diamond-shaped chambers resembling honeycombed structures, and could not be correlated to any known microorganism. These enigmatic subsurface filamentous fabrics were found within black ferromanganese deposits from a historic granite spring water tunnel in Porto city, Portugal (Miller *et al.*, 2012b), in lava tubes of Easter Island (Miller *et al.*, 2014) and in Spanish limestone caves as shown in Fig. 12.3. However, despite their efforts based on cultivation and molecular techniques, attempts to decipher the origin or to ascribe these structures to a definite microorganism have failed so far. These data exemplify how the research on

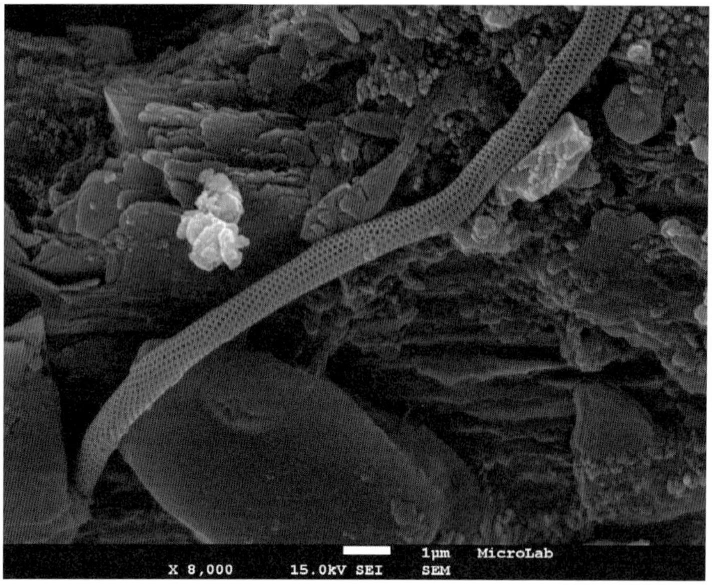

Figure 12.3 Reticulated filaments from Ardales Cave, Spain. FESEM image by A.Z. Miller and A. Dionísio.

subterranean environments can provide us with new microorganisms whose physiological functions are unknown. In addition, many of the new species described from caves are still waiting for a BAC screening.

A review on the biodiversity and distribution of bacteria in subterranean environments reveals the scant knowledge of this topic (Table 12.1). *Proteobacteria* represent the major portion of bacteria in caves when using molecular tools, but *Actinobacteria* constitute the majority of isolates (Groth *et al.*, 1999a; Barton *et al.*, 2007). In the last decade, actinobacteria have been shown to be abundant in caves (Groth *et al.*, 1999a; Schabereiter-Gurtner *et al.*, 2002a,b; Jurado *et al.*, 2005a,b,c, 2008, 2009; Cuezva *et al.*, 2009, 2012; Saiz-Jimenez *et al.*, 2011), suggesting that this is a habitat particularly favourable for this group of bacteria.

Groth and Saiz-Jimenez (1999) first stressed the importance of caves as a new habitat in which actinomycetes producing bioactive substances could be found. In a further study on Grotta dei Cervi, Porto Badisco, Italy, a high number of actinomycetes were isolated, including 163 *Streptomyces* (Groth *et al.*, 2001). Later, Herold *et al.* (2005) reported the production of cervimycin A–D by *Streptomyces tendae*, isolated from this cave. These novel polyketide glycosides have significant activity against multidrug-resistant staphylococci and vancomycin-resistant enterococci.

The data reported here indicates that caves, mines and any subterranean environment constitute an ecological niche with a solid potential for biodiversity studies. The prevailing conditions of high salinity, mineral concentrations and the stability of microenvironmental conditions over the years make these habitats of great interest and a target for in-depth study. In addition, many of these novel isolates from caves and mines represented rare actinomycetes. According Tawari and Gupta (2012) rare actinomycetes were the actinomycete strains whose isolation frequency is much lower than that of the streptomycete strains isolated by conventional methods.

Rare actinomycetes from caves and mines belong for example to the genera *Catelliglobosispora, Ferrimicrobium, Hoyosella, Thermobifida, Saccharothrix, and Fodinibacter*. Tawari and Gupta (2012) believe that with an intensive effort in isolating and screening rare genera of actinomycetes from unexplored environments can increase the likelihood of discovering new BACs. Subramani and Aalbersberg (2012) reported that the isolation of rare actinomycetes from underexplored diverse natural habitats resulted in the isolation of about 220 rare actinomycete genera of which more than 50 taxa produced 2500 bioactive compounds. Between 2007 and mid-2013, 80 new rare actinomycete species were isolated from marine habitats, grouped into 23 rare families, of which three were novel, and 20 novel genera. Similar or even higher figures could be obtained from a systematic and in-depth screening on caves, particularly those formed on gypsum, halite, and basalts, which have been hardly explored in our search for biodiversity. It must be also mentioned that mines and tunnels represent quasi unexplored environments where the nature of the mineral substratum can provide rare and unique bacteria. These artificial environments were drilled on ores and therefore the microbial diversity can be unique for the different types of minerals (pyrite, manganese, etc.).

Summary

We hope that studies presented in this chapter have optimistically convinced a number of researchers to take caves seriously as an untapped resource for novel microorganisms and drug discovery. However, challenges in isolation of microbiomes from soil, rock, water, speleothem samples of a number of unique cave environments and screening for antimicrobial production from the isolated cave microorganisms do exist. Caves, as an extreme habitat, possess taxing characteristics (physical, chemical, biological and ecological) for life to not just survive but thrive in (Barton *et al.*, 2007). Those that can thrive in this demanding home must have over time adjusted to selective pressures and the resulting adaptations may lead to a myriad of unique metabolic pathways to sustain life. Typically, caves are nutrient-limited, have almost constantly low temperature and high humidity with diverse and high concentration of minerals (Barton, 2006). Another compounding reason why caves are regards as nutrient poor environment is with limited to no light condition, these create less primary producers (less organic matter/inputs except in those caves with bats). With such demanding factors, microorganisms dwelling in caves are often slow growers and need to cooperate for survival. Choices of isolation media are vital to success in cave microorganisms' culture. Rich media at normal strength seldom yield the interesting oligotrophic and heterotrophic microbial community truly existing in these habitats. Often diluted version of isolation media is used for isolation, however, this depends on what types of caves and samples there are. To succeed in isolation of interesting cave microbiomes, we need to understand not only all of the nuances pertaining to factors that make cave an inhospitable place to live and thrive but also how microorganisms adapted and adopted a lifestyle in such harsh habitats. This is not a task for a faint of heart, but the outcome of discovering new species and related bioactive compounds is rewarding.

Acknowledgements

Our thanks go to Dr Ken Wagner of Thompson Rivers University for his time and assistance in proofreading this manuscript and to Dr Hazel Barton of the University of Akron for her insights and for sharing her latest and yet-to-be published work with us. NC was supported by TRU Research Innovation Funds. CSJ was supported by CSIC project 201230E125.

References

Alias-Villegas, C., Jurado, V., Laiz, L., Miller, A.Z., and Saiz-Jimenez, C. (2013a). *Nocardioides albertanoniae*, sp. nov., isolated from Roman catacombs. Int. J. Syst. Evol. Microbiol. *63*, 1280–1284.

Alias-Villegas, C., Jurado, V., Laiz, L., and Saiz-Jimenez, C. (2013b). *Sphingopyxis italica* sp. nov. isolated from Roman catacombs. Int. J. Syst. Evol. Microbiol. *63*, 2565–2569.

Banks, E.D., Taylor, N.M., Gulley, J., Lubbers, B.R., Giarrizzo, J.G., Bullen, H.A., Hoehler, T.M., and Barton, H.A. (2010). Bacterial calcium carbonate precipitation in cave environments: a function of calcium homeostasis. Geomicrobiol. J. 27, 444–454.

Barton, H.A. (2006). Introduction to cave microbiology: a review for the non-specialist. J. Cave Karst Stud. *68*, 43–54.

Barton, H.A., Taylor, N.M., Lubbers, B.R., and Pemberton, A.C. (2006). DNA extraction from low-biomass carbonate rock: an improved method with reduced contamination and the low-biomass contaminant database. J. Microbiol. Methods *66*, 21–31.

Barton, H.A., Taylor, N.M., Kreate, M.P., Springer, A.C., Oehrle, S.A., and Bertog, J.L. (2007). The impact of host rock geochemistry on bacterial community structure in oligotrophic cave environments. Int. J. Speleol. *36*, 93–104.

Bhullar, K., Waglechner, N., Pawlowski, A., Koteva, K., Banks, E.D., Johnston, M.D., Barton, H.A., and Wright,

G.D. (2012). Antibiotic resistance is prevalent in an isolated cave microbiome. PLoS One *7*, 1–11.

Chain, E., Florey, H.W., Hardner, A.D., Heatley, N.G., Jennings, M.A., Orr-Ewing, J., and Sanders, A.G. (1940). Penicillin as a chemotherapeutic agent. Lancet *239*, 226–228.

Cheeptham, N. (2013). Advances and challenges in studying cave microbial diversity. In Cave Microbiomes: A Novel Resource for Drug Discovery, N. Cheeptham, ed. (Springer, London, UK), pp. 1–34.

Cheeptham, N., Sadoway, T., Rule, D., Moote, P., Watson, K., Soliman, L., Azad, N., Donkor, K., and Horne, D. (2012). Cures from caves: volcanic cave actinomycetes and their potential in drug discovery. Int. J. Speleol. *42*, 35–47.

Cheng, J., Chen, W., Huo-Zhang, B., Nimaichand, S., Zhou, E.-M., Lu, X.-H., Klenk, H.-P., and Li, W.-J. (2013a) *Microlunatus cavernae* sp. nov., a novel actinobacterium isolated from Alu ancient cave, Yunnan, south-west China. Anton. Leeuw. *104*, 95–101.

Cheng, J., Zhang, Y.-G., Chen, W., Li, L., Zhang, D.-F., Wang, H.-F., Lu, X.-H., Duan, Y.-Q., and Li, W.-J. (2013b) *Saccharopolyspora cavernae* sp. nov., a novel actinomycete isolated from the Swallow Cave in Yunnan, south-west China. Anton. Leeuw. *104*, 837–843.

Contos, A.K., James, J.M., Heywood, B., Pitt, K., and Rogers, P. (2001). Morphoanalysis of bacterially precipitated subaqueous calcium carbonate from Weebubbie Cave, Australia. Geomicrobiol. J. *18*, 331–343.

Cuevas, C., and Francesch, A. (2009). Development of Yondelis (trabectedin, ET-743). A semisynthetic process solves the supply problem. Nat. Prod. Rep. *26*, 322–337.

Cuezva, S., Sanchez-Moral, S., Saiz-Jimenez, C., and Cañaveras, J.C. (2009). Microbial communities and associated mineral fabrics in Altamira Cave, Spain. Int. J. Speleol. 38, 83–92.

Cuezva, S., Fernandez-Cortes, A., Porca, E., Pasic, L., Jurado, V., Hernandez-Marine, M., Serrano-Ortiz, P., Cañaveras, J.C., Sanchez-Moral, S., and Saiz-Jimenez, C. (2012). The biogeochemical role of *Actinobacteria* in Altamira Cave, Spain. FEMS Microbiol. Ecol. *81*, 281–290.

De los Ríos, Bustillo, M.A., and Ascaso, C. (2011). Bioconstructions in ochreous speleothems from lava tubes on Terceira Island (Azores). Sediment. Geol. *236*, 117–128.

Dhanjal, S., Ruckmani, A., Cameotra, S.S., Pukall, R., Klenk, H.-P., and Mayilra, S. (2011a) *Yaniella fodinae* sp. nov., isolated from a coal mine. Int. J. Syst. Evol. Microbiol. *61*, 535–539.

Dhanjal, S., Kaur, I., Korpole, S., Schumann, P., Cameotra, S.S., Pukall, R., Klenk, H.-P., and Mayilraj, S. (2011b) *Agrococcus carbonis* sp. nov., isolated from soil of a coal mine. Int. J. Syst. Evol. Microbiol. *61*, 1253–1258.

Engel, A.S., Porter, M.L., and Kinkle, B.K. (2001). Ecological assessment and geological significance of microbial communities from Cesspool Cave, Virginia. Geomicrobiol. J. *18*, 259–274.

Everest, G.J., Curtis, S.M., De Leo, F., Urzì, C., and Meyers, P.R. (2013). *Kribbella albertanoniae* sp. nov., isolated from a Roman catacomb, and emended description of the genus *Kribbella*. Int. J. Syst. Evol. Microbiol. *63*, 3591–3596.

Feling, R.H., Buchanan, G.O., Mincer, T.J., Kauffman, C.A., Jensen, P.R., and Fenical, W. (2003). Salinosporamide A: a highly cytotoxic proteasome inhibitor from a novel microbial source, a marine bacterium of the new genus *Salinospora*. Angew. Chem. Int. Ed. *42*, 355–357.

Gonzalez, I., Laiz, L., Hermosin, B., Caballero, B., Incerti, C., and Saiz-Jimenez, C. (1999). Bacteria isolated from rock art paintings: the case of Atlanterra shelter (south Spain). J. Microbiol. Methods *36*, 123–127.

Groth, I., and Saiz-Jimenez, C. (1999). Actinomycetes in hypogean environments. Geomicrobiol. J. *16*, 1–8.

Groth, I., Vetermann, R., Schuetze, B., Schumann, P., and Saiz-Jimenez, C. (1999a). Actinomycetes in kartic caves of Northern Spain (Altamira and Tito Bustillo). J. Microbiol. Methods. *36*, 115–122.

Groth, I., Schumann, P., Schüetze, B., Augsten, K., Kramer, I., and Stackebrandt, E. (1999b). *Beutenbergia cavernae* gen. nov., sp. nov., an L-lysine-containing actinomycete isolated from a cave. Int. J. Syst. Evol. Microbiol. *49*, 1733–1740.

Groth, I., Schumann, P., Laiz, L., Sanchez–Moral S., Cañaveras, J.C., and Saiz-Jimenez, C. (2001). Geomicrobiological study of the Grotta dei Cervi, Porto Badisco, Italy. Geomicrobiol. J. *18*, 241–258.

Groth, I., Schumann, P., Schütze, B., Augsten, K., and Stackebrandt, E. (2002). *Knoellia sinensis* gen. nov., sp. nov. and *Knoellia subterranea* sp. nov., two novel actinobacteria isolated from a cave. Int. J. Syst. Evol. Microbiol. 52, 77–84.

Hamdali, H., Virolle, M.J., von Jan, M., Spröer, C., Klenk, H.-P., and Ouhdouch, Y. (2011). *Streptomyces youssoufiensis* sp. nov., isolated from a Moroccan phosphate mine. Int. J. Syst. Evol. Microbiol. *61*, 1104–1108.

Herold, K., Gollmick, F.A., Groth, I., Roth, M., Menzel, K.D., Möllmann, U., Gräfe, U., and Hertweck, C. (2005). Cervimycin A-D: a polyketide glycoside complex from a cave bacterium can defeat vancomycin resistance. Chemistry *11*, 5523–5530.

Hill, C.A., and Forti, P. (1997). Cave Minerals of the World, 2nd edition (National Speleological Society, Huntsville, AL).

Hose, L.D., Palmer, A.N., Palmer, M.V., Northup, D.E., Boston, P.J., and DuChene, H.R. (2000). Microbiology and geochemistry in a hydrogen-sulphide-rich karst environment. Chem. Geol. *169*, 399–423.

Jurado, V., Groth, I., Gonzalez, J.M., Laiz, L., and Saiz-Jimenez, C. (2005a). *Agromyces salentinus* sp. nov. and *A. neolithicus* sp. nov. Int. J. Syst. Evol. Microbiol. 55, 153–157.

Jurado, V., Groth, I., Gonzalez, J.M., Laiz, L., Schuetze, B., and Saiz-Jimenez, C. (2005b). *Agromyces italicus*, sp. nov., *Agromyces humatus*, sp. nov., and *Agromyces lapidis*, sp. nov., isolated from Roman catacombs. Int. J. Syst. Evol. Microbiol. 55, 871–875.

Jurado, V., Groth, I., Gonzalez, J.M., Laiz, L., and Saiz-Jimenez, C. (2005c). *Agromyces subbeticus* sp. nov., isolated from a cave in southern Spain. Int. J. Syst. Evol. Microbiol. 55, 1897–1901.

Jurado, V., Laurent, F., Boiron, P., Couble, A., Laiz, L., Gonzalez, J.M., Saiz-Jimenez, C., Pötter, G., Mouniée, D., Kroppenstedt, R.M., and Rodríguez-Nava, V. (2008). *Nocardia altamirensis* sp. nov., isolated from Altamira cave, Cantabria, Spain. Int. J. Syst. Evol. Microbiol. 58, 2210–2214.

Jurado, V., Kroppenstedt, R.M., Saiz-Jimenez, C., Klenk, H.-P., Mouniée, D., Laiz, L., Couble, A., Pötter, G., Boiron, P., and Rodríguez-Nava, V. (2009). *Hoyosella altamirensis* gen. nov., sp. nov., a new member of the order Actinomycetales isolated from a cave biofilm. Int. J. Syst. Evol. Microbiol. 59, 3105–3110.

Jurado, V., Laiz, L., Rodriguez-Nava, V., Boiron, P., Hermosin, B., Sanchez-Moral, S., and Saiz-Jimenez, C. (2010). Pathogenic and opportunistic microorganisms in caves. Int. J. Speleol. 39, 15–24.

Kelly, L.C., Cockell, C.S., Herrera-Belaroussi, A., Piceno, Y., Andersen, G., DeSantis, T., Brodie, E., Thorsteinsson, T., Marteinsson, V., Poly, F., and LeRoux, X. (2011). Bacterial diversity of terrestrial crystalline volcanic rocks, Iceland. Microb. Ecol. 62, 69–79.

Laiz, L., Groth, I., Gonzalez, I., and Saiz-Jimenez, C. (1999). Microbiological study of the dripping waters in Altamira cave (Santillana del mar, Spain). J. Microbiol. Methods 36, 129–138.

Lam, K.S. (2006). Discovery of novel metabolites from marine actinomycetes. Curr. Opin. Microbiol. 9, 245–251.

Lamprinou, V., Hernández-Mariné, M., Canals, T., Kormas, K., Economou-Amilli, A., and Pantazidou, A. (2011). Morphology and molecular evaluation of *Iphinoe spelaeobios* gen. nov., sp. nov. and *Loriellopsis cavernicola* gen. nov., sp. nov., two stigonematalean cyanobacteria from Greek and Spanish caves. Int. J. Syst. Evol. Microbiol. 61, 2907–2915.

Lamprinou, V., Skaraki, K., Kotoulas, G., Economou-Amilli, A., and Pantazidou, A. (2012). *Toxopsis calypsus* gen. nov., sp. nov. (*Cyanobacteria*, Nostocales) from cave 'Francthi', Peloponnese, Greece: a morphological and molecular evaluation. Int. J. Syst. Evol. Microbiol. 62, 2870–2877.

Lee, N.M., Meisinger, D.B., Aubrecht, R., Kovacik, L., Saiz-Jimenez,C., Baskar, S., Baskar, R., Liebl,W., Porter, M.L., and Engel, A.S. (2012). Caves and karst environments. In Life at Extremes: Environments, Organisms and Strategies for Survival, E.M. Bell, ed. (CAB International, Oxfordshire, UK), pp. 320–344.

Lee, S.D. (2013a). *Spelaeicoccus albus* gen. nov., sp. nov., an actinobacterium isolated from a natural cave. Int. J. Syst. Evol. Microbiol. 63, 3958–3963.

Lee, S.D. (2013b). *Humibacter antri* sp. nov., an actinobacterium isolated from a natural cave, and emended description of the genus *Humibacter.* Int. J. Syst. Evol. Microbiol. 63, 4315–4319.

Léveillé, R.J., Longstaffe, F.J., and Fyfe, W.P. (2002). Kerolite in carbon-rich speleothems and microbial deposits from basaltic caves, Kuai, Hawaii. Clay Clay Min. 50, 514–524.

Li, Y., Chen, F., Dong, K., and Wang, G. (2013). *Actinotalea ferrariae* sp. nov., isolated from an iron mine, and emended description of the genus *Actinotalea.* Int. J. Syst. Evol. Microbiol. 63, 3398–3403.

Liu, H., Wang, H., and Wang, G. (2012). *Intrasporangium chromatireducens* sp. nov., a chromate-reducing actinobacterium isolated from manganese mining soil, and emended description of the genus *Intrasporangium.* Int. J. Syst. Evol. Microbiol. 62, 403–408.

Lin, Y.B., Wang, X.Y., Li, H.F., Wang, N.N., Wang, H.X., Tang, M., and Wei, G.-H. (2011). *Streptomyces zinciresistens* sp. nov., a zinc-resistant actinomycete isolated from soil from a copper and zinc mine. Int. J. Syst. Evol. Microbiol. 61, 616–620.

Manivasagan, P., Venkatesan, J., Sivakumar, K., and Kim, S.K. (2013). Marine actinobacterial metabolites: current status and future perspectives. Microbiol. Res. 168, 311–332.

Margesin, R., Schumann, P., Spröer, C., and Gounot, A.-M. (2004). *Arthrobacter psychrophenolicus* sp. nov., isolated from an alpine ice cave Int. J. Syst. Evol. Microbiol. 54, 2067–2072.

Martin-Sanchez, P.M., Nováková, A., Bastian, F., Alabouvette, C., and Saiz-Jimenez, C. (2012). Two new species of the genus *Ochroconis, O. lascauxensis* and *O. anomala* isolated from black stains in Lascaux Cave, France. Fungal Biol. 116, 574–589.

Melim, L.A., Shinglman, K.M., Boston, P.J., Northup, D.E., Spilde, M.N., and Queen, J.M. (2001). Evidence for microbial involvement in pool finger precipitation, Hidden Cave, New Mexico. Geomicrobiol. J. 18, 311–329.

Melim, L.A., Northup, D.E., Spilde, M.N., Jones, B., Boston, P.J., and Bixby, R.J. (2008). Reticulated filaments in cave pool speleothems: microbe or mineral? J. Cave Karst Stud. 70, 135–141.

Miller, A.Z., Dionísio, A., Sequeira Braga, M.A., Hernández-Mariné, M., Afonso, M.J., Muralha, V.S.F., Herrera, L.K., Raabe, J., Fernandez-Cortes, A., Cuezva, S., *et al.* (2012a). Biogenic Mn oxide minerals coating in a subsurface granite environment. Chem. Geol. 322–323, 181–191.

Miller, A.Z., Hernandez-Marine, M., Jurado, V., Dionísio, A., Barquinha, P., Fortunato, E., Afonso, M.J., Chaminé, H.I., and Saiz-Jimenez, C. (2012b). Enigmatic reticulated filaments in subsurface granite. Environ. Microbiol. Rep. 4, 596–603.

Miller, A.Z., Pereira, M.F.C., Calaforra, J.M., Forti, P., Dionisio, A., and Saiz-Jimenez, C. (2014). Siliceous speleothems and associated microbe–mineral interactions from Ana Heva lava tube in Easter Island (Chile). Geomicrobiol. J. 31, 236–245.

Montano, E., and Henderson, L.O. (2013). Studies of antibiotic production by cave bacteria. In Cave Microbiomes: A Novel Resource for Drug Discovery, N. Cheeptham, ed. (Springer, London, UK), pp. 109–130.

Nakaew, N., Pathom-aree, W., and Lumyong, S. (2009a). First record of the isolation, identification and biological activity of a new strain of *Spirillospora albida* from Thai cave soil. Actinomycetologica 23, 1–7.

Nakaew, N., Pathom-aree, W., and Lumyong, S. (2009b). Generic diversity of rare actinomycetes from Thai cave soils and their possible use as new bioactive compounds. Actinomycetologica *23*, 21–26.

Nakaew, N., Sungthong, N., Yokota, A., and Lumyong, S. (2013). *Nonomuraea monospora* sp. nov., an actinomycete isolated from cave soil in Thailand, and emended description of the genus *Nonomuraea*. Int. J. Syst. Evol. Microbiol. *62*, 3007–3012.

Northup, D.E., Barns, S.M., Yu, L.E., Spilde, M.N., Schelble, R.T., Dano, K.E., Crossey, L.J., Connolly, C.A., Boston, P.J., Natvig, D.O., and Dahm, C.N. (2003). Diverse microbial communities inhabiting ferromanganese deposits in Lechuguilla and Spider Caves. Environ. Microbiol. *5*, 1071–1086.

Northup, D.E., Connolly, C.A., Trent, A., Peck, V.M., Spilde, M.N., Welbourn, W.C., and Natvig, D.O. (2004). The nature of bacterial communities in Four Windows Cave, El Malpais National Monument, New Mexico, USA. AMCS Bull. *19*, 119–125.

Northup, D.E., Melim, L.A., Spilde, M.N., Hathaway, J.J., Garcia, M.G., Moya, M., Stone, F.D., Boston, P.J., Dapkevicius, M.L., and Riquelme, C. (2011). Lava cave microbial communities within mats and secondary mineral deposits: Implications for life detection on other planets. Astrobiology *11*, 1–18.

Nováková, A., Hubka, V., Saiz-Jimenez, C., and Kolarik, M. (2012). *Aspergillus baeticus* sp. nov. and *Aspergillus thesauricus* sp. nov., two species in section *Usti* from Spanish caves. Int. J. Syst. Evol. Microbiol. *62*, 2778–2785.

Roldán, M., Ramírez, M., del Campo, J., Hernández-Mariné, M., and Komárek, J. (2013). *Chalicogloea cavernicola* gen. nov., sp. nov. (Chroococcales, Cyanobacteria), from low-light aerophytic environments: combined molecular, phenotypic and ecological criteria. Int. J. Syst. Evol. Microbiol. *63*, 2326–2333.

Rule, D., and Cheeptham, N. (2013). Effects of UV light on antimicrobial activity of cave actinomycete metabolites. Int. J. Speleol. *42*, 147–153.

Saiz-Jimenez, C. (2013). Cave Conservation: A Microbiologist's Perspective. In Cave Microbiomes: A Novel Resource for Drug Discovery, N. Cheeptham, ed. (Springer, London, UK), pp. 69–84.

Saiz-Jimenez, C., Cuezva, S., Jurado, V., Fernandez-Cortes, A., Porca, E., Benavente, D., Cañaveras, J.C., and Sanchez-Moral, S. (2011). Paleolithic art in peril: Policy and science collide at Altamira Cave. Science *334*, 42–43.

Saleh, A.M., El-Refai, H.A., Abdulla, H., Hanna, A.G., Hashem, A.M., and El-Refai, A.H. (2013). Bioprospecting of *Streptomyces* sp. MECO$_2$ from stones of an ancient Egyptian tomb with promising antitrichophyton activity under optimized cultivation conditions. World Appl. Sci. J. *24*, 1130–1140.

Schabereiter-Gurtner, C., Saiz-Jimenez, C., Piñar, G., Lubitz, W., and Rölleke, S. (2002a). Altamira cave paleolithic paintings harbour partly unknown bacterial communities. FEMS Microbiol. Lett. *211*, 7–11.

Schabereiter-Gurtner, C., Saiz-Jimenez, C., Piñar, G., Lubitz, W., and Rölleke, S. (2002b). Phylogenetic 16S rRNA analysis reveals the presence of complex and partly unknown bacterial communities in Tito Bustillo cave, Spain, and on its Paleolithic paintings. Environ. Microbiol. *4*, 392–400.

Schatz, A., Bugle, E., and Waksman, S.A. (1944). Streptomycin, a substance exhibiting antibiotic activity against Gram-positive and Gram-negative bacteria. Exp. Biol. Med. *55*, 66–69.

Schirmack, J., Mangelsdorf, K., Ganzert, L., Sand, W., Hillebrand-Voiculescu, A., and Wagner, D. (2013). *Methanobacterium movilense* sp. nov., a hydrogenotrophic, secondary alcohol utilizing methanogen from the anoxic sediment of the subsurface lake in Movile Cave, Romania. Int. J. Syst. Evol. Microbiol. *63*, doi: 10.1099/ijs.0.057224–0.

Shi, Z., Luo, G., and Wang, G. (2012). *Cellulomonas carbonis* sp. nov., isolated from coal mine soil. Int. J. Syst. Evol. Microbiol. *62*, 2004–2010.

Spear, J.R., Barton, H.A., Robertson, C.E., Francis, C.A., and Pace, N.R. (2007). Microbial community biofabrics in a geothermal mine adit. Appl. Environ. Microbiol. *73*, 6172–6180.

Spilde, M.N., Northup, D.E., Boston, P.J., Schelble, R.T., Dano, K.E., Crossey, L.J., and Dahm, C.N. (2005). Geomicrobiology of cave ferromanganese deposits: a field and laboratory investigation. Geomicrobiol. J. *22*, 99–116.

Subramani, R., and Aalbersberg, W. (2012). Marine actinomycetes: an ongoing source of novel bioactive metabolites. Microbiol. Res. *167*, 571–580.

Subramani, R., and Aalbersberg, W. (2013). Culturable rare Actinomycetes: diversity, isolation and marine natural product discovery. Appl. Microbiol. Biotechnol. *97*, 9291–9321.

Sugita, T., Kikuchi, K., Makimura, K., Urata, K., Someya, T., Kamei, K., Niimi, M., and Uehara, Y. (2005). *Trichosporon* species isolated from guano samples obtained from bat-inhabited caves in Japan. Appl. Environ. Microbiol. *71*, 7626–7629.

Tatar, D., Sazak, A., Guven, K., Cetin, D., and Sahin, N. (2013). *Amycolatopsis cihanbeyliensis* sp. nov., a halotolerant actinomycete isolated from a salt mine. Int. J. Syst. Evol. Microbiol. *63*, 3739–3743.

Tiwari, K., and Gupta, R.K. (2012). Rare actinomycetes: a potential storehouse for novel antibiotics. Crit. Rev. Biotechnol. *32*, 108–132.

Tiwari, K., and Gupta, R.K. (2013). Diversity and isolation of rare actinomycetes: an overview. Crit. Rev. Microbiol. *39*, 256–294.

Tomova, I., Lazarkevich, I., Tomova, A., Kambourova, M., and Vasileva-Tonkova, E. (2013). Diversity and biosynthetic potential of culturable aerobic heterotrophic bacteria isolated from Magura Cave, Bulgaria. Int. J. Speleol. *42*, 65–76.

Weidler, G.W., Dornmayr-Pfaffenhuemer, M., Gerbl, F.W., Heinen, W., and Stan-Lotter, H. (2007). Communities of archaea and bacteria in a subsurface radioactive thermal spring in the Austrian Central Alps, and evidence of ammonia-oxidizing Crenarchaeota. Appl. Environ. Microbiol. *73*, 259–270.

Williams, S.T., Locci, R., Beswick, A., Kurtböke, D.I., Kutzetsov, V.D., Le Monnier, F.J., Long, P.F., Maycroft, K.A., Palma, R.A., Petrolini, B., *et al.* (1993).

Detection and identification of novel actinomycetes. Res. Microbiol. *144*, 653–656.

Yamaç, M., Işik, K., and Şahin, N. (2011). Numerical classification of streptomycetes isolated from karstic caves in Turkey. Turkish J. Biol. *35*, 473–484.

Yücel, S., and Yamaç, M. (2010). Selection of *Streptomyces* isolates from Turkish karstic caves against antibiotic resistant microorganisms. Pak. J. Pharm. Sci. *23*, 1–6.

Zhang, L., An, R., Wang, J., Sun, N., Zhang, S., Hu, J., and Kuai, J. (2005). Exploring novel bioactive compounds from marine microbes. Curr. Opin. Microbiol. *8*, 276–281.

Zhou, J., Gu, Y., Zou, C., and Mo, M. (2007). Phylogenetic diversity of bacteria in an earth-cave in Guizhou province, southwest of China. J. Microbiol. *45*, 105–112.

Animal Venoms as Natural Sources of Antimicrobials

13

Ramar Perumal Samy, Seetharama Satyanarayanajois, Octavio L. Franco, Bradley G. Stiles and Ponnampalam Gopalakrishnakone

Abstract

A number of naturally occurring proteins/peptides exert antimicrobial activities reported throughout the literature, of which snake venoms (SV) represent a vast natural source of protein/peptides not thoroughly explored to date. Snake venoms represent rich sources of bioactive compounds, which are produced by venom glands located around the snake's jawbone. In this review, we focus more on the basis of antimicrobial potential within SV and further need to search for novel antibiotic prototypes. Several enzymes [i.e. phospholipase A_2 (PLA_2) (these are part of PLA_2) L-amino acid oxidase and metalloproteinase], as well as antimicrobial peptides (AMPs) such as cathelicidine and defensin, have been isolated by various groups from SV. Antimicrobial proteins/peptides work in various ways that include hydrolysing phospholipids on the bacterial surface. The presence of unusual amino acids and structure motifs in AMPs confer unique structural properties that contribute their specific mode of action. The ability of these active AMPs to act as multifunctional effectors such as signalling molecules and antibacterial agents makes them interesting candidates for structural and biological studies for prophylactic and therapeutic applications. In this review, we focus on the diversity and antimicrobial activities of various SV-derived molecules potentially useful as drug candidates for the pharmaceutical industry.

Introduction

Staphylococcus aureus causes a variety of infections from suppurative disease to food poisoning, pneumonia, eczema, scabies, sepsis, septic shock, skin and wound infection, etc. (Talan *et al.*, 2008). Methicillin-resistant *S. aureus* (MRSA) is a major cause of infection in health care, and with the emergence of vancomycin-resistant *S. aureus* (VRSA) there is clearly a serious problem that will extend well into the future (Komatsuzawa *et al.*, 2006; Appelbaum, 2006; Moran *et al.*, 2006, 2008; Loomba *et al.*, 2010; Talan *et al.*, 2011; Terp *et al.*, 2014). Furthermore, existing antibiotics fail to improve the mortality rate of multidrug-resistant Gram-negative *Burkholderia pseudomallei* that cause melioidosis (Leearasamee *et al.*, 1970; Dance, 1999; Schweizer, 2012; Limmathurotsakul *et al.*, 2014). In addition, multidrug-resistant (MDR) bacteria are generated by specific mutations on antibiotic targets, and acquired by transposon genes (Galbusera *et al.*, 2011), are one of the important healthcare concerns for long-term management. Antibiotic resistance has been of great concern during the last two decades due to extensive, often unwarranted use of antibiotics (Bonomo, 2000; Rubina and Jeyakumar, 2013). This is especially so for the antibiotics commonly used in treating *S. aureus*-infected surgical wounds and other skin-related infections, such as methicillin (Desbois *et al.*, 2010), vancomycin (Micek, 2007), glycopeptides (Kirby *et al.*, 2011), and β-lactams (Milheirico *et al.*, 2011), etc. However, the bacteria have also developed resistance to cationic antimicrobial peptides (AMPs) such as polylysine, protamine sulfate, human neutrophil (HNP-1) and polymyxins (Jones *et al.*, 1996; Ganz and Lehrer, 1997). Besides, more than 800 AMPs have already been identified in all species that include bacteria, fungi, amphibians, fish, and insects (Rozek *et al.*, 2000; Trabi *et al.*, 2001; Perez-Trallero and Iglesias, 2003; Ganz, 2003; Boman, 2003). However, AMPs generally possess broad antimicrobial activity but also haemolytic activity towards human erythrocytes. After six decades of widespread antibiotic use, bacterial pathogens have generally become increasingly resistant to many antimicrobial agents (WHO, 2001; Burke, 2000). This sobering reality should motivate many towards searching for novel bioactive compounds. Animal venoms naturally constitute one of the richest sources of pharmacologically active substances found in nature (Stiles *et al.*, 1991; Hancock and Lehrer, 1998). Current research highlights animal venoms as a platform for human drugs and translating toxins into therapeutics (King, 2011). Therefore, it is important to search and develop novel, safe and more effective antimicrobials with potent action that could effectively stunt the emergence of MDR pathogens.

Snake venom and its diverse molecules

Snakes are represented on earth today by at least 3150 species (Vidal *et al.*, 2007). The majority of these snakes are found in the superfamily Colubridea that includes the families of Elapidae (cobras, kraits, coral and sea snakes) and Viperidae (vipers and pit vipers) (Fry and Wuster, 2004). Snake venoms are a mixture of proteins and peptides (~90–95%), as well as the remainder being amino acids, nucleotides, lipids, carbohydrates and metallic elements (Russell, 1980; Tu, 1988; Heise *et al.*, 1995; Fry and Wuster, 2004). Snake venoms are an extremely rich source of biologically active proteins and peptides, with considerable clinical and medical potential (Fig. 13.1). As one example, vipers produce a complex mixture of many distinct proteins that number more than 100 (Fox and Serrano, 2005; Serrano *et al.*, 2005; Gomes *et al.*, 2007). These proteins include enzymes (serine proteinases, Zn2C-metalloproteases, L-amino acid oxidase, group II PLA_2) and proteins without enzymatic activity (disintegrins, C-type lectins, natriuretic peptides, myotoxins, cysteine-rich secretory protein (CRISP) toxins, nerve and vascular endothelium growth factors, cystatin, and Kunitz-type protease inhibitors. These molecules are produced by specialized glands around the snake's jawbone and evolutionarily related to salivary glands that possess toxins for subjugating prey (Kochva, 1987; de Oliveira Junior *et al.*, 2013). The molecular imaging tools of X-ray crystallography and nuclear magnetic resonance spectroscopy, key techniques towards developing rationally designed therapeutic agents (Bailey and Wilce, 2001).

Therapeutic values of snake venom molecules (SVM)

Animal venoms have a long history as being a rich source of medical treatments for various maladies (King, 2011, 2013). As just one example, snake venoms have been used in Ayurvedic medicine to treat arthritis and gastrointestinal ailments since the seventh century. In particular, cobra venom has been used to treat asthma, polio, multiple sclerosis, rheumatism and pain (Reid, 2007), During the modern era, major life-savers of novel venom-based drug discovery begins in 1970 with the development of the blockbuster anti-hypertensive drug captopril, based on a peptide from the venom of the Brazilian viper *Bothrops jaracaca* (Cushman *et al.*, 1991). Eptifibatidie (exenatide) and some of the best selling peptide drugs are derived from toxins in snake venom. Currently, animal (snake) toxins are the source of 10–20 medications that treat a myriad of maladies that include heart attack, diabetes, chronic pain and hypertension. Toxins are also the source of 10–15 clinical diagnostics for mostly haemostatic testing and wrinkle-smoothening cosmetics. Several agents from animal venoms such as the cone snail (ώ-conotoxin CVID used in medicine for chronic pain (Lewis and Garcia, 2003), sea anemone (Shk186) for autoimmune disease (phase Ia), spider (PcTx1) for pain/stroke (preclinical), centipede (Smlea) for pain (pre-clinical), additionally some of the disulfide rich venom peptides under preclinical trials. Clearly, the bites of certain snakes can be deadly, but these same venoms also contain diverse components of medical, biotechnological and pharmaceutical importance (Calvete *et al.*, 2007). The proteomic characterization of snake venom molecules, snake venomics, has

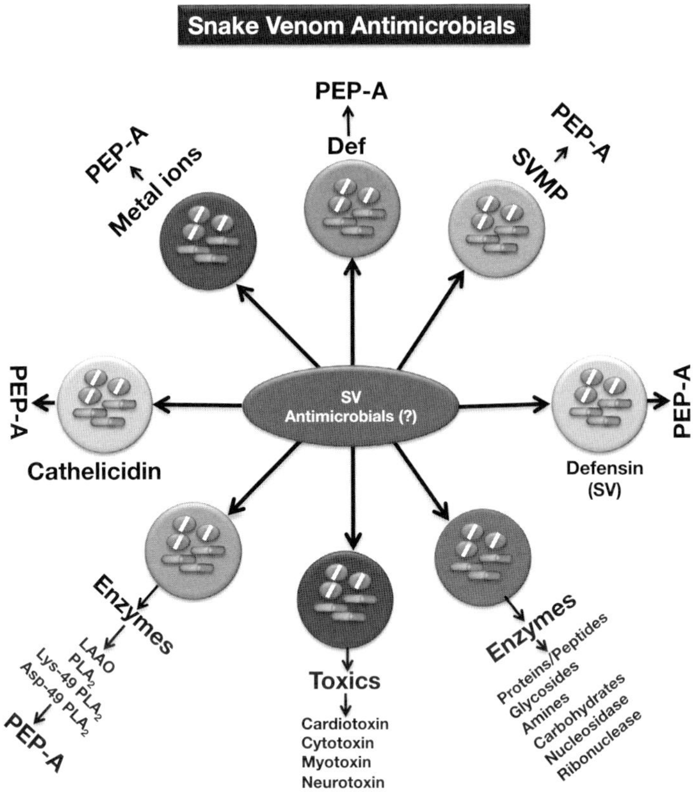

Figure 13.1 Snake venom contains a mixture of diverse enzymes that include phospholipase A_2, L-amino acid oxidase, nucleosidase, ribonuclease plus toxic (cardiotoxin, cytotoxin, myotoxin, neurotoxin) and non-toxic activities (antimicrobials) molecules considerably clinically and medically potential. Symbols denote: SV- snake venom antibiotics, CV-crude venom, SVMP-snake venom, PLA_2-phospholipase A_2 enzymes, Def-snake defensins, Dist-disintegrins, LAAO-L-amino acid oxidase, Pep-A –peptide antimicrobials, etc. A colour version of this figure is available in the plate section at the back of the book.

thus a number of potential benefits for basic research, clinical diagnosis and development of new research tools and drugs of potential clinical use (Calvete *et al.*, 2007). It is also relevant to obtain a deeper understanding of the evolution and biological effects of venoms, and to generate immunization protocols that elicit toxin-specific antibodies with greater specificity and effectiveness than conventional systems used for over a century (i.e. vaccination of horses with crude venom to generate anti-venom). However, the antimicrobial potential has not been fully explored, mainly because of our incomplete knowledge of the venom proteome and pharmacological properties of its many components, in particular those devoid of enzymatic activity (Georgieva *et al.*, 2008). Recently, more than 100 peptide-based drugs have reached the market (Craik *et al.*, 2013). Protein and peptide-based drugs are currently estimated at >40 billion US dollars in the pharmaceutical market (Danho *et al.*, 2009). This market share is growing much faster than that of other pharmaceuticals, and the success rates for bringing biologics to market are now about twice that of small molecule drugs (Vlieghe *et al.*, 2010). To promote peptide-based therapies, several exciting new technologies offer great potential for generating cyclic peptides (Katoh *et al.*, 2011). The inability of many current antibiotics to kill select pathogens affords society an opportunity to discover novel, antimicrobial strategies. We believe this opportunity is an important part of societies' fight to defeat multidrug resistance in the twenty-first century.

Antimicrobial potency of snake venom molecules

Antibiotics are a heterogeneous group of molecules, with various modes of action, produced by several organisms such as bacteria, fungi and others (Chopra *et al.*, 2002). The synthetic antimicrobials are produced by the pharmaceutical industry and display different mechanisms of action and a broad antibacterial spectrum (Harris and Thorarensen, 2004). Control of various pathogens was significantly increased by the introduction of sulfonamides and penicillin around 75 years ago (Brown, 2004). Because of increasing resistance among various pathogen populations, it is essential to discover novel antimicrobial agents from natural sources (Doley and Kini, 2009). Several venom compounds have been used for antimicrobial effects (de Oliveira Junior *et al.*, 2013; Georgieva *et al.*, 2008).

In vitro antimicrobial effects of SVM

Current advancement of proteomics has enabled venom complexity to become, less complex and involves identification of many proteins/peptides unknown just a few years ago. Several venom components such as enzymes (PLA_2, LAAO, metalloproteinase, disintegrins) and peptides (cathelicidine, defensin and others) were isolated from snake venoms. For instance, roughly 30 proteins have been identified with the molecular weight ranges of 7–110 kDa from venoms of *Bothrops cotiara* and *B. fonsecai* (Tashima *et al.*, 2008). These proteins and peptides are similar in structure but not identical to prey physiological systems (the venom proteins/peptides are able to better bind/exploit prey receptors for these toxins). On the other hand, envenomation is a process associated with a low incidence of bacterial infection due to antibacterial molecules in snake venoms that would protect the snake during feeding (De Lima *et al.*, 2005). Furthermore, having antimicrobial factors in venom would help protect this protein rich source from becoming contaminated or spoiled by proteolytic bacteria. Such an event would diminish venom potency for prey, thus making the snake 'weaker'.

Snake venom and antimicrobial activity

Several antimicrobial studies involving venom of snakes has been reported in the literature (Perumal Samy *et al.*, 2007, 2008, Sachidananda *et al.*, 2007; San *et al.*, 2010; Samy *et al.*, 2011a; Shebl *et al.*, 2012; Soares, 2012). Heavy oral and fang contamination of crotalid species contains a wide variety of potentially pathogenic bacteria (Talan *et al.*, 1991). Further studies demonstrate that venom have potent activity (MIC, less than or equal to 80 mg/ml) against *Staphylococcus, Pseudomonas, Enterobacter, Citrobacter, Proteus* and *Morganella* species of clinical isolates. Crotalid venoms are broadly active against aerobic Gram-negative and Gram-positive bacteria. These activities perhaps play an important role in the low incidence of infection after envenomation injuries (Talan *et al.*, 1991). Comparison of venoms bactericidal potency, elapid venom (*N. naja sputatrix*) shows greater activity than viperid-derived venoms of *V. russellii* or *C. adamanteus*. Electron microscopic examination reveals that extensive bacterial membrane (cell wall) damage caused by *N. naja sputatrix* venom (Stocker and Traynor, 1986). *Bothrops alternates* venom has more bactericidal effects against *E. coli* and *S. aureus* than *P. aeruginosa* and *Enterococcus faecalis*. More studies need to be done to identify the active component(s) of *B. alternates* venom. Even at the low concentrations assayed (0.312 mg/ml), this venom possesses significant bactericidal activity against *S. aureus* but it was ineffective against *E. coli* (Bustillo *et al.*, 2008). In our previous study, more than thirty snake venoms were tested for antibacterial activity by using a disc diffusion method at a 100 µg/ml concentration. Among the tested venoms, *Daboia russelli russelli, Crotalus adamanteus, Naja sumatrana, Pseudechis guttata, Agkistrodon halys, Acanthophis praelongus* and *D. russelli siamensis* displayed promising activity against four different pathogenic bacteria (Fig. 13.2). However, *D. russelli russelli* and *P. australis* venom exert the highest level of inhibitory activity against various bacteria that include *S. aureus, Burkholderia pseudomallei* (KHW/TES) *Proteus mirabilis, P. vulgaris, Enterobacter aerogenes, P. aeruginosa* and *E. coli* (Samy *et al.*, 2011b) while other venoms had relatively weak activity. However, there was a stronger

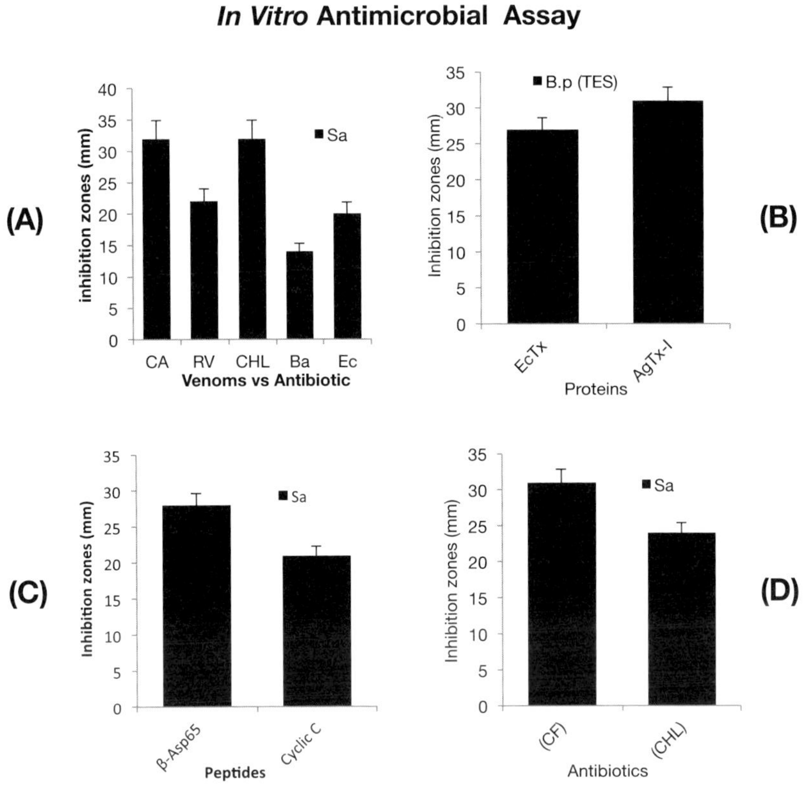

Figure 13.2 Antimicrobial effect of snake venoms, proteins (svPLA$_{2s}$) and venom derived synthetic peptides from natural sources of animal venoms specifically tested against Gram-negative and Gram-positive bacteria. (A) Antibacterial effect of snake venoms and its potency was compared within the venoms, (B) comparison of protein and peptides bioactivity, peptide showed higher antimicrobial action, (C) antimicrobial effect of peptides were equally effective compared with commercial antibiotics (chloramphenicol and ceftazidime).

inhibitory effect of Viperidae venoms with minimum inhibitory concentrations (MICs) of 20 μg/ml and 40 μg/ml of Elapidae venoms compared to the commercial antibiotics like chloramphenicol, streptomycin and penicillin, etc. (Perumal Samy *et al.*, 2006, 2007). Cumulative results suggest it worthwhile to further explore snake venoms as a source of novel antimicrobials to treat pathogens, perhaps those that are MDR.

Snake venom metalloproteases (SVMPs)

Metalloproteinases are abundant enzymes in crotalidae and viperidae snake venoms (Mazzi *et al.*, 2004; Perumal Samy *et al.*, 2008). Snake venom metalloproteinases (SVMPs) comprise a family of zinc-dependent enzymes that display a variety of biological properties that include haemorrhagic and necrosis can cause tissue damage (Mazzi *et al.*, 2004; Moura-da-Silva *et al.*, 2011). The structure and function of SVMPs are phylogenetically most closely related to mammalian ADAM (a disintegrin and metalloproteinase) and ADAMTS (ADAM with thrombospondin type-1 motif) family of proteins and together constitute the M12B clan of metalloendopeptidases (Lou *et al.*, 2005; Takeda *et al.*, 2012). The functional diversity of SVMPs is in part due to the structural organization of different combinations of catalytic, disintegrin, disintegrin-like and cysteine-rich domains, which causes SVMPs to be classified into three classes of precursor molecules (PI, PII and PIII) further divided into 11 subclasses, of which six belong to the PII group (Ramos and Selistre-de-Araujo, 2006; Markland and Swenson, 2013). Recent insights into snake biology with new technology (proteomics) allow high throughput screening of venoms (Vonk *et al.*, 2011), yielding attractive protein-based antimicrobials (de Oliveira Junior *et al.*, 2013). In our previous study, *Agkistrodon halys* metalloproteinase (AHM) was identified from (Chinese viper) snake venoms. The toxin consists of a single polypeptide chain, and the N-terminal amino acid sequence is similar to other venom metalloproteinases isolated from the Viperidae family. AHM has the most potent antibacterial activities against *S. aureus*, *Proteus vulgaris*, *P. mirabilis* and *MDR Burkholderia pseudomallei* strains of KHW at 120 a μM concentration. It also exerted diverse (*Proteus vulgaris* and *P. mirabilis*) bacteriostatic action (MIC 1.8–60 μM) that was less toxic (versus a 120 μM dose) to human macrophages (Perumal Samy *et al.*, 2008).

Antibacterial effect of amino acid oxidase (LAAO)

Snake venom L-amino acid oxidases (LAAOs) are flavoproteins that catalyse the oxidative deamination of L-amino acids, converting them into keto acids, ammonia and hydrogen peroxide (Du and Clemetson, 2002). The LAAOs are found in various snake venoms that include *B. alternatus* (Stabeli *et al.*, 2004) and *B. pirajai* (Izidoro *et al.*, 2006). LAAOs have become attractive molecules in biology and medicine due to their biological actions on various eukaryotic and prokaryotic cells (Kang *et al.*, 2011; Guo *et al.*, 2012). Snake venom LAAOs not only play an important role in venom toxicities, but also have anti-leishmania properties and anti-HIV; therefore, LAAOs have intrinsic effects not only on bacteria but also on protozoa and viruses (Du and Clemetson, 2002; Zuliani *et al.*, 2009; Fox *et al.*, 2013). Antibacterial activity of king cobra (*Ophiophagus hannah*) venom LAAO has been tested by broth dilution assay against a variety of clinical bacterial isolates such as *Staphylococcus aureus*, *S. epidermidis*, *Pseudomonas aeruginosa*, *Klebsiella pneumoniae* and *Escherichia coli*. King cobra LAAO is very effective in inhibiting Gram-positives (*S. aureus* and *S. epidermidis*), with a MIC of 0.78 μg/ml (0.006 μM) and 1.56 μg/ml (0.012 μM), respectively. The MICs are comparable to the MICs for antibiotics such as cefotaxime, kanamycin, tetracycline, vancomycin and penicillin) examined under the same conditions (Lee *et al.*, 2011). However, this same LAAO only moderately inhibited growth of *P. aeruginosa*, *K. pneumoniae* and *E. coli* with MIC ranges from 25 to 50 μg/ml (0.2–0.4 μM). The major LAAO of king cobra venom is known to be an unusual form of sv-LAAO and possesses unique structural and functional features that include thermal stability (Rodrigues *et al.*, 2009; Lee *et al.*, 2011). Additionally, the structures of inhibitor enzyme complexes provide ideal platforms for designing small potent inhibitors molecules useful for developing prototypes and lead compounds with potential therapeutic applications (Kang *et al.*, 2011). Bactericidal properties induced by LAAO from *B. atrox* venoms have also been studied to date (de Melo Alves Paiva *et al.*, 2011). Whereas, *B. mattogrosensis* (Bm) LAAO has remarkable antibacterial activity against both Gram-positive and Gram-negative bacteria, yet the BmLAAO1-BmLAAO3 were not cytolytic against macrophages as well as erythrocytes (Okubo *et al.*, 2012). In addition, structural analysis of BmLAAO-1 and BmLAAO-3 reveals proline-rich regions with BmLAAO-3 not as hydrophilic and cationic compared to BmLAAO-1. BmLAAO-3 has a net charge of +1, reducing its membrane attraction versus BmLAAO-1 (net charge +3). This latter protein's net charge is a well-known property that favours interaction with bacterial membranes through electrostatic interaction (Hancock and Rozek, 2002; Marcotte *et al.*, 2003). However, the peptide regions that are the key residues involved in antimicrobial activity are also structurally studied using theoretical models. The regions reported here represent promising candidates in the rational design of new antimicrobials that could be used to control resistant microorganisms (Okubo *et al.*, 2012).

svPLA$_2$s modulate antibacterial properties

Several snake venom phospholipase A$_2$ (svPLA$_2$) have been isolated and characterized from venoms of *Bothrops neuwiedi pauloensis* (Soares *et al.*, 2000; Rodriguez *et al.*, 2004), *B. jararacussu* (Roberto *et al.*, 2004a), *B. piraja* (Soares *et al.*, 2001a), *B. moojeni* (Stabeli *et al.*, 2004), and *Crotalus durissus terrificus* (Soares *et al.*, 2001b; Toyama *et al.*, 2003). Phospholipase A$_2$ (PLA$_2$) enzymes are among some of the best characterized components of snake venoms. PLA$_2$s catalyse the hydrolysis of glycerophospholipids at the sn-2 position, releasing lysophospholipids and fatty acids (Kini, 2003). The svPLA$_2$s are proteins belonging to two groups, I and II. Group I belongs to the Elapidae family, whereas those of group II are of Viperidae origins. These can be further subdivided into Asp49 PLA$_2$s which are catalytically active, and PLA$_2$ homologues, which possess most commonly a Lys49 residue and are devoid of catalytic activity (Arni and Ward, 1997; Lomonte *et al.*, 2009). A number of svPLA$_2$s that modulate antibacterial properties (Santamaria *et al.*, 2005; Perumal Samy *et al.*, 2010) include a Lys49/Asp49 protein from the myotoxic PLA$_2$ of *Bothrops asper* (Paramo *et al.*, 1998), basic PLA$_2$ of *Crotalus durissus terrificus* (Oliveira *et al.*, 2002), Lys49 PLA$_2$ of *B. atrox* (Nunez *et al.*, 2004), myotoxic PLA$_2$ of *B. neuwiedi pauloensis* (Rodrigues *et al.*, 2004), bactericidal PLA$_2$ from *Bungarus fasciatus* (Xu *et al.*, 2007), a basic svPLA$_2$ from *Agkistrodon halys* (Perumal Samy *et al.*, 2008) and peptides derived from the C-terminal region of myotoxic PLA$_2$ from crotalid venoms (Santamaria *et al.*, 2005). The svPLA$_2$ protein of *Echis carinatus* exert bactericidal and membrane damaging activities (Perumal Samy *et al.*, 2010). In addition, svPLA$_2$s have been shown to be inhibitory (bacteriostatic) or lethal (bactericidal) to Gram-positive bacteria *S. aureus* (Samy *et al.*, 2011a). In the case of a svPLA$_2$ isolated from *Crotalus durissus collilineatus* venom, the bactericidal effect was entirely dependent on its enzymatic activity (Toyama *et al.*, 2005). The effect of snake venoms (sv)PLA$_2$s towards Gram-positive and Gram-negative bacteria and their role in host defence against bacterial infection has been reported by Nevalainen *et al.* (2008). Various types of svPLA$_2$s and synthetic peptides derived from svPLA$_2$ can also possess anti-tumour activities. Whereas in the Viperidae family, svPLA$_2$s from *Vipera berus berus* (VBBPLA$_2$), *Vipera lebetina (VLPLA$_2$)* and *Naja naja oxiana* inhibits growth of soil inhabitant Gram-positive *Bacillus subtilis* at 500 µg/ml (36.2 µM) whereas the inhibitory effect was not due to its catalytic activity. For another Gram-positive, *S. aureus*, native sPLA$_2$ from *N. naja oxiana* inhibited the growth 42% but only slightly inhibited *B. subtilis*. None of the studied svPLA$_2$s was inhibitory to the growth of Gram-negative *E. coli* at 500 µg/ml (Samel *et al.*, 2013). Another study showed that myotoxic svPLA$_2$s induced bactericidal activity might be due to the common effect of PLA$_2$s (Santamaria *et al.*, 2005). Another recent study also reported that the *Porthidium nasutum* (Pn)PLA$_2$ possessed an antibacterial effect against *S. aureus* in a dose-dependent manner with minimum inhibitory concentration (MIC) of 32 µg/ml active against *E. coli* (Vargas *et al.*, 2012). *B. jararacussu* BthA-I-PLA$_2$ exerted antimicrobial activity (Roberto *et al.*, 2004b). Small peptide molecules designed and derivative from parent protein of svPLA$_2$ enzymes are bactericidal towards diverse species of Gram-positive bacteria (Nevalainen *et al.*, 2008). Bactericidal activity of PLA$_2$s involves recognition of anionic sites and enzymatic degradation of phospholipids in the membranes of target cells, preferentially of Gram-negative species (Koduri *et al.*, 2002).

Snake venom peptide as antimicrobials

Most svPLA$_2$s induce toxic effects to host cells, which makes them a challenge to harness for antimicrobial activities in mammals. These cysteine-rich proteins (~120 amino acids) possess similar structure and function. For instance, nearly 50% of the residues are packed into α-helices and only 10% are in β-sheets. The C-terminal region mainly determines the pharmacological, biological and antibacterial properties due to their enzyme action (Santamaria *et al.*, 2005; Samy *et al.*, 2011a). Designing of peptides are replacing by leucine (L) residues with tryptophan (W) in AMPs (I1WL5W and I4WL5W) by replacing Leu residues with two Trp residues at different position in the L-K6 peptide, tryptophan residues exhibited strong membrane-disruptive activity. Previous studies report that a short peptide (KKWRWWLKALAKK) was identified and designed from the original sequence of 115–129 within the C-terminal loop of *Bothrops asper* myotoxin II (Lys49 PLA$_2$). This peptide not only possessed antimicrobial activity against Gram-negative *Salmonella typhimurium* and Gram-positive *S. aureus* (thus reproducing the bactericidal action of the parent molecule), there were also less toxic effects on eukaryotic cells versus the holotoxin (Santamaria *et al.*, 2005). In one of our earlier studies, a variety of naturally occurring venom proteins display potent antimicrobial activity against *Burkholderia pseudomallei* (KHW/TES) and *S. aureus*. Further, a series of synthetic peptide variants designed from the original C-terminal sequence of PLA_2 and its Try→Trp substituted peptides (PIP-AMPs) were tested in an *in vitro* assay for bactericidal and cytolytic effects. Thus, these peptide variants retain very high bactericidal potency against both Gram-negative (*B. pseudomallei* KHW/TES) and Gram-positive (*S. aureus*) bacteria, yet were rather devoid of cytolytic effects towards human skin fibroblasts (HEPK) and macrophages (THP-1 and U-937 cells). Activity may be due to the number of tryptophan substitutions (Perumal Samy *et al.*, 2011). These molecules possess broad antimicrobial spectrum against a variety of bacteria by inserting into the cell membrane or binding to receptors (Fig. 13.3). Further, such peptides possess promise towards developing novel antimicrobials, especially those that could be useful for treating MDR pathogens that increase with time while our existing antimicrobial arsenal becomes less effective (Vaara *et al.*, 2008; Hadley and Hancock, 2010).

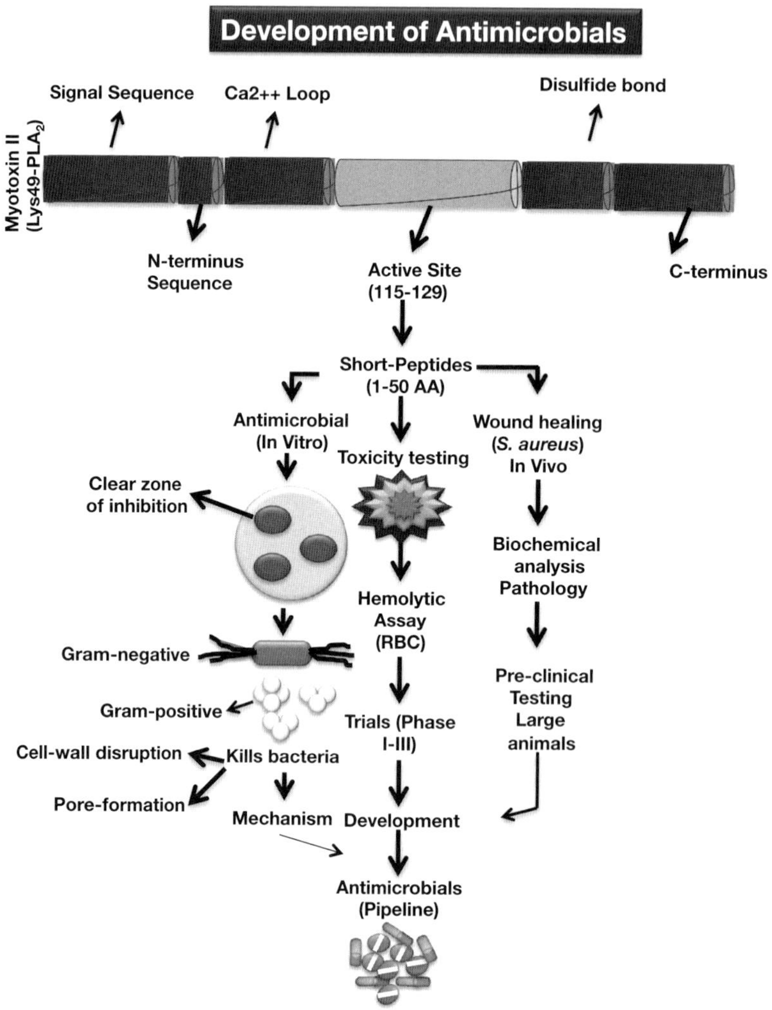

Figure 13.3 Flow chart represents the antimicrobial potency of snake components. Structure of a typical Lys49 PLA_2 and small synthetic molecules derived from the parent molecules derived of *B. asper* (Myotoxin II) leads for the possible drug candidates. A colour version of this figure is available in the plate section at the back of the book.

Snake cathelicidins

Cathelicidins are AMPs found in most vertebrates, including fish, birds and mammals, but not in amphibians or reptiles (although magainins are found in the African clawed toad). There are seven different cathelicidins in cattle, horse, pig, sheep and goat (Durr *et al.*, 2006). Only a single cathelicidin was found in human, rhesus monkey, mouse, rat and guinea pig (Bals *et al.*, 2001; Zhao *et al.*, 2001; Termen *et al.*, 2003; Durr *et al.*, 2006). Cathelicidin AMPs are released from their corresponding inactive precursors by proteolytic cleavage (Durr *et al.*, 2006). The cathelicidin family of proteins is characterized by the presence of highly conserved anionic cathelin like-domain at the N-terminus (Gennaro and Zanetti, 2000; Zanetti, 2004). Recently, cathelicidins from snake were identified and characterized. A cathelicidin-like AMP (cathelicidin-BF) has been identified from *Bungarus fasciatus* venom. As found with other cathelicidins, the precursor of cathelicidin-BF has a cathelin-like domain within the N-terminus and the mature cathelicidin-BF is located within the C-terminal sequence containing a typical acidic fragment between the cathelin-like domain and the C-terminus (Wang *et al.*, 2008). The acidic fragment is similar to acidic domains found in antimicrobial precursors from amphibians. The secondary structure of cathelicidin-BF showed an amphipathic α-helical conformation. Cathelicidin-BF effectively kills a variety of pathogens including drug-resistant bacteria and fungi. This molecule is also devoid of haemolytic and cytolytic action on mouse macrophage (RAW264) and human liver tumour ($HepG_2$) cells up to 400 mg/ml (Wang *et al.*, 2008). Similarly, cathelicidin-BF (BF-30) kills *P. aeruginosa* and *S. aureus* within 2 hours during *in vitro* testing (Zhou *et al.*, 2011).

In vivo antimicrobial efficacy of SVM

Innate immunity is the primary defence against infection, whether bacterial, fungal or viral. AMPs produced by animals (i.e. snake venoms) in many ways represent an innate, non-specific defence against microbial invaders. Clearly, these peptides have recently attracted important attention due to their relevance for treating some microbial-based diseases and also may be useful for future development of novel chemotherapeutic agents.

Wound healing potential of snake venom proteins

Viper venom induces diverse pathological changes such as oedema, haemorrhage, necrosis of skeletal muscle, as well as skin blistering and dermonecrosis (Gutierrez and Lomonte, 2003; Warrel, 2004). However, snake venom *B. asper* metalloproteinase (BaP1)-induced skin damage reproduces the pathological changes in a mouse ear model (Jimenez *et al.*, 2008). In this model the rapid angiogenic response leads to a partial revascularization, inflammation, granulation, and then successful re-epithelization within two weeks. This observation shows that the metalloproteinase BaP1 induces a series of effects similar to the skin of patient's envenomated by viperids. Study of the pathological alterations induced by this venom in skin is very clinically relevant (Jimenez *et al.*, 2008).

Among snake venom metalloproteinases (SVM), alternagin-C (ALT-C) is a disintegrin-like cysteine-rich protein identified from venom of the Brazilian snake *Bothrops alternates* (Selistre-de-Araujo *et al.*, 2005). A recent study demonstrates that ALT-C induces angiogenesis in mice, affecting fibroblasts and inflammatory cells, collagen types I and III, as well as TGF-alpha expression in a rat skin wound model (Sant'Ana *et al.*, 2011). Wounded animals were either treated with natrozol gel alone or 10, 60, and 100 ng of ALT-C for 1, 3, 5 and 7 days. ALT-C treatment increased the fibroblast density, collagen deposition and accelerated the inflammatory process, mostly in the ALT-C60 group. In addition, results indicate that ALT-C improves the wound healing process in rat skin (Sant'Ana *et al.*, 2011). Further, ALT-C and derived peptides (ALT-C PEP) were tested in a cutaneous wound model in rats. Wounds were treated with three different doses of ALT-C and ALT-C PEP and subsequently analysed, yielding induced vessel formation and modulated expression of growth factors such as vascular endothelial growth factor (VEGF) and fibroblast growth factor 1 (FGF1). Expression of VEGF was noticeably increased and detected up to 7 days after injury (Sant'Ana *et al.*, 2008) and there was induced angiogenesis (Ramos *et al.*, 2007). ALT-C (specific protein) also strongly activates protein kinase B phosphorylation, a signalling event involved in endothelial cell survival and angiogenesis. In human neutrophils, ALT-C has a potent chemotactic effect modulated by the intracellular signalling cascade characteristic of an integrin-activated pathway.

The alpha-2 and beta-1 integrins are major collagen receptors that play a vital role in adhesion of normal and tumour cells to the extracellular matrix (Selistre-de-Araujo *et al.*, 2005). Thus, ALT-C acts as a survival factor, promoting adhesion, migration and endothelial cell proliferation after binding to -2 and β-1 integrin on the cell surface. Thus, the ALT-C may be a potential candidate for developing a novel therapeutic strategy for skin wounds and tissue regeneration (Selistre-de-Araujo *et al.*, 2005; Durigan *et al.*, 2008).

An *in vivo* study reveals a dose–response effect of cathelicidin-BF (BF-30) at 0.75, 3, 12 mg/kg body weight per day in rats reduced colonization of *P. aeruginosa* in a burn and acute infection model. The peptide also prevents subsequent systemic infection and inflammation (Zhou *et al.*, 2011). In the earlier study (Perumal Samy *et al.*, 2007), experiments on the wound healing potential of snake venom proteins and peptides, a variety of $svPLA_2$ purified from crotalidae venom not only possess *in vitro* bactericidal effects but also *in vivo* activity in mice.

Furthermore, synthetic peptide derived from the parent $svPLA_2$ reproduced bactericidal (especially against Gram-positive *S. aureus*), as well as wound healing, activities (Samy *et al.*, 2011b). Peptide topically applied to skin wounds increases epithelialization, collagen synthesis and angiogenesis that then promotes the healing process. The peptide treated groups had at their wound site higher levels of macrophages, migration of cells and fibrocytes and macrophage-derived chemokine expression after two weeks compared to wound control. Snake venom derived peptides can promote healing of skin wounds with reduced toxicity towards normal human cells, while also retaining very high antimicrobial efficacy (Samy *et al.*, 2011a).

Wound healing and inflammatory cytokines

Wound healing is a complicated but well-organized biological process composed of diverse phases such as inflammation, proliferation and maturation (Singer and Clark, 1999; Martin, 1997; Reinke and Sorg, 2012). In the early stage of inflammation, different types of leucocytes like neutrophils and macrophages are recruited towards the wound sites. Neutrophil infiltration is a hallmark of the inflammatory process during wound healing, and presumed to have beneficial roles in the process by eradicating microbes and clearing cellular debris (Lin *et al.*, 2003; Ishida *et al.*, 2006). These cells eradicate microbes and provide cytokines and growth factors, which have a profound effect on successive proliferative events (Martin, 1997). However, impaired wound healing constitutes a major health problem around the world (Guo and DiPietro, 2010). Such wounds frequently enter a state of pathological inflammation due to delayed, incomplete and/or an uncoordinated healing process. Impaired wounds affect 3–6 million people in the United States, with patient's aged 65 years and older accounting for 85% of this event. As a result there are enormous health care expenditures estimated at more than $3 billion per year (Mathieu *et al.*, 2005; Menke *et al.*, 2007).

Impaired healing of wounds is caused by many complex factors like abnormal keratinocytes, fibroblast migration, proliferation, differentiation, apoptosis, abnormal macrophage polarization, impaired recruitment of mesenchymal stem cells (MSCs), endothelial progenitor cells (EPCs) and decreased vascularization (Xu *et al.*, 2013). Several of these cellular deficits have been linked to greater inflammation and pro-inflammatory cytokine production (Blakytny and Jude, 2006). Proinflammatory cytokines are elevated shortly after wounding, either in humans or animals, and include interleukin-1α (IL-1α), IL-1β, Il-6, IL-12, and tumour necrosis factor (TNF)-α (Wen *et al.*, 2006; Barrientos *et al.*, 2008, Hanel *et al.*, 2013). IL-1α and IL-1β have similar biological activities after binding to a common IL-1 receptor, and are considered essential mediators of tissue destruction in various inflammatory diseases such as septic shock and arthritis (Dinarello, 1996, 1997; Ishida *et al.*, 2006). In addition, abnormal cell responses and TNF-α play important roles in impaired healing of diabetic wounds (Tellechea *et al.*, 2010; Xu *et al.*, 2013). TNF-α regulates inflammation, immune response, migration and differentiation of white blood cells. TNF-α induces apoptosis via the caspase pathway, and it negatively regulates cell migration by increasing Smad 7 and inhibiting activation of Smad 2/3. TNF-α also induces nuclear factor κ (NF-κB) activation to enhance inflammatory responses. NF-κB is a heterodimeric transcription factor that plays a key role in inflammatory and immune responses (Karin and Delhase, 2000). Transcription of various inflammatory cytokines and CXC chemokines requires the activation of NF-κB, a major target of IL-1 and TNF-α mediated signal pathways (Ishida *et al.*, 2006). These cytokines (IL-1 and TNF-α) can induce the phosphorylation and subsequent degradation of IκB, followed by liberation of NF-κB and its nuclear translocation.

Structure–activity relationships of AMPs

Several short AMPs have been identified for the development of therapeutic agents. However, the structure–activity relationships of short AMPs have not been extensively studied to date (Oh *et al.*, 1999). As one example, a series of peptide analogues were designed from the C-terminal region of protein based on its amphipathic α-helical structure (Tossi *et al.*, 2000). Subsequent circular dichroism (CD) and nuclear magnetic resonance spectroscopy (NMR) studies also proved that the hybrid peptides adopt a largely helical conformation akin to the conformation of the parent molecule (Haney *et al.*, 2009; Sikorska *et al.*, 2009). The α-helical structure analogue of B1 (snake *Bungarus fasciatus*) shows that the peptide sequence of BF-15 (15 stands for number of amino acids (AA) provide an important contribution to the α-helical content of BF-30. Cathelicidin-BF15 is a 15-mer peptide derived from Cathelicidin BF-30 which is found in the venom of *B. fasciatus* and exhibit potent antimicrobial activity. The BF-15 peptide retains most of the potent antimicrobial activity of BF-30, with significantly reduced haemolytic action (Chen *et al.*, 2010). The α-helical and β-sheet structures are necessary for antimicrobial activity. Most peptides consisting of L-amino acids lost antifungal activity in the presence of heat-inactivated serum, while active diastereomers and scrambled peptide sequence (really, a scrambled sequence had antifungal activity) with a β-sheet structure retain antifungal activity under the same condition (Oh *et al.*, 1999). These peptides are much shorter than parent molecule (less toxic) and afford (cost) an attractive template for designing a novel class of AMPs (Koo *et al.*, 2008).

Mechanism of action employed by antibiotic peptides

Several proteins/peptides from natural venoms kill both Gram-negative and Gram-positive bacteria (Samy *et al.*, 2011b). For example, cathelicidin-BF15 and BF-30 peptides derived from *B. fasciatus* venom. The Cathelicidin BF-15 molecule has potent bactericidal activity against MDR bacteria with MICs of –64 μg/ml (2- to 16-fold more potent, so it takes more of shorter peptide to inhibit than BF-30) less haemolytic action (sentence is tough to understand). Furthermore, shorter peptide BF-15 specifically caused chaotic membrane morphology (i.e. membrane blebbing) and cell debris examined ultrastructurally, the antimicrobial effect of B1 is based on cytoplasmic membrane permeability (Chen *et al.*, 2010). Based on the evidence, AMPs may serve as safe and effective drug candidates against MDR microbes that resist existing antibiotic drugs. AMPs offer a broad spectrum of activity against a variety of microbial organisms (Varoga *et al.*, 2004). Several modes of action for these AMPs include membrane damage, pore formation, inhibition of nucleic acid and protein synthesis (Brogden, 2005). Furthermore, AMPs display their antibacterial effect via their cationic charge that via ionic interactions with cell membrane increases permeability, causing an efflux of intracellular contents and ultimately cell lysis (Izadpanah and Gallo, 2005). AMPs also show a different function with primary antimicrobial activity that includes tissue repair and remodelling (Bardan *et al.*, 2004). Most svPLA$_2$s penetrate the peptidoglycan envelope of Gram-positive bacteria and gain access to cell membrane phospholipids. These enzymes hydrolyse bacterial phospholipids in Gram-negatives only after disruption of the lipopolysaccharide (LPS)-rich layer. LPS is a complex molecule consisting of a fatty acid (lipid A), O-polysaccharide chain, and core sugar molecules inserted into the outer membrane of Gram-negative bacteria (Adrie and Pinsky, 2000; Hack and Zeerleder, 2001). Thus effects on these bacteria depend on factors that disrupt the outer membranes (Nevalainen *et al.*, 2008). Another proposed mechanism for the interaction between these peptides and bacterial membranes involves the 'carpet model', involving peptide-induced pore formation and detergent-like micelle formation (Papo *et al.*, 2006; Soman *et al.*, 2008). In addition, an earlier study also suggests that venom-derived peptides interact with LPS and lipid A from different Gram-negative bacteria or lipoteichoic acid (LTA) from Gram-positive bacteria, relying on a membrane-permeablizing mechanism to exert bactericidal effects (Paramo *et al.*, 1998).

Snake venoms as new sources of antibiotics

Venomous animals such as snakes, scorpions, marine snails and other animals have evolved for millions of years to namely secure food and defend against predators. The toxic components of venoms are among the most selective and high-affinity molecules found in nature (Boys *et al.*, 1960). There are over 100,000 venomous animal species on earth with more than 20 million toxins, of which 1000 of toxins have been studied in detail (Garza, 2013). Numerous other toxins are under clinical trials (des Ligneris and Grasset, 1936). Recently, new AMPs from natural sources (venom) have drawn attention. Since AMPs were initially identified in frog and insects in the 1980s, an additional 1200 AMPs have been elucidated to date (Park *et al.*, 2011). Among these AMPs are those isolated from snake venoms, displaying both antibacterial and antiviral activities (Rideiro *et al.*, 2007). Newly isolated derivatives of snake venom proteins such as crotamine and cathelicidine BF-15 exert broad-spectrum antibacterial activity (Oguiura *et al.*, 2011; Chen *et al.*, 2011; Zhou *et al.*, 2011). A number of published studies clearly indicate an increasing demand for antimicrobials from natural sources. The antimicrobial potency of snake venoms, involving biological activity of PLA_2s, LAAO and peptides, involves killing a variety of microorganisms (bacteria/fungi/virus), and in fact some of the unique proteins/peptides also possess wound healing potency in *a* mouse model for *S. aureus*-infected wounds (Samy *et al.*, 2011a). A majority of snake venom-derived AMPs are still in early development, thus relatively far from clinical studies to become an approved antimicrobial for human use. Ultimately, antimicrobial activity inherent in many animal venoms such as snake venoms can hopefully be translated into application and market opportunities in the near future to treat MDR pathogens.

Present challenges and development

Certain pathogens have become MDR, partly due to the over use of existing antibiotics (Xie *et al.*, 2003; Roos, 2004). However, these clinically challenging global health problems may decrease by discovering novel and safer antimicrobials necessary for staying at least one step ahead of these pathogens (Xie *et al.*, 2003). Therefore, new AMPs from animal (snake) venoms can possibly play an important role for treating especially children, elderly and those with otherwise compromised immune systems (Guardabassi and Kruse, 2003; Hancock and Diamond, 2000; Hujer *et al.*, 2004). Furthermore, technical advances that include introduction of high-throughput screening platforms (Vetter *et al.*, 2011) can greatly expedite future venom-based drug discoveries and help expand the growing pipeline of venom-derived drugs. Essentially, antimicrobial-based research to identify prototypical drugs from natural sources (i.e. snake venoms) represents important gold mines for the near and distant future.

Conclusion

An extremely wide range of animals have evolved venoms for predation and defence. Naturally occurring AMPs from some of these venoms are currently being explored as potential drugs. The major components of most venom are proteins/

peptides that are often protease-resistant due to their disulfide-rich architectures. Some of these toxins have become valuable tools and therapeutics due to their extremely high specificity and potency for particular molecular targets. Currently, there are six FDA-approved drugs derived from venom peptides or proteins in the market. Emerging trends have been identified by various groups, including an increasing demand by industry focused upon venom peptides. In particular, medicinal use of venom components seems quite appealing for engaging a wide array of molecular targets that go towards treating many clinical conditions/infections.

References

Adrie, C., and Pinsky. M.R. (2000). The inflammatory balance in human sepsis. Intensive Care Med. *26*, 364–375.

Appelbaum, P.C. (2006). The emergence of vancomycin-intermediate and vancomycin-resistant *Staphylococcus aureus*. Clin. Microbiol. Infect. Suppl. *1*, 16–23.

Arni, R.K., and Ward, R.J. (1996). Phospholipase A_2-a structural review. Toxicon *34*, 827–841.

Bailey, P., and Wilce, J. (2001). Venom as a source of useful biologically active molecules. Emerg. Med. (Fremantle) *13*(1), 28–36.

Bals, R., Lang, C., Weiner, D.J., Vogelmeier, C., Welsch, U., and Wilson, J.M. (2001). Rhesus monkey (*Macaca mulatta*) mucosal antimicrobial peptides are close homologues of human molecules. Clin. Diag. Lab. Immunol. *8*, 370–375.

Bardan, A., Nizet, V., and Gallo, R.L. (2004). Antimicrobial peptides and the skin. Expert Opin. Biol. Ther. *4*(4), 543–549.

Barrientos, S., Stojadinovic, O., Golinko, M.S., Brem, H., and Tomic-Canic, M. (2008). Growth factors and cytokines in wound healing. Wound Repair Regen. *16*(5), 585–601.

Blakytny, R., and Jude, E. (2006). The molecular biology of chronic wounds and delayed healing in diabetes. Diabetic Med. 23(6), 594–608.

Boman, H.G. (2003). Antibacterial peptides: basic facts and emerging concepts. J. Intern. Med. *254*, 197–215.

Bonomo, R.A. (2000). Multiple antibiotic-resistant bacteria in long-term-care facilities: an emerging problem in the practice of infectious diseases. Clin. Infect. Dis. *31*, 1414–1422.

Boys, F., Beamer, P., and Smith, H.M. (1960). Antibiotics in tests for toxicity of snake venom. JAMA *174*(3), 306–307.

Brogden, K.A. (2005). Antimicrobial peptides: pore formers or metabolic inhibitors in bacteria? Nat. Rev. Microbiol. 3(3), 238–250.

Brown, K. (2004). The history of penicillin from discovery to the drive to production. Pharm. Hist. (Lond) *34*, 37–43.

Burke, A. (2000). Antibiotic resistance. Med. Clinic. North Am. *84*, 1407–1429.

Bustillo, S., Laura, C., Merino, L.L., Acosta, O., de Kier Joffé, E.B., and Gorodner, J.O. (2008). Antimicrobial activity of *Bothrops alternatus* venom from the Northeast of Argentine. Microbiol. *50*(3–4), 79–82.

Calvete, J.J., Juárez, P., and Sanz, L. (2007). Snake venomics. Strategy and applications. J. Mass Spectrom. *42*(11), 1405–1414.

Chen, W., Yang, B., Zhou, H., Sun, L., Dou, J., and Qian, H., Huang, W., Mei, Y., and Han, J. (2011). Structure–activity relationships of a snake cathelicidin-related peptide, BF-15. Peptides. *32*(12), 2497–2503.

Chopra, I., Hesse, L., and O'Neill, A.J. (2002). Exploiting current understanding of antibiotic action for discovery of new drugs. Symp. Ser. Soc. Appl. Microbiol. *31*, 4S-15S.

Craik, D.J., Fairlie, D.P., Liras, S., and Price, D. (2013). The future of peptide-based drugs. Chem. Biol. Drug Des. *81*,136–147.

Cushman, D.W., and Ondetti, M.A. (1991). History of the design of captopril and related inhibitors of angiotensin converting enzyme. Hypertension *17*, 589–592.

Dance, D.A.B. (1996). Melioidosis. In Manson's Tropical Diseases, 20th Ed., G.C. Cook, ed. (WB. Saunders Co. Ltd, London, UK), pp. 925–930.

Danho, W., Swistok, J., Khan, W., Chu, X.J., Cheung, A., Fry, D., Sun, H., Kurylko, G., Rumennik, L., Cefalu, J., *et al.* (2009). Opportunities and challenges of developing peptide drugs in the pharmaceutical industry. Adv. Exp. Med. Biol. *611*, 467–469.

De Lima, C.D., Alvarez, A.P., Freitas, C., Oliveira, S.D., Oliveira, R.S., Dos Santos, C., Mendes Cabral, L., Rodrigues, C.R., and Castro, H.C. (2005). Snake venom: any clue for antibiotics and CAM? Evid. Based Complement. Alternat. Med. *2*(1), 39–47.

Desbois, A.P., Gemmell, C.G., and Coote, P.J. (2010). *In vivo* efficacy of the antimicrobial peptide ranalexin in combination with the endopeptidase lysostaphin against wound and systemic meticillin-resistant *Staphylococcus aureus* (MRSA) infections. Int. J. Antimicrob. Agents 35(6), 559–565.

Dinarello, C.A. (1996). Biologic basis for interleukin-1 in disease. Blood *87*, 209–2147.

Dinarello, C A. (1997). Interleukin-1. Cytokine Growth Factor Rev. *8*, 253–265.

Doley, R., and Kini, R.M. (2009). Protein complexes in snake venom. Cell. Mol. Life Sci. *66*, 2851–2871.

Du, X.Y., and Clemetson, K.J. (2002). Snake venom L-amino acid oxidases. Toxicon *40*, 659–665.

Durigan, J.L., Peviani, S.M., Russo, T.L., Delfino, G.B., Ribeiro, J.U., Cominetti, M.R., Selistre-de-Araujo, H.S., and Salvini, T.F. (2008). Effects of alternagin-C from *Bothrops alternatus* on gene expression and activity of metalloproteinases in regenerating skeletal muscle. Toxicon 52, 687–694.

Durr, U.H., Sudheendra, U.S., and Ramamoorthy, A. (2006). LL-37, the only human member of the cathelicidin family of antimicrobial peptides. Biochim. Biophys. Acta *1758*, 1408–1425.

Fox, J.W. (2013). A brief review of the scientific history of several lesser-known snake venom proteins: L-amino acid oxidases, hyaluronidases and phosphodiesterases. Toxicon. *62*, 75–82.

Fox, J.W., and Serrano, S.M.T. (2005). Snake toxins and hemostasis. Toxicon *45*, 951.

Fry, B.G., and Wuster, W. (2004). Assembling an arsenal: origin and evolution of the snake venom proteome inferred from phylogenetic analysis of toxin sequences. Mol. Biol. Evol. *21*(5), 870–883.

Galbusera, E., Renzoni, A., Andrey, D.O., Monod, A., and Barras, C., Tortora, P., Polissi, A., and Kelley, W. (2011). Site-specific mutation of *Staphylococcus aureus* VraS reveals a crucial role for the VraR-VraS sensor in the emergence of glycopeptide resistance. Antimicrob. Agents Chemother. *55*, 1008–1020.

Ganz, T. (2003). Defensins: antimicrobial peptides of innate immunity. Nat. Rev. Immunol. *3*, 710–720.

Ganz, T., and Lehrer, R.I. (1997). Antimicrobial peptides of leukocytes. Curr. Opin. Haematol. *4*, 53–58.

Garza, M.J. (2013). Snake venom used in medicine development. The Johns Hopkins News Letter – Science & Technology, 14th February (http://www.jhunewsletter.com).

Gennaro, R., and Zanetti, M. (2000). Structural features and biological activities of the cathelicidin-derived antimicrobial peptides. Biopolymers *55*, 31–49.

Georgieva, D., Arni, R.K., and Betzel, C. (2008). Proteome analysis of snake venom toxins: pharmacological insights. Expert Rev. Proteomics. *5*, 787–797.

Gomes, A., Choudhury, S.R., Saha, A., Mishra, R., Giri, B., Biswas, A.K., Debnath, A., and Gomes, A. (2007). A heat stable protein toxin (drCT-I) from the Indian viper (*Daboia russelli russelli*) venom having antiproliferative, cytotoxic and apoptotic activities. Toxicon *49*, 46–56.

Guardabassi, L., and Kruse, H. (2003). Overlooked aspects concerning development and spread of antimicrobial resistance. Expert Rev. Anti. Infect. Ther. *1*, 359–362.

Guo, C., Liu, S., Yao, Y., Zhang, Q., and Sun, M.Z. (2012). Past decade study of snake venom L-amino acid oxidase. Toxicon *60*, 302–311.

Guo, S., and DiPietro, L.A. (2010). Factors affecting wound healing. J. Dent. Res. *89*, 219–229.

Gutiérrez, J.M., and Lomonte, B. (2003). Efectos locales en el envenenamiento ofídico en América Latina. In Animais Peçonhentos no Brasil. Biología, Clínica e Terapêutica dos Accidentes, J.L.C. Cardoso, F.O.S. França, F.H. Wen, C.M. Málaque, and V. Haddad, eds. (Sarvier, São Paulo, Brasil), pp. 310–323.

Hack, C.E., and Zeerleder, S. (2001). The endothelium in sepsis: source of and a target for inflammation. Crit. Care Med. *29*, S21-S27.

Hadley, E.B., and Hancock, R.E. (2010). Strategies for the discovery and advancement of novel cationic antimicrobial peptides. Curr. Top. Med. Chem. *10*, 1872–1881.

Hancock, R.E.W., and Diamond, G. (2008). The role of antimicrobial peptides in innate host defense. Trends Microbiol. *8*, 402–410.

Hancock, R.E.W., and Lehrer, R. (1998). Cationic peptides: a new source of antibiotics. Trends Biotechnol. *16*, 82–88.

Hancock, R.E.W., and Rozek, A. (2002). Role of membranes in the activities of antimicrobial cationic peptides. FEMS Microbiol. Lett. *206*, 143–149.

Hänel, K.H., Cornelissen, C., Lüscher, B., and Baron, J.M. (2013). Cytokines and the skin barrier. Int. J. Mol. Sci. *14*, 6720–6745.

Haney, E.F., Hunter, H.N., Matsuzaki, K., and Vogel, H.J. (2009). Solution NMR studies of amphibian antimicrobial peptides: linking structure to function? Biochim. Biophys. Acta *1788*, 1639–1655.

Harris, C.R. (2004). Thorarensen A. Advances in the discovery of novel antibacterial agents during the year 2002. Curr. Med. Chem. *11*, 2213–2243.

Heise, P.J., Maxson, L.R., Dowling, H.G., and Hedges, S.B. (1995). Higher-level snake phylogeny inferred from mitochondrial DNA sequences of 12S rRNA genes. Mol. Biol. Evol. *12*, 259–265.

Hujer, A.M., Bethel, C.R., Hujer, K.M., and Bonomo, R.A. (2004). Antibiotic resistance in the institutionalized elderly. Clin. Lab. Med. *24*, 343–361.

Ishida, Y., Kondo, T., Kimura, A., Matsushima, K., and Mukaida, N. (2006). Absence of IL-1 receptor antagonist impaired wound healing along with aberrant NF-kB activation and a reciprocal suppression of TGF-beta signal pathway. J. Immunol. *176*, 5598–5606.

Izadpanah, A., and Gallo, R.L. (2005). Antimicrobial peptides. J. Am. Acad. Dermatol. *52*, 381–390; quiz 391–382.

Izidoro, L.F.M., Ribeiro, M.C., Souza, G.R.L., Sant'Ana, C.D., Hamaguchi, A., Homsi-Brandeburgo, M.I., Goulart, L.R., Beleboni, R.O., Nomizo, A., Sampaio, S.V., *et al.* (2006). Biochemical and functional characterization of an L-aminoacid oxidase isolated from *Bothrops pirajai* snake venom. Bioorg. Med. Chem. *14*, 7034–7043.

Jimenez, N., Escalante, T., Gutierrez, J.M., and Rucavado, A. (2008). Skin pathology induced by snake venom metalloproteinase: acute damage, revascularization, and re-epithelization in a mouse ear model. J. Investig. Dermatol. *128*, 2421–2428.

Jones, A.L., Beveridge, T.J., and Woods, D.E. (1996). Intracellular survival of *Burkholderia pseudomallei*. Infect. Immun. *64*, 782–790.

Kang, T.S., Georgieva, D., Genov, N., Murakami, M.T., Sinha, M., Kumar, R.P., Kaur, P., Kumar, S., Dey, S., Sharma, S., *et al.* (2011). Enzymatic toxins from snake venom: structural characterization and mechanism of catalysis. FEBS J. *278*, 4544–4576.

Karin, M., and Delhase, M. (2000). The IκB kinase (IKK) and NF-κB: key elements of pro-inflammatory signalling. Semin. Immunol. *12*, 85–98.

Katoh, T., Goto, Y., Reza, M.S., and Suga, H. (2011). Ribosomal synthesis of backbone macrocyclic peptides. Chem. Commun. (Camb) *47*, 9946–9958.

King, G. (2013). Venoms to drugs: translating venom peptides into therapeutics. Aust. Biochem. *44*, 1–8.

King, G.F. (2011). Venoms as a platform for human drugs: translating toxins into therapeutics. Expert Opin. Biol. Ther. *11*, 1469–1484.

Kini, R.M. (2003). Excitement ahead: structure, function and mechanism of snake venom phospholipase A_2 enzymes. Toxicon *42*, 827–840.
Kirby, A., Edwards, C., Broughton, C.M., and Williams, N.J. (2011). Glycopeptide and daptomycin resistance in community-associated MRSA in the UK. Infection *39*, 277–279.
Kochva, E. (1987). The origin of snakes and evolution of the venom apparatus. Toxicon *25*, 65–106.
Koduri, R.S., Gronroos, J.O., Laine, V.J., Le Calvez, C., Lambeau, G., Nevalainen, T.J., and Gelb, M.H. (2002). Bactericidal properties of human and murine groups I, II, V, X, and XII secreted phospholipases A_2. J. Biol. Chem. *277*, 5849–5857.
Komatsuzawa, H., Ouhara, K., Yamada, S., Fujiwara, T., Sayama, K., Hashimoto, K., and Sugai, M. (2006). Innate defenses against methicillin-resistant *Staphylococcus aureus* (MRSA) infection. J. Pathol. *208*, 249–260.
Koo, Y.S., Kim, J.M., Park, I.Y., Yu, B.J., Jang, S.A., Kim, K.S., Park, C.B., Cho, J.H., and Kim, S.C. (2008). Structure–activity relations of parasin I, a histone H2A-derived antimicrobial peptide. Peptides *29*, 1102–1108.
Lee, M.L., Tan, N.H., Fung, S.Y., and Sekaran, S.D. (2011). Antibacterial action of a heat-stable form of L-amino acid oxidase isolated from king cobra (*Ophiophagus hannah*) venom. Comp. Biochem. Physiol. C Toxicol. Pharmacol. *153*, 237–242.
Leelarasamee, A. (1998). *Burkholderia pseudomallei*: the unbeatable foe?. *Southeast* Asian J Trop. Med. Public Health *29*, 410–415.
Lewis, R.J., and Garcia, M.L. (2003). Therapeutic potential of venom peptides. Nature Rev. *2*, 790–802.
des Ligneris, M., and Grasset, E. (1936). Clinical experiments on the effect of african snake venoms on human cancer cases with or without concomitant deep therapy. Am. J. Cancer *26*, 512–520.
Limmathurotsakul, D., Wongsuvan, G., Aanensen, D., Ngamwilai, S., Saiprom, N., Rongkard, P., Thaipadungpanit, J., Kanoksil, M., Chantratita, N., Day, N.P.J., and Peacock, S.J. (2014). Melioidosis caused by *Burkholderia pseudomallei* in drinking water, Thailand, 2012. Emerg. Infect. Dis. *20*, 265–267.
Lin, Z.Q., Kondo, T., Ishida, Y., Takayasu, T., and Mukaida, N. (2003). Essential involvement of IL-6 in the skin wound-healing process as evidenced by delayed wound healing in IL-6-deficient mice. J. Leukoc. Biol. *73*, 713–721.
Lomonte, B., Angulo, Y., and Santamaría, C. (2003). Comparative study of synthetic peptides corresponding to region 115–129 in Lys49 myotoxic phospholipases A_2 from snake venoms. Toxicon *42*, 307–312.
Loomba, P.S., Taneja, J., and Mishra, B. (2010). Methicillin and vancomycin resistant *S. aureus* in hospitalized patients J. Glob. Infect. Dis. *2*, 275–283.
Lou, Z., Hou, J., Liang, X., Chen, J., Qiu, P., Liu, Y., Li, M., Rao, Z., and Yan, G. (2005). Crystal structure of a non-hemorrhagic fibrin(ogen)olytic metalloproteinase complexed with a novel natural tri-peptide inhibitor from venom of *Agkistrodon acutus*. J. Struct. Biol. *152*, 195–203.
Marcotte, I., Wegener, K.L., Lam, Y.H., Chia, B.C., de Planque, M.R., Bowie, J.H., Auger, M., and Separovic, F. (2003). Interaction of antimicrobial peptides from Australian amphibians with lipid membranes. Chem. Phys. Lipids *122*, 107–120.
Markland, F.S. Jr., and Swenson, S. (2013). Snake venom metalloproteinases. Toxicon. *62*, 3–18.
Martin, P. (1997). Wound healing-aiming for perfect skin regeneration. Science *276*, 75–81.
Mathieu, D., Linke, J.C., and Wattel, F. (2006). Non-healing wounds. In Handbook on Hyperbaric Medicine, D.E. Mathieu, ed. (Springer, Dordrecht, The Netherlands), pp. 401–427.
Mazzi, M.V., Marcussi, S., Carlos, G.B., Stabeli, R.G., Franco, J.J., Ticli, F.K., Cintra, A.C.O., Franca, S.C., Soares, A.M., and Sampaio, S.V. (2004). A new hemorrhagic metalloprotease from *Bothrops jararacussu* snake venom: isolation and biochemical characterization. Toxicon *44*, 215–223.
de Melo Alves Paiva, R., de Freitas Figueiredo, R., Antonucci, G.A., Paiva, H.H., de Lourdes Pires Bianchi, M., Rodrigues, K.C., Lucarini, R., Caetano, R.C., Linhari Rodrigues Pietro, R.C., Gomes Martins, C.H., *et al.* (2011). Cell cycle arrest evidence, parasiticidal and bactericidal properties induced by L-amino acid oxidase from *Bothrops atrox* snake venom. Biochimie *93*, 941–947.
Menke, N.B., Ward, K.R., Witten, T.M., Bonchev, D.G., and Diegelmann, R.F. (2007). Impaired wound healing. Clin. Dermatol. *25*, 19–25.
Micek, S.T. (2007). Alternatives to vancomycin for the treatment of methicillin-resistant *Staphylococcus aureus* infections. Clin. Infect. Dis. *45* (Supplement 3), S184-S190.
Milheirico, C., Portelinha, A., Krippahl, L., de Lencastre, H., and Oliveira, D.C. (2011). Evidence for a purifying selection acting on the β-lactamase locus in epidemic clones of methicillin-resistant *Staphylococcus aureus*. BMC Microbiol. *11*, 1–29.
Moran, G.J., and Talan, D.A. (2008). MRSA community-acquired pneumonia: should we be worried? Ann. Emerg. Med. *53*, 366–368.
Moran, G.J., Krishnadasan, A., Gorwitz, R.J., Fosheim, G.E., McDougal, K.L., Carey, R.B., and Talan, D.A. (2006). Methicillin-resistant *S. aureus* infections among patients in the emergency department. N. Engl. J. Med. *355*, 666–674.
Moura-da-Silva, A.M., Furlan, M.S., Caporrino, M.C., Grego, K.F., Portes-Junior, J.A., Clissa, P.B., Valente, R.H., and Magalhães, G.S. (2011). Diversity of metalloproteinases in *Bothrops neuwiedi* snake venom transcripts: evidences for recombination between different classes of SVMPs. BMC Genet. *1294*, 1–14.
Nevalainen, T.J., Graham, G.G., and Scott, K.F. (2008). Actions of secreted phospholipases A_2. Review. Biochim. Biophys. Acta *1781*, 1–9.
Núñez, V., Arce, V., Gutiérrez, J.M., and Lomonte, B. (2004). Structural and functional characterization of myotoxin I, a Lys49 phospholipase A_2 homologue from the venom of the snake *Bothrops atrox*. Toxicon *44*, 91–101.

Oguiura, N., Boni-Mitake, M., Affonso, R., and Zhang, G. (2011). *In vitro* antibacterial and hemolytic activities of crotamine, a small basic myotoxin from rattlesnake *Crotalus durissus*. J. Antibiot. *64*, 327–331.

Oh, J.E., Hong, S.Y., and Lee, K.H. (1999). Structure–activity relationship study: short antimicrobial peptides. J. Pept. Res. *53*, 41–46.

Okubo, B.M., Osmar, N., Silva, O.N., Migliolo, L., Gomes, D.G., Porto, W.F., Batista, C.L., Ramos, C.S., Holanda, H.H.S., Dias, S.C., *et al.* (2012). Evaluation of an antimicrobial L-amino acid oxidase and peptide derivatives from *Bothropoides mattogrosensis* pitviper venom. PLoS One 7, e33639.

Oliveira, D.G., Toyama, M.H., Novello, J.C., Beriam, L.O., and Marangoni, S. (2002). Structural and functional characterization of basic PLA_2 isolated from *Crotalus durissus terrificus* venom. J. Protein Chem. *21*, 161–168.

de Oliveira Junior, N.G., e Silva Cardoso, M.H., and Franco, O.L. (2013). Snake venoms: attractive antimicrobial proteinaceous compounds for therapeutic purposes. Cell. Mol. Life Sci. 70(24), 4645–4658.

Papo, N., Seger, D., Makovitzki, A., Kalchenko, V., Eshhar, Z., Degani, H., and Shai, Y. (2006). Inhibition of tumor growth and elimination of multiple metastases in human prostate and breast xenografts by systemic inoculation of a host defense-like lytic peptide. Cancer Res. *66*, 5371–5378.

Páramo, L., Lomonte, B., Pizarro-Cerdá, J., Bengoechea, J.A., Gorvel, J.P., and Moreno, E. (1998). Bactericidal activity of Lys49 and Asp49 myotoxic phospholipases A_2 from Bothrops asper snake venom-Synthetic Lys49 myotoxin II-(115–129)-peptide identifies its bactericidal region. Eur. J. Biochem. *253*, 452–461.

Park, S.C., Park, Y., and Hahm, K.S. (2011). The role of antimicrobial peptides in preventing multidrugresistant bacterial infections and biofilm formation. Int. J. Mol. Sci. *12*, 5971–5992.

Perez-Trallero, E., and Iglesias, L. (2003). Tetracyclines, sulfonamides and metronidazole. Enferm. Infecc. Microbiol. Clin. *21*, 520–529.

Perumal Samy, R., Pachiappan, A., Gopalakrishnakone, P., Thwin, M.M., Hian, Y.E., Chow, V.T., Bow, H., and Weng, J.T. (2006). *In vitro* antimicrobial activity of natural toxins and animal venoms tested against *Burkholderia pseudomallei*. BMC Infect. Dis. *20*, 6, 100.

Perumal Samy, R., Gopalakrishnakone, P., Thwin, M.M., Chow, T.K., Bow, H., Yap, E.H., and Thong, T.W. (2007). Antibacterial activity of snake, scorpion and bee venoms: a comparison with purified venom phospholipase A_2 enzymes. J. Appl. Microbiol. *102*, 650–659.

Perumal Samy, R., Gopalakrishnakone, P., Ho, B., and Chow, V.T. (2008). Purification, characterization and bactericidal activities of basic phospholipase A_2 from the venom of *Agkistrodon halys* (Chinese pallas). Biochimie *90*, 1372–1388.

Perumal Samy, R., Gopalakrishnakone, P., Bow, H., Puspharaj, P.N., and Chow, V.T.K. (2010). Identification and characterization of a phospholipase A_2 from the venom of the Saw-scaled viper: Novel bactericidal and membrane damaging activities. Biochimie *92*, 1854–1866.

Ramos, O.H., and Selistre-de-Araujo, H.S. (2006). Snake venom metalloproteases – structure and function of catalytic and disintegrin domains. Comp. Biochem. Physiol. C Toxicol. Pharmacol. *142*, 328–346.

Ramos, O.H., Terruggi, C.H., Ribeiro, J.U., Cominetti, M.R., Figueiredo, C.C., Bérard, M., Crepin, M., Morandi, V., and Selistre-de-Araujo, H.S. (2007). Modulation of *in vitro* and *in vivo* angiogenesis by alternagin-C, a disintegrin-like protein from *Bothrops alternatus* snake venom and by a peptide derived from its sequence. Arch. Biochem. Biophys. *461*, 1–6.

Reid, P.F. (2007). Alpha-cobratoxin as a possible therapy for multiple sclerosis: a review of the literature leading to its development for this application. Crit. Rev. Immunol. *27*, 291–302.

Reinke, J.M., and Sorg, H. (2012). Wound repair and regeneration. Eur. Surg. Res. *49*, 35–43.

Rideiro, P.H., Callejion, D.R., Baruffi, M.D., Sampaio, S.V., Nomizo, A., and Soares, A.M. (2007). L-amino acid oxidase isolated from *Bothrops moojeni* snake venom present selective cytotoxic activity against tumor. J. Venom. Anim. Toxins incl. Trop. Dis. *13*, 310.

Roberto, P.G., Kashima, S., Marcussi, S., Pereira, J., Astol.-Filho, S., Nomizo, A., Giglio, J.R., Fontes, M.R.M., Soares, A.M., and França, S.C. (2004a). Cloning and Identification of a complete cDNA coding for a bactericidal and antitumoral acidic phospholipase A_2 from *Bothrops jararacussu* Venom. Protein J. *23*, 273–285.

Roberto, P.G., Kashima, S., Soares, A.M., Chioato, L., Faca, V.M., Fuly, A.L., Astolfi-Filho, S., Pereira, J.O., and Franca, S.C. (2004b). Cloning and expression of an acidic platelet aggregation inhibitor phospholipase A_2 cDNA from *Bothrops jararacussu* venom gland. Protein Expr. Purif. *37*, 102–108.

Rodrigues, R.S., da Silva, J.F., Boldrini França, J., Fonseca, F.P., Otaviano, A.R., Henrique Silva, F., Hamaguchi, A., Magro, A.J., Braz, A.S., dos Santos, J.I., Homsi-Brandeburgo, M.I., *et al.* (2009). Structural and functional properties of Bp-LAAO, a new L-amino acid oxidase isolated from *Bothrops pauloensis* snake venom. Biochimie *91*, 490–501.

Rodrigues, V.M., Marcussi, S., Cambraia, R.S., de Araújo, A.L., Malta-Neto, N.R., Hamaguchi, A., Ferro, E.A., Homsi-Brandeburgo, M.I., Giglio, J.R., and Soares, A.M. (2004). Bactericidal and neurotoxic activities of two myotoxic phospholipases A_2 from *Bothrops neuwiedi pauloensis* snake venom. Toxicon *44*, 305–314.

Roos, K.L. (2004). Emerging antimicrobial-resistant infections. Arch. Neurol. *61*, 1512–1514.

Rozek, A., Friedrich, C.L., and Hancock, R.E. (2000). Structure of the bovine antimicrobial peptide indolicidin bound to dodecylphosphocholine and sodium dodecyl sulfate micelles. Biochemistry *39*, 15765–15774.

Rubina Lawrence, R., and Jeyakumar, E. (2013). Antimicrobial resistance: a cause for global concern. BMC Proc. 7 (Suppl. 3), S1, 1–14.

Russell, F.E. (1980). Venoms. In Snake Venom Poisoning, J.B. Lippincott, and G.M. Persol, eds. (Lippincott, Philadelphia, USA), pp. 139–234.

Sachidananda, M.K., Murar, S.K., and Channe, G.D. (2007). Characterization of an antibacterial peptide from Indian cobra (*Naja naja*) venom. J. Venom. Anim. Toxins incl. Trop. Dis. *13*, 446–461.

Samel, M., Vija, H., Kurvet, I., Künnis-Beres, K., Trummal, K., Subbi, J., Kahru, A., and Siigur, J. (2013). Interactions of PLA_2-s from *Vipera lebetina*, *Vipera berus berus* and *Naja naja oxiana* venom with platelets, bacterial and cancer cells. Toxins 5, 203–223.

Samy, R.P., Thwin, M.M., Chow, V.T., Bow, H., and Gopalakrishnakone, P. (2011a). Evaluation of antibacterial activity of proteins and peptides using a specific animal model for wound healing. Methods Mol. Biol. *716*, 245–265.

Samy, R.P., Stiles, B.G., Gopalakrishnakone, P., and Chow, V.T. (2011b). Antimicrobial proteins from snake venoms: Direct bacterial damage and activation of innate immunity against *Staphylococcus aureus* skin infection. Curr. Med. Chem. *18*, 5104–5113.

Samy, R.P., Gopalakrishnakone, P., Stiles, B.G., Girish, K.S., Swamy, S.N., Hemshekhar, M., Tan, K.S., Rowan, E.G., Sethi, G., and Chow, V.T. (2013). Snake venom phospholipases A(2): a novel tool against bacterial diseases. Curr. Med. Chem. *19*, 6150–6162.

San, T.M., Vejayan, J., Shanmugan, K., and Ibrahim, H. (2010). Screening Antimicrobial Activity of venoms from snakes commonly found in Malaysia. J. Appl. Sci. *10*, 2328–2332.

Santamaría, C., Larios, S., Angulo, Y., Pizarro-Cerda, J., Gorvel, J.P., Moreno, E., and Lomonte, B. (2005). Antimicrobial activity of myotoxic phospholipases A_2 from crotalid snake venoms and synthetic peptide variants derived from their C-terminal region. Toxicon *45*, 807–815.

Sant'Ana, E.M., Gouvêa, C.M., Nakaie, C.R., and Selistre-de-Araújo, H.S. (2008). Angiogenesis and growth factor modulation induced by alternagin C, a snake venom disintegrin-like, cysteine-rich protein on a rat skin wound model. Arch. Biochem. Biophys. *479*, 20–27.

Sant'Ana, E.M., Gouvêa, C.M., Durigan, J.L., Cominetti, M.R., Pimentel, E.R., and Selistre-de-Araújo, H.S. (2011). Rat skin wound healing induced by alternagin-C, a disintegrin-like, Cys-rich protein from *Bothrops alternatus* venom. Int. Wound. J. *8*, 245–252.

Schweizer, H.P. (2012). Mechanisms of antibiotic resistance in *Burkholderia pseudomallei*: implications for treatment of melioidosis. Future Microbiol. *7*, 1389–1399.

Selistre-de-Araujo, H.S., Cominetti, M.R., Terruggi, C.H., Mariano-Oliveira, A., De Freitas, M.S., Crepin, M., Figueiredo, C.C., and Morandi, V. (2005). Alternagin-C, a disintegrin-like protein from the venom of *Bothrops alternatus*, modulates alpha2beta1 integrin-mediated cell adhesion, migration and proliferation. Braz. J. Med Biol. Res.38, 1505–1511.

Serrano, S.M.T., Shannon, J.D., Wang, D., Camargo, A.C., and Fox, J.W. (2005). A multifaceted analysis of viperid snake venoms by two dimensional gel electrophoresis: an approach to understanding venom proteomics. Proteomics 5, 501–510.

Shebl, R.I., Mohamed, A.F., Ali, A.E., and Amin, M.A. (2012). Antimicrobial profile of selected snake venoms and their associated enzymatic activities. Br. Microbiol. Res. J. *2*, 251–263.

Sikorska, E., Greber, K., Rodziewicz-Motowidło, S., Szultka, L., Lukasiak, J., and Kamysz, W. (2009). Synthesis and antimicrobial activity of truncated fragments and analogs of citropin 1.1: The solution structure of the SDS micelle-bound citropin-like peptides. J. Struct. Biol. *168*, 250–258.

Singer, A.J., and Clark, R.A. (1999). Cutaneous wound healing. N. Engl. J. Med. *341*, 738–746.

Soares, A.M. (2012). Use of snake venom for biomedical researches and drug development. Open J. Biochem. Biotechnol. *1*, 1–4.

Soares, A.M., Guerra-Sa, R., Borja-Oliveira, C., Rodrigues, V.M., Rodrigues-Simioni, L., Rodrigues, V., Fontes, M., and Lomonte, B. (2000). Structural and functional characterization of BnSP-7, a Lys49 myotoxic phospholipase A_2 homologue from *Bothrops neuwiedi pauloensis* venom. Arch. Biochem. Biophys. *378*, 2, 201–209.

Soares, A.M., Andrião-Escarso, S.H., Bortoleto, R.K., Rodrigues-Simoni, L., Arni, R.K., Ward, R.J., Gutiérrez, J.M., and Giglio, J.R. (2001a). Dissociation of enzymatic and pharmacological properties of piratoxins-I and -III, two myotoxic phospholipases A_2 from *Bothrops pirajai* sanke venom. Arch. Biochem Biophys *387*, 188–196.

Soares, A.M., Mancin, A.C., Cecchini, A.L., Arantes, E.C., França, S.C., Gutiérrez, J.M., and Giglio, J.R. (2001b). Effects of chemical modifications of crotoxin B, the phospholipase A_2 subunit of crotoxin from *Crotalus durissus terrificus* snake venom, on its enzymatic and pharmacological activities. Int. J. Biochem. Cell Biol. *33*, 9, 877–888.

Soman, N.R., Lanza, G.M., Heuser, J.M., Schlesinger, P.H., and Wickline, S.A. (2008). Synthesis and characterization of stable fluorocarbon nanostructures as drug delivery vehicles for cytolytic peptides. Nano Lett. *8*, 1131–1136.

Stabeli, R.G., Marcussi, S., Guilherme, B.C., Rosemeire, C.L.R., and Pietro, H.S. (2004). Platelet aggregation and antibacterial effects of an L-amino acid oxidase purified from *Bothrops alternates* snake venom. Bioorg. Med. Chem. *12*, 2881–2886.

Stiles, B., Sexton, F., and Weinstein, S. (1991). Antibacterial effects of different snake venoms: purification and characterization of antibacterial proteins from *Pseudechis australis* (Australian king brown or mulga snake) venom. Toxicon *29*, 1129–1141.

Stocker, J.F., and Traynor, J.R. (1986). The action of various venoms on *Escherichia coli*. The J. Appl. Bacteriol. *61*, 383–188.

Takeda, S., Takeya, H., and Iwanaga, S. (2012). Snake venom metalloproteinases: structure, function and relevance to the mammalian ADAM/ADAMTS family proteins. Biochim. Biophys Acta *824*, 164–176.

Talan, D.A., Citron, D.M., Overturf, G.D., Singer, B., Froman, P., and Goldstein, E.J. (1991). Antibacterial activity of crotalid venoms against oral snake flora and other clinical bacteria. J. Infect. Dis. *164*, 195–198.

Talan, D.A., Moran, G.J., and Abrahamian, F.M. (2008). Severe sepsis and septic shock in the emergency department. Infect. Dis. Clin. North Am. *22*, 1–31.

Talan, D.A., Krishnadasan, A., Gorwitz, Fosheim, G.E., Limbago, B., Albrecht, V., and Moran, G.J. (2011). Comparison of *Staphylococcus aureus* from skin and soft tissue infections in US emergency department patients, 2004 and 2008. Clin. Infect. Dis. *53*, 144–149.

Tashima, A.K., Sanz, L., Camargo, A.C.M., Serrano, S.M.T., and Calvete, J.J. (2008). Snake venomics of the Brazilian pitvipers *Bothrops cotiara* and *Bothrops fonsecai*. Identification of taxonomy markers. J. Proteom. *71*, 473–485.

Tellechea, A.L.E., Veves, A., and Carvalho, E. (2010). Inflammatory and angiogenic abnormalities in diabetic wound healing: role of neuropeptides and therapeutic perspectives. Open Circul.Vasc. J. *3*, 43–55.

Termen, S., Tollin, M., Olsson, B., Svenberg, T., Agerberth, B., and Gudmundsson, G.H. (2003). Phylogeny, processing and expression of the rat Cathelicidin rCRAMP: a model for innate antimicrobial peptides. Cell. Mol. Life Sci. *60*, 536–549.

Terp, S., Krishnadasan, A., Bowen, W., Joo, J., Furoy, D., Chan, J., Moran, G., and Talan, D.A. (2014). Introduction of rapid methicillin-resistant *Staphylococcus aureus* polymerase chain reaction testing and antibiotic selection among hospitalized patients with purulent skin infections. Clin. Infect. Dis. 1–3.

Tossi, A., Sandri, L., and Giangaspero, A. (2000). Amphipathic, alpha-helical antimicrobial peptides. Biopolymers *55*, 4–30.

Toyama, M.H., Garcia de Oliveira, D., Beriam, L.O.S., Camillo Novello, J., Rodrigues-Simioni, L., and Marangoni, S. (2003). Structural, enzymatic and biological properties of new PLA_2 isoform from *Crotalus durissus terrificus* venom. Toxicon, 41, 1033–1038.

Toyama, M.H., Toyama, D.O., Joazeiro, P.P., Carneiro, E.M., Beriam, L.O., Marangoni, L.S., and Boschero, A.C. (2005). Biological and structural characterization of a new PLA_2 from the *Crotalus durissus collilineatus* venom. Protein J. *24*, 103–112.

Trabi, M., Schirra, H.J., and Craik, D.J. (2001). Three-dimensional structure of RTD-1, a cyclic antimicrobial defensin from Rhesus macaque leukocytes. Biochemistry *40*, 4211–4221.

Tu, A.T. (1988). Snake venoms: general background and composition In Venoms: Chemistry and Molecular Biology (John Willey and Sons, New York. NY), pp. 1–19.

Vaara, M., Fox, J., Loidl, G., Siikanen, O., Apajalahti, J., Hansen, F., Frimodt-Møller, N., Nagai, J., Takano, M., and Vaara, T. (2008). Novel polymyxin derivatives carrying only three positive charges are effective antibacterial agents. Antimicrob. Agents Chemother. *52*, 3229–3236.

Vargas, L.J., Londoño, M., Quintana, J.C., Rua, C., Segura, C., Lomonte, B., and Núñez, V. (2012). An acidic phospholipase A_2 with antibacterial activity from *Porthidium nasutum* snake venom. Comp. Biochem. Physiol. Part B *161*, 341–347.

Varoga, D., Pufe, T., Harder, J., Schroder, J.M., Mentlein, R., Meyer-Hoffert, U., Goldring, M.B., Tillmann, B., Hassenpflug, J., and Paulsen, F. (2005). Human betadefensin 3 mediates tissue remodeling processes in articular cartilage by increasing levels of metalloproteinases and reducing levels of their endogenous inhibitors. Arthritis Rheum. *52*, 1736–1745.

Vetter, I., Davis, J.L., Rash, L.D., Anangi, R., Mobli, M., Alewood, P.F., Lewis, R.J., and King, G.F. (2011). Venomics a new paradigm for natural products-based drug discovery. Amino Acids *40*, 15–28.

Vidal, N., Delmas, A.S., David, P., Cruaud., C., Couloux, A., and Hedges, S.B. (2007). The phylogeny and classification of caenophidian snakes inferred from seven nuclear protein-coding genes. C. R. Biol. *330*, 182–187.

Vlieghe, P., Lisowski, V., Martinez, J., and Khrestchatisky, M. (2010). Synthetic therapeutic peptides: science and market. Drug Discov. Today *15*, 40–56.

Vonk, F.J., Jackson, K., Doley, D., Madaras, F., Mirtschin, P.J., and Nicolas Vidal, N. (2011). Snake venom: from fieldwork to the clinic. Bioessays *33*, 269–279.

Wang, Y., Hong, J., Liu, X., Yang, H., Liu, R., Wu, J., Wang, A., Lin, D., and Lai, R. (2008). Snake Cathelicidin from *Bungarus fasciatus* Is a Potent Peptide Antibiotics. PLoS One *3*, e3217.

Wang, Y., Zhang, Z., Chen, L., Guang, H., Li, Z., Yang, H., Li, J., You, D., Yu, H., and Lai, R. (2011). Cathelicidin-BF, a snake cathelicidin-derived antimicrobial peptide, could be an excellent therapeutic agent for acne vulgaris. PLoS ONE *6*, 1–6.

Warrell, D.A. (2004). Snakebites in Central and South America: epidemiology, clinical features and clinical management. In The Venomous Reptiles of the Western Hemisphere, Vol II, J.A. Campbell, and W.W. Lamar, eds. (Comstock, Ithaca, NY), pp. 709–761.

Wen, Y., Gu, J., Li, S.L., Reddy, M.A., Natarajan, R., and Nadler, J.L. (2006). Elevated glucose and diabetes promote interleukin-12 cytokine gene expression in mouse macrophages. Endocrinology *147*, 2518–2525.

WHO (2001). WHO global strategy for containment of antimicrobial resistance (World Health Organization, Geneva, Switzerland) (Executive2001/WHO_CDS_CSR_DRS_2001.2a.pdf).

Xie, J.P., Yue, J., Xiong, Y.L., Wang, W.Y., Yu, S.Q., and Wang, H.H. (2003). *In vitro* activities of small peptides from snake venom against clinical isolates of drug-resistant *Mycobacterium tuberculosis*. Int. J. Antimicrob. Agents. *22*, 172–174.

Xu, C., Ma, D., Yu, H., Li, Z., Liang, J., Lin, G., Zhang, Y., and Lai, R. (2007). A bactericidal homodimeric phospholipases A_2 from *Bungarus fasciatus* venom. Peptides *28*, 969–973.

Xu, F., Zhang, C., and Graves, D.T. (2013). Abnormal cell responses and role of TNF-α in impaired diabetic wound healing. Biomed. Res. Int. *2013*, 754802.

Zanetti, M. (2004). Cathelicidins, multifunctional peptides of the innate immunity. J. Leukoc. Biol. *75*, 39–48.

Zhao, C., Nguyen, T., Boo, L.M., Hong, T., Espiritu, C., Orlov, D., Wang, W., Waring, A., and Lehrer, R.I.

(2001). RL-37, an alphahelical antimicrobial peptide of the rhesus monkey. Antimicrob. Agents Chemother. *45*, 2695–2702.

Zhou, H., Dou, J., Wang, J., Chen, L., Wang, H., Zhou, W., Li, Y., and Zhou, C. (2011). The antibacterial activity of BF-30 *in vitro* and in infected burned rats is through interference with cytoplasmic membrane integrity. Peptides *32*, 1131–1138.

Zuliani, J.P., Kayano, A.M., Zaqueo, K.D., Neto, A.C., Sampaio, S.V., Soares, A.M., and Stabeli, R.G. (2009). Snake venom L-amino acid oxidases: some consideration about their functional characterization. Protein Pept. Lett. *16*, 908–912.

New Targets for Antibacterial Compounds

14

Lynn L. Silver

Abstract

Resistance to antibacterials underwent a noticeable rise starting in the mid-1980s, at a time when natural product screening for antibacterial antibiotics had become less productive. The advent of genome sequencing of bacteria, starting in 1995, seemed to provide an answer to the resistance problem: find new gene products to target in order to find antibacterial agents that were different from previous classes and unlikely to be cross-resistant with them. This plan relied on the assumption that the limitation to discovery was a lack of novel targets. As the search for antibacterials over the past twenty years has been largely directed towards finding inhibitors of novel targets, and that search has been largely unproductive, it seems likely that the assumption that targets are rate limiting is wrong. This chapter will discuss the characteristics that define good targets, with an emphasis on recognizing the potential for rapid resistance development with single-gene targets. Until it is robustly demonstrated that combinations of single-targeted agents can prevent or retard resistance development (as is seen with tuberculosis, HIV and HCV) and a regulatory pathway is established to develop such combinations, it may be that we will have to depend on the successful targets that have already been found and exploited.

Introduction

This chapter summarizes developments in antibacterial discovery over the past 10 years, focusing on the validated targets of novel antimicrobial compounds. As will become apparent, these are limited. Describing antimicrobial targets as 'new' or 'unexploited' is a frustrating practice for old-time drug discoverers (including this author), as the history of antimicrobial discovery is one of >60 years of empirical discovery of compounds via 'kill-the-bug' screening without first choosing targets (White, 2012). The major classes of antibacterials are listed on a timeline in Fig. 14.1. Note that all of the antibacterial classes that were the progenitors of compounds currently in the clinic were discovered by 1987. The targets were, in almost all cases, discovered retrospectively and, the readout was generally cell death or growth inhibition. Thus all targets were 'exploited' for discovery. In the years from ~1960–1995, specific targets or pathways were chosen, based on physiology and microbial genetics using conditional mutants to pinpoint essentiality – but this was usually based on understanding of the well-studied bacterial species with well-defined genetic systems, such as *E. coli* or *S. aureus* or *B. subtilis*. Screening for inhibitors of these targets was often by whole-cell phenotypic assays that were designed to give a characteristic readout indicating that a specific target or pathway was inhibited (Silver, 2012b). With the advent of genome sequencing, many facets of the spectrum and homologies of prospective targets could be defined, although various genetic techniques are still necessary to prove or support essentiality. In actuality, few, if any, targets of interest nominated from genomic studies were novel or previously unknown. Screening strategies progressed from whole cell empirical and targeted phenotypic approaches to *in vitro* assays amenable to high throughput methodologies.

Target-based discovery has been the fashion in most therapeutic areas since the rise of

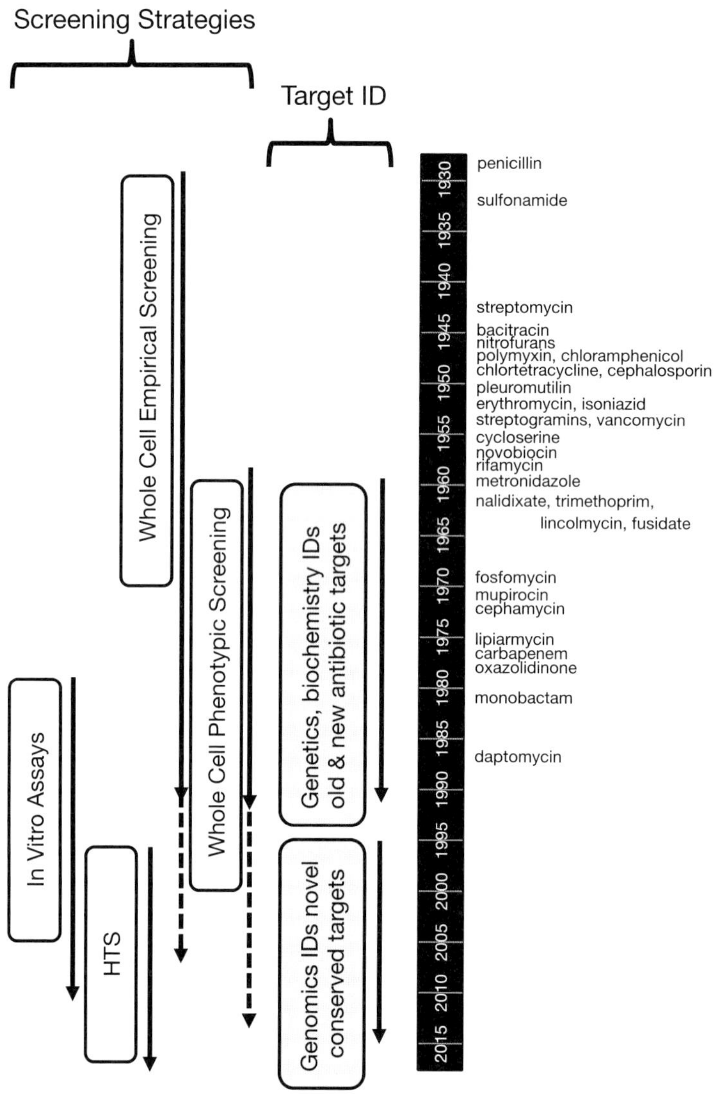

Figure 14.1 Timeline of dates of discovery of antibacterial classes. To the near left of the timeline are methods of target identification. Further left are the screening strategies used over time.

genomics. It has driven antibacterial discovery for that period – and unfortunately, that driving has generally led to the ditch. Note, in Fig. 14.1, that the more 'rational' approaches of the 1990s until the present have coincided with the (hopefully temporary) end of novel discovery. Even in other areas, including oncology and neurology, target-based discovery is being challenged by the re-emergence of phenotypic discovery. And as noted above, successful antibacterial discovery has been almost completely based on empirical and phenotypic screens. Thus, this chapter's title, New Targets for Antibacterial Compounds, is largely a misnomer. As discussed below, there may be novel compounds registered soon – but none from genomically identified targets.

Luckily, antibacterial discovery has not relied solely on finding new classes of molecules to hit new targets but has, instead, continued to turn out improved versions of the previously deployed compound classes which have been modified to overcome resistance or combined with inhibitors of resistance. As noted below, this approach has been productive and should continue to tide us over until new small molecule antibacterial classes or new paradigms of antibacterial therapy arrive.

Defining 'target'

In recent years, there have been many lists compiled, such as that in Table 14.1, of the qualities that define a good antibacterial target. Inhibitors of the targets in question, if such compounds can reach them, are those that should lead to cessation of growth or death of bacterial pathogens and should have a low probability of encountering pre-existing resistance. Thus, the main criteria are essentiality, potential useful antibacterial spectrum, lack of mammalian homologue, theoretical druggability, accessibility of location in the cell, and lack of cross-resistance with standard antibacterials (Pucci, 2001). It is conceivable that targets lacking some of those qualities might still be useful under some circumstances. Thus, in Table 14.1 are listed some caveats of selecting targets based on given criteria, and also some exceptions where targets apparently not meeting the criterion could 'pass'.

It is also important to predict or evaluate a low propensity for rapid selection of high level resistance upon challenge by the inhibitor. Indeed, while there are many proteins that might meet some of the criteria for identification as antibacterial targets, a key confounding point that may disqualify, or at least deprioritize, most of the apparent targets is that compounds inhibiting single enzyme targets have a high probability of selecting for single step resistance due to pre-existing mutations in the target gene (Silver, 2007, 2011; Silver and Bostian, 1993). In other therapeutic areas, including anti-viral, anti-tubercular, anti-parasitic and anti-cancer, such single target inhibitors are used in combinations that can overcome rapid resistance emergence due to target mutations. However, that has not been the case, thus far, in the treatment of more standard bacterial infections. While treating such infections with combinations to avoid rapid resistance development may well become established in the future, at present, a pathway for development of active

Table 14.1 Target qualification parameters

Quality	Identified	Caveat	Exceptions
Essentiality	Genetically: by lack of viable deletions; conditional mutations	Bypass mechanisms may be cryptic	If not essential, inhibitors might be useful in combinations. Possibility of 'virulence' targets
Spectrum sufficient for development	By sequence	Redundant or alternate functions may exist in certain species	Generally required to cover an indication, but newer regulations may allow very narrow, pathogen-specific drugs
Lack of mammalian homologue	By sequence	Similar functions with similar substrates may be inhibited by the same compounds	Small differences in sequence may allow differentiation;
Druggability	By biochemistry, enzymology, structure	Protein–protein interactions, shallow pockets may be difficult	Good screens of diverse libraries may find good inhibitors
Location	By sequence, biochemistry	Cytoplasmic targets, especially in Gram-negative bacteria, are hard to reach	Possibility of attaching smugglins – as long as these do not use inessential transport mechanisms
Lack of cross resistance with standard antibacterials	Empirically, structure		
Low propensity to select for resistance	Empirically, genetically [for bypass]	Single gene products may have a high propensity for selection of pre-existing single-step resistance mutations	Possibility of using such inhibitors in combination; possibility of iteratively optimizing to avoid and overcome resistance

compounds that *must* be given in combination has not been explored.

Many 'new' targets (those without previously validated inhibitors) have been proposed and programmes to discover inhibitors initiated – both in industry and academe. Efforts, based on genomics, to enumerate novel targets for a new generation of antibacterials have been presented and reviewed extensively, for example in (Abergel *et al.*, 2003; Allsop, 1998; Amini and Tavazoie, 2011; Black and Hodgson, 2005; Chalker and Lunsford, 2002; Chaudhuri *et al.*, 2009; Claverie *et al.*, 2002; Desnottes, 1996; Falconer and Brown, 2009; Forsyth *et al.*, 2002; Isaacson, 2002; Kim *et al.*, 2012; Loferer *et al.*, 2000; McDevitt and Rosenberg, 2001; Monaghan and Barrett, 2006; Payne *et al.*, 2004; Pucci, 2006, 2007; Sakharkar *et al.*, 2004; Trawick and Schilling, 2006; White and Kell, 2004; Yin *et al.*, 2004; Zalacain *et al.*, 2003). Thousands of papers document efforts towards discovering inhibitors of both broad and narrow spectrum targets. More recent reviews note the general failure to find good inhibitors of these targets and attempt to diagnose the obstacles to doing so (Gwynn *et al.*, 2010; Livermore, 2011; Overbye and Barrett, 2005; Payne *et al.*, 2007; Silver, 2011). Even with extensive efforts towards finding inhibitors of proposed targets, there are few new targets for which antibacterial inhibitors have been validated. As discussed below, validation of inhibitors, in this sense, means proof that the antibacterial activity – as measured by MIC – is due solely to inhibition of the targeted enzyme or function.

For monotherapeutic agents, essentiality is paramount. While this should be obvious, there have been inessential protein targets proposed which, if inhibited might lead to 'poison' formation, a dominant effect causing lethality. However, if the targets are not essential, then their loss by mutation or deletion of the structural gene would most probably lead to resistance at a high frequency (see ClpP discussion below).

Recently, in silico and bioinformatic triage of targets has been proposed – such as the antibacTR tool (Panjkovich *et al.*, 2014), which ranks targets in 74 Gram-negative pathogens on the basis of (1) their presence in a wide selection of pathogens, (2) sequence conservation, (3) essentiality, (4) possible redundancy and (5) similarity to human enzymes. A more global view of the application of informatics techniques to the discovery of new anti-infectives, including antibacterials (Hopkins *et al.*, 2011), takes into account the need to integrate methods including and beyond sequence analysis to predict essentiality, (beneficial) polypharmacology, chemical tractability, and the potential for rapid acquisition of drug resistance.

As noted above, 'location' is particularly important for anti-Gram-negative agents. The ability thus far of chemists rationally to confer bacterial-entry characteristics on *in vitro* active enzyme inhibitors is limited. It is generally possible to do so with Gram-positives, since the needed characteristics are generally similar to those for passive diffusion through most phospholipid bilayers. However, Gram-negatives are surrounded by two membranes with orthogonal sieving properties in addition to a wide variety of efflux pumps with overall broad substrate specificity [although varying by species]. To put it bluntly, having extensive lists of potential targets in the highly impermeable/actively effluxing lactose non-fermenters (*Pseudomonas*, *Acinetobacter*, *Burkholderia*, *Stenotrophomonas*) is largely useless as there is no clear way of rationally engineering/optimizing accumulation into those pathogens of initially impermeant enzyme inhibitors.

Resistance potential and how it influences target choice

In the discussion of target criteria above, the point was made that the potential for resistance selection by an inhibitor should be predicted or evaluated. While genetic and metabolic studies may predict that bypass (via mutation) of a specific function is possible, is it possible to predict other types of resistance, whether target-based or due to reduced accumulation? In fact, while it is not guaranteed, it would seem that with almost any single protein target, there is a very high probability that single mutational events can lead to changes in the target that reduce affinity sufficient to lead to high-level resistance. In other words, with single targets, the prediction, the null hypothesis, should be that single-step resistance *will* most likely occur. And these mutations will occur at some rate per generation in a population of bacteria BEFORE they are challenged with drug. There could be exceptions where the inhibitor has high affinity to target sites which cannot withstand change because their original essential function is too gravely compromised. It is generally thought that the fitness of a resistant mutant, its ability to grow at a normal rate and compete with wild type brethren, will determine whether such mutations will persist after initial selection, especially *in vivo*. But there is insufficient experimental foundation to allow useful predictions.

Because of the likelihood of single-step mutation to resistance, almost any single target that is defined via sequence is highly likely to be deprioritized a priori. As a consequence of this, and the recognition that the successful monotherapeutic systemic antibacterials are multitargeted or have targets that are the products of multiple genes or the action of those genes, most advances in antibacterials have focused on the same or related multitargets. This is a form of beneficial polypharmacology (Silver, 2012a). It is likely that the best targets for monotherapeutic agents are already in hand. But the possible utility of single target inhibitors may lie in combinations as discussed below. It is possible new multitargets may be discovered through informatics technology (Hopkins *et al.*, 2011).

Evaluation of any given target should be approached from a genetic point of view. One should try to predict the possibilities for single step resistance inherent in the specific target. Are there easy ways to evolve resistance by loss of a protein (as seen with PDF, below)? What kinds of changes to the target could be made to afford resistance? Even if this cannot be predicted based on structure, it must be assumed that target-based resistance (of a single enzyme target) will occur and thus this should be directly tested and dealt with. Although it is not clear if there is an absolute cut-off for an acceptable single-step resistance frequency, frequencies higher than 10^{-9} should probably be viewed as excessive. If no resistance is seen *in vitro* with a putative single target inhibitor, then the likelihood must be considered that the compound is hitting more than one target. This is not a bad thing for a drug, as long as the compound is non-toxic and effective; but if the goal is to rationally design and optimize single target inhibitors, then that goal will not have been met.

Validation of targets and inhibitors

Proposals for targets can arise from studies of conditional mutations, essentiality as defined by lack of knockouts in the gene, universal antisense scanning to identify essential genes (and transcripts). But each instance of the use of the chemical-biology method of defining targets by finding inhibitors first and deducing the target must be carefully vetted so that it is proven that inhibition of only that target is the proximal function that leads to cell death [or growth inhibition]. Similarly, when a compound is found to inhibit a specific candidate enzyme *in vitro*, this does not in any way prove that enzyme is the only possible target of the compound. Promiscuity/polypharmacology, especially with synthetic compounds, is a fact of life. If an inhibitor of a certain enzyme kills a bacterial strain, it does not follow that the bacterial strain dies as a result of inhibition of that enzyme.

In order to validate a target via an inhibitory compound, it is necessary for the inhibitor to have antibacterial activity (demonstrate an MIC) and for target-inhibition to be the sole initiator of the MIC (given that mediation may involve other players). If the target–inhibitor interaction is prohibited, are there any remaining phenotypes? The best way of showing this is to obtain resistance mutations that map in the structural gene for the proposed target; this may not be possible or might occur at a very low frequency – with other mutations occurring at higher frequency – leading to overexpression of the gene, bypass of the function, impermeability to or efflux of inhibitors. Of course, the existence of target mutations may be a Catch 22 and presage problems for development. Other methods for validating target/compound pairs have been reviewed (O'Neill and Chopra, 2004; Silver, 2011), and include measurement of inhibition of specific pathways of macromolecular synthesis, overexpression and underexpression of the putative target (leading to increased or decreased MIC, respectively) and comparisons with proteomic, transcriptomic and antisense arrays to identify patterns which indicate inhibition of certain gene products or pathways.

New compounds for old targets

The major monotherapeutic antibacterials have few targets, and the only enzymatic targets are the penicillin-binding proteins (PBPs), inhibited by the β-lactams, and the type II topoisomerases, DNA gyrase and topoisomerase IV (Topo IV), inhibited by the fluoroquinolones (FQs). Note that these drugs inhibit two or more essential proteins and, in so doing, lower drastically the potential for single-step high-level target-based resistance. Other 'old' targets are: ribosomal RNA, where multiple cistrons encode the rRNA in most pathogens; membranes, whose synthesis relies on the products of many genes and pathways; and intermediates in peptidoglycan synthesis such as Lipid II, the target of the glycopeptides. In all of these targets, the multiplicity of genes involved has been hypothesized to restrain resistance development. Additionally, these targets are all intimately involved in synthesis of macromolecular structures. The lack of metabolic targets in this list is notable and may reflect their single gene nature and the possibility that inhibitor-induced auxotrophs may be supplied with missing metabolites by the host. The possibilities of metabolic targets for novel antibacterials is discussed below. In the current section, the classic targets are discussed in terms of the compounds that inhibit them and the mechanisms of resistance that may affect them.

Topoisomerases

In the last decade of efforts towards discovery of new antibacterials, topoisomerase inhibitors are a, if not the, major classes under study. The two type II topoisomerases of bacteria (except mycobacteria) are DNA Gyrase, responsible for introducing supercoils into DNA ahead of the replication fork, and Topo IV, responsible for decatenating daughter chromosomes after replication. Both enzymes are essential for cell viability and are tetramers of type α2β2. The A subunits, GyrA and ParC (of Topoisomerase IV) are responsible for the catalytic breakage and rejoining of DNA strands in order to introduce supercoils. The B subunits, GyrB and ParE, include the ATP binding and ATPase domain which provides energy for the movement of the complex. Recent reviews cover the activity and general structure of the bacterial topoisomerases and prospects for new inhibitors (some discussed below) of the A and B subunits (Chan *et al.*, 2013; Pucci and Wiles, 2014; Tomasic and Peterlin, 2014).

The FQs such as ciprofloxacin and moxifloxacin (Fig. 14.2a and b) target the A subunits of two bacterial type II topoisomerases, GyrA and ParC, leading to formation of DNA–enzyme complexes which result in double-stranded breaks (demonstrated by drug-dependent DNA-cleavage reactions) that can lead to rapid killing (Drlica *et al.*, 2009). Both of these compounds have excellent activity against Gram-negatives, including *P. aeruginosa*, while, additionally, moxifloxacin has excellent anti-Gram-positive activity. While ciprofloxacin inhibits Gyrase at lower concentration than it does Topo IV, moxifloxacin has relatively balanced activity against the two enzymes. Due to this, ciprofloxacin selects resistance mutations in *gyrA* at significant frequency (~10^{-8}), and these lead to a rise in MIC to a level defined by the next susceptible target, Topoisomerase II. An additional mutation in *parC* can then lead to a much greater rise in MIC. Moxifloxacin's balanced inhibition, on the other hand, yields a very low probability of selecting mutations in a single step that significantly raise the MIC, since that would require changes in both targets simultaneously. Although varying with specific FQs and genera, it is generally the case that gyrase is the more sensitive FQ target in Gram-negatives while TopoIV is more sensitive in Gram-positives, especially *S. aureus*.

It is generally the case with inhibitors where single step resistance is seen, resistance at a significant frequency occurs at low multiples of the MIC *in vitro*, but is not seen at higher drug concentration. This may be due to the presence of a secondary target that is less sensitive to inhibition than the primary target – as is seen with FQs such as ciprofloxacin that are unbalanced inhibitors of Gyrase and Topo IV. The concentration at which single step mutations are no longer seen has been termed the mutant prevention concentration, or MPC (Dong *et al.*, 1999).

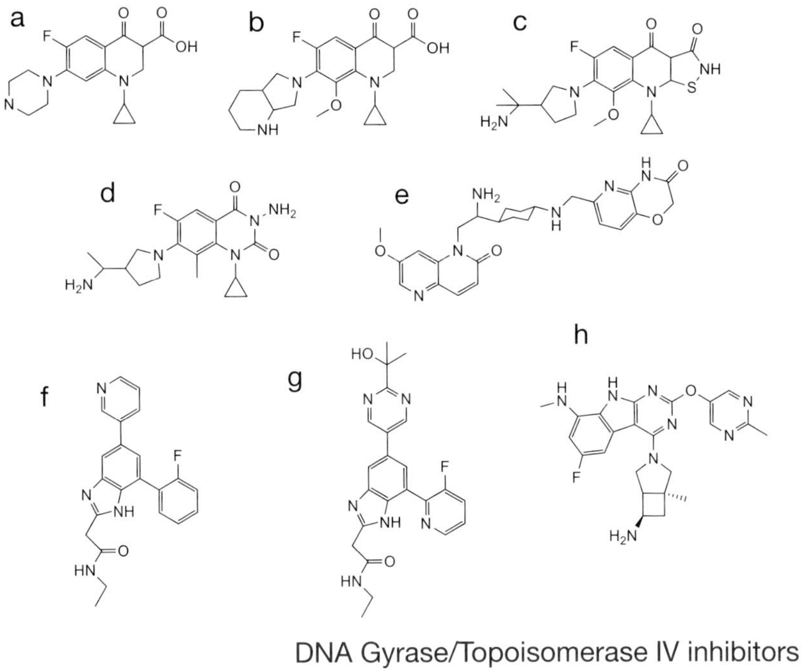

Figure 14.2 DNA Gyrase/Topoisomerase IV Inhibitors. (a) ciprofloxacin; (b) moxifloxacin; (c) ACH-702; (d) PD 0305970; (e) NBTI5463; (f) VRT-752586; (g) Vertex compound 4; (h) Trius C4.

Quinolone analogues, such as the isothiazoloquinolones were developed by Achillion, their lead being ACH-702 which has a mostly Gram-positive spectrum with balanced inhibition of Gyrase and Topo IV (Fig. 14.2c). The lowered Gram-negative activity is due to limitations in permeability and efflux susceptibility. ACH-702 remains highly active against ciprofloxacin-resistant *S. aureus* strains, although its MICs are raised above that of wild type, and it has a very low propensity ($< 10^{-10}$) for selecting additional resistance mutations (Pucci *et al.*, 2011). The compound showed efficacy in animal models, but its development as a systemic agent was halted due to very rapid metabolism. It may be used for ophthalmic infections (Pucci and Bush, 2013).

The quinazoline diones (QDs), are close analogues of FQs, not cross-resistant with the FQs but also leading to DNA-cleavage (Pan *et al.*, 2009). The 3-amino-QD PD0305970 (Fig. 14.2d), discovered by Pfizer, is active against FQ resistant strains and, interestingly selects for resistance mutations in the *gyrB* and *parE* genes (Huband *et al.*, 2007). Thus, although the activities of the QDs and FQs are similar, their binding, or at least resistance, mutations are at different sites. A class of inhibitors referred to as NBTIs (novel bacterial topoisomerase inhibitors) have varying structural motifs but all seem to bind to gyrase–DNA complexes, as do FQs, but do not form cleavage complexes and are not cross-resistant with FQs. They appear to be allosteric inhibitors of the DNA-dependent ATPase of the B subunits of the topoisomerases (Shapiro and Andrews, 2012). NBTIs have been described by GSK, Pfizer, Novexel, AstraZeneca, Actelion, and others. One of them, GSK2140944, whose structure has not been disclosed, has completed phase I trials and phase II dose ranging studies are planned. Most of the NBTIs have generally Gram-positive spectra, but NBTI5463 (Fig. 14.2e) from AstraZeneca has significant Gram-negative activity, including *P. aeruginosa* and shows *in vivo* activity in *P. aeruginosa* lung and thigh and *E. coli* thigh models.

The B-subunits of gyrase and Topo IV are important ‘new’ dual targets. While novobiocin, discovered in 1956, targets the GyrB subunit, it was unsuccessful in the clinic due to rapid resistance development and adverse events.

Novobiocin is a Gram-positive drug, as it is large and subject to efflux from Gram-negatives. Work by Vertex showed that novobiocin activity against TopoIV could be increased 20-fold by a single amino acid change in the protein (Bellon *et al.*, 2004). Thus, they reasoned that compounds might be found which could hit both targets in a balanced manner. Such dual targeting inhibitors, two aminobenzimidazoles, were disclosed (Grossman *et al.*, 2007; Mani *et al.*, 2006). The one, VRT-752586 (Fig. 14.2f), that had better potency and balanced inhibition of both targets was highly serum protein bound (albeit MICs in the presence of serum were still relatively low) and the compound, VRT-125853,with low serum binding was less well balanced. The spectrum was Gram-positive with activity against *H. influenzae* and *Moraxella catarrhalis*, as well. A recent paper from Vertex describes efforts towards reducing the serum binding and presents a compound (compound 4, Fig. 14.2g) that is active *in vivo* (Perola *et al.*, 2014). After the Vertex work, many other groups attacked the dual GyrB/ParE targets, mostly using high throughput *in vitro* screening. A number of molecules with similarities to the Vertex aminobenzimidazole leads have been reported by many groups. A different compound class, the pyrrolopyrimidines, with anti-Gram-negative activity and balanced activity against GyrB and ParE have been synthesize based on structural information and fragment-based drug design by the group at Trius (now Cubist) (Tari *et al.*, 2013; Trzoss *et al.*, 2013). Another class, the tricyclic GyrB/ParE (TriBE) inhibitors have been recently disclosed, with data given for representative compounds (such as compound C4, Fig. 14.2h), showing excellent broad spectrum activity and *in vivo* efficacy (mouse neutropenic thigh model with *E. coli*). The Trius work was done with attention paid to optimizing for balanced inhibition of both enzymes and, at the same time, antibacterial activity. Resistance rates are extremely low. This type of approach to this 'new' target gives some hope for the future of small molecule antibacterial discovery.

Penicillin-binding proteins (PBPs)

PBPs which encode essential peptidoglycan-transpeptidases are the prime antibacterial target. β-Lactams (BLs) which inhibit these enzymes have been a mainstay of antibacterial therapy since the discovery of penicillin and the development of many generations of penicillins, cephalosporins, monobactams and, finally, carbapenems. The PBP targets are present in all bacterial pathogens (although they are not involved in cell wall synthesis in *Chlamydia*), but the spectrum of any given compound is determined by its preference for various PBPs, its susceptibility to β-lactamases (BLAs) and, in Gram-negatives, its ability to enter the periplasm and avoid efflux. Since in each organism there are at least two essential PBPs, resistance does not arise through single-step target-based mutation. Instead, resistance has arisen via BLAs or the presence of modified BL-resistant, horizontally transmitted PBPs. As described below, initial resistance to BLs was by BLAs which have been classified into four main groups, the Ambler classes (Hall and Barlow, 2005). BLA inhibitors (BLIs) were first developed to combine with penicillins to overcome resistance. However, in the mid-1980s, methicillin-resistant-*S. aureus* (MRSA) appeared and its incidence rapidly increased (Deresinski, 2005). This resistance was due to the importation of a gene, *mecA*, likely originally from an environmental *Staphylococcus* (Wu *et al.*, 2001), that encodes a peptidoglycan transpeptidase activity, PBP2a, insensitive to BL inhibition, which can take over for the inhibited native PBP-transpeptidases. By the early 1990s, efforts to synthesize cephalosporins and carbapenems that could inhibit PBP2a were under way. This eventually resulted in the development of ceftaroline and ceftobiprole, both cephalosporins with anti-MRSA activity. The advent of MRSA also led to a major undertaking by pharmaceutical companies to find other compounds which were effective against MRSA and other increasingly problematic Gram-positive pathogens. This led to the development of the last novel antibacterial classes to be discovered (and arrive in the clinic), the oxazolidinones and daptomycin (see Fig. 14.1 and below).

Although a large variety of screens aimed at finding new cell wall inhibitors, including new classes of inhibitors of PBPs, have been deployed since the 1960s (Silver, 2012b), no developable non-BL inhibitors of PBPs have been found – until this past year. Such a class of non-BL inhibitors of PBPs, the oxadiazoles, have been described (O'Daniel *et al.*, 2014). They are active against MRSA, inactivating PBP2a, as well as other PBPs. The compounds, such as compound 3 (Fig. 14.3a) are 100% orally bioavailable, exhibit favourable toxicity profiles and pharmacokinetics and are active both IV and PO in animal models of MRSA infection.

Β-lactamases

Combinations of BLs and BLIs have been in use since the mid-1980s. These include amoxicillin-clavulanic acid (1985), ampicillin-sulbactam (1987) and piperacillin–tazobactam (1994), the BLs in each combination being penicillins. These BLIs are inhibitors, mainly, of Ambler Type A BLAs which have an active-site serine, are mostly plasmid-borne and are widely spread both among Gram-positives and Gram-negatives. These combinations have been successful, both medically and commercially. The specifics of the generations of BLs and the apparent co-evolution of BLAs has been reviewed (Bush, 2010; Jacoby and Munoz-Price, 2005) – but will be briefly summarized here. Early cephalosporins were relatively resistant to Type A BLAs and later generations of cephalosporins were designed to overcome hydrolysis by many of the Type A enzymes, without the addition of BLIs. However, Type A BLAs with extended spectra (extended spectrum BLs = ESBLs) that hydrolyse many cephalosporins have spread and other types of BLAs have become more prevalent and problematic, including the chromosomally encoded Type C BLA enzymes with serine at the active site. Additionally, BLAs resistant to the standard BLIs have become increasingly prevalent – thus possibly compromising the use of the penicillin/BLI combinations in use. Carbapenems, which are generally insensitive to the standard ESBLs and Type C BLs gained in usage, especially for enteric Gram-negatives. This most likely has led to the rise in problematic enzymes with potent hydrolytic activity against carbapenems, as well as the penicillins and cephalosporins. These include novel Type A enzymes, such as the KPCs (Bush, 2010), and the Type B (non-serine) metalloenzymes that can attack carbapenems, such as the recently described enzyme NDM-1 (Moellering, 2010). With the rise of carbapenem-hydrolysing BLAs in enteric bacteria, the spectre of untreatable carbapenem-resistant Enterobacteriaceae (CREs) has grown. Luckily, this trend had been recognized by drug developers in the early 2000s and new BLIs are in the pipeline for combination with a variety of BLs to combat BL-resistant Gram-negative infections. Avibactam (Fig. 14.3b), the diazabicyclooctane (DBO) developed by Novexel and now owned by AstraZeneca, that has good activity against most Type A and Type C BLs (Ehmann *et al.*, 2012), is under study in combination with ceftazidime, ceftaroline and aztreonam. Another DBO, MK7655 (Fig. 14.3c), is in development in combination with the imipenem-cilastatin (Livermore *et al.*, 2013). This very active area of research and development of new BLIs and BL/BLI combinations and their potential for staving off a post-antibiotic era, has been expertly and recently reviewed (Drawz *et al.*, 2014).

Ribosomal RNA

The many natural product-derived antibacterial protein synthesis inhibitors shown in Fig. 14.1 were discovered by empirical screening and later their mechanisms of action elucidated. Most of the developed protein synthesis inhibitors bind to ribosomal RNA rather than ribosomal proteins (although with certain inhibitors, such as streptomycin and spectinomycin, mutations in

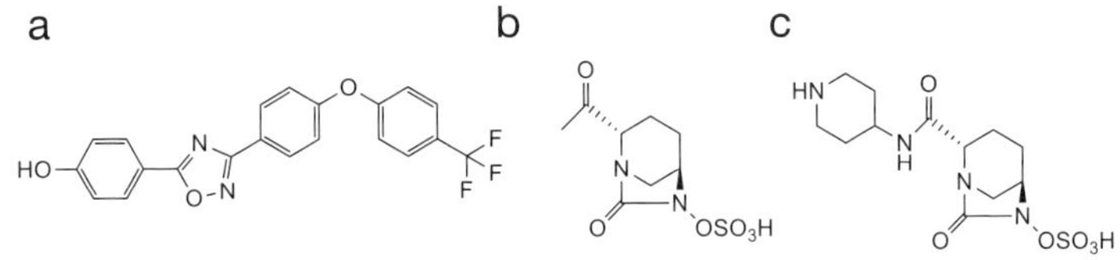

Figure 14.3 (a) O'Daniel compound 3; (b) avibactam; (c) MK7655.

proteins can disrupt binding and lead to resistance). Inhibitors binding to the 16 S RNA of the 30 S ribosomal subunit are the tetracyclines and aminoglycosides. Macrolides, lincosamides, chloramphenicol, pleuromutilins and streptogramins bind to 23S RNA of the 50S subunit and inhibit at or near the peptidyl transferase region. The newer, synthetic oxazolidinones also bind in that region. Resistance to these compounds is almost universally by various horizontally transmitted mechanisms of solute-specific efflux, methylation of the RNA target, or degradation of the antibiotic – but there are some exceptions. Mutational resistance in rRNA is very rarely seen in standard pathogens. Most of these pathogens have four or more copies of the rRNA genes. Changes in a single gene copy may give only minimal increases in MIC and multiple mutations are needed to afford high level resistance. Such multistep resistance to the oxazolidinones has been seen, albeit rarely, in the clinic with enterococci which have four or six copies of rRNA (Marshall *et al.*, 2002). Other pathogens have higher numbers of rRNA genes. On the other hand, *M. tuberculosis* has only one copy of the rRNA gene and hence is subject to single step high level resistance mutations in rRNA; *H. pylori* has two rRNA genes – and clarithromycin resistance in *H. pylori* does arise via mutation, since a single mutation, affecting 50% of the transcripts, does raise the MIC. The low target-based mutation potential of ribosomal RNA makes it an attractive antibacterial target.

While novel classes have not been brought forward recently, except for a new pleuromutilin (which class was discovered in 1950), there have been compounds in most classes which have been or are being developed to overcome resistance to the first generations of those classes. In the case of macrolides, the ketolides which have activity against many macrolide-resistant *S. pneumoniae* have been developed. Although telithromycin was registered, it met with toxicity problems that have minimized its use. A newer ketolide, solithromycin, being developed by Cempra, appears to have excellent activity against Gram-positives, including resistant *S. pneumoniae*, community-acquired MRSA and enterococci, and atypical organisms. Its structure could make it less prone to adverse events than is telithromycin. In the tetracycline class, semisynthetic derivatives, the glycylcyclines, showing good activity against many tetracycline-resistant pathogens, were investigated and tigecycline was developed by Wyeth/Pfizer. The pleuromutilins were discovered in 1950 but only relatively recently developed, by GSK, and that for topical use as retapamulin. There are phase II clinical trials in progress for a systemic pleuromutilin, BC3781, developed by Nabriva (Fig. 14.4a). Newer oxazolidinones, tedizolid and radezolid, are in clinical trials; they have activity against some oxazolidinone-resistant Gram-positive isolates and also appear to have lower toxicity potential.

Recently, the methodology for total synthesis of tetracycline, pioneered at Harvard (Charest *et al.*, 2005) and licensed by Tetraphase, has led to the production of thousands of novel tetracycline analogues that were previously inaccessible. Eravacycline is the first of these novel tetracyclines, a fluorocycline, being developed (Fig. 14.4b). It has broad-spectrum antibacterial activity including many MDR strains, but not including *P.*

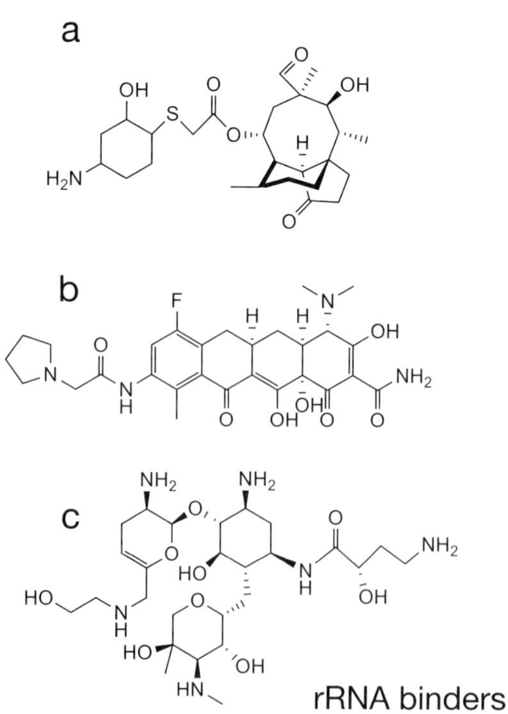

Figure 14.4 Ribosomal RNA targeting inhibitors (a) BC-3781; (b) eravacycline; (c) plazomicin.

aeruginosa; has successfully completed a phase II trial for IV dosing of intra-abdominal infection; and may have oral bioavailability (Solomkin *et al.*, 2014; Sutcliffe *et al.*, 2013). Aminoglycosides have been useful drugs, especially against Gram-negatives, but resistance has been rising and toxicities have limited their use. An aminoglycoside, plazomicin, designed to be insusceptible to almost all aminoglycoside modifying resistance enzymes, although not to resistance due to ribosomal modification, is being developed by Achaogen for the treatment of serious infections due to MDR Enterobacteriaceae including CRE and is soon to enter a phase III trial (Fig. 14.4c). Melinta Therapeutics (was Rib-X) has discovered by structure based design, a new class of synthetic ribosome-binding protein synthesis inhibitors (the RX-04 series) with good Gram-negative activity, including *P. aeruginosa* (Housman *et al.*, 2012). No specific candidate appears to have been selected for clinical development as yet, but the effort appears promising. The detailed structural studies of bacterial ribosomes should allow the discovery of further new classes of compounds with therapeutic potential.

Membranes

A number of useful antibacterials target the membrane, leading to depolarization and/or membranolysis. Polymyxin and colistin, natural products acting on Gram-negatives, are renally toxic and their use had been limited, but recent increases in problematic MDR Gram-negative infections due to *P. aeruginosa* and CREs have necessitated a rise in their use. Polymyxins are cationic cyclic peptides which bind to Lipid A of LPS, its basic groups displacing Mg^{2+} that neutralizes Lipid A phosphates, which leads to local LPS disruption, allowing penetration to, and possibly through, the cytoplasmic membrane. Lethality appears to result from multiple interactions with the membranes and, likely, cytoplasmic functions (Zhang *et al.*, 2000). Resistance to the polymyxins arises from changes to LipidA which reduce interaction with the phosphates (Fernández *et al.*, 2013). Daptomycin is an anionic lipopeptide with Ca^{2+}-dependent anti-Gram-positive activity. Its mechanism of action entails Ca^{2+}-dependent binding to the membrane, eventually leading to membrane depolarization (and ion release) which correlates with cell death (Silverman *et al.*, 2003). Resistance *in vitro* results from stepwise changes leading to increased MICs, some of which reduce initial binding; similar mutations are found in resistant clinical isolates. Daptomycin MICs are raised in vancomycin-intermediate *S. aureus* (VISA strains) which also arise through step-wise changes leading to thickened cell walls which are thought to present a physical barrier to penetration of drug to target (Cui *et al.*, 2006). Recently, PolyMedix had been developing an antibacterial cationic peptide-mimetic (actually an arylamide oligomer) called brilacidin, for systemic use against *S. aureus*, with potential for broad spectrum activity. The compound appears to act by binding to and depolarizing the membrane (Choi *et al.*, 2009). Resistance was not observed during 40 serial passages in the presence of drug. Phase II results in acute bacterial skin and skin structure infections with *S. aureus* or MRSA were positive and the compound promising (Brilacidin, 2013). Financial constraints led to the company's being bought by Cellceutics – and the fate of brilacidin is for the moment unknown. Antimicrobial peptides (AMPs) have been under study for many years as possible therapeutic agents but their development, particularly for systemic use, has not yet been successful due to problems of protein binding, poor pharmacokinetics and rapid clearance, metabolic instability, salt sensitivity and, most importantly, lack of selectivity for bacterial (or fungal) membranes leading to haemolysis and/or haemolysis. The peptide mimetics mentioned above have progressed farthest. Any further progress of AMPs will rely on optimization to overcome all the obstacles, probably simultaneously – as most programmes described in the literature indicate selection for potent antibacterial activity and counterselection for only a subset of the other obstacles.

Lipid II

Lipid I (GlcNAc-MurNAc(-pentapeptide)-P-P-undecaprenol) is the target of the glycopeptides, notably vancomycin. It is the product of the cytoplasmic Mur pathway steps of peptidoglycan synthesis and the substrate of the transglycosylases and transpeptidases which assemble the cell wall external to the cytoplasmic membrane. Lipid II as an antibacterial target has been recently reviewed (Schneider and Sahl, 2010; Silver, 2013). Various drugs and inhibitors bind to different parts of Lipid II, vancomycin binding to the terminal D-ala-D-ala of the pentapeptide portion. Vancomycin is active only against Gram-positive bacteria since it is excluded from the periplasm by the outer membrane. High level vancomycin resistance, seen almost exclusively in Enterococci, arose from import (by horizontal genetic transfer) of a novel substitute for part of the Mur pathway produces an altered pentapeptide, actually a depsipeptide, terminating in D-ala-D-lactate, for which vancomycin has very low affinity (Bugg *et al.*, 1991). As noted above, VISA strains, which may compromise therapy, have arisen through multiple changes. It should be noted that vancomycin resistance was not evident in the clinic for approximately 30 years after its introduction – so, it may be inferred that its target is a very good one, vis à vis low resistance potential. Two glycopeptides with lipophilic modifications, telavancin and oritavancin have activity against vancomycin-resistant enterococci (VRE). However, the label for telavancin, which has been registered, includes only vancomycin-susceptible enterococci. Oritavancin is still under development, so its labelling is as yet unknown.

Progress on single target inhibitors

Table 14.2 presents a list (not exhaustive, but extensive) of bacterial targets for which validated inhibitors have been reported over the last 10 years. While there have been inhibitors reported for many more bacterial enzymes nominated as antibacterial targets, only a subset of inhibitors have been shown to be antibacterial (have an MIC) due to inhibition of the chosen enzyme. In the table, the target and a reference to inhibitors are given. The fate of clinical candidates is included in a fourth column. Six targets are discussed below, five of which have inhibitors that are clinical candidates. The first three may be subject to resistance problems; the second three may be less subject to rapid resistance development.

Validated targets with clinical candidates where resistance may be problematic

If resistance can arise at significant frequency in the laboratory, will that compromise therapy? What frequency is 'significant'; what is the cut-off? These questions, unfortunately, are not yet addressable with certainty. Since the hypothesis, promoted in the early 1990s, that single target inhibitors would be inferior to multitarget inhibitors in the clinic due to a propensity for resistance development, no single target inhibitor had reached phase II in the clinic to challenge the hypothesis – until results of studies with GSK2251052.

The cautionary tale of AN3365/GSK2251052

In September of 2009, Anacor presented data at a national meeting on a representative oxaborole inhibitor, ABX, of the editing function of leucyl tRNA synthetase (LeuRS) which showed promising *in vitro* and *in vivo* activity against Gram-negative organisms, including *P. aeruginosa* (Freund *et al.*, 2009). The tRNA synthetases were recognized targets in the pre-genomic era, based on genetics, biochemistry and the existence of several antibiotics targeting them, including mupirocin, an inhibitor of isoleucyl tRNA synthetase. In November of 2009, it was announced that GSK was developing an Anacor LeuRS inhibitor, AN3365/GSK2251052 that had entered phase I trials (Fig. 14.5a). In February of 2012, a news release revealed that phase II trials of the compound had been suspended due to resistance arising during therapy. The details of the compound, trials and resistance emergence have been more recently presented. *In vitro* resistance frequencies were ≥ 10^{-8} and mutations mapped to a site in the editing function of LeuS (Hernandez *et al.*, 2013). *In vivo*, resistance arose between day 1 and day 2 of therapy during a phase IIb trial for complicated urinary tract infection (cUTI), in 3 of 14 patients treated with the test drug. Either single or double mutations in LeuS were found in the resistant isolates and the mutants exhibited MICs raised 32- to 2056-fold above baseline. On the other hand, no resistance was seen in a phase II trial of the compound in complicated intra-abdominal infection (cIAI). (Twynholm *et al.*, 2013). In hollow fibre *in vitro* infection models, based on conditions in the cUTI trial and using pre-treatment isolates, results mimicked what was seen *in vivo* (VanScoy *et al.*, 2013). The finding that potent single-target inhibitors appearing effective in animal models could lead rapidly to clinical resistance emergence during therapy was a wakeup call. While in this case, an *in vitro* model might have predicted such resistance emergence, it is difficult to rely on such information at present, since so few single-targeted compounds have been reported to have been run in such models, and there was no resistance seen in the cIAI trial. It seems necessary to establish standardized models to test for resistance emergence during therapy so that better clinical predictions can be made.

Peptidyl deformylase (Pdf)

Peptidyl deformylase (Pdf) is pre-genomic target proposed when methods were developed to stabilize and characterize the enzyme (Mazel *et al.*, 1994) In bacteria, the AUG initiation codon on mRNA is recognized by charged formyl-methionyl-tRNA^{fmet} – with formylation of the methionine, catalysed by formyl-methionine-transferase (Fmt), occurring on the methionine-charged tRNA. After translation, the formyl group is removed by Pdf and then, in many proteins, the deformylated N-terminal methionine is removed by methionine aminopeptidase, Map. It was shown (Mazel *et al.*, 1994) that Pdf could not be disrupted in wild type *E. coli*, but could be disrupted if Fmt was also deleted. Fmt is not essential for translation, although cells lacking it are less fit – but, if it is present then Pdf must be present to remove the formyl group. Apparently there are proteins which will not function correctly unless the fMet is removed While this target was taken up by almost every pharmaceutical company, it seemed obvious from the genetics that deletion of *fmt* could lead to resistance and should happen at a high rate. Indeed, inhibitors of Pdf select for resistance in *S. aureus*, *E. coli*, and *H. influenzae*, mainly via loss of Fmt. In *S. aureus*, such mutants appear to have reduced virulence (Lewandowski *et al.*, 2013). Over time and under selective pressure by inhibitor, *fmt* mutants pick up compensatory mutations. In *Salmonella enterica*, many compensatory mutants overexpress tRNA^{fmet} (Nilsson *et al.*, 2006). In *S. aureus*, most of the compensatory mutants are in *agrC* (Zorzet *et al.*, 2012). While the *fmt* and compensated *fmt agrC* double mutants grew slowly *in vitro*, they were found to grow normally in a neutropenic mouse thigh model. In *S. pneumoniae*, different from the other species mentioned, resistance mutations occur at lower frequency (10^{-8}) in *pdf*, the target itself (Margolis *et al.*, 2001) – and these have a normal growth rate. What will be the results in humans in the clinic? The development of at least two Pdf inhibitors was halted after phase I a number of years ago, however the Pdf inhibitor, GSK1322322 (Fig. 14.5b) is in phase II trials, with no results reported as yet.

LpxC

The first recognized LpxC inhibitor, L-573,655, a phenyl oxazoline hydroxamate (Fig. 14.5c) was found in phenotypic screen for inhibitors of LPS synthesis (present only in Gram-negative bacteria) and later shown to be an inhibitor of the second step of LipidA synthesis, UDP-3-O-(R-3-hydroxymyristoyl)-N-acetylglucosamine deacetylase (Onishi *et al.*, 1996). The gene for the enzyme was shown to be the previously recognized permeability locus, *envA* and was redubbed *lpxC* (Young *et al.*, 1995). The enzyme proved to be zinc containing (Hyland *et al.*, 1997). Optimization of the oxazoline hydroxamate series at Merck yielded L-161,240, which had potent activity against *E. coli* and showed *in vivo* activity (Fig. 14.5d). However, no significant anti-pseudomonal activity was obtained (Onishi *et al.*, 1996). British Biotech's compound, BB-78485 (Fig. 14.5e) demonstrated activity against the Enterobacteriacease, fastidious Gram-negatives and *Burkholderia cepacia*, but not *P. aeruginosa* (Clements *et al.*, 2002). Work at Chiron based on L-161,240, produced CHIR-090 (Fig. 14.5f), which was optimized for a broader spectrum, including *P. aeruginosa* (Anderson *et al.*, 2004; Kline *et al.*, 2002; McClerren *et al.*, 2005). Pfizer published its LpxC inhibitor series, of which several were shown to have *in vivo* efficacy in a murine neutropenic thigh *P. aeruginosa* model (Brown *et al.*, 2012). None of these produced clinical candidates. However, Achaogen's inhibitor, ACHN-975 (Fig. 14.5g), entered phase I. The

Table 14.2 Antibacterial targets 2004–2014 with validated antibacterial inhibitors

Target	Compound	References	Status
Fatty acid, lipid and membrane synthesis			
FabI	AFN-1252	Karlowsky *et al*. (2009)	Phase II
FabF	Platensimycin	Wang *et al*. (2006)	
FabF, FabH	Platencin	Wang *et al*. (2007)	
Acc AD	Pyrrolidinediones	Freiberg *et al*. (2004)	
AccC	Benzimidazoles	Cheng *et al*. (2009)	
	Pyridopyrimidine	Miller *et al*. (2009)	
CoaA	N-alkyl pantothenates	Choudhry *et al*. (2003)	
LpxC	L-161,240	Onishi *et al*. (1996)	
	CHIR090	McClerren *et al*. (2005)	
	ACHN975	Badal *et al*. (2013)	Phase I discontinued
LptD	POL7080	Srinivas *et al*. (2010)	Entering phase II
RNA and protein synthesis			
RNAP switch region	Myxopyronin	Moy *et al*. (2011) and Srivastava *et al*. (2012)	
	Ripostatin	Buurman *et al*. (2012)	
Ef-Tu	LFF-571	LaMarche *et al*. (2012)	Phase II
Pdf	BB-83698	Lofland *et al*. (2004)	Phase I discontinued
	LBM-415	McIntyre *et al*. (2005)	Phase I discontinued
	GSK1322322	O'Dwyer *et al*. (2013)	In phase II
LeuRS	AN3365	Hernandez *et al*. (2013)	Phase II discontinued
MetRS	REP8839	Ochsner *et al*. (2005)	On hold after phase I
DNA synthesis and substrates			
PolC	EMAIPU	Kuhl *et al*. (2005)	
PolE	Nargenicin	Painter *et al*. (2011)	
DnaB	Benzobisthiazoles	Li *et al*. (2013)	
LigA	Naphthyridine carboxamides	Surivet *et al*. (2012)	
	Adenosine analogues	Mills *et al*. (2011)	
GyrA/B	See text		
ParC,E	See text		
DHFR	Iclaprim	Schneider *et al*. (2003)	Post phase III; not approved
	Propargyl-based inhibitors	Frey *et al*. (2012)	
Tmk	Sulfonyl piperidines	Martínez-Botella *et al*. (2013)	
Cell wall synthesis and cell division			
MurI	Pyrazolopyrimidinediones	de Jonge *et al*. (2009)	
MraY	Muraymycin analogues	Tanino *et al*. (2011)	
MurG	Murgacil	Mann *et al*. (2013)	
GlmU	Aryl sulfonamides	Stokes S.S. *et al*. (2012)	
PBPs	Oxadiazoles	O'Daniel *et al*. (2014)	
TarG	Targocil	Wang *et al*. (2013)	

Table 14.2 (Continued)

Target	Compound	References	Status
FtsZ	PC190723	Stokes N.R. *et al.* (2012)	
MreB	Indole compound	Robertson *et al.* (2007)	
Proteases and signal peptidases			
ClpP	ADEP4	Conlon *et al.* (2013)	
SpsB	Arylomycin	Smith and Romesberg (2012) and Tan and Romesberg (2012)	
	Krisinomycin derivatives	Therien *et al.* (2012)	

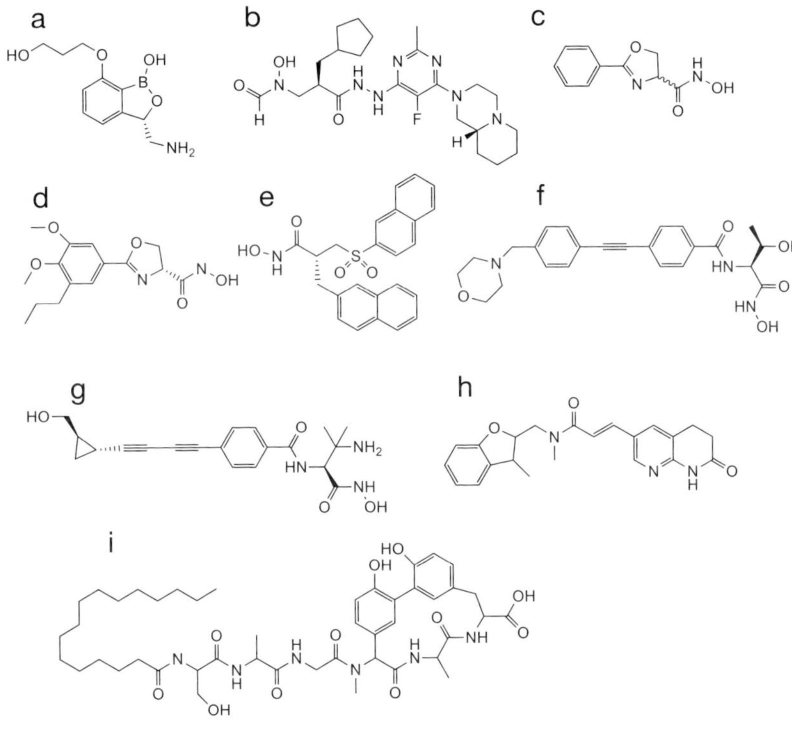

Figure 14.5 Single target inhibitors. (a) AN3365/GSK2251052; (b) GSK1322322 (c) L-573,655; (d) L-161,240; (e) BB78485; (f) CHIR-090; (g) ACHN-975; (h) AFN-1252; (i) arylomycin.

compound had excellent Gram-negative activity, including *P. aeruginosa*, and was active in an *in vivo* model of pneumonic plague (*Yersinia pestis*) (Badal *et al.*, 2013; Castanheira *et al.*, 2013; Heine *et al.*, 2013). Unfortunately, development of the molecule was halted when infusion site inflammation was seen, but further work on the series is envisaged (Achaogen, 2014).

Based on *in vitro* studies, a likely caveat for development of LpxC inhibitors is the potential for resistance development. Resistance to LpxC inhibitors has been shown in the past to arise via a bypass mutation in the *fabZ* gene which is a dehydratase involved in the synthesis of unsaturated fatty acids as well as by mutations in *lpxC* (Clements *et al.*, 2002). It is interesting that a missense mutation (leading to increased outer membrane permeability) in the *lpxC* (formerly *envA*) gene had been shown to be subject to an outside suppressor, *sefA1*, which was eventually identified as *fabZ* (Mohan *et al.*, 1994). The mechanism by which the *fabZ* mutation affords protection from

LpxC inhibitors and suppresses *lpxC* mutations was, until recently, unclear. Recent work, using CHIR090, shines light on this phenomenon (Zeng *et al.*, 2013). Single step mutations were selected in *E. coli* at 5× MIC of CHIR-090 at a frequency of 10^{-9} and subjected to a second round of selection at 50× MIC, yielding mutants at 10^{-7}. The first-round mutations mapped in *fabZ* and raised the MIC 60× over wild type. The second round of selection yielded mutations in *thrS* (the threonyl-tRNA synthetase gene) and led to an increase of 280-fold the wild type MIC. These mutations affected all the LpxC inhibitors tested, including a British Biotech compound, BB-78485 and L-161,240. The study by Zeng *et al.* (2013) led to the intriguing finding that resistance appeared to be due to a re-balancing of growth. This new work showed that the *fabZ* mutation led to a decrease of FabZ activity and also a decrease of LpxC activity (in the absence of inhibitor). This implies and indeed a variety of experiments support the idea that fatty acids and lipid A synthesis are co-regulated in order to maintain an optimum ratio between phospholipids and lipidA. In other words, the protection afforded by reduction of FabZ seems to be the rebalancing of the ratio. Sensitivity to LpxC inhibitors would then be due to unbalanced growth. The second round mutations in *thrS* leads to reduced protein synthesis and should serve to further restore homeostasis, albeit at a lower overall growth rate. There was no evidence that FabZ or ThrS were secondary targets of the inhibitors – but were instead acting as a bypass or rebalancing mechanism. This is supported by the effects of the mutations on compounds of different structures. There have been reports of spontaneous mutations in LpxC selected with L-161,240 and BB-78485, but they have not been reported for CHIR-090. Thus, there is evidence that resistance to LpxC inhibitors can arise that would be difficult to overcome by modifying the inhibitors. But it is possible that a sufficiently potent inhibitor might be difficult to suppress by restoration of homeostasis; furthermore, with a very potent inhibitor, the ultimate MIC reached in resistant mutants might still be attainable, i.e. the MPC might be low enough to reduce resistance emergence.

Single target/inhibitor pairs that might overcome resistance

There has been progress with target–inhibitor pairs that may overcome the 'single-target curse'. Previous experience in antibacterials and oncology has shown that certain inhibitors of single enzyme-targets can be iteratively optimized to overcome specific types of target-based resistance. This is true of iclaprim, an analogue of trimethoprim, which is active against both wild type and trimethoprim-resistant forms of its target dihydrofolate reductase (DHFR) in *S. aureus* (Schneider *et al.*, 2003). The BCL-ABL kinase is a prime target for treatment of chronic myeloid leukaemia. The first inhibitor, imatinib (Gleevec) revolutionized CML therapy, but resistant mutants arose, compromising therapy in some patients. Ariad went on to design a second-generation inhibitor, ponatinib (Iclusig), active against imatinib-resistant mutants (Zhou *et al.*, 2011). While developed successfully, it proved to have toxicities which have limited its clinical use. In the examples below, the specific target–inhibitor pairs may well lead to development of monotherapies with low resistance potential.

LptD

The Polyphor company's peptide β-hairpin mimetics, cyclized peptides based on the cationic antibacterial peptide protegrin-1 as a starting point, led to the clinical candidate POL7080, which was found to have potent activity and a narrow anti-Pseudomonal spectrum. Biochemical and genetic analysis showed that the compounds were not membranolytic but specifically target the LptD β-barrel outer membrane protein which plays a key role in the biogenesis of LPS (Srinivas *et al.*, 2010; Werneburg *et al.*, 2012). Thus, this was not a target-driven programme; rather serendipitous retroidentification, which might now be called chemical biology. The system of LPS biogenesis has only recently been worked out (Ruiz *et al.*, 2009). Interestingly, the frequency of resistance to POL7080 is very low, reported as $\leq 10^{-10}$ (Srinivas *et al.*, 2010). The three mutants isolated contained an 18 base pair tandem repeat in the lptD coding sequence. This may well be a sufficiently rare event to have low potential for clinically problematic resistance. While the structure of LptD is as yet undisclosed and the binding site for POL7080 has not been determined, it seems likely that, as it spans the outer membrane, LptD provides a prime location for an anti-*Pseudomonas* target, outside the periplasm, not subject to permeability or efflux problems. LptD is present in other Gram-negative pathogens and it is possible that peptides similar to POL7080 might be designed or selected which have broader spectra, or at least target other important pathogens such as *E. coli* and/or Klebsiella. Mutations in LptD, originally called *imp* in *E. coli*, were known to lead to greatly increased outer membrane permeability and there is evidence that the LptD inhibitors do synergize other antibacterials (Srinivas *et al.*, 2010). POL7080 has been shown to be active against *P. aeruginosa* infections *in vivo*, has completed phase I trials and is now being co-developed by Polyphor and Roche.

FabI

FabI, an enoyl-acyl carrier protein (ACP) reductase, is the target of the first-line antitubercular drug isoniazid and has become a target of interest in other pathogens (Lu and Tonge, 2008). Notably, FabI was the target of a screening and optimization campaign by GlaxoSmithKline (GSK) (Miller *et al.*, 2002). While GSK intended the inhibitors for activity against respiratory pathogens, especially *S. pneumoniae*, it was found, after the compounds were in hand, that they lacked activity against *S. pneumoniae*, an important respiratory pathogen that GSK wished to cover. In fact, *S. pneumoniae* uses a different enoyl-ACP reductase, FabK, which is insensitive to the inhibitors. Thus, the compound series, outlicensed to Affinium, was further optimized to produce a compound, AFN-1252 (Fig. 14.5h), with extremely potent anti-Staphylococcal activity (Karlowsky *et al.*, 2009). Recent work has shown that, while resistance mutations can arise in *fabI*, the ultimate MIC is sub-micromolar with no further MIC increases seen upon further selection *in vitro*. This may bode well for its development and low potential for rapid resistance development in the clinic (Yao *et al.*, 2013).

Signal peptidase

Signal peptidase 1 (SP1) has been recognized as a likely single-enzyme target (Paetzel *et al.*, 2000) and a natural product inhibitor, arylomycin (Fig. 14.5i) was described (Kulanthaivel *et al.*, 2004). More recent studies on arylomycin in the Romesberg laboratory have shown that the spotty spectrum it displays was due to the existence of naturally occurring target mutations in different genera (Smith *et al.*, 2010). Analogues broadening the spectrum to include the intrinsically resistant pathogens have been produced through total synthesis and analoging (Roberts *et al.*, 2011). Recently, RQx Pharmaceuticals was established in order to develop arylomycin derivatives and it has entered into a collaboration with Genentech to further broaden the spectrum of arylomycins to overcome the intrinsic resistance of the target due to specific mutations. Conceivably, this approach could lead to compounds which are themselves refractory to further resistance development.

Overcoming resistance with sufficient dosing

The concept of controlling resistance emergence by the use of proper dosing has been promoted by many, starting with Drlica who proposed the concept of MPC, as explained above, a drug concentration at which single step resistance mutations can no longer be selected (Dong *et al.*, 1999). Drusano has recommended this approach of dosing to inhibit resistance (Drusano, 2003) as have many other pharmacokinetic/pharmacodynamic practitioners. While MPC may be defined by the presence of a second drug target, as with FQs, it may also be due to the limit to which a specific target can change to prevent binding; that is, the affinity of drug to target may be lowered by mutation to raise the MIC, but it may still bind at higher concentration. Thus, for single targeted agents where resistance selection leads to a moderate MPC, it might be possible to dose with sufficiently high drug levels to reach an MPC (or the *in vivo* equivalent of MPC). This may be the case with fidaxomicin, the recently approved therapy for *C. difficile*. It is not systemically absorbed, and very high faecal concentrations, > 1000 µg/g, can be reached. *In vitro*, the resistance frequencies seen for fidaxomicin in *C. difficile* have been reported from $<7 \times 10^{-7}$ to $<1 \times 10^{-9}$, with an MPC of 16 µg/ml, which should be well covered at *in vivo* concentrations (Leeds *et al.*, 2013; Scott, 2013). However, with most antibacterials and in many (or most) body sites, it may be hard to reach the MPC due to limitations of solubility and/or toxicity.

The 'promise' of combinations

When discussing the potential resistance problems that may face single-target inhibitors, the possibility of developing two – or more – such inhibitors for use in combinations invariably arises. It is clear that therapy in other anti-infective areas has rapidly adopted the use of combination therapy. The use of combinations to retard the development of resistance to single step mutations to resistance is standard in HIV, HCV, *M. tuberculosis*, malaria, and now even targeted cancer therapy. Drugs for each of these are targeted to single gene products; mutations to resistance can (and do) appear in pre-treatment populations and are selected for clonal expansion. In fact, the use of combinations in these areas was a consequence of the available drugs rapidly selecting for resistance. But therapy for standard bacterial pathogens has favoured monotherapy. Where combination therapy is used for non-tubercular infections it is for empirical dosing to cover possible pathogens present, for (theoretical) synergy or additivity, or with the BL/BLI combinations pairing a standard drug with an inhibitor of a resistance mechanism. Would it be possible to develop two different antibacterials that, by themselves, are subject to resistance arising during therapy? Would it be ethical? It is possible that the efficacy – in reducing resistance potential – could be robustly demonstrated in animal models. But again, could such combinations of two new active chemical entities be tested in man? Still it seems very important to address the problem.

The BL/BLI approach could be broadened to the use of classical compounds combined with specific inhibitors of resistance or persistence. One area where this has been suggested is in the synergistic activity of various inhibitors, at concentrations below their MICs, to resensitize MRSA to β-lactams. Merck has published on a number of inhibitors which do this, including inhibitors if FtsZ (Tan *et al.*, 2012), signal peptidase 1 (SP1, product of the *spsB* gene in *S. aureus*) (Huber *et al.*, 2011; Therien *et al.*, 2012), TarG (Wang *et al.*, 2013), and MurG (Mann *et al.*, 2013). In fact, the latter two types of inhibitors were found in a natural product screen for synergists of imipenem against MRSA. One of the two SP1 inhibitors is actinocarbasin, which is related to arylomycin, discussed above.

An exciting combination approach to fight biofilm infections and persistors living in biofilms, has recently been proposed (Conlon *et al.*, 2013). It is known that the peptidase ClpP is not essential, but a specific compound, acyldepsipeptide 4 (ADEP4) can activate ClpP to become a poison, a highly active non-specific protease that can kill cells, including persistor cells. Since ClpP is not essential, its loss is strongly selected for. However, using ADPE4 in combination with rifampicin can lead to eradication of biofilms of *S. aureus in vitro* apparent sterilization in mouse thigh model of *S. aureus* infection.

Metabolic targets

Many *in silico* studies, especially those using flux balance analysis (FBA), have predicted sets of gene targets involved in metabolism that should be essential in various specific minimal or rich media *in vitro*, such as (Lee *et al.*, 2009) whose models involved many *S. aureus* strains. Two major questions arise about targets selected in this way: (1) what is the composition of 'media' components present in various human tissues that could feed cells in which these targets are inhibited and (2) what is the ability of given species to transport these components. Additionally, the resistance potential of any single enzyme target must be considered. An interesting study of potential metabolic targets in Salmonella (Becker *et al.*, 2006) found that most of the metabolic enzymes they evaluated are likely not good targets as there are many redundancies, and the host nutrient environment is apparently rich enough to support mutants deficient in those targets, as most of those tested *in vivo* are still virulent. Of the possibly good targets (that affect virulence) many are broad spectrum, but most are in pathways already targeted by antibiotics or previously plumbed for inhibitors. They conclude that are few new targets in unexploited pathways – however, they do point out that the 'old' targets might be the most attractive. The redundancies seen in many metabolic pathways led Lee *et al.* (2009), mentioned above, to note a number of enzyme pairs (many involved in the synthesis of amino-sugars) which are 'synthetic-lethal': while loss of one does not affect growth, loss of both is lethal. Ostensibly, the authors wish to infer that these synthetic–lethal pairs are good targets, either for inhibition by two compounds or, possible, by a single multitargeting compound. However, in most cases, such synthetic–lethal pairs would be subject to resistance if *either* target became resistant (unless susceptibility is dominant, as with formation of a poison complex – but this seems unlikely for pathways involving soluble intermediates and products). This is further explained in (Silver, 2012a).

Anti-virulence targets

The subject of anti-virulence targets will not be covered here. They have been hypothesized to reduce selective pressure for resistance – and that might be true if they affected only the host but had no effect on growth, colonization, adhesion, biofilm or other beneficial status that, if lost, would lead to a disadvantage relative to peers in the same niche. But that seems hard to ascertain without a validated inhibitor in hand. And, even with such an inhibitor, the mechanics of estimating resistance potential *in vitro* for a 'virulence' phenotype could be difficult without an engineered readout. It has been argued that even if resistant mutants appeared, they might not be spread significantly under drug selection (Allen *et al.*, 2014) and that paper presents good arguments and candidates for virulence targets. Aside from the question of resistance, such anti-virulence approaches might be viable – as some putative virulence inhibitors have been shown to rescue mice from virulent infection. One caveat is that the inhibitor might have direct antibacterial activity *in vivo*, either via the proposed target or due to off-target interactions only evident *in vivo*. Still, further work is worthwhile. With a validated inhibitor in hand, how would it be developed? Almost certainly, it would have to be developed, at least initially as a combination – with an established antibacterial. And, as discussed above, the use of a combination might retard resistance development.

Conclusions

There will be other types of targets and other types of therapeutic modalities that come to the fore in the future, such as targets that are required for virulence, useful systemic membrane-lytic antibacterial peptides and vaccines against problematic MDR pathogens such as MRSA and *P. aeruginosa*. In fact vaccines against those organisms are being actively explored. But for small molecule antibacterials, it is not clear that the way forward involves searching for unexploited targets. This is especially true for Gram-negative agents, for which the limiting factor is cell entry and retention. Research towards rationalizing chemistry for Gram-negative entry should greatly advance the field. As described here, there are some single targeted-agents which may avoid rapid resistance development, but that should not be assumed for all such targets and resistance should instead be expected and its potential to arise during therapy be evaluated as early as possible. Methodology for doing this, for relating *in vitro* rates and frequencies of resistance selection to what is seen *in vivo* during a fulminant infection, is not at all standardized, but is required for future progress in this area. A regulatory pathway to combination therapy of standard pathogens may eventually allow development of novel single-targeted agents despite their resistance potential (as is done with tuberculosis, HIV, HCV). But it must first be demonstrated that such combinations do indeed retard resistance development in robust *in vitro* and *in vivo* models. There may be other bacterial 'multitargets' which could be exploited, and both bio- and chemi-informatic approaches may uncover them. But it should be reasonably clear that genomics has not uncovered an array of new antibacterial targets.

References

Abergel, C., Coutard, B., Byrne, D., Chenivesse, S., Claude, J.-B., Deregnaucourt, C., Fricaux, T., Gianesini-Boutreux, C., Jeudy, S., Lebrun, R., *et al.* (2003). Structural genomics of highly conserved microbial genes of unknown function in search of new antibacterial targets. J. Struct. Func. Genomics *4*, 141–157.

Achaogen (2014). SEC – IPO S1 Document. p. 20. http://www.nasdaq.com/markets/ipos/filing.ashx?filingid=9291045

Allen, R.C., Popat, R., Diggle, S.P., and Brown, S.P. (2014). Targeting virulence: can we make evolution-proof drugs? Nat. Rev. Micro. *12*, 300–308.

Allsop, A.E. (1998). New antibiotic discovery, novel screens, novel targets and impact of microbial genomics. Curr. Opin. Microbiol. *1*, 530–534.

Amini, S., and Tavazoie, S. (2011). Antibiotics and the post-genome revolution. Curr. Opin. Microbiol. *14*, 513–518.

Anderson, N., Bowman, J., Erwin, A.L., Harwood, E.A., Kline, T., Mdluli, K., Ng, S.B., Pfister, K., Shawar, R., Wagman, A., *et al.* (2004). Antibacterial Agents. IPO WO 2004/062601 A2.

Badal, R., Hoban, D., Hackel, M., Bouchillon, S., Biedenbach, D., and Serio, A. (2013). *In vitro* activity of ACHN-975 against non-fermentative Gram-negative bacilli. Poster (E-614) presented at: 53rd Interscience Congress on Antimicrobial Agents and Chemotherapy (Denver, CO).

Becker, D., Selbach, M., Rollenhagen, C., Ballmaier, M., Meyer, T.F., Mann, M., and Bumann, D. (2006). Robust Salmonella metabolism limits possibilities for new antimicrobials. Nature *440*, 303–307.

Bellon, S., Parsons, J.D., Wei, Y., Hayakawa, K., Swenson, L.L., Charifson, P.S., Lippke, J.A., Aldape, R., and Gross, C.H. (2004). Crystal structures of *Escherichia coli* Topoisomerase IV ParE subunit (24 and 43 kilodaltons): a single residue dictates differences in novobiocin potency against Topoisomerase IV and DNA Gyrase. Antimicrob. Agents Chemother. *48*, 1856–1864.

Black, M.T., and Hodgson, J. (2005). Novel target sites in bacteria for overcoming antibiotic resistance. Adv. Drug Deliv. Rev. *57*, 1528–1538.

Brilacidin (2013). (PMX-30063) Antibiotic Fact Sheet. [http://files.shareholder.com/downloads/ABEA-4ITCYZ/1997024712x0x611205/FEA66C67-D02F-4FF2-B230-AA711F46ED7F/PMX-30063_Antibiotic_Fact_Sheet_Nov_2012_.pdf]

Brown, M.F., Reilly, U., Abramite, J.A., Arcari, J.T., Oliver, R., Barham, R.A., Che, Y., Chen, J.M., Collantes, E.M., Chung, S.W., *et al.* (2012). Potent inhibitors of LpxC for the treatment of Gram-negative infections. J. Med. Chem. *55*, 914–923.

Bugg, T.D.H., Wright, G.D., Dutka-Malen, S., Arthur, M., Courvalin, P., and Walsh, C.T. (1991). Molecular basis for vancomycin resistance in *Enterococcus faecium* BM4147: biosynthesis of a depsipeptide peptidoglycan precursor by vancomycin resistance proteins VanH and VanA. Biochemistry *30*, 10408–10415.

Bush, K. (2010). Bench-to-bedside review: the role of b-lactamases in antibiotic-resistant Gram-negative infections. Crit. Care *14*, 224.

Buurman, E.T., Foulk, M.A., Gao, N., Laganas, V.A., McKinney, D.C., Moustakas, D.T., Rose, J.A., Shapiro, A.B., and Fleming, P.R. (2012). Novel rapidly diversifiable antimicrobial RNA Polymerase switch region inhibitors with confirmed mode of action in *Haemophilus influenzae*. J. Bacteriol. *194*, 5504–5512.

Castanheira, M., Rhomberg, P.R., Dietrich, R., and Jones, R. (2013). Activity of the first-in-class LpxC inhibitor ACHN-975 and comparators tested against Gram-negative organisms including isolates with characterized resistance mechanisms. Poster (F-1228) presented at: 53rd Interscience Congress on Antimicrob. Agents Chemother. (Denver, CO).

Chalker, A.F., and Lunsford, R.D. (2002). Rational identification of new antibacterial drug targets that are essential for viability using a genomics-based approach. Pharmacol. Ther. *95*, 1–20.

Chan, P.F., Huang, J., Bax, B.D., and Gwynn, M.N. (2013). Recent developments in inhibitors of bacterial type IIA Topoisomerases. In Antibiotics: Targets, Mechanisms and Resistance, C.O. Gualerzi, L. Brandi, A. Fabbretti, and C.L. Pon, eds. (Wiley-VCH Verlag, Weinheim, Germany), pp. 263–297.

Charest, M.G., Lerner, C.D., Brubaker, J.D., Siegel, D.R., and Myers, A.G. (2005). A convergent enantioselective route to structurally diverse 6-deoxytetracycline antibiotics. Science *308*, 395–398.

Chaudhuri, R.R., Allen, A.G., Owen, P.J., Shalom, G., Stone, K., Harrison, M., Burgis, T.A., Lockyer, M., Garcia-Lara, J., Foster, S.J., *et al.* (2009). Comprehensive identification of essential *Staphylococcus aureus* genes using Transposon-Mediated Differential Hybridisation (TMDH). BMC Genomics *10*, 291.

Cheng, C.C., Shipps, G.W., Jr., Yang, Z., Sun, B., Kawahata, N., Soucy, K.A., Soriano, A., Orth, P., Xiao, L., Mann, P., *et al.* (2009). Discovery and optimization of antibacterial AccC inhibitors. Bioorg. Med. Chem. Lett. *19*, 6507–6514.

Choi, S., Isaacs, A., Clements, D., Liu, D., Kim, H., Scott, R.W., Winkler, J.D., and DeGrado, W.F. (2009). De novo design and *in vivo* activity of conformationally restrained antimicrobial arylamide foldamers. Proc. Natl. Acad. Sci. U.S.A. *106*, 6968–6973.

Choudhry, A.E., Mandichak, T.L., Broskey, J.P., Egolf, R.W., Kinsland, C., Begley, T.P., Seefeld, M.A., Ku, T.W., Brown, J.R., Zalacain, M., *et al.* (2003). Inhibitors of pantothenate kinase: novel antibiotics for staphylococcal infections. Antimicrob. Agents Chemother. *47*, 2051–2055.

Claverie, J.M., Monchois, V., Audic, S., Poirot, O., and Abergel, C. (2002). In search of new anti-bacterial target genes: a comparative/structural genomics approach. Comb. Chem. High Throughput Screen *5*, 511–522.

Clements, J.M., Coignard, F., Johnson, I., Chandler, S., Palan, S., Waller, A., Wijkmans, J., and Hunter, M.G. (2002). Antibacterial activities and characterization of novel inhibitors of LpxC. Antimicrob. Agents Chemother. *46*, 1793–1799.

Conlon, B.P., Nakayasu, E.S., Fleck, L.E., LaFleur, M.D., Isabella, V.M., Coleman, K., Leonard, S.N., Smith, R.D., Adkins, J.N., and Lewis, K. (2013). Activated ClpP kills persisters and eradicates a chronic biofilm infection. Nature *503*, 365–370.

Cui, L., Tominaga, E., Neoh, H.M., and Hiramatsu, K. (2006). Correlation between reduced daptomycin susceptibility and vancomycin resistance in vancomycin-intermediate *Staphylococcus aureus*. Antimicrob. Agents Chemother. *50*, 1079–1082.

Deresinski, S. (2005). Methicillin-Resistant *Staphylococcus aureus*: An evolutionary, epidemiologic, and therapeutic odyssey. Clin. Infect. Dis. *40*, 562–573.

Desnottes, J.F. (1996). New targets and strategies for the development of antibacterial agents. Trends Biotechnol. *14*, 134–140.

Dong, Y., Zhao, X., Domagala, J., and Drlica, K. (1999). Effect of fluoroquinolone concentration on selection of resistant mutants of *Mycobacterium bovis* BCG and *Staphylococcus aureus*. Antimicrob. Agents Chemother. *43*, 1756–1758.

Drawz, S.M., Papp-Wallace, K.M., and Bonomo, R.A. (2014). New b-lactamase inhibitors: a therapeutic renaissance in an MDR world. Antimicrob. Agents Chemother. *58*, 1835–1846.

Drlica, K., Hiasa, H., Kerns, R., Malik, M., Mustaev, A., and Zhao, X. (2009). Quinolones: Action and resistance updated. Curr. Topics Med. Chem. *9*, 981–998.

Drusano, G.L. (2003). Prevention of resistance: A goal for dose selection for antimicrobial agents. Clin. Infect. Dis. *36*, S42-S50.

Ehmann, D.E., Jahic, H., Ross, P.L., Gu, R.F., Hu, J., Kern, G., Walkup, G.K., and Fisher, S.L. (2012). Avibactam is a covalent, reversible, non-β-lactam β-lactamase inhibitor. Proc. Natl. Acad. Sci. U.S.A. *109*, 11663–11668.

Falconer, S.B., and Brown, E.D. (2009). New screens and targets in antibacterial drug discovery. Curr. Opin. Microbiol. *1*, 1.

Fernández, L., Álvarez-Ortega, C., Wiegand, I., Olivares, J., Kocíncová, D., Lam, J.S., Martínez, J.L., and Hancock, R.E.W. (2013). Characterization of the polymyxin B resistome of *Pseudomonas aeruginosa*. Antimicrob. Agents Chemother. *57*, 110–119.

Forsyth, R.A., Haselbeck, R.J., Ohlsen, K.L., Yamamoto, R.T., Xu, H., Trawick, J.D., Wall, D., Wang, L., Brown-Driver, V., Froelich, J.M., *et al.* (2002). A genome-wide strategy for the identification of essential genes in *Staphylococcus aureus*. Mol. Microbiol. *43*, 1387–1400.

Freiberg, C., Brunner, N.A., Schiffer, G., Lampe, T., Pohlmann, J., Brands, M., Raabe, M., Habich, D., and Ziegelbauer, K. (2004). Identification and characterization of the first class of potent bacterial acetyl-CoA carboxylase inhibitors with antibacterial activity. J. Biol. Chem. *279*, 26066–26073.

Freund, Y., Kimura, R., Baker, S.J., Mendes, R.E., Biedenbach, D.J., Simon, N.L., Demarsh, P.L., O'Leary, A., and Alley, M.R.K. (2009). *In vitro* activity and *in vivo* efficacy of (S)-3-(aminomethyl)benzo[c][1,2]oxaborol-1(3H)-ol (ABX): A Gram-negative antimicrobial. Poster (F1–1223a) presented at: 49th Interscience Congress on Antimicrobial Agents and Chemotherapy (San Francisco, CA).

Frey, K.M., Viswanathan, K., Wright, D.L., and Anderson, A.C. (2012). Prospective screening of novel antibacterial inhibitors of dihydrofolate reductase for mutational resistance. Antimicrob. Agents Chemother. *56*, 3556–3562.

Grossman, T.H., Bartels, D.J., Mullin, S., Gross, C.H., Parsons, J.D., Liao, Y., Grillot, A.L., Stamos, D., Olson, E.R., Charifson, P.S., *et al.* (2007). Dual targeting of GyrB and ParE by a novel aminobenzimidazole class of antibacterial compounds. Antimicrob. Agents Chemother. *51*, 657–666.

Gwynn, M.N., Portnoy, A., Rittenhouse, S.F., and Payne, D.J. (2010). Challenges of antibacterial discovery revisited. Ann. N.Y. Acad. Sci. *1213*, 5–19.

Hall, B.G., and Barlow, M. (2005). Revised Ambler classification of β-lactamases. J. Antimicrob. Chemother. *55*, 1050–1051.

Heine, H., Chuvala, L., Riggins, R., Hurteau, G., Drusano, G., Reyes, N., and Cirz, R.T. (2013). Efficacy of ACHN-975 in a murne pneumonic plague (*Yersinia pestis*) model. Poster (F-1230) presented at: 53rd Interscience Congress on Antimicrobial Agents and Chemotherapy (Denver, CO).

Hernandez, V., Crepin, T., Palencia, A., Cusack, S., Akama, T., Baker, S.J., Bu, W., Feng, L., Freund, Y.R., Liu, L., *et al.* (2013). Discovery of a novel class of boron-based antibacterials with activity against Gram-negative bacteria. Antimicrob. Agents Chemother. *57*, 1394–1403.

Hopkins, A.L., Bickerton, G.R., Carruthers, I.M., Boyer, S.K., Rubin, H., and Overington, J.P. (2011). Rapid analysis of pharmacology for infectious diseases. Curr. Topics Med. Chem. *11*, 1292–1300.

Housman, S.T., Sutherland, C., and Nicolau, D.P. (2012). *In vitro* evaluation of novel compounds against selected resistant *Pseudomonas aeruginosa* isolates. Antimicrob. Agents Chemother. *56*, 1646–1649.

Huband, M.D., Cohen, M.A., Zurack, M., Hanna, D.L., Skerlos, L.A., Sulavik, M.C., Gibson, G.W., Gage, J.W., Ellsworth, E., Stier, M.A., *et al.* (2007). *In vitro* and *in vivo* activities of PD 0305970 and PD 0326448, new bacterial gyrase/topoisomerase inhibitors with potent antibacterial activities versus multidrug-resistant gram-positive and fastidious organism groups. Antimicrob. Agents Chemother. *51*, 1191–1201.

Huber, J., Diez, M.T., Dorso, K., Genilloud, O., Hermes, J., Leavitt, P., Lee, S.H., Miesel, L., Ogawa, A., Onishi, H.R., *et al.* (2011). Elucidation of SpsB as the target of an acidic cyclic depsipeptide natural product carbapenem potentiator. Poster (F1–1352) presented at: 51st Interscience Conference on Antimicrobial Agents and Chemotherapy (Chicago, IL).

Hyland, S.A., Eveland, S.S., and Anderson, M.S. (1997). Cloning, expression, and purification of UDP-3-O-acyl-GlcNAc deacetylase from *Pseudomonas aeruginosa*: a metalloamidase of the lipid A biosynthesis pathway. J. Bacteriol. *179*, 2029–2037.

Isaacson, R.E. (2002). Genomics and the prospects for the discovery of new targets for antibacterial and antifungal agents. Curr. Pharm. Des. *8*, 1091–1098.

Jacoby, G.A., and Munoz-Price, L.S. (2005). The new β-Lactamases. N. Engl. J. Med. *352*, 380–391.

de Jonge, B.L.M., Kutschke, A., Uria-Nickelsen, M., Kamp, H.D., and Mills, S.D. (2009). Pyrazolopyrimidinediones are selective agents for

Helicobacter pylori that suppress growth through inhibition of glutamate racemase (MurI). Antimicrob. Agents Chemother. *53*, 3331–3336.

Karlowsky, J.A., Kaplan, N., Hafkin, B., Hoban, D.J., and Zhanel, G.G. (2009). AFN-1252, a FabI inhibitor, demonstrates a Staphylococcus-specific spectrum of activity. Antimicrob. Agents Chemother. *53*, 3544–3548.

Kim, H.U., Sohn, S.B., and Lee, S.Y. (2012). Metabolic network modeling and simulation for drug targeting and discovery. Biotechnol. J. *7*, 330–342.

Kline, T., Andersen, N.H., Harwood, E.A., Bowman, J., Malanda, A., Endsley, S., Erwin, A.L., Doyle, M., Fong, S., Harris, A.L., *et al.* (2002). Potent, novel *in vitro* inhibitors of the *Pseudomonas aeruginosa* deacetylase LpxC. J. Med. Chem. *45*, 3112–3129.

Kuhl, A., Svenstrup, N., Ladel, C., Otteneder, M., Binas, A., Schiffer, G., Brands, M., Lampe, T., Ziegelbauer, K., Rubsamen-Waigmann, H., *et al.* (2005). Biological characterization of novel Inhibitors of the gram-positive DNA polymerase IIIC enzyme. Antimicrob. Agents Chemother. *49*, 987–995.

Kulanthaivel, P., Kreuzman, A.J., Strege, M.A., Belvo, M.D., Smitka, T.A., Clemens, M., Swartling, J.R., Minton, K.L., Zheng, F., Angleton, E.L., *et al.* (2004). Novel lipoglycopeptides as inhibitors of Bacterial Signal Peptidase I. J. Biol. Chem. *279*, 36250–36258.

LaMarche, M.J., Leeds, J.A., Amaral, A., Brewer, J.T., Bushell, S.M., Deng, G., Dewhurst, J.M., Ding, J., Dzink-Fox, J., Gamber, G., *et al.* (2012). Discovery of LFF571: an investigational agent for *Clostridium difficile* infection. J. Med. Chem. *55*, 2376–2387.

Lee, D.S., Burd, H., Liu, J., Almaas, E., Wiest, O., Barabasi, A.L., Oltvai, Z.N., and Kapatral, V. (2009). Comparative genome-scale metabolic reconstruction and flux balance analysis of multiple *Staphylococcus aureus* genomes identify novel antimicrobial drug targets. J. Bacteriol. *191*, 4015–4024.

Leeds, J.A., Sachdeva, M., Mullin, S., Barnes, S.W., and Ruzin, A. (2013). *In vitro* selection, via serial passage, of *Clostridium difficile* mutants with reduced susceptibility to fidaxomicin or vancomycin. J. Antimicrob. Chemother. *69*, 41–44.

Lewandowski, T., Huang, J., Fan, F., Rogers, S., Gentry, D., Holland, R., DeMarsh, P., Aubart, K., and Zalacain, M. (2013). *Staphylococcus aureus* formyl-methionyl transferase mutants demonstrate reduced virulence factor production and pathogenicity. Antimicrob. Agents Chemother. *57*, 2929–2936.

Li, B., Pai, R., Aiello, D., Di, M., Barnes, M.H., Peet, N.P., Bowlin, T.L., and Moir, D.T. (2013). Optimization of a novel potent and selective bacterial DNA helicase inhibitor scaffold from a high throughput screening hit. Bioorg. Med. Chem. Lett. *23*, 3481–3486.

Livermore, D.M. (2011). Discovery research: the scientific challenge of finding new antibiotics. J. Antimicrob. Chemother. *66* 1941–1944.

Livermore, D.M., Warner, M., and Mushtaq, S. (2013). Activity of MK-7655 combined with imipenem against Enterobacteriaceae and *Pseudomonas aeruginosa*. J. Antimicrob. Chemother. *68*, 2286–2290.

Loferer, H., Jacobi, A., Posch, A., Gauss, C., Meier-Ewert, S., and Seizinger, B. (2000). Integrated bacterial genomics for the discovery of novel antimicrobials. Drug Disc. Today *5*, 107–114.

Lofland, D., Difuntorum, S., Waller, A., Clements, J.M., Weaver, M.K., Karlowsky, J.A., and Johnson, K. (2004). *In vitro* antibacterial activity of the peptide deformylase inhibitor BB-83698. J. Antimicrob. Chemother. *53*, 664–668.

Lu, H., and Tonge, P.J. (2008). Inhibitors of FabI, an enzyme drug target in the bacterial fatty acid biosynthesis pathway. Acc. Chem. Res. *41*, 11–20.

McClerren, A.L., Endsley, S., Bowman, J.L., Andersen, N.H., Guan, Z., Rudolph, J., and Raetz, C.R. (2005). A slow, tight-binding inhibitor of the zinc-dependent deacetylase LpxC of lipid A biosynthesis with antibiotic activity comparable to ciprofloxacin. Biochemistry *44*, 16574–16583.

McDevitt, D., and Rosenberg, M. (2001). Exploiting genomics to discover new antibiotics. Trends Microbiol. *9*, 611–617.

McIntyre, J., Castaner, J., and Martin, L. (2005). LBM-415 antibacterial peptide deformylase inhibitor. Drugs Fut. *30*, 23–28.

Mani, N., Gross, C.H., Parsons, J.D., Hanzelka, B., Muh, U., Mullin, S., Liao, Y., Grillot, A.-L., Stamos, D., Charifson, P.S., *et al.* (2006). *In vitro* characterization of the antibacterial spectrum of novel bacterial type II topoisomerase inhibitors of the aminobenzimidazole class. Antimicrob. Agents Chemother. *50*, 1228–1237.

Mann, P.A., Muller, A., Xiao, L., Pereira, P.M., Yang, C., Ho Lee, S., Wang, H., Trzeciak, J., Schneeweis, J., Dos Santos, M.M., *et al.* (2013). Murgocil is a highly bioactive staphylococcal-specific inhibitor of the peptidoglycan glycosyltransferase enzyme MurG. ACS Chem. Biol. *8*, 2442–2451.

Margolis, P., Hackbarth, C., Lopez, S., Maniar, M., Wang, W., Yuan, Z., White, R., and Trias, J. (2001). Resistance of *Streptococcus pneumoniae* to deformylase inhibitors Is due to mutations in *defB*. Antimicrob. Agents Chemother. *45*, 2432–2435.

Marshall, S.H., Donskey, C.J., Hutton-Thomas, R., Salata, R.A., and Rice, L.B. (2002). Gene dosage and linezolid resistance in *Enterococcus faecium* and *Enterococcus faecalis*. Antimicrob. Agents Chemother. *46*, 3334–3336.

Martínez-Botella, G., Loch, J.T., Green, O.M., Kawatkar, S.P., Olivier, N.B., Ann Boriack-Sjodin, P., and Keating, T.A. (2013). Sulfonylpiperidines as novel, antibacterial inhibitors of Gram-positive thymidylate kinase (TMK). Bioorg. Med. Chem. Lett. *23*, 169–173.

Mazel, D., Pochet, S., and Marliere, P. (1994). Genetic characterization of polypeptide deformylase, a distinctive enzyme of eubacterial translation. EMBO J. *13*, 914–923.

Miller, J.R., Dunham, S., Mochalkin, I., Banotai, C., Bowman, M., Buist, S., Dunkle, B., Hanna, D., Harwood, H.J., Huband, M.D., *et al.* (2009). A class of selective antibacterials derived from a protein kinase inhibitor pharmacophore. Proc. Natl. Acad. Sci. U.S.A. *106*, 1737–1742.

Miller, W.H., Seefeld, M.A., Newlander, K.A., Uzinskas, I.N., Burgess, W.J., Heerding, D.A., Yuan, C.C., Head, M.S., Payne, D.J., Rittenhouse, S.F., *et al.* (2002). Discovery of aminopyridine-based inhibitors of bacterial enoyl-ACP reductase (FabI). J. Med. Chem. *45,* 3246–3256.

Mills, S.D., Eakin, A.E., Buurman, E.T., Newman, J.V., Gao, N., Huynh, H., Johnson, K.D., Lahiri, S., Shapiro, A.B., Walkup, G.K., *et al.* (2011). Novel bacterial NAD+-dependent DNA ligase inhibitors with broad spectrum activity and antibacterial efficacy *in vivo*. Antimicrob. Agents Chemother. *55,* 1088–1096.

Moellering, R.C. (2010). NDM-1 – A cause for worldwide concern. N. Engl. J. Med. *363,* 2377–2379.

Mohan, S., Kelly, T.M., Eveland, S.S., Raetz, C.R., and Anderson, M.S. (1994). An *Escherichia coli* gene (FabZ) encoding (3R)-hydroxymyristoyl acyl carrier protein dehydrase. Relation to fabA and suppression of mutations in lipid A biosynthesis. J. Biol. Chem. *269,* 32896–32903.

Monaghan, R.L., and Barrett, J.F. (2006). Antibacterial drug discovery – then, now and the genomics future. Biochem. Pharmacol. *71,* 901–909.

Moy, T.I., Daniel, A., Hardy, C., Jackson, A., Rehrauer, O., Hwang, Y.S., Zou, D., Nguyen, K., Silverman, J.A., Li, Q., *et al.* (2011). Evaluating the activity of the RNA polymerase inhibitor myxopyronin B against *Staphylococcus aureus.* FEMS Microbiol. Lett. *319,* 176–179.

Nilsson, A.I., Zorzet, A., Kanth, A., Dahlström, S., Berg, O.G., and Andersson, D.I. (2006). Reducing the fitness cost of antibiotic resistance by amplification of initiator tRNA genes. Proc. Natl. Acad. Sci. U.S.A. *2103,* 6976–6981.

Ochsner, U.A., Young, C.L., Stone, K.C., Dean, F.B., Janjic, N., and Critchley, I.A. (2005). Mode of action and biochemical characterization of REP8839, a novel inhibitor of methionyl-tRNA synthetase. Antimicrob. Agents Chemother. *49,* 4253–4262.

O'Daniel, P.I., Peng, Z., Pi, H., Testero, S.A., Ding, D., Spink, E., Leemans, E., Boudreau, M.A., Yamaguchi, T., Schroeder, V.A., *et al.* (2014). Discovery of a new class of non-β-lactam inhibitors of penicillin-binding proteins with Gram-positive antibacterial activity. J. Amer. Chem. Soc. *136,* 3664–3672.

O'Dwyer, K., Hackel, M., Hightower, S., Hoban, D., Bouchillon, S., Qin, D., Aubart, K., Zalacain, M., and Butler, D. (2013). Comparative analysis of the antibacterial activity of a novel peptide deformylase inhibitor, GSK1322322. Antimicrob. Agents Chemother. *57,* 2333–2342.

O'Neill, A.J., and Chopra, I. (2004). Preclinical evaluation of novel antibacterial agents by microbiological and molecular techniques. Exp. Opin. Invest. Drugs *13,* 1045–1063.

Onishi, H.R., Pelak, B.A., Gerckens, L.S., Silver, L.L., Kahan, F.M., Chen, M.-H., Patchett, A.A., Galloway, S.M., Hyland, S.A., Anderson, M.S., *et al.* (1996). Antibacterial agents that inhibit lipid A biosynthesis. Science *274,* 980–982.

Overbye, K.M., and Barrett, J.F. (2005). Antibiotics: where did we go wrong? Drug Discov. Today *10,* 45–52.

Paetzel, M., Dalbey, R.E., and Strynadka, N.C.J. (2000). The structure and mechanism of bacterial type I signal peptidases: A novel antibiotic target. Pharmacol. Ther. *87,* 27–49.

Painter, R., Goetz, M.A., Arocho, M., Gill, C., Allocco, J., Nare, B., Powles, M.A., and Young, K. (2011). Elucidation of DnaE as the antibacterial target of the natural product, Nargenicin. Poster presented at: 52nd Annual Interscience Congress on Antimicrobial Agents and Chemotherapy (Chicago, IL).

Pan, X.-S., Gould, K.A., and Fisher, L.M. (2009). Probing the differential interactions of quinazolinedione PD 0305970 and quinolones with gyrase and Topoisomerase IV. Antimicrob. Agents Chemother. *53,* 3822–3831.

Panjkovich, A., Gibert, I., and Daura, X. (2014). antibacTR: dynamic antibacterial-drug-target ranking integrating comparative genomics, structural analysis and experimental annotation. BMC Genomics *15,* 36.

Payne, D.J., Gwynn, M.N., Holmes, D.J., and Rosenberg, M. (2004). Genomic approaches to antibacterial discovery. Methods Mol. Biol. *266,* 231–259.

Payne, D.J., Gwynn, M.N., Holmes, D.J., and Pompliano, D.L. (2007). Drugs for bad bugs: confronting the challenges of antibacterial discovery. Nat. Rev. Drug Discov. *6,* 29–40.

Perola, E., Stamos, D., Grillot, A.-L., Ronkin, S., Wang, T., LeTiran, A., Tang, Q., Deininger, D.D., Liao, Y., Tian, S.-K., *et al.* (2014). Successful application of serum shift prediction models to the design of dual targeting inhibitors of bacterial Gyrase B and Topoisomerase IV with improved *in vivo* efficacy. Bioorg. Med. Chem. Lett. *24,* 2177–2181.

Pucci, M.J. (2001). Overview of antibacterial target selection. In Current Protocols in Pharmacology. 31:A:13A.2:13A.2.1–13A.2.13.

Pucci, M.J. (2006). Use of genomics to select antibacterial targets. Biochem. Pharmacol. *71,* 1066–1072.

Pucci, M.J. (2007). Novel genetic techniques and approaches in the microbial genomics era: identification and/or validation of targets for the discovery of new antibacterial agents. Drugs R. D. *8,* 201–212.

Pucci, M.J., and Bush, K. (2013). Investigational antimicrobial agents of 2013. Clin. Microbiol. Rev. *26,* 792–821.

Pucci, M., and Wiles, J. (2014). Bacterial topoisomerase inhibitors: Quinolones and beyond. In Antimicrobials, F. Marinelli, and O. Genilloud, eds. (Springer, Berlin, Heidelberg, Germany), pp. 307–326.

Pucci, M.J., Podos, S.D., Thanassi, J.A., Leggio, M.J., Bradbury, B.J., and Deshpande, M. (2011). *In vitro* and *in vivo* profiles of ACH-702, an isothiazoloquinolone, against bacterial pathogens. Antimicrob. Agents Chemother. *55,* 2860–2871.

Roberts, T.C., Schallenberger, M.A., Liu, J., Smith, P.A., and Romesberg, F.E. (2011). Initial efforts toward the optimization of arylomycins for antibiotic activity. J. Med. Chem. *54,* 4954–4963.

Robertson, G.T., Doyle, T.B., Du, Q., Duncan, L., Mdluli, K.E., and Lynch, A.S. (2007). A Novel Indole Compound That Inhibits *Pseudomonas aeruginosa*

Growth by Targeting MreB Is a Substrate for MexAB-OprM. J. Bacteriol. *189*, 6870–6881.
Ruiz, N., Kahne, D., and Silhavy, T.J. (2009). Transport of lipopolysaccharide across the cell envelope: the long road of discovery. Nat. Rev. Microbiol. *7*, 677–683.
Sakharkar, K.R., Sakharkar, M.K., and Chow, V.T.K. (2004). A novel genomics approach for the identification of drug targets in pathogens, with special reference to *Pseudomonas aeruginosa*. In Silico Biol. *4*, 355–360.
Schneider, P., Hawser, S., and Islam, K. (2003). Iclaprim, a novel diaminopyrimidine with potent activity on trimethoprim sensitive and resistant bacteria. Bioorg. Med. Chem. Lett. *13*, 4217–4221.
Schneider, T., and Sahl, H.G. (2010). Lipid II and other bactoprenol-bound cell wall precursors as drug targets. Curr. Opin. Invest. Drugs *11*, 157–164.
Scott, L. (2013). Fidaxomicin: A review of its use in patients with *Clostridium difficile* infection. Drugs *73*, 1733–1747.
Shapiro, A.B., and Andrews, B. (2012). Allosteric inhibition of the DNA-dependent ATPase activity of *Escherichia coli* DNA gyrase by a representative of a novel class of inhibitors. Biochem. Pharmacol. *84*, 900–904.
Silver, L.L. (2007). Multi-targeting by monotherapeutic antibacterials. Nat. Rev. Drug Discov. *6*, 41–55.
Silver, L.L. (2011). Challenges of antibacterial discovery. Clin. Microbiol. Rev. *24*, 71–109.
Silver, L.L. (2012a). Polypharmacology as an emerging trend in antibacterial discovery. In Polypharmacology in Drug Discovery, J. Peters, ed. (Wiley, Hoboken, NJ), pp. 167–202.
Silver, L.L. (2012b). Rational approaches to antibacterial discovery: Pre-genomic directed and phenotypic screening. In Antibiotic Discovery and Development, T.J. Dougherty, and M.J. Pucci, eds. (Springer, New York), pp. 33–75.
Silver, L.L. (2013). Viable screening targets related to the bacterial cell wall. Ann. N.Y. Acad. Sci. *1277*, 29–53.
Silver, L.L., and Bostian, K.A. (1993). Discovery and development of new antibiotics: The problem of antibiotic resistance. Antimicrob. Agents Chemother. *37*, 377–383.
Silverman, J.A., Perlmutter, N.G., and Shapiro, H.M. (2003). Correlation of daptomycin bactericidal activity and membrane depolarization in *Staphylococcus aureus*. Antimicrob. Agents Chemother. *47*, 2538–2544.
Smith, P.A., and Romesberg, F.E. (2012). Mechanism of action of the arylomycin antibiotics and effects of signal peptidase I inhibition. Antimicrob. Agents Chemother. *56*, 5054–5060.
Smith, P.A., Roberts, T.C., and Romesberg, F.E. (2010). Broad-spectrum antibiotic activity of the arylomycin natural products is masked by natural target mutations. Chem. Biol. *17*, 1223–1231.
Solomkin, J.S., Ramesh, M.K., Cesnauskas, G., Novikovs, N., Stefanova, P., Sutcliffe, J.A., Walpole, S.M., and Horn, P.T. (2014). Efficacy and safety of two dose regimens of eravacycline versus ertapenem in adult community-acquired complicated intra-abdominal infections: Results of a Phase 2,randomized, double-blind study. Antimicrob. Agents Chemother. *58*, 1847–1854.
Srinivas, N., Jetter, P., Ueberbacher, B.J., Werneburg, M., Zerbe, K., Steinmann, J., Van der Meijden, B., Bernardini, F., Lederer, A., Dias, R.L., *et al.* (2010). Peptidomimetic antibiotics target outer-membrane biogenesis in *Pseudomonas aeruginosa*. Science *327*, 1010–1013.
Srivastava, A., Degen, D., Ebright, Y.W., and Ebright, R.H. (2012). Frequency, spectrum, and fitness costs of resistance to myxopyronin in *Staphylococcus Aureus*: Myxopyronin resistance has non-zero fitness cost. Antimicrob. Agents Chemother. *56*, 6250–6255.
Stokes, N.R., Baker, N., Bennett, J.M., Berry, J., Collins, I., Czaplewski, L.G., Logan, A., Macdonald, R., MacLeod, L., Peasley, H., *et al.* (2012). An improved small-molecule inhibitor of FtsZ with superior *in vitro* potency, drug-like properties and *in vivo* efficacy. Antimicrob. Agents Chemother. *57*, 317–325.
Stokes, S.S., Albert, R., Buurman, E.T., Andrews, B., Shapiro, A.B., Green, O.M., McKenzie, A., and Otterbein, L.R. (2012). Inhibitors of the acetyltransferase domain of N-acetylglucosamine-1-phosphate-uridylyltransferase/glucosamine-1-phosphate-acetyltransferase (GlmU). Part 2: Optimization of physical properties leading to antibacterial aryl sulfonamides. Bioorg. Med. Chem. Lett. *22*, 7019–7023.
Surivet, J.-P., Lange, R., Hubschwerlen, C., Keck, W., Specklin, J.-L., Ritz, D., Bur, D., Locher, H., Seiler, P., Strasser, D.S., *et al.* (2012). Structure-guided design, synthesis and biological evaluation of novel DNA ligase inhibitors with *in vitro* and *in vivo* anti-staphylococcal activity. Bioorg. Med. Chem. Lett. *22*, 6705–6711.
Sutcliffe, J.A., O'Brien, W., Fyfe, C., and Grossman, T.H. (2013). Antibacterial activity of eravacycline (TP-434), a novel fluorocycline, against hospital and community pathogens. Antimicrob. Agents Chemother. *57*, 5548–5558.
Tan, C.M., Therien, A.G., Lu, J., Lee, S.H., Caron, A., Gill, C.J., Lebeau-Jacob, C., Benton-Perdomo, L., Monteiro, J.M., Pereira, P.M., *et al.* (2012). Restoring methicillin-resistant *Staphylococcus aureus* susceptibility to b-lactam antibiotics. Sci. Transl. Med. *4*, 126ra135.
Tan, Y.X., and Romesberg, F.E. (2012). Latent antibiotics and the potential of the arylomycins for broad-spectrum antibacterial activity. Med. Chem. Comm. *3*, 916–925.
Tanino, T., Al-Dabbagh, B., Mengin-Lecreulx, D., Bouhss, A., Oyama, H., Ichikawa, S., and Matsuda, A. (2011). Mechanistic analysis of muraymycin analogues: A auide to the design of MraY Inhibitors. J. Med. Chem. *54*, 8421–8439.
Tari, L.W., Trzoss, M., Bensen, D.C., Li, X., Chen, Z., Lam, T., Zhang, J., Creighton, C.J., Cunningham, M.L., Kwan, B., *et al.* (2013). Pyrrolopyrimidine inhibitors of DNA gyrase B (GyrB) and topoisomerase IV (ParE). Part I: Structure guided discovery and optimization of dual targeting agents with potent, broad-spectrum enzymatic activity. Bioorg. Med. Chem. Lett. *23*, 1529–1536.

Therien, A.G., Huber, J.L., Wilson, K.E., Beaulieu, P., Caron, A., Claveau, D., Deschamps, K., Donald, R.G., Galgoci, A.M., Gallant, M., *et al.* (2012). Broadening the spectrum of b-lactam antibiotics through inhibition of signal peptidase type I. Antimicrob. Agents Chemother. *56*, 4662–4670.

Tomasic, T., and Peterlin, M.L. (2014). Prospects for developing new antibacterials targeting bacterial type IIA Topoisomerases. Curr. Top. Med. Chem. *14*, 130–151.

Trawick, J.D., and Schilling, C.H. (2006). Use of constraint-based modeling for the prediction and validation of antimicrobial targets. Biochem. Pharmacol. *71*, 1026–1035.

Trzoss, M., Bensen, D.C., Li, X., Chen, Z., Lam, T., Zhang, J., Creighton, C.J., Cunningham, M.L., Kwan, B., Stidham, M., *et al.* (2013). Pyrrolopyrimidine inhibitors of DNA Gyrase B (GyrB) and Topoisomerase IV (ParE), Part II: Development of inhibitors with broad spectrum, Gram-negative antibacterial activity. Bioorg. Med. Chem. Lett. *23*, 1537–1543.

Twynholm, M., Dalessandro, M., Barker, K., Scangarella-Oman, N., Li, G., Ingraham, K., Min, S., Jakielaszek, C., O'Dwyer, K., Holmes, D.J., *et al.* (2013). Termination of Phase II program due to emergence of resistance (EOR) on-therapy. Poster (G1251) presented at: 53rd Interscience Congress on Antimicrobial Agents and Chemotherapy (Denver, CO).

VanScoy, B.D., Bulik, C.C., Moseley, C., Scangarella-Oman, N.E., Miller, L.A., Bhavnani, S.M., and Ambrose, P.G. (2013). Hollow fiber infection model mimics both the time-to-resistance emergence and magnitude of *E. coli* resistance to GSK052 occurring in a Phase 2b clinical study. Paper presented at: 53rd Interscience Congress on Antimicrobial Agents and Chemotherapy (Denver, CO).

Wang, H., Gill, Charles, J., Lee, Sang, H., Mann, P., Zuck, P., Meredith, Timothy, C., Murgolo, N., She, X., Kales, S., Liang, L., *et al.* (2013). Discovery of wall teichoic acid inhibitors as potential anti-MRSA β-lactam combination agents. Chem. Biol. *20*, 272–284.

Wang, J., Soisson, S.M., Young, K., Shoop, W., Kodali, S., Galgoci, A., Painter, R., Parthasarathy, G., Tang, Y.S., Cummings, R., *et al.* (2006). Platensimycin is a selective FabF inhibitor with potent antibiotic properties. Nature *44*, 358–361.

Wang, J., Kodali, S., Lee, S.H., Galgoci, A., Painter, R., Dorso, K., Racine, F., Motyl, M., Hernandez, L., Tinney, E., *et al.* (2007). Discovery of platencin, a dual FabF and FabH inhibitor with *in vivo* antibiotic properties. Proc. Natl. Acad. Sci. U.S.A. *104*, 7612–7616.

Werneburg, M., Zerbe, K., Juhas, M., Bigler, L., Stalder, U., Kaech, A., Ziegler, U., Obrecht, D., Eberl, L., and Robinson, J.A. (2012). Inhibition of lipopolysaccharide transport to the outer membrane in *Pseudomonas aeruginosa* by peptidomimetic antibiotics. Chembiochem *13*, 1767–1775.

White, R.J. (2012). The early history of antibiotic discovery: empiricism ruled. In Antibiotic Discovery and Development (Springer, Heidelberg, Germany), pp. 3–31.

White, T.A., and Kell, D.B. (2004). Comparative genomic assessment of novel broad-spectrum targets for antibacterial drugs. Comp. Func. Genomics *5*, 304–327.

Wu, S.W., de Lencastre, H., and Tomasz, A. (2001). Recruitment of the *mecA* gene homologue of *Staphylococcus sciuri* into a resistance determinant and expression of the resistant phenotype in *Staphylococcus aureus*. J. Bacteriol. *183*, 2417–2424.

Yao, J., Maxwell, J.B., and Rock, C.O. (2013). Resistance to AFN-1252 arises from missense mutations in *Staphylococcus aureus* enoyl-acyl carrier protein reductase (FabI). J. Biol. Chem. *288*, 36261–36271.

Yin, D., Fox, B., Lonetto, M.L., Etherton, M.R., Payne, D.J., Holmes, D.J., Rosenberg, M., and Ji, Y. (2004). Identification of antimicrobial targets using a comprehensive genomic approach. Pharmacogenomics 5, 101–113.

Young, K., Silver, L.L., Bramhill, D., Cameron, P., Eveland, S.S., Raetz, C.R., Hyland, S.A., and Anderson, M.S. (1995). The *envA* permeability/cell division gene of *Escherichia coli* encodes the second enzyme of lipid A biosynthesis. UDP-3-O-(R-3-hydroxymyristoyl)-N-acetylglucosamine deacetylase. J. Biol. Chem. *270*, 30384–30391.

Zalacain, M., Biswas, S., Ingraham, K.A., Ambrad, J., Bryant, A., Chalker, A.F., Iordanescu, S., Fan, J., Fan, F., Lunsford, R.D., *et al.* (2003). A global approach to identify novel broad-spectrum antibacterial targets among proteins of unknown function. J. Mol. Microbiol. Biotech. *6*, 109–126.

Zeng, D., Zhao, J., Chung, H.S., Guan, Z., Raetz, C.R., and Zhou, P. (2013). Mutants resistant to LpxC inhibitors by rebalancing cellular homeostasis. J. Biol. Chem. *288*, 5475–5486.

Zhang, L., Dhillon, P., Yan, H., Farmer, S., and Hancock, R.E.W. (2000). Interactions of bacterial cationic peptide antibiotics with outer and cytoplasmic membranes of *Pseudomonas aeruginosa*. Antimicrob. Agents Chemother. *44*, 3317–3321.

Zhou, T., Commodore, L., Huang, W.S., Wang, Y., Thomas, M., Keats, J., Xu, Q., Rivera, V.M., Shakespeare, W.C., Clackson, T., *et al.* (2011). Structural mechanism of the Pan-BCR-ABL inhibitor ponatinib (AP24534): lessons for overcoming kinase inhibitor resistance. Chem. Biol. Drug Des. *77*, 1–11.

Zorzet, A., Andersen, J.M., Nilsson, A.I., Moller, N.F., and Andersson, D.I. (2012). Compensatory mutations in *agrC* partly restore fitness *in vitro* to peptide deformylase inhibitor-resistant *Staphylococcus aureus*. J. Antimicrob. Chemother. 1835–1842.

Novel Antimicrobial and Other Bioactive Metabolites Obtained from Silent Gene Clusters

15

Juan F. Martín and Paloma Liras

Abstract

Hundreds of secondary metabolite gene clusters have been found in the genome sequence of filamentous fungi, *Streptomyces* species and some rare actinomycetes (20–50 clusters per genome). However, the number of secondary metabolites found in the culture broths of each strain is much lower than the number of gene clusters in that particular strain. Many of the sequenced gene clusters are silent or very poorly expressed, and there is a high untapped potential for the discovery of novel antibiotics or other bioactive products. Classical strategies to trigger production of secondary metabolites rely on the use of nutritional stressing conditions such as phosphate starvation or ammonium limitation, and the addition of phosphate-precipitating or ammonium-trapping agents leads to formation of new products. Metal toxicity also triggers the production of novel secondary metabolites in some *Streptomyces* strains. Modification of the *rpsL* (for the ribosomal protein S12) or the *rpoB* (for the RNA polymerase β subunit) genes enhance the expression of silent or poorly expressed secondary metabolite gene clusters in several *Streptomyces* species. Addition of antibiotic biosynthesis 'remodelling compounds' re-directs the metabolic flux to obtain new antibiotics. Modern strategies to awaken sleeping clusters include the modification of wide domain regulators that control related or even disparate pathways. Alteration of butyrolactone receptor proteins or of an oligopeptide-binding protein results also in awakening of silent gene clusters in some producer strains. Co-cultivation of *Streptomyces* with other bacteria or fungi, involving contact between cells, frequently triggers silent gene clusters. The potential of these strategies to trigger the expression of silent clusters in poorly studied actinobacteria other than *Streptomyces* is very high.

In fungi, rearrangements of the chromatin structure by either directed mutations of some genes or by the use of 'chromatin modifiers', has provided several new compounds in each of the tested species. Particularly, deletion or overexpression of the *laeA* gene that influences chromatin rearrangement, modifies the transcription of some low-expression gene clusters and awakens the expression of some silent clusters providing new compounds. Expression of the *laeA* gene is increased by the fungal autoinducer 1,3-diaminopropane, what may trigger some silent clusters.

Examples of new molecules synthesized by the enzyme encoded by different silent or 'near-silent' gene clusters are provided in this article. Most of the novel products observed by HPLC or by bioassays still remain to be characterized chemically.

Silent and/or cryptic gene clusters

The availability of the genome sequence of numerous microorganisms, particularly soil-dwelling and marine microorganisms and some plant endophytes, has revealed the presence of hundreds of gene clusters, putatively encoding novel secondary metabolites (also named 'specialized metabolites', Martín *et al.*, 2000; Hopwood, 2013). This is particularly true in several high G+C actinobacteria (specially members of the genus *Streptomyces*), myxobacteria and filamentous fungi. Species of the genus *Streptomyces* contain large genomes, in the range 7–11 Mb. The largest *Streptomyces* genome known is that of *Streptomyces bingchenggensis* with a size of 11 Mb, encoding more than 10,000 proteins (Wang *et al.*, 2010). Most *Streptomyces* species contain between 20 and 30 gene clusters for secondary metabolites. The same is true for some myxobacteria that also have very large genomes.

In fungi a detailed analysis of the genome of *Aspergillus nidulans* revealed 53 gene clusters for secondary metabolites (von Döhren, 2009) and the genome of *Penicillium chrysogenum* harbours about 40 secondary metabolite gene clusters (van den Berg *et al.*, 2008).

However, the number of known secondary metabolites formed by these fungi is much lower than that of identified gene clusters. The same is true for *Streptomyces* and other actinomycetes. This means that many of the gene clusters are silent, i.e. they are not expressed or they are expressed to very low degree (near-silent).

In other cases the secondary metabolite gene clusters are expressed but the product(s) of the pathway are unknown. These compounds are seen after HPLC or UPLC separation but their chemical structure has not been determined yet. This second type of gene clusters are known as cryptic (or orphan) gene clusters (Gross, 2007). Some of those compounds will be identified chemically in the future. Since these cryptic clusters are expressed, they cannot be considered as silent clusters. We will refer in this article to silent and near-silent clusters but not to chemical characterization of cryptic compounds.

There is clear evidence that many secondary metabolites are 'communication molecules' (Davies, 2007; Yim *et al.*, 2007) that serve as signals for cross-talk: (i) between species (e.g. between *Streptomyces* and other bacteria or between filamentous fungi and bacteria) or, (ii) between cells of the same species (e.g. quorum sensing signals, see below). Antibiotics are a part of the many different types of specialized metabolites (Martín *et al.*, 2000; Keller *et al.*, 2005; Fischbach, 2009). Antibiotics, and other bioactive compounds, have received considerable attention because of their medical interest (Yim *et al.*, 2007; Fischbach and Walsh, 2009) and the mechanisms triggering their biosynthesis serve as a good model of the mechanisms that control expression of silent gene clusters.

Years ago Martín and Demain (1980) proposed that antibiotic production is triggered as a response to nutrients limitation (so called nutritional stress). We now know that, in addition to classical nutrient limitation of phosphate or nitrogen sources, there are several other mechanisms that trigger the onset of numerous secondary metabolites. These include: (a) metal toxicity (Haferburg *et al.*, 2009), (b) the presence of bacterial or fungal cell wall components (Rigali *et al.*, 2008), (c) attack by other bacteria or by their metabolites, and (d) cell to cell contact between fungi and *Streptomyces* in the soil. All these may be considered stress signals. We will describe strategies based on each of these mechanisms to trigger the expression of silent or near-silent clusters (Table 15.1).

Metabolite profiling to identify new polyketide synthases (PKS) or non-ribosomal peptide synthetases (NRPS) gene clusters

Many polyketide synthases (PKS) and non-ribosomal peptide synthetases (NRPS) clusters in microbial genomes can be identified by bioinformatics analysis (Medema *et al.*, 2010). An example of characterization of novel products is the use of mutants in the phoshopantethenyl transferase (PTTase) gene, an enzyme that modifies both PKSs and NRPSs by addition of the phophopantetheine molecule (García-Estrada *et al.*, 2008; Márquez-Fernández *et al.*, 2007). Since this PTTase activity is required for activation of enzymes involved in the biosynthesis of several secondary metabolites (including polyketide-derived pigments) comparison of metabolite profiles of the parental and the *ppt* mutant (named *cfwA* in *A. nidulans*) allows to characterize sets of polyketides and non-ribosomal peptides. This strategy allowed to identify the polyketide compounds emericellin, dehydroaustinol and shamixanthone (Márquez-Fernández *et al.*, 2007).

Table 15.1 Examples of novel compounds obtained from silent gene clusters using different strategies

Strategy	Compounds	Producer organism	Reference
Phosphate trapping	Nanaomycin	*Streptomyces rosa* sub. *notoensis*	Masuda *et al.* (1986)
Ammonium-trapping	Gilvocarcin	*Micromonospora chalcea*	Wei *et al.* (1982)
	Tetrocarcin A	*Micromonospora chalcea*	Tamaoki and Tomita (1982)
Blocked mutant in an antibiotic pathway	Emericellamide	*Aspergillus nidulans*	Chiang *et al.* (2008)
ropB mutants	Piperidamycins	*Streptomyces mauvecolor*	Hosaka *et al.* (2009)
	Neotrehalosediamine	*Bacillus subtilis*	Inaoka *et al.* (2011)
LAL-type regulators	Filipin	*Streptomyces avermitilis*	Santos-Aberturas *et al.* (2011)
	Stambomycins	*Streptomyces ambofaciens*	Laureti *et al.* (2011)
scbR2-γ-Butyrolactone receptors	Coelimycin P1	*Streptomyces coelicolor*	Pawlik *et al.* (2007)
			Gómez-Escribano *et al.* (2013)
Quorum-sensing modification	Thailandamycin	*Burkholderia thailandesis*	Ishida *et al.* (2010)
	Malleilactone	*Burkholderia pseudomallei*	Biggins *et al.* (2012)
Removal of oligopeptide-binding protein	Holomycin	*Streptomyces clavuligerus*	Fuente *et al.* (2002)
Challenging with bacteria	Holomycin	*Streptomyces clavuligerus*	Charusanti *et al.* (2012)
Removal of megaplasmids	Holomycin	*Streptomyces clavuligerus*	Álvarez-Álvarez *et al.* (2013b)
Challenging with bacteria (intimate contact)	Acyl-desferrioxamines	*Streptomyces coelicolor*	Trexler *et al.* (2013)
Co-cultivation with other organisms (fungi-bacteria)	Orsenillic acid Lecanoric acid F-9775A and F-9775B (Cathepsin K inhibitors)	*Aspergillus nidulans*	Schroechk *et al.* (2009)
Cross-pathway regulation	Aspyridones A, B	*Aspergillus nidulans*	Bergman *et al.* (2007)
	Asperfuranone	*Aspergillus nidulans*	Bergman *et al.* (2007)
	PR-toxin	*Penicillium chrysogenum*	Hidalgo *et al.* (2014)
Heterochromatin modification	Emodin	*Aspergillus nidulans*	Bok *et al.* (2009)
	Monodyctiophenone	*Aspergillus nidulans*	Swzewczyk *et al.* (2008)
	Asperthecin	*Aspergillus nidulans*	William *et al.* (2008)
	Calphostin	*Aspergillus nidulans*	William *et al.* (2008)
	Cladochrome	*Aspergillus nidulans*	William *et al.* (2008)
	Nygerone	*Aspergillus nidulans*	Henrikson *et al.* (2009)
	Sterigmatocystin	*Aspergillus nidulans*	Reyes-Domínguez *et al.* (2010)
LaeA: Chromatin structure reorganization	Terrequinone A	*Aspergillus nidulans*	Bok *et al.* (2006)

Other metabolites of unknown chemical structures formed from silent clusters have not been included in this table (see text).

Why is a silent gene cluster sleeping?

Increasing evidence indicates that there are at least three different genetic mechanisms responsible for the lack of expression of silent genes:

1. Some gene clusters are fully repressed by strong transcriptional negative regulators. These silent gene clusters may be awakened when external inducers or biochemical modifications of the metabolism of the producer cells decrease the affinity of the negative regulators.
2. A second class of silent gene clusters is clearly defective in their promoters or regulatory genes, e.g. the clavulanic acid gene cluster of *Streptomyces flavogriseus* (Álvarez-Álvarez *et al.*, 2013a). This cluster contains all clavulanic acid genes (known to be sufficient for clavulanic acid production in *Streptomyces clavuligerus*), including the SARP pathway-specific regulatory gene *ccaR* and the LysR-type regulatory gene *claR*, but the *S. flavogriseus* cluster could not be expressed when tested in nine culture media with different phosphate and nitrogen sources. Indeed, in *S. flavogriseus* the clavulanic acid cluster seems to be permanently silent under laboratory conditions due to a defective expression of some genes of the cluster. In the natural habitat *S. flavogriseus* might be able to express the clavulanic acid cluster when challenged by other organisms.
3. Finally, some of the fully silent clusters may be defective because they have lost key structural genes, i.e. they are 'degenerated clusters'. Also, is possible that some clusters are silent because the strain lacks the specific precursors for initiating or completing the antibiotic molecule.

Classical approaches to trigger low expression secondary metabolite gene clusters

Phosphate starvation stress

The biosynthesis of different types of secondary metabolites produced by microorganisms is regulated negatively by high phosphate concentrations (Martín, 1989). Expression of many antibiotic genes takes place only under phosphate limiting conditions (Martín, 2004). The two component PhoR–PhoP system, which responds to phosphate starvation controls antibiotic biosynthesis in *Streptomyces lividans* (Sola-Landa *et al.*, 2003). *S. lividans* mutants lacking PhoP produce large amounts of actinorhodin and undecyprodigiosin. However, not all *Streptomyces* species respond equally to phosphate, i.e. *S. coelicolor* ΔPhoP mutants do not respond as strongly as *S. lividans* mutants (Santos-Beneit *et al.*, 2009; Martín *et al.*, 2012b). In *Streptomyces natalensis* biosynthesis of polyene macrolides is extremely sensitive to phosphate and Δ*phoP* mutants are useful to increase the production of pimaricin (Mendes *et al.*, 2007).

The phosphate-trapping agents strategy

Optimal growth of *Streptomyces* species requires 10 to 30 mM inorganic phosphate. Since phosphate limitation is required for expression of numerous secondary metabolites biosynthetic genes (Martín, 2004; Mendes *et al.*, 2007; Martín *et al.*, 2012b) an obvious approach is to use phosphate-limited media (below 0.5 to 1.0 mM). This is possible when using defined culture media although, generally, in such media growth is poor and secondary metabolites are only synthesized in minute amounts. Complex media are more favourable for secondary metabolites production, due to the presence of nutritional factors and inducers, but often they contain excessive phosphate which does not allow the expression of phosphate-repressible genes. A strategy was developed at the Kitasato Institute (Tokyo) by Omura and co-workers to overcome this problem (Masuda *et al.*, 1986). They tested several phosphate trapping agents that chelate or absorb inorganic phosphate from the complex media. These chemicals include allophane, basic magnesium carbonate, aluminium oxide, silicon dioxide, aluminium silicate, acid clay, kaolin and calcium carbonate. The best of these compounds was allophane, an amorphous hydrogel formed by alumina gel and silica gel (Wada, 1989), that produces a large increase of the formation of the isochromane quinone antibiotic nanaomycin by *Streptomyces rosa* subsp. *notoensis*. Interestingly, allophane has a stimulatory effect on the production of several antibiotics by different *Streptomyces* species with the exception of the phosphate-containing antibiotic phosalacine produced by *Kitasatosporia phosalacinea*.

The effect of allophane on triggering secondary metabolite biosynthesis has been confirmed in other actinobacteria, including the vancomycin producer *Amycolatopsis orientalis* and the cephamycin producer *Amycolatopsis* (formerly *Nocardia*) *lactamdurans* (Masuda *et al.*, 1986; Leitao *et al.*, 1996). Allophane stimulated the expression of several genes of the cephamycin C gene cluster in the low producer *A. lactamdurans* resulting in higher activities of the isopenicillin N synthase and the late enzymes for methoxylation and carbamoylation of this antibiotic.

Nitrogen source stress

Nitrogen regulation of the biosynthesis of secondary metabolites is a well-known phenomenon (Aharonowitz, 1980; Omura and Tanaka, 1984). When different nitrogen sources are present simultaneously in the culture medium most microorganisms preferentially use ammonium ions. Ammonium represses the synthesis and inhibits the activity of amino acid transporters and aminotransferases required for amino acid utilization (Reuter and Wohlleben, 2007), and enzymes involved in the biosynthesis of nitrogen containing precursors of secondary metabolites, e.g. the α-aminoadipic acid moiety of β-lactam antibiotics or the pipecolic acid precursor of rapamycin, tacrolimus and related inmunosuppressants.

Ammonium ions are also known to repress the expression of genes involved in the biosynthesis of different metabolites. This repression is mediated by the global nitrogen metabolism regulator GlnR in bacteria (Reuter and Wohlleben, 2007), an orphan response regulator (Gross, 2007), although the GlnR-regulation transcriptional factors that control the expression of antibiotic gene clusters are still unclear. Similarly, ammonium ions also control expression of secondary metabolites genes in filamentous fungi (Haas *et al.*, 1995) although in this case transcription of the ammonium regulated genes is exerted by the global Nre regulator, a zinc-finger protein that acts as a positive regulator.

The ammonium ion-trapping strategy

Following the same strategy used to activate phosphate repressed gene clusters by phosphate trapping agents, Omura and Tanaka (1984) studied the action of several ammonium-trapping chemicals on the biosynthesis of different antibiotics. The tested ammonium trapping agents were magnesium phosphate and different natural zeolites. Magnesium phosphate is known to precipitate ammonium ions, without significant release of phosphate. Other ammonium ion trapping agents, such as calcium phosphate, sodium phosphotungstate and sodium urate were tested but they were not as efficient as magnesium phosphate. The later compound enhanced the production of several low produced macrolide antibiotics and other secondary metabolites by different actinobacteria from 2- to 7-fold (Omura and Tanaka, 1984).

Early applications of the ammonium ion-trapping agent magnesium phosphate were the triggering of the antitumour compounds gilvocarcin (Wei *et al.*, 1982) and tetrocarcin A (Tamaoki and Tomita, 1982) by *Micromonospora chalcea*. Since magnesium phosphate is not a good phosphate source, in this strategy the medium is supplemented with low concentrations of a soluble phosphate salt (e.g. potassium or sodium phosphate). In this way ammonium ion regulation and phosphate repression are bypassed, resulting in derepression of low expressing and sleeping genes.

Metal toxicity stress

Several metals, other than iron and magnesium, the so called 'trace metals' in culture media (including Zn, Mn, Cd, Co, Mo) are required in small concentrations for growth of many microorganisms. They are adequately provided as components of the complex nutrients or in the tap water, although in strictly defined medium their addition is necessary. The concentration of those trace metals is critical for the expression of secondary metabolites. Optimal levels of those metals are different for the production of each metabolite (e.g. a moderately high cobalt concentration is required to produce some cobalt-containing metabolites but excessive concentrations are usually toxic). In nature soil-dwelling microorganisms have to cope with a variety of metals in the soil, frequently as components originated from metal salts in the soil rocks. The metal contamination might be more complicated due to human disposal of metal salts used in a variety of chemical industries. When the metal concentration in a specific habitat is high, a metal toxicity stress response is triggered in the microbial population of the habitat. Interestingly, evidence is accumulating which indicates that this stress response triggers the awakening of sleeping gene clusters (Haferburg, *et al.*, 2009).

There are only scattered reports on the possibility to use heavy metals toxicity to trigger the awakening of sleeping clusters. Haferburg *et al.* (2009) reported that ten *Streptomyces* sp. strains isolated from an abandoned uranium mine and other soils rich in metal minerals contain sleeping gene clusters that are awakened by growth on cadmium or other heavy metals. These authors provide chemical evidence of new compounds being formed in metal supplemented cultures as observed in thin layer chromatography, and also of novel bioactive compounds as shown in bioassays using *Mycobacterium phlei* or *Staphylococcus aureus* as test strains. The cadmium concentration used to trigger these sleeping gene clusters was 3×10^{-4} M. This concentration is toxic for many *Streptomyces* species, although the strains isolated from metal-contaminated soils were resistant to at least 10-fold higher concentrations, probably due to the presence of additional metal detoxification mechanisms. The exact mechanisms by which heavy metals release expression of sleeping gene clusters is unknown but it may involve common metabolic SOS signals.

New products from remodelled pathways

Two different approaches within this strategy have been used, namely the use of pathway remodelling compounds and the utilization of blocked mutants in the pathway to produce new products.

Antibiotic biosynthesis remodelling compounds

This strategy is based on the well-known observation that different small molecular weight compounds act as 'antibiotic pathways remodelling compounds' (ARC) (Moore *et al.*, 2012). Many of these compounds may act as inducers of secondary metabolite gene clusters or as inhibitors of common pathways that will lead to 'reprogramming of antibiotic precursors metabolism' (Olano *et al.*, 2008).

A case study of this approach was the use of thousands of small molecules to search their ability to trigger the onset of secondary metabolites in *S. coelicolor* (Craney *et al.*, 2012). Nineteen of those compounds gave positive results. One family of positive compounds (ARC-2) were chemicals with structures resembling that of trichlosan, a known inhibitor of essential enoyl reductase activities involved in fatty acid synthesis. It is likely that these compounds trigger the early onset of secondary metabolites by saving acetyl-CoA and malonyl-CoA from fatty acid synthesis. The increased availability of these precursors may enhance the expression of poorly expressed gene clusters for specialized metabolite biosynthesis. The ARC-2 compounds triggered also the formation of new compounds in *Streptomyces peucetius* that are not formed in the absence of ARC-2.

Similarly, slowing the rate of protein synthesis with inhibitors may increase the production of non-ribosomal peptides by increasing the precursor amino acid availability. These protein synthesis inhibitors have to be added at subinhibitory concentrations or after the rapid (exponential) growth phase. Indeed, in early studies on non-ribosomal peptides it was observed that inhibition of protein synthesis after the growth phase in *Streptomyces antibioticus* increased the production of the non-ribosome-derived peptide-chromophore antibiotic actinomycin D (Katz and Weissbach, 1963). Also the inhibition of protein synthesis with cycloheximide in cells of *Penicillium chrysogenum* led to enhanced synthesis of the non-ribosomal tripeptide ACV, the first intermediate in penicillin biosynthesis (López Nieto *et al.*, 1985).

In general these ARC-mediated strategies may enhance the production of secondary metabolites from low-expression gene clusters but there is not enough evidence to know if they trigger silent gene clusters.

New products from knock-out mutants in a secondary metabolite pathway

Secondary metabolites are synthesized by complex biosynthetic pathways that frequently involve many enzymatic activities (Díez *et al.*, 1990; Martín *et al.*, 2000; Pel *et al.*, 2007; García-Estrada *et al.*, 2011; Hidalgo *et al.*, 2013). These enzymes are encoded by different ORFs of a gene cluster(s). The complexity of these pathways may be an advantage to generate new products when a late gene in the pathway is blocked. The metabolite profiling of cultures of blocked mutants may lead to novel products. Using this strategy Chiang *et al.* (2008) found in an *Aspergillus nidulans* mutant, a new product identified as emericellamide.

New products from engineered ribosome strains

Modification of some ribosome proteins or of the RNA polymerase β and ω subunits alter drastically the expression of secondary metabolites gene clusters.

Modification of the ribosomal protein S12 or the RNA polymerase β-subunit

The application of the modification of the genes *rpsL* or *rpoB*, encoding respectively the ribosomal protein S12 or the β-subunit of the RNA polymerase derives from the initial work of Ochi (Ochi *et al.*, 1997; Hu *et al.*, 2002) who observed that mutants of several species of *Streptomyces* resistant to streptomycin, kanamycin and other aminoglycosides, as well as mutants resistant to rifampicin were able to produce increased levels of antibiotics (Ochi and Hosaka, 2013). The term 'ribosome engineering' was coined to refer to the directed modification of the genes encoding ribosomal proteins or the *frr* gene that encodes the ribosome recycling-factor in order to improve the production of antibiotics (Hosaka *et al.*, 2006; Ochi, 2007).

These strategies have been applied to try to discover new products by awakening sleeping gene clusters. For this purpose clones of *Streptomyces* sp. isolated from nature that were unable to produce antibiotics, as detected by bioassay, were shown to gain the ability to produce antibiotics when mutated in the *rpsL* gene (e.g. the mutation K88R or K88E) or the mutations H437Y and H437R in the *rpoB* gene. One of the soil isolated actinomycetes identified as *Streptomyces mauvecolor*, that was unable to produce antibiotics when isolated, was altered by selecting antibiotic resistant clones that carried mutations in the *rpsL* and *rpoB* genes. The resulting antibiotic-resistant mutants produce the novel antibiotic piperidamycin (Hosaka *et al.*, 2009)

The evidence accumulated in the last years indicates that *rpoB* mutations, particularly H437Y and H437R, produce large increases in expression of different gene clusters including some silent clusters in *Streptomyces lividans, Streptomyces griseus, S. coelicolor, S. mauvecolor* and *Saccharopolysopora erythraea* (Ochi and Hosaka, 2013). Similarly *rpoB* mutants of *Bacillus subtilis* produce a novel compound, neotrehalosediamine that is not produced by the parental strain (Inaoka and Ochi, 2011). These results are explained by the ability of mutant RNA polymerase β-subunit to interact with the promoter of silent or poorly expressed genes. Another subunit of the RNA polymerase, the subunit ω encoded by the *rpoZ* gene, that interacts with the β-subunit of the core is required for optimal production of actinorhodin and probably other secondary metabolites in *S. coelicolor* (Santos-Beneit *et al.*, 2011).

Two RNA polymerases β-subunits are found in some actinomycetes

The important role of a modified RNA polymerase β-subunit on activating secondary metabolite gene cluster is supported by the discovery that the rare actinomycete *Nanomuraea* contains two *rpoB* genes that encode different RNA polymerase β subunits that are developmentally regulated (Talá *et al.*, 2009). The second *rpoB* gene, named *rpoB*(R), is expressed during the stationary phase and seems to be associated with expression of secondary metabolite gene clusters. This RNA polymerase β(R) form contains five amino acid substitutions when compared with the first one, *rpoB*(S), in the protein region related to rifampicin resistance. It is possible that in some actinomycetes evolution has developed an RNA polymerase (containing the β(R) subunit form) specific for expression of secondary metabolite genes whereas in others this role is played by specific transcriptional factors that modulate the affinity of the RNA polymerase for the promoter. A good possibility is to use the novel *rpoB*(R)-encoded RNA polymerase β form to awaken sleeping gene clusters in actinomycetes that lack the second *rpoB* gene.

Modification of wide-domain regulators and quorum sensing systems

The expression of silent or poorly expressed gene clusters is likely to be defective in their activation by wide domain or pathway-specific regulators. Higher expression of these regulatory genes is usually obtained when they are expressed under the control of an inducible promoter (see examples below). In uncharacterized clusters, in absence of information about the pathway-specific regulators, it is difficult to design strategies to modify them. However, wide-domain regulators that are conserved among many actinomycetes are amenable to genetic modification.

The DasR and LAL-type regulators

Two examples of this strategy to activate silent gene clusters have been reported in the last few years: (i) modification of the DasR regulator in *S. coelicolor*, and (ii) alteration of LAL-type regulators.

The DasR regulator, a member of the GntR family, is involved in signal transduction as a response to N-acetylglucosamine or chitin fragments resulting from the lysis of fungal cells or from the decay of dead arthropods (Rigali *et al.*, 2008). Chitin is a component of fungal cell walls and its presence in the actinobacteria habitat is taken as a signal for activation of secondary metabolism. The molecular mechanism involves binding of DasR to nucleotide sequences in the promoters of pathway-specific regulatory genes (e.g. *actII-orf4*) that in turn activate expression of the adjacent genes of the cluster.

Rigali *et al.* (2008) reported that a mutation in the *dasR* gene activates expression of a silent cluster (genes SCO6273 to SCO6288) that encodes a type II polyketide synthase that synthesized a novel polyketide.

It will be interesting to study if this strategy works in other *Streptomyces* species, since N-acetylglucosamine is known to activate secondary metabolite production in several *Streptomyces* species (Rigali *et al.*, 2008).

LAL-type (Large ATP-binding LuxR-like) regulators are associated to clusters of antibiotic biosynthesis genes, so they might be considered to be pathway-specific regulators. However, some of the best studied LAL regulators bind to several promoters within gene clusters for secondary metabolites (Santos-Aberturas *et al.*, 2011) and, therefore, may activate more than one gene cluster. The LAL-type PAS-Lux regulator PimM of *Streptomyces natalensis* pimaricin gene cluster resulted in activation of amphotericin B and filipin gene clusters in *Streptomyces nodosus* and *Streptomyces avermitilis*, respectively (Santos-Aberturas *et al.*, 2011).

Similarly, *Streptomyces ambofaciens* 23877 produces two antibiotics but contains other gene clusters that are silent under standard culture conditions. One of the silent gene clusters comprising 25 genes, encodes a deduced glycosylated macrolide with potential anti-tumour activity. This silent cluster includes a LAL-type regulator. Overexpression of this LAL regulator resulted in formation of four compounds (stambomycins A to D) that were not produced in the wild type parental strain (Laureti *et al.*, 2011).

Awakening of the *cpk* gene cluster by deletion of a regulatory gene

A type I polyketide synthetase cryptic cluster named *cpk* was found in the sequenced genome of *S. coelicolor* and its product has remained unknown for several years (Pawlik *et al.*, 2007). This cluster contains a putative regulatory gene *scbR2* that belongs to the *scb* family of gamma (γ)-butyrolactone receptor proteins (Takano, 2006). It was found that the ScbR2 protein binds to the promoter of the *scbO* gene and, therefore, it might be considered as a negative regulator of some genes of the *cpk* cluster

Gottelt and coworkers (2010) provided evidence showing that the deletion of the *scbR2* gene leads to the production of two novel compounds that are not produced in the parental *S. coelicolor* M145 strain. One of these is an antibacterial compound (abCpk) and the second is a yellow pigment, identified as coelimycin P1 (Gomez-Escribano *et al.*, 2012), a novel alkaloid containing a functionalized 1,5-oxathiocane moiety. Deletion and overexpression of *scbR2* confirmed that the *cpk* cluster is activated by the ScbR2 regulator (Gottelt *et al.*, 2010).

Activation of silent gene clusters by modification of the quorum-sensing system

Most studies on activation of silent clusters have been made in *Streptomyces* species and in some rare actinomycetes. However, some other bacteria are also able to produce secondary metabolites. The quorum-sensing system is a well-known communication system that coordinates population responses by intraspecific diffusible signals (Nishida *et al.*, 2007; Daniels *et al.*, 2004).

A new polyketide named thailandamycin was produced in *Burkholderia thailandesis* when its LuxR-like quorum sensing system was manipulated. From the analysis of the genome a secondary metabolite gene cluster encoding a hybrid PKS-NRPS was located. The cluster contains a single putative regulatory gene (*thaA*) that resembles LuxR quorum sensing regulators which contain a homoserine lactone binding sequence. Disruption of *thaA* in *B. thailandensis* triggers the production of the complex polyketide thailandamycin whose structure was determined (Ishida *et al.*, 2010). The mutant strain also produced a pigment that was absent in the parental strain. The ThaA protein acts, therefore, as a repressor of the thailandamycin gene cluster that is awakened when the gene encoding this repressor is deleted.

The pathogen *Burkholderia pseudomallei* genome is rich in cryptic natural products biosynthetic gene clusters. The *mal* gene cluster, highly conserved in the *pseudomallei* group, contains two large modular PKSs. The biosynthesis of these compounds is likely self-regulated. Indeed, replacement of the natural promoter of *malA* by a rhamnose inducible strong promoter resulted in activation of the cluster expression and formation of a compound identified as malleilactone, a polyketide-derived cytotoxic siderophore. This compound contains a γ-butyrolactone core that, as indicated above, is a signalling molecule. Both the possible quorum-sensing effect and the ability to sequester iron might affect the virulence in this organism (Biggins *et al.*, 2012).

Awakening of the antitumoral holomycin by deletion of oligopeptide-binding protein OppA2 and the removal of the megaplasmid pSCL4

In 2002 Fuente and coworkers reported one of the first examples of awakening a silent gene cluster. The *oppA2* gene of *S. clavuligerus*, located in the clavulanic acid gene cluster, encodes an oligopeptide-binding protein that has been crystallized and found to bind arginine and probably arginine-derived peptides (Mackenzie *et al.*, 2010). A mutant disrupted in the *oppA2* gene overproduces large amounts of the antitumoral holomycin (Fuente *et al.*, 2002; Lorenzana *et al.*, 2004). It seems that the oligopeptide-binding protein acts as a receptor of arginine-derived ligands and this OppA2–ligand complex represses the holomycin gene cluster (unrelated to clavulanic acid cluster) keeping it silent in the parental *S. clavuligerus* strain.

Interestingly, Álvarez-Álvarez *et al.* (2013b) have found that the high holomycin-producing strain lacks the 1.8 Mb megaplasmid pSCL4 which encodes 25 secondary metabolite gene clusters. Strains devoid of pSCL4 (by introducing a deletion of the partition *parA-parB*$_{\mathrm{pSCL4}}$ genes) produce 1000 μg holomycin per μg of cell DNA whereas an *oppA2*-deleted mutant carrying normal copy number of pSCL4 produces only 400 μg holomycin/μg DNA. When both mutations are present the production increases to 1600 μg holomycin per μg DNA. In summary, it seems that the awakening of the holomycin cluster responds to at least two genetic factors.

The available evidence indicates that a given silent gene cluster may be awakened by more than one mechanism. This is in agreement with the 'multifunctional activation' model proposed by Charusanti *et al.* (2012), who found in addition to the loss of the megaplasmid five other mutations in an holomycin-producing variant strain, some of which may contribute to trigger expression of the holomycin gene (see below).

The original *oppA2* holomycin overproducing strains were obtained during studies using protoplasts transformation procedures. It seems likely that the challenge of the wild-type *S. clavuligerus* strain with lysozyme and the stress produced by the cell wall damage resulted in mutation of genes leading to very low pSCL4 copy number. The loss of the pSCL4 megaplasmid by stressing conditions may be an important factor in the activation of chromosomal gene clusters. It is likely that the linear plasmid(s) encode(s) a repressor protein controlling the holomycin cluster expression.

New products obtained by alteration of cross-pathways regulation

Cross-pathways regulations are becoming better known in *Streptomyces* and filamentous fungi (Wang *et al.*, 2009; Bergmann *et al.*, 2010; Hidalgo *et al.*, 2013) Pleiotropic regulatory genes have been found that control distant unrelated gene clusters together with the cluster in which they are located. There are several examples of this strategy. Bergmann *et al.* (2007) studied a regulatory gene *apdR* located in a silent gene cluster found during bioinformatics analysis of the *A. nidulans* genome. The silent *apd* gene cluster encodes a putative hybrid PKS-NRPS product. When the *apdR* gene that encodes a Zn_2-cys_6 type regulator was expressed from a strong promoter the entire *apd* gene cluster was activated (as expected from a cis-acting regulator). In addition two novel NRPS-PKS compounds were triggered, that were absent in the parental *A. nidulans* strain. These compounds were identified as aspyridones A and B.

Later, Bergmann *et al.* (2010) applied the same principle to another regulatory gene, *scpR*, located in a cryptic NRPS gene cluster. When the *scpR* gene was expressed from the ethanol-inducible *alcA* promoter it resulted in the activation of a different gene cluster located in other chromosome, that encodes the asperfuranone biosynthesis pathway. The *scpR* gene activates the regulatory *afoA* gene of the asperfuranone cluster. Similarly, overexpression of the regulatory gene *afoA* of the asperfuranone gene cluster in *A. nidulans* lead to activation of this cluster (Chiang *et al.*, 2009).

Cross activation of the *Penicillium chrysogenum* PR-toxin gene cluster

Another example of cross-activation of unrelated pathways is the awakening of the PR-toxin gene cluster by *P. chrysogenum* mutants lacking the penicillin biosynthetic genes. The PR-toxin gene cluster has been characterized both in *Penicillium roqueforti* and *P. chrysogenum* by our group (Hidalgo *et al.*, 2013). This cluster contains (i) a 15 carbon atom terpene cyclase (aristolochene synthase) that forms aristolochene, (ii) several other enzymes that convert aristolochene to the final PR toxin and (iii) a MFS transport protein for secretion of the toxin.

In *P. chrysogenum* there is a silent gene cluster (now identified as the PR-toxin genes) which is silent under normal laboratory conditions used for penicillin production (van den Berg *et al.*, 2008). Following a study on the activation of certain genes in *P. chrysogenum* strains lacking the *pen* cluster (Harris *et al.*, 2009) we observed that a *P. chrysogenum* mutant lacking the region containing the *pen* cluster express the silent PR-toxin cluster, located in a different location, what results in the production of PR-toxin in this fungus (Hidalgo *et al.*, 2013). The gene in the *pen* cluster that acts as trigger for expression of the PR-toxin cluster is still under investigation.

Challenging by co-cultivation with other organisms

It is well known that many antibiotics induce expression of sets of genes, particularly antibiotic resistance genes. Several years ago different research groups studied the proteins induced by thiostrepton in *Streptomyces lividans* (Murakami *et al.*, 1989), *Amycolatopsis lactamdurans* and in other actinobacteria (Kumar and Martín, 1994). This peptide antibiotic enhances the expression of several *tip* genes (for thiostrepton-induced proteins). The role of the induced proteins is still unclear.

At subinhibitory concentrations, antibiotics act as 'communication signals' (Yim *et al.*, 2007;

Goh *et al.*, 2002) producing growth stress and triggering a variety of biochemical responses. There is evidence that the same effect may be produced by co-cultivation of a producer strain with other 'challenging' strains.

Challenging by co-cultivation with other organisms

The microbial interactions in the soil are extremely complex and the production of novel metabolites (Fischbach, 2000) in response to the interactions reflects several phenomena including: (i) precursor availability, (ii) antagonism as a fight for survival under nutrient limiting conditions, and (iii) specific communication signals, like quorum sensing, and signals for morphological differentiation.

Charusanti *et al.* (2012) reported a procedure to challenge *S. clavuligerus*, a well-known producer of different types of secondary metabolites including β-lactam antibiotics (Liras *et al.*, 2008), by co-cultivation with a pathogenic *Staphylococcus aureus* MRS (methicillin-resistant strain). After several passages of co-cultivation using this strategy a *S. clavuligerus* variant strain was isolated that produced a yellow compound, absent in the unchallenged wild-type *S. clavuligerus*, that had been previously characterized as holomycin (Fuente *et al.*, 2002). The challenge of *S. clavuligerus* with *S. aureus* also resulted in the triggering of three other compounds different from holomycin but of unknown structure. The mechanism by which these silent metabolites are awakened is unknown since sequencing of the genome of the holomycin-producing variant strains revealed four different mutations that may influence expression of the holomycin gene cluster (Charousanti *et al.*, 2012). In addition, these authors detected the loss of plasmid pSCL4 in the holomycin-producing strain but did not analyse its effect.

Intimate bacterial contacts

Mycolic acid-containing bacteria triggers the production of many new secondary metabolites in *Streptomyces* species. Several actinobacteria contains mycolic acids in their cell envelope. Several of these mycolic acid-containing bacteria, including *Tsukumurella pulmonis*, *Corynebacterium glutamicum* and *Rhodococcus erythropolis*, trigger the production of new secondary metabolites in *Streptomyces* species isolated from the soil. *T. pulmonis* was first observed to trigger the production of the pigmented antibiotics in *S. lividans* that usually does not produce pigments. Co-cultivation of *Streptomyces* with *T. pulmonis* changed the profile of secondary metabolites produced by a specific *Streptomyces* strain and resulted in the formation of new metabolites in 37% of the tested *Streptomyces* strains.

Interestingly, the induction is also exerted by co-cultivation of *Strepomyces* sp. strains with *C. glutamicum*. The alteration of the profiles of secondary metabolites, including the formation of entirely novel metabolites has been attributed to the presence of mycolic acids in the cell walls of the challenging 'inducer actinobacteria'. However, addition of pure mycolic acid to the *Streptomyces* cultures do not produce the same effect. An intimate contact between the 'inducer' and the 'producer' cells has been proposed (Ochi and Hosaka, 2013) but the real communication system and the signal molecules involved are still unknown.

The awakening of silent genes by interaction of two different actinobacteria has been supported by the analysis by microscale imaging mass spectrometry of the formation of new products by *S. coelicolor* in response to the interaction with five other *Streptomyces* species including *Streptomyces albus*, *Streptomyces viridochromogenes*, other *Streptomyces* sp. or *Amycolatopsis* species. Many metabolites not produced by the isolated *S. coelicolor* colonies were found when the colonies contacted colonies of the 'inducer (initiator) strains'. The sets of metabolites associated with *S. coelicolor* colonies were highly variable depending on the interacting partner and the growth time, but they were specific for each partner. The compounds observed include prodigionines, actinorhodin or acyl-desferroxiamines (Fig. 15.1), all of which are specifically encoded in the *S. coelicolor* genome and not present in the 'initiator strain' genome. The interactions with *Amycolatopsis* sp. AA4 and *Streptomyces* sp. E14 triggered the production of a set of new acyl-desferrioxamines not detected previously (Fig. 15.1) (Traxler *et al.*, 2013).

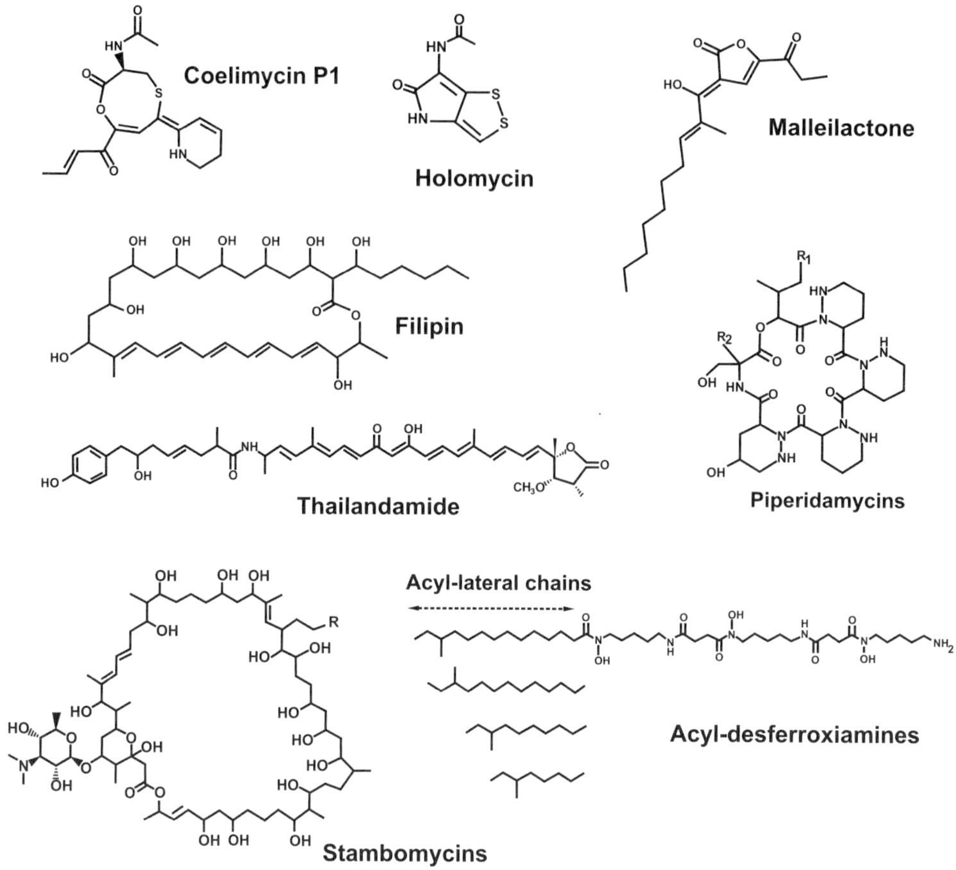

Figure 15.1 Chemical structure of representative compounds encoded by awaken gene clusters of Gram-positive and Gram-negative bacteria: Coelimycin P1 (*Streptomyces coelicolor*), Holomycin (*Streptomyces clavuligerus*), Filipin (*Streptomyces avermitilis*), Malleilactone (*Burkholderia pseudomallei*), Thailandamide (*Burkholderia thailandensis*), Piperidamycins (*Streptomyces mauvecolor*) Stambomycins (*Streptomyces ambofaciens*), Acyl-desferroxamines (*S. coelicolor*). For references see Table 15.1.

Intimate bacterial–fungal interactions

The molecular ecology of filamentous fungi is largely unknown (Barreiros *et al.*, 2012) despite the numerous studies of a few model fungi (Jami *et al.*, 2010). There are hundreds of species that contain thousands of secondary metabolites gene clusters. Genome sequencing reveals no redundancy of clusters in a single strain and only low conservation in different fungi. Many of these gene clusters are silent or very poorly expressed. Only about 10–20% of the gene clusters are expressed under standard laboratory conditions. Therefore, the filamentous fungi show a high untapped potential for the discovery of new bioactive products (Frisvad *et al.*, 2004).

It is known that in soil and also in other habitats, filamentous fungi interact with many bacteria; some of these interactions, as for example the attack by bacteria producing antifungal agents, cause a defence reaction in the fungi.

This type of interactions have been demonstrated by using genomics and microarrays studies to analyse the response of *Aspergillus nidulans* to the co-cultivation with species of 58 actinomycetes (Schroeckh *et al.*, 2009). Interaction with the actinomycetes triggered the expression in *A. nidulans* of a silent PKS gene cluster that encodes the polyketides orsenillic acid, lecanoric acid and the cathepsin K inhibitors F-9775A and F-9775B. Most interestingly, dialysis experiments showed that an intimate contact between hyphae of actinomycetes and *Aspergillus* is required to trigger the onset of the novel compounds. It seems, therefore, that the intergeneric communication signal is not diffusible and is probably associated to the actinomycetes cell walls.

Chromatin remodelling: epigenetic modifiers

Manipulation of the LaeA regulator and induction by diamines

The protein LaeA is a nuclear protein that resembles histone and arginine methyltransferases and has a functional S-adenosylmethionine-binding site as shown by mutation of this binding site. It was first reported in *A. nidulans* as a result of the observation that mutants in the *laeA* gene were unable to express the pathway specific activator of the aflatoxin pathway, *aflR* (Bok and Keller, 2004). *A. nidulans* mutants defective in LaeA showed reduced production of sterigmatocystin, penicillin and some other metabolites.

In *Aspergillus fumigatus laeA* mutants showed reduced production of gliotoxin. Thirteen of the known clusters encoding secondary metabolites in this fungus were up-regulated by LaeA but all of them were also expressed at low levels in the non-mutated strain (Perrin *et al.*, 2007). In contrast, in *A. nidulans* comparative metabolite profiling of cultures of the *laeA* mutant and the parental strain resulted in the characterization of a novel LaeA-dependent gene cluster that was identified as terrequinone A (Bok *et al.*, 2006).

P. chrysogenum laeA mutants showed a drastically reduced expression of the three penicillin biosynthesis genes *pcbAB*, *pcbC* and *penDE*. Interestingly, *laeA*-overexpressing transformants overproduce penicillin and also several other solvent-extractable and water-soluble compounds of still unknown structure (Kosalkova *et al.*, 2009). As in *A. nidulans* and *A. fumigatus* not all secondary metabolites of *P. chrysogenum* were regulated by LaeA; the production of roquefortine C was not affected by LaeA (Kosalkova *et al.*, 2009), presumably because the roquefortine C gene cluster is located in a chromosome region of low expression with heterochromatin structure transcriptionally restrictive (see below).

The LaeA regulator forms part of a three-component nuclear complex VeaB/veA/LaeA that in *A. nidulans* is involved in the light/darkness control of asexual differentiation (Bayram *et al.*, 2010). In *P. chrysogenum* LaeA also controls asexual sporulation and spore-pigment formation (Kosalkova *et al.*, 2009).

It is interesting that the effect of LaeA on the production of secondary metabolites in *P. chrysogenum* is reversed by the autoinducer 1,3-diaminopropane (Martín *et al.*, 2011, 2012a). This autoinducer was first characterized as a product of cultures of *P. chrysogenum* and *Acremonium chrysogenum* (Martín *et al.*, 2011). This inducer enhances the expression of the *laeA* gene, which in turn increases the expression of the penicillin biosynthesis genes. The effect of the autoinducer, mediated by LaeA, is also observed in the formation of a yellow pigment that is nearly absent in the parental strain. It is likely that this diamine triggers the production of silent or low-expressing secondary metabolite clusters also in other fungi.

Chromatin modifiers triggers expression of new clusters

The effect of LaeA on triggering the expression of secondary metabolite gene clusters suggested that LaeA acts by remodelling the chromatin structure (Keller *et al.*, 2005). Indeed, experiments on expression of the histone deacetylase gene, supported this hypothesis (Shwab *et al.*, 2007). Fungal histones can be modified by several reactions including methylation, deacetylation and phosphorylation. Other modifications such as ubiquitination and sumoylation are important for histone degradation or maintenance.

Methylation of histone H3 is regulated by the COMPASS protein complex. Inactivation of the gene *cclA* (for one protein of this complex), altered the histone methylation in *A. nidulans* and triggered the production of emodin, monodyctiphenone and some derivatives (Bok *et al.*, 2009). Also, alteration of the sumoylation *sumO* gene leads to a large increase in the synthesis of asperthecin and a decrease of austinol (Szewczyk *et al.*, 2008).

Taking into account the evidence of heterochromatin involvement in expression of secondary metabolite gene clusters some chemical compounds known to modify the chromatin structure have been used to trigger the synthesis of novel secondary metabolites. The best known 'epigenetic modifiers' include the histone deacetylase inhibitors suberoylanilide hydroxamate and the DNA-methyltransferase inhibitor 5′-azacitidine. In *Cladosporium cladosporioides* the addition of 5′-azacitidine triggered the production of oxylipins, and the addition of suberoylanilide hydroxamate led to the production of calphostin and cladochrome (Williams *et al.*, 2008). Similarly, in *A. niger* cultures, the addition of suberoylanilide hydroxamate triggered production of nygerone (Henrikson *et al.*, 2009).

In summary, the available evidence in different fungi suggests that 'epigenetics engineering' using chromatin modifiers is a good strategy to awaken silent gene clusters and to potentiate the transcription of very low expression clusters.

The arrangement in clusters of most secondary metabolites facilitate the chromatin-mediated transcriptional regulation of the expression of those clusters. This is particularly relevant when the cluster includes a specific regulatory gene that controls the activation of the other genes (e.g. *aflR* in the aflatoxin cluster).

An *A. nidulans* cc/A mutant (altered in the formation of the heterochromatin structure component proteins) shows increased expression of the sterigmatocystin, penicillin and terrequinone A clusters.

The sterigmatocystin gene cluster is silent during the *A. nidulans* rapid growth phase and contains chromatin rich in trimethylated H3 lysine9 (H3K9-trime) histone. In this growth phase the chromatin is associated with heterochromatin protein 1 (HepA). When the rapid growth ceases, the H3K9-trime content of chromatin decreases and instead the acetylated H3 increases and protein HepA is released (Reyes-Domínguez *et al.*, 2010). This chromatin arrangement is conserved at the edges of the cluster but not in genes outside the cluster. The nuclear protein LaeA changes the heteroprotein arrangement characteristic of the growth phase to that of the production phase, thus allowing expression of the serigmatocystin cluster.

It is likely that a similar mechanism controls the biosynthesis of several other gene clusters (but not all) in a given fungal species. Indeed, it is known that the expression of different gene clusters is highly dependent on the location in the fungal chromosomes (e.g. in subtelomeric or in chromosome internal regions) (Bok *et al.*, 2006). The chromatin arrangement in each of those chromosome regions explains the effect (or the lack of it) of LaeA.

Conclusions, remarks and perspectives

The genome sequences of a variety of actinobacteria, myxobacteria and filamentous fungi has revealed a large number of secondary metabolite gene clusters. Many of those clusters appear to be silent or near silent since the number of chemical products formed by those bacteria or fungi is much lower than the number of secondary metabolite gene clusters.

In addition to the classical strategies to trigger the expression of secondary metabolite gene clusters by nutrient limitation stresses or by resistance to antibiotics acting on the ribosomal protein S12 or on the RNA polymerase β subunit, several novel approaches have been tried in the last decades to awaken sleeping gene clusters. One of the strategies, the modification of wide-domain and pathway-specific regulators has been successful in bacteria. In fungi, cross-talk regulation between secondary metabolite pathway regulators has been manipulated to trigger expression of silent genes.

Another fruitful approach has been the challenging of putative secondary metabolite producer strains with competing bacteria or fungi. Co-cultivation 'challenging' of possible producer strains with other microorganisms has revealed a tremendous potential to awaken the expression of silent gene clusters. This aspect of microbial molecular ecology is extremely promising for the discovery of still sleeping gene clusters.

Furthermore, in fungi the modification of chromatin structure in the producer fungi by either directed mutations of genes affecting the heterochromatin structure or by addition of chromatin modifiers, is a very powerful tool to obtain new fungal products. Since hundreds of fungal species are still poorly studied, the untapped potential of fungal genomes to obtain new products is very high. Routine analysis of 'chromatin-modified' fungal cultures combined with the use of advanced mass spectrometry tools will uncover new secondary metabolites in the next decades, some of which may provide novel pharmacological activities to treat human diseases.

References

Aharonowitz, Y. (1980). Nitrogen metabolite regulation of antibiotic biosynthesis. Annu. Rev. Microbiol. *34*, 209–233.

Álvarez-Álvarez, R., Martínez-Burgo, Y., Pérez-Redondo, R., Braña, A.F., Martín, J.F., and Liras, P. (2013a). Expression of the endogenous and heterologous clavulanic acid cluster in *Streptomyces flavogriseus*: why a silent cluster is sleeping. Appl. Microbiol. Biotechnol. 97, 9451–9463.

Álvarez-Álvarez, R., Rodríguez-García, A., Martínez-Burgo, Y., Robles-Reglero, V., Santamarta, I., Pérez-Redondo, R., Martín, J.F., and Liras, P. (2013b). A 1.8-Mb-reduced *Streptomyces clavuligerus* genome: relevance for secondary metabolism and differentiation. Appl. Microbiol. Biotechnol. *98*, 2183–2195.

Barreiro, C., Martín, J.F., and García-Estrada C. (2012). Proteomics shows new faces for the old penicillin producer *Penicillium chrysogenum*. J. Biomed. Biotechnol. *2012*, 105–109.

Bayram, S., Bayram, O., Valerius, O., Park, H.S., Irniger, S., Gerke, J., Ni, M., Han, K.H., Yu, J.H., and Braus, G.H. (2010). LaeA control of velvet family regulatory proteins for light-dependent development and fungal cell-type specificity. PLoS Genet. *6*, e1001226.

van den Berg, M.A., Albang, R., Albermann, K., Badger, J.H., Daran, J.M., Driessen, A.J., García-Estrada, C., Fedorova, N.D., Harris, D.M., Heijne, W.H., *et al.* (2008). Genome sequencing and analysis of the filamentous fungus *Penicillium chrysogenum*. Nat. Biotechnol. *26*, 1161–1168.

Bergmann, S., Schümann, J., Scherlach, K., Lange, C., Brakhage, A.A., and Hertweck, C. (2007). Genomics-driven discovery of PKS–NRPS hybrid metabolites from *Aspergillus nidulans*. Nat. Chem. Biol. *3*, 213–217.

Bergmann, S., Funk, A.N., Scherlach, K., Schroeckh, V., Shelest, E., Horn, U., Hertweck, C., and Brakhage, A.A. (2010). Activation of a silent fungal polyketide biosynthesis pathway through regulatory cross talk with a cryptic nonribosomal peptide synthetase gene cluster. Appl. Environ. Microbiol. *76*, 8143–8149.

Biggins, J.B., Ternei, M.A., and Brady, S.F. (2012). Malleilactone, a polyketide synthase-derived virulence factor encoded by the cryptic secondary metabolome of *Burkholderia pseudomallei* group pathogens. J. Am. Chem. Soc. *134*, 13192–13195.

Bok, J.W., and Keller, N.P. (2004). LaeA, a regulator of secondary metabolism in *Aspergillus* spp. Eukaryot. Cell 3, 527e535.

Bok, J.W., Noordermeer, D., Kale, S.P., and Keller, N.P. (2006). Secondary metabolic gene cluster silencing in *Aspergillus nidulans*. Mol. Microbiol. *61*, 1636e1645.

Bok, J.W., Chiang, Y.M., Szewczyk, E., Reyes-Dominguez, Y., Davidson, A.D., Sánchez, J.F., Lo, H.C., Watanabe, K., Strauss, J., Oakley, B.R., Wang, C.C., and Keller, N.P. (2009). Chromatin-level regulation of biosynthetic gene clusters. Nat. Chem. Biol. *5*, 462–464.

Charusanti, P., Fong, N.L., Nagarajan, H., Pereira, A.R., Li, H.J., Abate, E.A., Su, Y., Gerwick, W.H., and Palsson, B.O. (2012). Exploiting adaptive laboratory evolution of *Streptomyces clavuligerus* for antibiotic discovery and overproduction. PLoS One. *7*, e33727.

Chiang, Y.-M., Szewczyk, E., Nayak, T., Davidson, A.D., Sanchez, J.F., Lo, H.C., Ho, W.Y., Simityan, H., Kuo, E., Praseuth, A., *et al.* (2008). Molecular genetic mining of the *Aspergillus* secondary metabolome: discovery of the emericellamide biosynthetic pathway. Chem. Biol. *15*, 527–532.

Chiang, Y.-M., Szewczyk, E., Davidson, A.D., Keller, N.P., Oakley, B.R., and Wang, C.C. (2009). A gene cluster containing two fungal polyketide synthases encodes the biosynthetic pathway for a polyketide, asperfuranone, in *Aspergillus nidulans*. J. Am. Chem. Soc. *131*, 2965–2970.

Craney, A., Ozimok, C., Pimentel-Elardo, S.M., Capretta, A., and Nodwell, J.R. (2012). Chemical perturbation of secondary metabolism demonstrates important links to primary metabolism. Chem. Biol. *19*, 1020–1027.

Daniels, R., Vanderleyden, J., and Michiels, J. (2004). Quorum sensing and swarming migration in bacteria. FEMS Microbiol. Rev. *28*, 261–289.

Davies, J. (2007). Small molecules: the lexicon of biodiversity. J. Biotechnol. *29*, 3–5.

Díez, B., Gutiérrez, S., Barredo, J.L., van Solingen, P., van der Voort, L.H., and Martín, J.F. (1990). The cluster of penicillin biosynthetic genes. Identification and characterization of the *pcbAB* gene encoding the alpha-aminoadipyl-cysteinyl-valine synthetase and linkage to the *pcbC* and *penDE* genes. J. Biol. Chem. *265*, 16358–16365.

von Döhren, H. (2009). A survey of nonribosomal peptide synthetase (NRPS) genes in *Aspergillus nidulans*. Fungal Genet. Biol. *46* Suppl 1, S45–52.

Fischbach, M.A. (2009). Antibiotics from microbes: converging to kill. Curr. Opin. Microbiol. *12*, 520–527.

Fischbach, M.A., and Walsh, C.T. (2009). Antibiotics for emerging pathogens. Science *325*, 1089–1093.

Frisvad, J.C., Smedsgaard, J., Larsen, T.O., and Samson, R.A. (2004). Mycotoxins, drugs and extrolites produced by species in *Penicillium* subgenus *Penicillium*. Stud. Mycol. *49*, 201–241.

Fuente, A. de la, Lorenzana, L.M., Martín, J.F., and Liras, P. (2002). Mutants of *Streptomyces clavuligerus* with disruptions in different genes for clavulanic acid biosynthesis produce large amounts of holomycin: possible cross-regulation of two unrelated secondary metabolic pathways. J. Bacteriol. *184*, 6559–6565.

García-Estrada, C., Ullán, R.V., Velasco-Conde, T., Godio, R.P., Teijeira, F., Vaca, I., Feltrer, R., Kosalková, K., Mauriz, E., and Martín, J.F. (2008). Post-translational enzyme modification by the phosphopantetheinyl transferase is required for lysine and penicillin biosynthesis but not for roquefortine or fatty acid

formation in *Penicillium chrysogenum*. Biochem. J. *415*, 317–324.
García-Estrada, C., Ullán, R.V., Albillos, S.M., Fernández-Bodega, M.Á., Durek, P., von Döhren, H., and Martín, J.F. (2011). A single cluster of coregulated genes encodes the biosynthesis of the mycotoxins roquefortine C and meleagrin in *Penicillium chrysogenum*. Chem. Biol. *18*, 1499–1512.
Goh, E.B., Yim, G., Tsui, W., McClure, J., Surette, M.G., and Davies, J. (2002). Transcriptional modulation of bacterial gene expression by subinhibitory concentrations of antibiotics. Proc. Natl. Acad. Sci. U.S.A. *99*, 17025–17030.
Gomez-Escribano, J.P., Song, L., Fox, D.J., Yeo, V., Bibb, M.J., and Challis, G.L. (2012). Structure and biosynthesis of the unusual polyketide alkaloid coelimycin P1, a metabolic product of the *cpk* gene cluster of *Streptomyces coelicolor* M145. Chem. Sci. *3*, 2716–2720.
Gottelt, M., Kol, S., Gomez-Escribano, J.P., Bibb, M., and Takano, E. (2010). Deletion of a regulatory gene within the *cpk* gene cluster reveals novel antibacterial activity in *Streptomyces coelicolor* A3(2). Microbiology *156*, 2343–2353.
Gross, H. (2007). Strategies to unravel the function of orphan biosynthesis pathways: recent examples and future prospects. Appl. Microbiol. Biotechnol. *75*, 267–277.
Haas, H., Bauer, B., Redl, B., Stöffler, G., and Marzluf, G.A. (1995). Molecular cloning and analysis of *nre*, the major nitrogen regulatory gene of *Penicillium chrysogenum*. Curr. Genet. *27*, 150–158.
Haferburg, G., Groth, I., Möllmann, U., Kothe, E., and Sattler, I. (2009). Arousing sleeping genes: shifts in secondary metabolism of metal tolerant actinobacteria under conditions of heavy metal stress. Biometals *22*, 225–234.
Harris, D.M., van der Krogt, Z.A., Klaassen, P., Raamsdonk, L.M., Hage, S., van den Berg, M.A., Bovenberg, R.A., Pronk, J.T., and Darans, J.M. (2009). Exploring and dissecting genome-wide gene expression responses of *Penicillium chrysogenum* to phenylacetic acid consumption and penicillinG production. BMC Genomics *10*, 10:75.
Henrikson, J.C., Hoover, A.R., Joyner, P.M., and Cichewicz, R.H. (2009). A chemical epigenetics approach for engineering the in situ biosynthesis of a cryptic natural product from *Aspergillus niger*. Org. Biomol. Chem. *7*, 435–438.
Hidalgo, P.I., Ullán, R.V., Albillos, S.M., Montero, O., Fernández-Bodega, M.A., García-Estrada, C., Fernández-Aguado, M., and Martín, J.F. (2013). Molecular characterization of the PR-toxin gene cluster in *Penicillium roqueforti* and *Penicillium chrysogenum*: Cross talk of secondary metabolite pathways. Fungal Genet. Biol. *62*, 11–24.
Hopwood, D.A. (2013). Imaging mass spectrometry reveals highly specific interactions between actinomycetes to activate specialized metabolic gene clusters. MBio. *4*, e00612–13.
Hosaka, T., Xu, J., and Ochi, K. (2006). Increased expression of ribosome recycling factor is responsible for the enhanced protein synthesis during the late growth phase in an antibiotic-overproducing *Streptomyces coelicolor* ribosomal *rpsL* mutant. Mol. Microbiol. *61*, 883–897.
Hosaka, T., Ohnishi-Kameyama, M., Muramatsu, H., Murakami, K., Tsurumi, Y., Kodani, S., Yoshida, M., Fujie, A., and Ochi, K. (2009). Antibacterial discovery in actinomycetes strains with mutations in RNA polymerase or ribosomal protein S12. Nat. Biotechnol. *27*, 462–464.
Hu, H., Zhang, Q., and Ochi, K. (2002). Activation of antibiotic biosynthesis by specified mutations in the *rpoB* gene (encoding the RNA polymerase beta subunit) of *Streptomyces lividans*. J. Bacteriol. *184*, 3984–3991.
Inaoka, T., and Ochi, K. (2011). Activation of dormant secondary metabolism neotrehalosadiamine synthesis by an RNA polymerase mutation in *Bacillus subtilis*. Biosci. Biotechnol. Biochem. *75*, 618–623.
Ishida, K., Lincke, T., Behnken, S., and Hertweck, C. (2010). Induced biosynthesis of cryptic polyketide metabolites in a *Burkholderia thailandensis* quorum sensing mutant. J. Am. Chem. Soc. *132*, 13966–13968.
Jami, M.S., Barreiro, C., García-Estrada, C., and Martín, J.F. (2010). Proteome analysis of the penicillin producer *Penicillium chrysogenum*: Characterization of protein changes during the industrial strain improvement. Mol. Cel. Proteomics *9*, 1182–1198.
Katz, E., and Weissbach, H. (1963). Incorporation of ^{14}C-labelled amino acids into actinomycin and protein in *Streptomyces antibioticus*. J. Biol. Chem. *238*, 666–675.
Keller, N.P., Turner, G., and Bennett, J.W. (2005). Fungal secondary metabolism – from biochemistry to genomics. Nat. Rev. Microbiol. *3*, 937–947.
Kosalková, K., García-Estrada, C., Ullán, R.V., Godio, R.P., Feltrer, R., Teijeira, F., Mauriz, E., and Martín, J.F. (2009).The global regulator LaeA controls penicillin biosynthesis, pigmentation and sporulation, but not roquefortine C synthesis in *Penicillium chrysogenum*. Bioquimie *91*, 214–225.
Kumar, C.V., and Martín, J.F. (1994). Thiostrepton induced proteins in *Streptomyces*, *Amycolatopsis* and *Nocardia* species. FEMS Microbiol. Lett. *118*, 107–112.
Laureti, L., Song, L., Huang, S., Corre, C., Leblond, P., Challis, G.L., and Aigle, B. (2011). Identification of a bioactive 51-membered macrolide complex by activation of a silent polyketide synthase in *Streptomyces ambofaciens*. Proc. Natl. Acad. Sci. U.S.A. *108*, 6258–6263.
Leitao, A.L., Enguita, F.J., Fuente, J.L., Liras, P., and Martín, J.F. (1996). Allophane increases the protein levels of several cephamycin biosynthesis enzymes in *Nocardia lactamdurans*. Microbiology *142*, 3399–3406.
Liras, P., Gomez-Escribano, J.P., and Santamarta, I. (2008). Regulatory mechanisms controlling antibiotic production in *Streptomyces clavuligerus*. J. Ind. Microbiol. Biotechnol. 35, 667–676.
López-Nieto, M.J., Ramos, F.R., Luengo, J.M., and Martín, J.F. (1985). Characterization of the biosynthesis *in vivo* of α-aminoadipyl-cysteinyl-valine in *Penicillium*

chrysogenum. Appl. Microbiol. Biotechnol. *22*, 343–351.

Lorenzana, L.M., Pérez-Redondo, R., Santamarta, I., Martín, J.F., and Liras, P. (2004). Two oligopeptide-permease-encoding genes in the clavulanic acid cluster of *Streptomyces clavuligerus* are essential for production of the β-lactamase inhibitor. J. Bacteriol. *186*, 3431–3438.

Mackenzie, A.K., Valegård, K., Iqbal, A., Caines, M.E., Kershaw, N.J., Jensen, S.E., Schofield, C.J., and Andersson, I. (2010). Crystal structures of an oligopeptide-binding protein from the biosynthetic pathway of the β-lactamase inhibitor clavulanic acid. J. Mol. Biol. *396*, 332–344.

Márquez-Fernández, O., Trigos, A., Ramos-Balderas, J.L., Viniegra-González, G., Deising, H.B., and Aguirre, J. (2007). Phosphopantetheinyl transferase CfwA/NpgA is required for *Aspergillus nidulans* secondary metabolism and asexual development. Eukaryot. Cell. *6*, 710–720.

Martín, J.F. (1989). Molecular mechanism for the control by phosphate of the biosynthesis of antibiotic and secondary metabolites. In Regulation of Secondary Metabolism in Actinomycetes, S. Shapiro, ed. (CRC Press, Boca Raton, FL), pp. 213–236.

Martín, J.F. (2004). Phosphate control of the biosynthesis of antibiotics and other secondary metabolites is mediated by the PhoR-PhoP system: an unfinished story. J. Bacteriol. *186*, 5197–5201.

Martín, J.F., and Demain, A.L. (1980). Control of antibiotic biosynthesis. Microbiol. Rev. *4*, 230–251.

Martín, J.F., Gutiérrez, S., and Aparicio, J.F. (2000). Secondary metabolites. In Encyclopedia of Microbiology, vol. 4, J. Lederberg, ed. (Academic Press, San Diego, CA, USA), pp. 213–236.

Martín, J., García-Estrada, C., Rumbero, A., Recio, E., Albillos, S.M., Ullan, R.V., and Martín, J.F. (2011). Characterization of an autoinducer of penicillin biosynthesis in *Penicillium chrysogenum*. Appl. Env. Microbiol. *77*, 5688–5696.

Martín, J., García-Estrada, C., Kosalková, K., Ullán, R.V., Albillos, S.M., and Martín, J.F. (2012a). The inducers 1,3-diaminopropane and spermidine produce a drastic increase in the expression of the penicillin biosynthetic genes for prolonged time, mediated by the *laeA* regulator. Fungal Genet. Biol. *49*, 1004–1013.

Martín, J.F., Santos-Beneit, F., Rodríguez-García, A., Sola-Landa, A., Smith, M.C., Ellingsen, T.E., Nieselt, K., Burroughs, N.J., and Wellington, E.M. (2012b). Transcriptomic studies of phosphate control of primary and secondary metabolism in *Streptomyces coelicolor*. Appl. Microbiol. Biotechnol. *95*, 61–75.

Masuma, R., Tanaka, Y., Tanaka, H., and Omura, S. (1986). Production of nanaomycin and other antibiotics by phosphate-depressed fermentation using phosphate-trapping agents. J. Antibiot. *39*, 1557–1564.

Medema, M.H., Trefzer, A., Kovalchuk, A., van den Berg, M., Müller, U., Heijne, W., Wu, L., Alam, M.T., Ronning, C.M., Nierman, W.C., *et al.* (2010). The sequence of a 1.8-mb bacterial linear plasmid reveals a rich evolutionary reservoir of secondary metabolic pathways. Genome Biol. Evol. *2*, 212–224.

Mendes, M.V., Tunca, S., Antón, N., Recio, E., Sola-Landa, A., Aparicio, J.F., and Martín, J.F. (2007). The two-component *phoR-phoP* system of *Streptomyces natalensis*: inactivation or deletion of *phoP* reduces the negative phosphate regulation of pimaricin biosynthesis. Metab. Eng. *9*, 217–227.

Moore, J.M., Bradshaw, E., Seipke, R.F., Hutchings, M.I., and McArthur, M. (2012). Use and discovery of chemical elicitors that stimulate biosynthetic gene clusters in *Streptomyces* bacteria. Methods Enzymol. *517*, 367–385.

Murakami, T., Holt, T.G., and Thompson, C.J. (1989). Thiostrepton-induced gene expression in *Streptomyces lividans*. J. Bacteriol. *171*, 1459–1466.

Nishida, H., Ohnishi, Y., Beppu, T., and Horinouchi, S. (2007). Evolution of gamma-butyrolactone synthases and receptors in *Streptomyces*. Environ. Microbiol. *9*, 1986–1994.

Ochi, K. (2007). From microbial differentiation to ribosome engineering. Biosci. Biotechnol. Biochem. *71*, 1373–1386.

Ochi, K., and Hosaka, T. (2013). New strategies for drug discovery: activation of silent or weakly expressed microbial gene clusters. Appl. Microbiol. Biotechnol. *97*, 87–98.

Ochi, K., Zhang, D., Kawamoto, S., and Hesketh, A. (1997). Molecular and functional analysis of the ribosomal L11 and S12 protein genes (*rplK* and *rpsL*) of *Streptomyces coelicolor* A3(2). Mol. Gen. Genet. *256*, 488–498.

Olano, C., Lombó, F., Méndez, C., and Salas, J.A. (2008). Improving production of bioactive secondary metabolites in actinomycetes by metabolic engineering. Metab. Eng. *10*, 281–292.

Omura, S., and Tanaka, Y. (1984). In Biological, Biochemical and Biomedical Aspects of Actinomycetes, L. Ortiz-Ortiz, L.F. Bojalil, and V. Yakoleff, eds. (Academic Press, Orlando, FL, USA), pp. 367–380.

Pawlik, K., Kotowska, M., Chater, K.F., Kuczek, K., and Takano, E. (2007). A cryptic type I polyketide synthase (*cpk*) gene cluster in *Streptomyces coelicolor* A3(2). Arch. Microbiol. *187*, 87–99.

Pel, H.J., de Winde, J.H., Archer, D.B., Dyer, P.S., Hofmann, G., Schaap, P.J., Turner, G., de Vries, R.P., Albang, R., Albermann, K., *et al.* (2007). Genome sequencing and analysis of the versatile cell factory *Aspergillus niger* CBS 513.88. Nat. Biotechnol. *25*, 221–231.

Perrin, R.M., Fedorova, N.D., Bok, J.W., Cramer, R.A., Wortman, J.R., Kim, H.S., Nierman, W.C., and Keller, N.P. (2007). Transcriptional regulation of chemical diversity in *Aspergillus fumigatus* by LaeA. PLoS Pathog. 3(4), e50.

Reuther, J., and Wohlleben, W. (2007). Nitrogen metabolism in *Streptomyces coelicolor*: transcriptional and post-translational regulation. J. Mol. Microbiol. Biotechnol. *12*, 139–146.

Reyes-Domínguez, Y., Bok, J.W., Berger, H., Shwab, E.K., Basheer, A., Gallmetzer, A., Scazzocchio, C., Keller, N., and Strauss, J. (2010). Heterochromatic marks are associated with the repression of secondary metabolism clusters in *Aspergillus nidulans*. Mol. Microbiol. *76*, 1376–1386.

Rigali, S., Titgemeyer, F., Barends, S., Mulder, S., Thomae, A.W., Hopwood, D.A., and van Wezel, G.P. (2008). Feast or famine: the global regulator DasR links nutrient stress to antibiotic production by *Streptomyces*. EMBO Rep. *9*, 670–675.

Schroeckh, V., Scherlach, K., Nützmann, H.W., Shelest, E., Schmidt-Heck, W., Schuemann, J., Martin, K., Hertweck, C., and Brakhage, A.A. (2009). Intimate bacterial–fungal interaction triggers biosynthesis of archetypal polyketides in *Aspergillus nidulans*. Proc. Natl. Acad. Sci. U.S.A. *106*, 14558–14563.

Santos-Aberturas, J., Payero, T.D., Vicente, C.M., Guerra, S.M., Cañibano, C., Martín, J.F., and Aparicio, J.F. (2011). Functional conservation of PAS-LuxR transcriptional regulators in polyene macrolide biosynthesis. Metab. Eng. *13*, 756–767.

Santos-Beneit, F., Rodríguez-García, A., Sola-Landa, A., and Martín, J.F. (2009). Cross-talk between two global regulators in *Streptomyces*: PhoP and AfsR interact in the control of *afsS*, *pstS* and *phoRP* transcription. Mol. Microbiol. 72, 53–68.

Santos-Beneit, F., Barriuso-Iglesias, M., Fernández-Martínez, L.T., Martínez-Castro, M., Sola-Landa, A., Rodríguez-García, A., and Martín, J.F. (2011).The RNA polymerase omega factor RpoZ is regulated by PhoP and has an important role in antibiotic biosynthesis and morphological differentiation in *Streptomyces coelicolor*. Appl. Environ. Microbiol. *77*, 7586–7594.

Sola-Landa, A., Moura, R.S., and Martín, J.F. (2003). The two-component PhoR-PhoP system controls both primary metabolism and secondary metabolite biosynthesis in *Streptomyces lividans*. Proc. Natl. Acad. Sci. U.S.A. *100*, 6133–6138.

Shwab, E.K., Bok, J.W., Tribus, M., Galehr, J., Graessle, S., and Keller, N.P. (2007). Histone deacetylase activity regulates chemical diversity in *Aspergillus*. Eukaryot. Cell. *6*, 1656–1664.

Szewczyk, E., Chiang, Y.M., Oakley, C.E., Davidson, A.D., Wang, C.C., and Oakley, B.R. (2008). Identification and characterization of the asperthecin gene cluster of *Aspergillus nidulans*. Appl. Environ. Microbiol. *74*, 7607–7612.

Takano, E. (2006). Gamma-butyrolactones: *Streptomyces* signaling molecules regulating antibiotic production and differentiation. Curr. Opin. Microbiol. *9*, 287–294.

Talà, A., Wang, G., Zemanova, M., Okamoto, S., Ochi, K., and Alifano, P. (2009). Activation of dormant bacterial genes by *Nonomuraea* sp. strain ATCC 39727 mutant-type RNA polymerase. J. Bacteriol. *191*, 805–814.

Tamaoki, T., and Tomita, F. (1982). Biosynthesis and production of tetrocarcin A, a new antitumor antibiotic in chemically defined medium. Agric. Biol. Chem. *46*, 1021–1026.

Traxler, M.F., Watrous, J.D., Alexandrov, T., Dorrestein, P.C., and Koiter, R. (2013). Interspecies interactions stimulate diversification of the *Streptomyces coelicolor* secreted metabolome. MBio. *20*, 4(4).

Yim, G., Wang, H.H., and Davies, J. (2007). Antibiotics as signalling molecules. Phil. Trans. R. Soc. Lond. B Biol. Sci. *362*, 1195–1200.

Wada, K. (1989). Allophane and imogolite. In Minerals in Soil Environments, 2nd edn. J.B. Dixon and S.B. Weed, eds. (Soil Science Society of America, Madison, WI) pp. 1051–1087.

Wang, L., Tian, X., Wang, J., Yang, H., Fan, K., Xu, G., Yang, K., and Tan, H.(2009). Autoregulation of antibiotic biosynthesis by binding of the end product to an atypical response regulator. Proc. Natl. Acad. Sci. U.S.A. *106*, 8617–8622.

Wang, X.J., Yan, Y.J., Zhang, B., An, J., Wang, J.J., Tian, J., Jiang, L., Chen, Y.H., Huang, S.X., Yin, M., *et al.* (2010). Genome sequence of the milbemycin-producing bacterium *Streptomyces bingchenggensis*. J. Bacteriol. *192*, 4526–4527.

Wei, T.T., Chain, J.A., Roller, P.P., Weiss, U., Stroshane, R.M., White, R.J., and Byrne, K.M. (1982). Detection of gilvocarcin antitumor complex by a biochemical induction assay (BIA). J. Antibiot. *35*, 529–532.

Williams, R.B., Henrikson, J.C., Hoover, A.R., Lee, A.E., and Cichewicz, R.H. (2008). Epigenetic remodeling of the fungal secondary metabolome. Org. Biomol. Chem. *6*, 1895–1897.

Combinatorial Biosynthesis for Antibiotic Discovery

16

Sung Ryeol Park and Yeo Joon Yoon

Abstract

The rapid and increasing prevalence of antibiotic-resistant pathogenic microorganisms has heightened the need for development of novel antibiotics. Natural products have traditionally been a prolific source of therapeutic agents for infectious diseases and other illnesses. However, as searching for new antibiotics with desired properties from natural products has become increasingly challenging, researchers have turned towards a genetic approach known as combinatorial biosynthesis to generate novel bioactive compounds for antibiotic discovery. In this chapter, we will review the past and current attempts and achievements to discover potentially novel antibiotic derivatives through combinatorial biosynthesis, and discuss its role and importance as well as its challenges, giving insights into the future trends and possibilities. In addition, multiple approaches, including heterologous expression and *in vitro* glycorandomization, coupled with the combinatorial biosynthesis technique will be described. Moreover, although we will focus on antibacterial compounds and their derivatives, several other related compounds will be mentioned to enlighten the concept of combinatorial biosynthesis.

Introduction

Since the discovery of the first antibacterial agent, penicillin, which is produced by the fungi *Penicillium notatum* and *Penicillium chrysogenum*, other nature-derived substances, including chloramphenicol and streptomycin, were identified sequentially. Through these breakthrough discoveries, a 'golden era' of antibiotic research lasted over 40 years (Demain and Sanchez, 2009). Although we believe that nature will continue to provide a vast number of fascinating sources for novel antibiotics, discovering innovative and nontoxic antibiotics from nature has become increasingly difficult, and also there is concern that the drug discovery potential remaining in existing libraries is diminishing (Bassetti *et al.*, 2013). In the most recent report, one of the major organizations, the Centers for Disease Control and Prevention (http://www.cdc.gov/drugresistance/threat-report-2013/), came to the conclusion that there are only a few potent drugs providing significant benefits over existing drugs.

Several strategies have been accepted as a means for the discovery of novel antibiotics. First, the classical screening of sources from nature, such as soil or marine microorganisms, fungi, plants, and invertebrates, have been used to excavate potent antibiotic compounds. Besides this traditional screening, chemical modification of natural product by semi-synthesis is an effective alternative as exemplified by clarithromycin and ketolides (Zhanel *et al.*, 2001). Despite the success of chemical modification or total synthesis of new natural product derivatives, many sites on the natural products are not susceptible to chemical derivatization, and total synthesis of structurally complex natural products is economically impractical in many cases. Another interesting way to generate potential antibiotic leads is to utilize a genetic engineering technique called combinatorial biosynthesis, which, theoretically, can create unlimited numbers of new compounds for drug assay and further development. In this chapter, we will define the concept of combinatorial biosynthesis and introduce its brief history. In addition, the past and current efforts for generating antibiotic analogues through combinatorial biosynthesis of the biosynthetic pathways of natural products, including polyketides, non-ribosomal peptides and aminoglycosides, will be described. Moreover, we will discuss the importance and challenges of the combinatorial biosynthesis technique to give an outlook for future prospects of antibiotic discovery.

Definition and brief history of combinatorial biosynthesis

Combinatorial biosynthesis can be described as the application of genetic manipulation to modify the biosynthetic machinery of natural products as a way to produce new and altered natural product-related compounds for drug discovery (Floss, 2006). Altering the biosynthetic pathway of natural products by deletion, mutation, and/or substitution may result in the synthesis of previously unknown natural-product analogues.

The concept of this approach was first introduced in 1985 by the Hopwood research group (1985) in their work on the production of 'hybrid antibiotics' through the transfer of biosynthetic genes between strains producing different members of the same chemical class of compounds. They were able to generate new antibiotics, mederrhodin and dihydrogranatirhodin, by transforming a partial actinorohodin gene cluster from *Streptomyces coelicolor* into the medermycin producer *Streptomyces* AM-7161 and the granaticin producer *Streptomyces violaceroruber*, respectively (Hopwood et al., 1985). These experiments facilitated the beginning of the combinatorial biosynthesis field. Almost 10 years later, the Khosla group (2007) were successful in altering the actinorhodin pathway by recombining biosynthetic genes from different clusters to create new compounds (McDaniel *et al.*, 1993). Further, it was demonstrated that engineering the biosynthetic gene cluster 6-deoxyerythronolide B (6-dEB), the core scaffold of erythromycin antibiotic, enables the generation of a number of novel molecules (Katz, 1997; Katz and McDaniel, 1999).

Combinatorial biosynthesis of polyketides

Definition of polyketides and polyketide synthase (PKS)

Most combinatorial biosynthetic approaches have been focused on genetic manipulation of genes encoding polyketide synthases (PKSs) and the related non-ribosomal peptide synthases (NRPSs). This technology allowed scientists to reprogram polyketide biosynthetic pathways and generate structural diversification that is difficult to obtain using conventional chemical approach because polyketide structures are complex and rich in stereocenters.

These PKSs are grouped into three types based on both their genetic structures and mode of biosynthesis (Fig. 16.1) (Shen, 2003). Type I PKSs (Fig. 16.1A) are multifunctional enzymes consisting of multiple modules, each of which controls the incorporation of the coenzyme A (CoA) thioester precursor unit into a polyketide backbone during chain elongation. A general pathway for type I polyketide biosynthesis begins when a starter unit is loaded onto the phosphopantotheinyl arm of the acyl carrier protein (ACP) of the loading module (module n). Next, the acyltransferase (AT) domain of the module n+1 transfers an extender unit to the ACP domain, whereas the β-ketoacyl synthase (KS) domain receives the starter unit from the preceding module. The KS domain then catalyses the decarboxylative condensation reaction between the ACP-bound extender unit and the starter unit. Each condensation cycle results in a β-keto group that is further processed by optional catalytic domains, including a ketoreductase (KR), a dehydratase (DH), and an enoyl reductase (ER), to a hydroxyl group, double bond, or single bond depending on the presence of reductive domains in each module. After this first assembly of the polyketide chain, additional extender units are incorporated sequentially, and the acyl chain elongation is terminated by a thioesterase (TE) domain. Fig. 16.2 shows the schematic organization of a modular type I PKS as exemplified by the 6-dEB synthase (DEBS) of *Saccharopolyspora erythraea*. The biosynthesis of 6-dEB has been a prototypical example for understanding the structure and function of the type I

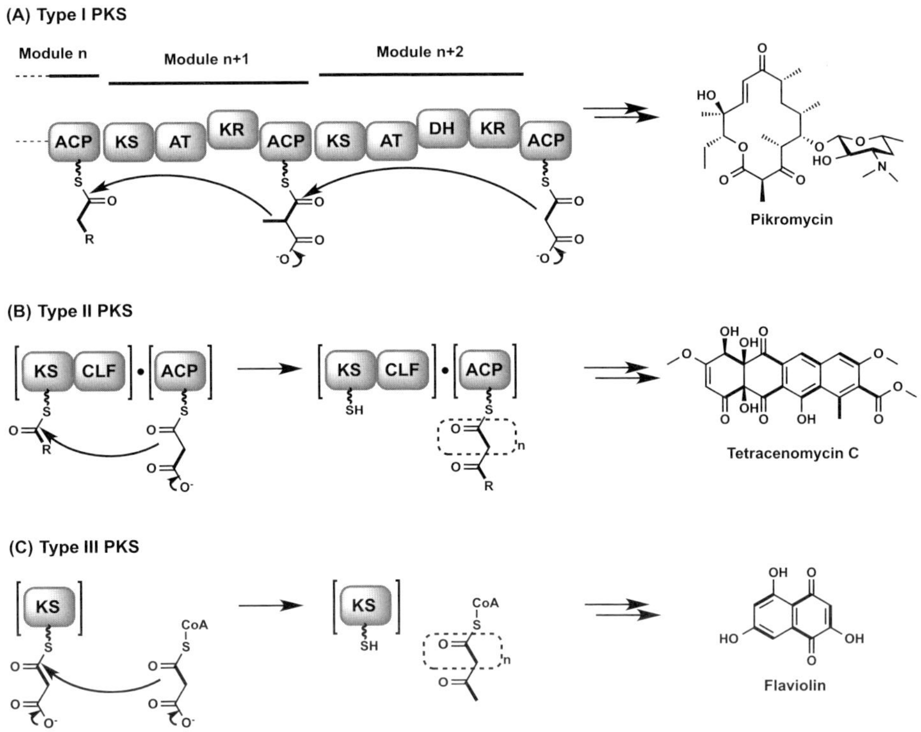

Figure 16.1 Organization and mechanisms of three different bacterial polyketide synthases (PKSs). (A) Type I PKS consists of multiple non-iterative modules. (B) Type II PKS consists of iteratively acting minimal KS and ACP domains for generating aromatic polyketides. Chain length factor (CLF), also known as KS_{β}, determines the number of iterative condensation steps. (C) Type III PKS consists of a single domain catalysing iterative condensation reaction. Abbreviations: KS, β-ketoacyl synthase; ACP, acyl carrier protein; AT, acyltransferase; KR, ketoreductase; DH, dehydratase.

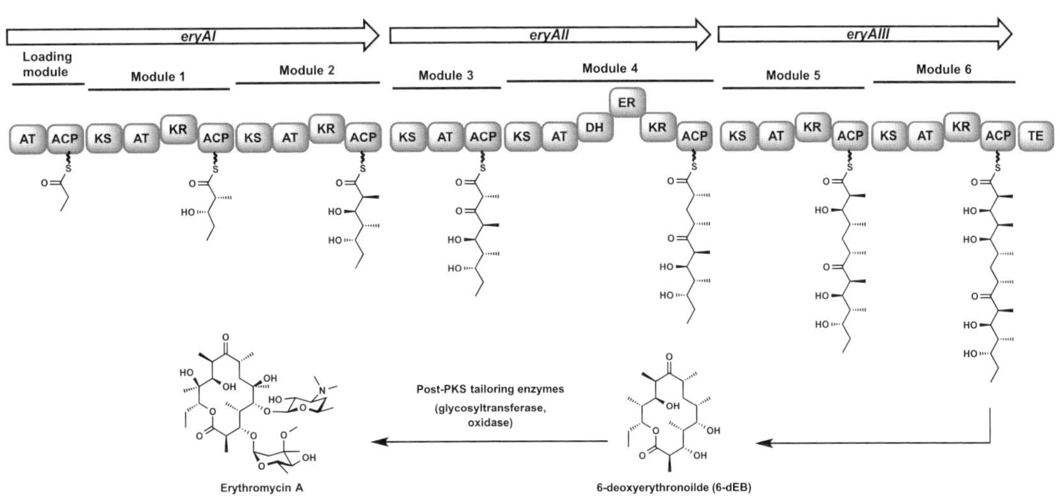

Figure 16.2 Scheme for erythromycin biosynthesis by type I PKS. The erythromycin PKS contains a loading module, six elongation modules, and a thioesterase (TE) domain. The *eryAI-III* genes encode 6-deoxyerythronolide B (6-dEB) synthase (DEBS), which is responsible for the biosynthesis of the polyketide backbone, 6-dEB, of the erythromycin antibiotic. Once the product of DEBS is released from the multienzyme by the TE domain, it is further elaborated by post-PKS tailoring enzymes. Abbreviation: ER, enoyl reductase.

PKSs that are required for assembling complex polyketides (Cortes *et al.*, 1990; Khosla *et al.*, 2007). Once the products of PKS are released from the multifunctional enzymes, they are further modified by post-PKS tailoring enzymes.

Unlike modular type I, type II PKSs (Fig. 16.1B) are complexes of monofunctional enzymes carrying a single set of iteratively acting activities for the biosynthesis of aromatic ring molecules (often polycyclic), such as actinorhodin (Hertweck *et al.*, 2007). The ACP-independent type III PKSs (Fig. 16.1C), also called as chalcone synthase-like PKSs, are structurally distinct from both type I and type II PKSs, and they can be found in both plants and a wide array of microbes. They generally synthesize mono- or bicyclic aromatic molecules by catalysing iterative condensation reactions (Yu *et al.*, 2012). Examples include chalcone and stilbene synthases of plant and the 1,3,6,8-tetrahydroxy naphthalene synthase (THNS) of *S. coelicolor*.

Combinatorial biosynthesis of type I polyketides

The erythromycin PKS genes, *eryAI-III* encoding DEBS (Fig. 16.2) (Cortes *et al.*, 1990; Khosla *et al.*, 2007) have been the most widely used model system for exploring the potential of modular type I PKSs to create new polyketides (Rodriguez and McDaniel, 2001). The initial achievements were obtained by changing the length of the polyketide chain, the composition of starter and extender units, and the extent of reduction of the polyketide scaffold and stereochemistry of carbon centres. These changes involve the deletion, site-specific mutagenesis, substitution or addition of individual domains or modules, or the generation of hybrid modules from different PKSs, thereby contributing to macrolide structural diversity. Several reports have shown that 'unnatural' triketide (Cortes *et al.*, 1995; Kao *et al.*, 1994a, 1995, 1996) and hexaketide (Kao *et al.*, 1995) lactones were produced by expressing engineered DEBS modules fused to the TE domain, indicating that the TE domain from the erythromycin pathway can mediate the release of chains with diverse lengths. Based on this finding, in later work, it was possible to produce the desired novel octaketides containing either an acetate or a propionate starter unit by the insertion of an extra module from the rapamycin PKS between the first two extension modules from the DEBS, which demonstrated that the individual modules of PKSs used combinatorially can generate novel polyketide libraries (Rowe *et al.*, 2001).

The incorporation of unnatural starter units by replacing the loading module also offers an alternative method to facilitate production of novel analogues of antibiotic polyketides. Since the structure of a starter unit of a polyketide is determined by the specificity of the AT domain of the loading module (Moore and Hertweck, 2001), exchange of the loading module with a distinct loading module from different PKS can lead to the structural diversification of the side chain originating from the starter unit. For example, by implanting the substrate-promiscuous loading module from the avermectin PKS of *Streptomyces avermitilis* into the place of the DEBS1 loading module, several novel erythromycins containing branched starter units were obtained (Marsden, 1998). In addition, the loading module from the oleandomycin or tylosin pathway was grafted on the DEBS loading module in the model PKS system DEBS1-TE (Wiesmann *et al.*, 1995). This hybrid PKS was overexpressed in the mutant *S. erythraea* bearing a deletion of DEBS genes, and the resulting strain produced a triketide lactone with an acetate- or a propionate-derived starter unit (Long *et al.*, 2002). As the AT domain also specifies different extender units and determines the structure of the side chain of a polyketide (Oliynyk *et al.*, 1996), a combinatorial biosynthesis approach has also been applied to alter the substrate specificity of the AT domain. When the methylmalonate-specific AT domain of DEBS was replaced with the ethylmalonate-specific AT domain of niddamycin, which is a 16-member polyketide with an ethyl side chain at C-6, the resulting mutant strain of *S. erythraea* was found to produce a novel compound, 6-desmethyl-6-ethylerythromycin A (Stassi *et al.*, 1998). Further, the desired new analogue of 6-deoxyerythronolide B, i.e. 6-desmethyl-6-deoxyerythronolide B, was obtained by alteration of the malonyl-specific motif in the AT domain through site-specific mutagenesis (Reeves *et al.*, 2001).

The degree of β-carbon processing, such as ketoreduction, can also be altered by combinatorial biosynthesis. For example, replacing sets of KR domains of DEBS with rapamycin PKS-derived domains, which would be impractical to achieve using conventional chemical synthesis, resulted in the formation of a library of more than 50 polyketides (McDaniel *et al.*, 1999). In addition, replacement of the amino acid residue tyrosine in the KR domain of DEBS3 through site-specific mutagenesis caused a loss of KR activity and led to the production of a novel keto derivative of 6-deoxyerythronolide (3-deoxy-3-oxo-6-dEB) (Reid *et al.*, 2003).

Moreover, this combinatorial biosynthetic principle can be extended to combine complete subunits from heterologous PKS clusters. The beginning or end of the pikromycin PKS subunit in *Streptomyces venezuelae* (Xue *et al.*, 1998, 2000) was replaced with the corresponding subunits from heterologous modular PKSs, tylosin PKS, and DEBS, resulting in the production of diverse macrolactone cores with 12- or 14-membered rings. The resulting macrolactones are further glycosylated and hydroxylated at multiple positions by the post-PKS tailoring enzymes, DesVII glycosyltransferase and PikC hydroxylase, leading to novel pikromycin derivatives (Yoon *et al.*, 2002). Combining synthetic genes encoding PKS modules further facilitated the generation of 'unnatural' products; genes which encode 14 modules from eight different PKS clusters were synthesized and assembled in 154 bimodular combinations, resulting in the production of 72 triketides in *Escherichia coli* (Menzella *et al.*, 2005).

Recently, many efforts to engineer fungal PKSs have been reported. Fungi produce a great variety of structurally diverse natural products. These fungal compounds are also of prime interest to the pharmaceutical industry as well as research groups due to their wide range of biological activities, including antibiotic, anticancer, and immunosuppressive activities. In addition, these compound are expected to provide rich sources for combinatorial biosynthesis for successful drug development (Chooi and Tang, 2012; Crawford and Townsend, 2010). Fungal polyketides are synthesized by iterative multidomain PKSs (Cox, 2007). Although the architecture of fungal PKSs is similar to a single module of bacterial type I PKSs, its iterative nature is analogous to bacterial type II PKSs. Particularly, non-reducing fungal PKSs contain starter unit ACP transacylase, KS, AT, product template, and ACT domains as minimal domains. Release of the polyketide chain from the PKS is normally catalysed by several different types of domains, including the TE, thioesterase/Claisen cyclase, and reductase (R) domains (Bailey *et al.*, 2007; Fujii *et al.*, 2001; Schroeckh *et al.*, 2009). One of the recent efforts in combinatorial biosynthesis of fungal PKSs is the generation of chimeric PKS by swapping the R domain from *Aspergillus niger* with the TE domains from *Aspergillus nidulans*, which enabled the production of new aromatic polyketides with carboxylic acids instead of the corresponding aldehydes (Yeh *et al.*, 2013). The results obtained from studies on the fungal PKS may open additional ways to generate desired 'unnatural' products as well as create biosynthetic novelty.

Combinatorial biosynthesis of type II polyketides

Fig. 16.3A illustrates the gene clusters of type II PKSs for the biosynthesis of actinorhodin and teteracenomycin from *S. coelicolor* and *Streptomyces glaucescens*, respectively. It was shown that the actinorhodin (*act*) minimal PKS catalyses the formation of an octaketide backbone (McDaniel *et al.*, 1994), whereas the tetracenomycin (*tcm*) minimal PKS catalyses the formation of a 20-carbon backbone (McDaniel *et al.*, 1995b; Shen *et al.*, 1995). Based on the fact that the *tcm* aromatase/cyclase (ARO/CYC) controls a C9-C14 cyclization in the *tcm* pathways, it was possible to generate a novel product RM77 by a rationally designed hybrid PKS composed of the *act* and *tcm* minimal PKSs (Fig. 16.3B) (McDaniel *et al.*, 1995b; Shen *et al.*, 1995). Likewise, expression of the *tcm* minimal PKS genes with the *actIII* encoding KR, which controls regiospecific reduction of the octaketide backbone, can produce decaketides RM20 and RM20b (Fig. 16.3C) (Fu *et al.*, 1994). Moreover, this hybrid PKS (*tcm* minimal PKS-*act* KR) was further used to combine with the griseusin-derived aromatase, which catalyses an aromatization of the first ring

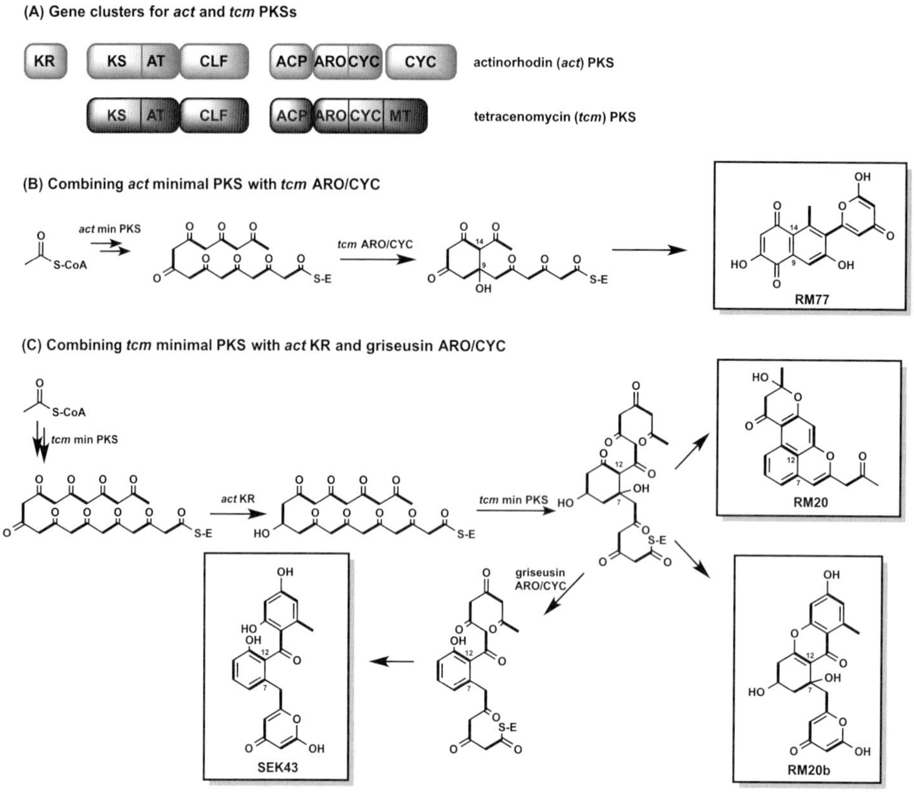

Figure 16.3 Biosynthetic pathways and gene clusters for type II aromatic PKSs. (A) Gene clusters for actinorhodin (*act*) and tetracenomycin (*tcm*) PKSs. (B) The *act*, *tcm*, and griseusin PKSs for combinatorial biosynthesis of type II polyketides. Abbreviations: CLF, chain length factor; ARO/CYC, aromatase/cyclase; MT, methyltransferase.

of a decaketide, leading to production of SEK43 as shown in Fig. 16.3C (McDaniel *et al.*, 1995a).

The *Streptomyces fradiae* Tü2717-derived urdamycins antibiotic belong to the angucycline group, the largest group of type II PKS-engineered natural products (Drautz *et al.*, 1986). Several new urdamycin analogues were generated typically through gene inactivation and/or gene overexpression experiments. Urdamycins I, J and K were generated by inactivation of *urdGT2* encoding a glycosyltransferase, which attaches D-olivose to an angucyclinone precursor (Faust *et al.*, 2000). In another report, a mutant lacking *urdR*, shown to be required for D-olivose formation, led to production of novel analogue urdamycin M carrying D-rhodinose. In addition, novel urdamycins N and O were produced from the chromosomal *urdGT1c*-deleted mutants of *S. fradiae* by overexpression of *urdGT1c*, which is predicted to encode a glycosyltransferase responsible for the linkage of the terminal L-rhodinose moiety (Hoffmeister *et al.*, 2000).

Combinatorial biosynthesis of type III polyketides

Combinatorial biosynthesis has also been performed for the production of plant-derived natural and unnatural products, which involve the type III PKSs. Most of therapeutic agents for bacterial disease have originated from microbial organisms, but the antibacterial properties of plant-derived compounds including flavonoids (Cushnie and Lamb, 2011) and stilbenes (Boulos *et al.*, 2013) synthesized by type III PKS are attracting increasing attention (Fig. 16.4). Despite the pharmaceutical and nutraceutical importance of plant-derived natural products, the availability of the valuable plant secondary metabolites is

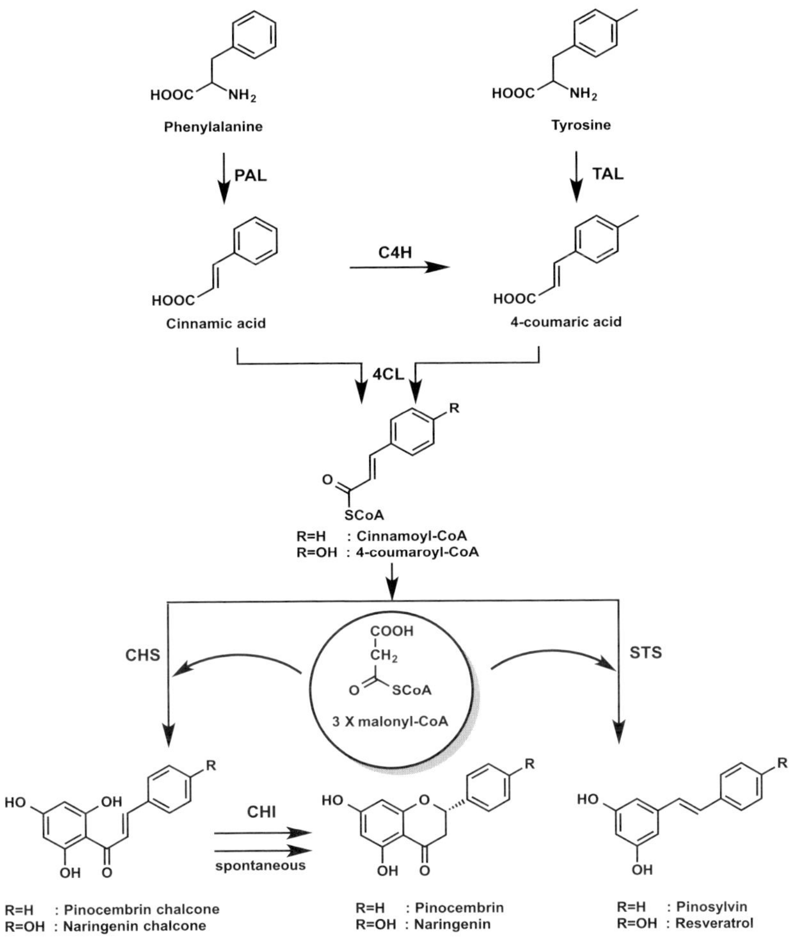

Figure 16.4 Biosynthetic pathways of type III polyketides, flavonoids (pinocembrin and naringenin) and stilbenes (pinosylvin and resveratrol). Enzyme abbreviations: PAL, phenylalanine ammonia lyase; TAL, tyrosine ammonia lyase; C4H, cinnamate-4-hydoxylase; 4CL, 4-coumarate/cinnamate: coenzyme A ligase; CHS, chalcone synthase; STS, stilbene synthase; CHI, chalcone isomerase.

limited by season- and region-dependent variations. Moreover, it is challenging to isolate and purify the single target compound from plant extracts due to the complexity of secondary metabolites from the plant. Therefore, reconstruction and optimization of plant-originated type III PKS pathways in microbial systems have been considered as a practical alternative strategy to tackle these problems.

Flavonoids represent a large and important group of plant natural products. The biosynthesis of the basic flavonoid-related molecules in plants starts from L-phenylalanine via the phenylpropanoid pathway (Fig. 16.4). Conversion of phenylalanine to chalcone is mediated by serial action of phenylalanine ammonia lyase (PAL); cinnamate-4-hydoxylase (C4H) and 4-coumarate/cinnamate: coenzyme A ligase (4CL); and chalcone synthase (CHS) (Weisshaar and Jenkins, 1998). Particularly, the type III PKS, CHS catalyses a condensation reaction of 4-coumaroyl-CoA (or cinnamoyl-CoA) with three molecules of malonyl-CoA to form chalcone, which serves as a gateway molecule to other diverse flavonoids (Fig. 16.4). The success of flavonoid production by an engineered *E. coli* depends on overcoming the difficulty of expression of cytochrome P450 enzyme C4H, which would not be efficiently expressed due to its instability and requirement for redox partner in bacteria, by employing bacterial 4CL enzyme from *S. coelicolor* instead of plant C4H. A hybrid flavonoid biosynthetic pathway from a combination of yeast, bacterial, and plant genes was expressed in a genetically engineered *E. coli*; as a result, the C4H step was bypassed in the phenylpropanoid pathway, and two flavonoids, naringenin and pinocembrin, were produced by *E. coli* (Hwang *et al.*, 2003). In another study, substitution of plant PAL and C4H with bacterial tyrosine ammonia lyase (TAL), which is known to deaminate both phenylalanine (PAL activity) and tyrosine (TAL activity), increased the yield of flavonoids in *E. coli* (Watts *et al.*, 2004). These results facilitated diverse opportunities of further expression and metabolic engineering of the biosynthetic pathway of plant-derived type III polyketides in bacteria such as *E. coli* (Leonard *et al.*, 2007; Miyahisa *et al.*, 2005), *S. cerevisiae* (Jiang *et al.*, 2005; Yan *et al.*, 2005), and *Streptomyces* (Park *et al.*, 2009, 2010). In a similar way, stilbene and its derivatives were synthesized by expressing PAL from *Rhodotorula rubra*, a 4CL from *Lithospermum erythrorhizon*, and the type III PKS stilbene synthase (STS) from *Arachis hypogaea*. Feeding tyrosine and phenylalanine to the culture of recombinant *E. coli* expressing the hybrid stilbene biosynthetic genes from different organisms led to production of resveratrol and pinocembrin, respectively. When other carboxylic acids were used as substrates, a number of non-natural stilbenes were produced (Horinouchi, 2009).

Although the type III PKSs were formerly believed to exist exclusively in plants, many type III PKSs have recently been found in microorganisms (Katsuyama and Ohnishi, 2012) since the first characterization of the bacterial type III PKS, 1,3,6,8-tetrahydroxynaphthalene synthase (THNS) RppA in 1999, which is involved in the biosynthesis of flaviolin and hexahydroxyperylenequinone (Funa *et al.*, 1999). In addition, type III PKSs are able to provide building blocks required for the biosynthesis of other natural products including kendomycin (Wenzel *et al.*, 2008) and the non-ribosomal peptide, balhimycin (Pfeifer, 2001). Therefore, mining more type III PKSs from nature and understanding the biochemistry of specific enzymes (or enzyme complex) will extend the boundary of combinatorial biosynthesis of polyketides with antibiotic activity.

Modification of polyketide structures through combinatorial biosynthesis of post-PKS enzymes

Most of early studies on combinatorial biosynthesis of polyketides pathways focused on the biosynthetic steps catalysed by PKS. However, a more productive way of generating biologically active antibiotic classes would be a post-PKS tailoring modification, which participates in the decoration of aglycone skeletons, catalysed by a suite of diverse enzymes, including oxidases, glycosyltransferases, halogenases and other modifying enzymes (Olano *et al.*, 2010; Rix *et al.*, 2002).

Glycosyltransferase

The sugar moieties attached to the core skeleton of polyketides by glycosyltransferase are necessary for biological activity and therefore contribute to a natural extension of combinatorial biosynthesis (Song *et al.*, 2013). Recently, a vast number of reports have provided increasing evidence to show the flexibility of glycosyltransferase towards the sugar donors and/or acceptors in the biosynthesis of secondary metabolites.

Erythromycin glycosyltranferase EryCIII, which attaches the D-desosamine moiety to 3-α-L-mycarosyl erythronolide B, has been known as a flexible glycosyltransferase that can transfer alternative sugar-donor substrates, including D-mycaminose and D-angolosamine (Schell *et al.*, 2008; Yuan *et al.*, 2005). EryCIII was able to accept the altered sugar acceptors, such as 3-*O*-(2′-*O*-methylrhamnosyl)-erythronolide B and 3-*O*-(2′,3′-bis-*O*-methylrhamnosyl)-erythronolide B as substrates to generate new compounds (Gaisser *et al.*, 2001). EryBV, another glycosyltransferase in the erythromycin pathway, is involved in transfer of L-mycarose to erythronolide B. EryBV is also a flexible glycosyltransferase that modifies erythronolide B with various sugar donors, including D-glucose and L-rhamnose (Zhang *et al.*, 2007). In addition, two new compounds, 3-*O*-rhamnosyl-6-deoxy-erythronolide B and 3-*O*-rhamnosyl-6-deoxy-8,8a-dihydroxyerythronolide B, were produced by expressing glycosyltranserase *oleG2* and cytochrome P450 gene *oleP* from the oleandomycin pathway in an engineered *S. erythraea*, which lacks the two glycosyltransferase genes, *eryCIII* and *eryBV*, and the native cytochrome P450 gene *eryF* (Gaisser *et al.*, 2002).

The pikromyin biosynthetic gene cluster of *S. venezuelae* contains one glycosyltransferase DesVII, which transfers D-desosamine to 12- and 14-membered macrolactone rings, 10-deoxymethynolide and narbonolide, to make YC-17 and narbomycin, respectively (Borisova *et al.*, 2004, 2006). A mutant strain of *S. venezuelae* designated as YJ003, in which the entire D-desosamine gene cluster was deleted, was used as a heterologous host to synthesize and transfer various unnatural sugars to 10-deoxymethynolide and narbonolide, along with the glycosyltransferase-auxiliary protein pair (DesVII-DesVIII) from the native pikromycin cluster; hybrid quinovose- and olivose-attached 10-deoxymethynolide and narbonolide were produced (Hong *et al.*, 2004). The same mutant was utilized as a platform in order to express different sugar biosynthetic cassettes and the genes encoding DesVII-DesVIII; seven hybrid narbomycin derivatives were produced. Among them, 3′-*O*-demethyl-D-chalcosyl narbonolide and L-rhamnosyl narbonolide displayed remarkably better antibacterial activity than narbomycin and the clinically relevant erythromycin (Han *et al.*, 2012). In a similar way, macrolide antibiotic YC-17 analogues containing unnatural sugars, instead of native D-desosamine, were produced by the *S. venezuelae* mutant YJ003. The structures of these metabolites were determined as D-quinovosyl-10-deoxymethynolide, L-rhamnosyl-10-deoxymethynolide, L-olivosyl-10-deoxymethynolide, and D-boivinosyl-10-deoxymethynolide. Especially, the YC-17 analogue with L-rhamnose exhibited better antibacterial activity against clinically isolated erythromycin-resistant pathogenic strains, as well as erythromycin-susceptible strains than the parent compound (Shinde *et al.*, 2013). Moreover, the introduction of the 3,5-epimerase gene from *Streptomyces antibioticus* Tu99, and the 4-aminotransferase gene from *Streptomyces* sp. KCTC0041BP to the derivative mutant of YJ003 resulted in the production of new hybrid compounds including 4-amino-4,6-dideoxy-L-glucosylated derivatives of YC-17, methymycin, novamethymycin, and pikromycin (Pageni *et al.*, 2008).

Tylosin, produced from *Streptomyces fradiae*, contains three deoxysugars, D-mycaminose, L-mycarose and D-mycinose, through post-PKS modification steps. The D-mycaminose can be linked to the 16-memebered tylactone by action of TylMII glycosyltransferase (Fish and Cundliffe, 1997). It has been shown that the TylMII transfers various sugar donors to the non-native aglycone narbonolide as well as its natural aglycone, tylactone (Gaisser *et al.*, 2000; Melançon *et al.*, 2005). For example, two new tylosin derivatives, 4′-deoxy-5-*O*-mycaminosyl-tylonolide and 4'-deoxy-20-dihydro-demycarosyltylosin, were produced from *S. fradiae* by expressing *nbmK* and *nbmJ*, which are required for deoxygenation at the C4 position during D-desosamine biosynthesis in the narbomycin pathway from *Streptomyces narbonensis*, together with *tylMII* (Butler *et al.*, 2002).

Dihydrochalcomycin, a 16-membered macrolide antibiotic with antimicrobial activity against Gram-positive and Gram-negative bacteria (Kim *et al.*, 1996), contains D-chalcose and D-mycinose. A mutant strain of *Streptomyces* sp. GERI-155, in which genes involved in D-chalcose biosynthesis are deleted, produced the novel macrolide 20-*O*-mycinosyl-dihydrochalconolide. This mutant was used further to express the D-desosamine biosynthetic gene cassette, resulting in the formation of the new macrolide derivative 5-*O*-desosaminyl-20-*O*-mycinosyl-dihydrochalconolide (Pageni *et al.*, 2009). Similarly, another mutant strain was constructed by deleting genes required for the biosynthesis of D-mycinose, D-chalcose, and some tailoring steps. When the UDP-xylose biosynthetic genes (*calS7*, *calS8*, and *calS9*) from the calicheamicin gene cluster was expressed in this strain, novel 5-*O*-xylosyl-chalconolide derivative was accumulated (Simkhada *et al.*, 2009).

OleG1 glycosyltransferase from *S. antibioticus* involves transfer of D-desosamine to the corresponding macrolactone in oleandomycin biosynthesis. OleG1 can attach the native sugar to the unnatural sugar donor, 3-*O*-L-mycarosyl erythronolide B, leading to a new erythromycin derivative (Doumith *et al.*, 1999). Another glycosyltransferase OleG2 is responsible for glycosylation of oleandolide with L-olivose. This glycosyltransferase is also able to transfer unnatural L-mycarose and L-rhamnose to the alternative aglycones, erythronolide B and 6-dEB (Doumith *et al.*, 1999).

The substrate promiscuity of several glycosyltransferases mentioned here has served as a good platform for efficient glycorandomization. Nonetheless, other stringent glycosyltransferases which limit enzymatic glycodiversifications need to be engineered. One recent report has shown that a library of oleandomycin glycosyltransferase OleDs is generated by error-prone PCR and screened through a fluorescence-based assay. Notably, some engineered OleDs selected exhibited enhanced catalytic efficiency and/or substrate flexibility (Gantt *et al.*, 2008; Williams *et al.*, 2007). Therefore, genetic engineering and evolution of glycosyltransferase can provide more opportunities to construct glycosyltransferases possessing broad substrate flexibility with potential for *in vivo* and *in vitro* glycodiversification.

Oxidase

Oxidation performed by diverse polyketide oxidases is also frequently found during post-PKS modifications leading to structural and biological diversity of polyketide compounds. Therefore, combinatorial biosynthesis of oxidase genes in polyketide pathways has been a useful approach for addition of important functional group to polyketide cores in order to generate novel products with potentially greater or altered biological activities (Rix *et al.*, 2002). EryK and PikC, cytochrome P450 monooxygenase from *S. erythraea* and *S. venezuelae*, respectively, have been extensively studied because of their flexibility. EryK accepts 14-membered precursors, i.e. erythromycins D and B, as natural substrates to generate erythromycins C3 and A4, respectively (Stassi *et al.*, 1993). It possesses flexible specificity and recognizes YC-17 and narbomycin, natural substrates of PikC from *S. venezuelae*, to give methymycin/neomethymycin and pikromycin, respectively (Lee *et al.*, 2004). It has also been shown that EryK converts 15-fluoro-6-deoxyerythronolide B to 15-fluoroerythromycin A (Desai *et al.*, 2004).

PikC is responsible for the biosynthesis of methymycin/neomethymycin and pikromycin from YC-17 and narbomycin, respectively (Betlach *et al.*, 1998; Xue *et al.*, 2004). PikC, with a more relaxed substrate specificity compared to EryK, is able to catalyse hydroxylation of diverse unnatural macrolides. An engineered *S. venezuelae* mutant carrying hybrid type I PKS genes produced multiple new macrolactones that are modified further by serial action of flexible glycosyltransferase DesVII and hydroxylase PikC, resulting in novel derivatives as previously discussed (Yoon *et al.*, 2002). Additionally, this flexible PikC accepted oleandomycin as an alternative substrate to introduce the hydroxyl group into the C4 position of elandomycin (Lee *et al.*, 2005). Further, it has been reported that 5-*O*-desosaminyl erythronolide A, a potent precursor of ketolides, the latest generation of antibiotic compounds derived from erythromycin A, was produced by the expressing *pikC* in the mutant of *S. erythraea* lacking of EryBV glycosyltransferase (Basnet *et al.*, 2008). Recent studies on X-ray structure analysis and site-specific mutagenesis based on crystal structures have facilitated the understanding of the diverse hydroxylation patterns of PikC and provided new information on the several important residues that are critical for the hydroxylation of native macrolides, as well as interaction with native sugar D-desosamine, for generation of novel hydroxylated derivatives (Fish and Cundliffe, 1997; Li *et al.*, 2009).

The gene cluster of oxytetracycline, a 19-carbon polyketide antibiotic produced by *Streptomyces rimosus*, possesses some oxygenases that have been used for combinatorial biosynthesis (Fig. 16.5). Five potentially novel compounds were detected in the fermentation broth of an engineered mutant of *S. rimosus* bearing a deletion of *oxyS* (formerly known as *otcC*), encoding an anhydrotetracycline oxygenase that catalyses a hydroxylation of the anthracycline structure (Peric-Concha, 2005). When *oxyL*, encoding a NADPH-dependent dioxygenase, was deleted in *S. rimosus*, the yield of the PW2 compound (Fig. 16.5) was increased while the production of

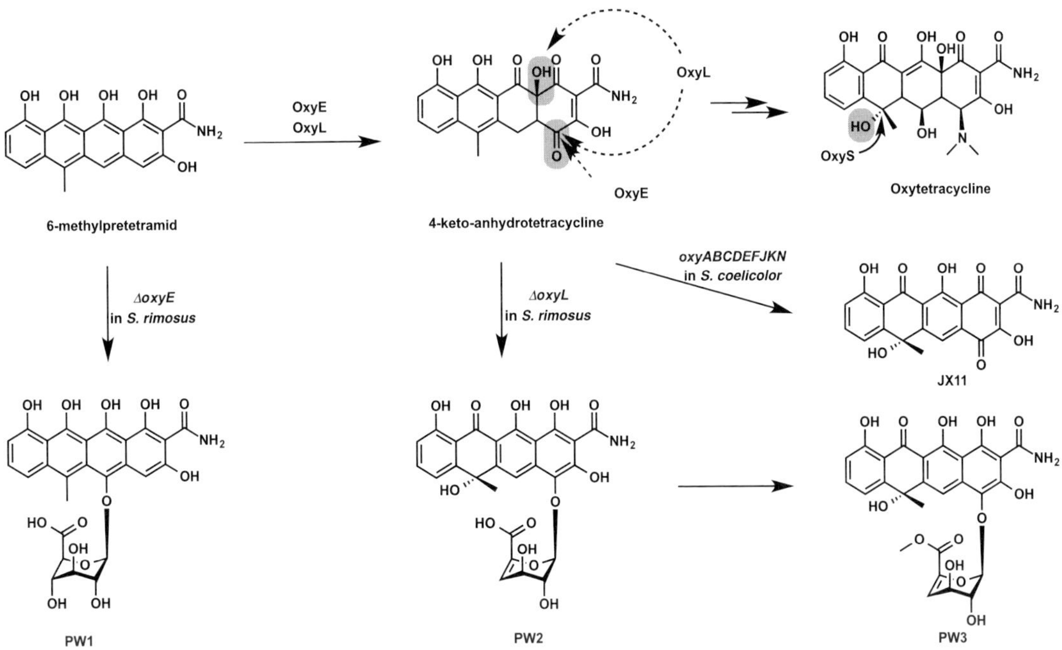

Figure 16.5 Biosynthetic pathway of oxytetracycline and structures of novel compounds generated through combinatorial biosynthesis and heterologous expression.

oxytetracycline was abolished. Interestingly, the PW2 placed in a solvent containing methanol was readily methylated, generating a more stable new product PW3 (Fig. 16.5) (Zhang *et al.*, 2008). Another example is the inactivation of *oxyE*, a gene encoding oxygenase during oxytetracycline biosynthesis in *S. rimosus*, leading to the emergence of a new metabolite PW1 (Fig. 16.5) (Wang *et al.*, 2009). The heterologous expression of *oxyE* in *S. coelicolor*, along with *oxyABCDFJKN* genes responsible for production of 6-methylpretetramid, yielded novel metabolite JX11 (Fig. 16.5) (Wang *et al.*, 2009).

Other decorating enzymes

Methylation, most often catalysed by methyltransferases that rely on the co-substrate *S*-adenosyl methionine as an electron-deficient methyl donor, is a common biotransformation in nature (Liscombe *et al.*, 2012). Therefore, methylation reactions can occur either at the polyketide core moiety or on sugar residues, either prior to or after the post-PKS modification mentioned above. This has facilitated the use of methyltransferase-encoding genes for the generation of novel antibiotic derivatives using combinatorial biosynthesis approaches (Olano *et al.*, 2010). When a substrate-relaxed methyltransferase from a rice plant was introduced into the construct containing hybrid stilbene synthetic genes from different organism, a collection of novel methylated stilbenes were generated (Katsuyama *et al.*, 2007). Further, when *rif-orf14* encoding methyltransferase was deleted from the pathway of rifamycin B, a prominent member of the ansamycin family of antibiotics, the mutant of rifamycin producer *Amycolatopsis mediterranei* accumulated 27-*O*-demethyl-rifamycin SV as the main product, along with a small amount of 27-*O*-demethyl-25-*O*-deacetyl-rifamycin SV (Xu *et al.*, 2003).

The enzymes involved in acylation of antibiotic compounds have also allowed the generation of diverse hybrid compounds (Olano *et al.*, 2010). Rif-Orf20, involving the *in vitro* acetylation of 27-*O*-demethyl-25-*O*-deacetyl-rifamycin SV to give 27-*O*-demethyl-rifamycin SV, has been found to have a certain degree of flexibility towards the acyl-CoA to be transferred, since it produced a propionyl ester derivative of 27-*O*-demethyl-25-*O*-deacetyl-rifamycin SV (Xiong *et al.*, 2005).

The utility of these post-tailoring enzymes, including glycosyltransferase, oxidase, methyltransferase, and acyltransferase, in combinatorial biosynthesis would provide a valuable tool for producing unnatural macrolides. Moreover, increasing knowledge of the broad substrate spectrum and X-ray crystal structure of the enzymes that participate in the polyketide biosynthesis will provide new insight into the combinatorial biosynthetic approaches for the generation of novel antibiotic compounds.

Combinatorial biosynthesis of non-ribosomal peptides

Like polyketides, non-ribosomal peptides are synthesized from simple building blocks, such as amino acids, by non-ribosomal peptide synthetases (NRPSs) (Fischbach and Walsh, 2006). The NRPSs are typically composed of a modular set of repeating enzyme domains to activate and incorporate specific amino acids into a growing chain in a particular order (Fig. 16.6A). Within each module of a typical NRPS, an adenylation (A) domain selects and activates a specific amino acid, a peptidyl carrier protein (PCP) domain, also known as a thiolation (T) domain, to which the activated amino acid is attached; it performs the propagation of the growing peptide chain, and a condensation (C) domain couples the amino acid to the nascent peptide chain. In addition, a variety of optional domains for methylation (M) or epimerization (E) that tailor individual amino acids may also be present, which increase the structural diversity of non-ribosomal peptides. As the NRPS is made of specific parts containing two or more enzyme domains that can be replaced with different domains from different pathways, it has been demonstrated that the number, content, and order of each individual domain can be genetically altered to generate novel peptide assembly machines producing novel antibiotic peptides.

Combinatorial biosynthesis of lipopeptides

Daptomycin (Fig. 16.6B), a cyclic lipopeptide antibiotic produced by *Streptomyces roseosporus*, was FDA approved in 2003 and marketed under the name Cubicin, for the treatment of infections caused by various Gram-positive pathogens (Kirkpatrick *et al.*, 2003). The core peptide of daptomycin is assembled by NRPS multienzyme subunits, DptA, DptBC, and DptD (Baltz, 2009; Mchenney *et al.*, 1998; Miao *et al.*, 2005). An initial attempt to produce novel active analogues has been made by expressing the biosynthetic genes of two related lipopeptides, A54154 and calcium-dependent antibiotic (CDA), in a *dptD*-deleted mutant of *S. roseosporus* (Miao *et al.*, 2006). A54145 is also a complex of cyclic lipopeptide antibiotics with 10-membered ring structures produced from *S. fradiae* (Boeck *et al.*, 1990). The A54154 gene cluster contains four NRPS genes, *lptA*, *lptB*, *lptC*, and *lptD*, whereas lipopeptide CDA derived from *S. coelicolor* has three (*cdaPS1*, *cdaPS2*, and *cdaPS3*) (Baltz *et al.*, 2005). Since the *dptD* gene has been replaced by introducing the *lptD* from A54154 gene cluster or *cdaPS3* from the CDA gene cluster to produce hybrid lipopeptide derivatives A21978C_{1-3}, containing isoleucine or tryptophan instead of native kynurenine, this aroused interest in demonstrating that the similar alteration of amino acid in tridecapeptides could be generated by heterologous NRPS module fusions. The amino acid exchange in the DptD module was further explored: as one example, the C-A (or C-A-T) domains from LptC (recognizing asparagine) were exchanged for the corresponding domains from DptD (recognizing kynurenine). The recombinant strain produced A21978C_{1-3} derivatives containing asparagine at the C13 position (Doekel *et al.*, 2008). Most of these derivatives displayed antibacterial activity against *Staphylococcus aureus*, even though they were slightly less potent than daptomycin (Baltz, 2009).

In addition, the possibility of coupling multi-domain substitutions with subunit exchanges, variations in lipid tails, and deletion of *dptI* gene encoding methyltransferase was investigated to generate multiple combinations. Five of the new compounds produced through multiple combinations were as active as daptomycin against Gram-positive test strains (Nguyen *et al.*, 2006). More recently, multiple A54145 derivatives were produced by expressing hybrid A54145

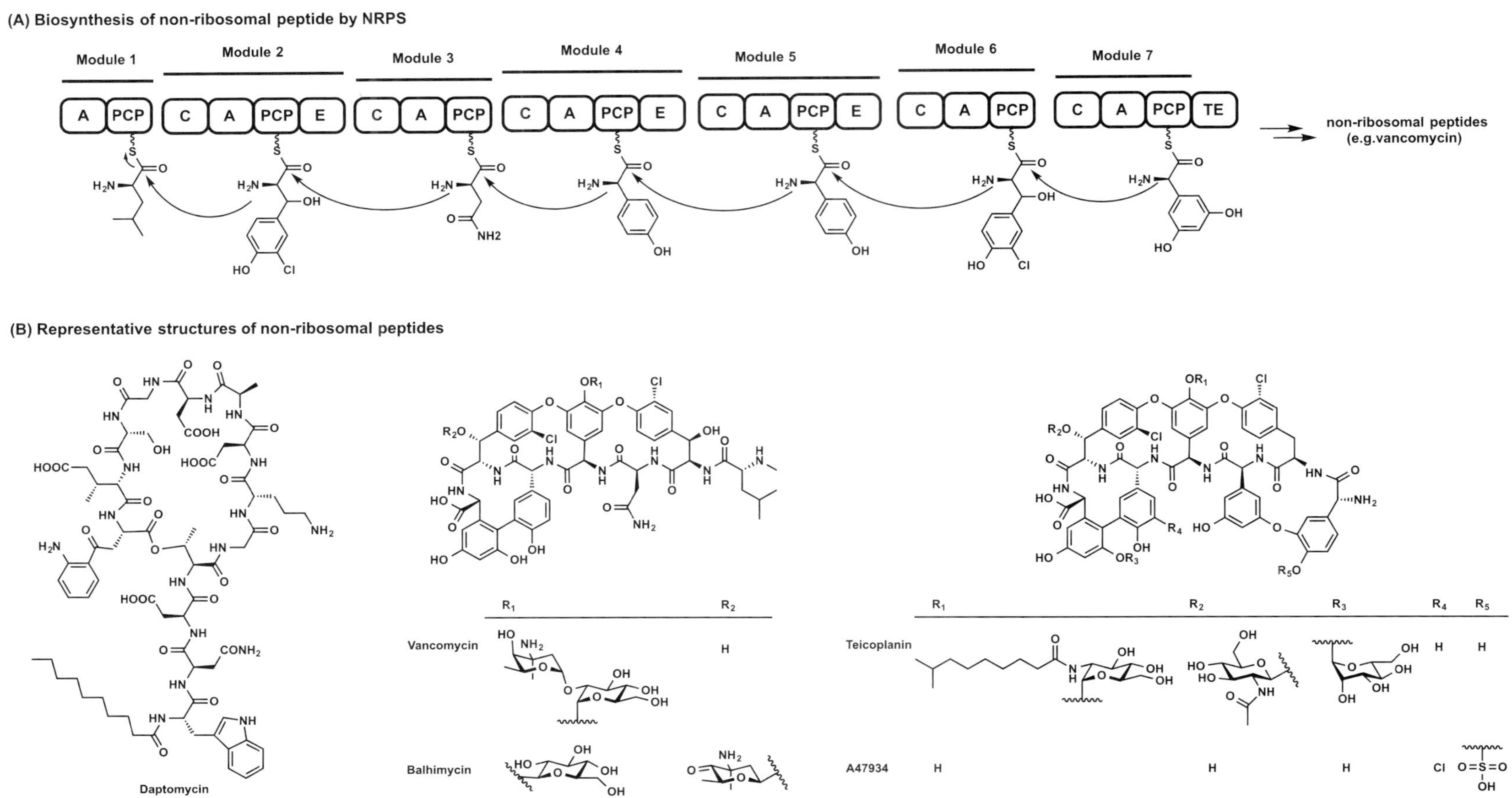

Figure 16.6 Organization and mechanisms of non-ribosomal peptide synthases (NRPSs) and structures of non-ribosomal peptides. (A) NRPS consists of a set of modular enzymes activating and incorporating specific amino acids to growing peptide backbone in the process. Abbreviations: A, adenylation; C, condensation; PCP, peptidyl carrier protein. (B) Examples of non-ribosomal peptides, daptomycin, vancomycin, balhimycin, teicoplanin, and A47934.

biosynthetic genes in the *lptI* methyltransferase gene-deleted *S. fradiae* mutant. λ-Red-mediated recombination was used to replace a single amino acid in single or multiple modules in the LptBC subunit of A54145 NRPS. A similar strategy was employed to substitute multiple amino acids in modules 2 to 4 or modules 2 to 8, with the corresponding modules from daptomycin. These exchanges resulted in production of 16 hybrid A54145 analogues from the recombinants. Interestingly, several ofthe A54145 derivatives were about 8-fold more active against vancomycin-susceptible *E. coli* imp-4213 strain than daptomycin (Nguyen *et al.*, 2010).

Combinatorial biosynthesis of glycopeptides

Glycopeptide antibiotics, including vancomycin and teicoplanin, are clinically important for the treatment of a variety of life-threatening infections caused by Gram-positive bacteria (Fig. 16.5B). They are structurally comprised of a heptapeptide scaffold synthesized by NRPS and modified through several sugar moieties. In addition, they are usually further embellished by chlorination, glycosylation, methylation, acylation, and/or sulfation (Donadio and Sosio, 2008). The glycopeptides inhibit bacterial growth in a late stage by binding to amino acids within the cell wall, thus interfering with peptidoglycan biosynthesis (Kohanski *et al.*, 2010). Since the sugar portions in glycopeptides are known to be important for their biological activity, efforts to explore the modifications of carbohydrate substituents of glycopeptide have been made in order to create novel glycopeptides over the past decades.

The vancomycin gene cluster in *Amycolatopsis orientalis* contains two genes encoding glycosyltransferase GtfE and GtfD that decorate the aglycone scaffolds, 4-hydroxyphenylglycine and glucosyl peptide with D-glucose and L-vancosamine, respectively (Losey *et al.*, 2001, 2002). Of the two vancomycin glycosyltransferases, GtfE is known to display a certain level of substrate promiscuity and attach different NDP-sugars to different algycones (Hubbard and Walsh, 2003; Losey *et al.*, 2001; Solenberg *et al.*, 1997). Diverse glycorandomized vancomycin derivatives were generated by *in vitro* reactions of flexible GtfE and vancomycin aglycone together with 23 natural and unnatural NDP-sugars as substrates. In addition, subsequent chemoselective enzymatic ligation (1,3-dipolar cycloaddition) led to discovery of a vancomycin derivative with remarkably enhanced antibacterial properties from the monoglycosylated vancomycin variant. Interestingly, the resulting derivative showed twice the activity of vancomycin against both *S. aureus* and *Enterococcus faecium* (Fu et al., 2003). An *in vitro* reaction containing D-xylose, desvancosaminyl vancomycin, and GtfE', a variant of GtfE with a single serine-for-proline substitution, resulted in production of a xylosyl-attached desvancosaminyl vancomycin. Another study showed that GtfE enabled the addition of D-glucose to an alternative teicoplanin-related heptapeptide core, A41030A, from *Streptomyces virginiae* and A47934 from *Streptomyces toyocaensis*, leading the generation of glucosyl-A41030A and glucosyl-A47934. The resulting hybrid glycopeptide showed antibacterial activity against *Micrococcus luteus*. This *in vitro* result prompted the *in vivo* production of the glucosyl-A47934 under the control of the constitutive *ermEp** promoter by introduction of the *gtfE'* gene into the A47934-producing *S. toyocaensis* (Solenberg *et al.*, 1997). It has also been reported that the hydroxyphenylglycine moiety of both vancomycin and teicoplanin aglycones was glycosylated by flexible GtfE, and the corresponding glycosylated variants could be further processed with 4-*epi*-vancosamine moiety by action of GtfD to generate vancomycin and teicoplanin derivatives with glucosyl- and 4-*epi*-vancosaminyl sugar (Losey *et al.*, 2002).

The acyltransferase from the teicoplanin or A40926 cluster has contributed to expand glycopeptide diversity. In addition to the *N*-acylation of 2-aminoglucosyl aglycones of vancomycin and teicoplanin, the enzyme was shown to be able to attach acyl groups to 6-aminoglucosyl- and glucosyl- scaffolds to generate glycopeptide variants. This allowed the production of new molecules through the combined action of acyltransferases and glycosyltranferases. The glycosyltransferase GtfB is first utilized to transfer a 6-aminoglucose moiety to an aglycone, followed by *N*-acylation and vancosaminylation by acyltransferase from the A40926 pathway and glycosyltransferase

GtfD from the vancomycin pathway, respectively, resulting in generation of a novel hybrid compound. An additional hybrid compound was produced by employing the serial action of teicoplanin glycosyltransferase GtfA, which adds the *N*-acetylglucosamine moiety to teicoplanin aglycon, and a set of A40926 acyltransferase and GtfD (Kruger *et al.*, 2005).

The balhimycin (Fig. 16.6B) producer *A. mediterranei* has been subjected to multiple gene inactivation experiments, yielding several new compounds. A mutant with an inactivation of the *bhaA* gene encoding halogenase or the *bgtfB* gene encoding glycosyltransferase produced non-chlorinated balhimycins or four different antibacterial balhimycin aglycones, respectively (Pelzer *et al.*, 1999; Puk *et al.*, 2002). A40926-producing strain, *Nonomuraea* sp. ATCC 39727, also has been used to produce novel glycopeptide antibiotics, such as the dechloromannosyl aglycone and a desoxy-derivative of A40926, devoid of the β-hydroxyl group on tyrosine moiety through inactivation of *orf8* encoding a putative hydroxylase. (Sosio *et al.*, 2003; Stinchi *et al.*, 2006). Additionally, a new derivative of A40926 with a deacetylated mannose was produced by inactivating the *dbv23* gene, which is responsible for adding the acetyl group to a mannose sugar moiety, from the A40926 pathway (Sosio and Donadio, 2006).

The antibiotic A47934 (Fig. 16.6B), produced by *Streptomyces toyocaensis*, is a unique glycopeptide possessing a non-glycosylated heptapeptide skeleton that is sulfated on the phenolic hydroxyl of the 4-hydroxy-L-phenylglycine residue by sulfotransferase StaL. The StaL was found to recognize teicoplanin scaffolds as substrates, thereby suggesting that the unique modification of glycopeptides via sulfation offers a potential strategy for the development of new glycopeptide antibiotics (Lamb *et al.*, 2006).

Combinatorial biosynthesis of aminoglycosides

The aminoglycoside antibiotics have been widely used to treat serious bacterial infections for decades, beginning with the first use of streptomycin as an effective antituberculosis agent (Jones *et al.*, 1944). They inhibit protein synthesis in the pathogenic bacteria by binding to the 16S rRNA subunit of the 30S bacterial ribosome and generating errors in translation (Magnet and Blanchard, 2005). These antibiotics are classified based on the position of their glycosidic linkages as follows: 4,5-disubstituted aminoglycosides include neomycin (Waksman and Lechevalier, 1949) and butirosin (Howells *et al.*, 1972), whereas 4,6-disubstituted aminoglycosides include kanamycin (Umezawa *et al.*, 1957) and tobramycin (Stark *et al.*, 1967). Mono-substituted aminoglycosides are exemplified by apramycin and hygromycin (Fig. 16.7).

Recently, the complete kanamycin biosynthetic pathway was characterized, which enabled facilitation of aminoglycoside antibiotic manipulation (Park *et al.*, 2011; Sucipto *et al.*, 2012). The findings that kanamycin biosynthesis is based on two parallel pathways governed by substrate flexibility of a glycosyltransferase allowed the use of other aminoglycoside biosynthetic enzymes in order to synthesize known and new highly active antibiotics. Introduction of seven butirosin genes (*btrG-btrH-btrI-btrJ-btrK-btrO-btrV*) that are required for biosynthesis and transfer of the 1-*N*-[*S*-4-amino-2-hydroxybutiric acid] (AHBA) side chain in the non-aminoglycoside-producing heterologous host of *S. venezuelae* engineered to biosynthesize kanamycin A, led to generation of 1-*N*-AHBA-kanamycin A, known as amikacin (Fig. 16.7). Further, co-expression of the set of butirosin genes in the *S. venezuelae* strain expressing genes for biosynthesis of 6′-deamino-6′-hydroxykanamyicn A (kanamycin X) resulted in the production of a new 1-*N*-AHBA-kanamycin X (Fig. 16.7). Surprisingly, compared with the currently used semisynthetic aminoglycosides amikacin, this new compound displayed enhanced activity against Gram-negative bacteria, including the amikacin-resistant clinical isolate of *Pseudomonas aeruginosa* (Park *et al.*, 2011). This combinatorial biosynthesis approach was further employed for tobramycin biosynthesis in the engineered *S. venezuelae*. When *Streptomyces tenebrarius*-derived two putative genes, *aprD3* and *aprD4*, which are predicted to be *S*-adenosylmethionine oxidoreductase and a dehydrogenase, respectively, are expressed in the *S. venezuelae* strain harbouring kanamycin B biosynthetic genes, the resulting strain produced nebramine as

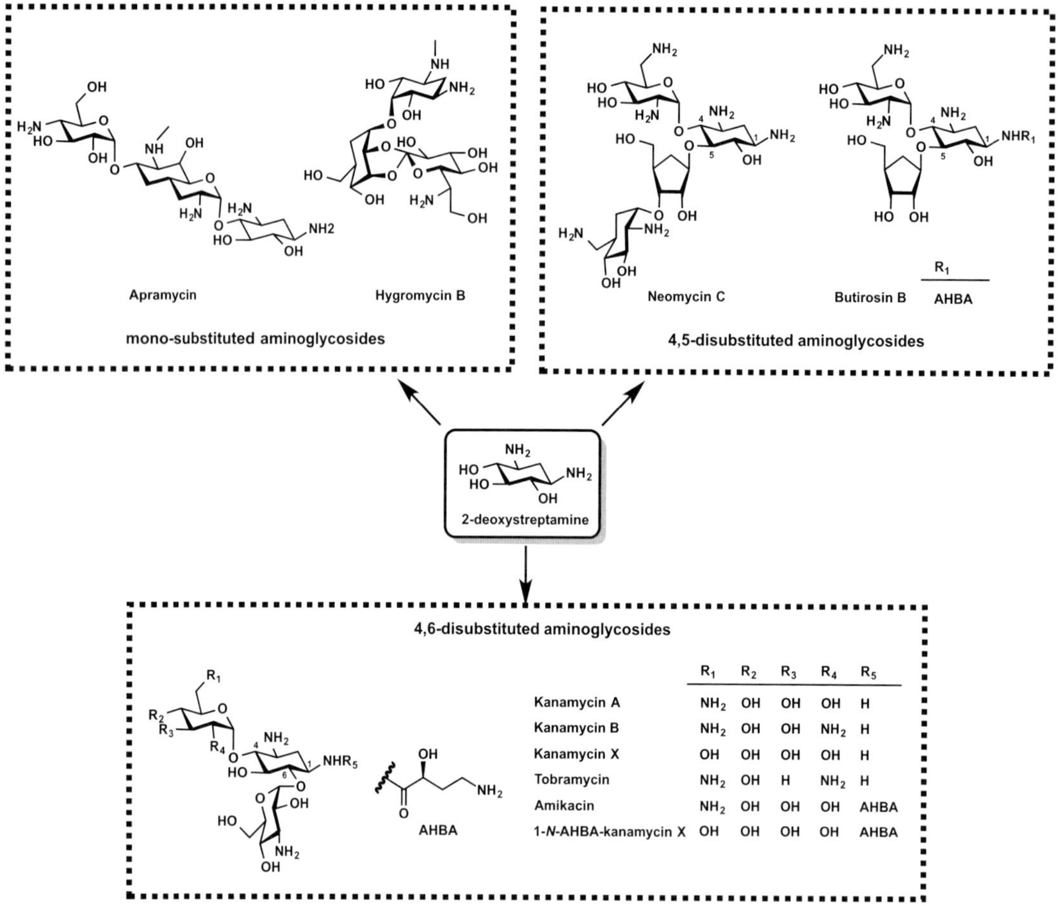

	R_1	R_2	R_3	R_4	R_5
Kanamycin A	NH_2	OH	OH	OH	H
Kanamycin B	NH_2	OH	OH	NH_2	H
Kanamycin X	OH	OH	OH	OH	H
Tobramycin	NH_2	OH	H	NH_2	H
Amikacin	NH_2	OH	OH	OH	AHBA
1-*N*-AHBA-kanamycin X	OH	OH	OH	OH	AHBA

Figure 16.7 Representative structures of the mono-, 4,5- and 4,6-disubstituted 2-deoxystreptamine-containing aminoglycosides.

well as 3′-deoxykanamycins A and C. In addition, exchange of the kanamycin glycosyltransferase gene *kanE* with the tobramycin gene *tobM2* from the kanamycin B biosynthetic gene cluster resulted in the generation of tobramycin from the engineered *S. venezuelae* containing *aprD3* and *aprD4* genes (Park *et al.*, 2011).

Heterologous expression of antibiotic biosynthetic genes

The increasing number of developed genetic tools and accumulated genetic knowledge have provided a number of novel antibiotic compounds through combinatorial biosynthesis of antibiotic genes in natural producers. However, this approach using native hosts has often encountered limitations, especially when applied to poor-growing or genetically unamenable organisms. In such cases, heterologous expression represents one solution to engineer secondary metabolite biosynthetic machinery (Ongley *et al.*, 2013).

Streptomyces and related actinomycetes producing the greatest number of pharmacologically relevant natural products as secondary metabolites have come to be known as robust heterologous hosts for the generation of new antibiotic compounds. *S. coelicolor* is one of the most well-known bacteria for such a purpose. The production of the antibiotic erythromycin aglycone along with its analogue, 8,8a-deoxyoleandolide was performed by expression of the entire DEBS in *S. coelicolor* (Wolpert *et al.*, 2008). In addition, the entire coumermycin A_1 biosynthetic pathway from *Streptomyces rishiriensis* was reconstituted via recombination technology and expressed in *S. coelicolor*. In this study, inactivation of the *couO* gene, encoding methyltransferase required for C-methylation of aminocoumarin moieties in coumermycin A_1, led to an accumulation of a C8-unsubstituted coumermycin A_1 derivative. When a halogenase gene *clo-hal* of the clorobiocin gene cluster was co-expressed with the coumermycin A_1 gene cluster in the heterologous strain, it resulted in two new hybrid antibiotics (Wolpert *et al.*, 2008). Similar to coumarin antibiotics, novobiocin (Steffensky *et al.*, 2000) and clorobiocin (Pojer *et al.*, 2002) gene clusters from *Streptomyces spheroides* and *Streptomyces roseochromogenes*, respectively, were also heterologously expressed in *S. coelicolor* (Eustáquio *et al.*, 2005). Heterologous expression of the modified novobiocin gene cluster with deletion of a methyltransferase gene *novO* resulted in the production of a C8'-unsubstituted novobiocin derivative. Expression of the clorobiocin-derived *clo-hal* gene in a *novO*-defective heterologous host led to the formation of two novclobiocin analogues carrying chlorine at the C8′ position (Flatman *et al.*, 2006). A novel, unglycosylated phenalinolactone derivative was generated by heterologous host *S. coelicolor* expressing a glycosyltransferase gene *plaA6* deficient-phenalinolactone gene cluster (Binz *et al.*, 2008).

Streptomyces lividans has also been subjected to generating antibiotic derivatives as a heterologous host. GE37468 (Selva *et al.*, 1997), produced from *Streptomyces* ATCC 55365, is a thiazolyl peptide antibiotic that potently inhibits ribosome protein synthesis in Gram-positive bacteria and malaria parasites (Bagley *et al.*, 2005). Structurally, GE37468 carries a unique β-methyl-δ-hydroxy-proline moiety, which results from the post-translational transformation of isoleucine 8 (Ile 8) by P450 monooxygenase GetJ. Inactivation of *getJ* by Red/ET recombination and heterologous expression of the modified gene cluster devoid of *getJ* in *S. lividans* led to the accumulation of a GE37468 analogue harbouring a monooxygenated Ile 8. The antibiotic activity of the GE37468 variant was decreased against methicillin-resistant *S. aureus*, showing the importance of the β-methyl-δ-hydroxy-proline residue for the GE37468's antibiotic activity (Young and Walsh, 2011).

Similarly, the pikromycin-producing *S. venezuelae* has recently been exploited for heterologous expression due to its relative rapid growth and ease of genetic manipulation, as compared to other actinomycetes (Xue and Sherman, 2001; Yoon *et al.*, 2002). In one study, the entire tylosin PKS gene cluster from *S. fradiae* was expressed in a *S. venezuelae* mutant, in which the native pikromycin PKS gene cluster was deleted, which led to the production of tylactone. The resulting macrolactone was further processed to produce desosaminyl-tylactone, probably by action of native glycosyltransferase DesVII (Jung *et al.*, 2006). As described in previous sections, the *S. venezuleae*-based heterologous expression system has also been used as a robust platform to reconstruct aminoglycoside genes (Park *et al.*, 2011) and type III polyketide biosynthetic genes (Park *et al.*, 2009) for generation of novel bioactive molecules.

In addition, *E. coli* has served as a desirable host for combinatorial biosynthesis due to its rapid growth, high efficiency of transformation, and well-developed genetic tools. Using the 6-dEB producing *E. coli* as a heterologous host (Pfeifer *et al.*, 2001) has enabled the generation of potent antibacterial polyketide erythromycins C and D by expression of the megalomycin gene cluster from *Micromonospora megalomicea*, as the source to catalyse the tailoring steps in erythromycin C biosynthesis from the 6-dEB intermediate (Peiru *et al.*, 2005). The production of the most bioactive erythromycin A was also accomplished via introduction of *S. erythraea* biosynthetic pathway in *E. coli*. Subsequently, two rationally designed analogues of erythromycin A were produced (Zhang *et al.*, 2010). More specifically, the production of an analogue of erythromycin D with a benzyl unit was obtained by exchanging the loading module of DEBS1 with the loading module from rifamycin PKS that is primed with a benzoate starter unit (Admiraal *et al.*, 2001). In similar fashion, the second module of DEBS3 was altered to accept a malonyl-CoA instead of methylmalonyl-CoA as extender unit, leading to production of a methyl group-deficient erythromycin A–D (Lu *et al.*, 1997). See Table 16.1 for the examples of

Table 16.1 Examples of heterologous expression of antibiotic biosynthetic genes

Compound	Native host	Heterologous host	References
Chalcomycin	*Streptomyces bikiniensis*	*S. fradiae*	Ward *et al.* (2004)
Chlorobiocin	*S. roseochromogenes*	*S. coelicolor*	Eustáquio *et al.* (2005)
Coumermycin A_1	*S. rishiriensis*	*S. coelicolor*	Wolpert *et al.* (2008)
Daptomycin	*S. roseosporus*	*S. lividans*	Penn *et al.* (2006)
Erythromycin	*S. erythraea*	*S. coelicolor*	Kao *et al.* (1994b), Rodriguez *et al.* (2004)
		E. coli	Pfeifer *et al.* (2001), Zhang *et al.* (2010)
GE37468	*Streptomyces* ATCC 55365	*S. lividans*	Bagley *et al.* (2005)
Kanamycin	*S. kanamyceticus*	*S. venezuelae*	Park *et al.* (2011)
Medermycin	*Streptomyces* sp. K73	*S. coelicolor*	Ichinose *et al.* (2003)
Midecamycin	*Streptomyces mycarofaciens*	*S. fradiae*	Rodriguez *et al.* (2004)
Novobiocin	*S. spheroides*	*S. coelicolor*	Eustáquio *et al.* (2005), Eustáquio *et al.* (2004)
Oleandomycin	*S. antibioticus*	*S. coelicolor*	Tang *et al.* (2000)
		S. lividans	Shah *et al.* (2000)
Oxytetracycline	*S. rimosus*	*Myxococcus xanthus*	Stevens et *al.* (2010)
Phenalinolactone	*Streptomyces* sp. Tü6071	*S. coelicolor*	Binz *et al.* (2008)
Pikromycin	*S. venezuelae*	*S. coelicolor*	Tang *et al.* (2000)
Tylosin	*S. fradiae*	*S. venezuelae*	Jung *et al.* (2006)

heterologous expression of antibiotic biosynthetic genes to generate antibiotic compounds including their derivatives.

Recently, a great number of antibiotic gene clusters have been continuously identified, and enzymes and intermediates in the antibiotic biosynthetic pathways have been characterized. In conjunction with the advent of expressing genes from biosynthetic pathways in heterologous organisms, combinatorial biosynthesis will be a viable approach for the generation of new antibiotic drugs.

Future perspectives in combinatorial biosynthesis: challenges and opportunities

As aforementioned, combinatorial biosynthesis has been extensively developed over almost three decades, since the initial research on the production of hybrid antibiotics was reported (Hopwood *et al.*, 1985). A wide range of studies on combinatorial biosynthesis have helped establish guidelines for the successful assembly of antibiotic synthetic machinery and developed valuable tools that can be used to build up a library of novel antibiotic compounds. However, at the same time, it has been realized that there are substantial obstacles to be overcome in combinatorial biosynthesis, even today, for design and expression of the biosynthetic pathways of antibiotic leads (Floss, 2006; Udwary, 2009).

The issue of relatively low production yield became one of challenges of the current combinatorial biosynthesis approach. One manner to solve this problem would be to construct a more efficient and robust production platform; it would be preferable to express hybrid antibiotic pathways in industrially optimized hosts (Rodriguez *et al.*, 2003). In a recent report, the industrial strain of *S. avermitilis* was developed into a versatile model host for the expression of biosynthetic gene clusters of secondary metabolites by deleting the large sub-telomeric region of the linear genome. The mutant strain of *S. avermitilis* did not produce any of the major native secondary metabolites, i.e. avermectins, filipins, and oligomycin. However, expression of foreign gene clusters encoding biosynthesis of the aminoglycoside antibiotic streptomycin (from *Streptomyces griseus*) and the β-lactam antibiotic cephamycin C (from *Streptomyces clavuligerus*) led to a higher production of the corresponding compounds, compared to those produced from original strains (Komatsu *et al.*, 2010). To confirm the feasibility of the engineered *S. avermitilis*, an additional 20 gene clusters for exogenous secondary metabolites, including polyketides, non-ribosomal peptides, and aminoglycosides, were cloned using Gibson's assembling method (Gibson *et al.*, 2009), and introduced into the model host of *S. avermitilis*. Almost all transformants harbouring the entire gene cluster produced metabolites and the productivity of some metabolites was approximately 2-fold higher than that obtained from the native-producing strains. Interestingly, silent biosynthetic gene clusters in the original host were also expressed in the genetically engineered mutant of *S. avermitilis* (Ikeda *et al.*, 2014; Komatsu *et al.*, 2013).

An additional issue is the limitation of DNA manipulation technologies for assembling large gene clusters combinatorially. To date, most successful attempts at combinatorial biosynthesis have relied upon the expression of a single pathway or a small unit of enzymes, due to the limitation of traditional DNA manipulating approaches, such as conventional ligation-based cloning methods using restriction digestion (Udwary, 2009). To overcome this hurdle, a number of revolutionary tools for rapid and efficient assembly of large gene cluster have been recently developed and it is now possible to transform tedious and time-consuming pathway reconstruction into routine laboratory work (See review (Cobb *et al.*, 2014)). The most prominent *in vitro* DNA assembly technique is the one-pot isothermal assembly, also known as the 'Gibson assembly'. In the process, splicing, gap-filling, and nick-sealing of multiple overlapping DNA fragments are conducted by serial action of exonuclease, polymerase, and DNA ligase in a single-tube reaction (Gibson *et al.*, 2009). Additional homology-based DNA assembly includes sequence- and ligase-independent cloning (SLIC) (Li and Elledge, 2012), polymerase incomplete primer extension (PIPE) (Klock *et al.*, 2008), uracil-specific excision reagent (USER) cloning (Bitinaite *et al.*, 2007), circular polymerase extension cloning (CPEC) (Quan and Tian, 2011), and site-specific recombination-based tandem assembly (SSRTA) (Zhang *et al.*, 2011). One example of *in vivo* DNA assembly, an alternative to the *in vitro* approaches, is the DNA assembler method with terminal overlap sequences using native homologous recombination machinery of *S. cerevisiae* (Shao *et al.*, 2009). The Red/ET recombination system also represents a major *in vivo* assembly technique, in which expression of the prophage-derived recombinase protein enhances the homologous recombination in *E. coli* (Zhang *et al.*, 1998). In addition to these homology-based methods, there are several ligation-based methods, including BioBrick assembly (Shetty *et al.*, 2008), golden gate assembly (Engler *et al.*, 2008), and methylation-assisted tailorable ends rational (MASTER) ligation (Chen *et al.*, 2013).

Other challenges include the issues of compatibility between enzymes in designed chimeric pathways, for example, differences in kinetics, localization within the cell, and substrate specificity. Although studies on the structure and mechanism of enzymes, including domains, modules and linkers capable of performing the spectrum of catalytic reaction in antibiotic biosynthetic pathways, have been continuously progressing, many questions still remain unanswered (Wong and Khosla, 2012).

Nevertheless, there are still bright spots and appealing routes to antibiotic discovery through combinatorial biosynthesis by expanding the boundary of combinatorial strategies. There are a large number of examples of *in vitro* combinatorial biosynthesis, although combinatorial biosynthesis can be understood as a special tool for the *in vivo* generation of novel compounds from the genetically amenable organism. One report has described a new *in vitro* combinatorial biosynthesis technique for the generation of large libraries of natural product analogues via a high-throughput, microarray-based approach (Kwon *et al.*, 2007). As a proof of concept, three type III PKSs, including 1,3,6,8-tetrahydroxynaphthalene synthase (THNS) from *S. coelicolor*, a mutant variant of THNS, and a chalcone synthase (CHS) from alfalfa, were contacted with 16 different precursors (acyl-CoA esters) and three post-PKS tailoring enzymes including soybean peroxidase, chloroperoxidase, or polyphenol oxidase: as a result, 144 potential combinations of enzyme reactions were obtained. This strategy was further adapted to a microarray-based synthesis, with a capacity of 448 individual reactions per slide, containing multicombinations of enzymes to enable high-throughput synthesis and derivatization of polyketide-derived products. Further, the same microarray format was used subsequently to screen bioactive products (Kwon *et al.*, 2007). Thus, the design of efficient *in vitro* biosynthesis coupled with high-throughput technology could serve as an efficient approach for the reconstruction of unnatural biosynthetic pathways to generate novel compounds displaying antibacterial activities.

In recent days, natural products derived from marine microorganisms or invertebrates have attracted much attention because their metabolic capacities continue to provide pharmacologically interesting activities (Blunt *et al.*, 2013; Leal *et al.*, 2012). Moreover, the increasing availability of multiorganism-derived pools of genomic DNA (metagenomic DNAs) may be a next key to accessing the desirable genetic sources for new antibiotics (Ghazanfar *et al.*, 2010; Schofield and Sherman, 2013). It has been estimated that up to 99% of microbes remain incapable of being cultured under standard laboratory conditions. Thus, metagenomics, which is a culture-independent analysis of a mixture of a microbial genome, has been developed to circumvent the difficulties and limitations associated with cultivation techniques. Further, the metagenomic studies have largely progressed with advances in development of efficient gene cloning vectors including bacterial artificial chromosomes (BACs) or fosmid/cosmids (Babcock *et al.*, 2007; Xu, 2006), which allow engineering and expressing large and complex DNA segments or genes. Thus, metagenomics coupled with the genetic engineering tools and heterologous expression systems described herein have the potential to open up an untapped pool of novel antibiotic molecules.

In addition to the increasing availability of prolific sources of antibiotic compounds, combination with the use of advanced techniques, such as genomic, proteomic, and/or bioinformatics analysis, will provide valuable information to identify biosynthetic gene clusters and increase the productivity of secondary metabolites and their derivatives generated by combinatorial biosynthesis. For example, genomics will play an important role in combinatorial biosynthesis, as it has already and rapidly become the routine approach in every biological research field. The development of cost-effective second generation DNA sequencing has easily enabled many researchers to mine interesting biosynthetic pathways and use the tremendous opportunity to apply combinatorial biosynthesis techniques for generation of new compounds exhibiting biological activity against antibiotic-resistant infectious bacteria. Of course, in the near term, further advances in genome sequencing (called third generation technologies) will be developed to further increase throughput, decrease cost, and reduce the time to obtain results (Donkor, 2013).

Further, modern tools of structural biology, X-ray crystallography and nuclear magnetic resonance (NMR) can help determine the three-dimensional structure of antibacterial target proteins, allowing scientists to visualize the location of the binding areas on the target proteins. This will enable researchers to design an effective 'key' drugs that can block the target and 'lock' bacterial proteins more effectively (Nicola and Abagyan, 2005).

These efforts to overcome the obstacles of current combinatorial biosynthesis will successfully contribute to developing a robust platform for discovery/creation of many presently unknown complex structures in sufficient quantities that can be tested for antibacterial activity. Although discovery or generation of new antibiotics is not easy, we need to continue to challenge ourselves and take action to find innovative and improved antibiotics. We believe the accumulated understating and highly developed techniques within and beyond combinatorial biosynthesis will give the scientific community as well as pharmaceutical companies a fresh starting point for the discovery of highly effective and nontoxic antibiotic agents in the future.

References

Admiraal, S.J., Walsh, C.T., and Khosla, C. (2001). The loading module of rifamycin synthetase is an adenylation–thiolation didomain with substrate tolerance for substituted benzoates. Biochemistry *40*, 6116–6123.

Babcock, D.A., Wawrik, B., Paul, J.H., McGuinness, L., and Kerkhof, L.J. (2007). Rapid screening of a large insert BAC library for specific 16S rRNA genes using TRFLP. J. Microbiol. Methods *71*, 156–161.

Bagley, M.C., Dale, J.W., Merritt, E.A., and Xiong, X. (2005). Thiopeptide antibiotics. Chem. Rev. *105*, 685–714.

Bailey, A.M., Cox, R.J., Harley, K., Lazarus, C.M., Simpson, T.J., and Skellam, E. (2007). Characterisation of 3-methylorcinaldehyde synthase (MOS) in *Acremonium strictum*: first observation of a reductive release mechanism during polyketide biosynthesis. Chem. Commun. *39*, 4053–4055.

Baltz, R. (2009). Biosynthesis and genetic engineering of lipopeptides in *Streptomyces roseosporus*. Methods Enzymol. *458*, 511–531.

Baltz, R.H., Brian, P., Miao, V., and Wrigley, S.K. (2005). Combinatorial biosynthesis of lipopeptide antibiotics in *Streptomyces roseosporus*. J. Ind. Microbiol. Biotechnol. *33*, 66–74.

Basnet, D.B., Park, J.W., and Yoon, Y.J. (2008). Combinatorial biosynthesis of 5-O-desosaminyl erythronolide A as a potent precursor of ketolide antibiotics. J. Biotechnol. *135*, 92–96.

Bassetti, M., Merelli, M., Temperoni, C., and Astilean, A. (2013). New antibiotics for bad bugs: where are we? Ann. Clin. Microbiol. Antimicrob. *12*, 22.

Betlach, M.C., Kealey, J.T., Betlach, M.C., Ashley, G.W., and McDaniel, R. (1998). Characterization of the macrolide P450 hydroxylase from *Streptomyces venezuelae* which converts narbomycin to picromycin. Biochemistry *37*, 14937–14942.

Binz, T.M., Wenzel, S.C., Schnell, H.-J., Bechthold, A., and Müller, R. (2008). Heterologous expression and genetic engineering of the phenalinolactone biosynthetic gene cluster by using Red/ET recombineering. Chembiochem. *9*, 447–454.

Bitinaite, J., Rubino, M., Varma, K.H., Schildkraut, I., Vaisvila, R., and Vaiskunaite, R. (2007). USERTM friendly DNA engineering and cloning method by uracil excision. Nucleic Acids Res. *35*, 1992–2002.

Blunt, J.W., Copp, B.R., Keyzers, R.A., Munro, M.H.G., and Prinsep, M.R. (2013). Marine natural products. Nat. Prod. Rep. *30*, 237–323.

Boeck, L.D., Papiska, H.R., Wetzel, R.W., Mynderse, J.S., Fukuda, D.S., Mertz, F.P., and Berry, D.M. (1990). A54145, a new lipopeptide antibiotic complex: discovery, taxonomy, fermentation and HPLC. J. Antibiot. (Tokyo) *43*, 587–593.

Borisova, S.A., Zhao, L., Melançon, C.E., Kao, C.-L., and Liu, H.-w. (2004). Characterization of the glycosyltransferase activity of DesVII: analysis of and implications for the biosynthesis of macrolide antibiotics. J. Am. Chem. Soc. *126*, 6534–6535.

Borisova, S.A., Zhang, C., Takahashi, H., Zhang, H., Wong, A.W., Thorson, J.S., and Liu, H.-w. (2006). Substrate specificity of the macrolide-glycosylating enzyme pair DesVII/DesVIII: opportunities, limitations, and mechanistic hypotheses. Angew. Chem. Int. Ed. *45*, 2748–2753.

Boulos, R.A., Man, N.Y.T., Lengkeek, N.A., Hammer, K.A., Foster, N.F., Stemberger, N.A., Skelton, B.W., Wong, P.Y., Martinac, B., Riley, T.V., *et al.* (2013). Inspiration from old dyes: Tris(stilbene) compounds as potent Gram-positive antibacterial agents. Chem. Eur. J. *19*, 17980–17988.

Butler, A., Bate, N., Kiehl, D., Kirst, H., and Cundliffe, E. (2002). Genetic engineering of aminodeoxyhexose biosynthesis in *Streptomyces fradiae*. Nat. Biotechnol. *20*, 713–716.

Chen, W.H., Qin, Z.J., Wang, J., and Zhao, G.P. (2013). The MASTER (methylation-assisted tailorable ends rational) ligation method for seamless DNA assembly. Nucleic Acids Res. *41*, e93.

Chooi, Y.-H., and Tang, Y. (2012). Navigating the fungal polyketide chemical space: from genes to molecules. J. Org. Chem. *77*, 9933–9953.

Cobb, R.E., Ning, J.C., and Zhao, H. (2014). DNA assembly techniques for next-generation combinatorial biosynthesis of natural products. J. Ind. Microbiol. Biotechnol. *41*, 469–477.

Cortes, J., Haydock, S.F., Roberts, G.A., Bevitt, D.J., and Leadlay, P.F. (1990). An unusually large multifunctional polypeptide in the erythromycin producing polyketide synthase of *Saccharopolyspora erythraea*. Nature *348*, 176–178.

Cortes, J., Wiesmann, K.E., Roberts, G.A., Brown, M.J., Staunton, J., and Leadlay, P.F. (1995). Repositioning of a domain in a modular polyketide synthase to promote specific chain cleavage. Science *268*, 1487–1489.

Cox, R.J. (2007). Polyketides, proteins and genes in fungi: programmed nano-machines begin to reveal their secrets. Org. Biomol. Chem. *5*, 2010–2026.

Crawford, J.M., and Townsend, C.A. (2010). New insights into the formation of fungal aromatic polyketides. Nat. Rev. Microbiol. *8*, 879–889.

Cushnie, T., and Lamb, A. (2011). Recent advances in understanding the antibacterial properties of flavonoids. Int. J. Antimicrob. Agents *38*, 99–107.

Demain, A.L., and Sanchez, S. (2009). Microbial drug discovery: 80 years of progress. J. Antibiot. (Tokyo) *62*, 5–16.

Desai, R.P., Rodriguez, E., Galazzo, J.L., and Licari, P. (2004). Improved Bioconversion of 15-Fluoro-6-deoxyerythronolide B to 15-fluoro-erythromycin A by overexpression of the *eryK* Gene in *Saccharopolyspora erythraea*. Biotechnol. Progress *20*, 1660–1665.

Doekel, S., Coëffet-Le Gal, M.F., Gu, J.Q., Chu, M., Baltz, R.H., and Brian, P. (2008). Non-ribosomal peptide synthetase module fusions to produce derivatives of daptomycin in *Streptomyces roseosporus*. Microbiology *154*, 2872–2880.

Donadio, S., and Sosio, M. (2008). Biosynthesis of glycopeptides: prospects for improved antibacterials. Curr. Top. Med. Chem. *8*, 654–666.

Donkor, E.S. (2013). Sequencing of bacterial genomes: principles and insights into pathogenesis and development of antibiotics. Genes *4*, 556–572.

Doumith, M., Legrand, R., Lang, C., Salas, J.A., and Raynal, M.C. (1999). Interspecies complementation in *Saccharopolyspora erythraea*: elucidation of the function of *oleP1*, *oleG1* and *oleG2* from the oleandomycin biosynthetic gene cluster of *Streptomyces antibioticus* and generation of new erythromycin derivatives. Mol. Microbiol. *34*, 1039–1048.

Drautz, H., Zähner, H., Rohr, J., and Zeeck, A. (1986). Metabolic products of microorganisms. 234. urdamycins, new angucycline antibiotics from *Streptomyces fradiae*. I. Isolation, characterization and biological properties. J. Antibiot. (Tokyo) *39*, 1657–1669.

Engler, C., Kandzia, R., and Marillonnet, S. (2008). A one pot, one step, precision cloning method with high throughput capability. PLoS One *3*, e3647.

Eustáquio, A.S., Gust, B., Li, S.-M., Pelzer, S., Wohlleben, W., Chater, K.F., and Heide, L. (2004). Production of 8′-halogenated and 8′-unsubstituted novobiocin derivatives in genetically engineered *Streptomyces coelicolor* strains. Chem. Biol. *11*, 1561–1572.

Eustáquio, A.S., Gust, B., Galm, U., Li, S.M., Chater, K.F., and Heide, L. (2005). Heterologous expression of novobiocin and clorobiocin biosynthetic gene clusters. Appl. Environ. Microbiol. *71*, 2452–2459.

Faust, B., Hoffmeister, D., Weitnauer, G., Westrich, L., Haag, S., Schneider, P., Decker, H., Künzel, E., Rohr, J., and Bechthold, A. (2000). Two new tailoring enzymes, a glycosyltransferase and an oxygenase, involved in biosynthesis of the angucycline antibiotic urdamycin A in *Streptomyces fradiae* Tü2717. Microbiology *146*, 147–154.

Fischbach, M.A., and Walsh, C.T. (2006). Assembly-line enzymology for polyketide and nonribosomal peptide antibiotics: logic, machinery, and mechanisms. Chem. Rev. *106*, 3468–3496.

Fish, S.A., and Cundliffe, E. (1997). Stimulation of polyketide metabolism in *Streptomyces fradiae* by tylosin and its glycosylated precursors. Microbiology *143*, 3871–3876.

Flatman, R.H., Eustaquio, A., Li, S.M., Heide, L., and Maxwell, A. (2006). Structure–activity relationships of aminocoumarin-type gyrase and topoisomerase IV inhibitors obtained by combinatorial biosynthesis. Antimicrob. Agents Chemother. *50*, 1136–1142.

Floss, H.G. (2006). Combinatorial biosynthesis—potential and problems. J. Biotechnol. *124*, 242–257.

Fu, H., McDaniel, R., Hopwood, D.A., and Khosla, C. (1994). Engineered biosynthesis of novel polyketides: stereochemical course of two reactions catalyzed by a polyketide synthase. Biochemistry 33, 9321–9326.

Fu, X., Albermann, C., Jiang, J., Liao, J., Zhang, C., and Thorson, J.S. (2003). Antibiotic optimization via *in vitro* glycorandomization. Nat. Biotechnol. *21*, 1467–1469.

Fujii, I., Watanabe, A., Sankawa, U., and Ebizuka, Y. (2001). Identification of Claisen cyclase domain in fungal polyketide synthase WA, a naphthopyrone synthase of *Aspergillus nidulans*. Chem. Biol. *8*, 189–197.

Funa, N., Ohnishi, Y., Fujii, I., Shibuya, M., Ebizuka, Y., and Horinouchi, S. (1999). A new pathway for polyketide synthesis in microorganisms. Nature *400*, 897–899.

Gaisser, S., Reather, J., Wirtz, G., Kellenberger, L., Staunton, J., and Leadlay, P.F. (2000). A defined system for hybrid macrolide biosynthesis in *Saccharopolyspora erythraea*. Mol. Microbiol. *36*, 391–401.

Gaisser, S., Lill, R., Wirtz, G., Grolle, F., Staunton, J., and Leadlay, P.F. (2001). New erythromycin derivatives from *Saccharopolyspora erythraea* using sugar *O*-methyltransferases from the spinosyn biosynthetic gene cluster. Mol. Microbiol. *41*, 1223–1231.

Gaisser, S., Lill, R., Staunton, J., Méndez, C., Salas, J., and Leadlay, P.F. (2002). Parallel pathways for oxidation of 14-membered polyketide macrolactones in *Saccharopolyspora erythraea*. Mol. Microbiol. *44*, 771–781.

Gantt, R.W., Goff, R.D., Williams, G.J., and Thorson, J.S. (2008). Probing the aglycon promiscuity of an engineered glycosyltransferase. Angew. Chem. Int. Ed. *47*, 8889–8892.

Ghazanfar, S., Azim, A., Ghazanfar, M.A., Anjum, M.I., and Begum, I. (2010). Metagenomics and its application in soil microbial community studies: biotechnological prospects. J. Anim. Plant Sci. *6*, 611–622.

Gibson, D.G., Young, L., Chuang, R.-Y., Venter, J.C., Hutchison, C.A., and Smith, H.O. (2009). Enzymatic assembly of DNA molecules up to several hundred kilobases. Nat. Methods *6*, 343–345.

Han, A.R., Shinde, P.B., Park, J.W., Cho, J., Lee, S.R., Ban, Y.H., Yoo, Y.J., Kim, E.J., Kim, E., Park, S.R., *et al.* (2012). Engineered biosynthesis of glycosylated derivatives of narbomycin and evaluation of their antibacterial activities. Appl. Microbiol. Biotechnol. *93*, 1147–1156.

Hertweck, C., Luzhetskyy, A., Rebets, Y., and Bechthold, A. (2007). Type II polyketide synthases: gaining a deeper insight into enzymatic teamwork. Nat. Prod. Rep. *24*, 162.

Hoffmeister, D., Ichinose, K., Domann, S., Faust, B., Trefzer, A., Dräger, G., Kirschning, A., Fischer, C., Künzel, E., Bearden, D., *et al.* (2000). The NDP-sugar co-substrate concentration and the enzyme expression level influence the substrate specificity of glycosyltransferases: cloning and characterization of deoxysugar biosynthetic genes of the urdamycin biosynthetic gene cluster. Chem. Biol. *7*, 821–831.

Hong, J.S.J., Park, S.R., Choi, C.Y., Sohng, J.K., and Yoon, Y.J. (2004). New olivosyl derivatives of methymycin/pikromycin from an engineered strain of of *Streptomyces venezuelae*. FEMS Microbiol. Lett. *238*, 391–399.

Hopwood, D.A., Malpartida, F., Kieser, H.M., Ikeda, H., Duncan, J., Fujii, I., Rudd, B.A., Floss, H.G., and Omura, S. (1985). Production of 'hybrid' antibiotics by genetic engineering. Nature *314*, 642–644.

Horinouchi, S. (2009). Combinatorial biosynthesis of plant medicinal polyketides by microorganisms. Curr. Opin. Chem. Biol. *13*, 197–204.

Howells, J.D., Anderson, L.E., Coffey, G.L., Senos, G.D., Underhill, M.A., Vogler, D.L., and Ehrlich, J. (1972).

Butirosin, a new aminoglycosidic antibiotic complex: bacterial origin and some microbiological studies. Antimicrob. Agents Chemother. *2*, 79–83.

Hubbard, B.K., and Walsh, C.T. (2003). Vancomycin assembly: nature's way. Angew. Chem. Int. Ed. Engl. *42*, 730–765.

Hwang, E.I., Kaneko, M., Ohnishi, Y., and Horinouchi, S. (2003). Production of plant-specific flavanones by *Escherichia coli* containing an artificial gene cluster. Appl. Environ. Microbiol. *69*, 2699–2706.

Ichinose, K., Ozawa, M., Itou, K., Kunieda, K., and Ebizuka, Y. (2003). Cloning, sequencing and heterologous expression of the medermycin biosynthetic gene cluster of *Streptomyces* sp. AM-7161: towards comparative analysis of the benzoisochromanequinone gene clusters. Microbiology *149*, 1633–1645.

Ikeda, H., Shin-ya, K., and Omura, S. (2014). Genome mining of the *Streptomyces avermitilis* genome and development of genome-minimized hosts for heterologous expression of biosynthetic gene clusters. J. Ind. Microbiol. Biotechnol. *41*, 233–250.

Jiang, H., Wood, K.V., and Morgan, J.A. (2005). Metabolic engineering of the phenylpropanoid pathway in *Saccharomyces cerevisiae*. Appl. Environ. Microbiol. *71*, 2962–2969.

Jones, D., Metzger, H., Schatz, A., and Waksman, S. (1944). Control of Gram-negative bacteria in experimental analysis by streptomycin. Science *100*, 103–105.

Jung, W.S., Lee, S.K., Hong, J.S.J., Park, S.R., Jeong, S.J., Han, A.R., Sohng, J.K., Kim, B.G., Choi, C.Y., Sherman, D.H., *et al.* (2006). Heterologous expression of tylosin polyketide synthase and production of a hybrid bioactive macrolide in *Streptomyces venezuelae*. Appl. Microbiol. Biotechnol. *72*, 763–769.

Kao, C., M., Luo, G., Katz, L., Cane, D.E., and Khosla, C. (1994a). Engineered biosynthesis of a triketide lactone from an incomplete modular polyketide synthase. J. Am. Chem. Soc. *116*, 11612–11613.

Kao, C.M., Katz, L., and Khosla, C. (1994b). Engineered biosynthesis of a complete macrolactone in a heterologous host. Science *265*, 509–512.

Kao, C.M., Luo, G., Katz, L., Cane, D.E., and Khosla, C. (1995). Manipulation of macrolide ring size by directed mutagenesis of a modular polyketide synthase. J. Am. Chem. Soc. *117*, 9105–9106.

Kao, C.M., Luo, G., Katz, L., Cane, D.E., and Khosla, C. (1996). Engineered biosynthesis of structurally diverse tetraketides by a trimodular polyketide synthase. J. Am. Chem. Soc. *118*, 9184–9185.

Katsuyama, Y., and Ohnishi, Y. (2012). Type III polyketide synthases in microorganisms. Methods Enzymol. *515*, 359–377.

Katsuyama, Y., Funa, N., and Horinouchi, S. (2007). Precursor-directed biosynthesis of stilbene methyl ethers in *Escherichia coli*. Biotechnol. J. *2*, 1286–1293.

Katz, L. (1997). Manipulation of modular polyketide synthases. Chem. Rev. *10*, 2557–2576.

Katz, L., and McDaniel, R. (1999). Novel macrolides through genetic engineering. Med. Res. Rev. *19*, 543–558.

Khosla, C., Tang, Y., Chen, A.Y., Schnarr, N.A., and Cane, D.E. (2007). Structure and mechanism of the 6-deoxyerythronolide B synthase. Annu. Rev. Biochem. *76*, 195–221.

Kim, S.D., Ryoo, I.J., Kim, C.J., Kim, W.G., Kim, J.P., Kong, J.Y., Koshino, H., Uramoto, M., and Yoo, I.D. (1996). GERI-155, a new macrolide antibiotic related to chalcomycin. J. Antibiot. (Tokyo) *49*, 955–957.

Kirkpatrick, P., Raja, A., LaBonte, J., and Lebbos, J. (2003). Daptomycin. Nat. Rev. Drug Discov. *2*, 943–944.

Klock, H.E., Koesema, E.J., Knuth, M.W., and Lesley, S.A. (2008). Combining the polymerase incomplete primer extension method for cloning and mutagenesis with microscreening to accelerate structural genomics efforts. Proteins *71*, 982–994.

Kohanski, M.A., Dwyer, D.J., and Collins, J.J. (2010). How antibiotics kill bacteria: from targets to networks. Nat. Rev. Microbiol. *8*, 423–435.

Komatsu, M., Uchiyama, T., Omura, S., Cane, D.E., and Ikeda, H. (2010). Genome-minimized *Streptomyces* host for the heterologous expression of secondary metabolism. Proc. Natl. Acad. Sci. U.S.A. *107*, 2646–2651.

Komatsu, M., Komatsu, K., Koiwai, H., Yamada, Y., Kozone, I., Izumikawa, M., Hashimoto, J., Takagi, M., Omura, S., Shin-ya, K., *et al.* (2013). Engineered *Streptomyces avermitilis* host for heterologous expression of biosynthetic gene cluster for secondary metabolites. ACS Synth. Biol. *2*, 384–396.

Kruger, R.G., Lu, W., Oberthür, M., Tao, J., Kahne, D., and Walsh, C.T. (2005). Tailoring of glycopeptide scaffolds by the acyltransferases from the teicoplanin and A-40,926 biosynthetic operons. Chem. Biol. *12*, 131–140.

Kwon, S.J., Lee, M.-Y., Ku, B., Sherman, D.H., and Dordick, J.S. (2007). High-throughput, microarray-based synthesis of natural product analogues via *in vitro* metabolic pathway construction. ACS Chem. Biol. *2*, 419–425.

Lamb, S.S., Patel, T., Koteva, K.P., and Wright, G.D. (2006). Biosynthesis of sulfated glycopeptide antibiotics by using the sulfotransferase StaL. Chem. Biol. *13*, 171–181.

Leal, M.C., Madeira, C., Brandão, C.A., Puga, J., and Calado, R. (2012). Bioprospecting of marine invertebrates for new natural products-a chemical and zoogeographical perspective. Molecules *17*, 9842–9854.

Lee, S.K., Basnet, D.B., Choi, C.Y., Sohng, J.K., Ahn, J.S., and Yoon, Y.J. (2004). The role of erythromycin C-12 hydroxylase, EryK, as a substitute for PikC hydroxylase in pikromycin biosynthesis. Bioorg. Chem. *32*, 549–559.

Lee, S.K., Basnet, D.B., Hong, J.S.J., Jung, W.S., Choi, C.Y., Lee, H.C., Sohng, J.K., Ryu, K.G., Kim, D.J., Ahn, J.S., *et al.* (2005). Structural diversification of macrolactones by substrate-flexible cytochrome P450 monooxygenases. Adv. Synth. Catal. *347*, 1369–1378.

Leonard, E., Lim, K., Saw, P., and Koffas, M. (2007). Engineering central metabolic pathways for high-level flavonoid production in *Escherichia coli*. Appl. Environ. Microbiol. *73*, 3822–3886.

Li, M.Z., and Elledge, S.J. (2012). SLIC: a method for sequence- and ligation-independent cloning. Methods Mol. Biol. *852*, 51–59.

Li, S., Ouellet, H., Sherman, D.H., and Podust, L.M. (2009). Analysis of transient and catalytic desosamine-binding pockets in cytochrome P450 PikC from *Streptomyces venezuelae*. J. Biol. Chem. *284*, 5723–5730.

Liscombe, D.K., Louie, G.V., and Noel, J.P. (2012). Architectures, mechanisms and molecular evolution of natural product methyltransferases. Nat. Prod. Rep. *29*, 1238–1250.

Long, P., Wilkinson, C., Bisang, C., Cortés, J., Dunster, N., Oliynyk, M., McCormick, E., McArthur, H., Mendez, C., Salas, J., *et al.* (2002). Engineering specificity of starter unit selection by the erythromycin-producing polyketide synthase. Mol. Microbiol. *43*, 1215–1225.

Losey, H., Peczuh, M., Chen, Z., Eggert, U., Dong, S., Pelczer, I., Kahne, D., and Walsh, C. (2001). Tandem action of glycosyltransferases in the maturation of vancomycin and teicoplanin aglycones: novel glycopeptides. Biochemistry *40*, 4745–4755.

Losey, H.C., Jiang, J., Biggins, J.B., Oberthür, M., Ye, X.Y., Dong, S.D., Kahne, D., Thorson, J.S., and Walsh, C.T. (2002). Incorporation of glucose analogs by GtfE and GtfD from the vancomycin biosynthetic pathway to generate variant glycopeptides. Chem. Biol. *9*, 1305–1314.

Lu, L., Thamchaipenet, A., Fu, H., Betlach, M., and Ashley, G. (1997). Biosynthesis of 2-Nor-6-deoxyerythronolide B by rationally designed domain substitution. J. Am. Chem. Soc. *119*, 10553–10554.

McDaniel, R., Ebert-Khosla, S., Hopwood, D.A., and Khosla, C. (1993). Engineered biosynthesis of novel polyketides. Science *262*, 1546–1550.

McDaniel, R., Ebert-Khosla, S., Hopwood, D.A., and Khosla, C. (1994). Engineered biosynthesis of novel polyketides: *actVII* and *actIV* genes encode aromatase and cyclase enzymes, respectively. J. Am. Chem. Soc. *116*, 10855–10859.

McDaniel, R., Ebert-Khosla, S., Hopwood, D.A., and Khosla, C. (1995a). Rational design of aromatic polyketide natural products by recombinant assembly of enzymatic subunits. Nature *375*, 549–554.

McDaniel, R., Hutchinson, C., and Khosla, C. (1995b). Engineered biosynthesis of novel polyketides: analysis of *tcmN* function in tetracenomycin biosynthesis. J. Am. Chem. Soc. *117*, 6805–6810.

McDaniel, R., Thamchaipenet, A., Gustafsson, C., Fu, H., Betlach, M., and Ashley, G. (1999). Multiple genetic modifications of the erythromycin polyketide synthase to produce a library of novel 'unnatural' natural products. Proc. Natl. Acad. Sci. U.S.A. *96*, 1846–1851.

Magnet, S., and Blanchard, J.S. (2005). Molecular insights into aminoglycoside action and resistance. Chem. Rev. *105*, 477–498.

Marsden, A.F. (1998). Engineering broader specificity into an antibiotic-producing polyketide synthase. Science *279*, 199–202.

Mchenney, M.A., Hosted, T.J., Dehoff, B.S., Rosteck, P.R.J., and Baltz, R.H. (1998). Molecular cloning and physical mapping of the daptomycin gene cluster from *Streptomyces roseosporus*. J. Bacteriol. *180*, 143–151.

Melançon, C.E., Yu, W.-l., and Liu, H.-w. (2005). TDP-mycaminose biosynthetic pathway revised and conversion of desosamine pathway to mycaminose pathway with one gene. J. Am. Chem. Soc. *127*, 12240–12241.

Menzella, H.G., Reid, R., Carney, J.R., Chandran, S.S., Reisinger, S.J., Patel, K.G., Hopwood, D.A., and Santi, D.V. (2005). Combinatorial polyketide biosynthesis by de novo design and rearrangement of modular polyketide synthase genes. Nat. Biotechnol. *23*, 1171–1176.

Miao, V., Coëffet-Legal, M.F., Brian, P., Brost, R., Penn, J., Whiting, A., Martin, S., Ford, R., Parr, I., Bouchard, M., *et al.* (2005). Daptomycin biosynthesis in *Streptomyces roseosporus*: cloning and analysis of the gene cluster and revision of peptide stereochemistry. Microbiology *151*, 1507–1523.

Miao, V., Coëffet-Le Gal, M.-F., Nguyen, K., Brian, P., Penn, J., Whiting, A., Steele, J., Kau, D., Martin, S., and Ford, R. (2006). Genetic engineering in *Streptomyces roseosporus* to produce hybrid lipopeptide antibiotics. Chem. Biol. *13*, 269–276.

Miyahisa, I., Kaneko, M., Funa, N., Kawasaki, H., Kojima, H., Ohnishi, Y., and Horinouchi, S. (2005). Efficient production of (2*S*)-flavanones by *Escherichia coli* containing an artificial biosynthetic gene cluster. Appl. Microbiol. Biotechnol. *68*, 498–504.

Moore, B., and Hertweck, C. (2001). Biosynthesis and attachment of novel bacterial polyketide synthase starter units. Nat. Prod. Rep. *19*, 70–99.

Nguyen, K.T., Ritz, D., Gu, J., Alexander, D., Chu, M., Miao, V., Brian, P., and Baltz, R.H. (2006). Combinatorial biosynthesis of novel antibiotics related to daptomycin. Proc. Natl. Acad. Sci. U.S.A. *103*, 7462–7467.

Nguyen, K.T., He, X., Alexander, D.C., Li, C., Gu, J.Q., Mascio, C., Van Praagh, A., Mortin, L., Chu, M., Silverman, J.A., *et al.* (2010). Genetically engineered lipopeptide antibiotics related to A54145 and daptomycin with improved properties. Antimicrob. Agents Chemother. *54*, 1404–1413.

Nicola, G., and Abagyan, R. (2005). Structure-based approaches to antibiotic drug discovery. Curr. Protoc. Microbiol. *chapter 17*, 1–10.

Olano, C., Méndez, C., and Salas, J.A. (2010). Post-PKS tailoring steps in natural product-producing actinomycetes from the perspective of combinatorial biosynthesis. Nat. Prod. Rep. *27*, 571–616.

Oliynyk, M., Brown, M., Cortés, J., Staunton, J., and Leadlay, P. (1996). A hybrid modular polyketide synthase obtained by domain swapping. Chem. Biol. *3*, 833–839.

Ongley, S.E., Bian, X., Neilan, B.A., and Müller, R. (2013). Recent advances in the heterologous expression of microbial natural product biosynthetic pathways. Nat. Prod. Rep. *30*, 1121–1138.

Pageni, B.B., Oh, T.J., Liou, K., Yoon, Y.J., and Sohng, J.K. (2008). Genetically engineered biosynthesis of macrolide derivatives including 4-amino-4,6-dideoxy-L-glucose from *Streptomyces venezuelae* YJ003-OTBP3. J. Microbiol. Biotechnol. *18*, 88–94.

Pageni, B.B., Oh, T.-J., and Sohng, J.K. (2009). Novel desosaminyl derivatives of dihydrochalcomycin from a genetically engineered strain of *Streptomyces* sp. Biotechnol. Lett. *31*, 1759–1768.

Park, J.W., Park, S.R., Nepal, K.K., Han, A.R., Ban, Y.-H., Yoo, Y.J., Kim, E.J., Kim, E.M., Kim, D., Sohng, J.K., *et al.* (2011). Discovery of parallel pathways of kanamycin biosynthesis allows antibiotic manipulation. Nat. Chem. Biol. *7*, 843–852.

Park, S., Yoon, J., Paik, J., Park, J., Jung, W., Ban, Y., Kim, E., Yoo, Y., Han, A., and Yoon, Y. (2009). Engineering of plant-specific phenylpropanoids biosynthesis in *Streptomyces venezuelae*. J. Biotechnol. *141*, 181–188.

Park, S.R., Paik, J.H., Ahn, M.S., Park, J.W., and Yoon, Y.J. (2010). Biosynthesis of plant-specific flavones and flavonols in *Streptomyces venezuelae*. J. Microbiol. Biotechnol. *20*, 1295–1299.

Peiru, S., Menzella, H.G., Rodriguez, E., Carney, J., and Gramajo, H. (2005). Production of the potent antibacterial polyketide erythromycin C in *Escherichia coli*. Appl. Environ. Microbiol. *71*, 2539–2547.

Pelzer, S., Süssmuth, R., Heckmann, D., Recktenwald, J., Huber, P., Jung, G., and Wohlleben, W. (1999). Identification and analysis of the balhimycin biosynthetic gene cluster and its use for manipulating glycopeptide biosynthesis in *Amycolatopsis mediterranei* DSM5908. Antimicrob. Agents Chemother. *43*, 1565–1573.

Penn, J., Li, X., Whiting, A., Latif, M., Gibson, T., Silva, C., Brian, P., Davies, J., Miao, V., Wrigley, S., *et al.* (2006). Heterologous production of daptomycin in *Streptomyces lividans*. J. Ind. Microbiol. Biotechnol. *33*, 121–128.

Peric-Concha, N. (2005). Ablation of the *otcC* gene encoding a post-polyketide hydroxylase from the oxytetracyline biosynthetic pathway in *Streptomyces rimosus* results in novel polyketides with altered chain length. J. Biol. Chem. *280*, 37455–37460.

Pfeifer, B.A., Admiraal, S.J., Gramajo, H., Cane, D.E., and Khosla, C. (2001). Biosynthesis of complex polyketides in a metabolically engineered strain of *E. coli*. Science *291*, 1790–1792.

Pfeifer, V. (2001). A polyketide synthase in glycopeptide biosynthesis. The biosynthesis of the non-proteinogenic amino acid (*S*)-3,5-dihyeroxyphenylglycine. J. Biol. Chem. *276*, 38370–38377.

Pojer, F., Li, S.M., and Heide, L. (2002). Molecular cloning and sequence analysis of the clorobiocin biosynthetic gene cluster: new insights into the biosynthesis of aminocoumarin antibiotics. Microbiology *148*, 3901–3911.

Puk, O., Huber, P., Bischoff, D., Recktenwald, J., Jung, G., Süssmuth, R.D., van Pée, K.H., Wohlleben, W., and Pelzer, S. (2002). Glycopeptide biosynthesis in *Amycolatopsis mediterranei* DSM5908: function of a halogenase and a haloperoxidase/perhydrolase. Chem. Biol. *9*, 225–235.

Quan, J., and Tian, J. (2011). Circular polymerase extension cloning for high-throughput cloning of complex and combinatorial DNA libraries. Nat. Protoc. *6*, 242–251.

Reeves, C.D., Murli, S., Ashley, G.W., Piagentini, M., Hutchinson, C.R., and McDaniel, R. (2001). Alteration of the substrate specificity of a modular polyketide synthase acyltransferase domain through site-specific mutations. Biochemistry *40*, 15464–15470.

Reid, R., Piagentini, M., Rodriguez, E., Ashley, G., Viswanathan, N., Carney, J., Santi, D.V., Hutchinson, C.R., and McDaniel, R. (2003). A model of structure and catalysis for ketoreductase domains in modular polyketide synthases. Biochemistry *42*, 72–79.

Rix, U., Fischer, C., Remsing, L.L., and Rohr, J. (2002). Modification of post-PKS tailoring steps through combinatorial biosynthesis. Nat. Prod. Rep. *19*, 542–580.

Rodriguez, E., and McDaniel, R. (2001). Combinatorial biosynthesis of antimicrobials and other natural products. Curr. Opin. Microbiol. *4*, 526–534.

Rodriguez, E., Hu, Z., Ou, S., Volchegursky, Y., Hutchinson, C.R., and McDaniel, R. (2003). Rapid engineering of polyketide overproduction by gene transfer to industrially optimized strains. J. Ind. Microbiol. Biotechnol. *30*, 480–488.

Rodriguez, E., Ward, S., Fu, H., Revill, W., McDaniel, R., and Katz, L. (2004). Engineered biosynthesis of 16-membered macrolides that require methoxymalonyl-ACP precursors in *Streptomyces fradiae*. Appl. Microbiol. Biotechnol. *66*, 85–91.

Rowe, C., Böhm, I., Thomas, I., Wilkinson, B., Rudd, B., Foster, G., Blackaby, A., Sidebottom, P., Roddis, Y., Buss, A., *et al.* (2001). Engineering a polyketide with a longer chain by insertion of an extra module into the erythromycin-producing polyketide synthase. Chem. Biol. *8*, 475–485.

Schell, U., Haydock, S.F., Kaja, A.L., Carletti, I., Lill, R.E., Read, E., Sheehan, L.S., Low, L., Fernandez, M.-J., Grolle, F., *et al.* (2008). Engineered biosynthesis of hybrid macrolide polyketides containing D-angolosamine and D-mycaminose moieties. Org. Biomol. Chem. *6*, 3315.

Schofield, M.M., and Sherman, D.H. (2013). Meta-omic characterization of prokaryotic gene clusters for natural product biosynthesis. Curr. Opin. Biotechnol. *24*, 1151–1158.

Schroeckh, V., Scherlach, K., Nützmann, H.W., Shelest, E., Schmidt-Heck, W., Schuemann, J., Martin, K., Hertweck, C., and Brakhage, A.A. (2009). Intimate bacterial–fungal interaction triggers biosynthesis of archetypal polyketides in *Aspergillus nidulans*. Proc. Natl. Acad. Sci. U.S.A. *106*, 14558–14563.

Selva, E., Montanini, N., Stella, S., Soffientini, A., Gastaldo, L., and Denaro, M. (1997). Targeted screening for elongation factor Tu binding antibiotics. J. Antibiot. (Tokyo) *50*, 22–26.

Shah, S., Xue, Q., Tang, L., Carney, J., Betlach, M., and McDaniel, R. (2000). Cloning, characterization and heterologous expression of a polyketide synthase and P450 oxidase involved in the biosynthesis of the antibiotic oleandomycin. J. Antibiot. (Tokyo) *53*, 502–508.

Shao, Z., Zhao, H., and Zhao, H. (2009). DNA assembler, an *in vivo* genetic method for rapid construction of biochemical pathways. Nucleic Acids Res. *37*, e16.

Shen, B. (2003). Polyketide biosynthesis beyond the type I, II and III polyketide synthase paradigms. Curr. Opin. Chem. Biol. *7*, 285–295.

Shen, B., Summers, R.G., Wendt-Pienkowski, E., and Hutchinson, C.R. (1995). The *Streptomyces glaucescens tcmKL* polyketide synthase and *tcmN* polyketide cyclase genes govern the size and shape of aromatic polyketides. J. Am. Chem. Soc. *117*, 6811–6821.

Shetty, R.P., Endy, D., and Knight, T.F. (2008). Engineering BioBrick vectors from BioBrick parts. J. Biol. Eng. *2*, 5.

Shinde, P.B., Han, A.R., Cho, J., Lee, S.R., Ban, Y.-H., Yoo, Y.J., Kim, E.J., Kim, E., Song, M.C., Park, J.W., *et al.* (2013). Combinatorial biosynthesis and antibacterial evaluation of glycosylated derivatives of 12-membered macrolide antibiotic YC-17. J. Biotechnol. *168*, 142–148.

Simkhada, D., Oh, T.-J., Pageni, B.B., Lee, H.C., Liou, K., and Sohng, J.K. (2009). Characterization of CalS9 in the biosynthesis of UDP-xylose and the production of xylosyl–attached hybrid compound. Appl. Microbiol. Biotechnol. *83*, 885–895.

Solenberg, P., Matsushima, P., Stack, D., Wilkie, S., Thompson, R., and Baltz, R. (1997). Production of hybrid glycopeptide antibiotics *in vitro* and in *Streptomyces toyocaensis*. Chem. Biol. *4*, 195–202.

Song, M.C., Kim, E., Ban, Y.-H., Yoo, Y.J., Kim, E.J., Park, S.R., Pandey, R.P., Sohng, J.K., and Yoon, Y.J. (2013). Achievements and impacts of glycosylation reactions involved in natural product biosynthesis in prokaryotes. Appl. Microbiol. Biotechnol. *97*, 5691–5704.

Sosio, M., and Donadio, S. (2006). Understanding and manipulating glycopeptide pathways: the example of the dalbavancin precursor A40926. J. Ind. Microbiol. Biotechnol. *33*, 569–576.

Sosio, M., Stinchi, S., Beltrametti, F., Lazzarini, A., and Donadio, S. (2003). The gene cluster for the biosynthesis of the glycopeptide antibiotic A40926 by *Nonomuraea* Species. Chem. Biol. *10*, 541–549.

Stark, W.M., Hoehn, M.M., and Knox, N.G. (1967). Nebramycin, a new broad-spectrum antibiotic complex. I. detection and biosynthesis. Antimicrob. Agents Chemother. *7*, 314–323.

Stassi, D., Donadio, S., Staver, M., and Katz, L. (1993). Identification of a *Saccharopolyspora erythraea* gene required for the final hydroxylation step in erythromycin biosynthesis. J. Bacteriol. *175*, 182–189.

Stassi, D., Kakavas, S., Reynolds, K., Gunawardana, G., Swanson, S., Zeidner, D., Jackson, M., Liu, H., Buko, A., and Katz, L. (1998). Ethyl-substituted erythromycin derivatives produced by directed metabolic engineering. Proc. Natl. Acad. Sci. U.S.A. *95*, 7305–7309.

Steffensky, M., Muhlenweg, A., Wang, Z.X., Li, S.M., and Heide, L. (2000). Identification of the novobiocin biosynthetic gene cluster of *Streptomyces spheroides* NCIB 11891. Antimicrob. Agents Chemother. *44*, 1214–1222.

Stevens, D., Henry, M., Murphy, K., and Boddy, C. (2010). Heterologous expression of the oxytetracycline biosynthetic pathway in *Myxococcus xanthus*. Appl. Environ. Microbiol. *76*, 2681–2683.

Stinchi, S., Carrano, L., Lazzarini, A., Feroggio, M., Grigoletto, A., Sosio, M., and Donadio, S. (2006). A derivative of the glycopeptide A40926 produced by inactivation of the beta-hydroxylase gene in *Nonomuraea* sp. ATCC39727. FEMS Microbiol. Lett. *256*, 229–235.

Sucipto, H., Kudo, F., and Eguchi, T. (2012). The last step of kanamycin biosynthesis: unique deamination reaction catalyzed by the α-ketoglutarate-dependent nonheme iron dioxygenase KanJ and the NADPH-dependent reductase KanK. Angew. Chem. Int. Ed. Engl. *51*, 3428–3431.

Tang, L., Fu, H., and McDaniel, R. (2000). Formation of functional heterologous complexes using subunits from the picromycin, erythromycin and oleandomycin polyketide synthases. Chem. Biol. *7*, 77–84.

Udwary, D.W. (2009). Natural product combinatorial biosynthesis: promises and realities. Nat. Prod. Chem. Drug Discov. *chapter 10*, 299–317.

Umezawa, H., Ueda, M., Maeda, K., Yagishita, K., Kondo, S., Okami, Y., Utahara, R., Osato, Y., Nitta, K., and Takeuchi, T. (1957). Production and isolation of a new antibiotic: kanamycin. J. Antibiot. (Tokyo) *10*, 181–188.

Waksman, S.A., and Lechevalier, H.A. (1949). Neomycin, a new antibiotic active against streptomycin-resistant bacteria, including tuberculosis organisms. Science *109*, 305–307.

Wang, P., Zhang, W., Zhan, J., and Tang, Y. (2009). Identification of OxyE as an ancillary oxygenase during tetracycline biosynthesis. Chembiochem. *10*, 1544–1550.

Ward, S., Hu, Z., Schirmer, A., Reid, R., Revill, W., Reeves, C., Petrakovsky, O., Dong, S., and Katz, L. (2004). Chalcomycin biosynthesis gene cluster from *Streptomyces bikiniensis*: novel features of an unusual ketolide produced through expression of the *chm* polyketide synthase in *Streptomyces fradiae*. Antimicrob. Agents Chemother. *48*, 4703–4712.

Watts, K.T., Lee, P.C., and Schmidt-Dannert, C. (2004). Exploring recombinant flavonoid biosynthesis in metabolically engineered *Escherichia coli*. Chembiochemistry *5*, 500–507.

Weisshaar, B., and Jenkins, G.I. (1998). Phenylpropanoid biosynthesis and its regulation. Curr. Opin. Plant Biol. *1*, 251–257.

Wenzel, S.C., Bode, H.B., Kochems, I., and Müller, R. (2008). A type I/type III polyketide synthase hybrid biosynthetic pathway for the structurally unique ansa compound kendomycin. Chembiochem. *9*, 2711–2721.

Wiesmann, K., Cortés, J., Brown, M., Cutter, A., Staunton, J., and Leadlay, P. (1995). Polyketide synthesis *in vitro* on a modular polyketide synthase. Chem. Biol. *2*, 583–589.

Williams, G.J., Zhang, C., and Thorson, J.S. (2007). Expanding the promiscuity of a natural-product glycosyltransferase by directed evolution. Nat. Chem. Biol. *3*, 657–662.

Wolpert, M., Heide, L., Kammerer, B., and Gust, B. (2008). Assembly and heterologous expression of the coumermycin A1 gene cluster and production of new derivatives by genetic engineering. Chembiochem. *9*, 603–612.

Wong, F., and Khosla, C. (2012). Combinatorial biosynthesis of polyketides-a perspective. Curr. Opin. Chem. Biol. *16*, 117–123.

Xiong, Y., Wu, X., and Mahmud, T. (2005). A homologue of the *Mycobacterium tuberculosis* PapA5 protein, Rif-Orf20, is an acetyltransferase involved in the biosynthesis of antitubercular drug rifamycin B by *Amycolatopsis mediterranei* S699. Chembiochem. *6*, 834–837.

Xu, J. (2006). Microbial ecology in the age of genomics and metagenomics: concepts, tools, and recent advances. Mol. Ecol. *15*, 1713–1731.

Xu, J., Mahmud, T., and Floss, H.G. (2003). Isolation and characterization of 27-O-demethylrifamycin SV methyltransferase provides new insights into the post-PKS modification steps during the biosynthesis of the antitubercular drug rifamycin B by *Amycolatopsis mediterranei* S699. Arch. Biochem. Biophys. *411*, 277–288.

Xue, Y., and Sherman, D.H. (2001). Biosynthesis and combinatorial biosynthesis of pikromycin-related macrolides in *Streptomyces venezuelae*. Metab. Eng. 3, 15–26.

Xue, Y., Zhao, L., Liu, H.-W., and Sherman, D.H. (1998). A gene cluster for macrolide antibiotic biosynthesis in *Streptomyces venezuelae*: architecture of metabolic diversity. Proc. Natl. Acad. Sci. U.S.A. 95, 12111–12116.

Xue, Y., Wilson, D., and Sherman, D.H. (2000). Genetic architecture of the polyketide synthases for methymycin and pikromycin series macrolides. Gene *245*, 203–211.

Xue, Y., Wilson, D., Zhao, L., Liu, H., and Sherman, D. (2004). Hydroxylation of macrolactones YC-17 and narbomycin is mediated by the *pikC*-encoded cytochrome P450 in *Streptomyces venezuelae*. Chem. Biol. 5, 661–667.

Yan, Y., Kohli, A., and Koffas, M.A.G. (2005). Biosynthesis of natural flavanones in *Saccharomyces cerevisiae*. Appl. Environ. Microbiol. *71*, 5610–5613.

Yeh, H.-H., Chang, S.-L., Chiang, Y.-M., Bruno, K.S., Oakley, B.R., Wu, T.-K., and Wang, C.C.C. (2013). Engineering fungal nonreducing polyketide synthase by heterologous expression and domain swapping. Org. Lett. *15*, 756–759.

Yoon, Y., Beck, B., Kim, B., Kang, H., Reynolds, K., and Sherman, D. (2002). Generation of multiple bioactive macrolides by hybrid modular polyketide synthases in *Streptomyces venezuelae*. Chem. Biol. 9, 203–214.

Young, T.S., and Walsh, C.T. (2011). Identification of the thiazolyl peptide GE37468 gene cluster from *Streptomyces* ATCC 55365 and heterologous expression in *Streptomyces lividans*. Proc. Natl. Acad. Sci. U.S.A. *108*, 13053–13058.

Yu, D., Xu, F., Zeng, J., and Zhan, J. (2012). Type III polyketide synthases in natural product biosynthesis. IUBMB Life *64*, 285–295.

Yuan, Y., Chung, H.S., Leimkuhler, C., Walsh, C.T., Kahne, D., and Walker, S. (2005). *In vitro* reconstitution of EryCIII activity for the preparation of unnatural macrolides. J. Am. Chem. Soc. *127*, 14128–14129.

Zhanel, G.G., Dueck, M., Hoban, D.J., Vercaigne, L.M., Embil, J.M., Gin, A.S., and Karlowsky, J.A. (2001). Review of macrolides and ketolides: focus on respiratory tract infections. Drugs *61*, 443–498.

Zhang, C., Fu, Q., Albermann, C., Li, L., and Thorson, J.S. (2007). The *in vitro* characterization of the erythronolide mycarosyltransferase EryBV and its utility in macrolide diversification. Chembiochem. *8*, 385–390.

Zhang, H., Wang, Y., Wu, J., Skalina, K., and Pfeifer, B.A. (2010). Complete biosynthesis of erythromycin A and designed analogs using *E. coli* as a heterologous host. Chem. Biol. *17*, 1232–1240.

Zhang, L., Zhao, G., and Ding, X. (2011). Tandem assembly of the epothilone biosynthetic gene cluster by *in vitro* site-specific recombination. Sci. Rep. *1*, 141.

Zhang, W., Watanabe, K., Cai, X., Jung, M.E., Tang, Y., and Zhan, J. (2008). Identifying the minimal enzymes required for anhydrotetracycline biosynthesis. J. Am. Chem. Soc. *130*, 6068–6069.

Zhang, Y., Buchholz, F., Muyrers, J.P., and Stewart, A.F. (1998). A new logic for DNA engineering using recombination in *Escherichia coli*. Nat. Genet. *20*, 123–128.

Lantibiotics and Other Antibacterial Peptides

17

Margherita Sosio and Stefano Donadio

Abstract

Lantibiotics and other ribosomally synthesized peptides have attracted increasing attention in the search for new chemical classes of antibacterial agents effective against multidrug-resistant pathogens. Several factors have probably contributed to this resurgent interest. Among them, the realization that many different classes of ribosomally synthesized peptides are produced by almost all taxa of bacteria; that lantibiotics, being ribosomally synthesized, are amenable to the generation of analogues by codon substitutions in the structural gene; and that, by binding to biosynthetic intermediates and not directly to enzymes, resistance in bacterial pathogens cannot be simply acquired by mutations in the target enzymes. In this chapter, we will review the chemistry, biosynthesis, occurrence and mechanism of action of lanthipeptides, with particular focus on lantibiotics. We will also present three lantibiotics that are currently under advanced preclinical or early clinical development, as well as some recently discovered compounds which represent improved variants of known lantibiotics.

Introduction

In the past decade or so, lantibiotics and other ribosomally synthesized peptides (RiPPs) have attracted increasing attention in the search for new chemical classes of antibacterial agents effective against multidrug-resistant pathogens. This renewed interest may seem a historical revenge for nisin, the prototype lantibiotic and actually the first antibiotic to be reported in the scientific literature, before penicillin (Rogers, 1928). Several factors have probably contributed to this resurgent interest. Among them, the realization that many different classes of RiPPS are produced by almost all taxa of bacteria (Arnison *et al.*, 2013); that these peptides, being ribosomally synthesized, are amenable to the generation of analogues by codon substitutions in the gene encoding the precursor peptide (Cotter *et al.*, 2006; Appleyard *et al.*, 2009; Cortes *et al.*, 2009); and that, by binding to biosynthetic intermediates and not directly to enzymes, resistance in bacterial pathogens cannot be simply acquired by mutations in the target enzymes (Breukink and de Kruijff, 2006).

The term 'lantibiotic', introduced in 1988 as a linguistic blend between *lan*thionine and an*tibiotic*, was meant to designate short, gene-encoded peptides that contain one or more lanthionine (Lan) or methyllanthionine (MeLan) residues (Schnell *et al.*, 1988). The name lanthionine derives from *lana*, the Latin word for wool, from which lanthionine was first extracted (Horn *et al.*, 1941). However, an increasing number of characterized (Me)Lan-containing peptides are devoid of any measurable antibacterial activity despite possessing other potent activities. Hence, in a general trend to switch from a bioactivity- to a structure-based nomenclature for natural products, the term 'lanthipeptides' has been introduced to designate any (Me)Lan-containing peptide, irrespective of its bioactivity (Arnison *et al.*, 2013). Given the widespread use of the term, the term lantibiotics will continue to be used to designate the subset of lanthipeptides endowed with antimicrobial activity.

A recent analysis listed >20 distinct different classes of RiPPs, many of which were unveiled in the last five years or so (Arnison *et al.*, 2013). We should also mention that, in this respect, thiopeptides represent the clinically most advanced class of RiPPS: LFF571 is a semisynthetic derivative of the thiopeptide GE2270 currently in phase 2 clinical studies for the treatment of *Clostridium difficile* infections (LaMarche *et al.*, 2012); while another derivative of GE2270 has completed a phase 1 clinical study as a topical treatment for acne (Donadio *et al.*, 2010). However, the focus of this chapter will be on lanthipeptides that are in the late stages of preclinical characterization or have entered clinical trials for the treatment of bacterial infections. Before describing in detail those compounds, we shall review some of the general features of lanthipeptides.

Chemistry, occurrence and biosynthesis of lanthipeptides

The post-translational modifications common to the lanthipeptides involve the dehydration of serine and threonine residues to form 2,3-didehydroalanine (Dha) and (Z)-2,3-didehydrobutyrine (Dhb), respectively. This is followed by the stereospecific intramolecular addition of a cysteine thiol onto Dha or Dhb to generate the thioether linkage present in Lan or MeLan residues, respectively (Fig. 17.1A). The latter residues are believed to have the *meso*-stereochemistry, although this has been established only for a subset of compounds (Chatterjee *et al.*, 2005). In recent years (Meindl *et al.*, 2010; Völler *et al.*, 2012; Wang and

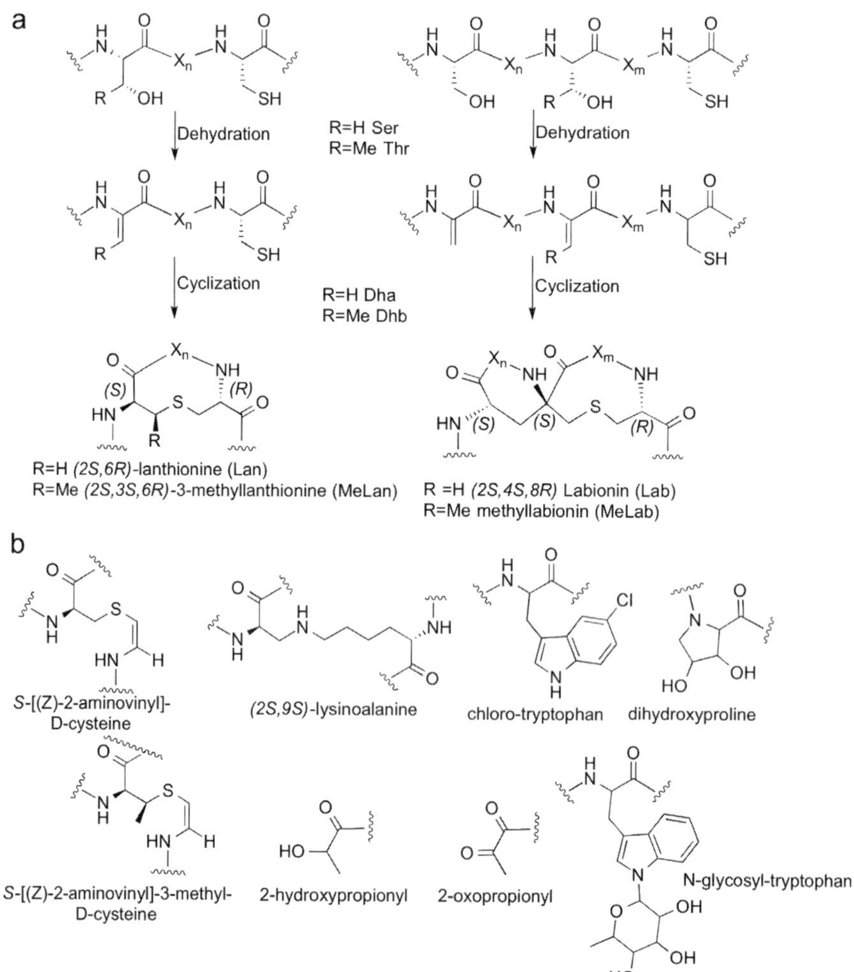

Figure 17.1 (a) Amino acid residues typical of lanthipeptides (Dha, Dhb, Lan, MeLan, Lab and MeLab) and their formation from serine (threonine) and cysteine residues. (b) Some other non-proteinogenic amino acids or deaminated residues found in lanthipeptides.

van der Donk, 2012; Iorio *et al.*, 2014), the repertoire of modified amino acids has been expanded by the discovery of the carbacyclic amino acid labionin (Lab) and its analogue methyl-labionin (MeLab), triamino acids consisting of a quaternary central carbon atom joined through a thioether and a methylene bridge to the *N*- and *C*-terminal residues, respectively (Fig. 17.1A). Each of the two Lab residues present in the labyrinthopeptins possess a *2S,4S,8R* configuration (Meindl *et al.*, 2010).

All post-translational modifications occur on a precursor peptide, from which the leader sequence is eventually removed by proteolytic cleavage. As the number of lanthipeptides increased, the original classification of lantibiotics as type A and type B on the basis of structural features proved limited. Accordingly, lanthipeptides are currently grouped into four different classes on the basis of biogenetic considerations. The synthesis of class I lanthipeptides requires separate dehydratase and cyclase enzymes, designated as LanB and LanC, respectively. In contrast, the biosynthesis of class II lanthipeptides is catalysed by the bifunctional enzyme LanM. Formation of class III and class IV lanthipeptides involves trifunctional synthetases that share an N-terminal phosphoSer/phosphoThr lyase domain and a central kinase domain, but differing in their C-terminal cyclization domains as class IV enzymes possess the same Zn-binding motif present in LanC and LanM. Interestingly, some but not all of the class III lanthipeptides contain (Me)Lab residues. However, the features of the precursor peptide and/or of the synthetase that lead to the formation of Lan or Lab residues in class III lanthipeptides are still under investigation.

Often, lanthipeptides belonging to the same class share common structural features. For example, class I compounds, exemplified by nisin (Fig. 17.2), are generally screw-shaped, elongated, flexible and amphipathic with an overall positive net charge, whereas class II lanthipeptides, for example mersacidin (Fig. 17.2), are generally rigid globular molecules. Unique within the class II molecules are two-component lantibiotics, such as lacticin 3147 and haloduracin, in which two peptides, each encoded by a separate structural gene and processed by a cognate enzyme, act synergistically to form the mature antibiotic (Knerr and van der Donk, 2012). So far, relevant antibacterial activities have been found only among members of classes I and II.

In addition to formation of the characteristic thioethers, lanthipeptides can undergo further post-translational modifications (Fig. 17.1B). Among them, the most common are oxidative decarboxylation of the C-terminus, β-hydroxylation of aspartate and lysinoalanine, formation of D-alanine and formation of α-ketoamide and α-hydroxyamide N-terminal caps (Chatterjee *et al.*, 2005). Other,

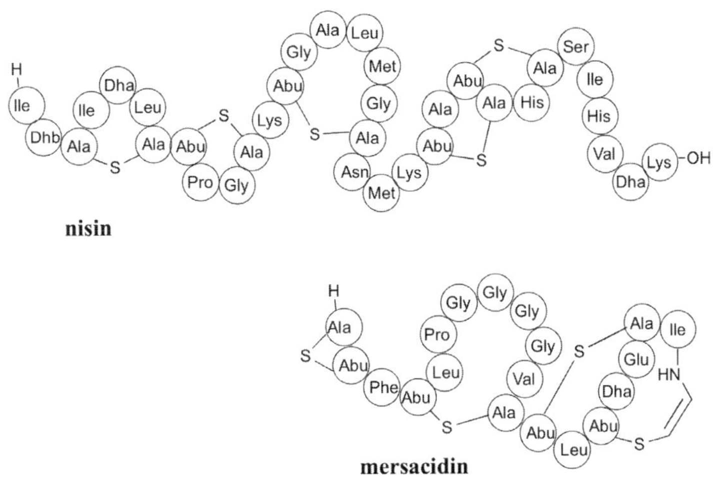

Figure 17.2 Structures of the prototype class I and class II lanthipeptides nisin and mersacidin, respectively.

less common modifications include *N*-glycosylation (Iorio *et al.* 2014), disulfide bond formation (Meindl *et al.*, 2010), as well as halogenation and proline hydroxylation, as described below.

Biosynthesis of lanthipeptides is directed by small, compact gene clusters characterized by the presence of the structural gene *lanA*, which encodes an *N*-terminal leader followed by the core region that undergoes the required post-translational modifications before the mature product is exported from the cell, coincident with, or followed by, leader peptide removal. The precise substrates for the other modifications have not been established in most cases, although they are likely to occur during or after lanthionine formation. In bacterial cells, other proteins participate in the processing and transport of the lanthipeptide, in regulating its biosynthesis and in conferring immunity to the producing strain.

Up to a decade ago or so, lanthipeptides were believed to be produced mostly by members of the class *Firmicutes* and, to a lesser extent, by the *Actinobacteria*. However, Lan-forming enzymes are encoded by members of the classes *Proteobacteria* and *Cyanobacteria*. As the number of RiPPs expands, lanthipeptide production may be found in additional microbial taxa (Velasquez and van der Donk, 2011).

Mechanism of action of lantibiotics

In this section, we will limit our analysis to the lanthipeptides with antibacterial activity, i.e. the lantibiotics. Other bioactivities reported for lanthipeptides include: inhibition of phospholipase A_2 by cinnamycin (Märki *et al.*, 1991); of angiotensin-converting enzyme by ancovenin (Kido *et al.*, 1983); morphogenic agents for aerial mycelium formation in *Streptomyces* strains such as SapB (Kodani *et al.*, 2004) and catenulipeptin (Wang and van der Donk, 2012); and efficacy against neuropathic pain in mouse models for labyrinthopeptin (Meindl *et al.*, 2010) and NAI-112 (Iorio *et al.*, 2014). We will also limit our analysis to the general properties of lantibiotics, while later sections will focus on selected compounds that are currently under advanced preclinical or early clinical investigation. For a recent overview of the biological activities of lantibiotics, the reader is referred to other recent reviews on the topic (Maffioli *et al.*, 2012; Arnison *et al.*, 2013).

The activity of lantibiotics against relevant bacterial pathogens can vary greatly and the elements contributing to potency are not fully understood (Schneider and Sahl, 2010). In any case, some compounds can exhibit potent activity against clinically relevant Gram-positive bacteria, including drug-resistant strains of *Staphylococcus, Streptococcus, Enterococcus,* and *Clostridium*. Lantibiotics are generally inactive against Gram-negative bacteria, although sporadic cases of activity against selected species have been reported. This is believed to be due to the outer membrane preventing access to the actual target on the outer side of the cytoplasmic membrane. Indeed, when the permeability of the outer membrane is increased by chelators, some Gram-negative bacteria can be sensitized to nisin (Delves-Broughton *et al.*, 1996).

In all cases that have been investigated, lantibiotics have been found to inhibit cell wall biosynthesis. In some cases, it has been demonstrated that this inhibition occurs through binding to the key cell wall intermediate lipid II (Hsu *et al.*, 2003; Hsu *et al.*, 2004; Breukink and de Kruijff, 2006). Remarkably, lantibiotics are active against vancomycin-resistant entercococci (VRE), strains that synthesize a lipid II intermediate modified in the terminal dipeptide of the pentapeptide moiety. As explained below, this feature is explained by the observation that binding of glycopeptides and lantibiotics involves different portions of lipid II.

Nisin (Fig. 17.2) represents the first and most studied lantibiotic, with a long history as food preservative in many countries. It acts by binding to the pyrophosphate moiety of the cell wall intermediate lipid II. The interaction primarily involves the formation of five hydrogen bonds between the pyrophosphate moiety of Lipid II and the A and B rings of nisin, which are conserved in several class I lanthipeptides. Once bound to Lipid II, a transmembrane orientation of nisin occurs involving insertion of its C-terminal part and a flexible hinge region leading to the formation of stable pores in a complex consisting of eight nisin and four lipid II molecules (Hsu *et al.*, 2004; Breukink and de Kruijff, 2006). This causes dissipation of the membrane potential, rapid efflux of small metabolites, and rapid cell lysis. Many class I lantibiotics contain the N-terminal rings involved in pyrophosphate caging but lack the C-terminal tail and do not form pores or form pores in a strain-specific fashion, dependent on the length of membrane phospholipids (Knerr and van der Donk, 2012). For example, binding of bovicin HC5 to Lipid II inhibits the pore-forming activity of nisin, presumably by sequestering most Lipid II binding sites (Paiva *et al.*, 2012).

Mersacidin, the prototype of class II lantibiotics (Fig. 17.2), also binds lipid II and inhibits transglycosylation in a Ca^{2+}-dependent manner (Böttiger *et al.*, 2009). Binding involves a conformational change likely to be caused by a movement around a hinge region located at residues 12 and 13. Mersacidin lacks the ability to form pores. For the two-component lantibiotics, the individual peptides display little or no activity alone, but afford potent and synergistic activity in combination (Knerr and van der Donk, 2012). The α-peptide of haloduracin shares the mersacidin-binding motif and interacts with Lipid II, where it benefits from the presence of the β peptide, which facilitates pore formation. The complex has been shown to consist of Lipid II:Halα:Halβ in 1:2:2 stoichiometry (Oman *et al.*, 2011).

Nisin is among the most potent lantibiotics, with minimal inhibitory concentration (MIC) values in the nanomolar range. Despite its potency, *in vivo* efficacy has been reported only for a *Streptococcus pneumoniae* mouse model (Goldstein *et al.*, 1998). While the compound was effective at low doses, its short half-life may have hindered further development. This might be due to nisin's intrinsic instability at physiological pH, which results in total loss of antibacterial activity (Chan *et al.*, 1989). Mutacin B-Ny266 has also MICs in the nanomolar range against several Gram-positive pathogens, and has demonstrated efficacy comparable to vancomycin against a methicillin-sensitive *S. aureus* strain in an intraperitoneal mouse infection model (Mota-Meira *et al.*, 2005).

Lantibiotics in development

In this section, we present a brief description of the lantibiotics that are part of the drug discovery pipeline of small pharmaceutical companies.

Mutacin 1140

The lantibiotic mutacin 1140 is produced by *Streptococcus mutans*, the major causative agent of dental caries (Hillman *et al.*, 1984). Its 2263-Da molecule accommodates four (Me)Lan-containing macrocyclic rings (Fig. 17.3). Mutacin 1140 belongs to the epidermin group of class I lanthipeptides, sharing rings A and B with nisin and containing a *C*-terminal *S*-[(*Z*)–2-aminovinyl]-D-cysteine, like other members of the epidermin group (Knerr and van der Donk., 2012). The 3-D structure of mutacin 1140 was determined in acetonitrile/water (Smith *et al.*, 2003). Like nisin and gallidermin, the thioether rings of mutacin 1140 maintain rigidity while the hinge region is quite flexible, forming a horseshoe-like shape.

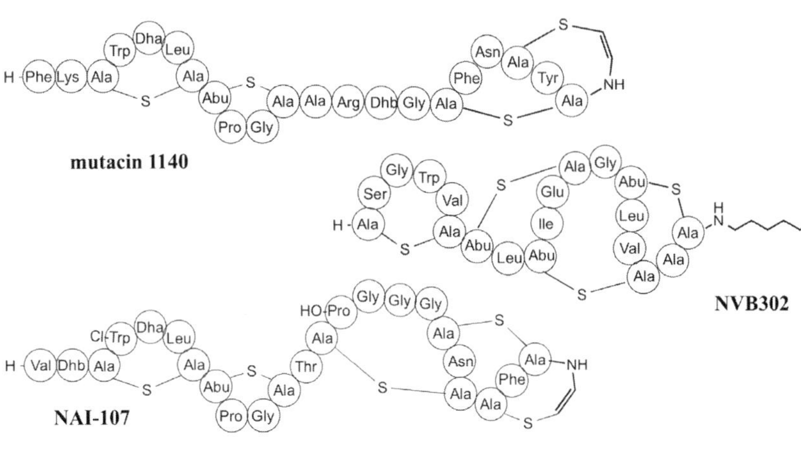

Figure 17.3 Structures of mutacin 1140, NVB302 and NAI-107.

Some interesting clinical applications have been proposed for mutacin 1140 (Hillman *et al.*, 2000). As the compound is effective against many strains of the same species, it has prompted interest by Oragenics (http://www.oragenics.com) for the development of the SMaRT (Streptococcus Mutans Replacement Therapy) therapy for the prevention of dental caries (Hillman *et al.*, 2000). A genetically modified strain of *S. mutans* with decreased virulence due to a mutation in the lactate dehydrogenase gene allows this beneficial bacterium to cause a therapeutic or health-enhancing effect in the oral cavity (Hillman *et al.*, 2007).

Currently, the lead compound MU 1140 is in preclinical development by Oragenics for the treatment of Gram-positive bacterial infections. MU 1140 has proven to be active *in vitro* against all Gram-positive bacteria against which it has been tested, including methicillin-resistant *Staphylococcus aureus* (MRSA), VREs and *Clostridium difficile* (Ghobrial *et al.*, 2010).

NVB302

An unusual feature of the class II lanthipeptides mersacidin and actagardine is that, despite being weakly active against Gram-positive bacteria *in vitro*, they show efficacy comparable to vancomycin in experimental models of infection (Malabarba *et al.*, 1990; Chatterjee *et al.*, 1992; Kruszewska *et al.*, 2004). This finding, together with the possibility of generating analogues by genetic engineering, has resulted in a renewed interest in this group of lantibiotics.

NVB302 is a carboxy amide derivative of deoxy-actagardine B, an actagardine variant produced by *Actinoplanes liguriae* carrying Leu-Val residues at position 15–16 instead of Val-Ile and devoid of the sulfoxide in the C-terminal ring (Fig. 17.3). NVB302 shows a general improvement in MIC against several Gram-positive bacteria (Appleyard *et al.*, 2009); it has *in vitro* activity against *C. difficile*, reduced activity against the intestinal bacteria *Bifidobacterium* and *Bacteroides* spp. and is stable in simulated gastric fluid (Appleyard *et al.*, 2009). Recently it has been demonstrated that NVB302 effectively eliminates vegetative *C. difficile* cells in an *in vitro* gut model, without displaying activity against spores (Crowther *et al.*, 2013). Re-infection by the *C. difficile* PCR ribotype 027 was not observed following successful treatment with NVB302 (Crowther *et al.*, 2013). This observation may offer clinical benefits with respect to a possible decreased likelihood of recurrence of *C. difficile* infection. In 2012, Novacta completed a phase 1 clinical study in healthy volunteers, in which NVB302 was found to be safe and well tolerated, with negligible systemic absorption of the drug and high concentrations recovered in the faeces (Crowther *et al.*, 2013).

NAI-107

NAI-107 is a lantibiotic produced by the actinomycete *Microbispora* sp. (Fig. 17.3). In addition to one MeLan and three Lan bridges, it carries a C-terminal S-[(Z)-2-aminovinyl]-D-cysteine and the lanthipeptide unique residues 5-chlorotryptophan and 3,4-dihydroxyproline (Maffioli *et al.*, 2014). As a result of halogenation and hydroxylation(s), as well as a spontaneous oxidation, NAI-107 is unusual among RiPPS as it is produced as a complex of related molecules, which differ for the presence of two, one or zero hydroxyl groups at Pro-14, by the presence/absence of a chlorine at Trp-4 and by the presence of a sulfoxide on the Lan (Maffioli *et al.*, 2014). NAI-107 was discovered after a high throughput screening of 120,000 microbial fermentation extracts (Castiglione *et al.*, 2008).

NAI-107 is active against multidrug resistant Gram-positive pathogens, including MRSA, VRE, penicillin-resistant *Streptococcus pneumoniae* and, unusual for a lantibiotic, also against some Gram-negative species (Jabés *et al.*, 2011). NAI-107 has been tested in several animal models of multidrug-resistant infections, in which it displays efficacy comparable or superior to the 'drugs of last resort' linezolid and vancomycin (Jabes *et al.*, 2011). As expected, NAI-107-resistant mutants were not observed during these studies.

To identify the molecular basis of its potency, the mode of action was analysed using whole cell and *in vitro* assays, as well as NMR titration experiments (Münch *et al.*, 2014). It has been observed that NAI-107 efficiently interferes with late stages of cell wall biosynthesis and induces accumulation of the soluble peptidoglycan precursor UDP-N-acetylmuramic acid-pentapeptide in the cytoplasm (Münch *et al.*, 2014). It can form complexes with any bactoprenol-pyrophosphate coupled precursor of the bacterial cell wall (Münch *et al.*, 2014). Titration experiments indicate that first a 1:1 stoichiometric complex occurs, which then transforms into a 2:1 (peptide: lipid II) complex, when excess peptide is added. Furthermore, lipid II and related molecules do not serve as anchor molecules for pore formation, but slow membrane depolarization is observed after NAI-107 treatment, which could contribute to killing of bacterial cells (Münch *et al.*, 2014). NAI-107 is currently under preclinical development by the US company Sentinella Pharmaceuticals for systemic treatment of Gram-positive infections (Fox, 2013).

Other recent, promising compounds. Geobacillin has been heterologously produced in *Escherichia coli* using a gene cluster derived from the thermophilic bacterium *Geobacillus thermodenitrificans* (Garg *et al.*, 2012). It contains seven thioether cross-links, including the A and B rings found in nisin that are important for lipid II binding. In comparison with nisin, it contains two additional thioether bridges at its C-terminus (Fig. 17.4). Nisin and geobacillin possess a similar antimicrobial activity and overall mode of action: binding to lipid II and pore formation (Garg *et al.*, 2014). Geobacillin is active against a wide range of Gram-positive bacteria, including VREs and *Bacillus anthracis*. Interestingly, in parallel experiments with nisin, geobacillin I demonstrated three-fold higher activity against *Streptococcus dysgalactiae*, a causative agents of bovine mastitis (Garg *et al.*, 2012). As predicted from the higher number of Lan residues cross-links and the thermophilic source, geobacillin I is more stable at pH 7 and 8 at both 37 and at 60°C than nisin (Garg *et al.*, 2012).

NAI-802 is a new lantibiotic produced by two *Actinoplanes* sp., along with minor amounts of its Ala(0)-congener (Simone *et al.*, 2013). It is closely related to the class II lantibiotic actagardine, sharing an identical 19-aa peptide and all four thioether bridges (Fig. 17.4). Interestingly, NAI-802 seems to combine the features of michiganin, a 3-Lan lantibiotic which carries one extra Ala and one extra Arg at the N- and C-termini (Holtsmark *et al.*, 2006), with the amino acid sequence and ring topology of actagardine. Because of the additional positive charge at physiological pH due to the C-terminal Arg leading to an overall higher positive charge than actagardine, NAI-802 is slightly more active than actagardine against staphylococci and streptococci (Simone *et al.*, 2013). Similarly to actagardine (Malabarba *et al.*, 1990), further improvement of NAI-802 antibacterial activity was achieved by adding one additional positive charge through conversion of the C-terminal carboxylate into the corresponding basic amide (Simone *et al.*, 2013).

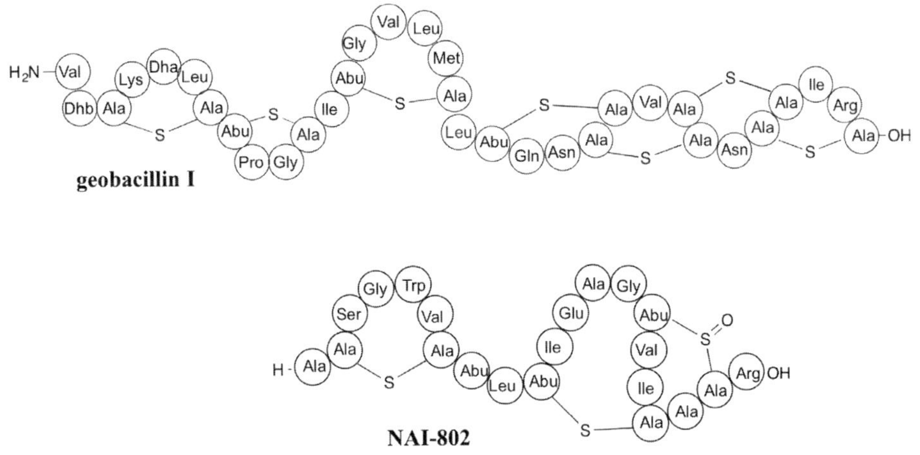

Figure 17.4 Structures of geobacillin I and NAI-802.

Outlook

The recent and increasing progress in the discovery and characterization of lanthipeptides with antibacterial activity has created interest in this class also for the treatment of infections in human and veterinary medicine, with a few small companies venturing into the uncharted territory of new chemical classes. In our opinion, the current threat of antibiotic resistance cannot be addressed without considering the introduction of new chemical classes not affected by existing resistance mechanisms. Hopefully, in the near future, the lantibiotics will represent one of several examples of new chemical classes of antibacterial agents under clinical development.

References

Appleyard, A.N., Choi, S., Read, D.M., Lightfoot, A., Boakes, S., Hoffmann, A., Chopra, I., Bierbaum, G., Rudd, B.A., Dawson, M.J., and Cortes, J. (2009). Dissecting structural and functional diversity of the lantibiotic mersacidin. Chem. Biol. *16*, 490–498.

Arnison, P.G., Bibb, M.J., Bierbaum, G., Bowers, A.A., Bugni, T.S., Bulaj, G., Camarero, J.A., Campopiano, D.J., Challis, G.L., Clardy, J., *et al.* (2013). Ribosomally synthesized and post-translationally modified peptide natural products: overview and recommendations for a universal nomenclature. Nat. Prod. Rep. *30*, 108–160.

Böttiger, T., Schneider, T., Martínez, B., Sahl, H.G., and Wiedemann, I. (2009). Influence of Ca^{2+} ions on the activity of lantibiotics containing a mersacidin-like lipid II binding motif. Appl. Environ. Microbiol. *75*, 4427–4434.

Breukink, E., and de Kruijff, B. (2006). Lipid II as a target for antibiotics Nat. Rev. Drug Discovery *5*, 321–323.

Castiglione, F., Lazzarini, A., Carrano, L., Corti, E., Ciciliato, I., Gastaldo, L., Candiani, P., Losi, D., Marinelli, F., Selva, E., and Parenti, F. (2008). Determining the structure and mode of action of microbisporicin, a potent lantibiotic active against multiresistant pathogens. Chem. Biol. *15*, 22–31.

Chan, W.C., Bycroft, B.W., Lian, L.Y., and Roberts, G.C.K. (1989). Isolation and characterization of two degradation products derived from the peptide antibiotic nisin. FEBS Lett. *252*, 29–36.

Chatterjee, C., Paul, M., Xie, L., and van der Donk, W.A. (2005). Biosynthesis and mode of action of lantibiotics. Chem. Rev. *105*, 633–684.

Chatterjee, S., Chatterjee, D.K., Jani, R.H., Blumbach, J., Ganguli, B.N., Klesel, N., Limbert, M., and Seibert, G. (1992). Mersacidin, a new antibiotic from *Bacillus*: *in vitro* and *in vivo* antibacterial activity. J. Antibiot. *45*, 839–845.

Cortes, J., Appleyard, A.N., Dawson, M.J., and David, A.H. (2009). Whole-cell generation of lantibiotic variants. Methods Enzymol. *458*, 559–574.

Cotter, P.D., Deegan, L.H., Lawton, E.M., Draper, L.A., O'Connor, P.M., Hill, C., and Ross, R.P. (2006). Complete alanine scanning of the two-component lantibiotic lacticin 3147: generating a blueprint for rational drug design. Mol. Microbiol. *62*, 735–747.

Crowther, G.S., Baines, S.D., Todhunter, S.L., Freeman, J., Chilton, C.H., and Wilcox, M.H. (2013). Evaluation of NVB302 versus vancomycin activity in an *in vitro* human gut model of *Clostridium difficile* infection. J. Antimicrob. Chemother. *68*, 168–176.

Delves-Broughton, J., Blackburn, P., Evans, R.J., and Hugenholtz, J. (1996). Applications of the bacteriocin, nisin. Antonie Van Leeuwenhoek *69*, 193–202.

Donadio, S., Maffioli, S., Monciardini, P., Sosio, M., and Jabes, D. (2010). Antibiotic discovery in the twenty-first century: current trends and future perspectives. J. Antibiot. *63*, 423–430.

Fox, J.L. (2013). Antimicrobial peptides stage a comeback. Nat. Biotechnol. *31*, 379–382.

Garg, N., Tang, W., Goto, Y., Nair, S.K., and van der Donk, W.A. (2012). Lantibiotics from *Geobacillus thermodenitrificans*. Proc. Natl. Acad. Sci. U.S.A. *109*, 5241–5246.

Garg, N., Oman, T.J., Andrew Wang, T.S., De Gonzalo, C.V., Walker, S., and van der Donk, W.A. (2014). Mode of action and structure–activity relationship studies of geobacillin I. J. Antibiot. *67*, 133–136.

Ghobrial, O., Derendorf, H., and Hillman, J.D. (2010). Pharmacokinetic and pharmacodynamic evaluation of the lantibiotic MU1140. J. Pharm. Sci. *99*, 2521–2528.

Goldstein, B.P., Wei, J., Greenberg, K., and Novick, R. (1998). Activity of nisin against *Streptococcus pneumoniae, in vitro,* and in a mouse infection model. J. Antimicrob. Chemother. *42*, 277–278.

Hillman, J.D., Johnson, K.P., and Yaphe, B.I. (1984). Isolation of *Streptococcus mutans* strain producing a novel bacteriocin. Infect. Immun. *44*, 141–144.

Hillman, J.D., Brooks, T.A., Michalek, S.M., Harmon, C.C., Snoep, J.L., and van Der Weijden, C.C. (2000). Construction and characterization of an effector strain of *Streptococcus mutans* for replacement therapy of dental caries. Infect. Immun. *68*, 543–549.

Hillman, J.D., Mo, J., McDonell, E., Cvitkovitch, D., and Hillman, C.H. (2007). Modification of an effector strain for replacement therapy of dental caries to enable clinical safety trials. J Appl Microbiol. *102*, 1209–1219.

Holtsmark, I., Mantzilas, D., Eijsink, V.G.H., and Brurberg, M.B. (2006). Purification, characterization, and gene sequence of michiganin A, an actagardine-like lantibiotic produced by the tomato pathogen *Clavibacter michiganensis* subsp. *michiganensis*. Appl. Environ. Microbiol. *72*, 5814–5821.

Horn, M.J., Jones, D.B., and Ringel, S.J. (1941). Isolation of a new sulfur-containing amino acid (lanthionine) from sodium carbonate-treated wool. J. Biol. Chem. *138*, 141–149.

Hsu, S.T., Breukink, E., Bierbaum, G., Sahl, H.G., de Kruijff, B., Kaptein, R., van Nuland, N.A., and Bonvin, A.M. (2003). NMR study of mersacidin and lipid II interaction in dodecylphosphocholine micelles: Conformational changes are a key to antimicrobial activity. J. Biol. Chem. *278*, 13110–13117.

Hsu, S.T., Breukink, E., Tischenko, E., Lutters, M.A., de Kruijff, B., Kaptein, R., Bonvin, A.M., and van Nuland, N.A. (2004). The nisin-lipid II complex reveals a pyrophosphate cage that provides a blueprint for novel antibiotics. Nat. Struct. Mol. Biol. *11*, 963–967.

Iorio, M., Sasso, O., Maffioli, S.I., Bertorelli, R., Monciardini, P., Sosio, M., Bonezzi, F., Summa, M., Brunati, C., Bordoni, R., *et al.* (2014). A glycosylated, labionin-containing lanthipeptide with marked antinociceptive activity. ACS Chem. Biol. 2013 Nov 20.

Jabés, D., Brunati, C., Candiani, G., Riva, S., Romanó, G., and Donadio, S. (2011). Efficacy of the new lantibiotic NAI-107 in experimental infections induced by multidrug-resistant Gram-positive pathogens. Antimicrob. Agents Chemother. *55*, 1671–1676.

Kido, Y., Hamakado, T., Yoshida, T., Anno, M., Motoki, Y., Wakamiya, T., and Shiba, T. (1983). Isolation and characterization of ancovenin, a new inhibitor of angiotensin I converting enzyme, produced by actinomycetes. J. Antibiot. *36*, 1295–1299.

Knerr, P.J., and van der Donk, W.A. (2012). Discovery, biosynthesis, and engineering of lantipeptides. Annu. Rev. Biochem. *81*, 479–505.

Kodani, S., Hudson, M.E., Durrant, M.C., Buttner, M.J., Nodwell, J.R., and Willey, J.M. (2004). The SapB morphogen is a lantibiotic-like peptide derived from the product of the developmental gene *ramS* in *Streptomyces coelicolor*. Proc. Natl. Acad. Sci. U.S.A. *101*, 11448–11453.

Kruszewska, D., Sahl, H.G., Bierbaum, G., Pag, U., Hynes, S.O., and Ljungh, A. (2004). Mersacidin eradicates methicillin-resistant *Staphylococcus aureus* (MRSA) in a mouse rhinitis model. J. Antimicrob. Chemother. *54*, 648–653.

LaMarche, M.J., Leeds, J.A., Amaral, A., Brewer, J.T., Bushell, S.M., Deng, G., Dewhurst, J.M., Ding, J., Dzink-Fox, J., Gamber, G., *et al.* (2012). Discovery of LFF571: an investigational agent for *Clostridium difficile* infection. J. Med. Chem. *55*, 2376–2387.

Maffioli, S.I., Monciardini, P., Sosio, M., and Donadio, S. (2012). New lantibiotics from natural and engineered strains. In Drug Discovery from Natural Products, O. Genilloud, and F. Vicente, eds. (RSC Publishing, Cambridge, UK), pp. 116–139.

Maffioli, S.I., Iorio, M., Sosio, M., Monciardini, P., Gaspari, E., and Donadio, S. (2014). Characterization of the congeners in the lantibiotic NAI–107 complex. J. Nat. Prod. 77, 79–84.

Malabarba, A., Pallanza, R., Berti, M., and Cavalleri, B. (1990). Synthesis and biological activity of some amide derivatives of the lantibiotic actagardine. J. Antibiot. *9*, 1089–1097.

Märki, F., Hanni, E., Fredenhagen, A., and van Oostrum, J. (1991). Mode of action of the lanthionine-containing peptide antibiotics duramycin, duramycin B and C, and cinnamycin as indirect inhibitors of phospholipase A_2. Biochem. Pharmacol. *42*, 2027–2035.

Meindl, K., Schmiederer, T., Schneider, K., Reicke, A., Butz, D., Keller, S., Gühring, H., Vértesy, L.,Wink, J., Hoffmann, H., *et al.* (2010). Labyrinthopeptins: a new class of carbacyclic lantibiotics. Angew. Chem. Int. Ed. *49*, 115–1154.

Mota-Meira, M., Morency, H., and Lavoie, M.C. (2005). *In vivo* activity of mutacin B-Ny266. J. Antimicrob. Chemother. *56*, 869–871.

Münch, D., Müller, A., Schneider, T., Kohl, B., Wenzel, M., Bandow, J.E., Maffioli, S., Sosio, M., Donadio, S., Wimmer, R., and Sahl, H.G. (2014). The lantibiotic NAI-107 binds to bactoprenol bound cell wall precursors and impairs membrane functions. J. Biol. Chem. *289*, 12063–12076.

Oman, T.J., Lupoli, T.J., Wang, T.S., Kahne, D., Walker, S., and van der Donk, W.A. (2011). Haloduracin α binds the peptidoglycan precursor lipid II with 2:1 stoichiometry. J. Am. Chem. Soc. *133*, 17544–17547.

Paiva, A.D., Irving, N., Breukink, E., and Mantovani, H.C. (2012). Interaction with lipid II induces conformational changes in bovicin HC5 structure. Antimicrob. Agents Chemother. *56*, 4586–4593.

Rogers L.A. (1928). The inhibiting effect of *Streptococcus lacti* on *Lactobacillus bulgaricus*. J. Bacteriol. *16*, 321–325.

Schneider, T., and Sahl, H.G. (2010). An oldie but a goodie – cell wall biosynthesis as antibiotic target pathway. Int. J. Med. Microbiol. *300*, 161–169.

Schnell, N., Entian, K.D., Schneider, U., Götz, F., Zahner, H., Kellner, R., and Jung, G. (1988). Prepeptide sequence of epidermin, a ribosomally synthesized antibiotic with four sulphide-rings. Nature *333*, 276–278.

Simone, M., Monciardini, P., Gaspari, E., Donadio, S., and Maffioli, S.I. (2013). Isolation and characterization of NAI-802, a new lantibiotic produced by two different *Actinoplanes* strains. J. Antibiot. *66*, 73–78.

Smith, L., Zachariah, C., Thirumoorthy, R., Rocca, J., Novak, J., Hillman, J.D., and Edison, A.S. (2003). Structure and dynamics of the lantibiotic mutacin 1140. Biochemistry *42*, 10372–10384.

Velásquez, J.E., and van der Donk, W.A. (2011). Genome mining for ribosomally synthesized natural products. Curr. Opin. Chem. Biol. *15*, 11–21.

Völler, G.H., Krawczyk, J.M., Pesic, A., Krawczyk, B., Nachtigall, J., and Süssmuth, R.D. (2012). Characterization of new class III lantibiotics-erythreapeptin, avermipeptin and griseopeptin from *Saccharopolyspora erythraea*, *Streptomyces avermitilis* and *Streptomyces griseus* demonstrates stepwise N-terminal leader processing. ChemBioChem. *13*, 1174–1183.

Wang, H., and van der Donk, W.A. (2012). Biosynthesis of the class III lantipeptide catenulipeptin. ACS Chem. Biol. *7*, 1529–1535.

Antiviral Compounds of Natural Origin

18

Patricia Veiga-Crespo, Miquel Viñas and Tomás G. Villa

Abstract

A great variety of plant- and medicinal herb-derived active principles have been traditionally known by the natural medicine of many civilizations. Recent advances in molecular biology have allowed us to characterize those active principles, to understand their mode of action and finally to include many of these compounds in the arsenal of antiviral drugs used by Western medicine. The higher number of resistant strains both in the bacterial and viral worlds makes mandatory the continuous search for new active principles to treat diseases. Many of the natural remedies used in medicine are now being characterized and will facility their use as regular chemotherapeutics.

Introduction

Antiviral compounds are drugs designed to treat viral infections. Normally, antivirals are specific and inhibit one or several steps in the viral replication cycle, interfering with the viral ability to copy itself and multiply. Therefore, together with antibacterial, antifungal, and antiparasitic compounds, they belong to the group of antimicrobials.

Viral illnesses encompass a wide range of diseases, from AIDS to influenza, hepatitis and herpes virus infections to various types of cancer. These pathologies are responsible for high medical costs around the world as well as an important part of patient morbidity.

Viruses, together with viroids, are the simplest life forms; they are at the frontier of life, as they are not able to replicate by themselves, but need to infect host cells and use their cellular machinery to their own advantage. This close relationship between hosts and viruses makes it difficult to design effective antiviral drugs, able not only to inhibit viral replication, but causing the lowest possible damage to the host cell.

New antivirals: production and development

The fundamental strategies against virus infections are threefold: vaccines, antiviral drugs and substances that stimulate the host's immune system.

Vaccines are the classical weapons in the fight against viruses, and are prepared either with attenuated versions of the virus or with inactive viral or subviral particles. This is a strategy usually with low undesirable side-effects, except for a few unfortunate cases, such as the isolated cases reported from vaccines against human papillomavirus (HPV). Vaccines are effective only when applied to uninfected people, but they are not useful in already infected patients. Additionally, this immune-therapy is more effective on stable virus, while vaccines are not very successful when dealing with viruses exhibiting high rates of mutation, such as the influenza virus and the human immunodeficiency virus (HIV).

In the past, there were no specific treatments for people infected with a virus, only palliative drugs to treat the symptoms while the illness followed its natural course. One of the major drawbacks in the development of antivirals has been the lack of precise knowledge about the virus's life cycle and viral genetics. For this reason, the first antiviral drugs did not start being developed until the

1960s. Viruses are a heterogeneous group, with different genetic material, RNA viruses (ssRNA, dsRNA), DNA viruses (dsDNA, ssDNA), with or without envelope, and with different capsid composition. This diversity notably increases the difficulty in designing specific drugs for these life forms. The advances in the knowledge of the molecular biology of viruses have allowed researchers to move a step forward in the design of more effective viral drugs.

Interferon molecules are the major immune system stimulators. In particular, α-interferon is a well-established drug for the treatment of hepatitis B and C (Burra, 2009) but other interferons are also utilized for this purpose.

Virus life cycle

Virus life cycles are complex and vary considerably among the different groups, but the fundamental steps include: (i) viral attachment to a host cell; (ii) penetration inside the cell and takeover of the host-cell machinery; (iii) replication of viral components using the host-cell machinery; (iv) assembly of viral components into complete viral particles; and, finally, (v) viral particle release.

Most of the antiviral drugs are designed to interfere one or more of these five steps, in order to interrupt the viral replication process and stop the infection from spreading by inhibiting the release of new viral progeny.

Viruses tend to undergo a high mutation rate, usually higher than that of the rest of living organisms, and this favours their becoming resistant to drugs or antiviral vaccines. For example, the seasonal influenza virus strains incorporate mutations in the neuraminidase protein (H274T), thus making it resistant to antivirals such as Tamiflu (Soundararajan *et al.*, 2009). In 1966, amantadine was approved for chemoprophylaxis of influenza A (H2N2) in the USA; in 1993, rimantadine was also approved. Both of them were recognized as adamantanes with activity against influenza A viruses but not against influenza B strains. In 2002, two other drugs could be administered against influenza: zanamivir and osealtamivir, approved in the USA in 1999 and 2000, respectively (CDC, 2002a). However, in the 2005–2006 seasonal influenza, the Health Alert not longer recommended the use of adamantanes, instead the use of neuraminidase inhibitors was recommended (CDC, 2002b).

The discovery of acyclovir, a nucleoside analogue of guanine, and its application in the treatment of herpesvirus infections during the last century, highlighted the beginning of a new therapeutical era, as this compound turned out to be highly specific and showed a low cytotoxicity (King, 1988). Another nucleoside analogue was, in 1987, the first antiviral approved for HIV treatment, zidovudine (AZT), as it clearly delayed the start of the illness. This compound is still used in multidrug antiretroviral regimes, in order to avoid the viral resistance to the treatment. Lamivudine is an antiviral, which acts against both HIV and HBV, also used in multidrug antiretroviral treatment. Many antiviral drugs have been developed to treat HIV and associated infections and, doubtless, they represent one of the main advances in the field of chemotherapy. For instance, fomivirsen is used in the treatment of cytomegalovirus (CMV)-caused retinitis in immune-compromised patients, including those with AIDS. It was licensed by the FDA, for CMV treatment, in August 1998 and, in fact, it represented the first antisense RNA antiviral approved by the FDA (Roush, 1997).

As it is the case with some drugs, some antiviral treatments are licensed to be used in Europe, but no in the USA. This is the case for Pleconaril, approved for treatment of picornavirus respiratory infections in Europe, but still in phase II of Clinical Trials in the USA (NCT00394914.2007, 2007).

Natural antiviral drugs

The search for new rational and effective drugs, together with the advances in molecular genetics, focused the attention of researchers in finding antiviral drugs in nature, for example, as reported by Lindequist *et al.* (2005) the presence of antiviral particles in mushrooms. In fact, Suzuki *et al.* (1989) had previously described the isolation of a protease inhibitor from shiitake mushrooms (*Lentinus edodes*) that was effective against HIV *in vitro*.

The strategies used to find natural antiviral drugs are diverse. One of the most commonly used, is the screening of a high number of plant extracts from different origins. For example,

Woo *et al.* (1997) tested more than 300 extracts, obtained by solvent fraction, from sixty plants and herbal medicines. The extracts were tested against Herpex simplex virus type I (HSV-1) and type II (HSV-2). From this study, they identified antiviral activity in several of the extracts, from the plants *Caragana sinica, Acer okamotoanum, Veratrum patulum* and *Osmunda japonica*.

Propolis, a substance manufactured by bees during the building of beehives, is another compound that displays antiviral activity against influenza, adenovirus, HIV and HSV (Amoros *et al.*, 1992; Huleihel and Isanu, 2002; Ito *et al.*, 2001; Schnitzler *et al.*, 2010; Serkedjieva *et al.*, 1992).

One of the current major hotspots is in the search for new anti-HIV drugs, in particular the finding of new viral inhibitors. The natural (+)-calanolide A is a non-nucleoside reverse transcriptase inhibitor of HIV replication. This drug was isolated from *Calophyllum lanigerum* trees in Malaysia and is produced by Sarawak MediChem for oral administration. This calanolide exhibits synergistic antiviral interactions with other nucleoside and non-nucleoside reverse transcriptase and protease inhibitors (Buckheit *et al.*, 2000). It was also found that dicaffeoylquinic acid is effective against HIV infection by acting on its integrase, an essential enzyme for HIV replication, active site (Robinson *et al.*, 1996). This drug, extracted from Chinese herbs, is now under clinical trials in phase I (Stellbrink, 2009).

The rapid emergence of HIV strains resistant to the drugs currently available implies that effective treatment therapies will require the use of a combination of drugs targeting different sites of the HIV life cycle. It is not a good strategy to increase the drug dose when dealing with drug-resistant viral strains. This strategy was applied in the fight against bacteria and is responsible for the drug-resistance problems that the worldwide healthcare systems are facing today. Another problem in the treatment of HIV infection is the undesirable side-effects caused by drugs, such as that found when using nucleoside analogues. Nucleoside analogues act by inhibiting the HIV reverse transcriptase and play a central role in the different therapeutical strategies used since the beginning of the AIDS epidemic, as a matter of fact, they are present in the majority of the pharmaceutical combination treatments currently used. However, long-term treatment with these compounds causes alterations in the mitochondrial polymerase, DNA polymerase γ (Anderson, 2010) and malfunction of this enzyme is related to severe illnesses, such as Parkinson disease.

The *in vitro* modification of natural compounds to use in a mix natural-synthetic strategy should be one of the main sources of successful future treatments. For example, polymenes are cationic polysaccharides synthesized by conjugation of various oligoamines to oxidized polysaccharides by reductive amination. One of these compounds, dextran-propan-1,3-diamine (DPD), showed a potent antiviral activity against HSV-1 and HSV-2 (Yudovin-Farber *et al.*, 2009). Acridone alkaloids obtained from *Citrus*, *Glycosmis* or *Severinia* plants (members of the *Rutaceae* family) were first compounds tested for their antimalarial activities *in vitro* and *in vivo* (Fujioka *et al.*, 1989), but later analyses showed the ability of these compounds, natural or synthetic, to act as antivirals (Sepulveda *et al.*, 2008). This kind of compounds is a good example of combinatorial synthesis, and of how this combined strategy can result in many new substances able to act against different viral targets, either DNA viruses, such as HSV, CMV, Epstein–Barr virus and adenovirus, or RNA ones, such as HIV or bovine viral diarrhoea virus. *Uncaria tomentosa* is a large woody vine native to the Amazon and Central American rainforest. This plant has been profusely used by indigenous people as a natural medicine. It shows immune-modulating, anti-inflammatory, cytotoxic, and antioxidant activities. The alkaloid extracted from *Uncaria tomentosa* displays two antiviral activities: reduction of the dengue virus-infected cell rates, and inhibition of cytokine activities, lowering TNF-α, IL-10 and IFN- α (Reis *et al.*, 2008).

Antivirals extracted from algae

When considering natural antivirals, a special attention must be paid to algae. The antibacterial activity of microalgae has been known for years, since Pratt and Fong reported such activity for *Chlorella vulgaris* as early as 1940 (Pratt and Fong, 1940). During the following years, several publications reported the antibacterial activity obtained from seaweed extracts. In 1958, Gerber *et al.* reported that algal polysaccharides had antiviral effects against mumps and influenza B viruses (Gerber *et al.*, 1958). Polysaccharides extracted from red algae were also able to inhibit the growth of virus such as HSV (Burkholder and Sharma, 1969). Witvrouw and De Clercq (1997) found, that sulfated polysaccharides from green algae were able to inhibit HIV replication. The best-known algal polysaccharides are the sulfated polysaccharides, such as fucoidan and carrageenans. They have been isolated from different green, brown and red algae as well as from cyanobacteria (Table 18.1 and Fig. 18.1). Carrageenans are mixtures of sulfated polysaccharides, extracted from different species of red seaweeds, which have a common structural backbone of D-galactose residues. These polysaccharides have a potent inhibitory effect over different HIV strains, in several human cell lines, and in peripheral blood lymphocytes (McClure *et al.*, 1992). These substances have mainly been tested against HIV, but they are also active against other viruses, such as HSV or influenza viruses. For example, three polysaccharides extracted from *Caulerpa brachypus* and *C. latum* showed a strong anti-HSV activity *in vitro*, being able to act at different replication stages (Lee *et al.*, 2004a). Diterpenes extracted from algae have also demonstrated their ability to inhibit HIV infections (Pereira *et al.*, 2004).

A spirulin-like substance extracted from *Arthrospira platensis* is able to act on CMV, HSV type 1, HSV type 6 or HIV viruses, but not against Epstein–Barr virus or influenza virus (Rechter *et al.*, 2006). So, these substances are strong candidates for broad-spectrum antiviral drugs. This broad spectrum is also observed in the extracts from *Spirulina maxima*: with antivirals effective against HSV, CMV, and HSV virus; whereas they are inactive against measles, sclerosing panencephalitis, and vesicular stomatitis virus, as well as poliovirus 1 (Hernandez-Corona *et al.*, 2002).

Since the beginning of human history, plants have been used as sources of a variety of natural remedies, and throughout the years many efforts have been made to find the active principles of their curative properties and antiviral agents.

Table 18.1 Antiviral extracts from algae

Species	Compound	Reference
Arthospira platensis	Spirulan-like molecules	Rechter *et al.* (2006)
Caulerpa brachypus and *C. latum*	Sulfated polysaccharides	Lee *et al.* (2004b)
Cochlodinium polykrikoides	Sulfated polysaccharides	Hasui M *et al.* (1995)
Dictyota menstrualis	Diterpenes	Pereira *et al.* (2004)
Fucus vesiculosus	Polysaccharides, fucoidan, polyphenols	Beress *et al.* (1993)
Microcystis aeruginosa	Aqueous extract	Nowotny *et al.* (1997)
Nostoc ellipsosporum	Cyanovirn-N	Boyd *et al.* (1997)
Nostoc ellipsosporum	Cyanovirin-N	Dey *et al.* (2000)
Nothogeniafastigiatia	Sulfated xylomannan	Damonte *et al.* (1994)
Oscillatoria spp. (TAU IL-76-12); *Scytonema* spp. (TAU SL-30-1-4); *Phormidum tenue* (TAU IL-144-1); *Ocscillatoria trichoides* (TAU IL-104-3-2); *Oscillatoria linnetica* (TAU NG 4-1-2)	Sulfolipids, sulfoglycolipids, hydrolysis products, synthetic derivates	Loya *et al.* (1998)
Peyssonelia sp.	Peyssonol A, Peyssonol B	Loya *et al.* (1998)
Porphyridium sp.	Sulfated polysaccharide	Huleihel and Arad (2001)
Sargassum horneri	Sulfated polysaccharide (fucose)	Hoshino *et al.* (1998)
Schizymenia pacifica	Polysaccharide	Nakashima *et al.* (1987)
Spirulina platensis	Calcium spirulin	Hayashi *et al.* (1996)
Spirulina platensis (now *Arthrospira platensis*)	Polysaccharide fraction	Ayehunie *et al.* (1998)

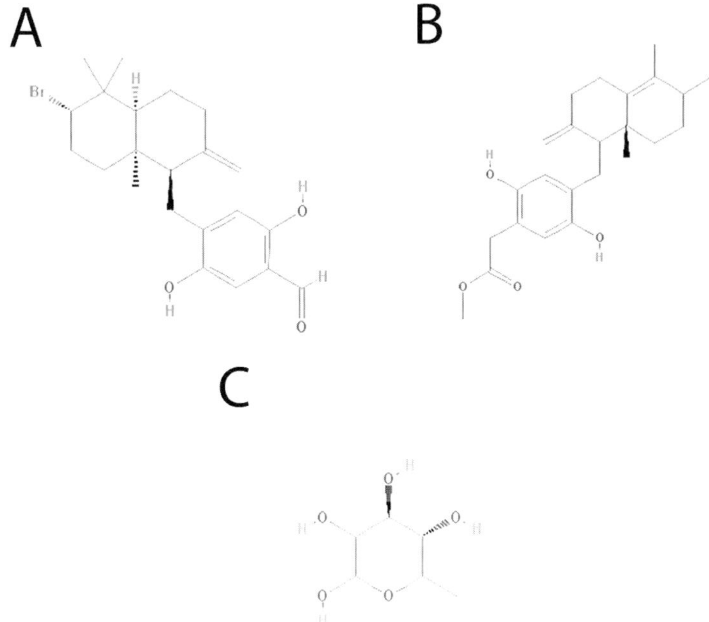

Figure 18.1. Structure of some antivirals extracted from algae. (A) peyssonol A; (B) peyssonol B; (C) fucose. A colour version of this figure is available in the plate section at the back of the book.

Many compounds from higher plants have been described that exert therapeutic activities, such as anti-thrombotic or anti-bacterial properties. One of the richest sources of natural clinical substances is the olive tree, *Olea europaea*. From its leaves it is possible to extract biophenols, capable of acting as antioxidants, antithrombotics, photoprotectives, or antimicrobials. One of the most abundant biophenols in olive tree leaves is oleuropein (Ole). Ole exhibits antiviral activity against Herpes mononucleosis, hepatitis virus, rotavirus, bovine rhinovirus, canine parvovirus, feline leukaemia virus, respiratory syncytial virus, parainfluenza type 3 virus, and salmonid haemorrhagic- septicaemia rhabdovirus (Fredickson, 2000; Ma *et al.*, 2001; Micol *et al.*, 2005).

The methanol-extracts from roots of *Byrsonima verbascifolia* possess a strong antiviral activity against HSV, but it is not active against poliovirus (Lopez *et al.*, 2001). These authors also penned an extensive review of potential antivirals in Colombian plants, and were able to find thirteen such compounds that inhibited HSV or poliovirus growth. Another noteworthy review was based in compounds obtained from Brazilian plants. In this study, *Arrabidaea samydoides* and *Callichlamys latifolia* contained compounds with the ability to act against Human herpes virus type 1, Vaccinia virus and murine encephalomyocarditis virus, whereas other plants analysed in their work showed activity against only one of the above mentioned viruses (Brandao *et al.*, 2010).

Pelargonium sidoides is a South African herbaceal plant whose root is commonly used in traditional medicine. It was shown that the aqueous root extract had a high activity against HSV. By mass spectrophotometry (LC-MS), the composition of the extract was shown to be predominantly coumarins, phenolic structures, flavonoids and catechin derivates (Schnitzler *et al.*, 2008a).

Essential oils have been commonly used in traditional medicine, so the antiviral effects of these essentials are being studied. For example, essential oils from dwarf-pine (*Pinus mugo*), anise (*Illicium verum*) and chamomile (*Matricaria recutita*) are able to inhibit the infection of HSV, because they are capable of inhibiting internalization of the viruses (Koch *et al.*, 2008). Additionally, *Melissa oficinalis* essential oils are able to alter the infectivity of the enveloped herpesvirus (Schnitzler *et al.*, 2008b). Essential oils from *Melaleuca alternifolia* and eucalyptus exhibited a high level of antiviral activity against HSV-1 and HSV-2, in viral suspension tests (Schnitzler *et al.*, 2001). Other essential oils with anti-HSV activity were those obtained from *Santolina insularis*, able to inhibit cell-to-cell transmission of HSV-1 and HSV-2 (De Logu *et al.*, 2000), and those from *Santalum album*, which were able to inhibit HSV-1 (Benencia and Courreges, 1999).

Other plant compounds, such as *Galanthus nivalis* agglutinin, have a high potential in antiviral drug development. This agglutinin, together with nelfinavir, was able to act, in a synergistic manner, against feline coronavirus (Hsieh *et al.*, 2010).

The list of potential antivirals from higher plants is high, with the number increasing everyday. The majority of these studies are inspired in the knowledge accumulated by traditional natural medicine. Oriental natural remedies, in particular, represent a mostly untapped source of compounds, many of which are unknown in western medicine. For example, wasabi (*Wasabia japonica*) is a widely used plant in traditional Japanese food and medicine. Ethanol extracts from its leaves have demonstrated high activity against influenza virus, inhibiting its replication (Kobayashi *et al.*, 2008).

The use of natural antivirals is also providing new antiviral compounds targeting different viral enzymes and replication stages. In the case of influenza virus, plants extracts are a source of neuroaminidase (NA) inhibitors. The Institute of Materia Medica from the Chinese Academy of Medical Science has tested more than 10,000 plant extracts, and found that *Elsholtzia rugulosa* (*Lamiaceae*) produces one of the most potent known inhibitors of the influenza virus neuraminidase enzyme.

Analysis of the extracts revealed that these inhibitors are flavonoids, in particular, apigenin, luteolin, apiin, galuteolin and luteolin 3'-glucuronyl acid methyl ester, and that their activity varies according to the number of –OH groups contained in their flavonoid β-ring (Liu *et al.*, 2008). Flavonoids (Alan and Wilkinson, 1997) are, without any doubt, one of the most interesting compounds due to their antiviral potential (Table 18.2 and Fig. 18.2). According its basic structure flavonoids can be classified into three different subclasses (Alan and Wilkinson, 1997): flavones, derived from 2-phenyl-1,4-benzopyrone; isoflavonoids, derived from 3-phenyl-1,4-benzopyrone; and neoflavonoids, derived from 3-phenylchromen-4-one.

Ethanol extracts from the *Chaenomeles sinensis* fruit, a traditional Chinese medicine for throat diseases, contain polyphenols, with the ability to act against influenza virus, by y suppressing NS2 protein synthesis (Sawai *et al.*, 2008). The hydrolysable tannins shephagenin A and B, extracted from the *Shepherdia argentea* leafs, showed anti-HIV-1 activity by inhibiting the viral reverse transcriptase (Yoshida *et al.*, 1996). Other polyphenolic-rich extracts from *Cistus incanus* showed a high activity against influenza virus, without negative side-effects or development of virus resistance (Ehrhardt *et al.*, 2007). *Acacia confuse* is indigenous to Taiwan and has been used as a traditional medicine, such as wound healing and anti-blood-stasis. A partially purified fraction, ACSB-M4, extracted with n-butanol-methanol displayed a high activity against hepatitis C virus

Table 18.2 Flavonoids with antiviral activity

Plant	Flavonoid	Reference
Achillea millefolium	Apigenin	Saeidnia *et al.* (2011)
Ginkgo biloba	Amentoflavone	Shuang-Cheng *et al.* (2001)
Silybum marianum	Silymarin	Ferenci *et al.* (2008)
Rhus spp.	Dihydrofisetin	Cushnie and Lamb (2005)
Alpinia officinarum	Galangine	Jullian *et al.* (2010)
Equisetum arvense	Luteolin	Liu *et al.* (2008)
Glycine max	Genistein	Vela *et al.* (2008)

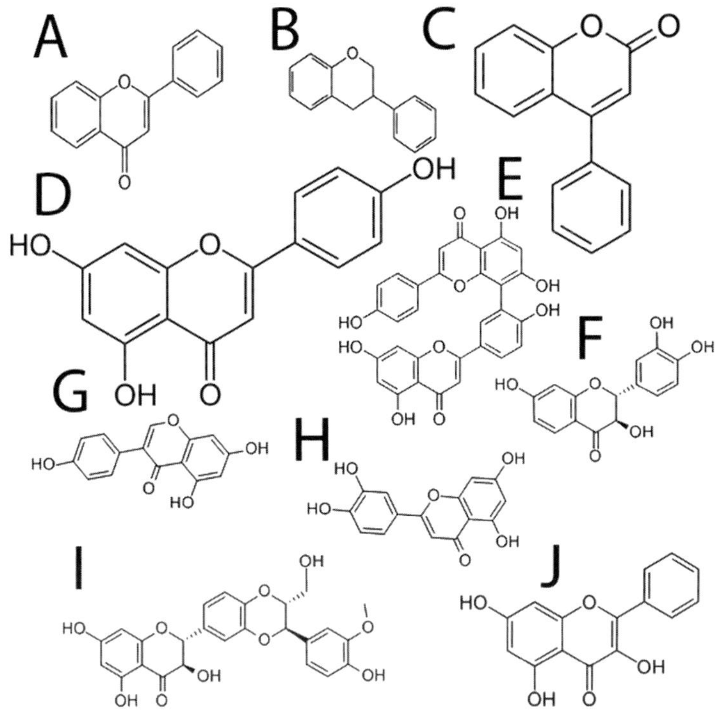

Figure 18.2 Structure of flavonoids. (A) 2-pheny-1,4-benzopyrone flavones basic structure. (B) 3-phenyl-1,4-benzopyrone isoflavonoids basic structure. (C) 4-phenyl-1,2-benzopyrone neoflavonoids basic structure. (D) apigenin. (E) amentoflavone. (F) dihydrofisetin. (G) genistein. (H) luteolin. (I) silymarin. (J) galangine.

(HCV), which is one of the major chronic pathogens worldwide (Lee *et al.*, 2011). *Scutellaria radix* major active ingredient, wogonin, is mainly active against hepatitis B virus (HBV). This plant has been commonly used in Chinese traditional medicine for centuries (Guo *et al.*, 2007). Curcumin (diferuloylmethane) is a polyphenolic compound widely used as a spice. It is derived from *Curcuma longa* and is able to reduce the yield of influenza virus in cell culture by inhibiting its haemagglutinin (Chen *et al.*, 2010).

More than 200 lignans with antiviral activity have so far been identified, indicating a widespread distribution of antivirals in the plant kingdom. Lignans isolated from *Larrea tridentata* are active against HIV (Gnabre *et al.*, 1996), whereas the lignans from *Rhinacanthus nasutus* are active against influenza virus. Different classes of Schirazins, lignans active against hepatitis virus, were extracted from *Kadsura matsudai* plants (Kernan *et al.*, 1997). Kadsurindutins A and B are dibenzocyclo-octane-lignans, isolated from *Kadsura induta*, and capable of inhibiting hepatitis B virus (Kuo *et al.*, 2001). Recently, two new lignans, as well as others already known, were recovered from *Illicium henryi*, and showed anti-Hepatitis B activity (Ma *et al.*, 2007). Justicidins represent another type of lignans, isolated from *Justicia procumbens* var. *leucantha*, and their related glycosides are active against virus such as vesicular stomatitis virus (VSV), and show low cytotoxicity in *in vitro* cultures of lung cells (Asano *et al.*, 1996).

Saponins are terpenoidic compounds, with a widespread distribution, that are able to act against a variety of viruses, with their different viral targets depending on their chemical structure. For example, saponins, isolated from *Anagallis arvensis*, act on HIV and polioviruses (Amoros *et al.*, 1987). Saponins are sugar-conjugates triterpenes of plant origin. Hence, the differences in their activity and in their viral target are mainly due to the nature of the conjugate. Oleanane represents a group of saponins that are able to inhibit DNA synthesis in viruses such as HSV-1; whereas the ursane group members inhibit HSV-1 viral capsid protein synthesis (Simoes *et al.*, 1999). Betulinic acid is a naturally occurring pentacyclic triterpenoid, which has anti-retroviral, anti-malarial, and anti-inflammatory properties. Today a semisynthetic drug derivative is by far one of the most promising compounds in the fight against HIV infections. This compound, called PA-457, was developed by Panacos Pharmaceuticals and is now under clinical trial evaluation (Butler, 2005).

As it is the case with synthetic drugs, natural extracts can also show undesirable side-effects such as cytotoxicity. For example, a liquid deodorized rosemary extract and an oily extract from a mixture of Provençal herbs were shown to be active against HIV infection, but both extracts showed a cytotoxic effect promoting DNA damage. This cytotoxic effect was eliminated when the active principle carnasol was removed from the samples (Aruoma *et al.*, 1996). However, the cytotoxicity of natural plant extracts is often lower than that from other drugs, such as Ribavirin, which is the only drug so far approved for the treatment of human respiratory syncytial virus (RSV) infections. This preparation shows a high toxicity and, hence, its use as a drug is very limited. Caffeoylquinic acids from *Schefflera heptaphylla* (L.) Frodin, can be very effective against RSV, since they have low undesirable side-effects (Li *et al.*, 2005).

The triterpene glycoside glycyrrhizic acid (GL) and its aglycone 18-β-glycyrrhetinic acid are the most intensively investigated bioactive compounds from liquorice roots (*Glycyrrhiza radix*). GL is able to protect human cells from influenza A virus infection, thus reducing virus uptake (Wolkerstorfer *et al.*, 2009) and also interferes with the early steps of Epstein–Barr virus infection (Lin *et al.*, 2008).

In summary, thousands of plant and herbal species and their antiviral potential activities have been studied so far, in order to unravel the nature of their active principles. All these studies have revealed a wide structural variety of compounds, including flavonoids, terpenoids, lignans, sulphides, polyphenolics, coumarins, saponins, furyl compounds, alkaloids, essential oils and others (Table 18.3).

Melaleuca alternifolia (tea tree) oil has an interesting antiviral activity against influenza A virus, with terpinen-4-ol, terpinoleene and alfa-terpineol being the main compounds responsible for this action (Garozzo *et al.*, 2011). In tea plants, hydrolysable tannins, such as strictinin and theogallin, are present as minor polyphenols. Strictin inhibits the replication of the Influenza A virus *in vitro*, and also inhibits the replication of influenza B virus and human parainfluenza virus type-1 (Saha *et al.*, 2010). Green tea, as well as other types of tea, is one of the major natural sources of active principles, with one of the main compounds being the catechins, in particular epigallocatechin-3-gallate (EGCG). ECCG displays anti-metastatic effects, and is also able to inhibit HIV, influenza and adenovirus infections (Weber *et al.*, 2003). Treatment of HBV with green tea extract (GTE), resulted in a surprising finding: EGCG was the major component of the extract, however the anti-HBV activity of EGCG was weaker than that of the complete extract, so the minor components were the most effective against this particular virus (Xu *et al.*, 2008).

In general, the above-mentioned antiviral activity is due to the polyphenol rosmarinic acid, and the low-molecular glycoside-forming compounds chlorogenic acid and caffeic acid and their derivatives. These phytochemicals are able to act at different sites in the virus life cycle, from genetic material replication to agglutination. One of the main subjects for further investigation must be the characterization of the exact molecular pathways these plant extracts act against the different viruses and the pharmacopoeia of natural principles they encompass.

Table 18.3 Main classes of plant antiviral compound and their principal targets

Antiviral	Target
Furyl compounds	DNA and RNA genomes
Alkaloids	DNA and virion proteins
Polysaccharides	Interferes with viral binding
Thiophenes	Membrane interactions
Flavonoid	RNA interactions
Terpenoids	Membrane interactions; DNA synthesis
Lignans	Different actions depending of the structure

Additional sources of natural antiviral compounds

Plants are usually our first thought when thinking about natural drugs, but higher vertebrates must also be considered. For example, camelids have been found to possess single-domain antigen-binding fragments, derived from heavy chain antibodies, with anti-polivirus activity (Thys *et al.*, 2010).

Chitosan is a linear polysaccharide, composed of randomly distributed β-(1-4)-linked D-glucosamine (deacetylated unit) and *N*-acetyl-D-glucosamine (acetylated unit), derived from the shells of shrimp and other crustaceans. It is obtained by deacetylation of chitin, which is the structural element in the exoskeleton of crustaceans, as well as in the cell walls of fungi.

Chitosan and its derivatives have been used in molecular and cell biology as gene deliverers. Chitosan oligomers are non-toxic and show good properties for gene delivery.

The role of chitosan in antiviral research has included both its use as a delivery vehicle and its truly antiviral effect. It has been established that chitosan strongly inhibits bacteriophages. Kochkina and Chirkov (2000) described the effect of chitosan fragments, with different degrees of polymerization as well as some derivatives, on *Bacillus thuringiensis* infected by phage 1–97A. Chitosan inhibits phage infection as well as inactivating phage particles. The extent of inhibition of phage infection is inversely dependent on the degree of polymerization of the chitosan fragments. On the other hand, the extent of inactivation of phage virulence is proportional to the degree of chitosan polymerization. In some cases, this compound has been used as a tool for virus removal from food. Su *et al.* (2009) investigated the effect of chitosan on three human enteric viral surrogates: murine norovirus 1 (MNV-1), feline calicivirus F-9 (FCV-F9), and (ssRNA) bacteriophage MS2 (MS2; using different formulations and drug concentrations, they obtained viral reductions ranging from 1.81 to 2.83 log PFU/ml, depending on the drug concentration. But the treatment had no effect on MNV-1 when the virus was found at high concentration. The authors claimed that their results in plaque reduction of human enteric virus surrogates by chitosan treatment, showed promise in its potential application as food treatment. Suckling mice injected with chitosan granules survived much longer than the control group, when challenged with foot-and-mouth disease virus (FMDV) and this anti-FMDV capacity was maintained for 1 week, although it was at its peak on the third day (Li, 2010).

Recently, it has been proposed that the antiviral activity of chitosan strongly depends on the structure and de-polymerization method used (Davydova *et al.*, 2011). This was analysed by determining the antiviral affect of low molecular derivatives of chitosan against the tobacco mosaic virus (TMV). The antiviral activity of LM chitosan significantly increased with the lowering of their polymerization degree. It was also revealed that the antiviral activity exhibited weakly depended on the degree of acetylation of the samples.

Influenza virus attaches to the host cell surface through the haemagglutinin (HA) protein on its surface, HA binds glycoligands, such as sialyllactose (SL). The synthesis and development of SL-containing chitosan compounds, either as water-soluble polymers or as functional fibres, have demonstrated their capability for viral adhesion and inhibition, as well as their use in decontamination. These syntheses were accomplished by grafting a lactoside-bearing and an aldehyde-functionalized aglycone to the amino groups of chitosan, or chitosan fibres, followed by enzymatic sialylation with sialyltransferase. The obtained water-soluble SL–chitosan conjugate bound HA with high affinity and effectively inhibited the viral attachment to host erythrocytes. Moreover, the SL-containing chitosan fibres efficiently removed the virus from an aqueous medium. The results, taken together, demonstrate the potential use of the new materials as virus adsorbents for the prevention and control of influenza (Davydova *et al.*, 2011).

Ganciclovir (GCV) is extensively used in the treatment of ocular viral infections. It is a toxic drug, since high doses results in bone marrow suppression and neutropenia. It has been shown that mucoadhesive nano-formulations for the topical ocular delivery of Ganciclovir including chitosan nanoparticles (GCV-NPs), resulted to be non-irritant and nontoxic (Akhter *et al.*, 2011). Thus, one can speculate on the future use of chitosan as an antiviral component of future drugs, alone or in combinations with many other molecules, both of natural or synthetic origin.

Microbial polysaccharides, such as chitosan, have received special attention as antiviral treatments over the years. The focus was on compounds such as sulfated polysaccharides, chitosan, chitin and curdlan, because of their ability to interfere with the HIV life cycle. Sulfated polysaccharides have been described as inhibitors of enveloped viruses, such as HSV, CMV, vesicular stomatitis virus and HIV (Baba *et al.*, 1988).

Curdlan is a linear polysaccharide composed of β-1,3-D-glucose units, produced by *Alcaligenes faecalis* and some strains of *Agrobacterium* (Gao *et al.*, 2008; Nikanishi *et al.*, 1976). This polymer and its derivatives show a strong anti-HIV activity, together with and anti-tumour activity. Curdlan sulfates possess different anti-HIV activities, with sulfated curdlan having a stronger action than sulfopropyl curdlan. Actually, the different antiviral and cytotoxic effects are due to their heterogenic structures. L-glycosyl-curdlan sulfates exhibit the highest anti-HIV activity, as well as the lowest cytotoxicity (Yoshida *et al.*, 1990).

Sulfated compounds frequently appear to exhibit higher antiviral activity, as is the case of the highly sulfated heparan sulfate, extracted from shrimp (*Penaeus brasiliensis*) heads. This molecule has a greater inhibitory effect than heparin against infections by dengue virus type 2 as well as Japanese encephalitis virus (Chen *et al.*, 2011).

Hence, the natural sources of antiviral compounds are still good sources for weapons in the fight against viral diseases as well as against resistant viruses. In depth studies on active principles, ways of action and potential chemical modifications are still needed in order to obtain higher efficiency and better antiviral activities.

Acknowledgements

The authors wish to thank Dr Angeles Sánchez-Pérez, The University of Sydney, Australia, for editing the English manuscript.

References

Akhter, S., Talegaonkar, S., Khan, Z.I., Jain, G.K., Khar, R.K., and Ahmad, F.J. (2011). Assessment of ocular pharmacokinetics and safety of Ganciclovir loaded nanoformulations. J. Biomed. Nanotechnol. *7*, 144–145.

Alan, D., and Wilkinson, A. (1997). Flavonoids (isoflavonoids and neoflavonoids). In UPAC Compendium of Chemical Terminology, IUAPC, ed. (Blackwell Scientific, Oxford, UK).

Amoros, M., Fauconnier, B., and Girre, R.L. (1987). *In vitro* antiviral activity of a saponin from Anagallis arvensis, Primulaceae, against herpes simplex virus and poliovirus. Antiviral Res. *8*, 13–25.

Amoros, M., Simoes, C.M., Girre, L., Sauvager, F., and Cormier, M. (1992). Synergistic effect of flavones and flavonols against herpes simplex virus type 1 in cell culture. Comparison with the antiviral activity of propolis. J. Nat. Prod. *55*, 1732–1740.

Anderson, K.S. (2010). A transient kinetic approach to investigate nucleoside inhibitors of mitochondrial DNA polymerase gamma. Methods *51*, 392–398.

Aruoma, O.I., Spencer, J.P., Rossi, R., Aeschbach, R., Khan, A., Mahmood, N., Munoz, A., Murcia, A., Butler, J., and Halliwell, B. (1996). An evaluation of the antioxidant and antiviral action of extracts of rosemary and Provencal herbs. Food Chem. Toxicol. *34*, 449–456.

Asano, J., Chiba, K., Tada, M., and Yoshii, T. (1996). Antiviral activity of lignans and their glycosides from Justicia procumbens. Phytochemistry *42*, 713–717.

Ayehunie, S., Belay, A., Baba, T.W., and Ruprecht, R.M. (1998). Inhibition of HIV-1 replication by an aqueous extract of Spirulina platensis (Arthrospira platensis). J. Acquir. Immune Defic. Syndr. 7–12.

Baba, M., Snoeck, R., Pauwels, R., and de Clercq, E. (1988). Sulfated polysaccharides are potent and selective inhibitors of various enveloped viruses, including herpes simplex virus, cytomegalovirus, vesicular stomatitis virus, and human immunodeficiency virus. Antimicrob. Agents Chemother. 32, 1742–1745.

Benencia, F., and Courreges, M.C. (1999). Antiviral activity of sandalwood oil against herpes simplex viruses-1 and -2. Phytomedicine *6*, 119–123.

Beress, A., Wassermann, O., Tahhan, S., Bruhn, T., Beress, L., Kraiselburd, E.N., Gonzalez, L.V., de Motta, G.E., and Chavez, P.I. (1993). A new procedure for the isolation of anti-HIV compounds (polysaccharides and polyphenols) from the marine alga Fucus vesiculosus. J. Nat. Prod. *56*, 478–488.

Boyd, M.R., Gustafson, K.R., McMahon, J.B., Shoemaker, R.H., O'Keefe, B.R., Mori, T., Gulakowski, R.J., Wu, L., Rivera, M.I., Laurencot, C.M., *et al.* (1997). Discovery of cyanovirin-N, a novel human immunodeficiency virus-inactivating protein that binds viral surface envelope glycoprotein gp120: potential applications to microbicide development. Antimicrob. Agents Chemother. *41*, 1521–1530.

Brandao, G.C., Kroon, E.G. dos Santos, J.R., Stehmann, J.R., Lombardi, J.A., and Braga de Oliveira, A. (2010). Antiviral activity of Bignoniaceae species occurring in the State of Minas Gerais (Brazil): part 1. Lett. Appl. Microbiol. *51*, 469–476.

Buckheit, R.W., Jr., Russell, J.D., Xu, Z.Q., and Flavin, M. (2000). Anti-HIV-1 activity of calanolides used in combination with other mechanistically diverse inhibitors of HIV-1 replication. Antiviral Chem. Chemother. *11*, 321–327.

Burkholder, P.R., and Sharma, G.M. (1969). Antimicrobial agents from the sea. Lloydia *32*, 466–483.

Burra, P. (2009). Hepatitis C. Seminars in liver disease *29*, 53–65.

Butler, M.S. (2005). Natural products to drugs: natural product derived compounds in clinical trials. Nat. Prod. Rep. *22*, 162–195.

CDC (2002a). 00240 – 1/14/2006 – CDC Recommends against the Use of Amantadine and Rimantadine for the Treatment or Prophylaxis of Influenza in the United States during the 2005–06 Influenza Season (Centers for Disease Control and Prevention, Atlanta, GA) (http://emergency.cdc.gov/HAN/han00240.asp).

CDC (2002b). Prevention and control of influenza: recommendations of the Advisory Committee on Immunization Practices. MMWR *51*, 23.

Chen, D.-Y., Shien, J.H., Tiley, L., Chiou, S.S., Wang, S.-Y., Chang, T.J., Lee, Y.-J., Kun-Wei, C., and Hsu, W.-L. (2010). Curcumin inhibits influenza virus infection and haemagglutination activity. Food Chem. *119*, 1346–1351.

Chen, J., Yamada, S., Hama, Y., Shetty, A.K., Kobayashi, T., Oda, H., Seiki, K., Kim, E., Kimura, T., Takahashi, N., *et al.* (2011). Unique heparan sulfate from shrimp heads exhibits a strong inhibitory effect on infections by dengue virus and Japanese encephalitis virus. Biochem. Biophys. Res. Commun. *412*, 136–142.

Cushnie, T.P., and Lamb, A.J. (2005). Antimicrobial activity of flavonoids. Int. J. Antimicrob. Agents. *26*, 343–356.

Damonte, E., Neyts, J., Pujol, C.A., Snoeck, R., Andrei, G., Ikeda, S., Witvrouw, M., Reymen, D., Haines, H., Matulewicz, M.C., *et al.* (1994). Antiviral activity of a sulphated polysaccharide from the red seaweed Nothogenia fastigiata. Biochem. Pharmacol. *47*, 2187–2192.

Davydova, V.N., Nagorskaia, V.P., Gorbach, V.I., Kalitnik, A.A., Reunov, A.V., Solov'eva, T.F., and Ermak, I.M. (2011). [Chitosan antiviral activity: dependence on structure and depolymerization method]. Prikl. Biokhim. Mikrobiol. *47*, 113–118.

De Logu, A., Loy, G., Pellerano, M.L., Bonsignore, L., and Schivo, M.L. (2000). Inactivation of HSV-1 and HSV-2 and prevention of cell-to-cell virus spread by Santolina insularis essential oil. Antiviral Res. *48*, 177–185.

Dey, B., Lerner, D.L., Lusso, P., Boyd, M.R., Elder, J.H., and Berger, E.A. (2000). Multiple antiviral activities of cyanovirin-N: blocking of human immunodeficiency virus type 1 gp120 interaction with CD4 and coreceptor and inhibition of diverse enveloped viruses. J Virol. *74*, 4562–4569.

Ehrhardt, C., Hrincius, E.R., Korte, V., Mazur, I., Droebner, K., Poetter, A., Dreschers, S., Schmolke, M., Planz, O., and Ludwig, S. (2007). A polyphenol rich plant extract, CYSTUS052, exerts anti influenza virus activity in cell culture without toxic side-effects or the tendency to induce viral resistance. Antiviral Res. *76*, 38–47.

Ferenci, P., Scherzer, T.M., Kerschner, H., Rutter, K., Beinhardt, S., Hofer, H., Schoniger-Hekele, M., Holzmann, H., and Steindl-Munda, P. (2008). Silibinin is a potent antiviral agent in patients with chronic hepatitis C not responding to pegylated interferon/ribavirin therapy. Gastroenterol. *135*, 1561–1567.

Fredickson, W.R. (2000). Method and composition for antiviral therapy with olive leaves. US Patent no 6,117,844.

Fujioka, H., Nishiyama, Y., Furukawa, H., and Kumada, N. (1989). *In vitro* and *in vivo* activities of atalaphillinine and related acridone alkaloids against rodent malaria. Antimicrob. Agents Chemother. *33*, 6–9.

Gao, F.P., Zhang, H.Z., Liu, L.R., Wang, Y.S., Jiang, Q., Yang, X.D., and Zhang, Q.Q. (2008). Preparation and physicochemical characteristics of self-assambled nanoparticles of deoxycholic acid modified carboxy–methyl-curdlan conjugates. Carbohydr. Polym. *71*, 606–613.

Garozzo, A., Timpanaro, R., Stivala, A., Bisignano, G., and Castro, A. (2011). Activity of Melaleuca alternifolia (tea tree) oil on Influenza virus A/PR/8: study on the mechanism of action. Antiviral Res. *89*, 83–88.

Gerber, P., Dutcher, J.D., Adams, E.V., and Sherman, J.H. (1958). Protective effect of seaweed extracts for chicken embryos infected with influenza B or mumps virus. Proc. Soc. Exp. Biol. Med. *99*, 590–593.

Gnabre, J.N., Ito, Y., Ma, Y., and Huang, R.C. (1996). Isolation of anti-HIV-1 lignans from *Larrea tridentata* by counter-current chromatography. J. Chromatogr., A *719*, 353–364.

Guo, Q., Zhao, L., You, Q., Yang, Y., Gu, H., Song, G., Lu, N., and Xin, J. (2007). Anti-hepatitis B virus activity of wogonin *in vitro* and *in vivo*. Antiviral Res. *74*, 16–24.

Hasui, M., Matsuda, M., Okutani, K., and Shigeta, S. (1995). *In vitro* antiviral activities of sulfated polysaccharides from a marine microalga (Cochlodinium polykrikoides) against human immunodefficiency virus and other enveloped viruses. Int. J. Biol. Macromol. *17*, 293–297.

Hayashi, T., Hayashi, K., Maeda, M., and Kojima, I. (1996). Calcium spirulan, an inhibitor of enveloped virus replication, from a blue-green alga Spirulina platensis. J. Nat. Prod. *59*, 83–87.

Hoshino, T., Hayashi, T., Hayashi, K., Hamada, J., Lee, J.B., and Sankawa, U. (1998). An antivirally active sulfated polysaccharide from Sargassum horneri (TURNER) C. AGARDH. Biol. Pharm. Bull. *21*, 730–734.

Hsieh, L.E., Lin, C.N., Su, B.L., Jan, T.R., Chen, C.M., Wang, C.H., Lin, D.S., Lin, C.T., and Chueh, L.L. (2010). Synergistic antiviral effect of Galanthus nivalis agglutinin and nelfinavir against feline coronavirus. Antiviral Res. *88*, 25–30.

Huleihel, M., and Arad, S. (2001). Effect of Porphyridium sp. polysaccharide on malignant cell transformation by Moloney murine sarcoma virus. Anticancer Res. *21*, 2073–2078.

Huleihel, M., and Isanu, V. (2002). Anti-herpes simplex virus effect of an aqueous extract of propolis. Israel Med. Ass. J. IMAJ *4*, 923–927.

Ito, J., Chang, F.R., Wang, H.K., Park, Y.K., Ikegaki, M., Kilgore, N., and Lee, K.H. (2001). Anti-AIDS agents.

48.(1) Anti-HIV activity of moronic acid derivatives and the new melliferone-related triterpenoid isolated from Brazilian propolis. J. Nat. Prod. *64*, 1278–1281.

Jullian, C., Alfaro, M., Zapata-Torres, G., and Olea-Azar, C. (2010). Inclusion Complexes of Cyclodextrins wiht Galangin: a Thermodynamic and Reactivity Study. J. Solution Chem. *39*, 1168–1177.

Kernan, M.R., Sendl, A., Chen, J.L., Jolad, S.D., Blanc, P., Murphy, J.T., Stoddart, C.A., Nanakorn, W., Balick, M.J., and Rozhon, E.J. (1997). Two new lignans with activity against influenza virus from the medicinal plant Rhinacanthus nasutus. J. Nat. Prod. *60*, 635–637.

King, D.H. (1988). History, pharmacokinetics, and pharmacology of acyclovir. J. Am. Acad. Dermatol. *18*, 176–179.

Kobayashi, T., Shimura, T., Araki, K., Ogawa, A., Mochida, Y., Suzuki, H., Suehiro, T., and Kuwano, H. (2008). Lymphoepithelial cyst of the pancreas: report of a case. Hepatogastroenterology *55*, 1107–1109.

Koch, C., Reichling, J., Kehm, R., Sharaf, M.M., Zentgraf, H., Schneele, J., and Schnitzler, P. (2008). Efficacy of anise oil, dwarf-pine oil and chamomile oil against thymidine-kinase-positive and thymidine-kinase-negative herpesviruses. J. Pharm. Pharmacol. *60*, 1545–1550.

Kochkina, Z.M., and Chirkov, S.N. (2000). [Effect of chitosan derivatives on the development of phage infection in cultured Bacillus thuringiensis]. Mikrobiologiia *69*, 266–269.

Kuo, Y.H., Li, S.Y., Huang, R.L., Wu, M.D., Huang, H.C., and Lee, K.H. (2001). Schizanrins [corrected] B, C, D, and E, four new lignans from Kadsura matsudai and their antihepatitis activities. J. Nat. Prod. *64*, 487–490.

Lee, J.B., Hayashi, K., Maeda, M., and Hayashi, T. (2004a). Antiherpetic activities of sulfated polysaccharides from Green Algae. Planta Med. *70*, 813–817.

Lee, J.B., Hayashi, K., Maeda, M., and Hayashi, T. (2004b). Antiherpetic activities of sulfated polysaccharides from Green Algae. Plant Med. *70*, 813–817.

Lee, J.C., Chen, W.C., Wu, S.F., Tseng, C.K., Chiou, C.Y., Chang, F.R., Hsu, S.H., and Wu, Y.C. (2011). Anti-hepatitis C virus activity of Acacia confusa extract via suppressing cyclooxygenase-2. Antiviral Res. *89*, 35–42.

Li, D. (2010). Chitosan can stop or postpone the death of the suckling mice challenged with foot-and-mouth disease virus. Virol. J. *7*, 125.

Li, Y., But, P.P., and Ooi, V.E. (2005). Antiviral activity and mode of action of caffeoylquinic acids from Schefflera heptaphylla (L.) Frodin. Antiviral Res. *68*, 1–9.

Lin, J.C., Cherng, J.M., Hung, M.S., Baltina, L.A., Baltina, L., and Kondratenko, R. (2008). Inhibitory effects of some derivatives of glycyrrhizic acid against Epstein–Barr virus infection: structure–activity relationships. Antiviral Res. *79*, 6–11.

Lindequist, U., Timo, H., Niedermeyer, J., and Julich, W.-D. (2005). The Pharmacological Potential of Mushrooms. Evid. Based Complement. Alternat. Med. *2*, 285–299.

Liu, A.L., Liu, B., Qin, H.L., Lee, S.M., Wang, Y.T., and Du, G.H. (2008). Anti-influenza virus activities of flavonoids from the medicinal plant Elsholtzia rugulosa. Planta Med. *74*, 847–851.

Lopez, A., Hudson, J.B., and Towers, G.H. (2001). Antiviral and antimicrobial activities of Colombian medicinal plants. J. Ethnopharmacol. *77*, 189–196.

Loya, S., Reshef, V., Mizrachi, E., Silberstein, C., Rachamim, Y., Carmeli, S., and Hizi, A. (1998). The inhibition of the reverse transcriptase of HIV-1 by the natural sulfoglycolipids from cyanobacteria: contribution of different moieties to their high potency. J. Nat. Prod. *61*, 891–895.

Ma, S.C., He, Z.D., Deng, X.L., But, P.P., Ooi, V.E., Xu, H.X., Lee, S.H., and Lee, S.F. (2001). *In vitro* evaluation of secoiridoid glucosides from the fruits of Ligustrum lucidum as antiviral agents. Chem. Pharm. Bull. *49*, 1471–1473.

Ma, W., Ma, X., Huang, H., Zhou, P., and Chen, D. (2007). Dibenzocyclooctane lignans from the stems of Kadsura induta and their antiviral effect on hepatitis B virus. Chem. Biodivers. *4*, 966–972.

McClure, M.O., Moore, J.P., Blanc, D.F., Scotting, P., Cook, G.M., Keynes, R.J., Weber, J.N., Davies, D., and Weiss, R.A. (1992). Investigations into the mechanism by which sulfated polysaccharides inhibit HIV infection *in vitro*. AIDS Res. Hum. Retroviruses *8*, 19–26.

Micol, V., Caturla, N., Perez-Fons, L., Mas, V., Perez, L., and Estepa, A. (2005). The olive leaf extract exhibits antiviral activity against viral haemorrhagic septicaemia rhabdovirus (VHSV). Antiviral Res. *66*, 129–136.

Nakashima, H., Kido, Y., Kobayashi, N., Motoki, Y., Neushul, M., and Yamamoto, N. (1987). Purification and characterization of an avian myeloblastosis and human immunodeficiency virus reverse transcriptase inhibitor, sulfated polysaccharides extracted from sea algae. Antimicrob. Agents Chemother. *31*, 1524–1528.

NCT00394914.2007, C.T.I. (2007). Effects of Pleconaril Nasal Spray on Common Cold Symptoms and Asthma Exacerbations Following Rhinovirus Exposure (Study P04295AM2) Clinical Trials gov. US National Institutes of Health.

Nikanishi, I., Kimura, K., Suzuki, T., Ishikawa, M., Banno, I., and Sakane, T. (1976). Demonstration of curdlan-type polysaccharide and some other B-1,3-glucan in mircroorganisms with aniline blau. J. Gen. Appl. Microbiol. *22*, 1–22.

Nowotny, A., Mentel, R., Wegner, U., Mundt, S., and Lindequist, U. (1997). Antiviral activity of an aqueous extract of the cyanobacterium Microcystis aeruginosa. Phytother. Res. *11*, 93–96.

Pereira, H.S., Leao-Ferreira, L.R., Moussatche, N., Teixeira, V.L., Cavalcanti, D.N., Costa, L.J., Diaz, R., and Frugulhetti, I.C. (2004). Antiviral activity of diterpenes isolated from the Brazilian marine alga Dictyota menstrualis against human immunodeficiency virus type 1 (HIV-1). Antiviral Res. *64*, 69–76.

Pratt, R., and Fong, J. (1940). Studies on Chlorella vulgaris. Further evidences that chlorella cells form a growth inhibiting substance. Am. J. Bot. 27, 341–436.

Rechter, S., Konig, T., Auerochs, S., Thulke, S., Walter, H., Dornenburg, H., Walter, C., and Marschall, M. (2006).

Antiviral activity of Arthrospira-derived spirulan-like substances. Antiviral Res. *72*, 197–206.

Reis, S.R., Valente, L.M., Sampaio, A.L., Siani, A.C., Gandini, M., Azeredo, E.L., D'Avila, L.A., Mazzei, J.L., Henriques, M., and Kubelka, C.F. (2008). Immunomodulating and antiviral activities of Uncaria tomentosa on human monocytes infected with Dengue Virus-2. Int. Immunopharmacol. *8*, 468–476.

Robinson, W.E., Jr., Cordeiro, M., Abdel-Malek, S., Jia, Q., Chow, S.A., Reinecke, M.G., and Mitchell, W.M. (1996). Dicaffeoylquinic acid inhibitors of human immunodeficiency virus integrase: inhibition of the core catalytic domain of human immunodeficiency virus integrase. Mol. Pharmacol. *50*, 846–855.

Roush, W. (1997). Antisense aims for a renaissance. Science *276*, 1192–1193.

Saeidnia, S., Gohari, A., Mokhber-Dezfuli, N., and Kiuchi, F. (2011). A review on phytochemistry and medicinal properties of the genus Achillea. Daru *19*, 173–186.

Saha, R.K., Takahashi, T., Kurebayashi, Y., Fukushima, K., Minami, A., Kinbara, N., Ichitani, M., Sagesaka, Y.M., and Suzuki, T. (2010). Antiviral effect of strictinin on influenza virus replication. Antiviral Res. *88*, 10–18.

Sawai, R., Kuroda, K., Shibata, T., Gomyou, R., Osawa, K., and Shimizu, K. (2008). Anti-influenza virus activity of *Chaenomeles sinensis*. Journal of ethnopharmacology *118*, 108–112.

Schnitzler, P., Schon, K., and Reichling, J. (2001). Antiviral activity of Australian tea tree oil and eucalyptus oil against herpes simplex virus in cell culture. DiePharmazie *56*, 343–347.

Schnitzler, P., Schneider, S., Stintzing, F.C., Carle, R., and Reichling, J. (2008a). Efficacy of an aqueous Pelargonium sidoides extract against herpesvirus. Phytomedicine *15*, 1108–1116.

Schnitzler, P., Schuchmacher, A., Astani, A., and Reichling, J. (2008b). Melissa officinalis oils affects infiectivity of enveloped herpes virus. Phytomedicine *15*, 734–740.

Schnitzler, P., Neuner, A., Nolkemper, S., Zundel, C., Nowack, H., Sensch, K.H., and Reichling, J. (2010). Antiviral activity and mode of action of propolis extracts and selected compounds. Phytother. Res.: PTR *24 Suppl 1*, S20–28.

Sepulveda, C.S., Fascio, M.L., Mazzucco, M.B., Palacios, M.L., Pellon, R.F., Garcia, C.C., D'Accorso, N.B., and Damonte, E.B. (2008). Synthesis and evaluation of N-substituted acridones as antiviral agents against haemorrhagic fever viruses. Antivir. Chem. Chemother. *19*, 41–47.

Serkedjieva, J., Manolova, N., and Bankova, V. (1992). Anti-influenza virus effect of some propolis constituents and their analogues (esters of substituted cinnamic acids). J. Nat. Prod. *55*, 294–302.

Shuang-Cheng, M.A., Pui-Hay, P., Eng-Choon, V., Yue-Hua, H., Spencer, H.-S., Song-Fong, L., and Rui-Chao, L. (2001). Antiviral Amentoflavone from Salaginella sinensis. Biol. Pharm. Bull. *24*, 311.

Simoes, C.M., Amoros, M., and Girre, L. (1999). Mechanism of antiviral activity of triterpenoid saponins. Phytother. Res.: PTR *13*, 323–328.

Soundararajan, V., Tharakaraman, K., Raman, R., Raguram, S., Shriver, Z., Sasisekharan, V., and Sasisekharan, R. (2009). Extrapolating from sequence – the 2009 H1N1 'swine' influenza virus. Nature Biotechnol. *27*, 510–513.

Stellbrink, H.J. (2009). Novel compounds for the treatment of HIV type-1 infection. Antivir. Chem. Chemother. *19*, 189–200.

Su, X., Zivanovic, S., and D'Souza, D.H. (2009). Effect of chitosan on the infectivity of murine norovirus, feline calicivirus, and bacteriophage MS2. J. Food Prot. *72*, 2623–2628.

Suzuki, H., Okubo, A., Yamazaki, S., Suzuki, K., Mitsuya, H., and Toda, S. (1989). Inhibition of the infectivity and cytopathic effect of human immunodeficiency virus by water-soluble lignin in an extract of the culture medium of Lentinus edodes mycelia (LEM). Biochem. Biophys. Res. Commun. *160*, 367–373.

Thys, B., Schotte, L., Muyldermans, S., Wernery, U., Hassanzadeh-Ghassabeh, G., and Rombaut, B. (2010). *In vitro* antiviral activity of single domain antibody fragments against poliovirus. Antiviral Res. *87*, 257–264.

Vela, E.M., Bowick, G.C., Herzog, N.K., and Aronson, J.F. (2008). Genistein treatment of cells inhibits arenavirus infection. Antiviral Res. *77*, 153–156.

Weber, J.M., Ruzindana-Umunyana, A., Imbeault, L., and Sircar, S. (2003). Inhibition of adenovirus infection and adenain by green tea catechins. Antiviral Res. *58*, 167–173.

Witvrouw, M., and De Clercq, E. (1997). Sulfated polysaccharides extracted from sea algae as potential antiviral drugs. Gen Pharmacol. *29*, 497–511.

Wolkerstorfer, A., Kurz, H., Bachhofner, N., and Szolar, O.H. (2009). Glycyrrhizin inhibits influenza A virus uptake into the cell. Antiviral Res. *83*, 171–178.

Woo, E.R., Kim, H.J., Kwak, J.H., Lim, Y.K., Park, S.K., Kim, H.S., Lee, C.K., and Park, H. (1997). Anti-herpetic activity of various medicinal plant extracts. Arch. Pharmacal Res. *20*, 58–67.

Xu, J., Wang, J., Deng, F., Hu, Z., and Wang, H. (2008). Green tea extract and its major component epigallocatechin gallate inhibits hepatitis B virus *in vitro*. Antiviral Res. *78*, 242–249.

Yoshida, T., Hatanaka, K., Uryu, T., Kaneko, Y., Suzuki, E., and Miyano, H. (1990). Synthesis and structural analysis of curdlan sulfate with a potent inhibitory effect *in vitro* of AIIDS virus infection. Macromolecules *23*, 3718–3722.

Yoshida, T., Ito, H., Hatano, T., Kurata, M., Nakanishi, T., Inada, A., Murata, H., Inatomi, Y., Matsuura, N., Ono, K., *et al.* (1996). New hydrolyzable tannins, shephagenins A and B, from Shepherdia argentea as HIV-1 reverse transcriptase inhibitors. Chem. Pharm. Bull. *44*, 1436–1439.

Yudovin-Farber, I., Gurt, I., Hope, R., Domb, A.J., and Katz, E. (2009). Inhibition of herpes simplex virus by polyamines. Antivir. Chem. Chemother. *20*, 87–98.

New Compounds with Antibacterial Activity

19

Patricia Veiga-Crespo, Angeles Sánchez-Pérez, Daniel Piso and Tomás G. Villa

Abstract

Infections caused by pathogenic microorganisms are among the major causes of global mortality and morbidity. The discovery and use of antibiotics revolutionized medicine as they constituted not only the first real therapeutic weapon against bacteria but also the possibility of eradication of bacterial disease in humans. However, the combination of antibiotic misuse and bacterial natural evolution rendered them less effective and resulted in the phenomenon known as antibiotic resistance. Although it is currently mandatory to search for new therapies, antibiotics still offer many potential uses, these range from enzybiotics to new nanotechnology-based delivery systems.

Introduction

In our opinion, nothing can be more moving and refreshing for the reader than re-reading Sir Howard Florey's classic article (Florey, 1945), published soon after the end of the Second World War, in which he summarized and gave his viewpoint on what we would now call the 'state of the art' on the 'use of micro-organisms for therapeutic purposes'. This idea was already old in 1945, as it had originated with Vuillemin's work (Vuillemin, 1890) in the previous century. It was him who coined the term 'antibiosis' to epitomize the tremendous fight taking place constantly in the microbial world regarding the interaction between predator and prey. Florey even included in his publication a quote from Vuillemin, referring to the new 'antibiosis' term and well worth of quoting here:

> The lion that springs on its prey and the serpent that poisons the wound before devouring its victim are not considered to be parasites. There is nothing equivocal about it, one creature destroys the life of another in order to sustain its own, the first being entirely active and the second entirely passive; one is in unrestricted opposition to the life of the other. The condition is so simple that it has never been named, but instead of being examined in isolation it can be viewed as a factor in more complex phenomena. For simplicity we shall refer to it as **antibiosis**; the active participant will be the antibiote.

The term antibiosis is almost identical to the word 'antibiotic' that we use today to describe a chemical compound produced by microorganisms that kills or inhibits the growth of other microorganisms (Waksman and Woodruff, 1942). The last three decades of the nineteenth century were indeed blessed with many papers concerning antibiosis. Pasteur and Joubert (Pasteur and Joubert, 1877) described the existence of antibiosis affecting *Bacillus anthracis*, a disease-producing microorganism. The following paragraph, quoted from Pasteur and Joubert, is also remarkable as one realizes that the grounds for the biological fight against disease-causing microorganisms were already being set in those early years of Microbiology:

> Neutral or slightly alkaline urine is an excellent medium for the bacteria. If the urine is sterile and the culture pure the

> bacteria multiply so fast that in the course of a few hours their filaments fill the fluid with a downy felt. But if when the urine is inoculated with these bacteria an aerobic organism, for example one of the 'common bacteria,' is sown at the same time, the anthrax bacterium makes little or no growth and sooner or later dies out altogether. It is a remarkable thing that the same phenomenon is seen in the body even of those animals most susceptible to anthrax, leading to the astonishing result that anthrax bacteria can be introduced in profusion into an animal, which yet does not develop the disease; it is only necessary to add some 'common bacteria' at the same time to the liquid containing the suspension of anthrax bacteria. These facts perhaps justify the highest hopes for therapeutics.

Cantani (1885) continued the work along these lines producing an interesting paper in which he demonstrated the elimination of *Mycobacterium tuberculosis* from the lungs of a seriously infected patient by 'insufflating' air containing a 'normal non-pathogenic bacterium'. An interesting application of this idea in those early years was the use of the so-called 'pyocyanase' in the treatment of a variety of serious diseases, at least at the level of animal models. Emmerich and co-workers (Emmerich, 1887), describe its use in the treatment of diseases caused by *Salmonella typhi, Staphylococcus, Corynebacterium diphtheriae* and *Yersinia pestis.* Despite its enormous therapeutic potential, the 'pyocyanase' technique felt gradually out of grace, mostly because the manufactured product was either inactive or not correctly prepared. However, this technique was still in use in several European countries during the mid-1930s. Thus, by 1890 the phenomenon of antibiosis was already well documented and was even used to treat animals and sometimes even humans. Additionally, in all of the studies, the investigators attributed the therapeutic biological action to some kind of chemical compound elaborated by the predatory bacterium (at that time, viruses were altogether unknown).

Another well-documented piece of work, also quoted in Florey's paper, concerns the research carried out by Lode (1903) who described the presence of what is now known as 'an active enzyme'. This enzyme was inactivated by high temperatures, produced by *Micrococcus tetragenus* and readily killed *Staphylococcus aureus* (a Gram-positive bacterium), but not Friedländer's bacillus (*Klebsiella pneumoniae*, a Gram-negative bacterium). Lode's publication probably represents the first description of an 'autolysin', a term used throughout the present chapter. It is worth noting here that, from the very beginning of microbiology, it was observed that Gram-positive and Gram-negative bacteria were differently affected by the above-mentioned active principles (enzymes) produced by other bacteria, with the former bacteria usually being far more sensitive to them that the latter.

By the turn of the twentith century the idea that the actinomycetes bacterial group could be of interest for the production of active principles against pathogenic bacteria was being actively pursued, but it was Greig-Smith, in 1917 (Greig-Smith, 1917), the first to demonstrate beyond the shadow of a doubt the antagonistic effects that this particular group of bacteria exert on other bacterial groups.

The pioneering work of Gosio, in 1896 (see Alsberg and Black 1913), demonstrated the role of fungi as producers of bioactive molecules against pathogenic bacteria He was able to prove that the broth in which *Penicillium* had been grown indeed contained a bioactive compound capable of inhibiting the growth of *B. anthracis*. This bioactive compound was never used *in vivo*, but it opened new avenues of research by demonstrating that eukaryotic microorganisms are capable of producing antibacterial principles. Long before Fleming's work, Vaudremer reported, in 1913 (Jennings, 1945), the use of the growth media from *Aspergillus fumigatus* in the treatment of tuberculosis. But it was Alexander Fleming who in 1929 set the precedent for the future use of bioactive principles produced by eukaryotic microorganisms (in this case *Penicillium notatun*) for the control of mainly Gram-positive bacteria, as the Gram-negative were more resistant to penicillin (Fleming, 1929). But it would take 11 years for the work by Chain and co-workers to be published, in 1940. They demonstrated the antibacterial effects of penicillin in *in vivo* tests in experimental animals, confirming the tremendous potential of penicillin as a therapeutic agent. Their publication also substantiated the original idea that antibiosis was mediated by chemical compounds generated by the predatory microorganism, and generally secreted to the culture media (Chain *et al.*, 1940). This represents the first example of a microbial active principle that could be used to combat a broad spectrum of bacterial infections in both humans and animals, due to its low toxicity for animal cells, hence having the potential to save millions of lives. The second such compound was isolated in Wakman's laboratory, after years of effort, and named streptomycin (Jones *et al.*, 1944). This molecule displayed relatively low mammalian cell toxicity (although it was capable of causing deafness after prolonged exposure). This novel antibiotic was isolated from *Streptomyces griseus* and exerted its effects not only on both on Gram-positive and Gran-negative bacteria, but even on intracellular bacterial infections, such as those caused by *Mycobacterium tuberculosis.* These antibiotic discoveries were recognized with Nobel Prizes for both Flemming and Waksman, the 1945 Physiology prize for Flemming and the 1952 Medicine prize for Waksman.

As it is commonly known, human mucosal membranes are reservoirs for some pathogenic bacteria, including those resistant to currently available antibiotics, for this reason no antibiotic-based therapies are used, due to the risk of increasing the bacterial antibiotic-resistance levels. In these circumstances, enzybiotic-based therapies are the best treatment alternatives to antibiotics. The name enzybiotic is a combination of the words 'enzyme' and 'antibiotic' and refers to either bacteriophages or their lysins, such as lysozymes, lysostaphin or proteases, that display strong antibacterial activity (Veiga-Crespo and Villa, 2010a). Enzybiotics have the ability to act in synergy with antimicrobial host defence peptides, produced shortly after microbial infection in order to neutralize the invading bacterial pathogens. These peptides primarily exert their lethal action on the bacterial cell wall, which consists of polymers not found in mammalian cells, hence not negatively affecting the host's homeostasis. To date, more than 800 antimicrobial peptides have been isolated, purified, and structurally and functionally characterized. Although this research area is still controversial, the peptides are starting to be used, in particular in the treatment of life-threatening infections involving multiresistant bacterial strains, not susceptible to classic antibiotic treatment.

It could be argued that, as it is the case with antibiotics, bacteria could develop lysin/lysozyme resistance but, as Borysowski and Górski (2010) pointed out, this is unlikely to occur, as lysins interact with bacterial cell wall components that are essential for their survival (Borysowski and Górski, 2010b). This is also supported by the results of independent studies on the effect of non-lethal concentrations of lysins on *S. pneumonia* and *B. cereus* strains. As reported by Loeffer *et al.* (2001) for *S. pneumoniae*, and Schuch *et al.* (2002) for *B. cereus*, neither bacteria developed resistance to the lysins they were exposed to.

Activation of bacterial autolysins

A little-known paper by Dubos in 1939 described the bactericidal effects of a compound associated with the protein fraction in the culture media of a soil *Bacillus.* This compound precipitated at a pH of 4.5 and was not only heat-resistant to 90 °C, but also protease-resistant. This active principle was capable of lysing all Gram-positive bacteria tested (Dubos, 1939), but did not have a deleterious effect on Gram-negative bacteria. Dubos concluded that the above-described active substance was not a protein, but associated with a protein, and it was the associated protein that determined the compound's solubility. The compound's lytic effect was maximal on pneumococci and staphylococci, as well as on unrelated Gram-positive spore-forming bacilli. Consequently it is likely that the cell lysis was indeed a secondary phenomenon, not directly caused by the compound, as it has been reported for other antiseptics (such as heavy metals, formaldehyde, iodine, bile salts, toluene, and acetone) (Wollman and Wollman, 1936). The 'bile solubility' test for pneumococci is a good example of this secondary lytic phenotype, since neither bile salts nor unsaturated fatty acids directly lyse pneumococcal cells. Bacterial lysis only occurs as a result of the bile disturbing and inhibiting essential metabolic functions, while the bacterial autolysins remain active, eventually causing the bacterial cell lysis.

In 1972 Gilpin and co-workers noted that a high salt concentration produced the activation of autolytic enzymes, in a ribitol teichoic acid-deficient mutant of *Staphylococcus aureus*, by causing the activation of a N-acyl-muramyl-L-Alanine amidase. This was an important finding, as it demonstrated that these autolysins from Gram-positive bacteria could be activated by increasing the salt concentration (Gilpin *et al.*, 1972). The effect of the high NaCl concentration appears to be similar to the action of chaotropic ions on hydrophobic areas of the molecule, causing some of the components of the cell wall to shrink and thus bringing together the enzymes and substrates. In other microorganisms such as *Bacillus subtilis*, this appears to foster an interaction between teichoic acids and autolysins that activates the autolysins and creates autolytic leakage points and bacterial cell death (Brown *et al.*, 1970).

Among the Gram-positive bacteria, *S. pneumoniae* merits a mention for its ability to activate its own autolysins. This bacterium is an important human pathogen capable of colonizing the respiratory tract, and it is a major cause of mortality, mainly in elderly patients. In addition, *S. pneumoniae* is commonly lysogenized by bacteriophages that, when induced to generate bacteriophage progeny, synthesize a phage lysin that activates the bacterial lytA lysin, thus causing the complete degradation of the bacterial cells (Frias *et al.*, 2009; Ronda-Lain *et al.*, 1977). These bacteriophages, as is the case for the majority of DNA-containing bacteriophages infecting a vast variety of bacteria, possess the 'holin-lysin' enzymatic duo. These two enzymes are necessary for the complete degradation of the bacterial peptidoglycan that precedes the release of the viral progeny. In fact, the genes encoding them are found in all known pneumococcal phages (Diaz *et al.*, 1996; Obregon *et al.*, 2003; Romero *et al.*, 2004) and the two enzymes must act in coordination. Holins generate non-specific lesions in the bacterial cytoplasmic membrane, facilitating the passage of lysin (which lacks both a typical N-terminal secretory signal sequence and transmembrane domains and hence cannot cross the membrane unaided) to its final destination in the bacterial cell. García and coworkers (1985) reported the phenotypical curing of *S. pneumoniae* treated with amidase, induced by the Dp-1 bacteriophage. In this manner, a phage-associated murein hydrolase activity (PAL), induced in an autolysis-defective mutant (lyt A minus) of *S. pneumoniae*, was able to complement the lyt mutation and rescue the bacterial mutant (Garcia *et al.*, 1985). This was further substantiated by Frias *et al* (2009), who demonstrated that the powerful amidase LytA, either alone or in combination with other enzymes, is activated during the pneumococcal phage-mediated lysis and that pneumococcal phages are able to use it to exit the host cell (Frias *et al.*, 2009).

The aforementioned best-characterized autolysin, the major pneumococcal LytA amidase, originated/evolved by the fusion of two protein domains, a N-terminal catalytic domain, and a C-terminal domain that is responsible for the binding to bacterial cell walls and contains six repeated units (for a comprehensive paper review see Garcia *et al.* (1994) and Lopez *et al.* (1997)). Biochemical analyses of truncated C-terminal domain mutants revealed that, to efficiently recognize the choline residues on the pneumococcal cell walls, the LytA amidase must contain at least four units. These units keep the enzyme tightly bound to the bacterial cell wall, in such a manner that the loss of either of the two extra repeats dramatically reduces the ability of the enzyme to bind onto the cell wall and results in an increased inhibition by either deoxycholate or choline (the N-terminal domain is unaffected by these treatments). Rationalizing the combinatorial module approach, Croux and colleagues (1993) were able to establish the role of the C-terminal domain of *Clostridium acetobutylicum* lysozyme (ATCC 824), by constructing a chimeric pneumococcal-clostridial cell wall lytic enzyme (Croux *et al.*, 1993). In this way, an active chimeric cell wall lytic enzyme was constructed by protein domain substitution between the major autolysins from *Cl. acetobutylicum* (ATCC 824) and *S. pneumoniae*. The chimeric enzyme, engineered by the fusion of the N-terminal domain of the pneumococcal LYTA amidase and the C-terminal domain of the clostridial LYC lysozyme, displayed an amidase activity capable of hydrolysing choline-containing clostridial cell walls with far higher efficiency than either of the original enzymes. These results open an alternative route in the creation of new antibacterial drugs, either by activation of the bacteria's native bacterial autolysins or by engineering new chimeric enzymes with novel and greater capabilities.

Another interesting phenomenon concerning autolysins and development of genetic transformation in Streptococcaceae is exemplified by the fact that during competence for genetic transformation, competence-induced pneumococci lyse non-competent cells of the same strain. Although this behaviour, on the surface, seems detrimental, it produces an increase in gene exchange under natural conditions. Lysis of the non-competent cells is mediated by the murein hydrolase CbpD and the autolysins LytA and LytC. The competent cells have a membrane-integrated protein (ComM) that minimizes the damaged caused by the CbpD, LytA and LytC autolysins. In addition, there is a second layer of protection (provided by the pneumococcal LiaFSR system), which senses the damage inflicted to the cell wall by the lysins (Eldholm *et al.*, 2010).

There are other species within the genus *Streptococcus* where regulation of autolysins is well documented. This is the case of the dental pathogen *S. mutans*, where an autolysin regulator (LytR) has been described. LytR plays a central role in cell division, as well as in cell wall metabolism, they are able to do this by regulating their autolysins (Chatfield *et al.*, 2005). Bioinformatic analyses on the LytR protein sequence showed that the deduced 407 amino acid sequence contained a typical LytR/CpsA/Psr family domain. Although its function is not completely understood, this domain is clearly important for Gram-positive bacteria, since it is present throughout its phylogenetic tree, as found in the pfam database. Indeed, this domain is present in almost 80 species of Gram-positive cocci and in *Mycobacterium tuberculosis* H37Rv, as well as Gram-positive bacilli and cocci such as *Bacillus anthracis* and *Staphylococcus aureus*, respectively. Thus, the LytR/CpsA/Psr domain is vital for bacteria and has survived through evolution in both pathogenic and non-pathogenic bacteria. For all these reasons, and because LytR is located outside of the cell, and it is postulated to be involved in the maintenance of cell-wall structure through autolysin regulation, it has been demonstrated that the LytR mutant displays increased autolysin activity. In *B. subtilis*, LytR has been described as an attenuator of autolysin transcription and, in *S aureus*, it has been shown that the absence of LytR increases cellular rates of autolysis (Brunskill and Bayles, 1996; Lazarevic *et al.*, 1992). In the latter, LytR is part of a regulatory system affecting the expression of the lrgA and lrgB genes, whose products are believed to be involved in the secretion of autolytic enzymes through the formation of bacteriophage's holin-like structures (Fujimoto *et al.*, 2000). Transcription of lytR is tightly controlled during the bacterial growth-phase, with the highest expression during the early- and mid-exponential periods, and a dramatic decline during the late-exponential phase (Chatfield *et al.*, 2005).

All taken together, it appears that autolysins can be activated by a variety of compounds, including acylated lipoteichoic acid (LPT). LPT is a putative endogenous regulator of autolysins that could mediate cellular lysis after treatment of the bacteria with detergents such as triton X-100, that produce the release of high amounts of LPT (Komatsuzawa *et al.*, 1994). The regulation of d-alanyl-lipoteichoic acid biosynthesis in *Streptococcus agalactiae* was studied by Poyart and co-workers in 2001 and involves a novel two-component regulatory system. The dlt operon of Gram-positive bacteria encompasses four genes (dltA, dltB, dltC, and dltD) that catalyse the incorporation of D-alanine residues into lipoteichoic acids (LTAs). *S. agalactiae*'s dlt operon, in addition to the dltA and dltD genes, includes two regulatory genes, designated dltR and dltS and located upstream of dltA. The dltR gene encodes a 224 amino acid long putative response regulator, belonging to the OmpR family of regulatory proteins. The dltS gene codes for a 395 amino acid long putative histidine kinase, thought to be involved in the sensing of environmental signals. This operon is mainly transcribed from the PdltR promoter, which directs synthesis of a 6.5 kb transcript encompassing dltR, dltS, dltA, dltB, dltC, and dltD. These genes, at least in Gram-positive bacteria, and most likely also in Gram-negative, define a key bacterial operon with the ability to modulate different acylated LPTs. This modulation results in either the activation or inhibition of the different autolysins associated with the cell wall in prokaryotic microorganisms and can give rise to apparently anomalous phenotypes, such as those with increased susceptibility, or resistance, to the cationic antimicrobial polypeptide colistin (Fig. 19.1).

Furthermore, McBride and Sonenshein propose that the D-alanylation of teichoic acids provides protection against antimicrobial peptides (such as nisin, gallidermin, polymyxin B and vancomycin) that could be essential for *C. difficile* to grow in the host (McBride and Sonenshein, 2011)

Another such example of a Gram-positive bacterium is *Enterococcus faecalis.* Some studies found that a small antibiotic peptide (termed AS-48), produced by a strain of the bacteria *Enterococcus faecalis* subsp. *liquefaciens,* had inhibitory action on both Gram-positive and Gram-negative bacteria and, in many cases, the phenotype involved cell lysis (Galvez *et al.*, 1989). As this phenotype was inhibited by cardiolipin, they suggested that the lytic phenotype was a secondary effect, due to the peptide inhibiting many cellular physiological functions, while keeping active the autolysins that take part in the normal process of cell wall elongation. The disequilibrium thus created would finally result in cell lysis. Often, as is the case with *Enterococcus hirae,* a bacteria with two autolytic enzymes (muramidase-1 and muramidase-2), this secondary lysis occurs because the autolysin is synthesized as a zymogen and, therefore, requires activation by a protease (Shockman, 1992). Muramidase-2, encoded by a different gene, is a very interesting enzyme displaying unusual features. This protein spans two to three separate domains, with a putative glycosidase-active site near the N-terminal end and six 45-amino-acid-long repeats at the C-terminal end. These repeats are presumed to be responsible for the protein's high-affinity binding to the insoluble peptidoglycan substrate. This enzyme's protease-mediated autolysin activation is accomplished by a different mechanism and facilitated by low molecular weight mediators.

The majority of Gram-positive bacteria contain polyanionic cell wall glycopolymers, which are covalently linked to either peptidoglycans (teichoic acid [WTA] or teichuronic acid) or membrane glycolipids (either lipoteichoic acid [LTA] or succinylated lipoglycans) (Weidenmaier and Peschel, 2008). Those compounds play different roles in the maintenance of Gram-positive cell envelopes, as well as in bacteria–host interactions (Reichmann and Grundling, 2011; Swoboda *et al.*, 2010; Xia *et al.*, 2010). Additionally, the cell wall polyanionic compounds have been involved in cation binding (Schirner *et al.*, 2009), hence playing an important role on the regulation of external pH and therefore on the activity of cell wall-associated enzymes such as autolysins.

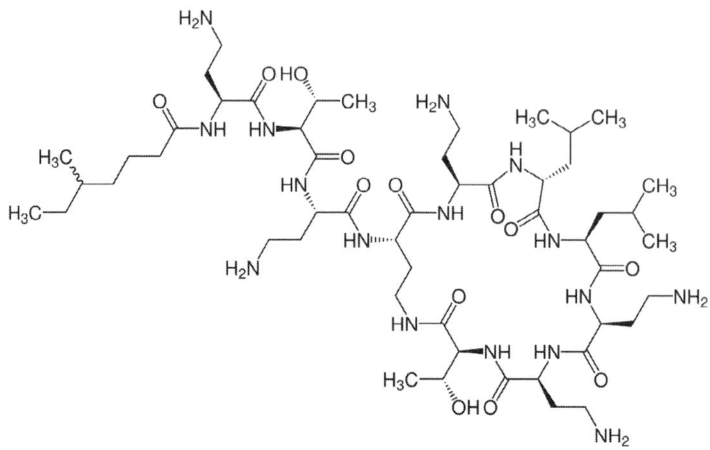

Figure 19.1 Primary structure of colistin, an enzyme produced by *Bacillus polymyxa* var. *colistinus*.

Calamita and co-workers provided evidence that the cell wall of *B. subtilis* is protonated during respiration, and that the resulting pH changes play a role in the regulation of autolysis (Calamita and Doyle, 2002; Calamita *et al.*, 2001). Subsequently, Biswas *et al.* (2012) showed that the cell wall teichoic acids (CWTA) contribute to the proton-binding capacity of *S. aureus* cell wall. They act like a cation exchanger, retaining protons in the cell wall and, in turn, modulating the autolytic activity of Gram-positive bacteria. Recent publications have shown that the CWTA present in the bacterial cell wall decrease the peptidoglycan-affinity of *S. aureus* autolysins, and that autolysin AtlA (the major *S. aureus* autolysin) binds to cell wall areas containing low amounts of CWTA (Frankel and Schneewind, 2012; Schlag *et al.*, 2010). However, the fine mechanisms involved in autolysin regulation, and mediated by the proton-motive force, remain unclear.

Lysostaphin

Since Nicolle's paper in 1907 stating that sterile filtrates of *Bacillus subtilis* had lytic effect on '*B. de Shiga, B. typhique and B. charbonneux*', bacteriolytic agents isolated from microorganisms has been an area of great interest to researchers. Fleming, about two decades later, widened this area to include a variety of bacteriolytic agents from other biological sources. These bacteriolytic agents included lysozyme, a most interesting compound with a diverse mode of action that was invaluable as a tool to study the bacterial cell wall structure. For more than one hundred years, microbiologist have obtained these lysing enzymes from filtrates of bacterial cultures, such as Actinomycetes (Waksman, 1947), *Streptococcus* and *Staphylococcus* (Richmond, 1959). In addition to extracellular staphylolytic agents, Ralston and co-workers reported the production of virolysin, a compound with an unique structure (Ralston *et al.*, 1957). The isolation of staphylolytic-specific agents was first reported by Schindler and Schuhardt (Schindler and Schuhardt, 1964). They also found that non-staphylococcal microorganisms, such as *Micrococcus lysodeikticus, Bacillus subtilis, Bacillus megaterium, Escherichia coli, Aerobacter aerogenes, Proteus morganii, Proteus rettgeri, Proteus vulgaris, Pseudomonas aeruginosa, Serratia marcescens, Lactobacillus plantarum, Gaffkya tetragena, Sarcina lutea, Corynebacterium diphtheria (gravis), C. diphtheriae (mitis), Streptococcus lactis, Streptococcus faecalis, Streptococcus pyogenes, Bordetella pertussis, Brucella abortus, Klebsiella pneumoniae, Pasteurella pestis, Pasteurella tularensis, Salmonella paratyphi, Salmonella schottmuelleri, Salmonella typhosa, Vibrio cholerae (Inaba), V. cholerae (Ogawa), Lysteria monocytogenes (type 1), Micrococcus lysodeikticus* and *Bacillus megaterium*, were not affected by these lytic agents.

The antibacterial enzyme lysostaphin, produced by a bacterium identified as *Staphylococcus simulans* biovar *staphylolyticus*, was first isolated in 1964 (Schuhardt and Schindler, 1964). Its main active agent is a 27-kDa zinc-containing glycylglycine endopeptidase (EC. 3.4.24.75; Fig. 19.2) enzyme, capable of cleaving the crosslinking pentaglycin bridges in the cell wall of staphylococci (Schuhardt and Schindler, 1964; Trayer and Buckley, 1970). This property makes lysostaphin a highly effective antimicrobial agent against both actively growing and quiescent bacteria (Wu *et al.*, 2003). This enzyme has also been shown to display a greater antibacterial activity than commonly used antibiotics, such as vancomycin (Yang *et al.*,

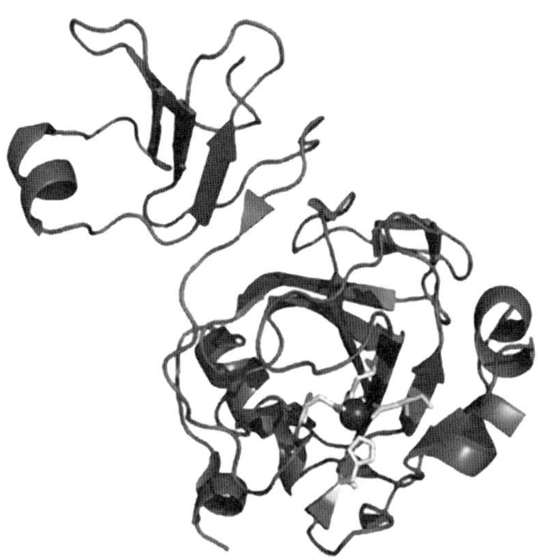

Figure 19.2 Crystal structure of the protein lysostaphin showing the active site. A colour version of this figure is available in the plate section at the back of the book.

2007). Lysostaphin is also effective against methicillin-resistant *S. aureus* (MRSA). Additionally, used in combination with other antimicrobials, such as cefazolin, clarithromycin, doxycycline, levofloxacin, linezolid and quinuprisitin/dalfopristin (Fig. 19.3), it has a synergistic effect on staphylococci (Dajcs *et al.*, 2000). It is worth quoting here some interesting properties of this staphylolytic enzyme which give a hint of what a new lytic agent might achieve: (i) it achieved bacterial lysis at a greater speed than virulent bacteriophages; (ii) this lytic factor was active against a wide variety of *S. aureus* strains, as well as 5 *S. epidermidis* strains; and (iii) there was no appreciable change in the lytic capacity of *S. aureus* preparations subsequent to their lysis. The possibility that lysostaphin was a type of lysozyme of microbial origin was rejected (Schindler and Schuhardt, 1964), as lysostaphin did not share the same antibacterial spectrum of action as lysozyme. Lysostaphin was found to be active against all tested *S. aureus* strains, independently of the differences found between strains, such as their ability to produce coagulase. Schindler and Schuhardt also used lysostaphin alone to treat mice infected with *S. aureus*. All these therapeutic studies were carried out on mice that were divided into four treatment groups and the therapy consisted on a single lysostaphin injection. The results obtained led the authors to conclude that lysostaphin could be a very valuable therapeutic agent in staphylococcal infections of both people and animals (Schuhardt and Schindler, 1964). The next advance was brought by Schaffner and colleagues (1967), who evaluated lysostaphin in *in vivo* experiments. They reached the conclusion that lysostaphin was superior to the penicillins in the treatment of experimental staphylococcal infections in mice, and this compound was even able to kill staphylococci inhabiting oxacillin-resistant abscesses (Schaffner *et al.*, 1967). Unfortunately, treatment with lysostaphin had the drawback that, although single doses of the compound had no deleterious effect, multiple doses generated precipitins and neutralizing antibodies. Even though, as early as 1968 it became clear that this new enzyme could represent an invaluable therapeutical tool in the treatment of methicillin-resistant staphylococci (Zygmunt *et al.*, 1968), provided that only a few doses of lysostaphin were used. In the above-mentioned study, all 16 methicillin-resistant *S. aureus* strains tested, were shown to be sensitive to lysostaphin's lytic activity, although they were resistant to tetracycline.

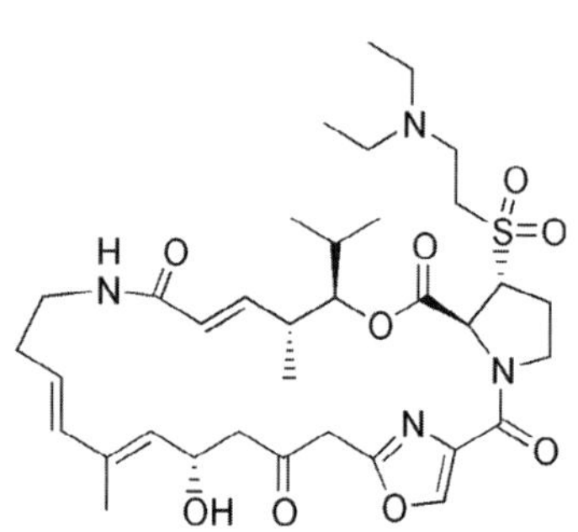

Figure 19.3 Structure of dalfopristin, an antibiotic that binds to the 23S rRNA component of the 50S ribosomal subunit.

Lysostaphin has even been used in food technology. Metcalf and Deibel (1969) used lysostaphin in the cheese-making industry. They spiked the raw materials used to make cheese with *S. aureus* and then treated them with lysostaphin. They found that addition of lysostaphin to the starting materials resulted in an initial decrease in bacterial contamination and that lysostaphin remained in the curd after separation of the whey (Metcalf and Deibel, 1969). However, the enzyme was unable to kill all the bacteria and a significant *S. aureus* population ultimately developed, even in the presence of lysostaphin, probably due to inactivation of the enzyme.

In addition to *Staphylococcus simulans* biovar *staphylolyticus*, other *Staphylococcus* strains are also able to produce lysostaphin, such as *S. epidermidis*, NRRL B-2628 (Iversen and Grov, 1973), thus suggesting that production of this staphylolytic compound is not species-specific but strain-specific. Lysostaphin was isolated from the culture supernatant of *S. epidermidis*, NRRL B-2628 (K-6-W1), by retention on the ion-exchange resin Amberlite IRC-50 and elution with a salt gradient. This released three different enzymes, identified as endo-β-N-acetylglucosaminidase, endopeptidase (glycinase) and N-acetylmuramyl-L-alanine amidase, respectively; the latter was did not produce bacteriolysis *per se*.

The presence of bacterial capsules in some pathogenic bacteria is thought to be a barrier against antibacterial compounds, thus preventing them from reaching their final targets. Despite this bacterial defence mechanism, lysostaphin was found to be able to penetrate and lyse both the encapsulated and non-encapsulated forms of *Staphylococcus aureus* cells, and both types of cells were lysed at the same rate (King *et al.*, 1980).

On the other hand, thus far four mechanisms have been identified that result in bacterial resistance to lysostaphin (Kusuma *et al.*, 2007; Shaw *et al.*, 2005), although some of them produce an increase in susceptibility to methicillin and/or other antibiotics, as well as a reduction in bacterial virulence (Kusuma *et al.*, 2007). These mechanisms of resistance include the gene FemA (the gene required for incorporation of the second and third glycine residues in the bacterial wall cross bridges), and its inactivation results in the formation of bridges, in the staphylococcal peptidoglycan, that contain a single glycine. As it name indicates, the lyrA gene is also involved in lysostaphin resistance (lyr stands for lysostaphin resistance). In this case, the resistance results from a modified cell wall envelope that, as is the case for FemA and FemB, can contain monoglycine cross bridges. The incorporation of amino acids other than glycine (usually serine, as in the lysostaphin-producing staphylococcal strain) in the peptidoglycan bridges can also lead to lysostaphin resistance, since the target of the enzyme is no longer present. Finally, lysostaphin resistance can also arise by mutations in the sspC gene, that encodes staphopain B, an inhibitor of the secreted protease (Shaw *et al.*, 2005).

An interesting array of applications for lysostaphin has been recently reviewed by Borysowski and Górski (Borysowski and Gorski, 2010a). They include elimination *of S. aureus* in nasal colonizations. Eradication of *S. aureus* colonization reduces the rate of infection. Owing to its rapid mode of action, the prophylactic use of lysostaphin in this application clearly supersedes antibiotic prophylaxis. In addition, interference with the host's normal microbiota is kept to a minimum; this is due to lysostaphin's specificity, targeting only glycine bridges. Special reference deserves the work of Kokai-Kun and co-workers, who studied the effect of lysostaphin on *S. aureus* nasal colonizations in rat models. These investigators found that a single dose of 0.5% lysostaphin, formulated as a cream and applied locally, eradicated the bacterium in *ca.* 93% of the animals. They also found that lowering the lysostaphin dose resulted in a lower eradication percentage, from 55% to 20%, a result that complies with one of basic principles of enzymology (Kokai-Kun *et al.*, 2003). The same authors reported, also in 2003 (Wu *et al.*, 2003), that lysostaphin was capable of disrupting not only *S. aureus* biofilms, but also those produced by *S. epidermidis*. They found that lysostaphin quickly, within minutes, killed *S. aureus* (with a minimum inhibitory concentration for 90% of the cells (MIC90) of 0.001–0.064 μg/ml) and that it was also effective against *S. epidermidis*, although at higher concentrations (MIC90: 12.5–64 μg/ml). Additionally, lysostaphin not only killed *S. aureus* in biofilms, but also broke down the extracellular matrix formed by this bacterium (only those biofilms made by lysostaphin-sensitive *S. aureus* were affected) in *in vitro*-formed biofilms on plastic surfaces, and did so when used at a reasonably low concentration (1 μg/ml).

These days it is becoming apparent that the use of lysostaphin to treat staphylococcal-biofilm-associated infections may be preferable to the use of antibiotics, such as vancomycin or clindamycin, since, as noted above, small doses of lysostaphin are enough to disrupt the staphylococcal biofilm thus obviating the need for surgical removal of the infected graft. Additionally, depending on the concentration of lysostaphin used, biofilm disruption can be accomplished within hours, instead of one or two days needed for the antibiotic treatment to produce the same effects. It has been reported that both oxacillin and vancomycin require at least 24 hours to achieve this purpose (Amorena *et al.*, 1999; Mack *et al.*, 2000).

Lysostaphin has been successfully used in eye drops as a prophylactic treatment against staphylococcal eye infections. McCormick *et al* (2006) tested the effectiveness of lysostaphin therapy in experimentally induced coagulase-negative *S. endophthalmitis* infections reporting a 4 log reduction in cfu/ml, as compared to untreated eye infections (McCormick *et al.*, 2006). Similarly, Balzli *et al.* (2010) showed that injection of lysostaphin into the rabbit aqueous humour resulted in the detection of anti-staphylococcal activity for at least two days after injection, whereas the eyes injected with gatifloxacin were no significantly different from the control ones, injected with saline solution (Balzli *et al.*, 2010). Recently, Satishkuma and co-workers (2013) have shown that lysostaphin can be used to coated meshes, maintaining its antimicrobial activity and, that in this manner, it can be used in the prevention of bacterial adhesion and colonization of the meshes used to surgically repair hernias (Satishkumar *et al.*, 2013).

Lysostaphin could potentially be used not only in the treatment of diseases, but even as a tool in the differential identification of *S. aureus* in a complex mixture of bacteria. In this manner, Potapova and co-workers (2013) investigated the potential of lysostaphin as a specific probe to identify staphylococci (Potapova *et al.*, 2013). They developed a novel two-step protocol that effectively differentiates *S. aureus* from *Escherichia coli*. As the procedure they developed is safe, it can potentially be used *in vivo* and it could represent a promising approach for diagnostic imaging of staphylococcal infections in clinical settings.

Fan *et al.* (2004) engineered transgenic goats, using an adenovirus vector containing the cloned lysostaphin gene, capable of expressing lysostaphin in their mammary glands. The idea behind this work was to protect the goats from developing mastitis. Unfortunately, the low persistence of the enzyme in the goat's mammary glands, together with the strong immune response that it provoked in the animals, made this approach totally unsuitable. The immune response, however, can be reduced to such an extent that lysostaphin could be directly injected to prevent the growth of *S. aureus*. This approach is facilitated by the work of Walsh and co-workers (2003), who developed the PEG conjugation method and demonstrated that PEGylation of lysostaphin, while diminishing its *S. aureus* killing activity, not only resulted in prolonged drug persistence in serum, but also substantially reduced antibody production (Walsh *et al.*, 2003).

Pangule and co-workers (2010) developed paint containing lysostaphin as an effective way of controlling bacterial populations on surfaces. Specifically, they incorporated conjugates of carbon nanotubes with lysostaphin into films to deliver bactericidal activity against *S. aureus* and *S. epidermidis*. These enzyme-based composites proved highly efficient at killing MRSA (>99% within 2 h) without causing the release of the enzyme into solution (Pangule *et al.*, 2010).

Lysozymes and related proteins

Lysozymes (EC 3.2.1.17) are enzymes widely found in animals and plants where they constitute one of the first barriers a pathogenic bacterium must overcome before it can establish an infection. Although the best-known antibacterial action of lysozyme is based on its enzymatic hydrolysis of murein, this protein can also destroy bacteria by non-enzymatic mechanisms, such as the activation of endogenous autolysins or by removing divalent cations from the surface of the bacterial cytoplasmic membrane, thus compromising its integrity. In fact, several reports suggest that these non-enzymatic mechanisms may be even more important than peptidoglycan hydrolysis (Masschalck and Michiels, 2003). Lysozymes are classified into several groups based on their DNA sequence (Masschalck and Michiels, 2003), similarity and structure (Jolles, 1996). These groups are denominated: c-type (chicken-type lysozyme), includes the stomach and insect lysozymes; goose-type lysozyme (g-type); plant lysozyme; bacterial lysozyme; and T4 phage-lysozyme (phage-type). c-Type lysozymes have been isolated from many vertebrates and also from insect species (Qasba and Kumar, 1997). Some of the characteristics of these lysozymes are worth noting here. While some stimulate the autolysin activity of bacteria (Iacono *et al.*, 1985), others, such as the human lysozyme and the hen egg white lysozyme (HEWL), show a bactericidal activity without actual peptidoglycan hydrolysis (Laible and Germaine, 1985). Additionally, the enzyme produced by the fungus *Chalaropsis punctulata* displays both lysozyme and β-1,4-*N*,6-*O*-diacetylmuramidase activity.

Animals usually produce more than one of these enzymes. In the human body, lysozyme is produced by cells of the immune system (polymorphonuclear granulocytes, monocytes and macrophages) and forms part of different biological fluids and tissues, including tears, urine, milk, saliva, liver, cartilage, and skin. Human lysozyme is similar to the c-type (HEWL) lysozymes and it plays a role in defence mechanisms against pathogenic bacteria. Ruminants, such as cows, deer or sheep, usually possess multiple lysozyme genes, including at least four that code for gastric lysozymes with digestive roles (Irwin, 1995; Irwin and Wilson, 1989). The recruitment of lysozymes as part of the digestive machinery appears to have occurred early in vertebrate evolution, and it could have been accomplished through gene duplication (Irwin, 1995). Four out of the ten lysozyme genes identified in ruminants are expressed in the stomach (in addition to two pseudogenes) and another four, non-monophyletic, are expressed in a variety of tissues. Ruminants are, therefore, an interesting group to study the evolution of this otherwise old gene (probably first produced by a bacteriophage) that encodes for lysozyme-like enzymes. In fact, the structural similarities between the lysozyme produced by the T4 phage and HEWL suggest that the two proteins have arisen by divergent evolution from a common precursor (Grutter *et al.*, 1983; Swanson *et al.*, 1985; Thunnissen *et al.*, 1995; Weaver *et al.*, 1984). Indeed, X-ray crystallography revealed the similarities between goose-type lysozyme (GEWL), T4 lysozyme, HEWL and a common ancestral precursor, even though their amino acid sequences appeared to be unrelated probably due to divergent evolution. It follows that lysozymes have undergone conformational changes, both at the global and local levels. So, throughout the evolution of these proteins, structural displacements can be observed as well as variations in the molecular weight that could account for the different activity of the lysozymes. Additionally, the glutamic acid residues of the active site are essential for lytic activity, but their positions can vary; for example, the glutamic acid is at position 73 in GEWL, position 35 in HEWL, and 11 in T4 lysozyme. In primates, the evolution of lysozyme is episodic, as this is not compatible with the neutral theory of evolution (Yang, 1998), it means that they have been evolving under negative pressure. Interestingly, the amino acid composition of α-lactalbumin, a protein found in cow's milk, is quite similar to HEWL. By comparing the amino acid sequences of HEWL, human lysozyme and α-lactalbumin, it has been postulated that all three share a common ancestral gene. The α-lactalbumin gene in the mammalian lineage appears to have evolved by duplication of an ancestral c-type gene, with the consequent loss of the lysozyme activity and the gain of a novel function (Prager and Wilson, 1988; Veiga-Crespo and Villa, 2010b).

Lysozymes are intriguing enzymes, since they not only have antibacterial activity but also behave as potent antivirals; they exhibit anti-inflammatory activity and yet are capable of arresting the growth of some tumours. They also possess immunomodulatory properties (Borysowski and Gorski, 2010b) and have been extensively used in humans, for decades now, to control the bacterial population either in pathogenic processes or in food science. Normally, lysozyme is able to destroy only Gram-positive bacteria, while Gram-negative bacteria tend to be resistant, due to the presence in them of an outer membrane. For the last few decades, lysozyme has been used, either alone or in combination with antibiotics, in the prophylaxis and treatment of different bacterial infections, including pharyngitis, tonsillitis, dysentery and wound infections (Sava, 1996). In recent years, a series of patents for lysozyme applications have been submitted and/or issued (Donovan, 2007). These include the use of lysozyme, formulated as a gel, for topical treatment of wounds, the use of different formulations of the enzyme in the treatment of acne, and the use of lysozyme in aerosol applications for the treatment of tracheitis, pneumonia, and faucitis (Borysowski and Gorski, 2010b). Lysozyme has also been used as a component of oral health products, such as mouthwash (Gil-Montoya *et al.*, 2008; Tenovuo, 2002).

The discovery of new types of lysozymes always brings to light new forms of the enzyme with potential medical use, as is the case for a new lysozyme isolated by Grinde in 1989 (Grinde, 1989). This author purified two lysozymes from rainbow trout kidney (types I and II) and subsequently tested them on eight bacterial strains causing clinical mastitis (staphylococci, streptococci and coliforms). He also compared this new enzyme to the antibacterial potential of chicken egg white lysozyme, lysostaphin and mutanolysin. The type II rainbow trout lysozyme decreased the number of live bacteria on all the bacterial strains tested, but was most efficient against staphylococci; thus mimicking the mode of action of lysostaphin.

The different lysozymes so far described display diverse bactericidal activity at different atmospheric pressures. Nakimbugwe and co-workers (2006) carried out a study comparing the bactericidal activity of six lysozymes both at atmospheric pressure and under high hydrostatic pressure (Nakimbugwe *et al.*, 2006). These authors found that, at normal atmospheric pressure, different Gram-positive bacteria displayed distinct sensitivity to the six lysozymes tested, but none of the Gram-negative bacteria was sensitive to any of the lysozymes. On the other hand, treatment under high hydrostatic pressure made several Gram-positive and Gram-negative bacteria more sensitive to one or more of the lysozymes. These results open a new array of applications in food preservation and in Food Science and Technology in general.

Yokogawa and colleagues, in 1974, coined the name mutanolysis to designate a lysozyme-type of enzyme capable of specifically degrading cariogenic streptococcal peptidoglycan (Yokogawa *et al.*, 1974). It was first purified, that same year, from the culture filtrate of *Streptomyces globisporus* 1829 and consisted of two main lytic enzymes with an isoelectric point close to pH 8.5 and 10, respectively; with the latter also acting as a proteolytic enzyme. Mutanolysin exhibited lytic or bactericidal activity against *Streptococcus mutans*, *Streptococcus salivarius*, *S. sanguis*, *S. mitis*, *Lactobacillus acidophilus*, and *Actinomyces viscosus*, all representing bacteria capable of causing dental caries (Hamada *et al.*, 1978; Linder *et al.*, 1983). The enzyme did not display bactericidal activity against either *S. aureus* or any of the Gram-negative strains tested (please note that this enzyme is not *muralysin*, a protein produced by some bacteriophages (see Barkulis *et al.*, 1964)). Mutanolysin's failure to destroy Gram-negative bacteria was a great disappointment at the time, as this enzyme was expected to be the panacea to cure all dental caries.

Mutanolysin was successfully used in the preparation of protoplasts from Group B streptococci, refractory to previously tested muralysins under physiological conditions (Calandra and Cole, 1980), and consequently used in the preparation of specific antigens from this streptococcal group. Mutanolysin has also been useful to decipher peptidoglycan structures from Gram-positive bacteria. Although its mode of action is similar to that of lysozyme, it exhibits significant differences that are advantageous in these structural studies (Fig. 19.4). For instance, the differential sensitivity of *B. anthracis* peptidoglycan to lysozyme and mutanolysin permitted the elucidation of its structure. It was found that 34% of the muramic acid residues in the peptidoglycan molecule contained unsubstituted amino groups, thereby providing an explanation for the resistance of these walls to lysozyme and, to a lesser extent, to mutanolysin (Zipperle *et al.*, 1984).

Infections in humans (as well as in animals, with rodents in particular) by group A streptococci and related bacteria, release peptidoglycan-polysaccharide polymers from the bacterial cell walls that produce acute joint inflammation, followed by a chronic, recurrent, erosive arthritis (Cromartie *et al.*, 1977). This pathology follows a clinical course characterized by repeated remissions and exacerbations, eventually progressing to ankylosis, with the final result of loss of function. This process involves activation of the complement, by both the classic and alternate pathways (Janusz *et al.*, 1987). Injection of peptidoglycan-polysaccharide polymers, obtained from these streptococci, readily induces arthritis in experimental animals, but this is not so when the injected polysaccharides are previously treated with mutanolysin. This information was used by Janusz and co-workers (1984) who treated the arthritic joints of experimental rats with mutanolysin. These authors found that, in their rat model, a single intravenous injection of mutanolysin, even if given at the peak of acute joint inflammation, was enough to prevent the development of recurrent, chronic erosive arthritis (Janusz *et al.*, 1984). The animals with arthritic symptoms also developed granulomatous lesions in the liver, spleen, and mediastinal lymph nodes. On the contrary, the animals treated with mutanolysin failed to show these symptoms, even when mutanolysin was injected 3 days after the disease had peaked. Janusz and colleagues (1986)

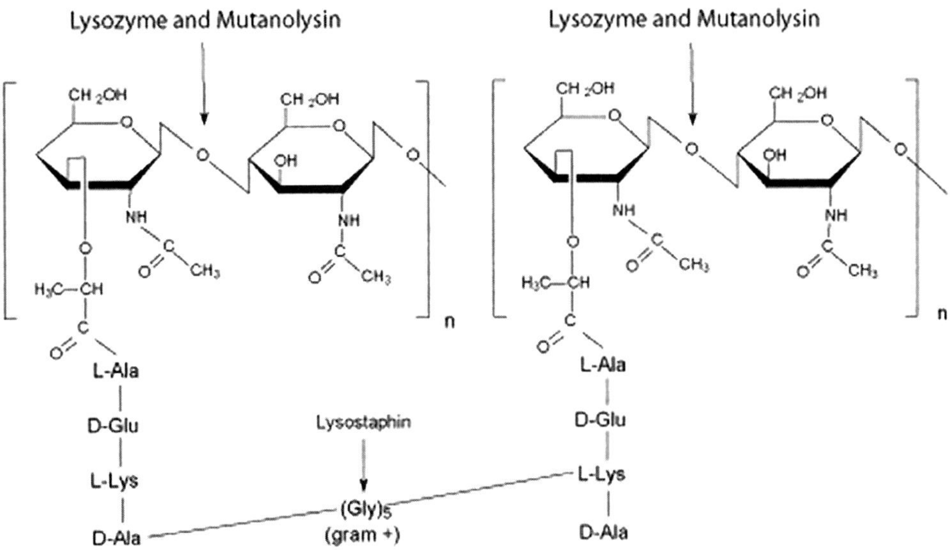

Figure 19.4 Regions in the *Staphylococcus aureus* peptidoglycan molecule that are susceptible to degradation by either lysozyme, mutanolysin, or lysostaphin.

proposed that mutanolysin abrogated arthritis by degrading peptidoglycan-derived polymers, present in the affected joints, to a size which was no longer able to induce chronic erosive arthritis, even though the polymers were still present in the limbs (Janusz *et al.*, 1986).

Production of mutanolysin by *S. globisporus* is stimulated by adding bacterial cell walls to the culture medium. This increase in bacteriolytic activity found in the supernatant correlates with *de novo* synthesis of mutanolysin, through increase in gene transcription. It is worth noting here that all bacterial cell walls tested were capable of eliciting the above-mentioned stimulation of lytic activity, including cell walls from archaea, which contained pseudomurein instead of murein (Bronneke and Fiedler, 1994). Glucose entirely abolished the stimulation caused by the bacterial cell walls, further inhibiting the low constitutive mutanolysin production.

It comes at no surprise that mutanolysin has played an essential role in unravelling the pathways used by *S. pneumoniae* to evade complement activation, hence avoiding its elimination and clearance from the blood stream. Crucial to this was the work by Angel and colleagues, in 1994, which revealed that pneumococci treated with 100 micrograms of mutanolysin released into the supernatant up to 94% of a specific C3-degrading activity from the pneumococcal surface, thus demonstrating that clinical and laboratory isolates of virulent pneumococci did degrade and inactivate soluble C3, hence precluding activation of complement fixation (Angel *et al.*, 1994).

Other proteins secreted and/or formed by surfaces/mucoses in the human body (other than the eyes, where lysozyme is one of the main proteins with antibacterial activity) include several polypeptides with striking antibacterial activity. *Pseudomonas aeruginosa*, a Gram-negative aerobic bacterium, is an opportunistic pathogen responsible for numerous human infections, including the chronic syndrome known as cystic fibrosis, with a high mortality index (Hoiby and Koch, 1990). This bacterium is associated with a strong biofilm formation (Singh *et al.*, 2000); to such an extent that abolition of this biofilm formation ability results in a dramatic reduction of symptoms and also makes the bacterium susceptible to antibiotic therapy. This is because the active principles reach more easily their targets in planktonic cells than in those 'captured' in biofilms, formed by bacteria immobilized on a surface and surrounded by an extracellular matrix they secrete (Costerton *et al.*, 1999). It is now clear that '*quorum sensing*' signals regulate this bacterial transition from a planktonic state to a sedentary way of life. Perhaps the best-studied *quorum sensing* mechanisms are those mediated by the acyl homoserine-lactone system, used by many different Gram-negative bacteria (Fuqua *et al.*, 2001), and the hormone-like (competence factors) peptides that mediate genetic transformation in Gram-positive cocci (Ephrussi-Taylor, 1951). In their natural environment, however, bacteria rarely encounter conditions similar to those found in batch cultures in the laboratory, but rather tend to form biofilms in arithmetical growth on surfaces, subjected to the influence of *quorum sensing* signals.

Several physical, biological and chemical factors can influence the *quorum sensing*-mediated biofilm growth, such as the hydrodynamic environment (Kirisits *et al.*, 2007), thus affecting biofilm thickness. This means that the *quorum sensing* signals, in the first place, will diffuse more easily into the bulk fluid and will therefore be washed away; and secondly, due to mass transfer problems, there will be a gradient of nutritional factors (Bollinger *et al.*, 2001). In these cases, the cells located deepest in the colony/biofilm, will find it harder to get nutrients than the outer cells. Iron availability can also modulate *quorum sensing*, while other nutritional components can delay its onset (Yarwood *et al.*, 2005)

The worldwide increase in bacterial resistance to conventional antibiotic therapy has prompted the search for new targets and strategies to fight recalcitrant microorganisms. The proteins secreted by mucosal surfaces are of potential use in this fight and include the so-called 'antimicrobial peptides' (AMPs), produced by vertebrates to ward off all kind of external pathogens (Gasset, 2010). The presence of antimicrobial activities in animal fluids was described as early as 1850, but it was not until the middle of the twentieth century (Skarnes and Watson, 1957) that some of these substances were finally isolated and tested against both Gram-positive and Gram-negative bacteria. This led to Bowman isolating and purifying, almost three decades later, the insect cecropin and attacin antibacterials (Hultmark *et al.*, 1983; Steiner *et al.*, 1981). In 1987, Zasloff succeeded in purifying the amphibian manganin, and this was continued the following year with the production of synthetic manganins (Zasloff *et al.*, 1988) and the isolation of human defensins (Ganz *et al.*, 1985; Zasloff, 1987; Zasloff *et al.*, 1988). Currently, more than 800 AMPs have been isolated, purified and characterized, and all form part of a battery of host defence peptides (for more detailed information, see Gasset, 2010) (Fig. 19.5). Defensins are small cysteine-rich cationic proteins, mainly found in invertebrates although they have also been isolated from invertebrates, as well as from plants (Pearce *et al.*, 2008; Thomma *et al.*, 2002). They always represent one of the first, if not the first, layer of defence in vertebrates against bacterial, fungal or even viral infections. Defensins are constituted of 18–45 amino acids, including 6–8 typically conserved cysteine residues. As reported, the immune system cells produce these types of peptides to help with the lysis of phagocytised bacteria. Most defensins act by binding to lipid targets in the microbial cell membrane, thus forming pore-like structures responsible for the efflux of essential ions and cellular nutrients, as is the case with fungal defensin plectasin that specifically binds to the bacterial lipid II (Schneider *et al.*, 2010).

Defensins are classified into three groups (α, β and θ). Alpha-defensin genes (DEFA) are mainly expressed in neutrophils and in NK cells, as well as in certain T-lymphocytes. Other defensins are found in Paneth cells in the small intestine and are directly involved in the control of intestinal bacteria. Beta-defensin genes (DEFB) are by far

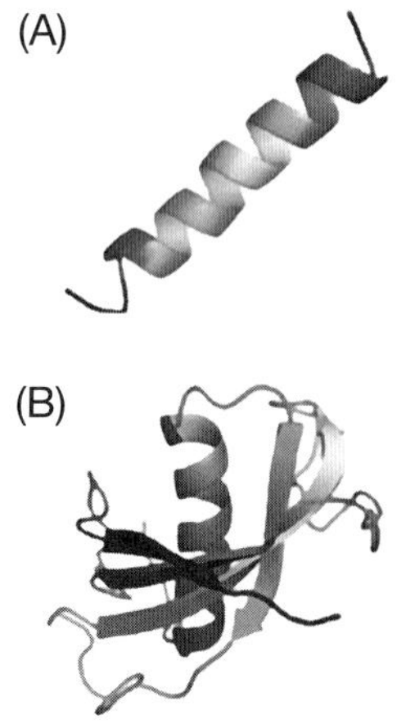

Figure 19.5 Schematic representation of two AMP polypeptides. (A) Magainin; (B) Protegin-1.

the most widely distributed, and are found in epithelial cells (in the tongue, skin, saliva, respiratory tract, etc.). Finally, θ-defensin genes (DEFT) are rare and only found in very low quantities in humans, although they are important in some monkeys, such as baboons (Garcia *et al.*, 2008). Plant defensins are also small basic peptides with a characteristic three-dimensional folding pattern that is stabilized by eight disulfide-linked cysteines. Up to date, more than 80 different plant defensins have been described. Some plants, for example *Arabidopsis thaliana*, contain at least 13 defensin genes that encode 11 proteins. As is the case in vertebrates, plant defensins inhibit the growth of a great variety of fungi, including many pathogenic species.

More than 500 AMPs have been described in amphibians, in either skin glands or the stomach mucosa (Gasset, 2010; Mangoni *et al.*, 2000; Rinaldi, 2002).

Magainins, isolated from the skin secretion of *Xenopus laevis*, are the prototype of the α-helical AMPs (Zasloff, 1987). Other amphibian α-helical AMPs include PGLa, fragments of proxenopsin and prolevitide hormones, bombinins, dermaseptins, adenoregulin, phylloxin, caerins, frenatins, maculatins, citropins, aureins, kassinateurins, and temporins (Gasset, 2010). These AMPs are about 10–30 amino acids long and half of them are C-terminally amidated. Most of these linear AMPs easily fold into amphipathic helical structures, adopt a single α-helix, and have a quite broad mode of action (Shai, 1999). Thus, they can kill both aerobic and anaerobic bacteria, multiresistant strains of human pathogens, moulds, yeasts, protozoa and, most interestingly, tumour cells.

The cathelicidin family is a group of antimicrobial peptides (AMP) found in a variety of mammalian species that are synthesized as preproproteins, with a highly conserved and inactive N-terminal domain, the cathelin domain, and a highly variable C-terminal domain, which exhibits antimicrobial activity. In humans, only one cathelicidin (CAP18) has been thus far identified (Gasset, 2010). The preprotein, after serine protease hydrolysis, generates the C-derived amphipathic fragment LL-37, containing 37 amino acids and two leucines at the N-terminus (Nagant *et al.*, 2012), producing the antibacterial activity through the formation of an α-helix upon contact with the lipids in the bacterial membrane.

It should be clear by now that evolution has selected these small universal peptides as perhaps the first, and best, line of defence for the producing organism, allowing it to keep pathogens under control. However, the exact mode of action of these enzymes still remains, in many cases, elusive. However the consensus is that they have either direct or indirect protective roles. As indicated above, the unique characteristics displayed by AMPs, such as amino acid composition, ionic charge and amphipaticity, allow them to act on the cell membranes of pathogenic organisms, damaging them and finally causing the microorganism's demise (Zhang and Falla, 2006).

The indirect antibacterial roles played by AMPs relate to their immunomodulatory abilities, including the modulation of pro-inflammatory signals, expression of hundreds of genes in different cells, and chemotaxis (Kolls *et al.*, 2008; Mookherjee and Hancock, 2007).

The direct antimicrobial action of AMPs starts, as pointed out by Gasset (2010), with the recognition of the pathogens and by using their positive charge to interact with the anionic polymers present in the pathogen's membrane. In some cases, this can be their final destination, but in other cases the membrane phase is only a transitory state and the AMP proceeds to an intracellular stage where it can reach other targets.

The therapeutic use of AMPs is still unclear and has been reviewed by Gordon in 2005 and, more recently, by Gasset in 2010. Frog-derived magainin 2 failed to gain FDA approval, in phase III, for the topical treatment of diabetic foot ulcers; porcine protegrin also failed in phase III, either as a mouth rinse or in aerosol form, for the treatment of ventilator-associated pneumoniae; bovine indolicidin failed in phase III for the purpose of controlling catheter-associated infections, but showed efficacy for the treatment of acne (phase IIb); histatins have showed high efficacy, in their formulation as mouth rinse, in patients with oral candidiasis (completed phase II); and, finally, human BPI failed, in phase III, its application to reduce mortality in paediatric meningococcaemia.

The AMPs, being peptides, are sensitive to

proteases and are indeed perfect substrates for trypsin and trypsin-like proteases. However, AMPS could be engineered to become resistant to proteases, by either amino acid replacement, introducing non-peptide bonds in the molecule, or by packaging them into liposomes or nanoparticles.

Apart from their antimicrobial actions, AMPs display other activities with potential medical uses. One of these is their ability to interfere with the viability of certain tumour cells (Mader *et al.*, 2006). This ability is also displayed by defensins and cecropins (Suttmann *et al.*, 2008).

Decapoda, such as the shrimp *Penaeus vannamei*, have been found to produce penaeidins, a novel family of antimicrobial peptides (Destoumieux *et al.*, 1997). So far three of these peptides have been isolated and all three displayed antibacterial activity, as well as antifungal activity, although they were more active on Gram-positive bacteria. Analyses of these peptides revealed a lack of homology to other antimicrobial peptides, hence representing a new antimicrobial family.

The common fruit fly, *Drosophila*, represents a new paradigm to study AMPs, their genetic organization and expression (Aparicio *et al.*, 2013). It has been reported that *Drosophila* uses either the Toll pathway, induced by Gram-positive bacteria and fungi, or the Imd pathway, induced by Gram-negative bacteria due to the presence of diaminopimelic acid, to regulate the expression of different AMP genes (Charroux and Royet, 2010). Once the infection is controlled, AMP expression is down-regulated to avoid deleterious immunopathological reactions (Aggarwal and Silverman, 2008).

The rise in antibiotic-resistant bacteria is a major concern in our society; particularly as we are running out of antibiotics to effectively treat the ever increasing number of resistant bacteria. It is now essential to find new bacterial targets where to direct the antimicrobial strategies. One of these potential new targets is the bacterial protein 'sortase' (Maresso and Schneewind, 2008).This enzyme catalyses the assembly of surface proteins and pili on the envelope of Gram-positive bacteria. It cleaves surface polypeptides and generates an acyl-enzyme intermediate that reacts with the neighbour's pilin, or adhesin molecules, to form either the complete pili or other structures on the bacterial surface. These proteins often play important roles in virulence, infection and colonization by pathogens and inhibition of these enzymes usually renders the bacteria unable to sustain an infection.

Laccase-type enzymes

Laccases (benzenediol: oxygen oxidoreductase, EC 1.10.3.2) are copper-containing enzymes (Fig. 19.6) that catalyse the oxidation of various aromatic substrates (such as guaiacol, 2,6-dimethoxyphenol or phenylene diamine), and this oxidation of the substrate is paired with the reduction of oxygen to water. Since their discovery, more than a century ago in *Rhus vernicifera*, laccases have been found in many plants, where they are involved in the synthesis of lignin and in the response to wounding. The work of Eggert in 1997 opened the possibility of using laccase-type enzymes as mediators to facilitate structural modifications of biomolecules, to either enhance their antibacterial activity or to produce new antibacterial compounds. He reported that the concentrated culture fluid of the wood-rotting

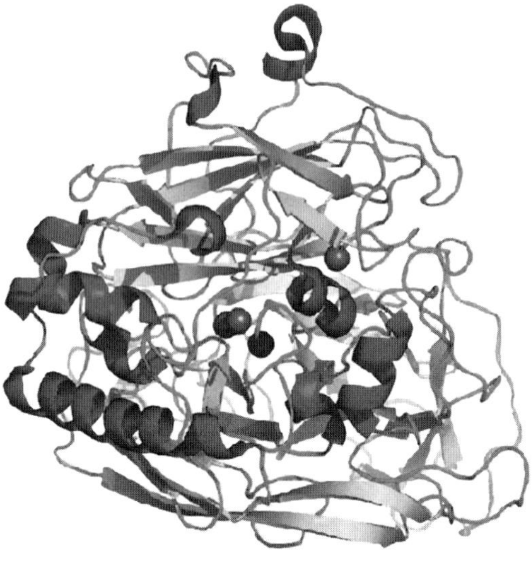

Figure 19.6 Laccase protein from *M. albomyces* displaying the four copper atoms present in the active site of the molecule. A colour version of this figure is available in the plate section at the back of the book.

basidiomycete *Pycnoporus cinnabarinus* exhibited biological activity against a variety of bacteria, mostly Gram-positive, including *Streptococcus* species (Eggert, 1997). The fungus produced the phenoxazinone derivative cinnabarinic acid, and this compound was formed after oxidation of the precursor, 3-hydroxyanthranilic acid, by laccases. This work was preceded by a publication from Agematu and co-workers (1993) that showed that penicillin X methyl ester could be transformed into three types of biologically active dimers by a laccase isolated from *Coriolus versicolor* (Agematu *et al.*, 1993).

Phenoloxidases (POs) are a group of copper proteins, including laccase, catecholase and tyrosinae, implicated in the immune defence mechanisms of invertebrates. In several insects and crustaceans, antibacterial substances are produced through the PO cascade, participating in the direct killing of invading microorganisms., This mechanism of action was first elucidated by Luna-Acosta and co-workers in 2011, working with the mollusc *Cassostrea gigas* (Luna-Acosta *et al.*, 2011). Grover and colleagues (2012) separately immobilized laccase and chloroperoxidase onto multiwalled carbon nanotubes, and subsequently mixed them with a commercial eco-friendly paint to generate biocatalytic coatings. The laccase-nanotube based paints killed more than 99% of *S. aureus* and *E. coli* cells and more than 98% *Bacillus cereus* and *Bacillus anthracis* spores (Grover *et al.*, 2012).

Recently, Rocasalbas and co-workers (2013) described the laccase-assisted formation of bioactive chitosan/gelatinhydrogel stabilized with plant polyphenols. The purpose was to assess the antibacterial and inhibitory effect of myeloperoxidase and collagenase, with the conclusion that these hydrogels showed relevant antibacterial activity against *Pseudomonas aeruginosa* and *S. aureus* (Rocasalbas *et al.*, 2013).

The use of laccase-type of enzymes has found unexpected applications, such as the synthesis of novel penicillins (hybridantibiotics) by biotransformation of the 'natural ones' with laccase from *Trametes* sp. Using this approach, eight novel penicillins were synthesized by heteromolecular reaction of either ampicillin or amoxicillin with 2,5-dihydroxybenzoic acid derivatives. These hybriantibiotics were even capable of inhibiting methicillin-resistant *S. aureus* strains and vancomycin-resistant enterococci (Mikolasch *et al.*, 2006). Following this line of thought, Mikolasch and co-workers, in 2007, synthesized 16 cephalosporins by amination of 2,5-dihydroxybenzoic acid derivatives (Mikolasch *et al.*, 2007). They also successfully used the laccase modification method to produce corollosporine, a substance with antimicrobial properties isolated from the fungus *Corollospora maritime* (Mikolasch *et al.*, 2008).

Antimicrobial metallopeptides

The most important classes of compounds exhibiting antibacterial activity include fluoroquinolones, Β-lactams, tetracyclins, aminoglycosides and macrolides, and their mode of action involves interference with bacterial cell wall synthesis, protein synthesis or nucleic acid synthesis. The latter is possible because the biochemical machinery differs from that of eukaryotic organisms (although partially shared by mitochondria and chloroplasts), and this difference has been extensively exploited in recent years for either the design or isolation of specific antibiotics. However, to date only a few of the vast number of potential nucleic acid targets in bacteria have been exploited. The current urgent need for novel antibacterial drugs, due to the appearance of bacterial resistance to new antibiotics, requires that we find new targets within the bacterial nucleic acid metabolism, to design new antibiotics capable of controlling pathogenic species.

It is well known that human saliva contains non-immune proteins with relevant antimicrobial (against both bacteria and fungi) activities (Melino *et al.*, 2006), that contribute to the defence system in the oral cavity. Histatin 5, a member of the histatin family of proteins, is a proteolytic product of histatin 3 and displays the highest antimicrobial activity against several yeasts, such as *Candida glabrata*, *Candida krusei*, *Saccharomyces cerevisiae* and *Cryptococcus neoformans* (Melino *et al.*, 2006; Tsai and Bobek, 1997). Histatin acts at low concentrations and its activity is pH-dependent. This protein also shows antibacterial activity against bacterial species such as *S. mutans, S. mitis* and *Porfiromonas gingivalis* (MacKay *et al.*, 1984). Melino and co-workers (2006) clearly demonstrated that these oligopeptides readily bound DNA in gel retardation assays, and displayed nuclease activity in the presence of copper, zinc and ascorbate. They were the first to suggest a novel use for these compounds as biotechnological tools (Melino *et al.*, 2006).

Gunderson and colleagues (2009) found that a small oligopeptide, WRWYCR, exhibited antibacterial activity because it interfered with the resolution of the Holliday junctions, thus directly hampering the normal process of homologous recombination in bacteria (Gunderson *et al.*, 2009). Joyner *et al.* (2013) recently reported a modified version of this oligopeptide, their metallo-oligopeptide contained both the complex Cu^{2+}-glycyl-glycylhistidine (Cu-GGH) and the sequence WRWYCR and displayed antimicrobial activity against a variety of pathogenic bacteria.

Fish are a good example of the effectiveness of metallopeptidases in the control the pathogenic bacteria (Tvete and Haugan, 2008). The antibacterial compounds in fish mucus include lysozyme, lectins, peptides, and at least three different antibacterial proteins, among them is metalloprotease C, a compound responsible for killing up to 74% of the bacteria.

Non-lysostaphin proteases, protease inhibitors and phospholipases

Proteases are hydrolytic enzymes capable of breaking the peptide bonds of proteins, and they represent *ca.* 60% of all currently commercialized enzymes worldwide. They have many applications, including in the pharmaceutical industry. Plants produce a variety of proteins that are involved in the defence against pathogens; these involve not only proteases, but also protease inhibitors. As some plant protease inhibitors also exhibit antimicrobial effects, both proteases and inhibitors are considered in the same group. The wide distribution of proteases among plants, animals, and microorganisms demonstrates that they are essential for living organisms, playing important physiological roles in a wide variety of biological processes (Rao *et al.*, 1998).

Phospholipases

A2 type phospholipases (EC 3.1.1.14) and related proteases achieve their antibacterial activity by degrading phosphatidylethanolamine, the major component of bacterial membranes. The phospholipases secreted by mammalians are currently classified into four groups (I, II, V, and X). Group II sPLA2 has been further divided into five subtypes (IIA, IIC, IID, IIE, and IIF), based on their primary structure. Group IIA PLA2 is one of the key enzymes involved in inflammation, and it regulates the synthesis of both arachidonic acid and lysophospholipids (Karray *et al.*, 2011; Russell *et al.*, 2013). The structural features of human group IIA PLA2 are unusual; they exhibit a highly cationic nature and contain a high number of positively charged residues. The highly cationic nature of the enzyme, together with its cellular localization, helps explain its antibacterial action. As a matter of fact, these enzymes are densely packed in the areas of the body susceptible to external bacterial colonization, such as the small intestine (Paneth cells), prostate and eyes (tears).

As early as 1979, Elsbach and co-workers were able to isolate and purify a potent bactericidal/permeability-increasing protease complex, from rabbit polymorphonuclear leucocytes (Elsbach *et al.*, 1979). In fact, the complex was later resolved into two antibacterial proteins, a polypeptide that increased the permeability of the envelope of susceptible Gram-negative bacteria, and a phospholipase A2. While the Gram-positive bacteria were insensitive to these proteins, the Gram-negative, such as *E. coli* and *Salmonella typhimurium*, were highly sensitive to them. Later on, Beers and colleagues (2002) found that the antibacterial properties of the human phospholipase A2 damaged the membrane of Gram-positive bacteria. This was accomplished by the high pI of these proteins, that generated a large excess of cationic residues on the enzyme surface, thus allowing it to penetrate through the bacterial cell wall and reach the cell membrane (Beers *et al.*, 2002). However, the ability of these enzymes to cause bacterial cell membrane degradation can have deleterious effects on host's cells. In this line, Winkler and Daugherty (1989) reported the presence of unusually high levels of phospholipase activity associated with the growth of *Rickettsia prowazeckii* on L929 cells (Winkler and Daugherty, 1989).

Karray and co-workers unequivocally proved the protective role of phospholipase A2 in the small intestine of chickens. The enzyme displayed a potent bactericidal activity against Gram-positive bacteria, but was ineffective against Gram-negative (Karray *et al.*, 2011). ChPLA2-IIA bactericidal activity was found to be dependent on the presence of calcium ions and this activity was potentiated by lysozyme. Similar data were obtained for the intestinal phospholipase A2 obtained from the common stingray *Dasyatis pastinaca* (Ben Bacha *et al.*, 2012).

Animal venoms are probably one of nature's richest sources of pharmacologically active substances, with a high potential for novel antibacterial drugs. Vargas *et al.* (2012) were able isolate both basic and acidic phospholipases A2 from the venom of the snake *Porthidium nasutum* venom. They also obtained an additional acidic phospholipase A2 that displayed indirect haemolytic activity in agarose erythrocyte-egg yolk gels, as well as exhibiting bactericidal activity against *Staphylococcus aureus* (Vargas *et al.*, 2012).

Serine proteases

Serine proteases (or serine endopeptidases) are a large group of proteases involved in diverse biological roles, such as digestive enzymes (e.g. chymotrypsin, trypsin, elastase), blood clotting factors (i.e. Factors 10 (X), 11 (XI), and 12 (XII), thrombin, plasmin, etc.), complement factors (i.e. C1r and C1s, etc.). As a general rule, they are produced by both prokaryotic and eukaryotic organisms and exhibit a high antibacterial activity. These enzymes cleave peptide bonds in proteins, in which serine serves as the nucleophilic amino acid at the (enzyme's) active site, and are classified into two categories (chymotrypsin-like (trypsin-like) or subtilisin-like) based on their structure.

Tsuji and co-workers (2001) purified a serine protease, from pupal extracts of *Sarcophaga peregrina* (flesh fly), with a molecular mass of 26 kDa, that exhibited potent antibacterial activity. They suggested that residues 6 to 18 of peptide 9 form an amphiphilic α-helix, under hydrophobic conditions, with a N-terminal basic loop that enabled the enzyme to interact, in a barrel manner, with acidic phospholipids in the bacterial membranes. This resulted in leakage of low molecular weight components through the bacterial membrane and the outer membrane of Gram-negative bacteria, a mechanism of action similar to that of phospholipase A2 (Tsuji *et al.*, 2001).

Mammal's milk is rich in lactoferrin, a member of the lactotransferrin family of non-haem iron-binding glycoproteins. This protein is also found at high concentrations in other exocrine secretions, where it plays a major role in mucosal defence. Bhinami *et al* (1999) reported that human and bovine lactoferrins, containing different degrees of iron saturation (9–97%), exhibited antibacterial activity (Bhimani *et al.*, 1999). These authors spiked the drinking water they gave to their experimental mice with 2% lactoferrin, and managed to reduce their mice's kidney infections by 40–60%. This treatment also reduced the number of viable bacteria by a factor of 5–12; suggesting a potential use of lactoferrins as natural antibacterial proteins. The mechanism of action of this protein was finally elucidated when lactoferricin B was characterized. Pepsin-treated bovine lactoferrin released a 25-residue long amphipathic peptide with the ability to interact with the bacterial membrane and produce its antimicrobial action (Schibli *et al.*, 1999). The antibacterial effect of lactoferrin was further established by the work of Sanchez and Watts (1999), who demonstrated that addition of ferrin to certain antibiotics, such as novobiocin, enhanced their action and resulted in the control of Gran-negative bacteria (Sanchez and Watts, 1999).

Recent work revealed that lactoferrin has a proteolytic activity and attenuates the pathogenic potential of some bacteria, such as *Haemophilus influenzae*, by cleaving and hydrolysing two bacterial colonization factors. This also classified lactoferrin as part of a serine protease family containing a 'catalytic diad', formed by either Ser259 or Lys73, analogous to a number of bacterial serine proteases (Hendrixson *et al.*, 2003). Mutations of either Ser259 or Lys73 resulted in a dramatic decrease in the enzyme's proteolytic activity and hence its antibacterial activity.

Mammalian serine proteases could help control *Mycobacterium tuberculosis* infections. A contribution to this field is the work of Reece and co-workers (2010), who found that host's serine proteases played an important role in the control of *M. tuberculosis* in hypoxic lung granulomas (Reece *et al.*, 2010). These granulomas contained large amounts of the serine protease cathepsin G, as well as clade B serine protease inhibitors (serpins). Reece's data suggest that serine protease activity acts as a protective mechanism within hypoxic regions of lung granulomas and represents a potential new strategy for the treatment of tuberculosis. The serine proteases released by infiltrated neutrophils could, however, inactivate surfactant protein D, by cleaving it within a conserved subregion of the carbohydrate recognition domain, thus causing deleterious effects on the overall host's defence mechanisms in lung infections (Hirche *et al.*, 2004). In addition, an excess of serine proteases (particularly cathepsins) could degrade human beta-defensins 2 and 3, also favouring bacterial growth (Taggart *et al.*, 2003).

Plants can also produce serine proteases in their fight against pathogenic bacteria. Siritapetawee and co-workers (2012) isolated a 48-kDa serine protease from the latex of the jackfruit (*Artocarpus heterophyllus*) which was active against both *Pseudomonas aeruginosa* and clinical isolates of the yeast *Candida albicans* (Siritapetawee *et al.*, 2012).

Bacteria are good serine protease producers, including serine proteases that possess antibacterial activity. Apart from the lysostaphin endopeptidase, dealt with earlier on in this chapter, soil bacteria such as *Lysobacter enzymogenes* have a remarkable ability to lyse not only other bacteria (earning the above species its generic name) but also soil nematodes. These metalloendopeptidases are capable of lysing *Arthrobacter globiformis*, *Micrococcus luteus* and *Staphylococcus aureus* cells, thus making them potentially very useful as antimicrobial agents (Supuran *et al.*, 2002).

Inhibitors of serine proteases can also show antimicrobial activity per se. One such an example is trappin-2, an inhibitor of neutrophil serine proteases, that displays antimicrobial activity against the bacteria *Pseudomonas aeruginosa*, *Staphylococcus aureus*, *Klebsiella pneumoniae*, *Haemophilus influenzae*, *Streptococcus pneumoniae*, *Branhamella catarrhalis* as well as against the pathogenic fungi *Aspergillus fumigatus* and *Candida albicans* (Baranger *et al.*, 2008).

A different aspect of the serine protease inhibitors group is their ability to modulate the bacterial signal endopeptidases by acylation with penems. The signal peptidases are, in fact, serine proteases responsible for the proteolytic removal of the N-terminal amino acid signal peptide in all kind of organisms; hence this approach targets an essential mechanism able to inhibit and control undesirable bacterial growth. Signal peptidases are really a novel class of serine proteases which generally utilize a serine and a lysine to form a unique *catalytic dyad*, instead of the prototypical Ser-His-Asp catalytic triad, for peptide hydrolysis (Sung and Dalbey, 1992).

Cysteine proteases

Polymorphonuclear leucocytes (PMNs) play an essential role in host defence against bacterial infection, and they produce several compounds that include cytotoxic and antibacterial polypeptides. Most of these proteins are stored in cytoplasmic granules and are released under the appropriate stimuli; cysteine proteases can be one of the components involved in antibacterial activity. Levy and co-workers described, in 1993, the isolation of two protein isoforms of an antibacterial 15-kDa protein that appeared to belong to a novel family of leucocyte proteins, with strong LPS-binding and protease-inhibitory activities (Buhling *et al.*, 2004; Levy *et al.*, 1993).

Cathepsins can also be included here. The term 'cathepsin', was coined in 1920 and stands for 'lysosomal proteolytic enzyme' (Buhling *et al.*, 2004). As indicated in Bühling's review, in addition to cysteine proteases, this term also includes serine proteases (cathepsins A and G) and aspartic proteases (cathepsins D and E). The cysteine proteases (cathepsins B, C, H, L and S) appear to have diverged from the rest early in eukaryotic evolution. Some of these genes encoding novel enzymes (e.g. cathepsins F, K, W and X) could have emerged by gene duplication. Cathepsins are mainly monomeric proteins with two separate domains (an α-helix and a β-barrel-like, respectively) and their active sites (Cys-25 and His-159 amino acids) are located between both domains and are highly homologous to plant papain.

The antibacterial activity of some of these enzymes is not a direct result of the mammalian cysteine protease, but is achieved through the inhibition of a bacterial cysteine protease essential for bacterial growth. The three human cystatins (S, SA and SN) are multifunctional proteins with marked cysteine protease inhibitory activity, and are capable of inhibiting the growth of *Porphyromonas gingivalis*, a bacteria associated with periodontal diseases (Baron *et al.*, 1999).

Plants, in particular tropical plants, are good producers of cysteine proteases. Some examples include *Carica papaya* (papain, chymopapain, caricain, glycil endopeptidase), *Ananas comosus* (fruit bromelain, stem bromelain, ananain, comosain), and *Ficus glabrata* (ficin). These enzymes are of considerable commercial importance, due to their strong proteolytic activity against a broad range of protein substrates and because they are active over a broad range of pH and temperature. According to the MEROPS database, cysteine proteases are divided into ten clans (CA, CD, CE, CF, CH, CL, CM, CN, CO, and C) and plant CPs are currently classified into five of groups (CA, CD, CF, CO, and CE). Most plant CPs belong to the C1 family, also known as the papain family (clan CA) (Feijoo-Siota and Villa, 2011; Rawlings *et al.*, 2010).

Cysteine proteases are mainly used in food science, with applications not only in bakery, the brewing industry, food processing, and in the production of fish protein hydrolysates, but also they are currently being used in pharmaceutical and medical applications. They have been used to treat gastrointestinal parasite infestations, as they degrade the worm's cuticle (Mueller and Mechler, 2005). Papain is the main ingredient of Papacarie, a gel used for chemomechanical dental caries removal. This gel effectively breaks the bonds in collagen fibrils in the dentine, thus avoiding the use of the traditional rotary cutting tools (Kotb *et al.*, 2010)

Non-enzymatic antibacterials: glycans

Glycans can display antibacterial activity in two different ways: either as isolated polysaccharides or forming part of an O-glycosylated peptide. Perhaps the best example of the latter pertains to the insect immunity system and the inducible antibacterial peptide diptericin, a compound that encompasses two O-glycans that are necessary for its biological activity (Bulet *et al.*, 1995). These authors described that the larvae of *Phormia terranovae*, when challenged with bacteria, produce an 82 amino acid long peptide. This polypeptide spans a N-terminal proline-rich region and a central glycine- rich domain. It also contains two threonine O-substitutions, one in the proline-rich region (residue 10) and one in the glycine-rich domain (residue 54). Removal of the O-glycans entirely abolishes the antibacterial activity of the compound.

Paragons of the first group are the polysaccharides synthesized by bacterial lytic transglycosylases (Pfeffer *et al.*, 2012). These enzymes act on peptidoglycan, with the same substrate specificity as lysozyme, but apart from displaying hydrolase activity, they further cut the peptidoglycan to generate an 1,6-anydromuramylmolecule. Because of the autolytic potential they display, these enzymes must be tightly regulated within the producing cells. Pfeffer and coworkers (2012) identified three areas of action for these enzymes, they involve: the modification of substrate, membrane association and complex formation, and the production of proteinaceous inhibitors. These areas represent potential new targets for designing novel drugs to control bacterial growth. Hence, by using transglycosylases isolated from the pathogenic bacteria's cell wall, it is possible to synthesize new polysaccharides that mimic host's carbohydrates, thus resulting in the pathogen's demise.

Glycosaminoglycans are negatively charged polysaccharides consisting of a repeating disaccharide unit; they include the biologically-active compounds: heparin, heparan sulfate, chondroitin sulfate, hyaluronan, dermatan sulfate and keratan sulfate. Glycans can either be directly used or be part of (i) small-molecule drugs, such as streptomycin, erythromycin A, doxorubicin or digoxin; (ii) low molecular weight inhibitors of influenza neuraminidase, such as Zanamivir (Relenza™, modified by the addition, at C-4 position, of a bulky guanidino side chain from a well-known neuraminidase inhibitor, 2-deoxy-2,3-dehydro-N-acetyl-neuraminic acid) with an increased affinity for influenza neuraminidase, but without affecting the host's neuraminidases; (iii) therapeutic glycoproteins that include erythropoietin, cytokines, antibodies, glycosyltransferases, and glycosidases as well as other compounds with antibacterial activity; (iv) therapeutic approaches to metabolic diseases; (v) therapeutic applications of glycosaminoglycans; (vi) formulation of vaccine components, either for microbial vaccines (e.g. against *Haemophilus influenzae* type b); (vii) compounds capable of blocking glycan recognition in some diseases. As it is widely known, many microorganisms bind to mammalian tissues by recognizing specific glycan receptors on their surface, and this could be used to our advantage by designing small soluble glycans that mimic such receptors, therefore blocking the microorganism's initial point of entry into the host's cells and preventing disease from occurring.

Laminins are a family of heterotrimeric extracellular matrix glycoproteins, found in the basement membrane of tissues, and composed of alpha, beta, and gamma chains (Senyurek *et al.*, 2010), as well as glycopeptides derived from human laminin. They exhibit a dose-dependent antimicrobial activity against both Gram-positive and Gram-negative bacteria, carried out by permeabilizing the bacterial membrane. From these data, we can conclude that the different extracellular matrix components are capable of protecting the tissues they enclose from the attack of invading pathogens, and indeed constitute an important part of the host defence mechanism.

Plants also produce bioglycans, such as the sulphated polysaccharide found in *Tridax procumbens*, with anticoagulant, antiherpetic and antibacterial activities (Naqash and Nazeer, 2011).

Antibacterial nanoparticles

Metal nanoparticles were first used to carry active principles, such as antibiotics, by Tom and co-workers (2004) who used gold nanoparticles as a delivery system for ciprofloxacin. This was successful and they found that most of the bound molecules could be released over an extended period of time (Tom *et al.*, 2004). A modification of this use of gold nanoparticles was reported by Nath and colleagues (2008) who described a high-throughput antimicrobial susceptibility assay, based on the concanavalin-induced self-assembly of dextran-coated gold nanoparticles. Their assay is very quick, it yields results within three hours, and can be used for the high-throughput screening of samples during epidemics to expedite clinical decisions (Nath *et al.*, 2008).

Gold nanoparticles, due to their physicochemical properties, can be applied to a variety of biological situations, but they must first be nucleated to the desired particle size and shape. They must then be coated with the active principle and targeted to the appropriate site of action, where they would be activated to accomplish their role, such as carry out different therapeutic applications, in diagnosis, or implement their antibacterial activity.

Gold nanoparticles range from very tiny spheres, with a diameter of 4 nm, obtained by using a strong reductant such as borohydride, to large 100–200 nm nanoshells (Gu *et al.*, 2003; Pissuwan *et al.*, 2010). They are stable and represent a good option as drug delivery systems; additionally, nanoparticles can be specifically targeted to particular areas in the body and this considerably increases their effectiveness and selectivity. In fact, there is mounting evidence that a variety of antibiotics, such as ciprofloxacin, streptomycin, gentamicin, neomycin, ampicillin, kanamycin and 5-fluorouracil, can be easily packaged into gold nanoparticles. In this manner, Gu and co-workers (2003) conjugated vancomycin to gold nanoparticles to successfully inhibit vancomycin-resistant *Enterococcus faecium* and *Enterococcus faecalis*. Their results demonstrated that the activity of vancomycin increased by a factor of 50 as a result of its conjugation to the nanoparticle (Gu *et al.*, 2003). This approach proved also successful against Gram-negative bacteria, although this required an increase in the antibiotic doses (to twice as much as those required to treat Gram-positive bacteria), probably due to the presence of an additional barrier (the outer membrane) in Gram-negative microorganisms.

Novel strategies are continually emerging in the design of gold nanoparticles with enhanced antibacterial activity. One such case is the use of amino-substituted pyrimidines linked to gold nanoparticles, although these compounds are inactive as antibiotics, they are effective antibacterials by sequestering magnesium or calcium ions that in turn disrupt the bacterial cell membrane, finally causing bacterial death (Zhao *et al.*, 2010).

Lysozyme is perhaps the paragon of the above mention enzymes, for a variety of reasons including the fact that it can be easily crystallized. This enzyme spans 129 amino acids, including three phenylalanines, three tyrosines, six tryptophans and one histidine. Two types of crystalline structures can be obtained for lysozyme (orthorhombic or tetragonal), depending on the crystallization temperature.

Silver nanoparticles exhibit unique antibacterial properties that make them ideal candidates for medical applications because, if they are appropriately prepared, they can be used as antimicrobials (Kemp *et al.*, 2009). Currently metal nanoparticles are being considered for biological uses, because of their unique properties as drug delivery systems. Silver (and metals in general) does not normally display an antibacterial activity, but has many uses in medicine, such as in eye drops to prevent neonatal blindness caused by neisserial species, and in the manufacture of catheters. Their antibacterial activity is exerted by inhibiting HS-containing enzymes and thus altering bacterial membrane permeability, and it is this characteristic that has been exploited in a wide variety of applications, such as in the manufacture of medical catheters and prostheses, in textiles and in water treatment.

References

Agematu, H., Tsuchida, T., Kominato, K., Shibamoto, N., Yoshioka, T., Nishida, H., Okamoto, R., Shin, T., and Murao, S. (1993). Enzymatic dimerization of penicillin X. J. Antibiot. *46*, 141–148.

Aggarwal, K., and Silverman, N. (2008). Positive and negative regulation of the Drosophila immune response. BMB reports *41*, 267–277.

Alsberg, C.L., and Black, O.F. (1913). Contribution to the study of maize deterioration, biochemical and toxicological investigations of *Penicillium puberulum and Penicillium stoloniferum*. Bull. Burl. Anim. Ind. US Dept. Agr. *270*, 1–47.

Amorena, B., Gracia, E., Monzon, M., Leiva, J., Oteiza, C., Perez, M., Alabart, J.L., and Hernandez-Yago, J. (1999). Antibiotic susceptibility assay for *Staphylococcus aureus* in biofilms developed *in vitro*. J. Antimicrob. Chemother. *44*, 43–55.

Angel, C.S., Ruzek, M., and Hostetter, M.K. (1994). Degradation of C3 by *Streptococcus pneumoniae*. J. Infect. Dis. *170*, 600–608.

Aparicio, R., Neyen, C., Lemaitre, B., and Busturia, A. (2013). dRYBP contributes to the negative regulation of the Drosophila Imd pathway. PloS One *8*, e62052.

Balzli, C.L., McCormick, C.C., Caballero, A.R., and O'Callaghan, R.J. (2010). Sustained anti-staphylococcal effect of lysostaphin in the rabbit aqueous humor. Curr. Eye Res. *35*, 480–486.

Baranger, K., Zani, M.L., Chandenier, J., Dallet-Choisy, S., and Moreau, T. (2008). The antibacterial and antifungal properties of trappin-2 (pre-elafin) do not depend on its protease inhibitory function. FEBS J. *275*, 2008–2020.

Barkulis, S.S., Smith, C., Boltralik, J.J., and Heymann, H. (1964). Structure of Streptococcal Cell Walls. Iv. Purification and properties of Streptococcal phage muralysin. J Bio. Chem. *239*, 4027–4033.

Baron, A.C., Gansky, S.A., Ryder, M.I., and Featherstone, J.D. (1999). Cysteine protease inhibitory activity and levels of salivary cystatins in whole saliva of periodontally diseased patients. J. Periodont. Res. *34*, 437–444.

Beers, S.A., Buckland, A.G., Koduri, R.S., Cho, W., Gelb, M.H., and Wilton, D.C. (2002). The antibacterial properties of secreted phospholipases A2: a major physiological role for the group IIA enzyme that depends on the very high pI of the enzyme to allow penetration of the bacterial cell wall. J. Bio. Chem *277*, 1788–1793.

Ben Bacha, A., Abid, I., and Horchani, H. (2012). Antibacterial properties of intestinal phospholipase A2 from the common stingray Dasyatis pastinaca. Appl. Biochem. Biotechnol. *168*, 1277–1287.

Bhimani, R.S., Vendrov, Y., and Furmanski, P. (1999). Influence of lactoferrin feeding and injection against systemic staphylococcal infections in mice. J. Appl. Microbiol. *86*, 135–144.

Biswas, R., Martinez, R.E., Gohring, N., Schlag, M., Josten, M., Xia, G., Hegler, F., Gekeler, C., Gleske, A.K., Gotz, F., *et al.* (2012). Proton-binding capacity of *Staphylococcus aureus* wall teichoic acid and its role in controlling autolysin activity. PloS One 7, e41415.

Bollinger, N., Hassett, D.J., Iglewski, B.H., Costerton, J.W., and McDermott, T.R. (2001). Gene expression in *Pseudomonas aeruginosa*: evidence of iron override effects on quorum sensing and biofilm-specific gene regulation. J. Bacteriol. *183*, 1990–1996.

Borysowski, J., and Gorski, A. (2010a). Anti-staphylococcal lytic enzymes. In Enzybiotics: Enzymes as Drugs and Therapeutics, P. Veiga-Crespo, and T.G. Villa, eds (Wiley and Sons, Hoboken, NJ), pp. 149–172.

Borysowski, J., and Gorski, A. (2010b). Enzybiotics and their potential applications in medicine. In Enzybiotics: Enzymes as Drugs and Therapeutics, P. Veiga-Crespo and T.G. Villa, eds (Wiley and Sons Inc, Hoboken, NJ), pp. 1–26.

Bronneke, V., and Fiedler, F. (1994). Production of bacteriolytic enzymes by Streptomyces globisporus regulated by exogenous bacterial cell walls. Appl. Envioron. Microbiol. *60*, 785–791.

Brown, W.C., Fraser, D.K., and Young, F.E. (1970). Problems in purification of a *Bacillus subtilis* autolytic enzyme caused by association with teichoic acid. Biochim. Biophys. Acta *198*, 308–315.

Brunskill, E.W., and Bayles, K.W. (1996). Identification and molecular characterization of a putative regulatory locus that affects autolysis in *Staphylococcus aureus*. J. Bacteriol. *178*, 611–618.

Buhling, F., Waldburg, N., Reisenauer, A., Heimburg, A., Golpon, H., and Welte, T. (2004). Lysosomal cysteine proteases in the lung: role in protein processing and immunoregulation. Eur. Respir. J *23*, 620–628.

Bulet, P., Hegy, G., Lambert, J., van Dorsselaer, A., Hoffmann, J.A., and Hetru, C. (1995). Insect immunity. The inducible antibacterial peptide diptericin carries two O-glycans necessary for biological activity. Biochemistry *34*, 7394–7400.

Calamita, H.G., and Doyle, R.J. (2002). Regulation of autolysins in teichuronic acid-containing *Bacillus subtilis* cells. Mol. Microbiol. *44*, 601–606.

Calamita, H.G., Ehringer, W.D., Koch, A.L., and Doyle, R.J. (2001). Evidence that the cell wall of *Bacillus subtilis* is protonated during respiration. PNAS *98*, 15260–15263.

Calandra, G.B., and Cole, R.M. (1980). Lysis and protoplast formation of group B streptococci by mutanolysin. Infect. Immun. *28*, 1033–1037.

Cantani, A. (1885). Un tentativo di batterioterapia. Gior. Int. Sci. Med. *7*, 493.

Chain, E., Florey, H.W., Gardner, A.D., Heatley, N.G., Jennings, M.A., Orr-Ewing, J., and Sanders, A.G. (1940). Penicillin as a chemotherapeutic agent. Lancet *236*, 226–228.

Charroux, B., and Royet, J. (2010). Drosophila immune response: From systemic antimicrobial peptide production in fat body cells to local defense in the intestinal tract. Fly *4*, 40–47.

Chatfield, C.H., Koo, H., and Quivey, R.G., Jr. (2005). The putative autolysin regulator LytR in *Streptococcus mutans* plays a role in cell division and is growth-phase regulated. Microbiol. *151*, 625–631.

Costerton, J.W., Stewart, P.S., and Greenberg, E.P. (1999). Bacterial biofilms: a common cause of persistent infections. Science *284*, 1318–1322.

Cromartie, W.J., Craddock, J.G., Schwab, J.H., Anderle, S.K., and Yang, C.H. (1977). Arthritis in rats after systemic injection of streptococcal cells or cell walls. J. Exp. Med. *146*, 1585–1602.

Croux, C., Ronda, C., Lopez, R., and Garcia, J.L. (1993). Role of the C-terminal domain of the lysozyme of Clostridium acetobutylicum ATCC 824 in a chimeric pneumococcal-clostridial cell wall lytic enzyme. FEBS Lett. *336*, 111–114.

Dajcs, J.J., Hume, E.B., Moreau, J.M., Caballero, A.R., Cannon, B.M., and O'Callaghan, R.J. (2000). Lysostaphin treatment of methicillin-resistant *Staphylococcus aureus* keratitis in the rabbit. IOVS *41*, 1432–1437.

Destoumieux, D., Bulet, P., Loew, D., Van Dorsselaer, A., Rodriguez, J., and Bachere, E. (1997). Penaeidins, a new family of antimicrobial peptides isolated from the shrimp Penaeus vannamei (Decapoda). J. Bio. Chem *272*, 28398–28406.

Diaz, E., Munthali, M., Lunsdorf, H., Holtje, J.V., and Timmis, K.N. (1996). The two-step lysis system of pneumococcal bacteriophage EJ-1 is functional in gram-negative bacteria: triggering of the major pneumococcal autolysin in *Escherichia coli*. Mol. Microbiol. *19*, 667–681.

Donovan, D.M. (2007). Bacteriophage and peptidoglycan degrading enzymes with antimicrobial applications. Recent Pat. Biotechnol. *1*, 113–122.

Dubos, R.J. (1939). Studies on a bactericidal agent extracted from a soil Bacillus: I. Preparation of the agent. Its activity *in vitro*. J. Exp. Med. *70*, 1–10.

Eggert, C. (1997). Laccase-catalyzed formation of cinnabarinic acid is responsible for antibacterial activity of Pycnoporus cinnabarinus. Microbiol. Res. *152*, 315–318.

Eldholm, V., Gutt, B., Johnsborg, O., Bruckner, R., Maurer, P., Hakenbeck, R., Mascher, T., and Havarstein, L.S. (2010). The pneumococcal cell envelope stress-sensing system LiaFSR is activated by murein hydrolases and lipid II-interacting antibiotics. J. Bacteriol. *192*, 1761–1773.

Elsbach, P., Weiss, J., Franson, R.C., Beckerdite-Quagliata, S., Schneider, A., and Harris, L. (1979). Separation and purification of a potent bactericidal/permeability-increasing protein and a closely associated phospholipase A2 from rabbit polymorphonuclear leukocytes. Observations on their relationship. J. Bio. Chem *254*, 11000–11009.

Emmerich, R. (1887). Die Heilung des Milzbrandes. Arch f Hyg, Berlin *6*, 442–501.

Ephrussi-Taylor, H. (1951). Genetic aspects of transformations of pneumococci. Cold Spring Harbor Symposia on Quantitative Biology *16*, 445–456.

Fan, W., Plaut, K., Bramley, A.J., Barlow, J.W., Mischler, S.A., and Kerr, D.E. (2004). Persistency of adenoviral-mediated lysostaphin expression in goat mammary glands. J. Dairy Sci. *87*, 602–608.

Feijoo-Siota, L., and Villa, T.G. (2011). Native and biotechnologically engineered plant proteases wiht industrial applications.Food Bioprocess. Tech. *4*, 1066–1088.

Fleming, A. (1929). On the antibacterial action of cultures of a Penicillium with special reference to their use in the isolation of B. influenza. Br. J. Exper. Path. *10*, 226–237.

Florey, H.W. (1945). Use of micro-organisms for therapeutic purposes. Br. Med. J. *2*, 635–642.

Frankel, M.B., and Schneewind, O. (2012). Determinants of murein hydrolase targeting to cross-wall of *Staphylococcus aureus* peptidoglycan. J. Bio. Chem. *287*, 10460–10471.

Frias, M.J., Melo-Cristino, J., and Ramirez, M. (2009). The autolysin LytA contributes to efficient bacteriophage progeny release in *Streptococcus pneumoniae*. J. Bacteriol. *191*, 5428–5440.

Fujimoto, D.F., Brunskill, E.W., and Bayles, K.W. (2000). Analysis of genetic elements controlling *Staphylococcus aureus* lrgAB expression: potential role of DNA topology in SarA regulation. J. Bacteriol. *182*, 4822–4828.

Fuqua, C., Parsek, M.R., and Greenberg, E.P. (2001). Regulation of gene expression by cell-to-cell communication: acyl-homoserine lactone quorum sensing. Ann. Rev. Genet. *35*, 439–468.

Galvez, A., Valdivia, E., Martinez, M., and Maqueda, M. (1989). Effect of peptide AS-48 on Enterococcus faecalis subsp. liquefaciens S-47. Antimicrob. Agents Chemother. *33*, 641–645.

Ganz, T., Selsted, M.E., Szklarek, D., Harwig, S.S., Daher, K., Bainton, D.F., and Lehrer, R.I. (1985). Defensins. Natural peptide antibiotics of human neutrophils. J. Clin. Invest. *76*, 1427–1435.

Garcia, A.E., Osapay, G., Tran, P.A., Yuan, J., and Selsted, M.E. (2008). Isolation, synthesis, and antimicrobial activities of naturally occurring theta-defensin isoforms from baboon leukocytes. Infect. Immun. *76*, 5883–5891.

Garcia, E., Garcia, J.L., Ronda, C., Garcia, P., and Lopez, R. (1985). Cloning and expression of the pneumococcal autolysin gene in *Escherichia coli*. Mol. Gen. Genet. *201*, 225–230.

Garcia, J.L., Diaz, E., Romero, A., and Garcia, P. (1994). Carboxy-terminal deletion analysis of the major pneumococcal autolysin. J. Bacteriol. *176*, 4066–4072.

Gasset, M. (2010). Membrane-targeted enzybiotics. In Enzybiotics: Enzymes as Drugs and Therapeutics, P. Veiga-Crespo, and T.G. Villa, eds. (Wiley and Sons, Hoboken, NJ).

Gil-Montoya, J.A., Guardia-Lopez, I., and Gonzalez-Moles, M.A. (2008). Evaluation of the clinical efficacy of a mouthwash and oral gel containing the antimicrobial proteins lactoperoxidase, lysozyme and lactoferrin in elderly patients with dry mouth – a pilot study. Gerodontology *25*, 3–9.

Gilpin, R.W., Chatterjee, A.N., and Young, F.E. (1972). Autolysis of microbial cells: salt activation of autolytic enzymes in a mutant of *Staphylococcus aureus*. J. Bacteriol. *111*, 272–283.

Greig-Smith, R. (1917). Contributions to our knowledge of soil fertility, XV. Proc. Linn. Soc. NSW *42*, 162–166.

Grinde, B. (1989). A lysozyme isolated from rainbow trout acts on mastitis pathogens. FEMS Microbiol. Lett. *51*, 179–182.

Grover, N., Borkar, I.V., Dinu, C.Z., Kane, R.S., and Dordick, J.S. (2012). Laccase- and chloroperoxidase-nanotube paint composites with bactericidal and sporicidal activity. Enzyme Microb. Technol. *50*, 271–279.

Grutter, M.G., Weaver, L.H., and Matthews, B.W. (1983). Goose lysozyme structure: an evolutionary link between hen and bacteriophage lysozymes? Nature *303*, 828–831.

Gu, H., Ho, P.L., Tsang, K.W., Wang, L., and Xu, B. (2003). Using biofunctional magnetic nanoparticles to capture vancomycin-resistant enterococci and other gram-positive bacteria at ultralow concentration. J. Am. Chem. Soc. *125*, 15702–15703.

Gunderson, C.W., Boldt, J.L., Authement, R.N., and Segall, A.M. (2009). Peptide wrwycr inhibits the excision of several prophages and traps holliday junctions inside bacteria. J. Bacteriol. *191*, 2169–2176.

Hamada, S., Torii, M., Kotani, S., Masuda, N., Ooshima, T., Yokogawa, K., and Kawata, S. (1978). Lysis of *Streptococcus mutans* cells with mutanolysin, a lytic enzyme prepared from a culture liquor of Streptomyces globisporus 1829. Arch. Oral Biol. *23*, 543–549.

Hendrixson, D.R., Qiu, J., Shewry, S.C., Fink, D.L., Petty, S., Baker, E.N., Plaut, A.G., and St Geme, J.W., 3rd (2003). Human milk lactoferrin is a serine protease that cleaves Haemophilus surface proteins at arginine-rich sites. Mol. Microbiol. *47*, 607–617.

Hirche, T.O., Crouch, E.C., Espinola, M., Brokelman, T.J., Mecham, R.P., DeSilva, N., Cooley, J., Remold-O'Donnell, E., and Belaaouaj, A. (2004). Neutrophil serine proteinases inactivate surfactant protein D by cleaving within a conserved subregion of the carbohydrate recognition domain. J. Bio. Chem. *279*, 27688–27698.

Hoiby, N., and Koch, C. (1990). Cystic fibrosis. 1. *Pseudomonas aeruginosa* infection in cystic fibrosis and its management. Thorax 45, 881–884.

Hultmark, D., Engstrom, A., Andersson, K., Steiner, H., Bennich, H., and Boman, H.G. (1983). Insect immunity. Attacins, a family of antibacterial proteins from Hyalophora cecropia. EMBO J. *2*, 571–576.

Iacono, V.J., Zove, S.M., Grossbard, B.L., Pollock, J.J., Fine, D.H., and Greene, L.S. (1985). Lysozyme-mediated aggregation and lysis of the periodontal microorganism Capnocytophaga gingivalis 2010. Infect. Immun. *47*, 457–464.

Irwin, D.M. (1995). Evolution of the bovine lysozyme gene family: changes in gene expression and reversion of function. J. Mol. Evol. *41*, 299–312.

Irwin, D.M., and Wilson, A.C. (1989). Multiple cDNA sequences and the evolution of bovine stomach lysozyme. J. Bio. Chem. *264*, 11387–11393.

Iversen, O.J., and Grov, A. (1973). Studies on lysostaphin. Separation and characterization of three enzymes. Eur. J. Biochem./FEBS *38*, 293–300.

Janusz, M.J., Chetty, C., Eisenberg, R.A., Cromartie, W.J., and Schwab, J.H. (1984). Treatment of experimental erosive arthritis in rats by injection of the muralytic enzyme mutanolysin. J. Exp. Med. *160*, 1360–1374.

Janusz, M.J., Esser, R.E., and Schwab, J.H. (1986). *In vivo* degradation of bacterial cell wall by the muralytic enzyme mutanolysin. Infect. Immun. *52*, 459–467.

Janusz, M.J., Eisenberg, R.A., and Schwab, J.H. (1987). Effect of muralytic enzyme degradation of streptococcal cell wall on complement activation *in vivo* and *in vitro*. Inflammation *11*, 73–85.

Jennings, M.A. (1945). Activity of Helvolic Acid against *Mycobacterium tuberculosis*. Nature *156*, 633–633.

Jolles, P. (1996). From the discovery of lysozyme to the characterization of several lysozyme families. Exs *75*, 3–5.

Jones, D., Metzger, H.J., Schatz, A., and Waksman, S.A. (1944). Control of Gram-negative bacteria in experimental animals by streptomycin. Science *100*, 103–105.

Joyner, J.C., Hodnick, W.F., Cowan, A.S., Tamuly, D., Boyd, R., and Cowan, J.A. (2013). Antimicrobial metallopeptides with broad nuclease and ribonuclease activity. Chem. Commun. *49*, 2118–2120.

Karray, A., Ali, Y.B., Gargouri, Y., and Bezzine, S. (2011). Antibacterial properties of chicken intestinal phospholipase A2. Lipids Health Dis. *10*, 4.

Kemp, M.M., Kumar, A., Clement, D., Ajayan, P., Mousa, S., and Linhardt, R.J. (2009). Hyaluronan- and heparin-reduced silver nanoparticles with antimicrobial properties. Nanomedicine *4*, 421–429.

King, B.F., Biel, M.L., and Wilkinson, B.J. (1980). Facile penetration of the *Staphylococcus aureus* capsule by lysostaphin. Infect. Immun. *29*, 892–896.

Kirisits, M.J., Margolis, J.J., Purevdorj-Gage, B.L., Vaughan, B., Chopp, D.L., Stoodley, P., and Parsek, M.R. (2007). Influence of the hydrodynamic environment on quorum sensing in *Pseudomonas aeruginosa* biofilms. J. Bacteriol. *189*, 8357–8360.

Kokai-Kun, J.F., Walsh, S.M., Chanturiya, T., and Mond, J.J. (2003). Lysostaphin cream eradicates *Staphylococcus aureus* nasal colonization in a cotton rat model. Antimicrob. Agents Chemother. *47*, 1589–1597.

Kolls, J.K., McCray, P.B., Jr., and Chan, Y.R. (2008). Cytokine-mediated regulation of antimicrobial proteins. Nature Rev. Immunol. *8*, 829–835.

Komatsuzawa, H., Suzuki, J., Sugai, M., Miyake, Y., and Suginaka, H. (1994). Effect of combination of oxacillin and non-β-lactam antibiotics on methicillin-resistant *Staphylococcus aureus*. The Journal of antimicrobial chemotherapy 33, 1155–1163.

Kotb, R.M., Abdella, A.A., El Kateb, M.A., and Ahmed, A.M. (2010). Clinical evaluation of papacarie in primary teeth. J. Clin. Pediatr. Dent. *34*, 117–123.

Kusuma, C., Jadanova, A., Chanturiya, T., and Kokai-Kun, J.F. (2007). Lysostaphin-resistant variants of *Staphylococcus aureus* demonstrate reduced fitness *in vitro* and *in vivo*. Antimicrob. Agents Chemother. *51*, 475–482.

Laible, N.J., and Germaine, G.R. (1985). Bactericidal activity of human lysozyme, muramidase-inactive lysozyme, and cationic polypeptides against Streptococcus sanguis and Streptococcus faecalis: inhibition by chitin oligosaccharides. Infect. Immun. *48*, 720–728.

Lazarevic, V., Margot, P., Soldo, B., and Karamata, D. (1992). Sequencing and analysis of the *Bacillus subtilis* lytRABC divergon: a regulatory unit encompassing the structural genes of the N-acetylmuramoyl-L-alanine amidase and its modifier. J. Gen. Microbiol. *138*, 1949–1961.

Levy, O., Weiss, J., Zarember, K., Ooi, C.E., and Elsbach, P. (1993). Antibacterial 15-kDa protein isoforms (p15s) are members of a novel family of leukocyte proteins. J. Bio. Chem. *268*, 6058–6063.

Linder, L., Andersson, C., Sund, M.L., and Shockman, G.D. (1983). Protoplast formation and localization of enzymes in *Streptococcus mitis*. Infect. Immun. *40*, 1146–1154.

Lode, A. (1903). Centralbl. f. Bakteriol. Abt Orig *33*, 196.

Loeffler, J.M., Nelson, D., and Fischetti, V.A. (2001). Rapid killing of *Streptococcus pneumoniae* with a bacteriophage cell wall hydrolase. Science *294*, 2170–2172.

Lopez, R., Garcia, E., Garcia, P., and Garcia, J.L. (1997). The pneumococcal cell wall degrading enzymes: a modular design to create new lysins? Microb Drug Resist 3, 199–211.

Luna-Acosta, A., Saulnier, D., Pommier, M., Haffner, P., De Decker, S., Renault, T., and Thomas-Guyon, H. (2011). First evidence of a potential antibacterial activity involving a laccase-type enzyme of the phenoloxidase system in Pacific oyster *Crassostrea gigas* haemocytes. Fish Shellfish Immun. *31*, 795–800.

McBride, S.M., and Sonenshein, A.L. (2011). The dlt operon confers resistance to cationic antimicrobial peptides in *Clostridium difficile*. Microbiol. *157*, 1457–1465.

McCormick, C.C., Dajcs, J.J., Reed, J.M., Marquart, M.E., and O'Callaghan, R.J. (2006). The effectiveness of lysostaphin therapy for experimental coagulase-negative Staphylococcus endophthalmitis. Curr. Eye Res. *31*, 225–230.

Mack, D., Rohde, H., Dobinsky, S., Riedewald, J., Nedelmann, M., Knobloch, J.K., Elsner, H.A., and Feucht, H.H. (2000). Identification of three essential regulatory gene loci governing expression of *Staphylococcus epidermidis* polysaccharide intercellular adhesin and biofilm formation. Infect. Immun. *68*, 3799–3807.

MacKay, B.J., Denepitiya, L., Iacono, V.J., Krost, S.B., and Pollock, J.J. (1984). Growth-inhibitory and bactericidal effects of human parotid salivary histidine-rich polypeptides on *Streptococcus mutans*. Infect. Immun. *44*, 695–701.

Mader, J.S., Smyth, D., Marshall, J., and Hoskin, D.W. (2006). Bovine lactoferricin inhibits basic fibroblast growth factor- and vascular endothelial growth factor165-induced angiogenesis by competing for heparin-like binding sites on endothelial cells. Am. J. Pathol. *169*, 1753–1766.

Mangoni, M.L., Rinaldi, A.C., Di Giulio, A., Mignogna, G., Bozzi, A., Barra, D., and Simmaco, M. (2000). Structure–function relationships of temporins, small antimicrobial peptides from amphibian skin. Eur. J. Biochem./FEBS 267, 1447–1454.

Maresso, A.W., and Schneewind, O. (2008). Sortase as a target of anti-infective therapy. Pharmacol. Rev. *60*, 128–141.

Masschalck, B., and Michiels, C.W. (2003). Antimicrobial properties of lysozyme in relation to foodborne vegetative bacteria. Crit. Rev. Microbiol. *29*, 191–214.

Melino, S., Gallo, M., Trotta, E., Mondello, F., Paci, M., and Petruzzelli, R. (2006). Metal-binding and nuclease activity of an antimicrobial peptide analogue of the salivary histatin 5. Biochemistry *45*, 15373–15383.

Metcalf, R.H., and Deibel, R.H. (1969). *Staphylococcus aureus* response to lysostaphin in some fermented foods. Appl. Microbiol. *17*, 63–67.

Mikolasch, A., Niedermeyer, T.H., Lalk, M., Witt, S., Seefeldt, S., Hammer, E., Schauer, F., Gesell, M., Hessel, S., Julich, W.D., *et al.* (2006). Novel penicillins synthesized by biotransformation using laccase from Trametes spec. Chem. Pharm. Bull. *54*, 632–638.

Mikolasch, A., Niedermeyer, T.H., Lalk, M., Witt, S., Seefeldt, S., Hammer, E., Schauer, F., Gesell

Salazar, M., Hessel, S., Julich, W.D., *et al.* (2007). Novel cephalosporins synthesized by amination of 2,5-dihydroxybenzoic acid derivatives using fungal laccases II. Chem. Pharm. Bull. *55*, 412–416.

Mikolasch, A., Hessel, S., Salazar, M.G., Neumann, H., Manda, K., Gordes, D., Schmidt, E., Thurow, K., Hammer, E., Lindequist, U., *et al.* (2008). Synthesis of new N-analogous corollosporine derivatives with antibacterial activity by laccase-catalyzed amination. Chem. Pharm. Bull. *56*, 781–786.

Mookherjee, N., and Hancock, R.E. (2007). Cationic host defence peptides: innate immune regulatory peptides as a novel approach for treating infections. Cell. Mol. Life Sci. *64*, 922–933.

Mueller, M.S., and Mechler, E. (2005). Medicinal plants in tropical countries. Traditional use-Experience-facts (Thieme, Stuttgart, Germany).

Nagant, C., Pitts, B., Nazmi, K., Vandenbranden, M., Bolscher, J.G., Stewart, P.S., and Dehaye, J.P. (2012). Identification of peptides derived from the human antimicrobial peptide LL-37 active against biofilms formed by *Pseudomonas aeruginosa* using a library of truncated fragments. Antimicrob. Agents Chemother. *56*, 5698–5708.

Nakimbugwe, D., Masschalck, B., Atanassova, M., Zewdie-Bosuner, A., and Michiels, C.W. (2006). Comparison of bactericidal activity of six lysozymes at atmospheric pressure and under high hydrostatic pressure. Int. J. Food Microbiol.*108*, 355–363.

Naqash, S.Y., and Nazeer, R.A. (2011). Anticoagulant, antiherpetic and antibacterial activities of sulphated polysaccharide from Indian medicinal plant Tridax procumbens L. (Asteraceae). Appl. Biochem. Biotechnol. *165*, 902–912.

Nath, S., Kaittanis, C., Tinkham, A., and Perez, J.M. (2008). Dextran-coated gold nanoparticles for the assessment of antimicrobial susceptibility. Anal. Chem. *80*, 1033–1038.

Obregon, V., Garcia, J.L., Garcia, E., Lopez, R., and Garcia, P. (2003). Genome organization and molecular analysis of the temperate bacteriophage MM1 of *Streptococcus pneumoniae*. J. Bacteriol. *185*, 2362–2368.

Pangule, R.C., Brooks, S.J., Dinu, C.Z., Bale, S.S., Salmon, S.L., Zhu, G., Metzger, D.W., Kane, R.S., and Dordick, J.S. (2010). Antistaphylococcal nanocomposite films based on enzyme–nanotube conjugates. ACS Nano. *4*, 3993–4000.

Pasteur, L., and Joubert, J.F. (1877). Charbon et septicémie. C. r. Séanc. Acad. Sci. 85, 101–115.

Pearce, G., Yamaguchi, Y., Munske, G., and Ryan, C.A. (2008). Structure-activity studies of AtPep1, a plant peptide signal involved in the innate immune response. Peptides *29*, 2083–2089.

Pfeffer, J.M., Moynihan, P.J., Clarke, C.A., Vandenende, C., and Clarke, A.J. (2012). Control of lytic transglycosylase activity within bacterial cell walls. In Bacterial Glycomics: Current Research, Technology and Applications, C.W. Reid, S.M. Twine and A.N. Reid, eds (Ottawa, Canada: Caister Academic Press).

Pissuwan, D., Cortie, C.H., Valenzuela, S.M., and Cortie, M.B. (2010). Functionalised gold nanoparticles for controlling pathogenic bacteria. Trends Biotechnol. *28*, 207–213.

Potapova, I., Eglin, D., Laschke, M.W., Bischoff, M., Richards, R.G., and Moriarty, T.F. (2013). Two-step labeling of *Staphylococcus aureus* with lysostaphin-azide and DIBO-alexa using click chemistry. J. Microbiol. Methods *92*, 90–98.

Prager, E.M., and Wilson, A.C. (1988). Ancient origin of lactalbumin from lysozyme: analysis of DNA and amino acid sequences. J. Mol. Evol. *27*, 326–335.

Qasba, P.K., and Kumar, S. (1997). Molecular divergence of lysozymes and alpha-lactalbumin. Crit. Rev. Biochem. Mol. *32*, 255–306.

Ralston, D.J., Lieberman, M., Baer, B., and Krueger, A.P. (1957). Staphylococcal virolysin, a phage-induced lysin; its differentiation from the autolysis of normal cells. J. Gen. Physiol. *40*, 791–807.

Rao, M.B., Tanksale, A.M., Ghatge, M.S., and Deshpande, V.V. (1998). Molecular and biotechnological aspects of microbial proteases. Microbiol. Mol. Biol. Rev. *62*, 597–635.

Rawlings, N.D., Barrett, A.J., and Bateman, A. (2010). MEROPS: the peptidase database. Nucl. Acids Res.*38*, D227–233.

Reece, S.T., Loddenkemper, C., Askew, D.J., Zedler, U., Schommer-Leitner, S., Stein, M., Mir, F.A., Dorhoi, A., Mollenkopf, H.J., Silverman, G.A., *et al.* (2010). Serine protease activity contributes to control of Mycobacterium tuberculosis in hypoxic lung granulomas in mice. J. Clin. Invest. *120*, 3365–3376.

Reichmann, N.T., and Grundling, A. (2011). Location, synthesis and function of glycolipids and polyglycerolphosphate lipoteichoic acid in Gram-positive bacteria of the phylum Firmicutes. FEMS Microbiol. Lett. *319*, 97–105.

Richmond, M.H. (1959). Lytic enzymes of *Staphylococcus aureus* 524 Biochim. Biophys. Acta. *31*, 564–565.

Rinaldi, A.C. (2002). Antimicrobial peptides from amphibian skin: an expanding scenario. Curr. Opin. Chem. Biol. *6*, 799–804.

Rocasalbas, G., Francesko, A., Tourino, S., Fernandez-Francos, X., Guebitz, G.M., and Tzanov, T. (2013). Laccase-assisted formation of bioactive chitosan/gelatin hydrogel stabilized with plant polyphenols. Carbohydr. Polymers 92, 989–996.

Romero, P., Lopez, R., and Garcia, E. (2004). Genomic organization and molecular analysis of the inducible prophage EJ-1, a mosaic myovirus from an atypical pneumococcus. Virology 322, 239–252.

Ronda-Lain, C., Lopez, R., Tapia, A., and Tomasz, A. (1977). Role of the pneumococcal autolysin (murein hydrolase) in the release of progeny bacteriophage and in the bacteriophage-induced lysis of the host cells. J. Virol. *21*, 366–374.

Russell, A.B., LeRoux, M., Hathazi, K., Agnello, D.M., Ishikawa, T., Wiggins, P.A., Wai, S.N., and Mougous, J.D. (2013). Diverse type VI secretion phospholipases are functionally plastic antibacterial effectors. Nature *496*, 508–512.

Sanchez, M.S., and Watts, J.L. (1999). Enhancement of the activity of novobiocin against *Escherichia coli* by lactoferrin. J. Dairy Sci. *82*, 494–499.

Satishkumar, R., Sankar, S., Yurko, Y., Lincourt, A., Shipp, J., Heniford, B.T., and Vertegel, A. (2013). Evaluation of the antimicrobial activity of lysostaphin-coated hernia repair meshes. Antimicrob. Agents Chemother. *55*, 4379–4385.

Sava, G. (1996). Pharmacological aspects and therapeutic applications of lysozymes. Exs *75*, 433–449.

Schaffner, W., Melly, M.A., Hash, J.H., and Koenig, M.G. (1967). Lysostaphin: an enzymatic approach to staphylococcal disease. I. *In vitro* studies. Yale J. Biol. Med. *39*, 215–229.

Schibli, D.J., Hwang, P.M., and Vogel, H.J. (1999). The structure of the antimicrobial active center of lactoferricin B bound to sodium dodecyl sulfate micelles. FEBS Lett. *446*, 213–217.

Schindler, C.A., and Schuhardt, V.T. (1964). Lysostaphin: A new bacteriolytic agent for the Staphylococcus. PNAS *51*, 414–421.

Schirner, K., Marles-Wright, J., Lewis, R.J., and Errington, J. (2009). Distinct and essential morphogenic functions for wall- and lipo-teichoic acids in *Bacillus subtilis*. EMBO J. *28*, 830–842.

Schlag, M., Biswas, R., Krismer, B., Kohler, T., Zoll, S., Yu, W., Schwarz, H., Peschel, A., and Gotz, F. (2010). Role of staphylococcal wall teichoic acid in targeting the major autolysin Atl. Mol. Microbiol. *75*, 864–873.

Schneider, T., Kruse, T., Wimmer, R., Wiedemann, I., Sass, V., Pag, U., Jansen, A., Nielsen, A.K., Mygind, P.H., Raventos, D.S., *et al.* (2010). Plectasin, a fungal defensin, targets the bacterial cell wall precursor Lipid II. Science *328*, 1168–1172.

Schuch, R., Nelson, D., and Fischetti, V.A. (2002). A bacteriolytic agent that detects and kills Bacillus anthracis. Nature *418*, 884–889.

Schuhardt, V.T., and Schindler, C.A. (1964). Lysostaphin therapy in mice Infected with *Staphylococcus aureus*. J. Bacteriol. *88*, 815–816.

Senyurek, I., Klein, G., Kalbacher, H., Deeg, M., and Schittek, B. (2010). Peptides derived from the human laminin alpha4 and alpha5 chains exhibit antimicrobial activity. Peptides *31*, 1468–1472.

Shai, Y. (1999). Mechanism of the binding, insertion and destabilization of phospholipid bilayer membranes by alpha-helical antimicrobial and cell non-selective membrane-lytic peptides. Biochim. Biophys. Acta *1462*, 55–70.

Shaw, L.N., Golonka, E., Szmyd, G., Foster, S.J., Travis, J., and Potempa, J. (2005). Cytoplasmic control of premature activation of a secreted protease zymogen: deletion of staphostatin B (SspC) in *Staphylococcus aureus* 8325-4 yields a profound pleiotropic phenotype. J. Bacteriol. *187*, 1751–1762.

Shockman, G.D. (1992). The autolytic ('suicidase') system of Enterococcus hirae: from lysine depletion autolysis to biochemical and molecular studies of the two muramidases of Enterococcus hirae ATCC 9790. FEMS Microbiol. Lett. *79*, 261–267.

Singh, P.K., Schaefer, A.L., Parsek, M.R., Moninger, T.O., Welsh, M.J., and Greenberg, E.P. (2000). Quorum-sensing signals indicate that cystic fibrosis lungs are infected with bacterial biofilms. Nature *407*, 762–764.

Siritapetawee, J., Thammasirirak, S., and Samosornsuk, W. (2012). Antimicrobial activity of a 48-kDa protease (AMP48) from Artocarpus heterophyllus latex. Eur. Rev. Med. Pharmacol. Sci. *16*, 132–137.

Skarnes, R.C., and Watson, D.W. (1957). Antimicrobial factors of normal tissues and fluids. Bacteriol. Rev. *21*, 273–294.

Steiner, H., Hultmark, D., Engstrom, A., Bennich, H., and Boman, H.G. (1981). Sequence and specificity of two antibacterial proteins involved in insect immunity. Nature *292*, 246–248.

Sung, M., and Dalbey, R.E. (1992). Identification of potential active-site residues in the *Escherichia coli* leader peptidase. J. Bio. Chem. *267*, 13154–13159.

Supuran, C.T., Scozzafava, A., and Clare, B.W. (2002). Bacterial protease inhibitors. Med. Res. Rev. *22*, 329–372.

Suttmann, H., Retz, M., Paulsen, F., Harder, J., Zwergel, U., Kamradt, J., Wullich, B., Unteregger, G., Stockle, M., and Lehmann, J. (2008). Antimicrobial peptides of the Cecropin-family show potent antitumor activity against bladder cancer cells. BMC Urology *8*, 5.

Swanson, R., Weaver, L.H., Remington, S.J., Matthews, B.W., and Baldwin, T.O. (1985). Crystals of luciferase from Vibrio harveyi. A preliminary characterization. J. Bio. Chem. *260*, 1287–1289.

Swoboda, J.G., Campbell, J., Meredith, T.C., and Walker, S. (2010). Wall teichoic acid function, biosynthesis, and inhibition. ChemBioChem. *11*, 35–45.

Taggart, C.C., Greene, C.M., Smith, S.G., Levine, R.L., McCray, P.B., Jr., O'Neill, S., and McElvaney, N.G. (2003). Inactivation of human beta-defensins 2 and 3 by elastolytic cathepsins. J. Immunol. *171*, 931–937.

Tenovuo, J. (2002). Clinical applications of antimicrobial host proteins lactoperoxidase, lysozyme and lactoferrin in xerostomia: efficacy and safety. Oral Dis. *8*, 23–29.

Thomma, B.P., Cammue, B.P., and Thevissen, K. (2002). Plant defensins. Planta *216*, 193–202.

Thunnissen, A.M., Rozeboom, H.J., Kalk, K.H., and Dijkstra, B.W. (1995). Structure of the 70-kDa soluble lytic transglycosylase complexed with bulgecin A. Implications for the enzymatic mechanism. Biochemistry *34*, 12729–12737.

Tom, R.T., Suryanarayanan, V., Reddy, P.G., Baskaran, S., and Pradeep, T. (2004). Ciprofloxacin-protected gold nanoparticles. Langmuir *20*, 1909–1914.

Trayer, H.R., and Buckley, C.E., 3rd (1970). Molecular properties of lysostaphin, a bacteriolytic agent specific for *Staphylococcus aureus*. J. Bio. Chem. *245*, 4842–4846.

Tsai, H., and Bobek, L.A. (1997). Studies of the mechanism of human salivary histatin-5 candidacidal activity with histatin-5 variants and azole-sensitive and -resistant Candida species. Antimicrob. Agents Chemother. *41*, 2224–2228.

Tsuji, Y., Aoyama, T., Takeuchi, K., Homma, K., Takahashi, H., Nakajima, Y., Shimada, I., and Natori, S. (2001). Identification and characterization of an antibacterial peptide of the 26-kDa protease of *Sarcophaga peregrina* with antibacterial activity. J. Biochem. *130*, 313–318.

Tvete, T., and Haugan, K. (2008). Purification and characterization of a 630 kDa bacterial killing metalloprotease (KilC) isolated from plaice Pleuronectes platessa (L.), epidermal mucus. J. Fish Dis. *31*, 343–352.

Vargas, L.J., Londono, M., Quintana, J.C., Rua, C., Segura, C., Lomonte, B., and Nunez, V. (2012). An acidic phospholipase A(2) with antibacterial activity from Porthidium nasutum snake venom. Comparative biochemistry and physiology Part B, Biochem. Mol. Biol. *161*, 341–347.

Veiga-Crespo, P., and Villa, T.G. (2010a). Advantages and disadvantages in the use of antibiotics or phages as therapeutic agents. In Enzybiotics: Antibiotic Enzymes as Drugs and Therapeutics, P. Veiga-Crespo and T.G. Villa, eds (John Wiley and Sons, Hoboken, NJ, USA), pp. 27–58.

Veiga-Crespo, P., and Villa, T.G. (2010b). Phylogeny of enzybiotics. In Enzybiotics: Enzymes as Drugs and Therapeutics, P. Veiga-Crespo and T.G. Villa, eds (John Wiley and Sons, Hoboken, NJ), pp. 75–106.

Vuillemin (1890). Antibiosis. Compte Rendu de l'Assoc Française pour l'Avancem des Sci *ii*, 526 (OED).

Waksman, S.A. (1947). Antibiotics and tuberculosis: a microbiologic approach. J. Am. Med. Soc. *135*, 478–485.

Waksman, S.A., and Woodruff, H.B. (1942). Selective antibiotic action of various substances of microbial origin. J. Bacteriol. *44*, 373–384.

Walsh, S., Shah, A., and Mond, J. (2003). Improved pharmacokinetics and reduced antibody reactivity of lysostaphin conjugated to polyethylene glycol. Antimicrob. Agents Chemother. *47*, 554–558.

Weaver, L.H., Grutter, M.G., Remington, S.J., Gray, T.M., Isaacs, N.W., and Matthews, B.W. (1984). Comparison of goose-type, chicken-type, and phage-type lysozymes illustrates the changes that occur in both amino acid sequence and three-dimensional structure during evolution. J. Mol. Evol. *21*, 97–111.

Weidenmaier, C., and Peschel, A. (2008). Teichoic acids and related cell-wall glycopolymers in Gram-positive physiology and host interactions. Nature Rev. Microbiol. *6*, 276–287.

Winkler, H.H., and Daugherty, R.M. (1989). Phospholipase A activity associated with the growth of Rickettsia prowazekii in L929 cells. Infect. Immun. *57*, 36–40.

Wollman, E., and Wollman, E.A. (1936). Recherches sur le phenomene de Twort-D'Herelle (bacteriophagie ou autolyse heredo-contagieuse, quatriem memoire. Annals Institute Pasteur *56*, 137–164.

Wu, J.A., Kusuma, C., Mond, J.J., and Kokai-Kun, J.F. (2003). Lysostaphin disrupts *Staphylococcus aureus* and *Staphylococcus epidermidis* biofilms on artificial surfaces. Antimicrob. Agents Chemother. *47*, 3407–3414.

Xia, G., Kohler, T., and Peschel, A. (2010). The wall teichoic acid and lipoteichoic acid polymers of *Staphylococcus aureus*. International J. Med. Microbiol. *300*, 148–154.

Yang, X.Y., Li, C.R., Lou, R.H., Wang, Y.M., Zhang, W.X., Chen, H.Z., Huang, Q.S., Han, Y.X., Jiang, J.D., and You, X.F. (2007). *In vitro* activity of recombinant lysostaphin against *Staphylococcus aureus* isolates from hospitals in Beijing, China. J. Med. Microbiol. *56*, 71–76.

Yang, Z. (1998). Likelihood ratio tests for detecting positive selection and application to primate lysozyme evolution. Mol. Biol. Evol. *15*, 568–573.

Yarwood, J.M., Volper, E.M., and Greenberg, E.P. (2005). Delays in *Pseudomonas aeruginosa* quorum-controlled gene expression are conditional. PNAS *102*, 9008–9013.

Yokogawa, K., Kawata, S., Nishimura, S., Ikeda, Y., and Yoshimura, Y. (1974). Mutanolysin, bacteriolytic agent for cariogenic streptococci: partial purification and properties. Antimicrob. Agents Chemother. *6*, 156–165.

Zasloff, M. (1987). Magainins, a class of antimicrobial peptides from Xenopus skin: isolation, characterization of two active forms, and partial cDNA sequence of a precursor. PNAS *84*, 5449–5453.

Zasloff, M., Martin, B., and Chen, H.C. (1988). Antimicrobial activity of synthetic magainin peptides and several analogues. PNAS *85*, 910–913.

Zhang, L., and Falla, T.J. (2006). Antimicrobial peptides: therapeutic potential. Expert Opin. Pharmacother, *7*, 653–663.

Zhao, Y., Tian, Y., Cui, Y., Liu, W., Ma, W., and Jiang, X. (2010). Small molecule-capped gold nanoparticles as potent antibacterial agents that target Gram-negative bacteria. J. Am. Chem. Soc. *132*, 12349–12356.

Zipperle, G.F., Jr., Ezzell, J.W., Jr., and Doyle, R.J. (1984). Glucosamine substitution and muramidase susceptibility in Bacillus anthracis. Can. J. Microbiol. *30*, 553–559.

Zygmunt, W.A., Harrison, E.F., Browder, H.P., and Tavormina, P.A. (1968). Comparative inhibition of methicillin-resistant strains of *Staphylococcus aureus* by lysostaphin and other antibiotics. Appl. Microbiol. *16*, 1174–1178.

Use of Antibiotic Core Structures to Generate New and Useful Macrolide Antibiotics

20

Prabhavathi Fernandes

Abstract

Macrolides, such as erythromycin, are safe and effective antibiotics that are broadly used in adults and children. Clarithromycin and azithromycin were introduced in the 1990s and became popular because of improved gastrointestinal tolerability and pharmacokinetics but they exhibit cross resistance with erythromycin. Intense efforts in many pharmaceutical companies led to the third-generation macrolide, telithromycin, a 3-keto-aglycone of clarithromycin with an extended 11,12-carbamate side chain. Telithromycin was invented using two lessons learned from older macrolides: (i) additional side chains on the macrolide ring, as found on 16-membered macrolides, can gain activity against resistant organisms, and (ii) an aglycone with a keto group as noted in picromycin, instead of erythromycin's cladinose can retain antibacterial activity. Although active against drug-resistant bacteria, telithromycin failed in the clinic due to serious adverse events. Solithromycin is a fourth-generation macrolide, a fluoroketolide, in late stage development that is differentiated from telithromycin in lacking the latter's pyridine-containing side chain, which interacts with human nicotinic acetylcholine receptors to produce treatment-limiting side-effects, and in having a third binding region in the bacterial ribosome to help overcome resistance. Solithromycin is in late-stage clinical development. Molecular mode of action studies could lead to even more potent molecules in the future, perhaps with uses beyond antiinfectives.

Introduction

Antibiotics were among the first drugs to be discovered and many classes of antibiotics targeting different structures of the bacteria have been developed. Most successful antibiotics are natural products, or are derivatives of natural products (Fernandes, 2006). Many antibiotics are complex molecules and are difficult for chemists to synthesize, but have been used to make semisynthetic analogues to overcome disadvantages in safety, bioavailability or spectrum of activity when compared to the parent molecule. The success of natural products as antibiotics is attributed to their complexity, their core structures, multiple sugars, and alkyl or aryl side chains. This complexity allows for multiple binding sites to the bacterial target, thereby decreasing the ability of bacteria to become resistant by mere single step, simple point mutations. These molecules have not followed the medicinal chemistry rules called the Lipinski's Rule of Five (Lipinski, 2001) that require small, simple synthetic molecules to be active, soluble and orally bioavailable. Most natural products do not fit the Rule of Five but are still orally bioavailable and well distributed, including penetration into cells.

These natural products have evolved within the producing organism in an environment where they must co-exist with other bacteria and microorganisms. Thus, antibiotics are secondary metabolites that provide a survival advantage to the antibiotic-producing organism. Selectivity, therefore, is governed by Nature, in that the producing organism is not killed while competing microorganisms in the surrounding milieu are killed. In order for an antibiotic to be used in humans or animals, it must be selectively active against bacteria while not interfering with eukaryotic cells of the host.

We human beings have derived the benefit of evolution of antibiotics in the environment through millions of years. However, in recent years, multidrug resistant bacteria have become increasingly common (CDC, 2013). Have we exhausted our resources or is there room to find new antibiotics? This author believes that Nature has an inexhaustible supply of compounds, provided we know how to find them. This chapter will focus on continued efforts in the modification of a popular class of antibiotics, the macrolides, and point to lessons learned from Nature.

History of macrolide class of antibiotics

Erythromycin was isolated at Eli Lilly Laboratories from *Saccharopolyspora erythraea* (previously called *Streptomyces erythreus*) growth from soil samples obtained from the Philippines in 1949 and was marketed in 1952. During the following two decades many new antibiotics were discovered, including the cephalosporins and other antibiotic classes that inhibit protein synthesis. Some of the marketed antibiotics that belong to classes of bacterial protein synthesis inhibitors are listed in Table 20.1.

Some protein synthesis inhibitors target 50S RNA and the others, for example erythromycin, target the 23S RNA. Those that target the 50S RNA, for example the aminoglycoside gentamicin, are bactericidal and have potent activity against both Gram-negative bacteria, including *Pseudomonas aeruginosa* and Gram-positive bacteria such as staphylococci. Another major class, the tetracyclines, also binds to the 50S RNA and has broad-spectrum activity. Erythromycin and its analogues, on the other hand, bind to the 23S RNA at the exit of the peptide synthesis tunnel, impeding the exit of the nascent peptide chain and rendering erthromycin bacteriostatic. Other antibiotics that target the 23S RNA are lincomycin and its analogue clindamycin, chloramphenicol, and the synthetic antibacterial agent linezolid, an oxazolidinone. Lincomycin and its analogue clindamycin were also discovered at Eli Lilly and had potent activity against Gram-positive bacteria including intestinal anaerobic microflora. Clindamycin binds to the peptide tunnel, but reaches close to the catalytic peptidyl-transferase centre (PTC). It is known to induce erythromycin resistance. Chloramphenicol and its analogues bind to the PTC. The *cfr* gene, whose product, the Cfr methyltransferase, modifies a 23S rRNA residue in the PTC, confers resistance to chloramphenicol. Chloramphenicol has potent activity against a broad spectrum of pathogens and is well distributed intracellularly, but its use is limited because it can result in rare cases of aplastic anaemia. Linezolid is not derived from natural products but was designed and made synthetically. It was the first oral drug to be approved for serious infections caused by methicillin resistant *Staphylococcus aureus* (MRSA) and enterococci. Linezolid also

Table 20.1 Some antibiotics that inhibit protein synthesis

Antibiotic class	RNA subunit target	Primary antibiotics	Examples of others in the class
Aminoglycosides	50S	Streptomycin, gentamicin	Tobramycin, amikacin
Tetracyclines	50S	Tetracycline	Doxycycline, minocycline
Chloramphenicol	23S	Chloramphenicol	Thiamphenicol, florphenicol
Clindamycin	23S	Lincomycin	Clindamycin
Streptogramins	23S	Streptogramins A and B	Pristinamycin
Macrolides	23S	Erythromycin	Clarithromycin, azithromycin, roxithromycin
Oxazolidinones*	23S	Linezolid	Tedizolid

*Not natural product derived.

targets the 23S RNA, where it binds in the PTC and inhibits protein synthesis (LaMarre *et al.*, 2013). Interestingly, the Cfr methyltransferase also confers resistance to linezolid, indicating that it binds to the same site as chloramphenicol. Thus, Cfr-mediated C8 methylation of A2503 of the 23S RNA makes bacteria resistant to all of the peptidyl transferase site-targeting antibiotics. Compounds that bind the PTC site in the bacterial ribosome become liabilities for human mitochondrial protein synthesis inhibition because this site is also present in the ribosome of mitochondria (Leach *et al.*, 2007; Nagiec *et al.*, 2005). Thus, many natural antibiotics and the man-made antibiotic, linezolid, target the 23S RNA, in the region of the peptide synthesis tunnel (Fig. 20.1). Because of its vulnerability, this region of the ribosome is known as the 'Achilles heel' of bacteria. All of these antibiotics have remained useful and are used broadly in treating infections.

Erythromycin and its activity

Erythromycin and other macrolides are mostly bacteriostatic and are not considered useful for treating serious or life-threatening respiratory tract infections. However, because of its oral bioavailability, spectrum, efficacy and safety, erythromycin was widely used for a broad variety of simple respiratory tract infections. A common reason for visiting a physician is a respiratory tract infection and erythromycin and its analogues have become the go-to antibiotic for treating mild pneumonia, sinusitis, bronchitis and other respiratory tract infections. When introduced to the clinic, erythromycin had the appropriate spectrum of activity for respiratory tract infections, targeting Gram-positive, Gram-negative and atypical bacteria that frequently occur alone or in combination during infections. Among these pathogens, *Streptococcus pneumoniae* (pneumococcus) (Gram-positive diplococcus) is the most prevalent pathogen. Other organisms such as *Hemophilus influenzae*, a small Gram-negative cocco-bacillus, *Moraxella catarrhalis*, a Gram-negative diplococcus, *Staphylococcus aureus*, a Gram-positive coccus often seen in clusters, *Streptococcus pyogenes* (Group A, beta-haemolytic streptococcus), a Gram-positive coccus in chains, or atypical bacteria that are neither Gram-positive or Gram-negative, such as *Legionella pneumophila*, *Mycoplasma pneumoniae*, and *Chlamydophila pneumoniae* could all be pathogens in the respiratory tract. Macrolides are unique in having an optimum spectrum: sufficiently broad for targeting these pathogens, but usually not too broad so as to disrupt intestinal and other microflora that could result in *Clostridium difficile* pseudomembranous enterocolitis, which is known to be a serious disease.

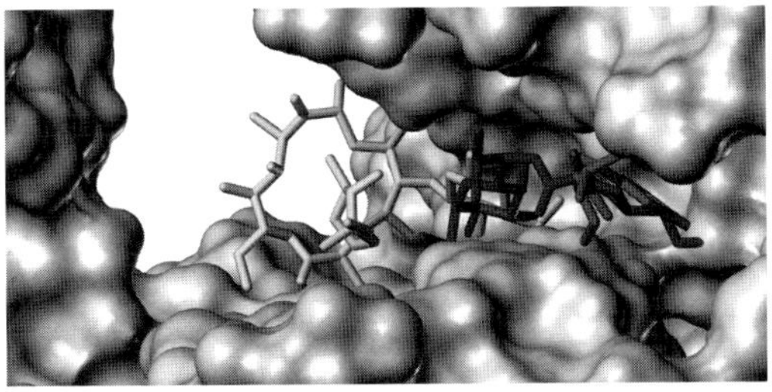

Figure 20.1 Computer modelled representation of the peptide synthesis tunnel in 23S RNA. Yellow: erythromycin at the exit of the peptide tunnel. Red: chloramphenicol at the peptidyl transferase centre. Blue: clindamycin. A colour version of this figure is available in the plate section at the back of the book.

Being generally safe and orally bioavailable, erythromycin was also developed for use in paediatric patients. Erythromycin and related macrolides, in addition to their antibacterial properties have strong anti-inflammatory properties thereby preventing collateral cytokine damage from inflammation in any infection and augmenting their overall efficacy. It was also noted that although blood concentrations were low, erythromycin distributed well into tissues, achieved high intracellular concentrations, and demonstrated activity against bacteria in the lysosomal and cytoplasmic compartments. Intracellular activity is clearly an advantage of this class of antibiotics.

Imperfections of erythromycin and development of second generation macrolides

Erythromycin is known for low oral bioavailability and for gastrointestinal contractions resulting in nausea and vomiting. It is not stable in acidic media, as is present in the stomach milieu. Numerous salt formulations were developed, such as erythromycin estolates and erythromycin ethyl succinate, all of which increased bioavailability, but with undesirable increased liver toxicity (Viluksela *et al.*, 1988). Erythromycin is known to form 8,9-anydro-6,9-hemiketal in acid as noted in (Fig. 20.2). The anhydrous form has no antibacterial activity, and has been thought to be responsible for the gastrointestinal side-effects.

Many new macrolides were developed, such as roxithromycin and flurithromycin (Ying and Tang, 2010; Asaka *et al.*, 2003) that had improved pharmacokinetics, but none of them demonstrated clear clinical advantage over erythromycin and erythromycin remained the leader except in some regions of the world. In the 1980s, the second generation of macrolides presented better gastrointestinal tolerability. The two successful products were clarithromycin (Taisho Pharmaceuticals, Abbott Laboratories) and azithromycin (Pliva, Pfizer) (Fig. 20.3). Both molecules had modifications that protected the macrocyclic ring from acid conversion to the inactive hemiketal form. In clarithromycin the macrocyclic ring was modified by the addition of 6-*O*-methyl, and in azithromycin the ring was expanded by the addition of nitrogen, making azithromycin the first azalide. These changes contributed to better bioavailability and lower gastrointestinal adverse effects, but the spectrum of activity remained mostly unchanged.

Erythromycin, clarithromycin and azithromycin bind to domain V of the 23S RNA via the N-dimethyl of their desosamine sugars to the nucleosides at A2058 and A2059. Since the activity is similar between the first and second generation macrolides, hospital laboratories test erythromycin as representative of the class, except against *H. influenzae* where azithromycin could have greater potency because of its ability to penetrate the Gram-negative membrane because of its lower lipophilicity.

Clarithromycin and azithromycin reached sales of over $1 billion soon after marketing

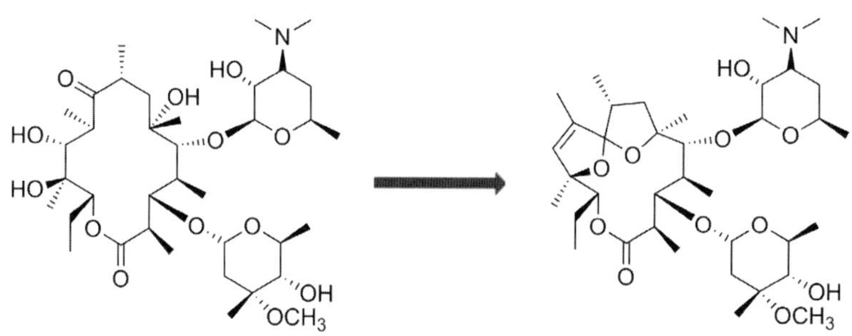

Erythromycin

8,9- Anhydro-6,9-hemiketal of Erythromycin

Figure 20.2 Degradation of erythromycin to 8,9-anhydro-6,9-hemiketal.

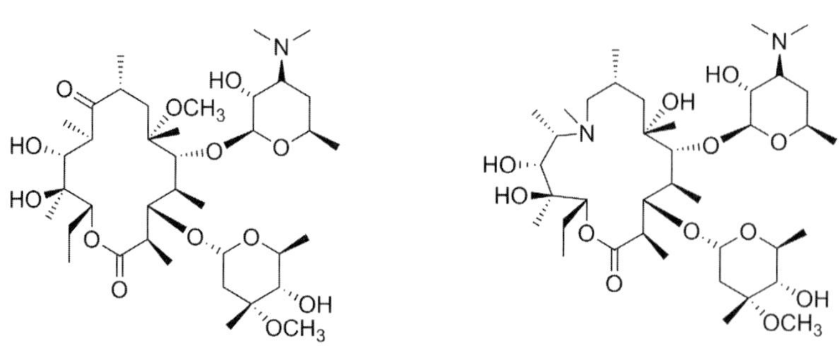

Clarithromycin

Azithromycin

Figure 20.3 Structures of clarithromycin and azithromycin.

approval, characterizing them as blockbuster antibiotics and in the late 2000s these macrolides reached combined sales of about $5 billion. Widespread use has its consequences, namely, selection of resistant strains. Resistance to erythromycin, azithromycin and clarithromycin can be constitutive or inducible and results from the methylation of nucleotides A2058 and A 2059 in Domain V of the 23S RNS through the action of the genes *ermA* and *ermB*. Lower level resistance could also occur through the *mef* efflux pump as well as in ribosomal protein mutations, such as L22 mutations. In the last decade, macrolide resistance has increased and the broad spectrum fluoroquinolones, such as levofloxacin and moxifloxacin, overtook the macrolides in the market, in spite of their lower safety profiles.

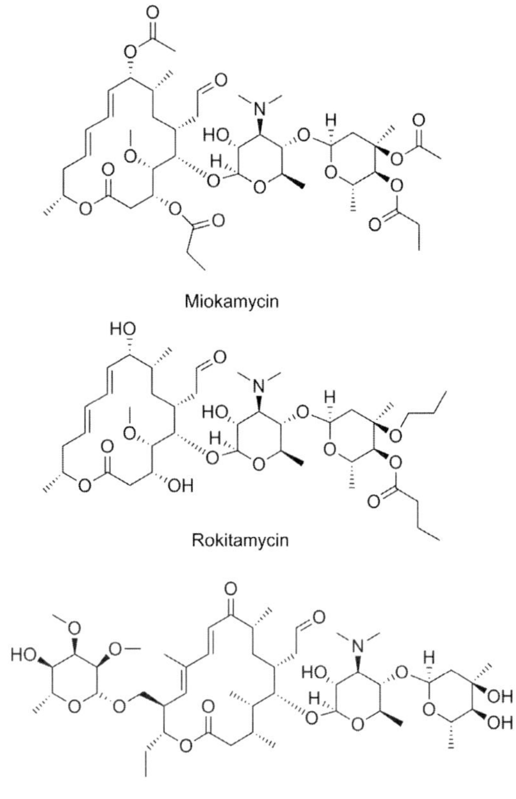

Figure 20.4 Structures of 16-membered macrolides demonstrating extended side chains.

The third-generation macrolides

Many large pharmaceutical companies began to address resistance to the older macrolides and in the 1980s scientists at Abbott laboratories synthesized erythromycin analogues that could reach further into the peptide tunnel to gain a second foothold on the ribosome to overcome antibiotic resistance. This step was based on a fundamental lesson from Nature, wherein the 16-membered ring macrolides, such a miokamycin, rokitamycin and tylosin, overcame resistance to erythromycin by way of their extended chains attached to the desosamine and by additional sugar moiety on the macrolide ring as in the case of tylosin (Fig. 20.4). These molecules have demonstrated improved activity against macrolide-resistant strains *in vitro*, but have metabolically unstable side chains *in vivo* (Sunazuka *et al.*, 2002). Although they inhibited macrolide-resistant bacteria, the 16-membered macrolides were not successful as products for human therapeutic use because of their poor pharmacokinetics, but they are used in veterinary medicine.

Their activity against macrolide-resistant bacteria was a key attribute and the lesson used here was put to use by making a series of clarithromycin analogues with an 11, 12-carbamate side chain; these analogues of clarithromycin had activity against constitutively resistant staphylococci (Fig. 20.5 and Table 20.2) (Fernandes *et al.*, 1989). However, these compounds did not have sufficient potency to be progressed further.

Designing the ketolides (third generation macrolides)

At Hoecht-Roussel laboratories, the length of the 11,12-carbamate side chain was increased by adding heterocycle moieties, namely, imidazole and pyridine. The resulting molecule was large, and limited bacterial penetration and hence activity. Next, Hoecht-Roussel scientists made further progress by substitution of the cladinose sugar at position 3 of the macrolide ring with a 3-keto group (Fig. 20.6).

It was already known from Nature, that the cladinose sugar on erythromycin was not essential for activity, as was seen in pikromycin (Fig.

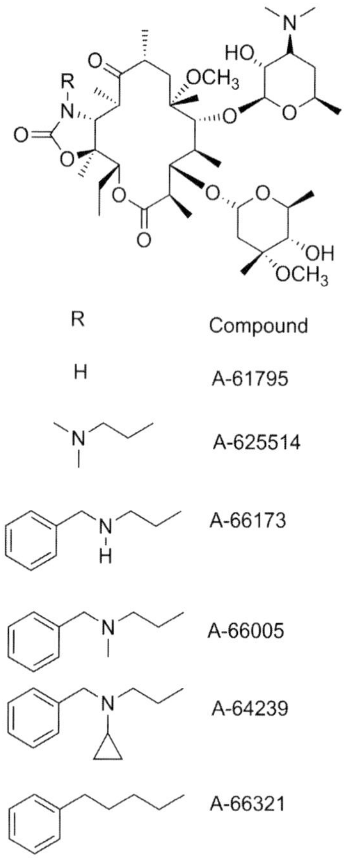

Figure 20.5 The 11, 12-carbamate analogues of clarithromycin that had activity against macrolide resistant bacteria (Fernandes *et al.*, 2008).

Table 20.2 Activities of 11, 12- carbamate analogues of clarithromycin against macrolide-susceptible and macrolide-resistant *Streptococcus pyogenes* strains (Fernandes *et al.*, 1989)

	MIC (µg/ml) against *S. pyogenes* strains			
Compound	EES61	2548	2707	930
A-61795	0.12	1	0.06	64
A-62514	0.12	4	0.25	64
A-66173	0.06	0.25	0.03	4
A-66005	0.03	0.25	0.03	4
A-64239	0.12	0.5	0.12	8
A-66321	0.06	0.5	0.06	4

20.7) a naturally occurring macrolide discovered in 1950. Pikromycin is a 14-membered macrolide that is an aglycone at position 3 with a 3-keto group instead. The result of this discovery programme was a molecule named telithromycin and the new macrolide class thus created was called ketolide to distinguish it from the older macrolides.

Telithromycin has activity against bacterial strains that have *erm* resistance and binds to the ribosome at Domain II, specifically nucleotide 752A, in addition to the primary A2058/2059 site in the peptide synthesis tunnel (Asaka *et al.*, 2003). The compound was well tolerated and showed superior efficacy, allowing the treatment of moderate to moderately severe community-acquired bacterial pneumonia (CABP) (Shain and Amsden, 2002). Telithromycin received regulatory approval in 2000 in Europe and in 2003 in the United States based on the positive data from large clinical trials. It was approved only as oral tablets for multiple respiratory tract infections, including simple infections such as sinusitis and bronchitis (Shi *et al.*, 2005). There was a great deal of pharmaceutical and medical interest in telithromycin (brand name, Ketek), and it was initially used extensively. However, adverse events were soon noted. Many of these adverse events were serious and included reversible loss of visual accommodation, loss of consciousness, exacerbation of myasthenia gravis and liver failure. These effects were collectively called the 'Ketek effect' (Jarvis, 2008). Visual accommodation changes had been noted in clinical trials, but were overlooked by regulatory agencies. Although there were regulatory concerns about the conduct of the phase 3 clinical trials, the serious side-effects were not noted prior to approval. Soon after its approval, it was withdrawn from use for all indications except in the case of severe CABP. However, because of the major adverse events associated with death, it is not used.

Other ketolides

The development of telithromycin and its projected sales encouraged many pharmaceutical companies to develop new ketolides. Only a few highlights are provided here and the readers can refer to several reviews on the large body of work that was done to find additional ketolides (Ying and Tang, 2010; Asaka *et al.*, 2003, Katz and Ashley, 2005). The goal of each of these programmes was to add or extend a side chain to gain a

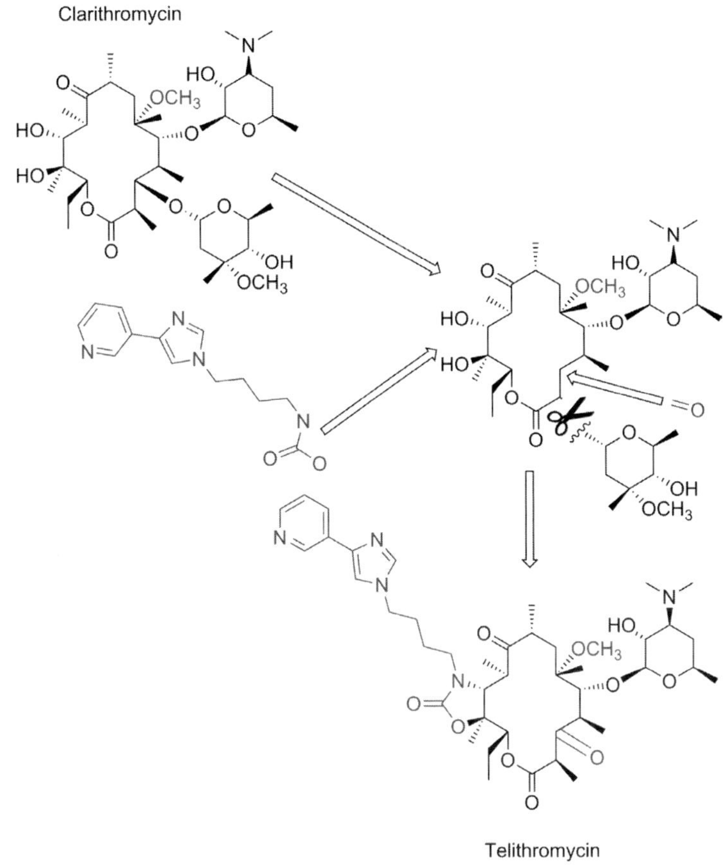

Figure 20.6 Design of telithromycin- a ketolide. A colour version of this figure is available in the plate section at the back of the book.

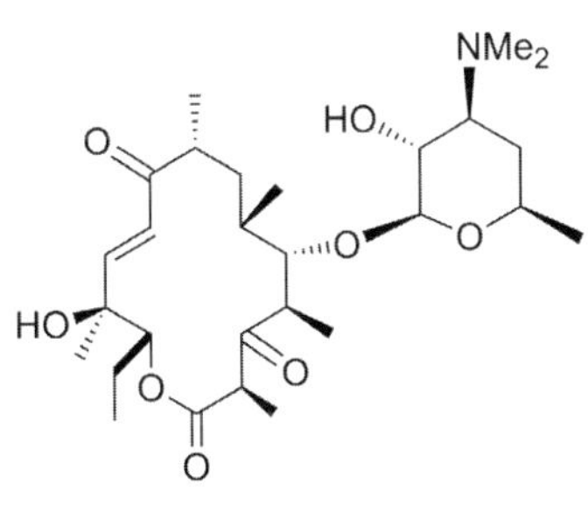

Figure 20.7 Pikromycin: an old macrolide that is descladinose and with a 3-keto group as in telithromycin.

second binding site. Like telithromycin, the 11,12- carbamate analogues were common. Other sites used for structural modifications were the C6 of the macrolide ring, and the hydroxyl of the cladinose; in the latter scenario, the structures were still macrolides but not ketolides. Scientists at Abbott Laboratories synthesized ketolides with the side chain at the C6 position of the macrolide ring. One of these molecules, called ABT-773 progressed into phase 1 trials, but the development was stopped. This product was licensed by Advanced Life Sciences, and was developed further as cethromycin (Rafie *et al.*, 2010) (Fig. 20.8). It was developed only as oral tablets at a dose of 300 mg for use in cases of simple pneumonia (Severity Class PORT I) (English *et al.*, 2012). It was not non-inferior to clarithromycin at a dose of 250 mg in the clinical studies and was not licensed by the Food and Drug Administration (FDA). Higher doses of cethromycin were limited by dose related gastrointestinal motility and abdominal cramps. Cethromycin retained the *in vitro* activity of telithromycin, but *in vivo* it bound albumin as well as α-1 acid glycoprotein in plasma (approximately 98% bound) and therefore its bioavailability became limited. In addition, it induced CYP3A4 during treatment, which further decreased blood levels by increasing the metabolism of cethromycin and limiting its bioavailability during therapy. An important feature of cethromycin was that no visual or other side-effects were noted.

Many analogues were made where the cladinose sugar at position 3 was retained, and the 3′-OH of the cladinose modified to give compounds that had antibacterial activity similar to that of telithromycin, the side chain provided an additional binding site in the peptide synthesis tunnel to overcome resistance. One such compound made by Pfizer is shown in Fig. 20.9. Pliva scientists described azithromycin analogues with an extended side chain at C-6 (Fig. 20.10), while Kosan made an extended C-6 side chain containing ketolide which was unusual in also having a C-13 propyl group (Fig. 20.11). Shorter linkers and unusual linkers were also tested by Glaxo as shown in Fig. 20.12. These compounds and many other ketolides are discussed in several reviews (Ying and Tang, 2010; Asaka *et al.*, 2003, Katz and Ashley, 2005). Through these studies, it was noted that the 4-atom linker at 11, 3 and 6 position is optimal for activity.

Pfizer had a large programme where several approaches were taken (Magee *et al.*, 2009). Key features were to reduce CYP3A4 interaction and metabolism of the molecule by hepatic microsomes and also to reduce inhibition of the hERG channel, a predictor of effects on the human cardiac Q-T interval, a feature noted in older macrolides as well as in telithromycin. Pfizer's programme focused on making a novel series of macrolides called the azetidinyl ketolides. In the Pfizer programme, attached to the azetidinyl

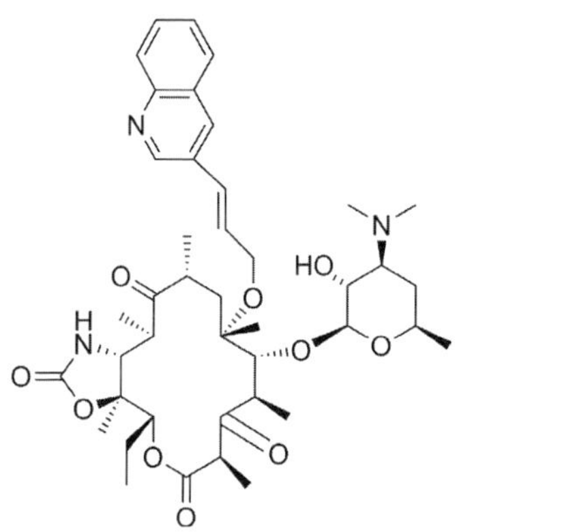

Figure 20.8 Cethromycin. A ketolide with a side chain at the C6 position.

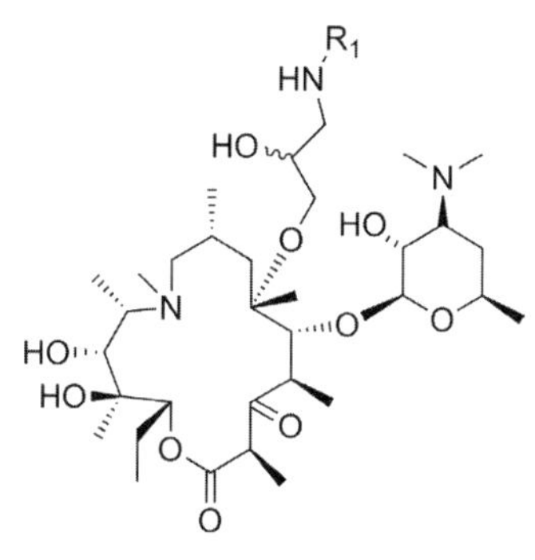

Figure 20.10 A ketolide with an azithromycin ring.

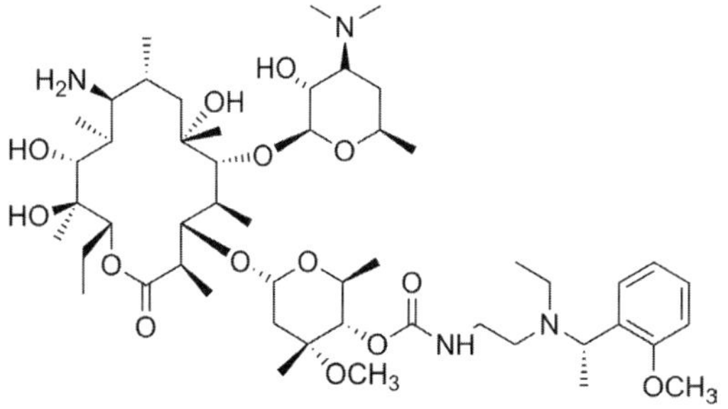

CP-544372

Figure 20.9 Macrolide with an extended side chain at the C-3 position.

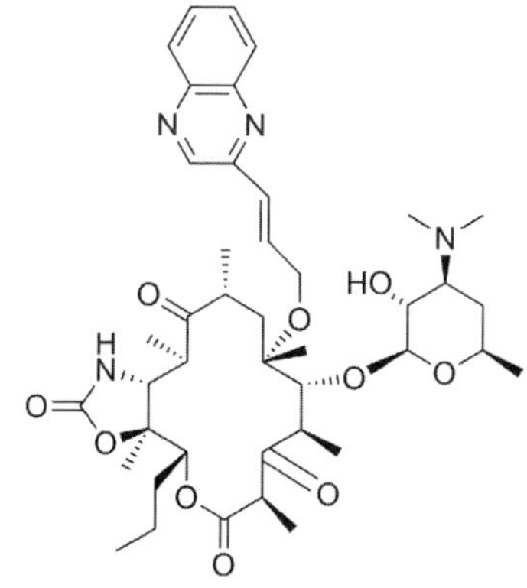

Figure 20.11 Macrolides with the extended chain at the C-6 position, and with an ethyl at the C-13 position.

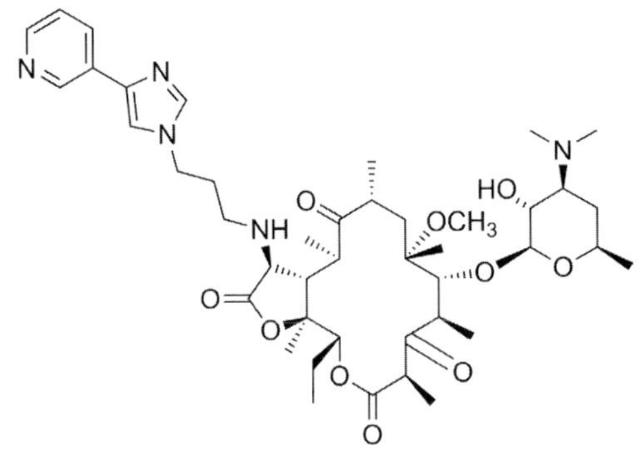

GW708408

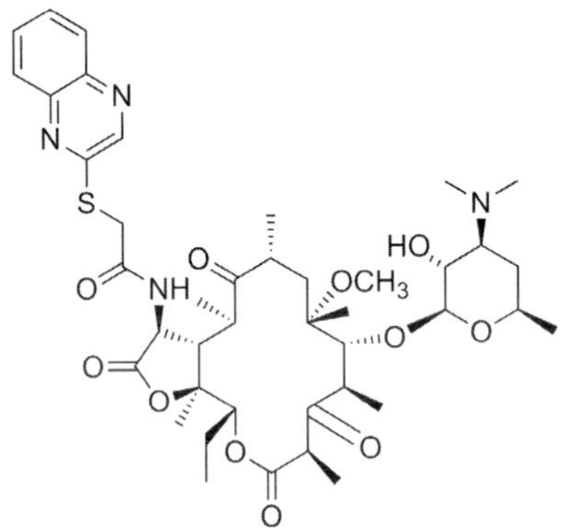

GW 773546

Figure 20.12 Examples of ketolides with shorter linkers in the side chain (GW 708408) and variations in the linker (GW 773546).

template were structures that ranged from a simple benzyl group to various complex naphthyridines. It was noted that the bicyclic fused ring, like a benzimidazole or a naphthryridine, had potency related advantages. A benzylic methylene bridge was used to connect the aryl ring to the azetidinyl nitrogen, with an ethyl substitution used to block metabolism by CYP3A4. The diastereomer shown in Fig. 20.13 was tested in phase 1 clinical trials. Interestingly this compound underwent metabolism in humans even though it was designed to avoid CYP metabolism. It was found that the compound was metabolized by a cytoplasmic enzyme, aldehyde oxidase which is located in the cytosol fraction rather than in the microsomal fraction with the CYP enzymes and was therefore missed during the preclinical screening. In order to avoid both the CYP3A4 and the aldehyde oxidase metabolism, the 3-hydroxy

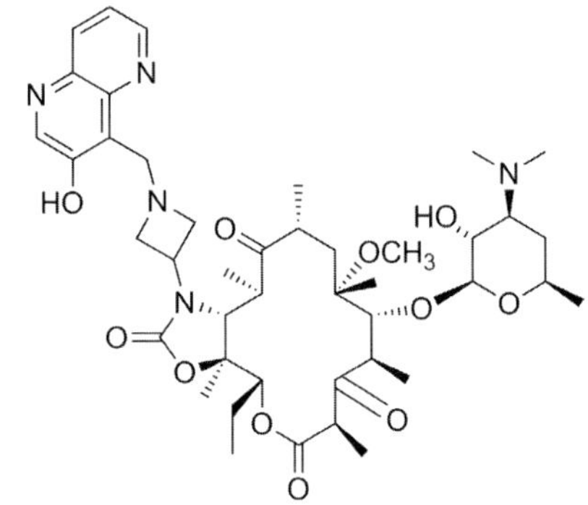

Figure 20.13 Azetidinyl ketolide that is not subject to CYP3A metabolism.

[1,8] naphthyridininyl analogue was made. This molecule was also taken into clinical trials, but no further information is known to this author (Betts *et al.*, 2010). This author's experience and understanding is that drug must have a route for excretion from the body and blocking all routes to metabolism may not result in favourable outcomes.

A few smaller companies continued to work on ketolides. Enanta Pharmaceuticals produced bicyclolides by bridging hydroxyl groups by esterification, which led to the bicyclolide derivative, EDP-420 (Sato *et al.*, 2011). EDP-420 is modithromycin, and was licensed to Shionogi for clinical development, but was later returned to Enanta Pharmaceuticals (Fig. 20.14). This molecule has a very long half-life of 17–18 h and it also has an inactive metabolite 9-keto-EDP-420 (Jiang and Or, 2009). The Shionogi scientists also synthesized their own ketolides with an

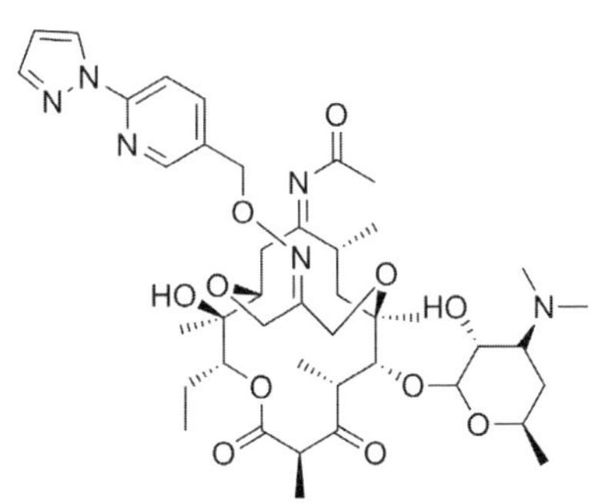

Figure 20.14 Bridge macrolides/ketolide – Bicyclolide – Modithromycin.

11,12-carbamate which are 9-oxime derivatives but these compounds did not have improved activity.

Many other series were made and many companies were involved with trying to make a better third generation macrolide. None of these compounds have activity that is notably better than that of telithromycin and none are in clinical development. Details of these compounds as well as many other compounds are described in the reviews cited above.

None of these products progressed past early clinical development because of intolerability or lack of potency. In addition, all macrolides in development were under the 'Ketek safety shadow'. It was thought that it would be necessary to show in clinical trials that adverse events like those that occurred with telithromycin would not occur with any new macrolide or ketolide. In order to ensure that the side-effects seen with telithromycin would not occur, the trials would have to be in the tens of thousands of patients, and this is not feasible. Soon after the FDA advisory meeting on telithromycin in late 2006, all major pharmaceutical companies stopped working on newer macrolides and ketolides. Thus, the safety image of macrolides as a class was tarnished by the experience with telithromycin and physicians were left with using the older second-generation macrolides, azithromycin and clarithromycin.

The hurdle for new macrolide development after telithromycin

The FDA changed regulatory guidance for antibacterial development following the troubled regulatory review around telithromycin's approval. In the proposed draft guidance approval for the use of antibiotics for treating respiratory tract infections, it would first be used for treating moderate to moderately severe CABP (PORT classes II–IV) with no allowance of PORT class I patients in the double blind, comparator controlled trials. PORT class I patients could be studied only in superiority trials compared to placebo, but these are not very feasible because of the large studies needed. In addition, microbial isolation and early response endpoints in efficacy that were within 10% non-inferiority margins of the comparator were required. These requirements made the development of just an oral formulation difficult, since PORT class III and PORT class IV patients frequently require intravenous dosing. Intravenous formulations have not been successfully developed for most macrolides owing to their cardiac Q-T effect as well as severe pain at the site of injection and phlebitis.

Improving efficacy and retaining safety

In 2007, it became apparent that development of a new macrolide could only be possible if the mechanistic reason for the severe adverse events associated with telithromycin was determined. These adverse events were hitherto not noted with macrolides. Ketolides, including telithromycin, are macrolides with a keto group at C-3 replacing the cladinose sugar found in older, safe macrolides. The other change in telithromycin relative to older macrolides is the 11,12-carbamate side chain. When focusing on the side chain, the presence of an imidazole ring was noted. This ring is found in other drugs, such as the imidazole antifungal drugs that are relatively safe and therefore it is unlikely that the imidazole group could be the cause of the severe adverse events seen with telithromycin. It was noted that the imidazole ring is labile *in vivo*, and related this metabolism to lower than expected efficacy when telithromycin was tested at a 500 mg oral dose while azithromycin, which is 4–8 times less potent *in vitro*, was effective against susceptible bacteria at 500 mg per dose. It is noted that in clinical trials with telithromycin, doses of 800 mg per day were necessary to demonstrate efficacy in clinical trials. Next, the focus was shifted to the second heterocyclic moiety, the pyridine, at the end of the 11, 12-carbamate side chain of telithromycin. Pyridine is an analogue of nicotine and drugs made for the central nervous system (CNS) area often contain pyridine. It was taken into consideration that CNS drugs are administered in very small amounts, a range of 2–10 mg is generally administered, while antibiotics are administered as hundreds of milligrams, which was 800 mg per day in the case of telithromycin. Therefore, it was considered worthwhile to hold the pyridine moiety responsible for the adverse events seen with telithromycin and it was decided that a ketolide without a pyridine in the side chain would be worth evaluating.

Characterization of the mechanism of toxicity of telithromycin

The visual side-effects and the curare-like side-effects at the neuromuscular junction suggested that nicotinic acid acetylcholine (nACh) receptors may be the target for adverse events seen for telithromycin. Led by Dr Daniel Bertrand, the activity of telithromycin was explored against the nACh receptors (Bertrand *et al.*, 2010). These receptors are pentameric receptors that are made of a combination of subunits, called α, β, δ and ε. Using cloned human receptors expressed in *Xenopus* oocytes, Dr Bertrand and his colleagues showed that telithromycin inhibited key receptors that are found in the eye, the muscle, the liver and in the frontal cortex of the brain. The specific receptors inhibited include the α7 and α3β4 nACh receptors, which occur in the ciliary ganglion of the eye, the blocking of which could result in changes in visual accommodation. Blocking the α3β2 andαβδε receptors at the neuromuscular junction could exacerbate myasthenia gravis and result in the reported curare-like effect of telithromycin. Blocking the α7 receptor at the end of the vagus nerve in the liver and on inflammatory cells would prevent the protective effect of the vagus nerve on inflammatory injury to the liver (Wang *et al.*, 2003). Telithromycin does not penetrate into the brain; therefore it was necessary to explore the reason for the loss of consciousness. The α4β2 nACh receptors present in the frontal cortex of the brain were blocked by the imidazole-pyridine side chain fragment. This side chain metabolite had not been reported, but it is known that the compound cleaves at the imidazole with the potential for the release of the pyridine moiety. This small molecule with a pyridine could enter the CNS and inhibit the receptors in the frontal cortex resulting in loss of consciousness. Older macrolides and macrolides without pyridine moieties do not inhibit the ACh receptors significantly. Also, it was noted that the antifungal voriconazole, which causes visual disturbances similar to telithromycin, also blocks the α3β4 receptors. Voriconazole's structure has a free pyrimidine, unlike other azole antifungals, which have not been associated with visual disturbances, and a pyrimidine could act like the pyridine at the nACh receptors (Bertrand *et al.*, 2011). A key feature of the visual side-effects was that it occurred more often in young women after administration of telithromycin than in male patients, resulting from the progesterone sensitization of nACh receptors (Bertrand *et al.*, 2010). Thus, 'Ketek effects' can more correctly be called 'Pyridine effects'.

Modifications may have altered the mode of action

Professor A. Mankin and his team at University of Illinois in Chicago made the unusual observation (Kannan *et al.*, 2012) that although the new ketolides are more potent at inhibiting bacteria, they are not potent inhibitors of peptide synthesis. On the contrary, while looking at total peptide synthesis, it can be said that they are poor inhibitors of peptide synthesis. How could this be true? Mankin and his team showed that certain peptides are mistranslated and that these mistranslated peptides are likely those of essential proteins. The incomplete blockage of the exit of the peptide tunnel could be related to the lack of the cladinose ring, leaving enough space for the peptides to escape from the tunnel. Such mistranslated peptides could be responsible for the bactericidal action of ketolides against some bacteria, unlike the strictly bacteriostatic activity of the older macrolides. These findings could lead to designing more potent ketolides that mistranslate additional essential proteins thereby resulting in toxicity to the bacterial cell.

The fourth generation macrolide, solithromycin

Solithromycin (CEM-101, OP-1068) was selected from a collection of macrolides synthesized at Optimer Pharmaceuticals that was licensed by Cempra, Inc. (Pereira and Fernandes, 2011) (Fig. 20.15).

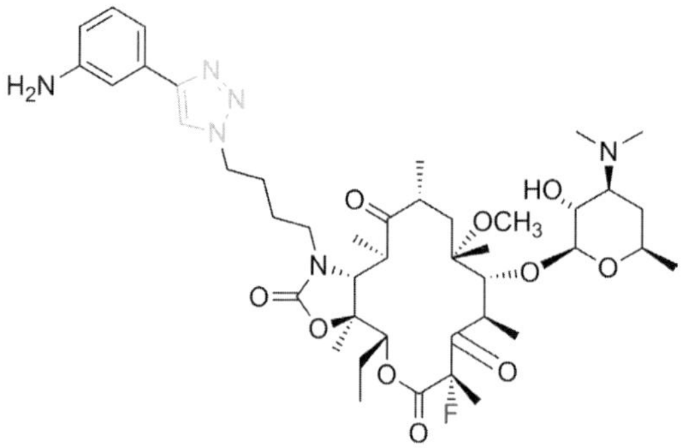

Figure 20.15 Structure of solithromycin. The side chain contains a 1,2,3, triazole (green), and an aminophenyl (blue) instead of the imidazole and pyridine found in telithromycin. Solithromycin also has a 2F (red). A colour version of this figure is available in the plate section at the back of the book.

Solithromycin is a ketolide, but is different from other ketolides in that it has a fluorine at the 2 position. Solithromycin is very potent against pneumococcus and other bacteria and has activity against bacteria that have developed resistance to older macrolides, including those resistant to telithromycin. We investigated the effect of the fluorine at the 2 position by making a solithromycin analogue without the fluorine called OP-1055. As noted in Table 20.3, fluorine is essential for activity against resistant strains. It was also noted that adding a fluorine to telithromycin did not confer the additional activity seen in solithromycin. Thus, it was concluded that there was cooperative activity between the solithromycin side chain and the 2-fluorine. The side chain of solithromycin contains a 1,2,3-triazole instead of the metabolically labile imidazole in telithromycin. Instead of the pyridine as in telithromycin, the side chain of solithromycin contains a terminal aminophenyl moiety. We considered the aminophenyl as a potential liability, but noted that when bound to the side chain it did not form free aniline. Importantly, solithromycin was shown to be non-mutagenic in mutagenicity assays, such as the Ames test with and without activation by the S9 fraction and in several other studies *in vitro* and in animals. No trace of free aniline was found in any *in vitro* or *in vivo* studies.

Intracellular activity

Macrolides are known to concentrate in tissues and cells and are used to treat infections with intracellular pathogens, such as pneumonia caused by *Legionella pneumophila.* Solithromycin has been shown to accumulate in the epithelial lining fluid in the lung and macrophages in the lung in human subjects (Rodvold *et al.,* 2012). Using macrophages *in vitro,* it was shown that solithromycin accumulates intracellularly to a higher degree than azithromycin and clarithromycin (Lemaire *et al.,* 2009; Mallegol *et al.,* 2013). In these studies it was also shown to have potent activity against intracellular bacteria like *Legionella pneumophila, Listeria monocytogenes* and *Neisseria gonorrhoeae.* This study demonstrated that solithromycin reached significant concentrations in the cytoplasm where *Listeria* is located and in the phagolysosome where *Legionella* replicates. The complete perfusion of solithromycin into tissues and cells prevents bacteria from escaping the effects of solithromycin.

Structure and activity

The superior activity of solithromycin *in vitro* was confirmed in many microbiological studies (Farrell *et al.,* 2010; Fernandes *et al.,* 2011a; Golparian *et al.,* 2012; Putnam *et al.,* 2010, 2011; Roblin *et al.,* 2010). Solithromycin was active against resistant bacteria carrying *ermA, ermB, ermA* and *ermB,* and L22 mutations (McGhee *et al.,* 2010). Investigators at Berkeley and at the University of Chicago have studied the mechanism of action of solithromycin. X-ray crystallographic studies were conducted on co-crystals of solithromycin and *E. coli* ribosomes. These studies showed multiple interacting regions that could be responsible for conferring high potency, with multiple binding sites enabling activity against resistant strains (Llano-Sotelo *et al.,* 2010). In addition to the traditional interaction at Domain V at nucleotide A2058 and A2959, extensive interaction was noted at Domain II with multiple interactions at the aminophenyl as well as the 1,2,3 triazole region. The 2-F interacts with the base of the peptide tunnel. Thus, solithromycin was shown to interact with three regions of the ribosome explaining its activity against strains resistant to the third generation macrolides and ketolides that have only two binding sites. Resistance is

Table 20.3 Role of 2-fluorine atom on activity of solithromycin and telithromycin against streptococci

		Solithromycin		Des-F-solithromycin		Telithromycin		2-F-Telithromycin	
Species and phenotype[a]	N	MIC_{50} (μg/ml)	MIC_{90} (μg/ml)	MIC_{50} (μg/ml)	MIC_{90} (μg/ml)	MIC_{50} (μg/ml)	MIC_{90} (μg/ml)	MIC_{50} (μg/ml)	MIC_{90} (μg/ml)
S. pneumoniae Pen-S Tel-S Mac-S	10	0.03	0.03	0.03	0.03	0.03	0.03	0.03	0.03
S. pneumoniae Pen-R Tel-I/R Mac-R	12[b]	0.5	0.5	4	4	2	8	1	8
S. pneumoniae Pen-S Tel-S Mac-R	23	0.03	0.03	0.06	0.06	0.06	0.06	0.03	0.06
S. pyogenes Tel-S Mac-S	10	0.03	0.03	0.03	0.03	0.03	0.03	0.03	0.03
S. pyogenes Tel-R Mac-R	10	0.125	0.25	2	2	32	32	32	32
S. pyogenes Tel-S Mac-R	30	0.03	0.03	0.125	0.25	0.125	0.5	0.25	0.5

[a]Clinical isolates differing in their penicillin (Pen), telithromycin (Tel), and macrolide (Mac) susceptibility.
[b]All the Mac-R strains of *S. pneumoniae* carried a constitutive *erm*B gene.

rare, even with multiple transfers in subinhibitory concentrations of solithromycin (McGhee *et al.*, 2010) possibly because of the need for multiple ribosomal regions to mutate in order to block binding at multiple sites.

Solithromycin has been shown to be selective for bacterial protein synthesis and does not inhibit protein synthesis in mammalian cells. Traditionally macrolides have been selective and have not inhibited protein synthesis in mitochondria since they do not extend to or bind the PTC site. In bacterial protein synthesis studies, macrolides have been demonstrated to block newly synthesized peptides from exiting the peptide tunnel. However when tested with ketolides, peptides continue to be made, although many peptides may be mistranslated and thus results in abnormal peptides and proteins as noted above (Kannan *et al.*, 2012).

The effect of Fluorine at C-2 on the structure of the macrolide core

Early in development of solithromycin it was noted that solithromycin can exist in the keto form, while telithromycin could exist in either of the two forms, the enol form or the keto form. Extensive work by Dr S. Matsa (unpublished work, contracted by Cempra, Inc.) demonstrated that in solution telithromycin exists in the enol form because the keto group is unprotected, while solithromycin exists as a true ketolide because the 2-F prevents the keto group from enolization (Fig. 20.16). Thus the name ketolides to the entire class of macrolides without a cladinose is

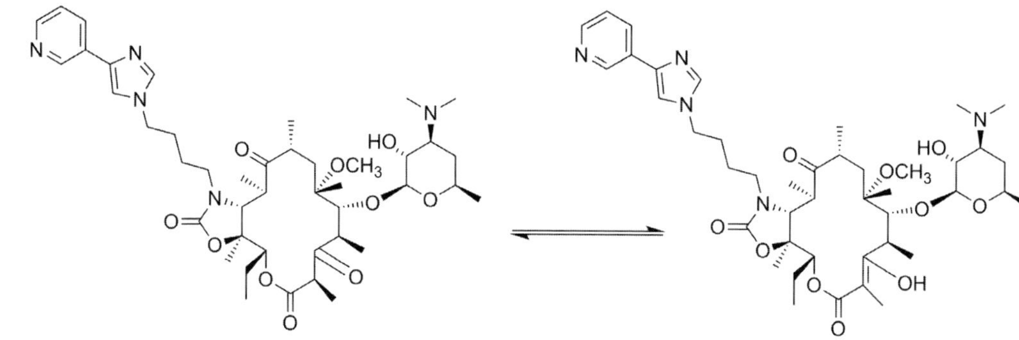

Figure 20.16 Conversion of a ketolide to the more stable enol form of telithromycin.

somewhat of a misnomer as the older ketolides are really 'enolides' and not ketolides. Because of the importance of the fluorine both to the structure and the activity of solithromycin, we have placed it in a superclass, as a fluoroketolide. It is possible that the fluorine provides conformational rigidity to the macrocyclic ring. Thus, solithromycin is a fluoroketolide that is both structurally and biologically different from telithromycin.

Clinical profile of solithromycin

Solithromycin has shown good oral bioavailability and has the best oral bioavailability and pharmacokinetics compared to the older approved macrolides, including azithromycin (Table 20.4). The pharmacokinetic and pharmacodynamic drivers of activity *in vivo* are AUC (Area under the curve) for macrolides.

Solithromycin inhibits CYP3A4 like all macrolides, but it does not induce CYP3A4 like cethromycin. It has limited protein binding, 82% bound to albumin, with no significant binding to α1-acid glycoprotein as is seen with the older macrolides. Unlike older macrolides, solithromycin is most active at an acidic pH because it is more soluble at acidic pH. When tested in phase 1 studies at doses of up to 1600 mg, little nausea and few gastrointestinal effects were noted. In multidose phase 1 studies (Still *et al.*, 2011) some drug accumulation was noted as is seen with older macrolides and a loading dose of 800 mg with a maintenance dose of 400 mg once a day dosing was selected by PK/PD modelling studies (Okusanya *et al.*, 2012). Solithromycin has shown no food effects and patients can be administered solithromycin with or without food. It was shown to be well tolerated and effective in a phase 2 study in CABP (Oldach *et al.*, 2013). An intravenous formulation has been developed and systemic safety in the intravenous phase 1 study was demonstrated (Fernandes *et al.*, 2011b; Oldach *et al.*, 2012). Both the oral and intravenous formulations are in phase 3 clinical development in two global trials for CABP. Solithromycin has also been tested in a phase 2 trial against gonorrhoea where it was shown to have 100% efficacy in all culture proven gonorrhoea cases (Hook *et al.*, 2013).

Anti-inflammatory effect

In addition to antibacterial activity, macrolide antibiotics are noted for their anti-inflammatory effects. These effects have translated into efficacy in patients with congestive obstructive pulmonary disease (COPD) where the macrolides are also used to treat pneumonia. In COPD patients, macrolides are used to re-sensitize patients to steroids such that the steroid dose can be lowered (Barnes and Adcock, 2009). This interesting effect has been thoroughly studied by Professor Peter Barnes and his team at Imperial College, London. These investigators have noted the up-regulation of the HDAC2 promoter (histone deacetylase 2 promoter) which results in increased HDAC2 and the inhibition of cytokine release, such as MMP9 and IL-8 (Kobayashi *et al.*, 2013a,b). Solithromycin was shown to have strong anti-inflammatory properties in their

Table 20.4 Comparative pharmacokinetic parameters. (a) Absolute oral bioavailability; (b) Key pharmacokinetic parameters of azithromycin, solithromycin and telithromycin

a)

Macrolide	Bioavailability
Clarithromycin[a]	~50%
Azithromycin[b]	38%
Telithromycin[c]	57%
Solithromycin[d]	67%

b) Mean pharmacokinetic parameters following single oral dose

		Azithromycin 500 mg ($n=12$)[b]	Azithromycin 1200 mg ($n=12$)[b]	Solithromycin (800 mg) (n = 5)[e]	Telithromycin (800 mg) ($n=18$)[c]
C_{MAX}	(μg/ml)	0.4	0.7	1.3	1.9
$t_{1/2}$	(h)	-	40	6.7	7.2
AUC_{0-24}	(μg•h/ml)	2.6	6.8	13.7	8.3

[a]Chu *et al.* (1992); [b]Foulds *et al.* (1990); [c]Perret *et al.* (2002); [d]Oldach *et al.* (2012); [e]Still *et al.* (2011).

assays and it is expected to be studied in COPD patients in the future.

Broad clinical potential of solithromycin

Solithromycin has been well tolerated to date and it has been differentiated from telithromycin both chemically and biologically. It is being tested in paediatric populations and a paediatric suspension is under development. If successful, solithromycin would be the first antibiotic that has been developed in oral, intravenous and paediatric suspension formulations in 25 years. Development of these formulations will allow flexibility of dosing and use in all populations. Its safety has been demonstrated in long term toxicology studies as well as in developmental toxicology studies. In studies to date, no Q-T effect has been noted. Solithromycin has also been shown to have activity against pathogenic bacteria that cause infections in pregnancy, such as Group B beta-haemolytic streptococcus (Picinelli *et al.*, 2013) and *Ureaplasma urealytium* (Waites *et al.*, 2009). Solithromycin has also been demonstrated to penetrate well into the amniotic fluid and the fetus of pregnant ewe (Kellan *et al.*, 2014), a recognized model system for human pregnancy. Solithromycin has shown *in vitro* activity against *Plasmodium* species and also *in vivo* efficacy against mouse malaria (Wittlin *et al.*, 2012). Considering the multiple use potential for solithromycin and its tolerability, it could be the next antibiotic to have broad use potential for many indications in all populations.

Non-antibiotic uses of macrolides

Motilin agonist – motilides

It is said that one man's meat is another man's poison. The adverse event of erythromycin, i.e. gastrointestinal motility, that causes nausea and vomiting, is used therapeutically to induce stomach contractions in diabetic gastroparesis (Camilleri, 2007). Erythromycin is effective when injected intravenously in inducing stomach contractions in the diabetic gastroparesis patients. However, repeated administration can result in desensitization. The pain on injection of erythromycin and the damage to the veins is also a limiting factor. It has been shown that erythromycin mimics human motilin, which is the natural peptide agonist for the motilin receptor (Xu *et al.*, 2005). When activated, the motilin receptor is responsible for contraction of the stomach wall, resulting in stomach emptying and the initiation of the migratory motor complex down the intestinal tract. A few companies have developed analogues of erythromycin to improve activity upon oral administration and remove antibacterial properties (Sanger *et al.*, 2009) (Fig. 20.17).

To date no compound has been successfully developed, either due to lack of efficacy or tolerability, one of the challenges being that the half-life of the molecule must be short in order to prevent continued stimulation of the motilin receptor in the stomach wall. By training, chemists focus on making very potent compounds. In the case of the motilin receptor, as is true for other G protein coupled receptors, highly potent compounds can cause receptor internalization with resulting desensitization (Mitselos *et al.*, 2008). The cladinose sugar of erythromycin is considered essential and thus ketolides are not expected to induce stomach contractions, although cethromycin was known to cause severe gastrointestinal contractions at high doses. With the increasing need for treating diabetes patients, it is possible that in the future motilides, a name coined for macrolides that act as motilin agonists could be designed.

Anti-inflammatory macrolides

As noted above, macrolides are known to have anti-inflammatory properties. This activity has aided rapid recovery of patients, since much of the pain and discomfort that patients feel could be due to inflammation caused by the infection. Macrolides are also used in cystic fibrosis, late stage asthma and COPD patients to decrease inflammation and to lower the dose of steroids. There is ongoing work in some companies to explore making macrolides without antibacterial properties to allow chronic administration without the liability of resistance development.

GnRH antagonist

The gonadotropins, luteinizing hormone (LH) and follicular stimulating hormone (FSH), play key roles in the process of human reproduction. LH-releasing hormone (LHRH) antagonists inhibit the function of gonads and are used to treat disorders such as endometriosis, uterine fibroids, precocious puberty, as well as cancers of the breast and prostate. Injectable complex peptides are used for treating patients and an oral, non-peptidic LHRH antagonist is needed. An analogue of erythromycin A was identified as a LHRH antagonist from a high throughput screen at Abbott laboratories (Randolph *et al.*, 2004). LHRH antagonism was achieved with modifications of the cladinose. Interestingly, the lead molecule that was identified was a 11,12-cyclic carbamate of clarithromycin (6-*O*-methyl erythromycin) with known antibacterial properties described above. This lead was optimized to make descladinose derivatives to yield the 5-(S) methyloxazikidin-2-one carbamate substitution at position 3 of the macrolide ring (Fig. 20.18). This lead molecule bound human LHRH receptors with approximately 1 nM affinity *in vitro*. When tested orally and intravenously in rats it showed dose dependent inhibition of LH in castrated male rats. The descladinose analogue had the advantage of oral bioavailability. No further development work has been reported on this compound. However, work in the area of LH antagonists continues and

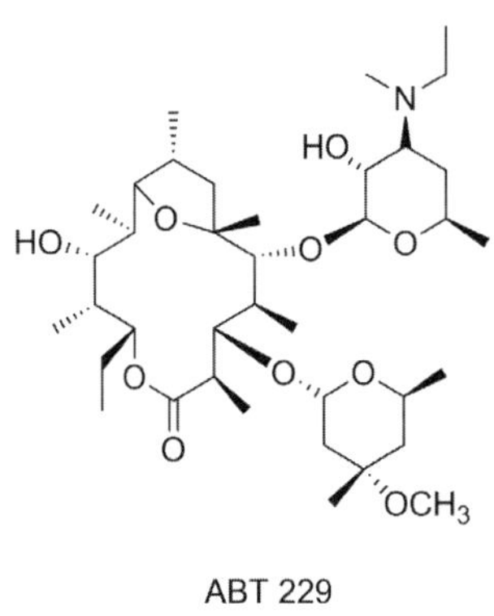

ABT 229

Figure 20.17 A macrolide optimized to be a motilin agonist.

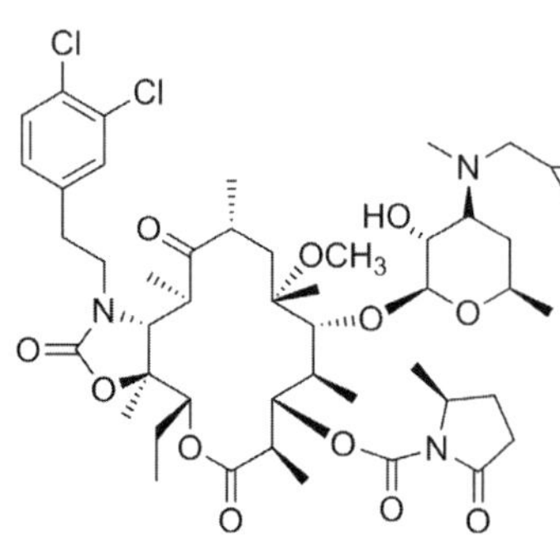

Figure 20.18 A macrolide optimized to be a LHRH Agonist (Randolph *et al.*, 2004).

someday a macrolide may emerge as an effective drug in the LH area.

Conclusion

The story of macrolide antibiotics started with the discovery of erythromycin. Extensive numbers of macrolides have been synthesized, of which, only a handful have reflected the safety and efficacy of the first macrolide, erythromycin. However, nurturing nature's molecule has led to newer macrolides, such as solithromycin, that could continue and expand the utility of this historically popular antibiotic class. In the future, scientists could learn from the new macrolides and ketolides and expand beyond antibacterial uses, perhaps even using macrolides to induce the correction of the mistranslation of mutant proteins in other diseases, that could correct diseases caused by misfolded proteins (Kannan *et al.*, 2012), such as the CFTR in cystic fibrosis or in cancers arising from misfolded P53 tumour suppressor protein. Thus, the research may have just begun in expanding the uses of macrolide antibiotics.

Acknowledgements

I thank my colleagues at Cempra, Inc., as well as at Abbott and Optimer Pharmaceuticals, for their collaborative work and discussions. I also thank Dr Kara Keedy and Dr Lovy Pradeep for their comments, Dr Sunil Matsa for providing the computer model of the peptide tunnel with bound antibiotics, and Dr David Pereira for help in drawing all the structures and reviewing the manuscript.

References

Asaka, T., Manaka, A., and Sugiyama, H. (2003). Recent developments in macrolide antimicrobial research. Curr. Top. Med. Chem. 3, 961–989.

Barnes, P.J., and Adcock, I.M. (2009). Glucocorticoid resistance in inflammatory diseases. Lancet 373, 1905–1917.

Bertrand, D., Bertrand, S., Neveu, E., and Fernandes, P. (2010). Molecular characterization of off-target activities of telithromycin: a potential role for nicotinic acetylcholine receptors. Antimicrob. Agents Chemother. *54*, 5399–5402.

Bertrand, D., Bertrand, S., Pereira, D., and Fernandes, P. (2011). Side-effects of telithromycin and voriconazole are attributable to nicotinic acetylcholine receptor interactions. Abstr. 51stIntersci. Conf. Antimicrob. Agents Chemother. abstr A2–588. Sept 17 – 20) – Chicago, IL.

Betts, A., Finegan, S., Hanna, D., King, B., Klichinsky, M., Lin. J., O'Donnel. J., Ripp, S., and Van Deusen., J. (2010). Effective clinical dose selection of the ketolide PF-04287881 through application of a mechanism-based pharmacokinetic/pharmacodynamic (PK/PD) model. Poster (A1–693) presented at: Interscience Conference on Antimicrobial Agents and Chemotherapy.

Camilleri M. (2007). Clinical practice. Diabetic gastroparesis. N. Engl. J. Med. *356*, 820–829.

CDC (Centers for Disease Control and Prevention) (2013). Antibiotic resistance threats in the United States (Centers for Disease Control and Prevention). http://www.cdc.gov/drugresistance/threat-report-2013/pdf/ar-threats-2013–508.pdf

Chu, S.Y., Deaton, R., and Cavanaugh, J. (1992). Absolute bioavailability of clarithromycin after oral administration in humans. Antimicrob. Agents Chemother. *36*, 1147–1150.

English, M.L., Fredericks, C.E., Milanesio, N.A., Rohowsky, N., Xu, Z.Q., Jenta, T.R., Flavin, M.T., and Eiznhamer, D.A. (2012). Cethromycin versus clarithromycin for community-acquired pneumonia: comparative efficacy and safety outcomes from two double-blinded, randomized, parallel-group, multicenter, multinational noninferiority studies. Antimicrob. Agents Chemother. *56*, 2037–2047.

Farrell, D.J., Sader, H.S., Castanheira, M., Biedenbach, D.J., Rhomberg, P.R., and Jones, R.N. (2010). Antimicrobial characterization of CEM-101 activity against respiratory tract pathogens including multidrug-resistant pneumococcal serogroup 19A isolates. Int. J. Antimicrob. Agents *35*, 537–543.

Fernandes, P. (2006). Antibacterial discovery and development – the failure of success? Nature Biotechnol. *24*, 1497–1503.

Fernandes, P., Pereira, D., Jamieson, B., and Keedy, K. (2011a). Solithromycin. Macrolide antibiotic. Drugs Fut. *36*, 751–758.

Fernandes, P., Pereira, D., Wu, S., and Slee, A. (2011b). Intravenous formulation of solithromycin, a painless macrolide antibiotic in a rabbit intravenous injection model. Abstract (A2–036) presented at: 51st Interscience Conference on Antimicrobial Agents and Chemotherapy.

Fernandes, P.B., Baker, W.R., Freiberg, L.A., Hardy, D.J., and McDonald, E.J. (1989). New macrolides active against *Streptococcus pyogenes* with inducible and constitutive type of macrolide-lincosamide-streptogramin B resistance. Antimicrob. Agents Chemother. 33, 78–81.

Foulds, G., Shepard, R.M., and Johnson, R.B. (1990). The pharmacokinetics of azithromycin in human serum and tissues. J. Antimicrob. 25, 73–82.

Golparian, D., Fernandes, P., Ohnishi, M., Jensen, J.S., and Unemo, M. (2012). *In vitro* activity of the new fluoroketolide solithromycin (CEM-101) against a large collection of clinical *Neisseria gonorrhoeae* isolates and international reference strains including those with various high-level antimicrobial resistance – potential treatment option for gonorrhea? Antimicrob. Agents Chemother. *56*, 2739–2742.

Hook III, E.W., Jamieson, B.D., Oldach, D., Harbison, H., Whittington, A., and Fernandes, P. (2013). A Phase II, dose ranging study to evaluate the efficacy and safety of single-dose oral solithromycin (CEM-101) for treatment of patients with uncomplicated urogenital gonorrhea. Presented at: 20th Meeting of the International Society of Sexually Transmitted Diseases Research and 14th World Congress of the International Union of Sexually Transmitted Infections, abstract, O02.5. July 14–17), Vienna, Austria.

Jarvis, L.M. (2008). The Ketek effect. Chem. Eng. News *86*, 861–862.

Jiang, L.J., and Or, Y.S. (2009). Pharmacokinetics of EDP-420 after multiple oral doses in healthy adult volunteers and in a bioequivalence study. Antimicrob. Agents Chemother. *53*, 3218–3225.

Kannan, K., Vázquez-Laslop, N., and Mankin, A.S. (2012). Selective protein synthesis by ribosomes with a drug-obstructed exit tunnel. Cell. *15*, 508–520.

Katz, L., and Ashley, G.W. (2005). Translation and protein synthesis: Macrolides. Chem Rev. *105*, 499–527.

Keelan, J.A., Kemp, M.W., Payne, M.S., Johnson, D., Stock, S.J., Saito, M., Fernandes, P., and Newnham, J.P. (2014). Maternal administration of solithromycin, a new, potent, broad-spectrum fluoroketolide antibiotic, achieves fetal and intraamniotic antimicrobial protection in a pregnant sheep model. Antimicrob. Agents Chemother. *58*, 447–454.

Kobayashi, Y., Wada, H., Rossios, C., Takagi, D., Charron, C., Barnes, P.J., and Ito, K. (2013a). A novel macrolide/fluoroketolide, solithromycin (CEM-101), reverses corticosteroid insensitivity via phosphoinositide 3-kinase pathway inhibition. Br. J. Pharmacol. *169*, 1024–1034.

Kobayashi, Y., Wada, H., Rossios, C., Takagi, D., Higaki, M., Mikura, S., Goto, H., Barnes, P.J., and Ito, K. (2013b). A novel macrolide solithromycin exerts superior anti-inflammatory effect via NF-κB inhibition. J. Pharmacol. Exp. Ther. *345*, 76–84.

LaMarre, J., Mendes, R.E., Szal, T., Schwarz, S., Jones, R.N., and Mankin, A.S. (2013). The genetic environment of the cfr gene and the presence of other mechanisms account for the very high linezolid resistance of *Staphylococcus epidermidis* isolate 426–3147L. Antimicrob. Agents Chemother. *57*, 1173–1179.

Leach, K.L., Swaney, S.M., Colca, J.R., McDonald, W.G., Blinn, J.R., Thomasco, L.M., Gadwood, R.C., Shinabarger, D., Xiong, L., and Mankin, A.S.(2007). The site of action of oxazolidinone antibiotics in living bacteria and in human mitochondria. Mol. Cell. *26*, 393–402.

Lemaire, S., Van Bambeke, F., and Tulkens, P.M. (2009). Cellular accumulation and pharmacodynamic evaluation of the intracellular activity of CEM-101, a novel fluoroketolide, against *Staphylococcus aureus, Listeria monocytogenes,* and *Legionella pneumophila* in human THP-1 macrophages. Antimicrob. Agents Chemother. *53*, 3734–3743.

Lipinski, C.A., Lombardo, F., Dominy, B.W., and Feeney, P.J. (2001). Experimental and computational approaches to estimate solubility and permeability in drug discovery and development settings. Adv. Drug Deliv. Rev. *46*, 3–26.

Llano-Sotelo, B., Dunkle, J., Klepacki, D., Zhang, W., Fernandes, P., Cate, J.H., and Mankin, A.S. (2010). Binding and action of CEM-101, a new fluoroketolide antibiotic that inhibits protein synthesis. Antimicrob. Agents Chemother. *54*, 4961–4970.

McGhee, P., Clark, C., Kosowska-Shick, K.M., Nagai, K., Dewasse, B., Beachel, L., and Appelbaum, P.C. (2010). *In vitro* activity of CEM-101 against *Streptococcus pneumoniae* and *Streptococcus pyogenes* with defined macrolide resistance mechanisms. Antimicrob. Agents Chemother. *54*, 230–238.

Magee, T.V., Ripp, S.L., Li, B., Buzon, R.A., Chupak, L., Dougherty, T.J., Finegan, S.M., Girard, D., Hagen, A.E., Falcone, M.J., *et al.* (2009). Discovery of azetidinylketolides for the treatment of susceptible and multidrug resistant community-acquired respiratory tract infections. J. Med. Chem. *52*, 7446–7457.

Mallegol, J., Fernandes, P., Seah, C., Guyard, C., and Melano, R.G. (2013). Determination of *in vitro* activity of solithromycin at different pHs and its intracellular activity tested against clinical isolates of *Neisseria gonorrhoeae* from a laboratory collection. Antimicrob. Agents Chemother. *57*, 4322–4328.

Mitselos, A., VandenBerghe, P., Peeters, T.L., and Depoortere, I. (2008). Differences in motilin receptor desensitization after stimulation with motilin or motilides are due to alternative receptor trafficking. Biochem. Pharmacol. *75*, 1115–1128.

Nagiec, E.E., Wu, L., Swaney, S.M., Chosay, J.G., Ross, D.E., Brieland, J.K., and Leach, K.L. (2005). Oxazolidinones inhibit cellular proliferation via inhibition of mitochondrial protein synthesis. Antimicrob. Agents Chemother. *49*, 3896–3902.

Okusanya, O., Bhavnani, S.M., Forrest, A., Bulik, C.C., Oldach, D., Fernandes, P., and Ambrose, P.G. (2012). Population pharmacokinetic (PPK) and pharmacokinetic–pharmacodynamic (PK–PD) target attainment (TA) analyses for solithromycin (SOL, CEM-101) to support intravenous (IV) dose selection in patients with community-acquired bacterial pneumonia (CABP). Abstract (A1269) presented at: 52nd Interscience Conference on Antimicrobial Agents and Chemotherapy.

Oldach, D., Jamieson, B., Clark, K., Keedy, K., and Fernandes, P. (2012). Systemic safety profile of solithromycin in Phase 1 & 2 clinical trials. Abstract (A1286) presented at: 52nd Interscience Conference on Antimicrobial Agents and Chemotherapy.

Oldach, D., Clark, K., Schranz, J., Das, A., Craft, J.C., Scott, D., Jamieson, B.D., and Fernandes, P. (2013). A randomized, double-blind, multi-center, phase 2 study comparing the efficacy and safety of oral solithromycin (CEM-101) to oral levofloxacin in the treatment of patients with community-acquired bacterial pneumonia. Antimicrob. Agents Chemother. *57*, 2526–2534.

Pereira, D., and Fernandes, P. (2011). Synthesis and antibacterial activity of novel 4-aryl-[1,2,3]-triazole containing macrolides. Bioorg. Med. Chem. Lett. *21*, 510–513.

Perret, C., Lenfant, B., Weinling, E., Wessels, D.H., Scholtz, H.E., Montay, G., and Sultan, E. (2002). Pharmacokinetics and absolute oral bioavailability of an 800-mg oral dose of telithromycin in healthy young and elderly volunteers. Chemother. *48*, 217–213.

Piccinelli, G., Fernandes, P., Bonfanti, C., Caccuri, F., Caruso, A., and De Francesco, M.A. (2013). *In vitro* activity of solithromycin against erythromycin – resistant *Streptococcus agalactiae*. Antimicrob. Agents Chemother. Antimicrobial. Agents Chemother. *58*, 1693–1698.

Putnam, S.D., Castanheira, M., Moet, G.J., Farrell, D.J., and Jones, R.N. (2010). CEM-101, a novel fluoroketolide: antimicrobial activity against a diverse collection of Gram-positive and Gram-negative bacteria. Diag. Microbiol. Infect. Dis. *66*, 393–401.

Putnam, S.D., Sader, H.S., Farrell, D.J., Biedenbach, D.J., and Castanheira, M. (2011). Antimicrobial characterisation of solithromycin (CEM-101), a novel fluoroketolide: activity against staphylococci and enterococci. Int. J. Antimicrob. Agents *37*, 39–45.

Rafie, S., MacDougall C., and James, C.L. (2010). Cethromycin: a promising new ketolide antibiotic for respiratory infections. Pharmacotherapy *30*, 290–303.

Randolph, J.T., Waid, P., Nichols, C., Sauer, D., Haviv, F., Diaz, G., Bammert, G., Besecke, L.M., Segreti, J.A., Mohning, K.M., *et al.* (2004). Nonpeptide luteinizing hormone-releasing hormone antagonists derived from erythromycin A: design, synthesis, and biological activity of cladinose replacement analogues. J. Med. Chem. *47*, 1085–1097.

Roblin, P.M., Kohlhoff, S.A., Parker, C., and Hammerschlag, M.R. (2010). *In vitro* activity of CEM-101, a new ketolide antibiotic, against *Chlamydia trachomatis* and *Chlamydia* (*Chlamydophila*) *pneumoniae*. Antimicrob. Agents Chemother. *54*, 1358–1359.

Rodvold, K.A., Gotfried, M.H., Still, J.G., Clark, K., and Fernandes, P. (2012). Comparison of plasma, epithelial lining fluid, and alveolar macrophage concentrations of solithromycin (CEM-101) in healthy adult subjects. Antimicrob. Agents Chemother. *56*, 5076–5081.

Sanger, G.J., Westaway, S.M., Barnes, A.A., Macpherson, D.T., Muir, A.I., Jarvie, E.M., Bolton, V.N., Cellek, S., Näslund, E., Hellström, P.M., *et al.* (2009). GSK962040: a small molecule, selective motilin receptor agonist, effective as a stimulant of human and rabbit gastrointestinal motility. Neurogastroenterol. Motil. *21*,657–664.

Sato, T., Tateda, K., Kimura, S., Iwata, M., Ishii, Y., and Yamaguchi, K. (2011). *In vitro* antibacterial activity of modithromycin, a novel 6,11-bridged bicyclolide, against respiratory pathogens, including macrolide-resistant Gram-positive cocci. Antimicrob. Agents Chemother. *55*, 1588–1593.

Shain, C.S., and Amsden, G.W. (2002). Telithromycin: the first of the ketolides. Ann Pharmacother. *36*, 452–464.

Shi, J., Montay, G., and Bhargava, V.O. (2005). Clinical pharmacokinetics of telithromycin, the first ketolide antibacterial. Clin Pharmacokinet. *44*, 915–934.

Still, J.G., Schranz, J., Degenhardt, T.P., Scott, D., Fernandes, P., Gutierrez, M.J., and Clark, K. (2011). Pharmacokinetics of solithromycin (CEM-101) after single or multiple oral doses and effects of food on single-dose bioavailability in healthy adult subjects. Antimicrob. Agents Chemother. *55*, 1997–2003.

Sunazuka, T., Oumra, S., Iwasaki, S., and Omura, S. (2002). Chemical modification of macrolides. In Macrolide Antibiotics. Chemistry Biology and Practice, 2nd edition, S. Omura, ed. (Academic Press, San Diego, CA), pp. 99–180.

Viluksela, M., Hanhijärvi, H., Husband, R.F., Kosma, V.M., Collan, Y., and Männistö, P.T. (1988). Comparative liver toxicity of various erythromycin derivatives in animals. J. Antimicrob. Chemother. *21* (Suppl. D), 9–27.

Waites, K.B., Crabb, D.M., and Duffy, L.B. (2009). Comparative *in vitro* susceptibilities of human mycoplasmas and ureaplasmas to a new investigational ketolide, CEM-101. Antimicrob. Agents Chemother. *53*, 2139–2141.

Wang, H., Yu, M., Ochani, M., Amella, C.A., Tanovic, M., Susarla, S., Li, J.H., Wang, H., Yang, H., Ulloa, L., *et al.* (2003). Nicotinic acetylcholine receptor alpha7 subunit is an essential regulator of inflammation. Nature *421*, 384–388.

Wittlin, S., Ekland, E., Craft, J.C., Lotharius, J., Bathurst, I., Fidock, D.A., and Fernandes, P. (2012). *In vitro* and *in vivo* activity of solithromycin (CEM-101) against *Plasmodium* species. Antimicrob. Agents Chemother. *56*, 703–707.

Xu, L., Depoortere, I., Vertongen, P., Waelbroeck, M., Robberecht, P., and Peeters, T.L. (2005). Motilin and erythromycin-A share a common binding site in the third transmembrane segment of the motilin receptor. Biochem. Pharmacol. *70*, 879–887.

Ying, L., and Tang, D. (2010). Recent advances in the medicinal chemistry of novel erythromycin-derivatized antibiotics. Curr. Top. Med. Chem. *10*, 1441–1469.

Antibiotics in the Pipeline

Hyunjun Park and Michael Thomas

21

Abstract

The enormous success of using antibiotics to save lives is being challenged by the steady evolution of resistance to these antibiotics by the targeted bacteria. This resistance has led to predictions that we will soon be returning to the pre-antibiotic era whereby even the most minor infection has the potential to have severe consequences on the infected person. Fortunately, there has been a steady increase in the efforts to combat this rise in antibiotic resistance. This chapter discusses antibiotics for treating bacterial infections that were in the clinical pipeline at the beginning of 2014. While a list of the drugs that have recently been approved for clinical use or are in various phases of clinical trials is provided, a select few are highlighted in more detail to show how drug development itself is evolving to generate the next generation of antibiotics so the pre-antibiotic era stays in the past.

Introduction

The threat of bacterial resistance to antibiotics and the ensuring need for a robust pipeline of new drugs has been extensively reviewed in numerous journals as well as in previous chapters in this book. We have listed the small molecule and peptide antibacterial drug candidates that are currently or have recently completed phases I-III clinical trials that the authors were able to identify from journal articles and the websites of pharmaceutical companies and regulatory agencies (Tables 21.1–21.3; Butler *et al.*, 2013; Coates and Halls, 2012; Fjell *et al.*, 2012; Fox, 2013; Jabes, 2011; Pucci and Bush, 2013; de Souza Mendes and de Souza Antunes, 2013). A complete list of web pages used is provided at the end of the references section. It is not the intent of this chapter to systematically discuss the attributes of each drug listed in these tables. Rather, we focus our discussion on a select subset of these drugs that best exemplify how some of the most salient challenges in antibiotic development are being addressed. This approach may ensure the specific examples discussed provide relevance past just success in clinical trials.

The challenges associated with antibiotic development include, but are not limited to, access to the drug target, host toxicity, drug efflux, persister cells, and limited molecular targets. Also discussed are noteworthy innovative approaches to antibacterial therapy. These include the use of prodrugs, combination therapy, altered binding to established targets, and inhibition of new targets. Discussion on specific structural elements was accentuated wherever functional improvements or desirable features of a drug could be attributed to them. Some molecules are discussed in more than one section, as they represent significant improvement in those areas over existing options.

Overcoming challenges associated with antibiotic development

Access to the drug target

Antibiotics can encounter several barriers prior to gaining access to their cellular target, especially if the target is intracellular. For drugs that target the cell wall biosynthetic machinery of Gram-positive bacteria, the permeable peptidoglycan layer

Table 21.1 Antibiotics in phase III clinical trials

Class	Name	Spectrum	Indications	Company
Ethambutols	SQ109	+,–	TB, *H. pylori*	Sequella, Infectex
Fluoroquinolones	Avarofloxacin (**12**)	+,–	IBS, CABP, ABSSSI	Furiex Pharmaceuticals
	Delafloxacin (**13**)	+,–	ABSSSI, uncomplicated gonorrhoea	Melinta Therapeutics
	Finafloxacin	+,–	Acute Otitis Media, *H. pylori*, UTI	MerLion Pharmaceuticals
	Gatifloxacin	+	TB (approved for RI)	Bristol-Myers Squibb
	Moxifloxacin	+	TB (approved for RI)	Global Alliance for TB Drug Development, Bayer
	Zabofloxacin (**32**)	+,–	CAP	Dong Wha Pharm
Lipoglycopeptides	Dalbavancin (**1**)	+	ABSSSI	Durata Therapeutics
	Oritavancin (**2**)	+	ABSSSI	The Medicines Company
Lipopeptides	Surotomycin	+	CDAD	Cubist Pharmaceuticals
Macrolides	Solithromycin (**9**)	+,–	CAP	Cempra
Nitroimidazoles	Delamanid (**22**)	+	TB	Otsuka Pharmaceuticals
Oxazolidinones/ quinolones hybrid	Cadazolid	+	CDAD	Actelion Pharmaceuticals
Peptide	Pexiganan acetate	+,–	Diabetic foot ulcer	Dipexium Pharmaceuticals
Quinolones	Nemonoxacin (**8**)	+,–	CAP (completed in Taiwan and China)	TaiGen Biotechnology
	Ozenoxacin (**7**)	+	Impetigo acne, topical	Ferrer Internacional, Maruho Co
Tetracyclines	Eravacycline (**11**)	+,–	cIAI, cUTI	Tetraphase Pharmaceuticals
	Omadacycline (**10**)	+,–	ABSSSI (CABP phase II, UTI phase I)	Paratek
Thiosemi-carbazones	Perchlozone	+	TB	Pharmasyntez
β-lactam/β-lactamase inhibitor	Ceftazidime (**24**)/ avibactam (**29**)	+,–	cIAI, UTI	AstraZeneca
	Ceftolozane (**23**)/ tazobactam (**28**)	+,–	cUTI, cIAI, (HABP/VABP phase II)	Cubist Pharmaceuticals

Abbreviations: +, Gram-positive bacteria; -, Gram-negative bacteria; ABSSSI, acute bacterial skin and skin structure infection; CABP, community-acquired bacterial pneumonia; CAP, community-acquired pneumonia; CDAD, *Clostridium difficile*-associated diarrhoea; cIAI, complicated intra-abdominal infection; cUTI, complicated urinary tract infection; *H. pylori*, *Helicobacter pylori*; HABP, hospital-acquired bacterial pneumonia; HAP, hospital-acquired pneumonia; IBS, irritable bowel syndrome; RI, respiratory infection; TB, tuberculosis; UTI, urinary tract infection; VABP, ventilator-associated bacterial pneumonia.

usually presents less of a challenge for access to the drug target compared to the multicomponent barrier system of Gram-negative bacteria (Denyer and Maillard, 2002; Lambert, 2002). However, even when Gram-positive organisms are the target, alterations in the cell wall composition and structure that possibly leads to affinity trapping of antibiotics has been shown to confer varying levels of resistance, as seen in several clinically isolated *Staphylococcus aureus* strains with intermediate resistance of vancomycin (Boyle-Vavra *et al.*, 2001; Cui *et al.*, 2000; Hiramatsu, 2001; Peschel *et al.*, 2000; Pfeltz *et al.*, 2000; Sieradzki and Tomasz, 1997). Therefore, efficient access to the correct drug target is demanded of all known antibiotic drugs regardless of class or mechanisms of action.

Several antibiotics in the drug pipeline that belong to two different classes of drugs are noteworthy for the modifications that enable them to have better access to their targets. For one class,

Table 21.2 Antibiotics in phase II clinical trials

Class	Name	Spectrum	Indications	Company
Actinonins	GSK1322322 (**20**)	+,–	cSSSI	GlaxoSmithKline
Aminoglycosides	Plazomicin (**31**)	+,–	UTI, pyelonephritis	Achaogen
Benzodiazepines	AFN-1252 (**19**)	+	ABSSSI	Affinium Pharmaceuticals
Fluoroquinolones	WCK-2349 (prodrug of WCK-771)	+,–	MRSA and Gram-negatives	Wockhardt
	WCK-771	+,–	MRSA and Gram-negatives	Wockhardt
Fusidic acids	CEM-192	+	ABSSSI (completed), prosthetic joint infections	Cempra
Glycopeptide/ cephalosporin hybrid	TD-1792	+	SSTI, HAP	Theravance, R-Pharm
N-Chlorotaurines	Auriclosene	+,–	Urinary catheter blockage and encrustation, adenoviral conjunctivitis; impetigo asymptomatic bacteriuria	Novabay Pharmaceuticals
Nitroimidazoles	PA-824 (**15**)	+	TB (in combination with moxifloxacin and pyrazinamide, bedaquiline and clofazimine or pyrazinamide)	Global Alliance for TB Drug Development, Chiron
Oxazolidinones	Posizolid	+	TB	AstraZeneca
	Radezolid	+,–	uSSSI, CAP	Melinta Therapeutics
	Sutezolid	+	TB	Sequella
	Tedizolid phosphate (**6**)	+,–	HABP/VABP (ABSSSI, cSSTI approved 2014)	Cubist Pharmaceuticals
Peptides	Brilacidin	+,–	ABSSSI	Cellceutix Corporation
	DPK-060	+,–	External otitis, atopic dermatitis	Pergamum AB
	SGX942	N/A	Immunomodulator – ABSSSI in combination with antibiotic	Soligenix
	LL-37	+,–	Chronic leg ulcer	Pergamum AB
	LTX-109	+,–	Impetigo (nasal decolonization in phase I)	Lytix Biopharma
	OP-145	+,–	Otitis (completed)	OctoPlus
Pleuromutilins	BC-3781	+,–	ABSSSI, CABP	Nabriva Therapeutics
Polychlorophen-oxyphenols	CG400549 (**18**)	+	ABSSSI	Crystal Genomics
Tetracyclines	Sarecycline	+	Acne, rosacea (completed)	Warner Chilcott
Thiazoles	LFF571	+	CDAD	Novartis
β-lactam/β-lactamase inhibitor	Ceftaroline (**25**)/ avibactam (**29**)	+,–	MRSA	AstraZeneca
	Imipenem/ cilastatin/MK-7655	+,–	UTI, cIAI	Merck & Co.

Abbreviations: +, Gram-positive bacteria; –, Gram-negative bacteria; ABSSSI, acute bacterial skin and skin structure infection; CABP, community-acquired bacterial pneumonia; CAP, community-acquired pneumonia; CDAD, *Clostridium difficile*-associated diarrhoea; cIAI, complicated intra-abdominal infection; cSSSI, complicated skin and skin structure infection; cSSTI, complicated skin and soft tissue infection; HABP, hospital-acquired bacterial pneumonia; HAP, hospital-acquired pneumonia; MRSA, methicillin-resistant *Staphylococcus aureus*; SSTI, skin and soft tissue infection; TB, tuberculosis; uSSSI, uncomplicated skin and skin structure infection; UTI, urinary tract infection; VABP, ventilator-associated bacterial pneumonia.

Table 21.3 Antibiotics in phase I clinical trials

Class	Name	Spectrum	Indications	Company
Bicyclic macrolide	EDP-788	+	MRSA	Enanta Pharmaceuticals
bis-Benzimidazoles	SMT-19969	+	CDAD	Summit Corporation
Cephems	S-649266	+,–	Gram-negatives	Shionogi/GSK
Fluoroquinolones	DS-8587	+,–	Phase I	Daiichi Sankyo
	KPI-10	+,–	Phase I	Kalidex
Lantibiotic	NVB302	+	CDAD	Novacta Biosystems
Oxazolidinones	LCB01–0371	+	Phase I	LegoChem Biosciences
	MRX-I	+	Phase I (completed)	MicuRx Pharmaceuticals
Peptides	POL7080	–	Phase I	Polyphor
Pleuromutilins	BC-7013	+,–	SSSI, acne, eye and wound infections	Nabriva
Porphyrins	Exeporfinium chloride	+	MRSA topical	Destiny Pharma
Quinolones	KRP-AM1977X(oral)	+,–	Phase I (completed)	Kyorin Pharmaceutical Co.
	KRP-AM1977Y (intravenous)	+,–	Phase I	Kyorin Pharmaceutical Co.
Sulfactams	BAL30072 (**4**)	+,–	Gram-negatives	Basilea Pharmaceutica
β-lactam/β-lactamase inhibitor	Aztreonam (**26**)/avibactam (**29**)	+,–	Phase I	AstraZeneca
	Biapenem (**27**)/RPX7009 (**30**)	+,–	Phase I	Rempex
Unknown	ACHN-975 (**21**)	–	Phase I	Achaogen
	GSK-214094		Phase I	GlaxoSmithKline

Abbreviations: +, Gram-positive bacteria; –, Gram-negative bacteria; CDAD, *Clostridium difficile*-associated diarrhoea; MRSA, methicillin-resistant *Staphylococcus aureus*; SSSI, skin and skin structure infection.

the structural modifications to the next generation of glycopeptide antibiotics allow for the enhancement of their presence near the drug target in Gram-positive bacteria. For the second class, the drug candidate exploits the iron-siderophore uptake system to gain access to the cell wall biosynthesis machinery of Gram-negative organisms.

Lipoglycopeptides: Dalbavancin (**1**, phase III, Durata Therapeutics), oritavancin (**2**, phase III, The Medicines Company), and telavancin (**3**, Vibativ®, approved in 2009, Theravance) are semisynthetic lipoglycopeptide drugs, structurally similar to vancomycin and are used to combat infections by multidrug-resistant Gram-positive bacteria (Fig. 21.1). These molecules contain a heavily cross-linked heptapeptide core that binds to the C-terminal D-alanyl-D-alanine of the growing peptidoglycan chains, inhibiting both the transpeptidation and transglycosylation reactions during cell wall biosynthesis. The lipophilic modifications to the cross-linked heptapeptide core is hypothesized to anchor these molecules into the bacterial cell membrane, increasing the effective molarity of these molecules relative to their molecular target, thereby increasing their efficacy. Furthermore, oritavancin (**2**) and telavancin (**3**) insertion into the cell membrane causes disruption of the membrane structure, resulting in membrane permeability and disruption of membrane potential, making the antibiotics bacteriocidal (Belley *et al.*, 2009; Higgins *et al.*, 2005).

Trojan horse strategy: Access to the cell wall biosynthetic machinery of Gram-negative bacteria, a molecule must be able to traverse across the outer membrane. For small hydrophilic molecules, such as many β-lactams, passive diffusion through outer membrane porins enable access to the periplasm (Yoshimura and Nikaido, 1985). As such, mutations leading to the modification or loss of porins have been widely associated with antibiotic resistance (Kumar and Schweizer,

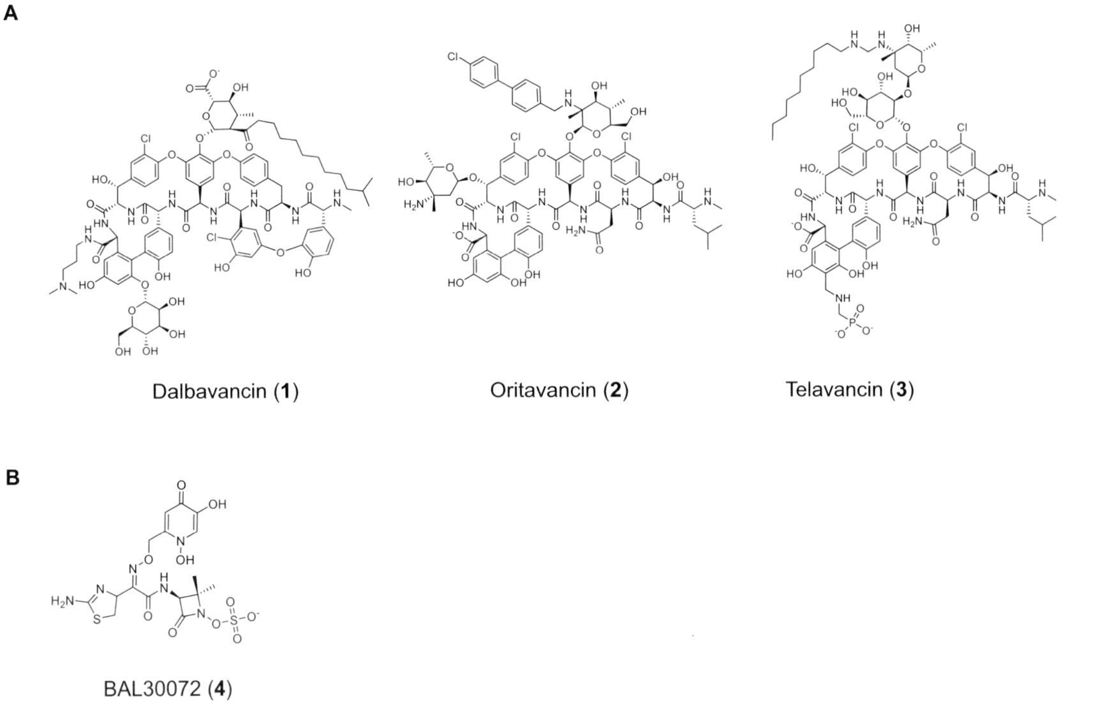

Figure 21.1 Access to the drug target. (A) Structures of lipoglycopeptides. The lipophilic side chains anchor these molecules to the cell membrane. Oritavancin and telavancin are also discussed in Reduced Toxicity and Targeting Persister Cells, respectively. (B) Structure of the siderophore sulfactam, BAL30072. The molecule is proposed to be taken up by the siderophore uptake system.

2005). One mechanism for overcoming this outer membrane barrier is to trick the bacterium into transporting the molecule across the outer membrane through one of its many outer membrane transporters. One such target is the outer membrane transporters involved in the uptake of Fe^{3+}-siderophore complexes. Siderophores are iron-chelating molecules made by many bacteria to acquire ferric iron from the environment. Because Fe^{3+}-siderophore complexes are actively transported into the periplasm by transporters with broad substrate specificity, and many bacteria have outer membrane transporters for their own siderophores as well as others (Schalk *et al.*, 2012), the system is an attractive target for drug discovery.

Nature has already taken advantage of the Fe^{3+}-siderophore transport system in delivering alternative cargo into competing cells, by producing siderophore conjugated microcins (Vassiliadis *et al.*, 2010) and sideromycins (Möllmann *et al.*, 2009). Artificial efforts to mimic this strategy have recently been reviewed (Page, 2013). This 'Trojan Horse' strategy has the benefit of not only exploiting an efficient transport system to enable the drug to penetrate the outer membrane barrier, but resistance arising from disruption of this transport system is likely to lead to reduced fitness since the cells lose the ability to access the needed iron. Even with the positive attributes, BAL30072 (**4**) is the only candidate currently in clinical trials (Fig. 21.1, phase I, Basilea Pharmaceutica). BAL30072 (**4**) is a monocyclic β-lactam conjugated to a dihydropyridone siderophore moiety, which is proposed to facilitate its uptake. The potent activity of BAL30072 (**4**) against multidrug-resistant Gram-negative pathogens is attributed to its ability to bind penicillin-binding proteins 1a, 1b, and 3 with high affinity, thereby inhibiting cell wall biosynthesis (Page *et al.*, 2010).

Reduced toxicity

The development of an antibiotic is always hampered by the negative impact of unwanted side-effects, such as liver toxicity, ototoxicity, and neurotoxicity. While there is a need to reduce these side-effects, the problem of multidrug-resistant pathogens compounded by the dearth of viable treatment options for Gram-negative bacterial infections has led to the reconsideration of old therapeutics. For example, polymyxin B (**5a** and **5b**) has emerged as a frontline therapy for treating urinary tract infections, meningitis, blood infections and eye infection, even after their abandonment by most countries approximately 30 years ago due to common nephrotoxicity and neurotoxicity issues (Fig. 21.2; Falagas and Kasiakou, 2005; Walsh and Wencewicz, 2013). Thus, modifications of 'old' drugs to remove unwanted side-effects or designing new drugs that do not contain moieties with known side-effect issues is a priority.

This section provides examples of structural modifications belonging to new antibiotics that allows for a better toxicity profile over the previous generation molecule of the same respective class. In most cases the novel structural component of interest is correlated with lower toxicity and serves as the basis for hypotheses for the reduced toxicity.

Oxazolidinones: Linezolid is the first and currently only member of the new class of synthetic drugs known as oxazolidinones to be approved by the Food and Drug Administration (FDA). Although it has been a valuable treatment option since its approval in 2000, the potential side-effects of prolonged linezolid use include reversible myelosuppression, lactic acidosis, optic and peripheral neuropathies, and inhibition of monoamine oxidase (Narita *et al.*, 2007; Renslo, 2010). Research into the molecular basis of these side-effects has linked the inhibition of mitochondrial protein synthesis by linezolid with myelosuppression and neuropathy. The inhibition of mitochondrial protein synthesis and monoamine oxidase has been correlated with the *S*-configuration at C-5 and the

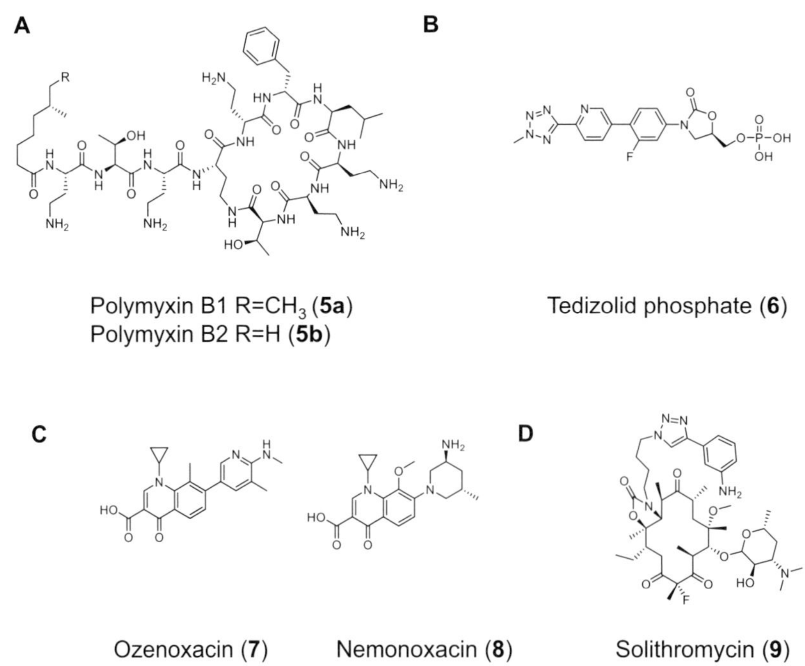

Figure 21.2 Reduced toxicity. (A) Structure of the re-emerging antibiotic, polymyxin B. (B) Structure of the oxazolidinone, tedizolid phosphate. The C-5-hydroxymethyl substituent is responsible for improved efficacy, allowing for lower dosage leading to improved toxicity profile. Also discussed in Prodrugs. (C) Structures of non-fluorinated quinolones. The lack of fluorination at the C-6 position has been hypothesized to result in lower toxicity. (D) Structure of ketolide, solithromycin. Solithromycin lacks the pyridine moiety present in telithromycin, which may cause adverse effects by binding to human nicotinic acetylcholine receptors.

C-ring of the molecule, respectively. The structure-toxicity and structure–function relationships of various oxazolidinones have been extensively reviewed (Barbachyn and Ford, 2003; Locke *et al.*, 2010; Renslo, 2010). As identical parts of the molecule have been implicated in both the desired as well as the adverse effects of oxazolidinones, modifications to them as present in the next generation of this class of drugs is expected to confer altered efficacy as well as reducing the side-effects.

Of the four next generation oxazolidinones currently in clinical trials, tedizolid phosphate (**6**) is the furthest along (phase III, Cubist Pharmaceuticals). The dephosphorylated drug, tedizolid, is the active drug and consists of a 1-(4-((2-hetero)-pyridin-5-yl)-oxazolidinone with a C-5-hydroxymethyl substituent, which is responsible for improved efficacy compared to linezolid (Im *et al.*, 2011). Because of the improved efficacy, tedizolid phosphate can be taken once daily (at 200 mg), compared to twice daily (600 mg) for linezolid. The altered chemical structure and reduced amounts needed for its biological activity, provides a much improved toxicity profile compared to linezolid. At these reduced doses, tedizolid phosphate (**6**) has not been shown to cause hematologic abnormalities or inhibit monoamine oxidase (Fig. 21.2; Das *et al.*, 2014; Urbina *et al.*, 2013). Additionally, the phosphorylated version of the drug covers up the C5 hydroxymethyl group, further decreasing monoamine oxidase inhibition. The prodrug aspects of tedizolid phosphate (**6**) are discussed in more detail below.

<u>Lipoglycopeptides:</u> As discussed above, telavancin (**3**) is a lipoglycopeptide containing a decylaminoethyl group attached to the heptapeptide core. In addition to the lipophilic side chain, telavancin also contains a phosphonomethyl aminomethyl group. This negatively charged side chain improves the absorption, distribution, metabolism, and excretion of telavancin, reducing the nephrotoxicity observed with vancomycin (Dunbar *et al.*, 2008; Leadbetter *et al.*, 2004; Zhanel *et al.*, 2010).

<u>Non-fluorinated quinolones:</u> The quinolones have been an extremely successful class of synthetic antibiotics since their introduction in the 1960s. There are now multiple generations of quinolone derivatives with various modifications to its basic scaffold (Walsh and Wencewicz, 2013). However, the rise of resistant bacterial strains as well as the toxic side-effects have been a challenge for many fluorinated quinolones (Lipsky and Baker, 1999). This is particularly true for topically formulated halogenated quinones, which can cause issues with photoallergic and photoirritation reactions (Oliveira *et al.*, 2000). The recent development of ozenoxacin (**7**) for impetigo (phase III, Ferrer Internacional), a non-fluorinated quinolone that lacks these side-effects, is a promising direction. In addition to ozenoxacin (**7**), the non-fluorinated quinolone nemonoxacin (**8**, phase III, TaiGen Biotechnology) is being developed for the treatment of community-acquired pneumonia (Fig. 21.2). This quinolone is missing a fluorination at the C-6 position and it has been hypothesized that the lack of this fluorination may explain the reduced toxic side-effects (Barry *et al.*, 2001).

<u>Ketolides:</u> Ketolides are semisynthetic derivatives of the 14-membered macrolide erythromycin A that retain the D-desosamine sugar but lack the L-cladinose from the C-3 position of the macrolide ring. Furthermore, the C-3 hydroxy is oxidized to a keto group (Zhanel *et al.*, 2002). Telithromycin was the first member of the ketolides to be used in the clinic and its side-effects include reversible visual disturbance and hepatotoxicity, likely due to the pyridine moiety binding to the nicotinic acetylcholine receptors (Bertrand *et al.*, 2010). The fluoroketolide solithromycin (**9**, CEM-101, Cempra) is currently in phase III clinical trials for the treatment of moderate to moderately severe community-acquired bacterial pneumonia (Fig. 21.2). In addition to being fluorinated at the C-2 position of the macrolide, solithromycin (**9**) lacks the pyridine moiety that causes problems with binding to the nicotinic acetylcholine receptors, and likely explains the greater efficacy, increased metabolic stability, and lower toxicity of the antibiotic relative to telithromycin (Still *et al.*, 2011).

Evasion of efflux pumps

The phenomenon of antibiotic resistance by active efflux was first reported for tetracycline (Ball *et al.*, 1980; Levy and McMurry, 1978). Since then efflux pumps have come to be known to play increasingly larger roles in the spread of bacterial multidrug resistance (Butaye, 2003). There are five families of bacterial drug efflux pumps: the adenosine triphosphate-binding cassette superfamily, the major facilitator superfamily (MFS), the multidrug and toxic compound extrusion family, the small multidrug resistance family, and the resistance-nodulation-cell division (RND) super family (Li and Nikaido, 2004). Given that efflux pumps are major contributors to multidrug resistance, efflux pump inhibitors (EPIs) have been investigated as potential adjuvants to conventional antibiotics, although none to date are in clinical trials. Reader are referred to several reviews for comprehensive coverage of multidrug efflux pump mediated drug resistance in bacteria (Denyer and Maillard, 2002; Kumar and Schweizer, 2005; Li and Nikaido, 2009; Nikaido and Pagès, 2012; Schweizer, 2012) and bacterial EPIs (Mahamoud *et al.*, 2007; Pagès and Amaral, 2009; Zechini and Versace, 2009).

Currently in the pipeline, two tetracycline derivatives that evade the tetracycline-specific efflux pumps belonging to the MFS transporters and two new fluoroquinolones that evade members of the MFS, are in clinical trials. Also, the recent report of an AcrAB (of the RND super family) inhibitor is discussed in this section.

Tetracycline derivatives: Efflux and ribosomal protection are two major mechanisms that can give rise to bacterial resistance of tetracyclines (Thaker *et al.*, 2010). These mechanisms can be found together in one organism or separately, and generally efflux is more common in Gram-negatives, whereas ribosomal protection is more common in Gram-positives (Draper *et al.*, 2013; Roberts, 1996). Plasmid-encoded Tet proteins are the tetracycline-specific pumps that are primarily responsible for the resistance to tetracyclines by efflux and have been studied extensively (Levy, 1992).

Omadacycline (**10**, PTK0796, Paratek Pharmaceuticals), a broad-spectrum antibiotic, is the first member of a new semisynthetic subset of tetracycline derivatives called aminomethylcyclines to have entered clinical trials (currently in phase III) (Fig. 21.3; Nelson and Levy, 2011). Although the mechanism by which omadacycline (**10**) is able to evade efflux pumps is unknown, the presence of the efflux pump Tet(K) in *S. aureus* RN4250 did not affect its IC_{50} (Draper *et al.*, 2013). In a separate study, omadacycline (**10**) was shown to be active against the Gram-positive pathogens *S. aureus*, *Enterococcus faecalis*, *Enterococcus faecium*, *Streptococcus pneumoniae*, and β-haemolytic streptococci carrying genes encoding the efflux pumps Tet(K) and Tet(L). In the same report, the MIC of omadacycline (**10**)

A

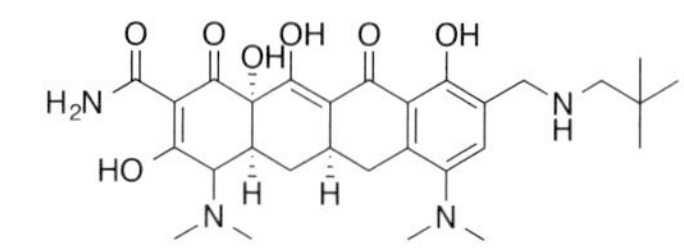

Omadacycline (**10**)

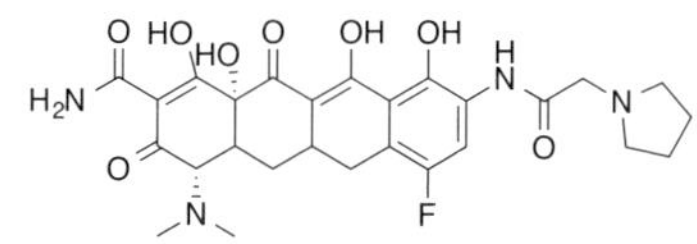

Eravacycline (**11**)

B

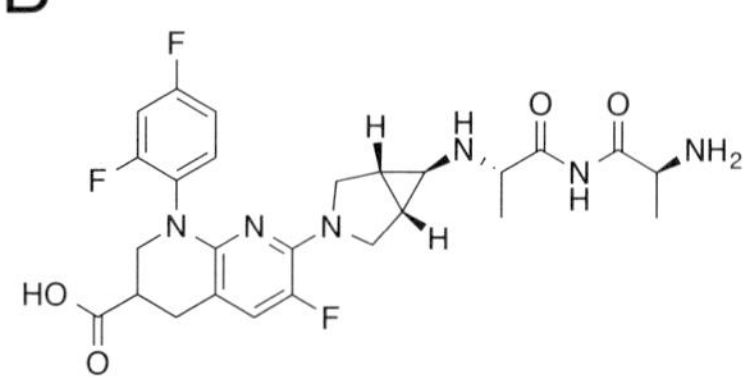

Avarofloxacin (**12**)

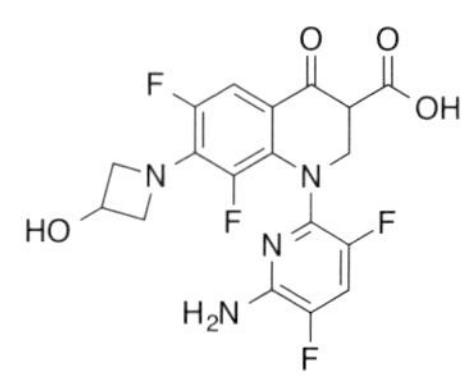

Delafloxacin (**13**)

Figure 21.3 Evasion of efflux pumps. (A) Structures of tetracycline derivatives. These molecules were shown to be active in the presence of tetracycline-specific efflux pumps. Also discussed in Altered Binding to Established Targets. (B) Structures of fluoroquinolones that remain active in the presence of multidrug efflux pumps. Also discussed in Multitarget Therapy.

for strains of *Escherichia coli* carrying the efflux gene *tet*(A) was lower than that of tetracycline or doxycycline (Macone *et al.*, 2014). Also of interest is omadacycline (**10**) is also evades ribosome modification protection seen with some tetracycline derivatives.

Eravacycline (**11**, TP-434, Tetraphase Pharmaceuticals), a broad-spectrum antibiotic, is the first member of a fluorocyclines to have entered clinical trials (currently in phase III) (Fig. 21.3; Sutcliffe *et al.*, 2013). Fluorocyclines are synthetic derivatives of tetracyclines that are substituted at the C-7 and C-9 positions of the tetracyclic core (Xiao *et al.*, 2012). A study showed that MIC of eravacycline (**11**) is unaffected against *E. coli* expressing the genes encoding efflux pumps Tet(K) or Tet(B), and increased 4-fold in cells expressing Tn1721-associated Tet(A). It was noted that the increased eravacycline (**11**) MIC of 0.25μg/mL against the latter is still 4-fold lower than that of tigecycline (Grossman *et al.*, 2012). The presence of NorA and MepA efflux pumps in *S. aureus* had no effect on the MIC of eravacycline (**11**) (Sutcliffe *et al.*, 2013).

Fluoroquinolones: Avarolfloxacin (**12**, JNJ-Q2, Furiex) and delafloxacin (**13**, RX-3341, Melinta Therapeutics) are two fluoroquinolones currently in phase III clinical trials (Fig. 21.3). These antibiotics are being evaluated for treating acute bacterial skin and skin structure infections. The MIC of avarolfloxacin (**12**) was not affected by the overexpression of *nor*A, encoding a multidrug efflux pump, in methicillin-resistant or methicillin-susceptible *S. aureus* (Morrow *et al.*, 2011). A personal communication cited in a study of delafloxacin (**13**) activity (Remy *et al.*, 2012) claims that delafloxacin (**13**) activity is unaffected by efflux pumps such as NorA, B, and C that contribute to quinolone resistance.

Efflux pump inhibitor (EPI)

Because of the importance of efflux pumps in conferring bacterial resistance to antibiotics, molecules to inhibit them have been under development for some time. There have been reports of EPIs that are in pre-clinical stages of development (Yoshida *et al.*, 2007), most recently MBX2319 (**14**, Microbiotix, Inc.), a pyranopyridine that inhibits the RND class AcrAB-TolC efflux pump in *Enterobacteriaceae* (Fig. 21.4; Opperman *et al.*, 2014). The efflux pump inhibited by MBX2319 (**14**) comprises three parts: integral membrane multidrug efflux transporter (AcrB), outer membrane channel (TolC), and periplasmic protein adapter (AcrA). AcrB uses the proton motive force to pump out the antibiotic molecules that act as substrate through TolC to the exterior of the cell (Nikaido and Takatsuka, 2009). MBX2319 (**14**) was shown to inhibit the efflux activity of AcrAB, likely by competitive inhibition of the substrate binding to AcrB (Opperman *et al.*, 2014). It also was shown to potentiate the antibacterial activities of molecules that are known substrates of AcrAB, including fluoroquinolones and β-lactams (Opperman *et al.*, 2014).

Targeting persister cells

Persister cells are a subset of the normal bacterial population that become dormant through one of many redundant pathways. The original discovery of persister cells dates back to 1944. In one of the early experiments with penicillin, Dr Joseph discovered that addition of penicillin to *Staphylococcus* cultures does not result in complete killing of all bacterial cells; instead a small fraction survives, giving rise to progeny that remain susceptible to penicillin (Bigger, 1944). Because persister cells generally have a high tolerance for antibiotics, they are a major concern in the treatment of chronic infections and the development of antibiotic resistance (Lewis, 2010).

Conventional drugs have shown virtually no effect against persister cells under standard

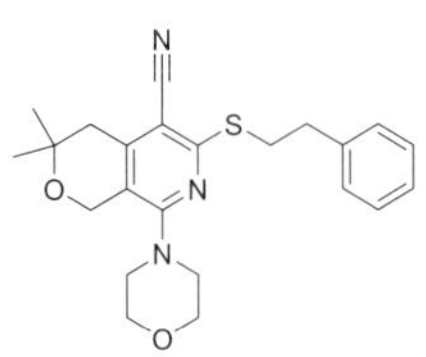

MBX2319 (**14**)

Figure 21.4 Efflux pump inhibitor. Structure of the efflux pump inhibitor in preclinical development. MBX2319 inhibits the multidrug efflux pump AcrAB-TolC, likely by competitive inhibition of the substrate binding to AcrB, an integral membrane protein.

regimen (Fauvart *et al.*, 2011), although there have been reports of persister eradication when they are used in combination with specific metabolites (Allison *et al.*, 2011) or increased dissolved oxygen (Grant *et al.*, 2012). Here, two drug candidates in the pipeline that have shown activity against persisters are discussed, as well as the recent report of an acyldepsipeptide antibiotic, which forces dormant cells to self-digest by hijacking their ClpP protease.

PA-824: PA-824 (**15**, Global Alliance for TB Drug Development) is a synthetic bicyclic nitroimidazole derivative currently in phase II clinical trials for the treatment of multidrug-resistant *Mycobacterium tuberculosis*. Intracellular activation of PA-824 (**15**) from the pro-drug form requires a deazaflavin dependent nitroreductase (Manjunatha *et al.*, 2006), and will be discussed in detail at a later section (Fig. 21.5). PA-824 (**15**) shows antibacterial activity towards both actively replicating and non-replicating *M. tuberculosis* (Barry *et al.*, 2004). The mechanism of action of PA-824 (**15**) is very complex and discrete modes are believed to be in play for its activity against cells in different phases of growth. Although the exact molecular mechanism is yet to be elucidated, actively replicating cells is expected to be killed by the inhibition of cell wall mycolic acid biosynthesis (Manjunatha *et al.*, 2009). However, as other mycolic acid synthesis inhibitors fail to kill non-replicating *M. tuberculosis*, the same mechanism is unlikely to be responsible for the efficacy of PA-824 (**15**) towards non-replicating cells. It has been shown that conversion of PA-824 (**15**) into des-nitroimidazole by a deazaflavin dependent nitroreductase generates nitric oxide in cells, which is correlated with the killing of hypoxic non-replicating *M. tuberculosis*, but does not seem to be a major factor under aerobic conditions (Manjunatha *et al.*, 2009; Singh *et al.*, 2008).

Oritavancin: As discussed above, oritavancin (**2**) is a lipoglycopeptide, a subclass of glycopeptides characterized by a lipophilic side chain that anchors the molecule to the bacterial membrane. This 4′-chlorobiphenylmethyl group has also been shown to cause the disruption of membrane potential and to increase permeability in exponential-phase[7] and stationary phase[8] Gram-positive organisms. Taken together with the fact that oritavancin (**2**) and vancomycin can both inhibit cell wall biosynthesis but only the former retains activity against stationary phase cells *in vitro*, it has been proposed that the disruption of membrane integrity is responsible for the killing of cells in the stationary phase (Belley *et al.*, 2009).

ADEP4: Acyldepsipeptide (ADEP) antibiotic ADEP4 (**16**) has not yet entered clinical trials, but a recent report of its efficacy towards persisters (Conlon *et al.*, 2013) shows promise for further development as well as illustrates an interesting new strategy for battling persisters (Fig. 21.5). The discovery and report of *in vitro* activity against several Gram-positive organisms of a group of ADEPs in 1985 (Michel and Kastner, 1985) led to the investigation of their *in vivo* potency, mechanism of action, and structural optimization (Brötz-Oesterhelt *et al.*, 2005). Derivatives of the natural product showed superior activity compared to linezolid for the treatment of rodent bacterial infections, and ADEPs were found to

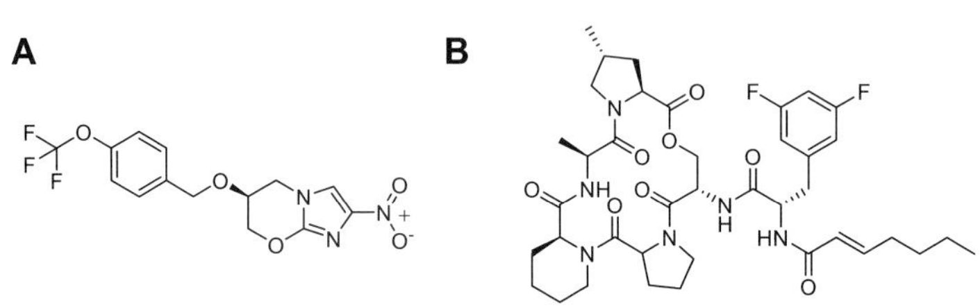

PA-824 (**15**) ADEP4 (**16**)

Figure 21.5 Targeting persister cells. (A) Structure of PA-824. Generation of nitric oxide on activation of the drug by a deazaflavin-dependent nitroreductase is hypothesized to eradicate persister cells. Also discussed in Prodrugs. (B) Structure of acyldepsipeptide antibiotic ADEP4. Binds and activates ATP-independent peptidase activity of ClpP protease, leading to extensive protein degradation and cell death.

cause ATP-independent peptidase activity of ClpP protease (Brötz-Oesterhelt *et al.*, 2005). Furthermore, structures of ClpP in complex with ADEPs have been reported (Lee *et al.*, 2010), as well as the finding that FtsZ, a mature protein, is a major target of ClpP bound to ADEP4 (**16**) (Sass *et al.*, 2011), a potent derivative. These data have led Conlon *et al.* to investigate whether ADEP4 (**16**) can stimulate the peptidase activity of ClpP in persisters, leading to cell death. They have found that ADEP4 (**16**) causes extensive protein degradation and ultimately death in persisters. Also, in combination with rifampicin that acts on the null *clpP* mutants resistant to ADEP4 (**16**), ADEP4 (**16**) is able to sterilize biofilm and eradicate infection (Conlon *et al.*, 2013).

Altered binding to established targets and inhibition of new targets

With only a handful of major bacterial targets utilized by all antibiotics, the field has historically been considered 'a target-poor therapeutic arena' (Walsh and Wencewicz, 2013). The major bacterial pathways targeted by conventional antibiotics are: (1) cell wall biosynthesis, (2) protein biosynthesis, (3) DNA replication and repair, (4) folate biosynthesis, and (5) membrane integrity (Walsh, 2003; Walsh and Wencewicz, 2013). Because many different drugs target one of these pathways and sometimes even the identical enzyme in a particular pathway, resistance to multiple drugs can arise as the result of a single mutation.

Strategies to circumvent the resistance problem arising from the bacterial modification of drug targets include compensatory modification of the antibiotic molecule to adapt to the remodelled binding site or to increase affinity for secondary binding sites, and development of drugs with entirely new targets. Two tetracycline derivatives currently in clinical trials take the former approach to battle ribosomal protection in tetracycline resistant bacteria. Also discussed in this section are antibiotics in various stages of clinical trials with new targets including fatty acid biosynthesis, ATP synthase, peptide deformylase, and lipopolysaccharide biosynthesis.

Altered binding to established targets: As discussed above, omadacycline and eravacycline are the first members of the tetracycline analogue subgroups aminomethylcyclines and fluorocyclines, respectively, to have entered clinical trials. These two drugs are able to evade various tetracycline-specific efflux pumps as well as continue to function in the presence of ribosomal protection proteins. Omadacycline (**10**) was shown to be active in the presence of Tet(M), Tet(O), and Tet(S), which interact with the ribosome and promote the release of bound tetracycline (Draper *et al.*, 2013; Macone *et al.*, 2014). It has been proposed that the evasion of ribosomal protection by omadacycline involves a unique way of binding to the ribosome that prevents Tet proteins from interacting with the ribosome (Draper *et al.*, 2013). Eravacycline activity was also shown to be unaffected by the presence of Tet(M) (Grossman *et al.*, 2012; Sutcliffe *et al.*, 2013).

Inhibition of new targets: To overcome the challenges inherent in having only a few targets, active research is being conducted to establish novel ones. A recent report on the use of a bacteriophage lysin to identify a species-specific cell-wall glycopolymer in *Bacillus anthracis* as a novel target is of particular interest, for its applicability to find targets for which the inhibitors can be narrow-spectrum, potent, and potentially not induce high-level resistance (Schuch *et al.*, 2013). In addition to the five major bacterial pathways targeted by traditional antibiotics, several more candidates are being tested by drugs in the pipeline.

Bedaquiline (**17**, Sirturo®, Janssen Pharmaceutica and Johnson & Johnson) is the first member of the diarylquinoline class of antibiotics to be approved by the FDA (Fig. 21.6). Its indications include the treatment of multidrug-resistant pulmonary tuberculosis as part of a combination therapy when other drugs are unavailable (Pucci and Bush, 2013). It was found that bedaquiline (**17**) resistant mycobacteria had mutations in the genes encoding the F_0 subunit of the ATP synthase (Andries *et al.*, 2005). Confirmation of its mechanism of action, via disruption of energy metabolism, came with the finding that there is a dose-dependent decrease of cellular ATP in *M. tuberculosis*, decreased sensitivity to the drug when the genes encoding the targets are overexpressed, and finally, *in vitro* demonstration of an amine analogue of bedaquiline (**17**) binding to the c-subunit (part of the F_0 subunit) of the ATP

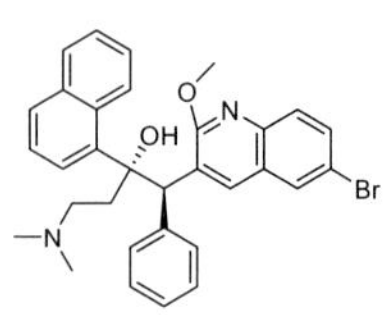

Bedaquiline (**17**)

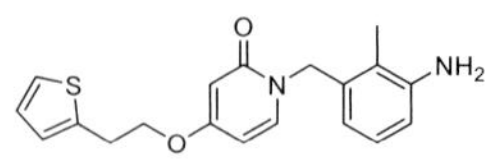

CG400549 (**18**)

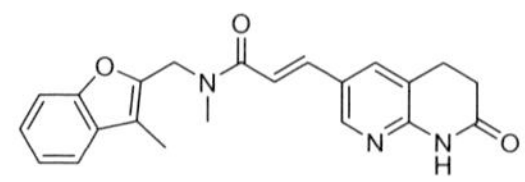

AFN-1252 (**19**)

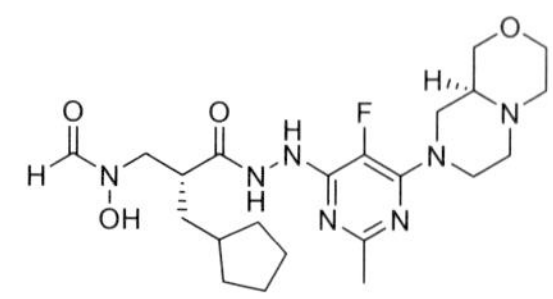

GSK1322322 (**20**)

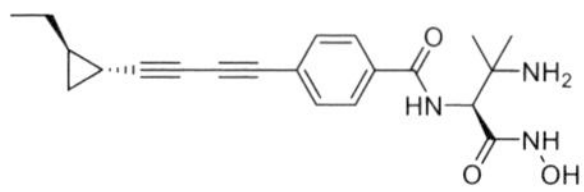

ACHN-975 (**21**)

Figure 21.6 Altered binding to established targets and inhibition of new targets. Structures of antibiotic molecules that bind to new targets. Bedaquiline targets the c-subunit of the ATP synthase; CG400549 and AFN-1252 inhibit fatty acid biosynthesis by inactivating FabI; GSK1322322 inhibits peptide deformylase; ACHN-975 inhibits zinc-dependent deacetylase LpxC, which catalyses the rate-limiting step in lipopolysaccharide biosynthesis.

synthase (Koul *et al.*, 2007). The use of the ATP synthase as a target in antibacterial therapy is an exciting new development that opens the possibility of finding more targets in bacterial energy metabolism.

Bacterial fatty acid biosynthesis pathway is a validated but arguably under-exploited target for antibiotics (Lu and Tonge, 2008). Antibiotics such as isoniazid and triclosan target a particular enzyme in this pathway, namely, FabI, which is an NAD-dependent enoyl reductase that saturates the double bond in a growing fatty acid chain as the final step in each round of elongation. Although FabI has been shown to be essential for cell survival and has been validated as a drug target, interest in it diminished with the discovery that not all pathogens carry a homologue of FabI. However, with the spread of multidrug-resistant bacteria coupled with efforts to develop narrow-spectrum antibiotics for various reasons, FabI inhibitors are receiving renewed attention. CG400549 (**18**, Crystal Genomics) and AFN-1252 (**19**, Affinium Pharmaceuticals) are two FabI inhibitor drugs currently in phase II of clinical trials for major cutaneous abscesses and acute bacterial skin and skin structure infections, respectively (Fig. 21.6). Both are synthetic molecules identified in a high-throughput screening programmes by Crystal Genomics and GlaxoSmithKline, respectively, showing potent activity against clinically isolated staphylococcal strains (Bogdanovich *et al.*, 2007; Kaplan *et al.*, 2012; Karlowsky *et al.*, 2009; Park *et al.*, 2007; Yum *et al.*, 2007).

Prokaryotic protein synthesis typically begins with an *N*-formyl-methioninyl-transfer RNA. In bacteria and possibly in eukaryotic organelles, the *N*-formyl group is removed from the protein by a peptide deformylase (PDF). With the discovery of the essentiality of PDFs to bacterial survival (Mazel *et al.*, 1994), the enhancement of their *in vitro* stability by adding divalent cations during purification (Groche *et al.*, 1998), the mechanism of action of natural product antibiotic actinonin as a PDF inhibitor (Chen *et al.*, 2000), and the presence of genes encoding PDFs in all sequenced bacterial genomes (Guilloteau *et al.*, 2002), PDF inhibitors have been under intense investigation for use as antimicrobial therapeutics (Aubart and Zalacain, 2006; Leeds and Dean, 2006; Yuan and White, 2006). GSK1322322 (**20**, GlaxoSmithKline) is a PDF inhibitor belonging to the hydrazide family, currently in phase II clinical trials and has been shown to be active against a wide range of pathogens, including some Gram-negative pathogens (Fig. 21.6; O'Dwyer *et al.*, 2013). Interestingly, GSK1322322 (**20**) was also shown to prevent the growth of *S. aureus* at concentrations 8- to 32-fold below its MIC (Butler *et al.*, 2014).

Lipopolysaccharide (LPS) biosynthesis pathway has been considered a highly attractive target in Gram-negative bacteria for decades (Walsh and Wencewicz, 2013). Currently in the pipeline there are two drug candidates in phase I of clinical trials that target this pathway in multidrug-resistant *Pseudomonas aeruginosa*. ACHN-975 (**21**, Achaogen) is a synthetic molecule that inhibits the zinc-dependent deacetylase LpxC, which catalyses the rate-limiting step in LPS biosynthesis, the deacetylation of the acyl-UDP-GlcNAc (Fig. 21.6; Pucci and Bush, 2013; Walsh and Wencewicz, 2013). POL7080 (structure unavailable, Polyphlor) is a synthetic peptide derived from β-hairpin mimetics of protegrin I, targeting the outer membrane β-barrel protein LptD, involved in exporting LPS molecules across the periplasm (Pucci and Bush, 2013; Walsh and Wencewicz, 2013).

Innovative antibiotic therapy

Prodrugs

The focus on prodrugs as a viable direction for antibiotic development has returned after a nearly 50-year absence. Prodrugs must be metabolically activated to exhibit their biological activities. Their successful use as antibiotics is best exemplified by the antituberculosis drugs isoniazid and ethionamide. Both of these prodrugs are modified by metabolic enzymes of *M. tuberculosis* to generate the active form of the drugs that bind to NADH, forming a cofactor–drug adduct. These adducts subsequently bind to, and inhibit, the function of InhA, a key enzyme in mycolic acid biosynthesis (Banerjee *et al.*, 1994). Prodrugs provide potential advantages over standard drugs (Lewis, 2013). First, in many cases the conversion of the prodrug to its biologically active form requires the enzymatic activity of the targeted microorganism, therefore there is a reduced likelihood of host toxicity. Second, the impact of multidrug efflux pumps is reduced by the prodrug being enzymatically modified once it is transported into the cell. Finally, the prodrug form may have enhanced bioavailability relative to the active drug form. As noted recently (Lewis, 2013), the movement away from prodrugs may be explained, at least in part, by unknowingly removing these molecules from the screens used to identify new drug leads. Adjustments to these screens have the potential to identify new prodrug leads.

Tedizolid phosphate: As noted above, oxazolidinones are a new class of synthetic drugs that have had extensive SAR studies performed. Tedizolid phosphate (**6**, previously known as TR-701, torezolid phosphate), a member of this drug family, recently passed phase III trials for the treatment of acute bacterial skin and skin structure infections, which are most commonly caused by *S. aureus* and *Streptococcus pyogenes*. At the end of 2013 the Food and Drug Administration accepted the New Drug Application by Cubist Pharamceuticals for treatment of these infections. Cubist is also investigating tedizolid phosphate use for the treatment of hospital-acquired/ventilator-associated bacterial pneumonia, with phase III trials beginning in 2014.

Tedizolid phosphate (**6**) is a prodrug that requires catalytic dephosphorylation to generate the active form of the drug, which binds to the 50S ribosomal subunit, inhibiting translation (Locke *et al.*, 2009). Tedizolid phosphate (**6**) is significantly different than other oxazolidinone derivatives in having reduced impact on the activity of monoamine oxidases while the phosphate improves the water solubility issue that arises with structurally related molecules and bioavailability (Shaw and Barbachyn, 2011).

Delamanid (OPC-67683) and PA-824: Recently it was proposed that the ideal antibiotic for treating bacterial infections is a prodrug that is inactive until it enters the bacterial cell where is enzymatically activated by a bacterial enzyme(s) (Lewis, 2013). Once activated, the reactive drug would have more than one target and would evade efflux mechanisms. Additionally, the reactive drug would be functional against microorganisms that are metabolically active or inactive. Two recently developed prodrugs fit this ideal antibiotic. Delaminid (**22**) and PA-824 (**15**) are both bicyclic nitroimidazoles that are known to inhibit mycolic acid biosynthesis of metabolically active *M. tuberculosis* (Figs 21.5 and 21.7; Mukherjee

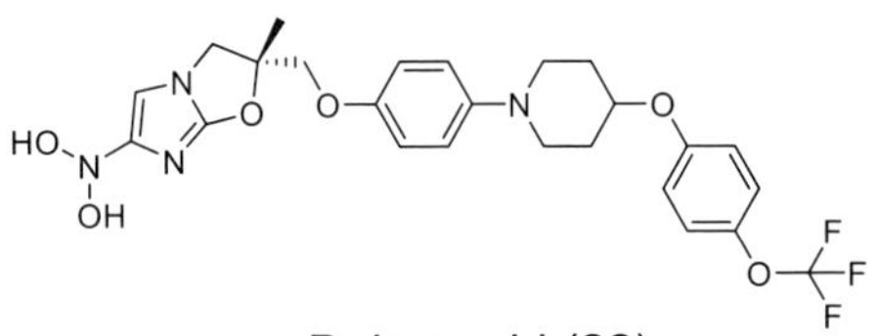

Delamanid (**22**)

Figure 21.7 Prodrugs. Structure of the prodrug delamanid. The prodrug is reductively activated by deazaflavin-dependent reductase, releasing nitric oxide and the active drug.

and Boshoff, 2011). The mechanism for this inhibition has yet to be determined. Interestingly, this inhibition of mycolic acid biosynthesis is unlikely to be the reason for killing of non-replicating *M. tuberculosis* because mycolic acid turnover is not thought to be essential at that stage (Manjunatha *et al.*, 2006). The key to understanding this activity was the finding that PA-824 (**15**) is reductively activated by enzymes that use deazaflavin F_{420} as a cofactor. The reduction of the prodrug results in the release of reactive nitrogen species, one of which is nitric oxide. This nitric oxide is thought to then target cytochromes oxidases and other targets. A similar mechanism of reductive activation of delamanid (**22**) is thought to occur. Thus, both drugs fit well with the proposed ideal antibiotic scheme. The European Medicines Agency's Committee for Medicinal Products for Human Use recently recommended conditional approval of the use of delamanid (**22**) for treatment of drug-resistant pulmonary tuberculosis. However, additional studies by Otsuka Pharmacuetical Co. were required for an understanding of the long-term benefits and safety. PA-824 (**15**) is currently in phase II development by the Global Alliance for TB Drug Development (Diacon *et al.*, 2012).

Combination therapies

The use of combination therapies have been looked at favourably for treating bacterial infections due to their potential to increase drug efficacy while also decreasing the emergence of antibiotic resistance. In fact, combination therapy is the norm for treating *M. tuberculosis* infections due to the high frequency of resistance during monotherapy. In the case of treating *M. tuberculosis*, the combination therapy of isoniazid (targeting mycolic acid biosynthesis) and rifampicin (targeting RNA polymerase) is a classic example of combination therapy where the components target different pathways. Other combinations include: (1) antibiotic plus an inhibitor of enzymes that inactivate the antibiotic, (2) an antibiotic plus an inhibitor of the efflux pump that would remove the antibiotic, and (3) an antibiotic plus an inhibitor of mechanisms for the cell to respond to the action of the antibiotic.

β-lactam/β-lactamase inhibitor combinations: There is a rich history of successful use of combination therapy when β-lactams are used to target cell wall biosynthesis. In the combination therapy, a β-lactamase inhibitor is added in combination of the β-lactam to ensure the efficacy of the antibiotic. For example, Augmentin® is an enormously successful example of this where amoxicillin is combined with clavulanate, with the latter inhibiting the β-lactamase resistance enzyme. The success of Augmentin® (GlaxoSmithKline) and other similar combinations (Timentin® [GlaxoSmithKline], Zosyn® [Wyeth], Unasyn® [Pfizer]) has led to next generation combinations that are at each phase of clinical trials. For example, ceftolozane (**23**)/tazobactam (**28**) (Cubist Pharmaceuticals) and ceftazidime (**24**)/avibactam (**29**) (AstraZeneca) combinations are currently in phase III clinical trials, while ceftaroline (**25**)/avibactam (**29**) (AstraZeneca) combinations are in phase II clinical trials and aztreonam (**26**)/avibactam (**29**) (AstraZeneca) and biapenem (**27**)/RPX7009 (**30**) (Rempex) are in phase I trials (Fig. 21.8).

Aminoglycoside/β-lactam combinations: Plazomicin (**31**, ACHN-490; Achaogen) is a next generation synthetically derived aminoglycoside (based on sisomicin from bacterial genus *Micromonospora*) that was developed to evade clinically relevant modifying enzymes (Fig. 21.8). It is currently in phase III clinical trials for the treatment of carbapenem-resistant *Enterobacteriaceae* (CRE). While it is currently in clinical trials as a monotherapy, treatment of CRE commonly involves combination therapy. Plazomicin (**31**), which targets the ribosome, has been found to have synergistic activity with daptomycin (targeting cell membrane function) (Lin *et al.*, 2010) along with many other β-lactam antibiotics (e.g.

A

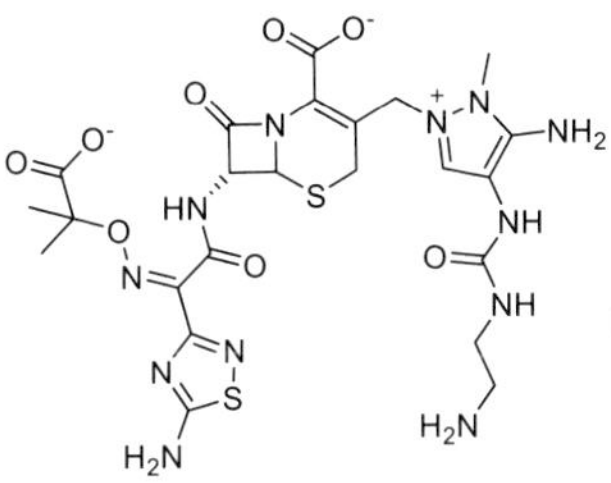

Ceftolozane (**23**)

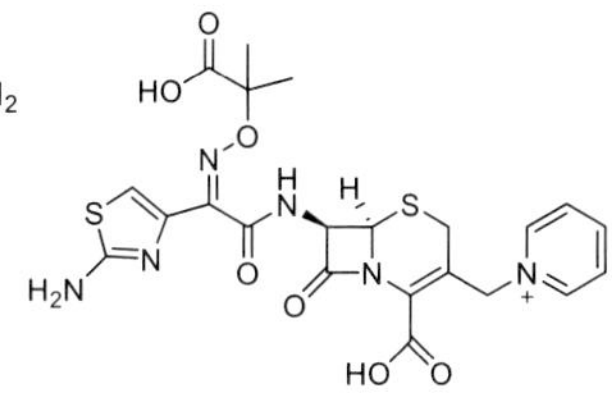

Ceftazidime (**24**)

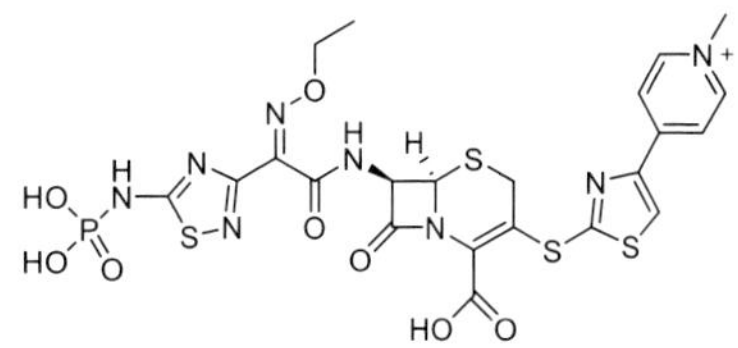

Ceftaroline (**25**)

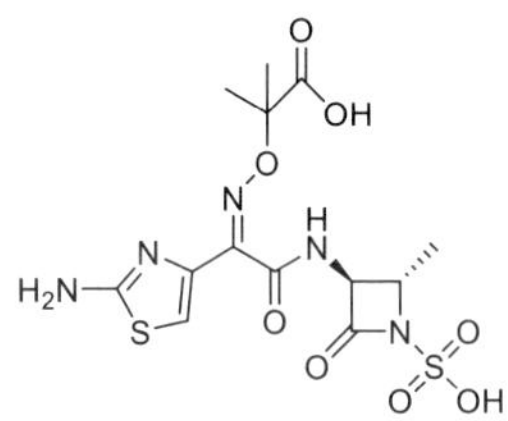

Aztreonam (**26**)

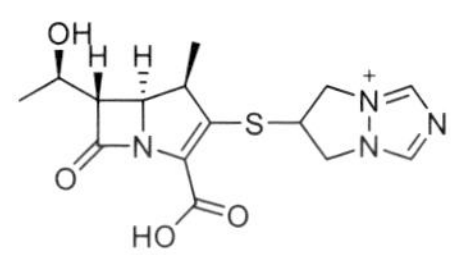

Biapenem (**27**)

B

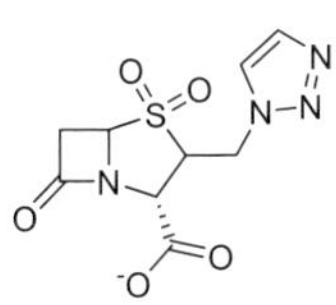

Tazobactam (**28**)

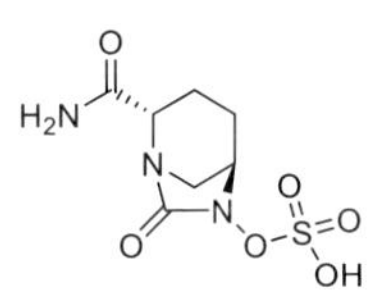

Avibactam (**29**)

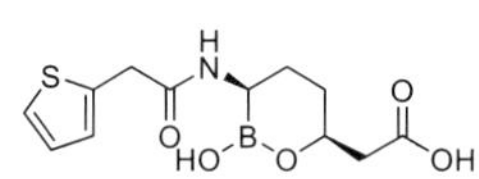

RPX7009 (**30**)

C

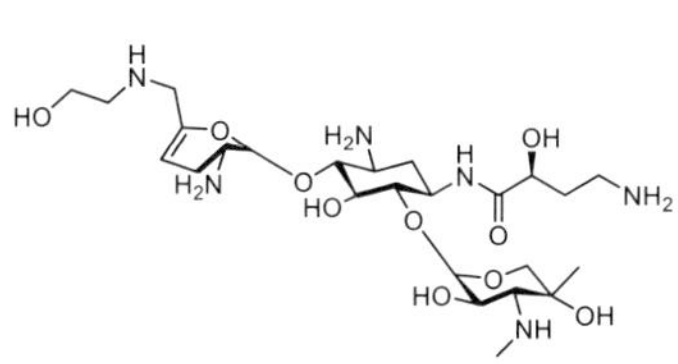

Plazomicin (**31**)

Figure 21.8 Combination therapies. (A) Structures of β-lactams in combination therapy. (B) Structures of β-lactamases in combination therapy. (C) Structure of plazomicin. Plazomicin targets the ribosome and has shown synergistic effects with daptomycin as well as β-lactam antibiotics.

deftobiprole, defepime, doripenem, imimpenem) (Zhanel *et al.*, 2012). Thus, there is the potential to use plazomicin (**31**) in combination therapy to target both ribosome function and cell wall biosynthesis/function.

Multitarget therapies

Utilizing multiple targets to combat bacterial infections has only recently begun to receive attention, although it has traditionally been the default in other therapeutic areas such as anti-HIV treatment and cancer chemotherapy. Multitarget therapy can be categorized by three different approaches: (1) drug cocktails that combine two or more separate agents, (2) multicomponent drugs comprising two or more agents formulated as one entity, and (3) multiple ligands that are single compounds that interact with two or more targets (Morphy and Rankovic, 2005). Drug cocktails and multicomponent drugs are collectively referred to as combinatorial therapy and is discussed above. Multiple ligands form a separate class of multitarget therapy and quinolones that inhibit both DNA gyrase as well as topoisomerase IV is a representative example (East and Silver, 2013).

The antibiotic multiple ligands identified so far utilize the same structural features to bind to the ligand-binding sites of two or more target molecules, whereas in other therapeutic areas, it is common to have multiple fused or conjugated structural motifs with affinities to different targets (East and Silver, 2013; Morphy and Rankovic, 2005). The former strategy is feasible when the targets have structurally similar binding sites, which may also be their closely related substrate binding sites. While the existing antibiotic multiple ligands were not designed to have multiple targets from the development stage and this property has been determined afterwards, a recent review covers current research that now explicitly try to exploit antibacterial multiple ligands in several pathways including DNA replication, cell wall biosynthesis, lipopolysaccharide biosynthesis, fatty acid biosynthesis, and folate biosynthesis (East and Silver, 2013).

Currently in clinical trials, three fluoroquinolones – delafloxacin (**13**) (Remy *et al.*, 2012), avarolfloxacin (**12**) (Morrow *et al.*, 2011), and

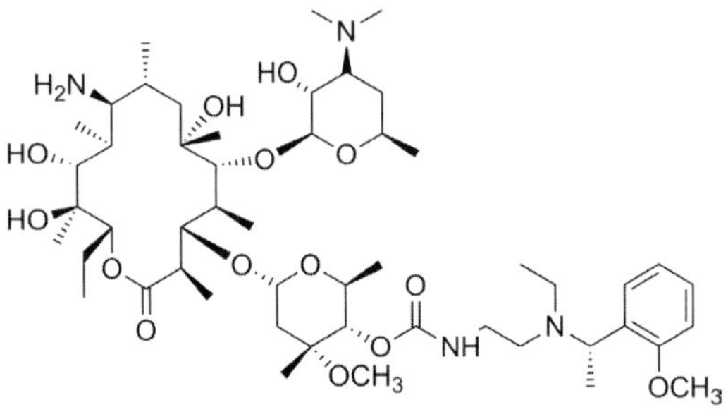

CP-544372

Figure 21.9 Multitarget therapies. Structure of zabofloxacin. Zabofloxacin inhibits type II topoisomerases GyrA and ParC.

zabofloxacin (**32**, Dong Wha Pharm; Kwon *et al.*, 2006) have been identified to have two molecular targets within the cell (Figs 21.3 and 21.9). These and other fluoroquinolones are able to inhibit the activity of the two related bacterial type II topoisomerases: DNA gyrase (GyrA) and topoisomerase IV (ParC). While the A-subunit of the DNA gyrase and the C-subunit of topoisomerase IV are targeted by the above as well as other fluoroquinolones, the B- and E-subunits of these proteins, respectively, are both ATPases and provide further opportunities for the development of multiple ligands (East and Silver, 2013).

Conclusion

The various efforts to overcome the many challenges in antibiotic development offer a modest degree of hope in a situation some have dubbed 'the perfect storm' (Patel and Bonomo, 2013; Shlaes, 2010). Despite lapses in funding for antibiotics research and development in big pharmaceutical companies, a few new classes of antibiotic molecules with new modes of action are currently in clinical trials. However, considering the historical rate at which the development and spread of bacterial resistance eventually renders obsolete any given agent, what we have in the pipeline is unlikely to be enough to sustain the modern lifestyle and medical practices. Importantly, there are no recently approved drugs that are effective against Gram-negative organisms. As a result, there are many instances where no treatment options are available (Boucher *et al.*, 2013).

Plazomicin (**31**), eravacycline (**11**), along with new β-lactam/β-lactamase inhibitor combinations show promise in addressing this issue (Pucci and Bush, 2013), but it cannot be stressed enough that more needs to be done.

We have seen progress in regulation with the passing of the Generating Antibiotic Incentives Now (GAIN) Act (included in the Food and Drug Administration Safety and Innovation Act) in 2012. Among other things the GAIN Act resulted in the formation of the Antibacterial Drug Development Task Force within the FDA to develop and revise guidance related to antibacterial drug development (http://www.fda.gov/NewsEvents/Newsroom/PressAnnouncements/ucm320643.htm). The new law also mandates the FDA to designate some antibiotics as qualified infectious disease products, which provides incentives such as eligibility for fast-track status and extended exclusivity rights (Fox, 2013). However, a draft guidance published in 2013 by the FDA to stimulate the development of new antibiotics was criticized for its emphasis on proving superiority, which may not be adequate when bacteria have developed resistance to existing drugs (http://www.raps.org/focus-online/news/news-article-view/article/4419/phrma-to-fda-antibiotic-guidance-focuses-too-much-on-superiority.aspx). Others have argued that antibiotics should be approved based on microbiological endpoints rather than clinical endpoints to reduce the costs and time associated with clinical trials (Coates and Halls, 2012). Clearly, streamlining the antibiotics testing procedure and approval system while maintaining appropriate safety standards is not an easy task and demands further attention of scientists and policy makers.

Other resources: list of webpages used to identify antibiotics in the pipeline (accessed: 3. 10. 2014)

Achaogen: http://www.achaogen.com/pipeline-overview/
Actelion Pharmaceuticals: http://www1.actelion.com/en/scientists/development-pipeline/phase-3/index.page
Affinium Pharmaceuticals: http://www.afnm.com/programs/afn-1252.htm
AstraZeneca: http://www.astrazeneca.com/Research/Our-pipeline-summary
Basilea Pharmaceutica: http://www.basilea.com/Development/Pipeline/
Cellceutix Corporation: http://cellceutix.com/pipeline/
Cempra: http://www.cempra.com/products/pipeline/
ClinicalTrialsFeeds.org: http://clinicaltrialsfeeds.org/clinical-trials/show/NCT01842516
ClinicalTrials.gov: http://clinicaltrials.gov/show/NCT02020434
Crystal Genomics: http://crystalgenomics.com/clinical/main.html
Cubist Pharmaceuticals: http://www.cubist.com/products/
Daiichi Sankyo: http://www.daiichisankyo.com/rd/pipeline/index.html
Destiny Pharma: http://www.destinypharma.com/pipeline.shtml
Dipexium Pharmaceuticals: http://www.dipexiumpharmaceuticals.com/index.html
Dong Wha Pharm: http://www.dong-wha.co.kr/english/rnd/rnd02_03.asp
Durata Therapeutics: http://www.duratatherapeutics.com/dalbavancin/clinical-trials
Enanta Pharmaceuticals: http://www.enanta.com/research/pipeline/
Ferrer Internacional: http://www.ferrer.com/Innovation_Innovacion-Pipeline-de-proyectos-ENG
Furiex Pharmaceuticals: http://www.furiex.com/pipeline/discoverydevelopment-pipeline/fluoroquinolone/
Global Alliance for TB Drug Development: http://www.tballiance.org/portfolio/
Kyorin Pharmaceutical Co.: http://www.kyorin-gr.co.jp/en/ir/pdf/201111kp.pdf
Lytix Biopharma: http://www.lytixbiopharma.com/pipeline/
Médecins Sans Frontières: http://www.msfaccess.org/content/msf-responds-ema-approval-new-tuberculosis-drug-delamanid
Melinta Therapeutics: http://www.melinta.com/our_pipeline.php
MeLion Pharmaceuticals: http://www.merlionpharma.com/?q=node/16
Merk & Co: http://www.merck.com/research/pipeline/home.html
MicuRx Pharmaceuticals: http://www.micurx.com/d1.htm
Nabriva Therapeutics: http://www.nabriva.com/programs/pipeline/
NovaBay Pharmaceuticals: http://novabay.com/technology/pipeline
Novacta Biosystems: http://www.novactabio.com/careers.php
Novartis: http://www.novartis.com/downloads/investors/reports/novartis-annual-report-2013-en.pdf
OctoPlus: http://www.octoplus.nl/files/2113/5543/4711/Dr._Reddys_Offer_Memorandum.pdf
Parak Pharmaceuticals: http://www.paratekpharm.com/
Pergamum: http://www.pergamum.com/programs/
Pharmasyntez: http://pharmasyntez.com/en/node/5

Polyphor: http://www.polyphor.com/products/pol7080
PR Newswire Association: http://www.prnewswire.com/news-releases/kalidex-introduces-novel-fluoroquinolone-kpi-10-at-the-52nd-interscience-conference-on-antimicrobial-agents-and-chemotherapy-icaac-168871396.html
Sequella: http://www.sequella.com/pipeline/index.html
Shionogi: http://www.shionogi.co.jp/en/company/pmrltj0000000u4v-att/e_kaihatsu.pdf
Soligenix: http://www.soligenix.com/prod_thera_sgx942.shtml
Summit Corporation: http://www.summitplc.com/programmes/c-difficile/
TaiGen Biotechnology: http://www.taigenbiotech.com/Product/2–1
Tetraphase Pharmaceuticals: http://tphase.com/pipeline/overview
The Medicines Company: http://www.themedicinescompany.com/page/pipeline
Theravance: http://www.theravance.com/pipeline
Wockhardt: http://www.wockhardt.com/how-we-touch-lives/new-drug-discover.aspx
Working Group on New TB Drugs: http://www.newtbdrugs.org/project.php?id=153

References

Allison, K.R., Brynildsen, M.P., and Collins, J.J. (2011). Metabolite-enabled eradication of bacterial persisters by aminoglycosides. Nature *473*, 216–220.

Andries, K., Verhasselt, P., Guillemont, J., Göhlmann, H.W.H., Neefs, J.-M., Winkler, H., Van Gestel, J., Timmerman, P., Zhu, M., Lee, E., *et al.* (2005). A diarylquinoline drug active on the ATP synthase of *Mycobacterium tuberculosis*. Science *307*, 223–227.

Aubart, K., and Zalacain, M. (2006). Peptide deformylase inhibitors. Prog. Med. Chem. *44*, 109–143.

Ball, P.R., Shales, S.W., and Chopra, I. (1980). Plasmid-mediated tetracycline resistance in *Escherichia coli* involves increased efflux of the antibiotic. Biochem. Biophys. Res. Commun. *93*, 74–81.

Banerjee, A., Dubnau, E., Quemard, A., Balasubramanian, V., Um, K., Wilson, T., Collins, D., de Lisle, G., and Jacobs, W. (1994). inhA, a gene encoding a target for isoniazid and ethionamide in *Mycobacterium tuberculosis*. Science *263*, 227–230.

Barbachyn, M.R., and Ford, C.W. (2003). Oxazolidinone structure–activity relationships leading to linezolid. Angew. Chem. Int. Ed. Engl. *42*, 2010–2023.

Barry, A.L., Fuchs, P.C., and Brown, S.D. (2001). *In vitro* activities of three nonfluorinated quinolones against representative bacterial isolates. Antimicrob. Agents Chemother. *45*, 1923–1927.

Barry, C., Boshoff, H., and Dowd, C. (2004). Prospects for clinical introduction of nitroimidazole antibiotics for the treatment of tuberculosis. Curr. Pharm. Des. *10*, 3239–3262.

Belley, A., Neesham-Grenon, E., McKay, G., Arhin, F.F., Harris, R., Beveridge, T., Parr, T.R., and Moeck, G. (2009). Oritavancin kills stationary-phase and biofilm *Staphylococcus aureus* cells *in vitro*. Antimicrob. Agents Chemother. *53*, 918–925.

Belley, A., McKay, G. a, Arhin, F.F., Sarmiento, I., Beaulieu, S., Fadhil, I., Parr, T.R., and Moeck, G. (2010). Oritavancin disrupts membrane integrity of *Staphylococcus aureus* and vancomycin-resistant enterococci to effect rapid bacterial killing. Antimicrob. Agents Chemother. *54*, 5369–5371.

Bertrand, D., Bertrand, S., Neveu, E., and Fernandes, P. (2010). Molecular characterization of off-target activities of telithromycin: a potential role for nicotinic acetylcholine receptors. Antimicrob. Agents Chemother. *54*, 5399–5402.

Bigger, J. (1944). Treatment of staphylococcal infections with penicillin by intermittent sterilisation. Lancet *244*, 497–500.

Bogdanovich, T., Clark, C., Kosowska-Shick, K., Dewasse, B., McGhee, P., and Appelbaum, P.C. (2007). Antistaphylococcal activity of CG400549, a new experimental FabI inhibitor, compared with that of other agents. Antimicrob. Agents Chemother. *51*, 4191–4195.

Boucher, H.W., Talbot, G.H., Benjamin, D.K., Bradley, J., Guidos, R.J., Jones, R.N., Murray, B.E., Bonomo, R.A., and Gilbert, D. (2013). 10 x '20 Progress – development of new drugs active against Gram-negative bacilli: an update from the Infectious Diseases Society of America. Clin. Infect. Dis. *56*, 1685–1694.

Boyle-Vavra, S., Labischinski, H., Ebert, C.C., Ehlert, K., and Daum, R.S. (2001). A spectrum of changes occurs in peptidoglycan composition of glycopeptide-intermediate clinical *Staphylococcus aureus* isolates. Antimicrob. Agents Chemother. *45*, 280–287.

Brötz-Oesterhelt, H., Beyer, D., Kroll, H., Endermann, R., Ladel, C., Schroeder, W., Hinzen, B., Raddatz, S., Paulsen, H., Henninger, K., *et al.* (2005). Dysregulation of bacterial proteolytic machinery by a new class of antibiotics. Nat. Med. *11*, 1082–1087.

Butaye, P. (2003). Mobile genes coding for efflux-mediated antimicrobial resistance in Gram-positive and Gram-negative bacteria. Int. J. Antimicrob. Agents *22*, 205–210.

Butler, D., Chen, D., O'Dwyer, K., Lewandowski, T., Aubart, K., and Zalacain, M. (2014). Potent sub-MIC effect of GSK1322322 and other peptide deformylase inhibitors on *in vitro* growth of *Staphylococcus aureus*. Antimicrob. Agents Chemother. *58*, 290–296.

Butler, M.S., Blaskovich, M.A., and Cooper, M.A. (2013). Antibiotics in the clinical pipeline in 2013. J. Antibiot. (Tokyo). *66*, 571–591.

Chen, D.Z., Patel, D.V., Hackbarth, C.J., Wang, W., Dreyer, G., Young, D.C., Margolis, P.S., Wu, C., Ni, Z.J., Trias, J., *et al.* (2000). Actinonin, a naturally occurring antibacterial agent, is a potent deformylase inhibitor. Biochemistry 39, 1256–1262.

Coates, A.R.M., and Halls, G. (2012). Antibiotics in phase II and III clinical trials. In Handbook of Experimental Pharmacology, A.R.M. Coates, ed. (Springer, Berlin, Germany), pp. 167–183.

Conlon, B.P., Nakayasu, E.S., Fleck, L.E., LaFleur, M.D., Isabella, V.M., Coleman, K., Leonard, S.N., Smith,

R.D., Adkins, J.N., and Lewis, K. (2013). Activated ClpP kills persisters and eradicates a chronic biofilm infection. Nature *503*, 365–370.

Cui, L., Murakami, H., Kuwahara-Arai, K., Hanaki, H., and Hiramatsu, K. (2000). Contribution of a thickened cell wall and Its glutamine nonamidated component to the vancomycin resistance expressed by *Staphylococcus aureus* Mu50. Antimicrob. Agents Chemother. *44*, 2276–2285.

Das, D., Tulkens, P.M., Mehra, P., Fang, E., and Prokocimer, P. (2014). Tedizolid phosphate for the management of acute bacterial skin and skin structure infections: safety summary. Clin. Infect. Dis. *58 Suppl 1*, S51–7.

Denyer, S.P., and Maillard, J.Y. (2002). Cellular impermeability and uptake of biocides and antibiotics in Gram-negative bacteria. Symp. Ser. Soc. Appl. Microbiol. 35S-45S.

Diacon, A.H., Dawson, R., du Bois, J., Narunsky, K., Venter, A., Donald, P.R., van Niekerk, C., Erondu, N., Ginsberg, A.M., Becker, P., *et al.* (2012). Phase II dose-ranging trial of the early bactericidal activity of PA-824. Antimicrob. Agents Chemother. *56*, 3027–3031.

Draper, M.P., Weir, S., Macone, A., Donatelli, J., Trieber, C.A., Tanaka, S.K., and Levy, S.B. (2013). The mechanism of action of the novel aminomethylcycline antibiotic omadacycline. Antimicrob. Agents Chemother. *58*, 1279–1283.

Dunbar, L.M., Tang, D.M., and Manausa, R.M. (2008). A review of telavancin in the treatment of complicated skin and skin structure infections (cSSSI). Ther. Clin. Risk Manag. *4*, 235–244.

East, S.P., and Silver, L.L. (2013). Multitarget ligands in antibacterial research: progress and opportunities. Expert Opin. Drug Discov. *8*, 143–156.

Falagas, M.E., and Kasiakou, S.K. (2005). Colistin: the revival of polymyxins for the management of multidrug-resistant gram-negative bacterial infections. Clin. Infect. Dis. *40*, 1333–1341.

Fauvart, M., De Groote, V.N., and Michiels, J. (2011). Role of persister cells in chronic infections: clinical relevance and perspectives on anti-persister therapies. J. Med. Microbiol. *60*, 699–709.

Fjell, C.D., Hiss, J. a, Hancock, R.E.W., and Schneider, G. (2012). Designing antimicrobial peptides: form follows function. Nat. Rev. Drug Discov. *11*, 37–51.

Fox, J.L. (2013). Antimicrobial peptides stage a comeback. Nat. Biotechnol. *31*, 379–382.

Grant, S.S., Kaufmann, B.B., Chand, N.S., Haseley, N., and Hung, D.T. (2012). Eradication of bacterial persisters with antibiotic-generated hydroxyl radicals. Proc. Natl. Acad. Sci. U.S.A. *109*, 12147–12152.

Groche, D., Becker, A., Schlichting, I., Kabsch, W., Schultz, S., and Wagner, A.F. (1998). Isolation and crystallization of functionally competent *Escherichia coli* peptide deformylase forms containing either iron or nickel in the active site. Biochem. Biophys. Res. Commun. *246*, 342–346.

Grossman, T.H., Starosta, A.L., Fyfe, C., O'Brien, W., Rothstein, D.M., Mikolajka, A., Wilson, D.N., and Sutcliffe, J.A. (2012). Target- and resistance-based mechanistic studies with TP-434, a novel fluorocycline antibiotic. Antimicrob. Agents Chemother. *56*, 2559–2564.

Guilloteau, J.-P., Mathieu, M., Giglione, C., Blanc, V., Dupuy, A., Chevrier, M., Gil, P., Famechon, A., Meinnel, T., and Mikol, V. (2002). The crystal structures of four peptide deformylases bound to the antibiotic actinonin reveal two distinct types: a platform for the structure-based design of antibacterial agents. J. Mol. Biol. *320*, 951–962.

Higgins, D.L., Chang, R., Debabov, D.V., Leung, J., Wu, T., Krause, K.M., Sandvik, E., Hubbard, J.M., Kaniga, K., Schmidt, D.E., *et al.* (2005). Telavancin, a multifunctional lipoglycopeptide, disrupts both cell wall synthesis and cell membrane integrity in methicillin-resistant *Staphylococcus aureus*. Antimicrob. Agents Chemother. *49*, 1127–1134.

Hiramatsu, K. (2001). Vancomycin-resistant *Staphylococcus aureus*: a new model of antibiotic resistance. Lancet Infect. Dis. *1*, 147–155.

Im, W.B., Choi, S.H., Park, J.-Y., Choi, S.H., Finn, J., and Yoon, S.-H. (2011). Discovery of torezolid as a novel 5-hydroxymethyl-oxazolidinone antibacterial agent. Eur. J. Med. Chem. *46*, 1027–1039.

Jabes, D. (2011). The antibiotic R&D pipeline: an update. Curr. Opin. Microbiol. *14*, 564–569.

Kaplan, N., Albert, M., Awrey, D., Bardouniotis, E., Berman, J., Clarke, T., Dorsey, M., Hafkin, B., Ramnauth, J., Romanov, V., *et al.* (2012). Mode of action, *in vitro* activity, and *in vivo* efficacy of AFN-1252, a selective antistaphylococcal FabI inhibitor. Antimicrob. Agents Chemother. *56*, 5865–5874.

Karlowsky, J.A., Kaplan, N., Hafkin, B., Hoban, D.J., and Zhanel, G.G. (2009). AFN-1252, a FabI inhibitor, demonstrates a *Staphylococcus*-specific spectrum of activity. Antimicrob. Agents Chemother. *53*, 3544–3548.

Koul, A., Dendouga, N., Vergauwen, K., Molenberghs, B., Vranckx, L., Willebrords, R., Ristic, Z., Lill, H., Dorange, I., Guillemont, J., *et al.* (2007). Diarylquinolines target subunit c of mycobacterial ATP synthase. Nat. Chem. Biol. *3*, 323–324.

Kumar, A., and Schweizer, H.P. (2005). Bacterial resistance to antibiotics: active efflux and reduced uptake. Adv. Drug Deliv. Rev. *57*, 1486–1513.

Kwon, A.-R., Min, Y.-H., Ryu, J.-M., Choi, D.-R., Shim, M.-J., and Choi, E.-C. (2006). *In vitro* and *in vivo* activities of DW-224a, a novel fluoroquinolone antibiotic agent. J. Antimicrob. Chemother. *58*, 684–688.

Lambert, P. a (2002). Cellular impermeability and uptake of biocides and antibiotics in Gram-positive bacteria and mycobacteria. Symp. Ser. Soc. Appl. Microbiol. 46S-54S.

Leadbetter, M.R., Adams, S.M., Bazzini, B., Fatheree, P.R., Karr, D.E., Krause, K.M., Lam, B.M.T., Linsell, M.S., Nodwell, M.B., Pace, J.L., *et al.* (2004). Hydrophobic vancomycin derivatives with improved ADME properties: discovery of telavancin (TD-6424). J. Antibiot. (Tokyo). *57*, 326–336.

Lee, B.-G., Park, E.Y., Lee, K.-E., Jeon, H., Sung, K.H., Paulsen, H., Rübsamen-Schaeff, H., Brötz-Oesterhelt,

H., and Song, H.K. (2010). Structures of ClpP in complex with acyldepsipeptide antibiotics reveal its activation mechanism. Nat. Struct. Mol. Biol. *17*, 471–478.

Leeds, J.A., and Dean, C.R. (2006). Peptide deformylase as an antibacterial target: a critical assessment. Curr. Opin. Pharmacol. *6*, 445–452.

Levy, S.B. (1992). Active efflux mechanisms for antimicrobial resistance. Antimicrob. Agents Chemother. *36*, 695–703.

Levy, S.B., and McMurry, L. (1978). Plasmid-determined tetracycline resistance involves new transport systems for tetracycline. Nature *276*, 90–92.

Lewis, K. (2010). Persister cells. Annu. Rev. Microbiol. *64*, 357–372.

Lewis, K. (2013). Platforms for antibiotic discovery. Nat. Rev. Drug Discov. *12*, 371–387.

Li, X.-Z., and Nikaido, H. (2004). Efflux-mediated drug resistance in bacteria. Drugs *64*, 159–204.

Li, X.-Z., and Nikaido, H. (2009). Efflux-mediated drug resistance in bacteria: an update. Drugs *69*, 1555–1623.

Lin, G., Ednie, L.M., and Appelbaum, P.C. (2010). Antistaphylococcal activity of ACHN-490 tested alone and in combination with other agents by time-kill assay. Antimicrob. Agents Chemother. *54*, 2258–2261.

Lipsky, B.A., and Baker, C.A. (1999). Fluoroquinolone toxicity profiles: a review focusing on newer agents. Clin. Infect. Dis. *28*, 352–364.

Locke, J.B., Hilgers, M., and Shaw, K.J. (2009). Novel ribosomal mutations in *Staphylococcus aureus* strains identified through selection with the oxazolidinones linezolid and torezolid (TR-700). Antimicrob. Agents Chemother. *53*, 5265–5274.

Locke, J.B., Finn, J., Hilgers, M., Morales, G., Rahawi, S., G.C.K., Picazo, J.J., Im, W., Shaw, K.J., and Stein, J.L. (2010). Structure–activity relationships of diverse oxazolidinones for linezolid-resistant *Staphylococcus aureus* strains possessing the cfr methyltransferase gene or ribosomal mutations. Antimicrob. Agents Chemother. *54*, 5337–5343.

Lu, H., and Tonge, P.J. (2008). Inhibitors of FabI, an enzyme drug target in the bacterial fatty acid biosynthesis pathway. Acc. Chem. Res. *41*, 11–20.

Macone, A.B., Caruso, B.K., Leahy, R.G., Donatelli, J., Weir, S., Draper, M.P., Tanaka, S.K., and Levy, S.B. (2014). *In vitro* and *in vivo* antibacterial activities of omadacycline, a novel aminomethylcycline. Antimicrob. Agents Chemother. *58*, 1127–1135.

Mahamoud, A., Chevalier, J., Alibert-Franco, S., Kern, W.V., and Pagès, J.-M. (2007). Antibiotic efflux pumps in Gram-negative bacteria: the inhibitor response strategy. J. Antimicrob. Chemother. *59*, 1223–1229.

Manjunatha, U.H., Boshoff, H., Dowd, C.S., Zhang, L., Albert, T.J., Norton, J.E., Daniels, L., Dick, T., Pang, S.S., and Barry, C.E. (2006). Identification of a nitroimidazo-oxazine-specific protein involved in PA-824 resistance in *Mycobacterium tuberculosis*. Proc. Natl. Acad. Sci. U.S.A. *103*, 431–436.

Manjunatha, U., Boshoff, H.I., and Barry, C.E. (2009). The mechanism of action of PA-824: Novel insights from transcriptional profiling. Commun. Integr. Biol. *2*, 215–218.

Mazel, D., Pochet, S., and Marlière, P. (1994). Genetic characterization of polypeptide deformylase, a distinctive enzyme of eubacterial translation. EMBO J. *13*, 914–923.

Michel, K.H., and Kastner, R.E. (1985). AS4556 antibiotics and process for production thereof (US Patent No 4492650).

Möllmann, U., Heinisch, L., Bauernfeind, A., Köhler, T., and Ankel-Fuchs, D. (2009). Siderophores as drug delivery agents: application of the 'Trojan Horse' strategy. Biometals *22*, 615–624.

Morphy, R., and Rankovic, Z. (2005). Designed multiple ligands. An emerging drug discovery paradigm. J. Med. Chem. *48*, 6523–6543.

Morrow, B.J., Abbanat, D., Baum, E.Z., Crespo-Carbone, S.M., Davies, T.A., He, W., Shang, W., Queenan, A.M., and Lynch, A.S. (2011). Antistaphylococcal activities of the new fluoroquinolone JNJ-Q2. Antimicrob. Agents Chemother. *55*, 5512–5521.

Mukherjee, T., and Boshoff, H. (2011). Nitroimidazoles for the treatment of TB: past, present and future. Future Med. Chem. 3, 1427–1454.

Narita, M., Tsuji, B.T., and Yu, V.L. (2007). Linezolid-associated peripheral and optic neuropathy, lactic acidosis, and serotonin syndrome. Pharmacotherapy *27*, 1189–1197.

Nelson, M.L., and Levy, S.B. (2011). The history of the tetracyclines. Ann. N.Y. Acad. Sci. *1241*, 17–32.

Nikaido, H., and Pagès, J.-M. (2012). Broad-specificity efflux pumps and their role in multidrug resistance of Gram-negative bacteria. FEMS Microbiol. Rev. *36*, 340–363.

Nikaido, H., and Takatsuka, Y. (2009). Mechanisms of RND multidrug efflux pumps. Biochim. Biophys. Acta *1794*, 769–781.

O'Dwyer, K., Hackel, M., Hightower, S., Hoban, D., Bouchillon, S., Qin, D., Aubart, K., Zalacain, M., and Butler, D. (2013). Comparative analysis of the antibacterial activity of a novel peptide deformylase inhibitor, GSK1322322. Antimicrob. Agents Chemother. *57*, 2333–2342.

Oliveira, H.S., Gonçalo, M., and Figueiredo, A.C. (2000). Photosensitivity to lomefloxacin. A clinical and photobiological study. Photodermatol. Photoimmunol. Photomed. *16*, 116–120.

Opperman, T.J., Kwasny, S.M., Kim, H.-S., Nguyen, S.T., Houseweart, C., D'Souza, S., Walker, G.C., Peet, N.P., Nikaido, H., and Bowlin, T.L. (2014). Characterization of a novel pyranopyridine inhibitor of the AcrAB efflux pump of *Escherichia coli*. Antimicrob. Agents Chemother. *58*, 722–733.

Page, M.G.P. (2013). Siderophore conjugates. Ann. N.Y. Acad. Sci. *1277*, 115–126.

Page, M.G.P., Dantier, C., and Desarbre, E. (2010). *In vitro* properties of BAL30072, a novel siderophore sulfactam with activity against multiresistant Gram-negative bacilli. Antimicrob. Agents Chemother. *54*, 2291–2302.

Pagès, J.-M., and Amaral, L. (2009). Mechanisms of drug efflux and strategies to combat them: challenging the efflux pump of Gram-negative bacteria. Biochim. Biophys. Acta *1794*, 826–833.

Park, H.S., Yoon, Y.M., Jung, S.J., Kim, C.M., Kim, J.M., and Kwak, J.-H. (2007). Antistaphylococcal activities of CG400549, a new bacterial enoyl-acyl carrier protein reductase (FabI) inhibitor. J. Antimicrob. Chemother. *60*, 568–574.

Patel, G., and Bonomo, R.A. (2013). 'Stormy waters ahead': global emergence of carbapenemases. Front. Microbiol. *4*, 48.

Peschel, A., Vuong, C., Otto, M., and Gotz, F. (2000). The D-alanine residues of *Staphylococcus aureus* teichoic acids alter the susceptibility to vancomycin and the activity of autolyticeEnzymes. Antimicrob. Agents Chemother. *44*, 2845–2847.

Pfeltz, R.F., Singh, V.K., Schmidt, J.L., Batten, M.A., Baranyk, C.S., Nadakavukaren, M.J., Jayaswal, R.K., and Wilkinson, B.J. (2000). Characterization of passage-selected vancomycin-resistant *Staphylococcus aureus* strains of diverse parental backgrounds. Antimicrob. Agents Chemother. *44*, 294–303.

Pucci, M.J., and Bush, K. (2013). Investigational antimicrobial agents of 2013. Clin. Microbiol. Rev. *26*, 792–821.

Remy, J.M., Tow-Keogh, C.A., McConnell, T.S., Dalton, J.M., and Devito, J.A. (2012). Activity of delafloxacin against methicillin-resistant *Staphylococcus aureus*: resistance selection and characterization. J. Antimicrob. Chemother. *67*, 2814–2820.

Renslo, A.R. (2010). Antibacterial oxazolidinones: emerging structure–toxicity relationships. Expert Rev. Anti. Infect. Ther. *8*, 565–574.

Roberts, M.C. (1996). Tetracycline resistance determinants: mechanisms of action, regulation of expression, genetic mobility, and distribution. FEMS Microbiol. Rev. *19*, 1–24.

Sass, P., Josten, M., Famulla, K., Schiffer, G., Sahl, H.-G., Hamoen, L., and Brötz-Oesterhelt, H. (2011). Antibiotic acyldepsipeptides activate ClpP peptidase to degrade the cell division protein FtsZ. Proc. Natl. Acad. Sci. U.S.A. *108*, 17474–17479.

Schalk, I.J., Mislin, G.L.A., and Brillet, K. (2012). Structure, function and binding selectivity and stereoselectivity of siderophore-iron outer membrane transporters. (Elsevier).

Schuch, R., Pelzek, A.J., Raz, A., Euler, C.W., Ryan, P.A., Winer, B.Y., Farnsworth, A., Bhaskaran, S.S., Stebbins, C.E., Xu, Y., *et al.* (2013). Use of a bacteriophage lysin to identify a novel target for antimicrobial development. PLoS One *8*, e60754.

Schweizer, H.P. (2012). Understanding efflux in Gram-negative bacteria: opportunities for drug discovery. Expert Opin. Drug Discov. *7*, 633–642.

Shaw, K.J., and Barbachyn, M.R. (2011). The oxazolidinones: past, present, and future. Ann. N.Y. Acad. Sci. *1241*, 48–70.

Shlaes, D.M. (2010). Antibiotics: The Perfect Storm (Springer, Berlin, Germany).

Sieradzki, K., and Tomasz, A. (1997). Inhibition of cell wall turnover and autolysis by vancomycin in a highly vancomycin-resistant mutant of *Staphylococcus aureus*. J. Bacteriol. *179*, 2557–2566.

Singh, R., Manjunatha, U., Boshoff, H.I.M., Ha, Y.H., Niyomrattanakit, P., Ledwidge, R., Dowd, C.S., Lee, I.Y., Kim, P., Zhang, L., *et al.* (2008). PA-824 kills nonreplicating *Mycobacterium tuberculosis* by intracellular NO release. Science *322*, 1392–1395.

De Souza Mendes, C., and de Souza Antunes, A. (2013). Pipeline of known chemical classes of antibiotics. Antibiotics *2*, 500–534.

Still, J.G., Schranz, J., Degenhardt, T.P., Scott, D., Fernandes, P., Gutierrez, M.J., and Clark, K. (2011). Pharmacokinetics of solithromycin (CEM-101) after single or multiple oral doses and effects of food on single-dose bioavailability in healthy adult subjects. Antimicrob. Agents Chemother. *55*, 1997–2003.

Sutcliffe, J.A., O'Brien, W., Fyfe, C., and Grossman, T.H. (2013). Antibacterial activity of eravacycline (TP-434), a novel fluorocycline, against hospital and community pathogens. Antimicrob. Agents Chemother. *57*, 5548–5558.

Thaker, M., Spanogiannopoulos, P., and Wright, G.D. (2010). The tetracycline resistome. Cell. Mol. Life Sci. *67*, 419–431.

Urbina, O., Ferrández, O., Espona, M., Salas, E., Ferrández, I., and Grau, S. (2013). Potential role of tedizolid phosphate in the treatment of acute bacterial skin infections. Drug Des. Devel. Ther. *7*, 243–265.

Vassiliadis, G., Destoumieux-Garzón, D., Lombard, C., Rebuffat, S., and Peduzzi, J. (2010). Isolation and characterization of two members of the siderophore-microcin family, microcins M and H47. Antimicrob. Agents Chemother. *54*, 288–297.

Walsh, C. (2003). Antibiotics : actions, origins, resistance (ASM Press, Washington, DC).

Walsh, C.T., and Wencewicz, T.A. (2013). Prospects for new antibiotics: a molecule-centered perspective. J. Antibiot. *67*, 7–22.

Xiao, X.-Y., Hunt, D.K., Zhou, J., Clark, R.B., Dunwoody, N., Fyfe, C., Grossman, T.H., O'Brien, W.J., Plamondon, L., Rönn, M., *et al.* (2012). Fluorocyclines. 1. 7-fluoro-9-pyrrolidinoacetamido-6-demethyl-6-deoxytetracycline: a potent, broad spectrum antibacterial agent. J. Med. Chem. *55*, 597–605.

Yoshida, K.-I., Nakayama, K., Ohtsuka, M., Kuru, N., Yokomizo, Y., Sakamoto, A., Takemura, M., Hoshino, K., Kanda, H., Nitanai, H., *et al.* (2007). MexAB-OprM specific efflux pump inhibitors in *Pseudomonas aeruginosa*. Part 7: highly soluble and *in vivo* active quaternary ammonium analogue D13–9001, a potential preclinical candidate. Bioorg. Med. Chem. *15*, 7087–7097.

Yoshimura, F., and Nikaido, H. (1985). Diffusion of β-lactam antibiotics through the porin channels of *Escherichia coli* K-12. Antimicrob. Agents Chemother. *27*, 84–92.

Yuan, Z., and White, R.J. (2006). The evolution of peptide deformylase as a target: contribution of biochemistry, genetics and genomics. Biochem. Pharmacol. *71*, 1042–1047.

Yum, J.H., Kim, C.K., Yong, D., Lee, K., Chong, Y., Kim, C.M., Kim, J.M., Ro, S., and Cho, J.M. (2007). *In vitro* activities of CG400549, a novel FabI inhibitor, against recently isolated clinical staphylococcal strains in Korea. Antimicrob. Agents Chemother. *51*, 2591–2593.

Zechini, B., and Versace, I. (2009). Inhibitors of multidrug resistant efflux systems in bacteria. Recent Pat. Antiinfect. Drug Discov. *4*, 37–50.

Zhanel, G.G., Walters, M., Noreddin, A., Vercaigne, L.M., Wierzbowski, A., Embil, J.M., Gin, A.S., Douthwaite, S., and Hoban, D.J. (2002). The ketolides: a critical review. Drugs *62*, 1771–1804.

Zhanel, G.G., Calic, D., Schweizer, F., Zelenitsky, S., Adam, H., Lagacé-Wiens, P.R.S., Rubinstein, E., Gin, A.S., Hoban, D.J., and Karlowsky, J.A. (2010). New lipoglycopeptides: a comparative review of dalbavancin, oritavancin and telavancin. Drugs *70*, 859–886.

Zhanel, G.G., Lawson, C.D., Zelenitsky, S., Findlay, B., Schweizer, F., Adam, H., Walkty, A., Rubinstein, E., Gin, A.S., Hoban, D.J., *et al.* (2012). Comparison of the next-generation aminoglycoside plazomicin to gentamicin, tobramycin and amikacin. Expert Rev. Anti. Infect. Ther. *10*, 459–473.

Index

B

C

E

F

T

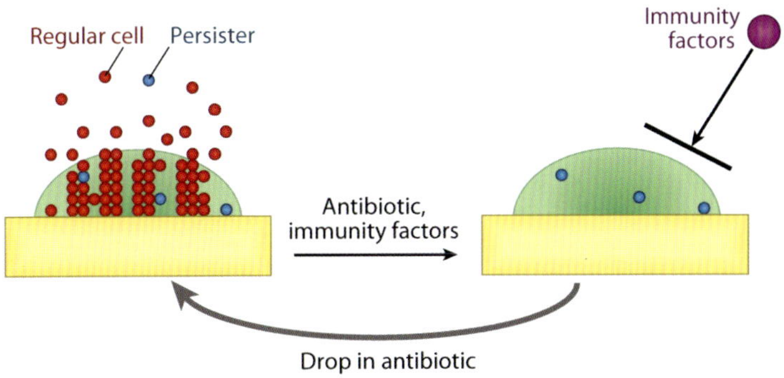

Figure 6.1 A model of relapsing biofilm infection (Lewis, 2001).

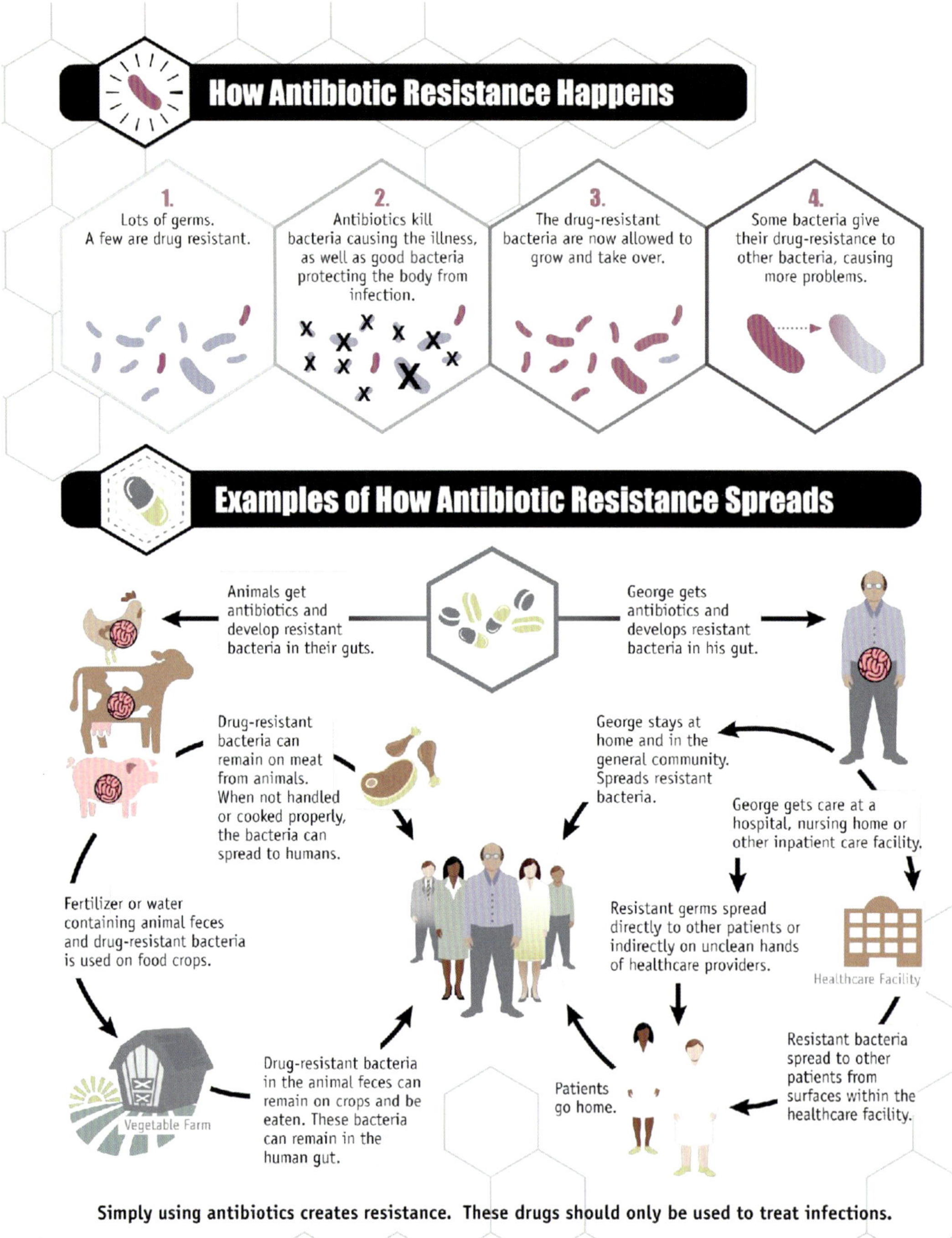

Figure 8.1 How antibiotic resistance happens, and examples of how antibiotic resistance spreads. Frieden, T. (2013). Antibiotic resistance threat in the United States, 2013. CDC (Washington, DC, Centers for Disease Control and Prevention).

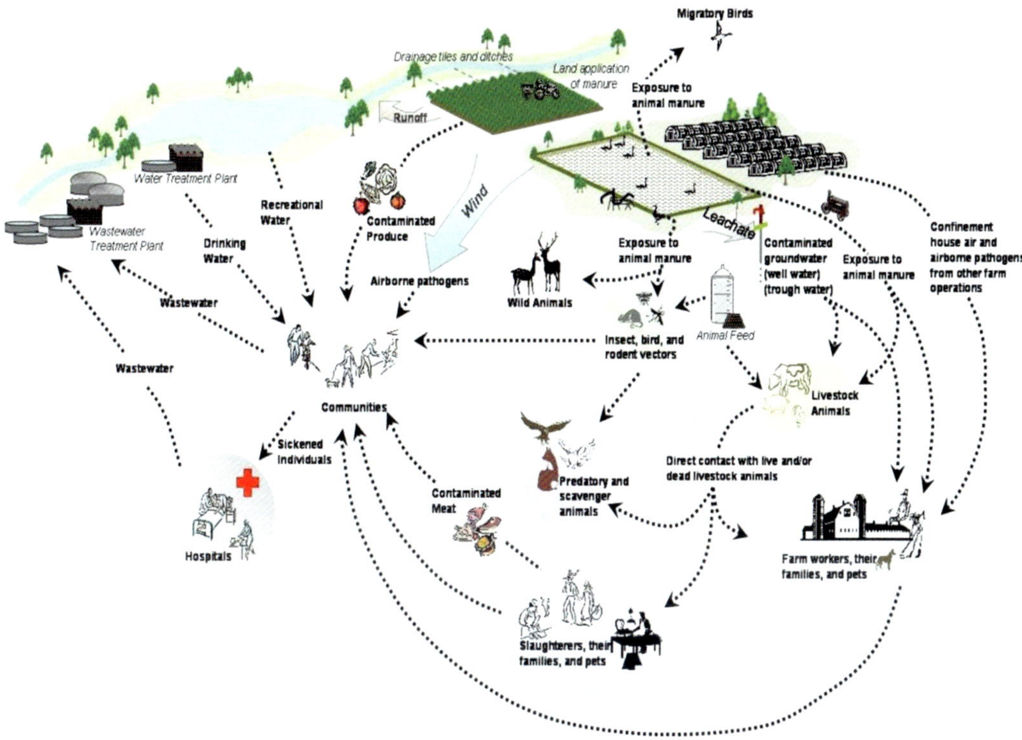

Figure 8.2 Movement of pathogens: an ecological perspective (Rogers *et al.*, 2005).

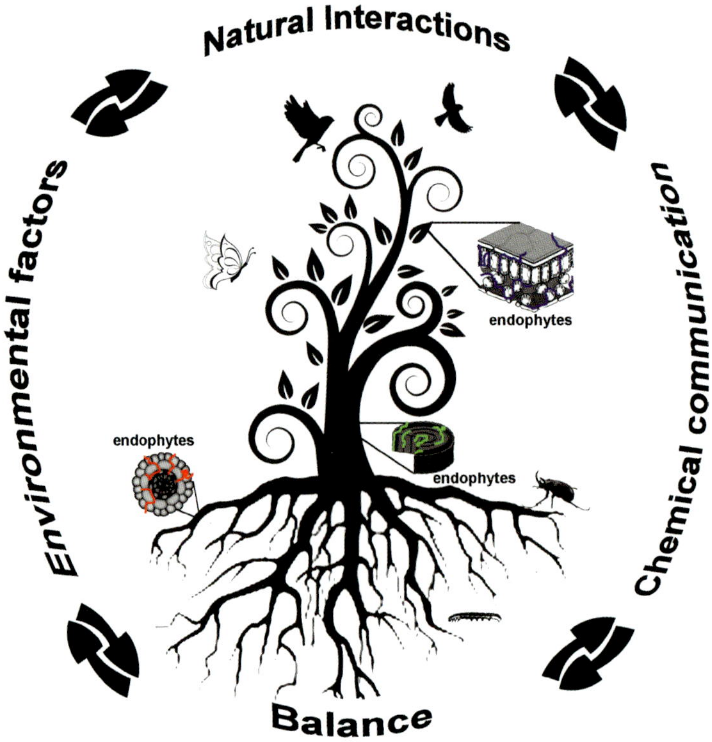

Figure 10.1 Depiction of interactions among the host plant with endophytes and diverse environmental factors.

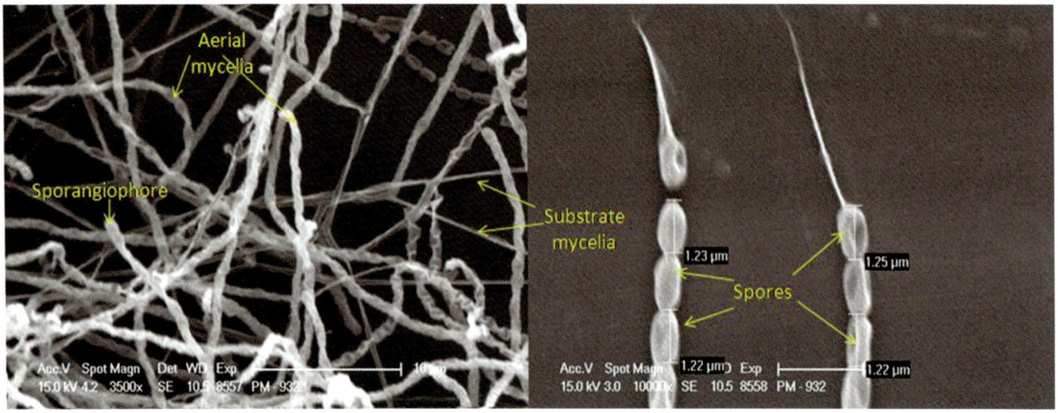

Figure 11.1 Scanning electron microscopy of *Streptomycetes* strain (PM-932) from a hot spring at Himachal Pradesh, India.

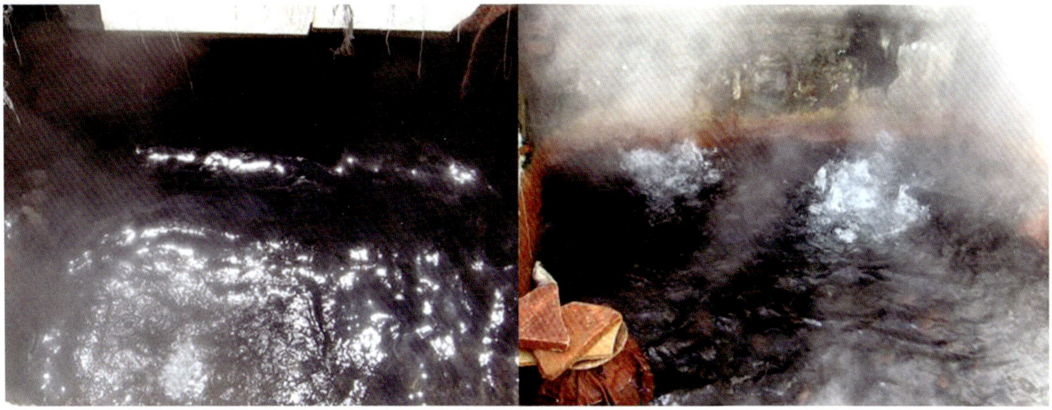

Figure 11.3 Hot spring site at Himachal Pradesh (A).

Figure 11.4 Hot spring site at Himachal Pradesh (B).

Figure 12.1 Yellow colonizations on the walls of a gypsum cave (Sorbas, Spain).

Figure 12.2 Helmcken Falls Cave, British Columbia, Canada. Volcanic caves have a great potential for biodiversity studies. Image by Kent Watson.

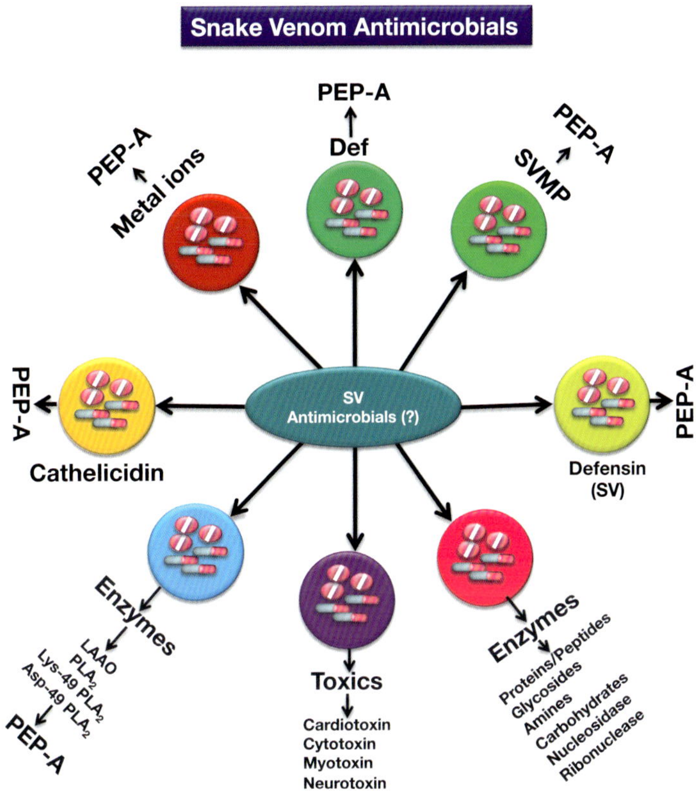

Figure 13.1 Snake venom contains a mixture of diverse enzymes that include phospholipase A2, L-amino acid oxidase, nucleosidase, ribonuclease plus toxic (cardiotoxin, cytotoxin, myotoxin, neurotoxin) and non-toxic activities (antimicrobials) molecules considerably clinically and medically potential. Symbols denote: SV- snake venom antibiotics, CV-crude venom, SVMP-snake venom, PLA2-phospholipase A2 enzymes, Def-snake defensins, Dist-disintegrins, LAAO-L-amino acid oxidase, Pep-A –peptide antimicrobials, etc.

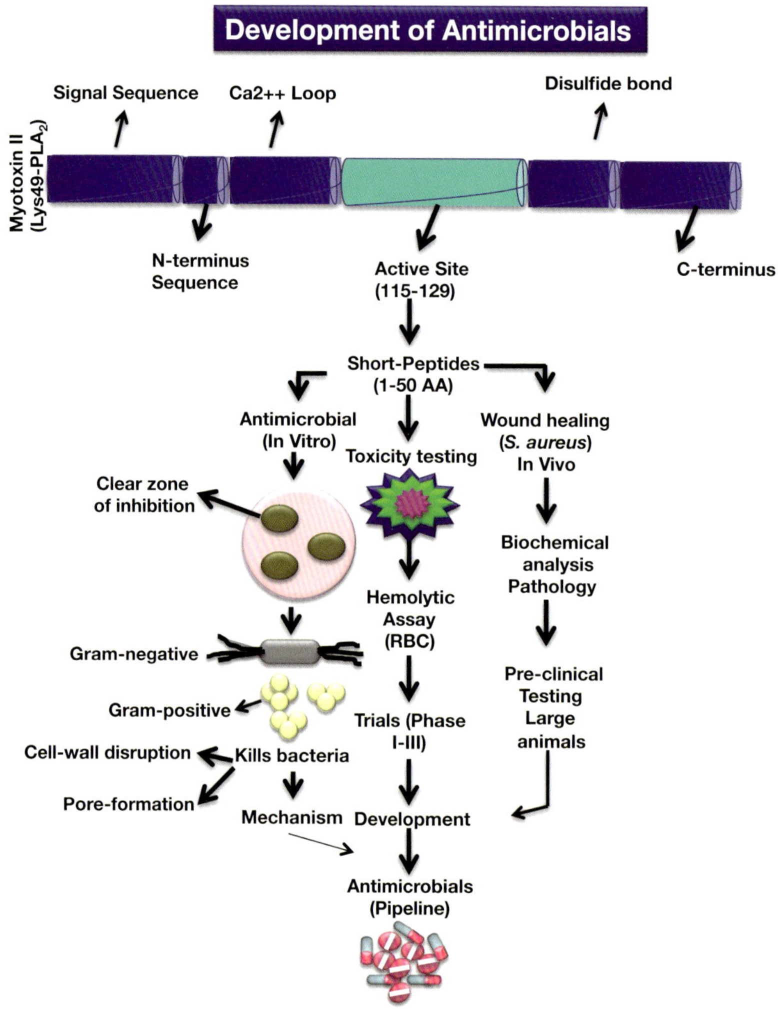

Figure 13.3 Flow chart represents the antimicrobial potency of snake components. Structure of a typical Lys49 PLA2 and small synthetic molecules derived from the parent molecules derived of *B. asper (*Myotoxin II) leads for the possible drug candidates.

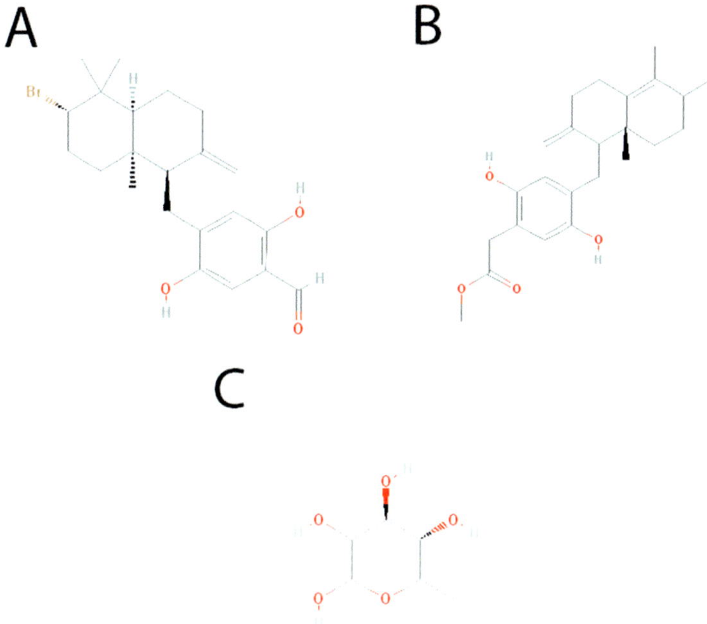

Figure 18.1. Structure of some antivirals extracted from algae. (A) peyssonol A; (B) peyssonol B; (C) fucose.

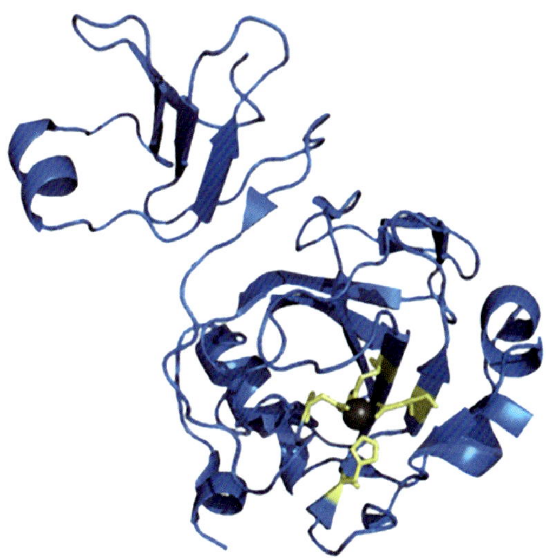

Figure 19.2 Crystal structure of the protein lysostaphin showing the active site.

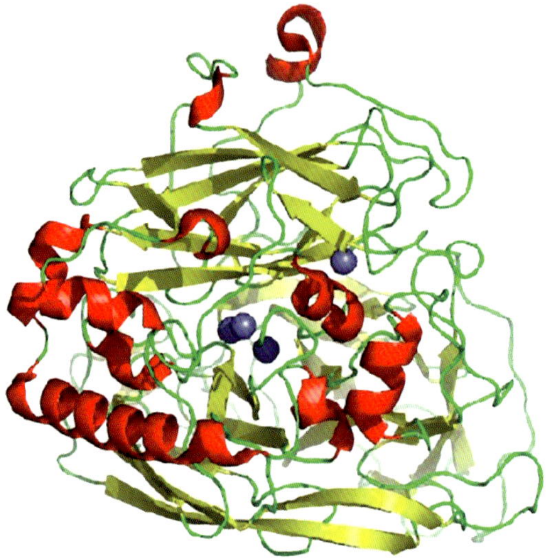

Figure 19.6 Laccase protein from *M. albomyces* displaying the four copper atoms present in the active site of the molecule.

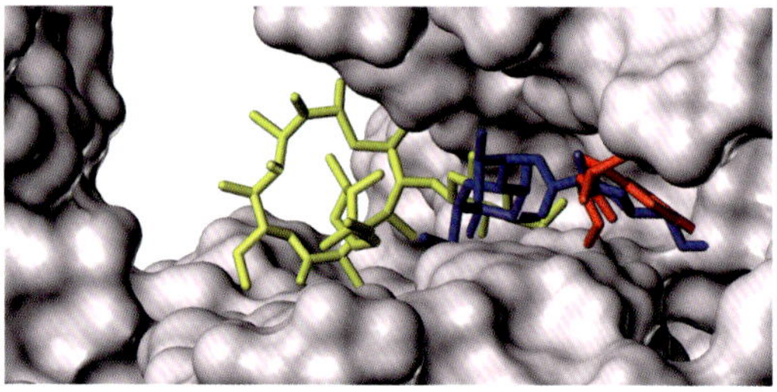

Figure 20.1 Computer modelled representation of the peptide synthesis tunnel in 23S RNA. Yellow: erythromycin at the exit of the peptide tunnel. Red: chloramphenicol at the peptidyl transferase centre. Blue: clindamycin.

Clarithromycin

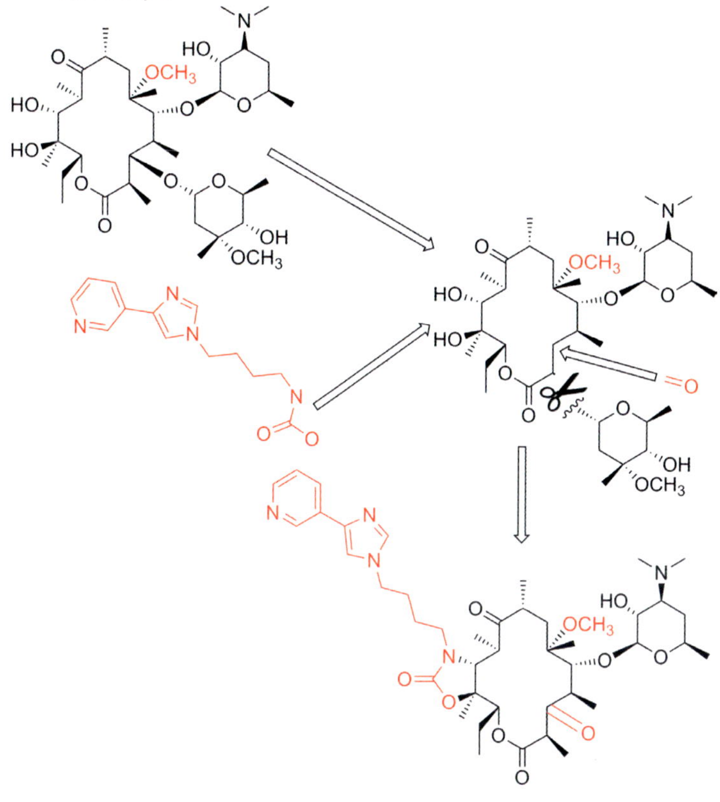

Telithromycin

Figure 20.6 Design of telithromycin- a ketolide.

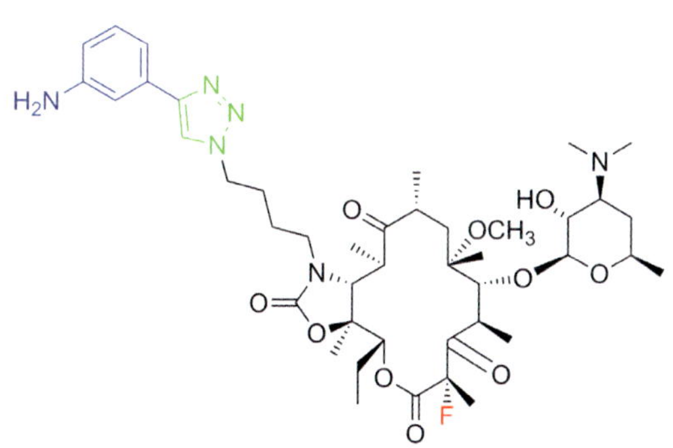

Figure 20.15 Structure of solithromycin. The side chain contains a 1,2,3, triazole (green), and an aminophenyl (blue) instead of the imidazole and pyridine found in telithromycin. Solithromycin also has a 2F (red).